# Lecture Notes in Computer Science 12178

More information about this series at http://www.springer.com/series/7407

Luca Pulina · Martina Seidl (Eds.)

# Theory and Applications of Satisfiability Testing – SAT 2020

23rd International Conference
Alghero, Italy, July 3–10, 2020
Proceedings

 Springer

*Editors*
Luca Pulina (iD)
University of Sassari
Sassari, Italy

Martina Seidl (iD)
Johannes Kepler University of Linz
Linz, Austria

ISSN 0302-9743          ISSN 1611-3349  (electronic)
Lecture Notes in Computer Science
ISBN 978-3-030-51824-0          ISBN 978-3-030-51825-7  (eBook)
https://doi.org/10.1007/978-3-030-51825-7

LNCS Sublibrary: SL1 – Theoretical Computer Science and General Issues

This Springer imprint is published by the registered company Springer Nature Switzerland AG
The registered company address is: Gewerbestrasse 11, 6330 Cham, Switzerland

# Preface

This volume contains the papers presented at the 23rd International Conference on Theory and Applications of Satisfiability Testing (SAT 2020) held during July 3–10, 2020. Originally planned to be held in Alghero, Italy, it was not possible to have an onsite event due to of COVID-19. Instead SAT went online and was organized as a virtual event.

SAT is the premier annual meeting for researchers focusing on the theory and applications of the propositional satisfiability problem, broadly construed. Aside from plain propositional satisfiability, the scope of the meeting includes Boolean optimization, including MaxSAT and pseudo-Boolean (PB) constraints, quantified Boolean formulas (QBF), satisfiability modulo theories (SMT), and constraint programming (CP) for problems with clear connections to Boolean reasoning. Many hard combinatorial problems can be tackled using SAT-based techniques, including problems that arise in formal verification, artificial intelligence, operations research, computational biology, cryptology, data mining, machine learning, mathematics, etc. Indeed, the theoretical and practical advances in SAT research over the past 25 years have contributed to making SAT technology an indispensable tool in a variety of domains. SAT 2020 welcomed scientific contributions addressing different aspects of SAT interpreted in a broad sense, including (but not restricted to) theoretical advances (such as exact algorithms, proof complexity, and other complexity issues), practical search algorithms, knowledge compilation, implementation-level details of SAT solvers and SAT-based systems, problem encodings and reformulations, applications (including both novel application domains and improvements to existing approaches), as well as case studies and reports on findings based on rigorous experimentation.

SAT 2020 received 69 submissions, comprising 52 long papers, 11 short papers, and 6 tool papers. Each submission was reviewed by at least three Program Committee members. The reviewing process included an author response period, during which the authors of the submitted papers were given the opportunity to respond to the initial reviews for their submissions. To reach a final decision, a Program Committee discussion period followed the author response period. External reviewers supporting the Program Committee were also invited to participate directly in the discussion for the papers they reviewed. This year, most submissions received a meta-review, summarizing the discussion that occurred after the author response and an explanation of the final recommendation. In the end, the committee decided to accept a total of 36 papers; 25 long, 9 short, and 2 tool papers.

In addition to presentations on the accepted papers, the scientific program of SAT announced two invited talks by:

- Georg Gottlob, Oxford University, UK, and TU Wien, Austria
- Aaarti Gupta, Princeton University, USA

SAT 2020 hosted various associated events coordinated by our workshop and competition chair Florian Lonsing. This year, three workshops were affiliated with SAT.

- Pragmatics of SAT Workshop, organized by Matti Järvisalo and Daniel Le Berre
- QBF Workshop, organized by Hubie Chen and Friedrich Slivovsky
- Model Counting Workshop, organized by Johannes Fichte and Markus Hecher

Also the workshops were organized as virtual events before or after the main conference. The results of the following four competitive events were announced at SAT:

- SAT Competition 2020, organized by Thomas Balyo, Marijn Heule, Markus Iser, Matti Järvisalo, and Martin Suda
- MaxSAT Evaluation 2020, organized by Fahiem Bacchus, Jeremias Berg, Matti Järvisalo, and Ruben Martins
- QBFEVAL 2020, organized by Luca Pulina, Martina Seidl, and Ankit Shukla
- Model Counting 2020, organized by Markus Hecher and Johannes K. Fichte

Last, but not least, we thank everyone who contributed to making SAT 2020 a success. In particular, we thank our workshop chair Florian Lonsing and our publication chair Laura Pandolfo. We are indebted to the Program Committee members and the external reviewers, who dedicated their time to review and evaluate the submissions to the conference. We thank the authors of all submitted papers for their contributions, the SAT association for their guidance and support in organizing the conference, and the EasyChair conference management system for facilitating the submission and selection of papers as well as the assembly of these proceedings. Last but not least we thank Springer for supporting the Best Paper Awards.

May 2020

Luca Pulina
Martina Seidl

# Organization

## Program Committee

| | |
|---|---|
| Fahiem Bacchus | University of Toronto, Canada |
| Olaf Beyersdorff | Friedrich Schiller University Jena, Germany |
| Armin Biere | Johannes Kepler University Linz, Austria |
| Nikolaj Bjorner | Microsoft, USA |
| Maria Luisa Bonet | Universitat Politècnica de Catalunya, Spain |
| Sam Buss | University of California, San Diego, USA |
| Florent Capelli | Université de Lille, France |
| Pascal Fontaine | Université de Liège, Belgium |
| Marijn Heule | Carnegie Mellon University, USA |
| Alexey Ignatiev | Monash University, Australia |
| Mikolas Janota | INESC-ID/IST, University of Lisbon, Portugal |
| Jie-Hong Roland Jiang | National Taiwan University, Taiwan |
| Jan Johannsen | Ludwig Maximilian University of Munich, Germany |
| Matti Järvisalo | University of Helsinki, Finland |
| Benjamin Kiesl | CISPA Helmholtz Center for Information Security, Germany |
| Oliver Kullmann | Swansea University, UK |
| Daniel Le Berre | CNRS, Université d'Artois, France |
| Florian Lonsing | Stanford University, USA |
| Ines Lynce | INESC-ID/IST, University of Lisbon, Portugal |
| Vasco Manquinho | INESC-ID/IST, University of Lisbon, Portugal |
| Felip Manyà | IIIA-CSIC, Spain |
| Joao Marques-Silva | ANITI, France |
| Ruben Martins | Carnegie Mellon University, USA |
| Kuldeep S. Meel | National University of Singapore, Singapore |
| Alexander Nadel | Intel, Israel |
| Aina Niemetz | Stanford University, USA |
| Jakob Nordstrom | University of Copenhagen, Denmark, and Lund University, Sweden |
| Luca Pulina | University of Sassari, Italy |
| Markus N. Rabe | Google, USA |
| Roberto Sebastiani | University of Trento, Italy |
| Martina Seidl | Johannes Kepler University Linz, Austria |
| Natasha Sharygina | Università della Svizzera Italiana, Switzerland |
| Laurent Simon | Labri, Bordeaux Institute of Technology, France |
| Friedrich Slivovsky | Vienna University of Technology, Austria |
| Stefan Szeider | Vienna University of Technology, Austria |

| Ralf Wimmer | Concept Engineering GmbH, Albert-Ludwigs-Universität Freiburg, Germany |
| Christoph M. Wintersteiger | Microsoft, UK |

## Additional Reviewers

Abraham, Erika
Asadi, Sepideh
Bendík, Jaroslav
Berg, Jeremias
Blinkhorn, Joshua
Brown, Christopher
Böhm, Benjamin
Carbonnel, Clément
Chen, Hubie
Devriendt, Jo
Dodaro, Carmine
England, Matthew
Fleury, Mathias
Franzén, Anders
Gaspers, Serge
Ge-Ernst, Aile
Gebser, Martin
Gocht, Stephan
Hendrian, Diptarama
Hermann, Miki
Hyvärinen, Antti
Kaufmann, Daniela
Kochemazov, Stepan
Kokkala, Janne I.

Korhonen, Tuukka
Levy, Jordi
Marescotti, Matteo
Mengel, Stefan
Morgado, Antonio
Ordyniak, Sebastian
Otoni, Rodrigo
Paxian, Tobias
Peitl, Tomáš
Preiner, Mathias
Razgon, Igor
Rebola Pardo, Adrian
Risse, Kilian
Scheder, Dominik
Semenov, Alexander
Sherratt, David
Soos, Mate
Strozecki, Yann
Torán, Jacobo
Wallon, Romain
Winterer, Felix
Zaikin, Oleg
Zamir, Or
Zunino, Roberto

# Contents

# Sorting Parity Encodings by Reusing Variables

Leroy Chew[✉] and Marijn J. H. Heule

Computer Science Department, Carnegie Mellon University, Pittsburgh, PA, USA
lchew@andrew.cmu.edu

**Abstract.** Parity reasoning is challenging for CDCL solvers: Refuting a formula consisting of two contradictory, differently ordered parity constraints of modest size is hard. Two alternative methods can solve these reordered parity formulas efficiently: binary decision diagrams and Gaussian Elimination (which requires detection of the parity constraints). Yet, implementations of these techniques either lack support of proof logging or introduce many extension variables.

The compact, commonly-used encoding of parity constraints uses Tseitin variables. We present a technique for short clausal proofs that exploits these Tseitin variables to reorder the constraints within the DRAT system. The size of our refutations of reordered parity formulas is $\mathcal{O}(n \log n)$.

## 1  Introduction

Modern SAT solving technology is based on Conflict Driven Clause Learning (CDCL) [12]. The resolution proof system [16] has a one-to-one correspondence [15] with CDCL solving. In practice, however, the preprocessing techniques used in modern solvers go beyond what can be succinctly represented in a resolution proof. As a consequence, when we need to present verifiable certificates of unsatisfiable instances, resolution is not always sufficient. Extended Resolution (ER) [20] is a strong proof system that can polynomially simulate CDCL and many other techniques. However, ER is not necessarily the most useful system in practice, as we also want to minimise the degree of the polynomial simulation.

The DRAT proof system [7] is polynomially equivalent to ER [8]. Yet most practitioners favour DRAT due to its ability to straightforwardly simulate known preprocessing and inprocessing techniques. DRAT works by allowing inference to go beyond preserving models and instead preserves only satisfiability.

In this paper, we demonstrate DRAT's strengths on a particular kind of unsatisfiable instances that involve parity constraints. Formulas with parity constraints have been benchmarks for SAT for decades. The Dubois family encodes the refutation of two contradictory parity constraints over the same variables using the same variable ordering. Urquhart formulas [21] encode a modulo two sum of the degree of each vertex of a graph, the unsatisfiability comes from an assertion that this sum is odd, a violation of the Handshake Lemma. The Parity family from Crawford and Kearns [3] takes multiple parity instances on a set of variables and combines them together. For these problems, practical solutions have been studied using Gaussian elimination, equivalence reasoning, binary decision diagrams and other approaches [5,6,9–11,13,18,19,22].

© Springer Nature Switzerland AG 2020
L. Pulina and M. Seidl (Eds.): SAT 2020, LNCS 12178, pp. 1–10, 2020.
https://doi.org/10.1007/978-3-030-51825-7_1

Extracting checkable proofs in a universal format has been another matter entirely. While it is believed that polynomial size circuitry exists to solve these problems, actually turning them into proofs could mean they may only be "short" in a theoretical polynomial-size sense rather than a practical one. Constructing a DRAT proof of parity reasoning has been investigated theoretically [14], but no implementation exists to actually produce them nor is it clear whether the size is still reasonable to be useful in practice.

There has been some investigation into looking at DRAT without the use of extension variables. DRAT$^-$, which is DRAT without extension variables, is somewhere in between resolution and ER in terms of power. Several simulation results for DRAT$^-$ [2], show that it is a powerful system even without the simulation of ER. A key simulation technique was the elimination and reuse of a variable, which we use to find short DRAT$^-$ proofs of a hard parity formula.

The structure of parity constraints can be manipulated by reusing variables and we exploit the associativity and commutativity of the parity function. We demonstrate this on formulas similar to the `Dubois` family except the variables now appear in a random order in one parity constraint. We show how to obtain DRAT proofs of size $\mathcal{O}(n \log n)$ without using additional variables. Our method can also be used to produce ER proofs of similar size with new variables.

## 2    Preliminaries

In propositional logic a literal is a variable $x$ or its negation $\overline{x}$, a clause is a disjunction of literals and a *Conjunctive Normal Form* (CNF) is conjunction of clauses. A unit clause is a clause containing a single literal. We denote the negation of literal $l$ as $\overline{l}$ (or $\neg l$). The variable corresponding to literal $l$ is var($l$). If $C$ is a clause, then $\overline{C}$ is the conjunction of the negation of the literals in $C$ each a unit clause. In this paper, we treat clauses/formulas as unordered and not containing more than one copy of each literal/clause respectively.

*Unit propagation* simplifies a conjunctive normal form $F$ by building a partial assignment and applying it to $F$. It builds the assignment by satisfying any literal that appears in a unit clause. Doing so may negate opposite literals in other clauses and result in them effectively being removed from that clause. In this way, unit propagation can create more unit clauses and can keep on propagating until no more unit clauses remain or the empty clause is reached. We denote that the empty clause can be derived by unit propagation applied to CNF $F$ by $F \vdash_1 \bot$. Since unit propagation is an incomplete but sound form of logical inference this is a sufficient condition to show that $F$ is a logical contradiction.

*The DRAT proof system.* Below we define the rules of the DRAT proof system. Each rule modifies a formula by either adding or removing a clause while preserving satisfiability or unsatisfiability, respectively.

**Definition 1 (Asymmetric Tautology (AT) [7]).** *Let $F$ be a CNF formula. A clause $C$ is an asymmetric tautology w.r.t. $F$ if and only if $F \wedge \overline{C} \vdash_1 \bot$.*

Asymmetric tautologies are also known as RUP (reverse unit propagation) clauses. The rules ATA and ATE allow us to add and eliminate AT clauses. ATA steps can simulate resolution steps and weakening steps.

$$\frac{F}{F \wedge C} \;(\text{ATA: } C \text{ is AT w.r.t. } F) \qquad \frac{F \wedge C}{F} \;(\text{ATE: } C \text{ is AT w.r.t. } F)$$

**Definition 2 (Resolution Asymmetric Tautology (RAT))** [7]). *Let $F$ be a CNF formula. A clause $C$ is a resolution asymmetry tautology w.r.t. $F$ if and only if there exists a literal $l \in C$ such that for every clause $\bar{l} \vee D \in F$ it holds that $F \wedge \overline{D} \wedge \overline{C} \vdash_1 \bot$.*

The rules RATA and RATE allow us to add and eliminate RAT clauses. RATA can be used to add new variables that neither occur in $F$ or anywhere else. This can be used to simulate extension steps in ER.

$$\frac{F}{F \wedge C} \;(\text{RATA: } C \text{ is RAT w.r.t. } F) \qquad \frac{F \wedge C}{F} \;(\text{RATE: } C \text{ is RAT w.r.t. } F)$$

## 3 A parity contradiction based on random orderings

In this section we will detail the main family of formulas investigated in this work. These formulas will be contradictions expressing both the parity and non-parity on a set of variables.

We define the parity of propositional literals $a, b, c$ as follows

$$\text{xor}(a, b, c) := (\overline{a} \vee \overline{b} \vee \overline{c}) \wedge (\overline{a} \vee b \vee c) \wedge (a \vee \overline{b} \vee c) \wedge (a \vee b \vee \overline{c})$$

Let $X = \{x_1, \ldots, x_n\}$, and let $\sigma$ be a bijection between literals on $X$, that preserves negation ($\sigma(\neg l) = \neg \sigma(l)$). Let $e$ denote the identity permutation on the literals of $X$. Let $T = \{t_1, \ldots, t_{n-3}\}$. We define $\text{PARITY}(X, T, \sigma)$ as

$$\text{xor}(\sigma(x_1), \sigma(x_2), t_1) \wedge \bigwedge_{j=1}^{n-4} \text{xor}(t_j, \sigma(x_{j+2}), t_{j+1}) \wedge \text{xor}(\overline{t}_{n-3}, \sigma(x_{n-1}), \sigma(x_n))$$

This formula is satisfiable if and only if the total parity of $\{\sigma(x_i) \mid x_i \in X\}$ is 1. The $T$ variables act as Tseitin variables and whenever the formula is satisfied $t_{i+1}$ is the sum modulo two of $\sigma(x_1), \ldots, \sigma(x_{i+2})$. The final clauses, $\text{xor}(\overline{t}_{n-3}, \sigma(x_{n-1}), \sigma(x_n))$ thus are satisfied when the sum of $t_{n-3}, \sigma(x_{n-1})$ and $\sigma(x_n)$ is 1 mod 2. Suppose we pick $\sigma$ so that there is some $i \in [n]$ such that $\sigma(x_j)$ is a negative literal if and only if $j = i$. Let $T' = \{t'_1, \ldots, t'_{n-3}\}$ be another set of Tseitin variables. $\text{PARITY}(X, T, \sigma) \wedge \text{PARITY}(X, T', e)$ is false as it states the parity of $X$ is true but also states it false. However the permutation $\sigma$ obfuscates the similarities between the two PARITY parts of the formula.

Were $\sigma(x_j) = x_j$ for all $j \neq i$ and $\sigma(x_i) = -x_i$ then these formulas would be equivalent to the Dubois formulas and a linear proof could be made by inductively deriving clauses that express $t'_j = t_j$ for $j < i - 1$ and then $t'_j \neq t_j$ for $j \geq i - 1$. This will always allow us to derive a contradiction. While a linear

| ATA | RATE | RATA | ATE |
|---|---|---|---|
| $\overline{q} \vee a \vee b \vee c$ | d $\overline{p} \vee a \vee b$ | $\overline{p} \vee c \vee b$ | d $\overline{q} \vee a \vee b \vee c$ |
| $\overline{q} \vee \overline{a} \vee \overline{b} \vee c$ | d $\overline{p} \vee \overline{a} \vee \overline{b}$ | $\overline{p} \vee \overline{c} \vee \overline{b}$ | d $\overline{q} \vee \overline{a} \vee \overline{b} \vee c$ |
| $\overline{q} \vee a \vee \overline{b} \vee \overline{c}$ | d $p \vee a \vee \overline{b}$ | $p \vee c \vee \overline{b}$ | d $\overline{q} \vee a \vee \overline{b} \vee \overline{c}$ |
| $\overline{q} \vee \overline{a} \vee b \vee \overline{c}$ | d $p \vee \overline{a} \vee b$ | $p \vee \overline{c} \vee b$ | d $\overline{q} \vee \overline{a} \vee b \vee \overline{c}$ |
| $q \vee \overline{a} \vee b \vee c$ | d $\overline{p} \vee q \vee c$ | $\overline{p} \vee q \vee a$ | d $q \vee \overline{a} \vee b \vee c$ |
| $q \vee a \vee \overline{b} \vee c$ | d $\overline{p} \vee \overline{q} \vee \overline{c}$ | $\overline{p} \vee \overline{q} \vee \overline{a}$ | d $q \vee a \vee \overline{b} \vee c$ |
| $q \vee a \vee b \vee \overline{c}$ | d $p \vee q \vee \overline{c}$ | $p \vee q \vee \overline{a}$ | d $q \vee a \vee b \vee \overline{c}$ |
| $q \vee \overline{a} \vee \overline{b} \vee \overline{c}$ | d $p \vee \overline{q} \vee c$ | $p \vee \overline{q} \vee a$ | d $q \vee \overline{a} \vee \overline{b} \vee \overline{c}$ |

**Fig. 1.** DRAT steps required for Lemma 1, d denotes a deletion step.

resolution proof is known for the Dubois, it is unknown what the size of the shortest resolution proof of $\text{PARITY}(X, T, \sigma) \wedge \text{PARITY}(X, T', e)$ for a random $\sigma$. While the Dubois family are a special case of these formulas, these formulas are in turn a special case of the Tseitin graph formulas [20] where the vertex degree is always 3.

When $\sigma \neq e$ we still have a contradiction due to the commutativity of the parity function. However such a straightforward DRAT proof becomes obstructed by the disarranged ordering. This permutation also makes these formulas hard for CDCL solvers (see Section 4). We will show that $\text{PARITY}(X, T', e)$ can be efficiently reordered. Afterwards a short resolution proof arises. This brings us to our main theoretical result.

**Theorem 1.** *Let* $X = \{x_1, \ldots, x_n\}$, $T = \{t_1, \ldots, t_n\}$, $T' = \{t'_1, \ldots, t'_n\}$ *and let* $\sigma$ *be a bijection between literals on* $X$, *that preserves negation. Suppose there is some* $i \in [n]$ *such that* $\sigma(x_j)$ *is a negative literal if and only if* $j = i$. *The 3-CNF* $\text{PARITY}(X, T, \sigma) \wedge \text{PARITY}(X, T', e)$ *has a proof of size* $O(n \log n)$.

In the remainder of this section we will prove Theorem 1. We begin with an essential lemma that uses the DRAT rules to perform elementary adjacent swaps on literals in xor constraints.

**Lemma 1.** *Suppose we have a CNF* $F$ *and two sets of xor clauses* $\text{xor}(a, b, p)$ *and* $\text{xor}(p, c, q)$, *where variable* $p$ *appears nowhere in* $F$. *We can infer*

$$\frac{F \wedge \text{xor}(a, b, p) \wedge \text{xor}(p, c, q)}{F \wedge \text{xor}(b, c, p) \wedge \text{xor}(p, a, q)}$$

*in a constant number of DRAT steps without adding new variables.*

*Proof.* The idea is to eliminate variable $p$ so that we define $q$ directly as the parity of $a, b, c$ using eight "ternary xor" clauses. Each of these clauses can be added directly via ATA. We can now remove (using RATE) all clauses that contain variable $p$. These steps are equivalent to performing Davis-Putnam (DP) resolution [4] on variable $p$.

What we are left with is that two levels of parity have been replaced with one level of ternary parity. We can reverse the above steps to get us two levels of parity yet again, but we can swap $a$ and $c$ (since they appear symmetrically in our "ternary xor" clauses). We re-use the eliminated $p$ to now mean the xor of $b$ and $c$ using RATA. Finally, we remove the "ternary xor" clauses using ATE. □

Note that here elimination is required only because we want to re-use the variable $p$. We can also show a similar step in ER without the elimination steps, introducing the "ternary xor" clauses immediately with resolution. We can introduce the four $\mathrm{xor}(p', b, c)$ extension clauses for $p'$, and by resolving them with the ternary clauses on $b$ we get eight intermediate clauses which can resolve with each other on $c$ to get the remaining four $\mathrm{xor}(p', a, q)$ clauses. Without ATA and RATA steps, this process involves 50% more addition steps, but since it contains no deletion steps we have 25% fewer steps in total. ER proofs allow us to keep lemmas without deletion so where we have more than two parity constraints we may wish to reuse derived parity clauses. On the other hand DRAT$^-$ keeps the number of variables and clauses fixed when searching when we do not know the structure of formula and can stop the search size from growing.

**Sorting the input literals.** We can switch the two parity inputs using Lemma 1 in a constant number of proof steps. Furthermore the technique in DRAT does not require any additional extension variables and since the number of addition and deletion steps in Lemma 1 is the same, the working CNF does not change in size. Sorting using adjacent variables requires $\Theta(n^2)$ swaps.

Let us ignore the variables $x_{n-1}$ and $x_n$ and the clauses that include them as special cases. We can take $\mathrm{xor}(x_1, x_2, t_1) \wedge \bigwedge_{i=1}^{n-4} \mathrm{xor}(t_i, x_{i+2}, t_{i+1})$ as the definition of $t_{n-3}$ in circuit form, using the $X$ variables as input gates and the $t_i$ variables as xor ($\oplus$) gates. This circuit is a tree with linear depth. the distance between two input nodes is linear in the worst-case, which is why we get $\Omega(n^2)$ many swaps. However Lemma 1 allows us even more flexibility, we can not only rearrange the input variables but the Tseitin variables.

For example if we have $\mathrm{xor}(t_i, x_{i+2}, t_{i+1})$ and $\mathrm{xor}(t_{i-1}, x_{i+1}, t_i)$ clauses we can eliminate $t_i$ so that $t_{i+1}$ is defined as the parity of $x_{i+1}$, $x_{i+2}$, and $t_{i-1}$. However

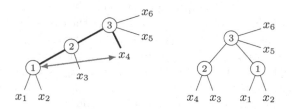

**Fig. 2.** Swapping the position of an internal node to balance the tree.

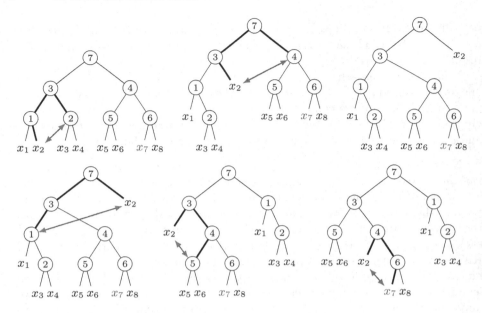

**Fig. 3.** Moving two literals next to each other. Top: moving $x_2$ up to the source of the tree. Bottom: moving $x_2$ down to swap with $x_7$

we can now redefine $t_i$ as the xor of $x_{i+1}$, $x_{i+2}$ (using xor($t_i, x_{i+1}, x_{i+2}$)) and $t_{i+1}$ as the xor of $t_i$ and $t_{i-1}$ (using xor($t_{i+1}, t_i, t_{i-1}$)). See Figure 2 for an example and notice how we change the topology of the tree.

In $\lfloor \frac{n}{2} \rfloor$ many swaps we can change our linear depth tree into a tree that consists of a two linear branches of depth at most $\lceil \frac{n}{2} \rceil$ joined at the top by an xor. This means that using a divide and conquer approach, we can turn this tree in a balanced binary tree of $\lceil \log_2 n \rceil$ depth in $\mathcal{O}(n \log n)$ many steps.

The purpose of a log depth tree structure is to allow leaf-to-leaf swapping from both ends of the the tree without having to do a linear number of swaps, in fact we can do arbitrary leaf swaps in $\mathcal{O}(\lceil \log n \rceil)$ many individual steps. This is done by pushing a variable up its branch to the source node of the tree and pushing it back down another branch to its destination as in Figure 3. Then we can reverse the steps with the variable being swapped out. The resulting tree even retains the position of all other nodes.

Note that we also have the variables $x_n$ and $x_{n-1}$ that only appear in the clauses of xor($\bar{t}_{n-3}, x_{n-1}, x_n$). Suppose the two children of $t_{n-3}$ in its definition circuit are $a$ and $b$, in other words xor($t_{n-3}, a, b$) are the clauses currently defining $t_{n-3}$. Without loss of generality suppose we want to swap $x_{n-1}$ with $a$.

The clauses of xor($\bar{t}_{n-3}, x_{n-1}, x_n$) are exactly the same as the clauses of xor($t_{n-3}, x_{n-1}, \bar{x}_n$). Using Lemma 1 we can eliminate $t_{n-3}$ and gain eight clauses that represent that $\bar{x}_n$ is the ternary xor of $a, b$ and $x_{n-1}$. Then we can reverse the steps but instead swap the positions of $x_{n-1}$ and $a$.

In this way we can introduce $x_{n-1}$ or $x_n$ into the tree and swap it with any leaf. Once again we only require $\mathcal{O}(\log n)$ many applications of Lemma 1 to completely swap the position of $x_{n-1}$ or $x_n$ with any leaf.

**Arriving at the empty clause.** The total number of leaf-to-leaf swaps we are required to perform is bounded above linearly so we stay within $\mathcal{O}(n \log n)$ many steps. We can now undo the balanced tree into a linear tree in (we reverse what we did to balance it) keeping within an $\mathcal{O}(n \log n)$ upper bound.

Recall that we performed a sort on the variables in $\text{PARITY}(X, T', e)$ thereby transforming it into $\text{PARITY}(X, T', \sigma')$ with $\text{var}(\sigma'(x)) = \text{var}(\sigma(x))$, resulting in the formula $\text{PARITY}(X, T, \sigma) \wedge \text{PARITY}(X, T', \sigma')$. Thus the final part of the proof now involves refuting a formula equivalent to one of the Dubois formulas.

We create a proof that inductively shows equivalence or non-equivalence between variables $t_j \in T$ and the $t'_j \in T'$ starting from $j = 1$ to $j = n - 3$. If there is an even number of instances $i$, $1 \leq i \leq j + 1$ where $\sigma'(x_i) \neq \sigma(x_i)$ we derive $(t'_j \vee \bar{t}_j)$ and $(\bar{t'}_j \vee t_j)$. If there are an odd number of instances $i$, $1 \leq i \leq j + 1$ where $\sigma'(x_i) \neq \sigma(x_i)$ we instead derive $(\bar{t'}_j \vee \bar{t}_j)$ and $(t'_j \vee t_j)$.

Whichever case, we can increase $j$ with the addition (ATA) of six clauses. We can think of this as working via DP resolution in a careful order: $\sigma(x_{j+1})$, $t_{j-1}$, $t'_j$ from $j = 1$ to $n - 3$ in increasing $j$ (and treat $\sigma(x_1)$ as $t_0$).

Finally, when $j = n - 3$, we have either already exceeded the single value $i$ such that $\sigma'(x_i) \neq \sigma(x_i)$, or it appears in $n - 1$ or $n$. Either way, we can add the four clauses $(\sigma(x_{n-1}) \vee \sigma(x_n))$, $(\neg\sigma(x_{n-1}) \vee \sigma(x_n))$, $(\sigma(x_{n-1}) \vee \neg\sigma(x_n))$, $(\neg\sigma(x_{n-1}) \vee \neg\sigma(x_n))$ then the two unit clauses $(\sigma(x_n))$ and $(\neg\sigma(x_n))$ and finally the empty clause. This final part of the refutation uses $\mathcal{O}(n)$ many ATA steps.

## 4    Evaluation

The formulas we ran experiments on are labelled $\mathbf{rpar}(n, g)$. Which represent $\text{PARITY}(X, T, \sigma_{(n,g)}) \wedge \text{PARITY}(X, T', e)$ using the DIMACS format. The parameter $n$ is the number of input variables and a random number generator $g$. The CNF uses variables $X = \{1, \ldots, n\}$, $T = \{n + 1, \ldots, 2n - 3\}$, $T' = \{2n - 2, \ldots, 3n - 6\}$, $e$ is the identity permutation, and $\sigma_{(n,g)}$ is a random permutation based on $g$, where one random literal $i_{n,g}$ is flipped by $\sigma$.

We ran a program rParSort that generated an instance $\mathbf{rpar}(n, \text{rnd}_s)$ based on a seed $s$ and also generated a DRAT proof based on Theorem 1. We compare the size of our proofs by ones produced by the state-of-the-art SAT solver CaDiCaL [1] (version 1.2.1) and the tool EBDDRES [17] (version 1.1). The latter solves the instance using binary decision diagrams and turns the construction into an ER proof. These ER proofs can easily be transformed into the DRAT format as DRAT generalizes ER. Proof sizes (in the number of DRAT steps, i.e. lines in the proof) are presented and compared in Figure 4.

rParSort proofs remained feasible for values as large as $n = 4000$ with proofs only being 150MB due to the $\mathcal{O}(n \log n)$ upper bound in proof lines. We believe

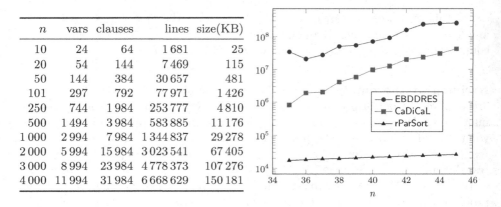

| $n$ | vars | clauses | lines | size(KB) |
|---|---|---|---|---|
| 10 | 24 | 64 | 1 681 | 25 |
| 20 | 54 | 144 | 7 469 | 115 |
| 50 | 144 | 384 | 30 657 | 481 |
| 101 | 297 | 792 | 77 971 | 1 426 |
| 250 | 744 | 1 984 | 253 777 | 4 810 |
| 500 | 1 494 | 3 984 | 583 885 | 11 176 |
| 1 000 | 2 994 | 7 984 | 1 344 837 | 29 278 |
| 2 000 | 5 994 | 15 984 | 3 023 541 | 67 405 |
| 3 000 | 8 994 | 23 984 | 4 778 373 | 107 276 |
| 4 000 | 11 994 | 31 984 | 6 668 629 | 150 181 |

**Fig. 4.** rParSort proof sizes for $\mathtt{rpar}(n, \mathsf{rnd}_{53})$ formulas (left). Comparisons of average (of 10) proof sizes on $n \in \{35, \ldots, 45\}$ (right).

leading coefficient is also kept small by number of factors such as the proof lines being width 4 and only 16 being needed per swap step.

CaDiCaL showed difficulty for modest values of $n$. While proofs with less than $10^6$ lines are common for $n = 35$, the size and running time grows exponentially and by $n = 41$ proofs are larger than $10^7$ lines. CaDiCaL times out using a 5000 seconds limit on some instances with $n = 46$ and on most instances with $n \geq 50$.

The size of proofs produced by EBDDRES appears to grow slower compared to CDCL, which is not surprising as BDDs can solve the formulas in polynomial time. However, as can be observed in Figure 4, the ER proofs are actually bigger for small $n$. The extracted DRAT proofs (converted from the ER proofs) are large: the average proof with $n \geq 35$ had more than $10^7$ lines. This means that this BDD-based approach is not practical to express parity reasoning in DRAT.

## 5   Conclusion

We have shown that through manipulating existing encoding variables DRAT can take advantage of the commutativity of xor definitions via Lemma 1. Our proof generator is capable of producing reasonable-sized proofs for instances with tens of thousands of variables, while state-of-the-art SAT solvers without xor detection and Gaussian elimination, such as CaDiCaL, can only solve instances up to about 60 variables. Although these formulas are also doable for BDD-based approaches, the resulting proofs are too big for practical purposes.

The DRAT proofs are in the fragment of DRAT$^-$, where the number of variables stays fixed, which is of potential benefit to the checker. If we are not concerned with the introduction of new variables, our DRAT proofs can easily be made into ER proofs with only a 50% increase in addition steps (and the introduction of new variables). This is an alternative approach that may prove useful in other settings where elimination of a variable is not so easy.

*Acknowledgements.* The authors thank Armin Biere for his useful comments on an earlier draft. This work has been support by the National Science Foundation (NSF) under grant CCF-1618574.

# References

1. Biere, A.: CaDiCaL at the SAT Race 2019 (2019)
2. Buss, S., Thapen, N.: DRAT proofs, propagation redundancy, and extended resolution. In: International Conference on Theory and Applications of Satisfiability Testing. pp. 71–89. Springer (2019)
3. Crawford, J.M., Kearns, M.J., Schapire, R.E.: The minimal disagreement parity problem as a hard satisfiability problem (1994)
4. Davis, M., Putnam, H.: A computing procedure for quantification theory. Journal of the ACM **7**, 210–215 (1960)
5. Han, C.S., Jiang, J.H.R.: When Boolean satisfiability meets Gaussian elimination in a simplex way. In: Madhusudan, P., Seshia, S.A. (eds.) Computer Aided Verification. pp. 410–426. Springer, Berlin, Heidelberg (2012)
6. Heule, M., van Maaren, H.: Aligning CNF- and equivalence-reasoning. In: Hoos, H.H., Mitchell, D.G. (eds.) Theory and Applications of Satisfiability Testing. pp. 145–156. Springer, Berlin, Heidelberg (2005)
7. Järvisalo, M., Heule, M.J.H., Biere, A.: Inprocessing rules. In: Gramlich, B., Miller, D., Sattler, U. (eds.) Automated Reasoning. pp. 355–370. Springer, Berlin, Heidelberg (2012)
8. Kiesl, B., Rebola-Pardo, A., Heule, M.J.H.: Extended resolution simulates DRAT. In: Galmiche, D., Schulz, S., Sebastiani, R. (eds.) Automated Reasoning - 9th International Joint Conference, IJCAR 2018, Held as Part of the Federated Logic Conference, FloC 2018, Oxford, UK, July 14–17, 2018, Proceedings. Lecture Notes in Computer Science, vol. 10900, pp. 516–531. Springer (2018)
9. Laitinen, T., Junttila, T., Niemelä, I.: Extending clause learning DPLL with parity reasoning. In: Proceedings of the 2010 Conference on ECAI 2010: 19th European Conference on Artificial Intelligence. pp. 21–26. IOS Press, NLD (2010)
10. Laitinen, T., Junttila, T., Niemela, I.: Equivalence class based parity reasoning with DPLL(XOR). In: Proceedings of the 2011 IEEE 23rd International Conference on Tools with Artificial Intelligence. pp. 649–658. ICTAI '11, IEEE Computer Society, USA (2011)
11. Li, C.M.: Equivalent literal propagation in the DLL procedure. Discrete Applied Mathematics **130**(2), 251–276 (2003), the Renesse Issue on Satisfiability
12. Marques Silva, J.P., Lynce, I., Malik, S.: Conflict-driven clause learning SAT solvers. In: Handbook of Satisfiability. IOS Press (2009)
13. Ostrowski, R., Grégoire, É., Mazure, B., Saïs, L.: Recovering and exploiting structural knowledge from CNF formulas. In: Van Hentenryck, P. (ed.) Principles and Practice of Constraint Programming - CP 2002. pp. 185–199. Springer, Berlin, Heidelberg (2002)
14. Philipp, T., Rebola-Pardo, A.: DRAT proofs for XOR reasoning. In: Michael, L., Kakas, A. (eds.) Logics in Artificial Intelligence. pp. 415–429. Springer International Publishing, Cham (2016)
15. Pipatsrisawat, K., Darwiche, A.: On the power of clause-learning SAT solvers as resolution engines. Artificial Intelligence **175**(2), 512–525 (2011)

16. Robinson, J.A.: Theorem-proving on the computer. Journal of the ACM **10**(2), 163–174 (1963)
17. Sinz, C., Biere, A.: Extended resolution proofs for conjoining BDDs. In: Grigoriev, D., Harrison, J., Hirsch, E.A. (eds.) Computer Science – Theory and Applications. pp. 600–611. Springer, Berlin, Heidelberg (2006)
18. Soos, M.: Enhanced Gaussian elimination in DPLL-based SAT solvers. In: Berre, D.L. (ed.) POS-10. Pragmatics of SAT. EPiC Series in Computing, vol. 8, pp. 2–14. EasyChair (2012)
19. Soos, M., Nohl, K., Castelluccia, C.: Extending SAT solvers to cryptographic problems. In: Kullmann, O. (ed.) Theory and Applications of Satisfiability Testing - SAT 2009. pp. 244–257. Springer, Berlin, Heidelberg (2009)
20. Tseitin, G.C.: On the complexity of derivations in propositional calculus. In: Slisenko, A.O. (ed.) Studies in Mathematics and Mathematical Logic, Part II, pp. 115–125 (1968)
21. Urquhart, A.: Hard examples for resolution. Journal of the ACM **34**(1), 209–219 (1987)
22. Warners, J.P., van Maaren, H.: A two-phase algorithm for solving a class of hard satisfiability problems. Operations Research Letters **23**(3), 81–88 (1998). ISSN 0167-6377

# Community and LBD-Based Clause Sharing Policy for Parallel SAT Solving

Vincent Vallade[1](✉), Ludovic Le Frioux[2], Souheib Baarir[1,3], Julien Sopena[1,4], Vijay Ganesh[5], and Fabrice Kordon[1]

[1] CNRS, LIP6, UMR 7606, Sorbonne Université, Paris, France
vincent.vallade@lip6.fr
[2] LRDE, EPITA, Le Kremlin-Bicêtre, France
[3] Université Paris Nanterre, Nanterre, France
[4] DELYS Team, Inria, Paris, France
[5] University of Waterloo, Waterloo, ON, Canada

**Abstract.** Modern parallel SAT solvers rely heavily on effective clause sharing policies for their performance. The core problem being addressed by these policies can be succinctly stated as "the problem of identifying high-quality learnt clauses". These clauses, when shared between the worker nodes of parallel solvers, should lead to better performance. The term "high-quality clauses" is often defined in terms of metrics that solver designers have identified over years of empirical study. Some of the more well-known metrics to identify high-quality clauses for sharing include clause length, literal block distance (LBD), and clause usage in propagation.

In this paper, we propose a new metric aimed at identifying high-quality learnt clauses and a concomitant clause-sharing policy based on a combination of LBD and community structure of Boolean formulas. The concept of community structure has been proposed as a possible explanation for the extraordinary performance of SAT solvers in industrial instances. Hence, it is a natural candidate as a basis for a metric to identify high-quality clauses. To be more precise, our metric identifies clauses that have low LBD and low community number as ones that are high-quality for applications such as verification and testing. The community number of a clause $C$ measures the number of different communities of a formula that the variables in $C$ span. We perform extensive empirical analysis of our metric and clause-sharing policy, and show that our method significantly outperforms state-of-the-art techniques on the benchmark from the parallel track of the last four SAT competitions.

## 1 Introduction

The encoding of complex combinatorial problems as Boolean satisfiability (SAT) instances has been widely used in industry and academy over the last few decades. From AI planning [19] to cryptography [26], modern SAT solvers have demonstrated their ability to tackle huge formulas with millions of variables

© Springer Nature Switzerland AG 2020
L. Pulina and M. Seidl (Eds.): SAT 2020, LNCS 12178, pp. 11–27, 2020.
https://doi.org/10.1007/978-3-030-51825-7_2

and clauses. This is instinctively surprising since SAT is an NP-complete problem [12]. The high-level view of a SAT solver algorithm is the successive enumeration of all possible values for each variable of the problem until a solution is found or the unsatisfiability of the formula is concluded. What makes SAT solving applicable to large real-world problems is the conflict-driven clause learning (CDCL) paradigm [24].

During its search, a CDCL solver is able to learn new constraints, in the form of new implied clauses added to the formula, which allows it to avoid the exploration of large parts of the search space. In practice too many clauses are learnt and a selection has to be done to avoid memory explosion. Many heuristics have been proposed, in the sequential context, to reduce the database of learnt clauses. Such methods of garbage collection are usually quite aggressive and are based on measures, such as the literal block distance (LBD), whose aim is to quantify the usefulness of clauses [4].

The omnipresence of many-core machines has led to considerable efforts in parallel SAT solving research [7]. There exist two main classes of parallel SAT strategies: a cooperative one called divide-and-conquer [32] and a competitive one called portfolio [15]. Both rely on the use of underlying sequential worker solvers that might share their respective learnt clauses. Each of these sequential solvers has a copy of the formula and manages its own learnt clause database. Hence, not all the learnt clauses can be shared and a careful selection must be made in order for the solvers to be efficient. In state-of-the-art parallel solvers this filtering is usually based on the LBD metric. The problem with LBD is its locality, indeed a clause does not necessarily have the same LBD value within the different sequential solvers' context.

In this work we explore the use of a more global quality measure based on the community structure of each instance. It is well-known that SAT instances encoding real-world problems expose some form of modular structure which is implicitly exploited by modern CDCL SAT solvers. A recurring property of industrial instances (as opposed to randomly-generated ones) is that some variables are more constraint together (linked by more clauses). We say that a group of variables that have strong link with each other and few links with the rest of the problem form a community (a type of cluster over the variable-incidence graph of Boolean formulas). A SAT instance may contain tens to thousand of communities.

**Contributions.** The primary contributions of this paper are the following:

- Based on statistics gathered during sequential SAT solver's executions on the benchmark from the SAT competition 2018, we study the relationship between LBD and community, and we analyse the efficacy of LBD and community as predictive metrics of the usefulness of newly learnt clauses.
- Based on this preliminary analysis, we propose to combine both metrics to form a new one and to use it to implement a learnt clause sharing policy in the parallel SAT solving context.

- We implement our new sharing strategy in the solver (P-MCOMSPS [22]) winner
  of the last parallel SAT competition in 2018, and evaluate our solver on the
  benchmark from the SAT competition 2016, 2017, 2018, and 2019. We show
  that our solver significantly outperforms competing solvers over this large and
  comprehensive benchmark of industrial application instances.

**Paper Structure.** The remainder of the paper is structured as follows: we introduce the basic concepts necessary to understand our work in Sect. 2. Section 3
presents preliminary analysis on LBD usage to provide intuition and motivate
our work. Section 4 explores combination of LBD and COM measures to detect
useful learnt clauses. Solvers and experimental results are presented in Sect. 5.
Section 6 surveys some related works and Sect. 7 concludes this paper.

## 2   Preliminaries

This section introduces useful definitions, notations, and concepts that will be
used in the remaining of the paper. It is worth noting that we consider the context
of complete SAT solving, and thus we focus on the well-known conflict-driven
clause learning (CDCL) algorithm [24]. For details on CDCL SAT algorithm we
refer the reader to [9].

### 2.1   Boolean Satisfiability Problem

A *Boolean variable* is a variable that has two possible values: *true* or *false*. A
*literal* is a Boolean variable or its negation (NOT). A *clause* is a finite disjunction
(OR) of literals. A *conjunctive normal form (CNF) formula* is a finite conjunction
(AND) of clauses. In the rest of the paper we use the term formula to refer to
CNF formula. Moreover, clauses are represented by the set of their literals, and
formulas by the set of their clauses.

   For a given formula $F$, we define an *assignment* of variables of $F$ as a function
$\mathcal{A} : \mathcal{V} \rightarrow \{true, false\}$, where $\mathcal{V}$ is the set of variables appearing in $F$. A clause
is satisfied when at least one of its literals is evaluated to *true*. A formula is
satisfied if all its clauses are evaluated to *true*. A formula is said to be *satisfiable*
(SAT) if there is at least one assignment that makes it *true*; it is reported *unsatisfiable* (UNSAT) otherwise. The Boolean satisfiability (SAT) problem consists in
determining if a given formula is SAT or UNSAT.

### 2.2   Literal Block Distance

Literal block distance (LBD) [4] is a positive integer, that is used as a learnt
clause quality metric in almost all competitive sequential CDCL-like SAT solvers.
The LBD of a clause is the number of different decision levels on which variables
of the clause have been assigned. Hence, the LBD of a clause can change over
time and it can be (re)computed each time the clause is fully assigned.

## 2.3   Community

It is well admitted that real-life SAT formulas exhibit notable "structures", explaining why some heuristics such as VSIDS [27] or phase saving [30], for example, work well. One way to highlight such a structure is to represent the formula as a graph and analyze its shape. A structure of interest in this paper is the so-called *community structure* [1]. Let us have a closer look on this latter.

An *undirected weighted graph (graph, for short)* is a pair $G = (N, w)$, where $N$ is the set of nodes of $G$, and $w : N \times N \to \mathbb{R}^+$ is the associated weight function which should be commutative.

The *variable incident graph (VIG)* [1] of a formula $F$ is a graph whose nodes represent variables of $F$, and there exists an edge between two variables iff they shared appearance in at least a clause. Hence, a clause $C$ results in $\binom{|C|}{2}$ edges. Thus, to give the same importance to each clause, edges have a weight: $w(x, y) = \sum_{\substack{C \in F \\ x,y \in C}} 1/\binom{|C|}{2}$.

The *community detection* of a graph is usually captured using the modularity metric [28]. The modularity function $\mathcal{Q}(G, P)$ (see Eq. 1), takes a graph $G$ and a partition $P = \{P_1, \ldots, P_n\}$ of nodes of $G$. It evaluates the density of the connection of the nodes within a part relatively to the density of the entire graph *w.r.t.* a random graph with the same number of nodes and the same degree.

$$\mathcal{Q}(G, P) = \sum_{P_i \in P} \frac{\sum_{x,y \in P_i} w(x, y)}{\sum_{x,y \in N} w(x, y)} - \left( \frac{\sum_{x \in P_i} deg(x)}{\sum_{x \in N} deg(x)} \right)^2 \tag{1}$$

The *modularity* of $G$ is the maximal modularity, for any possible partition $P$ of its nodes: $Q(G) = max\{\mathcal{Q}(G, P) \mid P\}$, and ranges over $[0, 1]$.

Computing the modularity of a graph is an NP-hard problem [11]. However, there exists greedy and efficient algorithms, returning an approximated lower bound for the modularity of a graph, such as the *Louvain method* [10].

In the remaining of the paper we use the Louvain method (with a precision $\epsilon = 10^{-7}$) to compute communities. Graphs we consider are VIG of formulas already simplified by the `SatElite` [13] preprocessor.

**Community Value of a Clause.** We call COM the number of communities on which a clause span: we work on the VIG of the problem, so we consider communities of variables. Each variable belongs to a unique community (determined by the Louvain algorithm). To compute the COM of a clause, we consider the variables corresponding to the literals of the clause and we count the number of distinct communities represented by these variables.

# 3   Measures and Intuition

Our main goal is to improve the overall performances of parallel SAT solving. One way to do that is to study the quality/impact of shared information between the underling SAT engines (sequential SAT solvers) in a parallel strategy.

Since determining, a priori, the usefulness of information is already a challenging task in a sequential context, we can easily imagine the hardness of this guessing in the parallel setting.

In almost all competitive parallel SAT solvers, the sharing is limited to particular forms of learnt clauses (*unit* clauses, clauses with an $LBD \leq i$ [8,22], double touched clauses [5]). We propose here to focus our study on LBD and evaluate the impact of sharing clauses with particular values.

## 3.1   Sequential SAT Solving, Learnt Clauses and LBD

The starting point of our analysis is to evaluate the usage of learnt clauses in the two main components of sequential solvers, namely, the *unit propagation* and the *conflict analysis* procedures. Performing this analysis in a sequential setting makes perfect sense since parallel solvers launch multiple sequential solvers.

To conduct this study, we run `Maple-COMSPS` [23] on the main track benchmark from the SAT competition in 2018[1] with a 5000 s timeout. Figure 1 depicts our observations. The x-axis shows the percentage of the mean number of learnt clauses (considering all learnt clauses of the whole benchmark). The y-axis corresponds to the cumulative usage percentage. Hence, the curve with dots depicts, for different LBDs (from 1 to 9), the usage of these clauses in unit propagation. The curve with triangles highlights the same information but for conflict analysis.[2]

In both curves, the key observation concerns the inflection located around LBD = 4. The impact of clauses with LBD $\leq$ 4 can be considered as very positive: 60% of usage in unit propagation, and 40% in conflict analysis, while representing only 3% of the total number of learnt clauses. Moreover, clauses with an LBD > 4 do not bring a significant added value when considering their quantities.

Based on these results, 4 appears to be a good LBD value for both limiting the rate of shared clauses and maximizing the percentage of tracked events in a parallel context.

## 3.2   A First Parallel Sharing Strategy

To assess our previous observation, we developed a strategy implementing an *LBD-based clauses sharing*: clauses learnt with an LBD below a predetermined threshold are shared. We then operated this strategy with LBD = 4, but also

---

[1] http://sat2018.forsyte.tuwien.ac.at.

[2] On the contrary of this qualitative clause study, benchmarks presented in Sects. 3.2 and 5.2 do not have any logs.

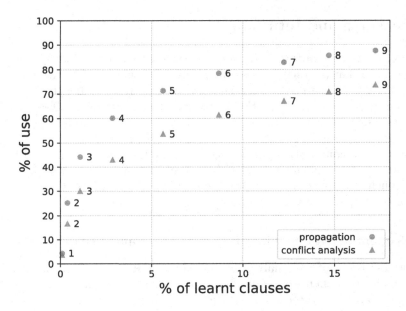

**Fig. 1.** Usage of learnt clause considering LBD

with the surrounding values (LBD = 3 and LBD = 5). Our strategy has been integrated into `Painless` [21][3] using `Maple-COMSPS` as a sequential solver engine and a portfolio parallelization strategy. The different solvers we compare are:

- `P-MCOMSPS`, a portfolio SAT solver, winner of the parallel track of the SAT competition 2018, is used as a reference. It implements a sharing strategy based on incremental values for LBD [22].
- `P-MCOMSPS-L`$\langle n \rangle$, our new portfolio solver, with LBD $\leq n$.

All solvers were processed on a 12-core Intel Xeon CPU at 2.40 GHz and 62 GB of RAM running Linux 4.9. The solvers have been compiled with the version 9.2.1 of *GCC*. They have been launched with 12 threads, a 61 GB memory limit, and a wall clock-time limit of 5000 s (as for the SAT competitions). Table 1 presents our measures on the SAT 2017, SAT 2018, and SAT 2019 competition benchmarks.[4] The shaded cells indicate which solver has the best results for a given benchmark. It shows that the strategy based on an LBD $\leq 4$ is not as efficient as we could expect. In particular, we note a large instability depending on the sets of instances to be treated.

---

[3] A framework to implement parallel solvers, thus allowing a fair comparison between strategies.

[4] In Table 1, PAR-$k$ is the penalized average runtime, counting each timeout as $k$ times the running time cutoff. The used value for $k$ in the yearly SAT competition is 2.

**Table 1.** Performances comparison for several values of LBD (3, 4, and 5). `P-MCOMSPS` is used as a reference.

| Solvers | 2017 | | 2018 | | 2019 | |
|---|---|---|---|---|---|---|
| | PAR-2 | # solved instances | PAR-2 | # solved instances | PAR-2 | # solved instances |
| P-MCOMSPS | **355h42** | **237** | 430h13 | 258 | **380h03** | **273** |
| P-MCOMSPS-L3 | 361h27 | 234 | 420h53 | 263 | 393h04 | 269 |
| P-MCOMSPS-L4 | 356h44 | **237** | **411h14** | **265** | 391h38 | 269 |
| P-MCOMSPS-L5 | 369h27 | 229 | 415h52 | 264 | 389h27 | 269 |

We believe these results are due to the fact that the LBD is too related to the local state of the solver engines. The intuition we investigate in this paper is that the LBD metrics must be strengthened with more global information.

## 4 Combining LBD and Community for Parallel SAT Solving

As previously stated, we need a metric independent of the local state of a particular solver engine. Structural information about the instance to be solved can be useful. For instance, in a portfolio solver, structural information can be shared among the solvers working on the same formula.

In this paper, we focus on the community structure exhibited by (industrial) SAT instances. The metrics (COM, defined in Sect. 2.3) derived from this structure has been proven to be linked with the LBD in [29]. Besides, it has been used to improve the performances of sequential SAT solving via a preprocessing approach in [2].

This section shows that communities are good candidates to provide the needed global information. To do so, we use the same protocol as the one used in the previous section: *(i)* studying data on sequential SAT solving to exhibit good candidates for the parameter values, and then *(ii)* check its efficiency in a parallel context. We completed the logs extracted from the sequential experiments presented in Sect. 3 by information on communities and studied the whole package as follows.

### 4.1 LBD Versus Communities

First, we studied the relationship between LBD and COM values thanks to two heatmaps (see Fig. 2). The left part (Fig. 2a) shows, for each LBD value, the distribution of COM values. The right part (Fig. 2b) shows, for each COM value, the LBD values distribution. For example, in Fig. 2a, we observe that $\approx 65\%$ of the clauses with LBD = 1 span on one community (COM = 1), $\approx 20\%$ of them span on two communities (COM = 2), etc. From these figures, we can conclude two important statements:

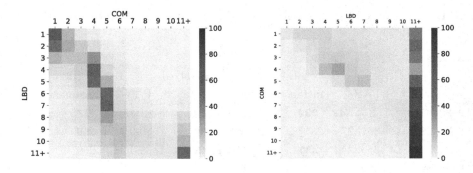

(a) COM distribution for various LBD    (b) LBD distribution for various COM

**Fig. 2.** Heatmap showing the distribution between COM and LBD

- from Fig. 2a, warm zones on the diagonal for low LBD values indicate a sort of correlation between the LBD and COM values (a result that has been already presented in [29]). Looking closer, we observe that a significant part of the clauses does not follow this correlation: the warm zones remain below 65%, and mainly range below 25%. Hence, the COM metrics appears to be a good candidate to refine the clauses already selected using LBD;
- from Fig. 2b, the COM values are almost uniformly distributed all over the LBD values. We conclude that using COM as the only selecting metrics is misleading (we assessed this with several experiments with parallel solvers that are not presented in this paper).

## 4.2 Composing LBD and Communities

As previously stated, COM is a good additional criterion to LBD. We thus need to discover the good couples of values that maximize the usage of shared clauses, while maintaining a reasonable size.

First, we note from Fig. 1 that clauses with LBD $\leq$ 3 represent 1.1% of the total clauses for a percentage of use rising up to 44.2%. Thus, there is no benefit to further filter those clauses. Besides, this first set of clauses is not sufficient since Table 1 reports that the parallel solver for LBD = 3 never wins (this is mainly due to a lack of shared clauses). Therefore, we look for the set of clauses that can be added to LBD $\leq$ 3 to improve performances. Based on the observations made in Sect. 3, we believe the best candidate is a subset of LBD = 4 or LBD = 5. To identify the good couple(s), we reused our previous experimentation protocol and tracked data for both LBD and COM.

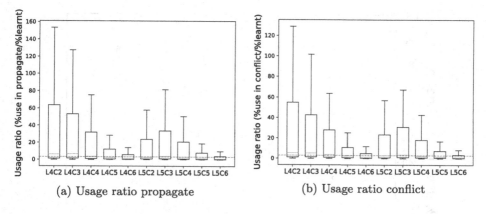

(a) Usage ratio propagate    (b) Usage ratio conflict

**Fig. 3.** Efficiency of LBD and COM combination. The blue dotted line corresponds to the average and the orange lines to the median. (Color figure online)

To capture the usefulness of learnt clauses, we define the *usage ratio* as follows: the ratio of the percentage of use (propagation or conflict analysis) to the percentage of learnt clauses (among those of a given formula). For instance, in a given formula, a clause with a *usage ratio* of 10 is used 10 times more than the average use of all learnt clauses. By extension, the *usage ratio of a set of clauses* is the average of the usage ratio of all its clauses.

The resulting data is displayed as box-plot diagrams in Fig. 3. These diagrams integrate the distribution of the usage ratio for all the formulas in the SAT competition 2018. A box-plot denoted $L\langle x\rangle C\langle y\rangle$ corresponds to the set of clauses with LBD = $x$ and COM = $y$.

Our immediate observation (it confirms our intuition) is that, for a fixed value of LBD, the usage ratio varies heavily considering different COM values. Secondly, we discern that clauses with LBD = 4 have a global better usage ratio than those with LBD = 5. The third, and critical, observation allows us to extract the best promising configurations. Actually, L4C2 and L4C3 configurations have the most impressive usage ratio: in 25% (the $3^{rd}$ quartile) of the treated instances, clauses within these configurations have a usage ratio greater than 50 in the unit propagation (Fig. 3a), 40 in the conflict analysis (Fig. 3b) and extends to very high values (up to 150 in propagation and 130 in conflict analysis). Moreover, the median of propagation usage ratio for L4C2 and L4C3 (6.0 and 6.5, respectively) are twice as big as the mean of the entire LBD = 4 and LBD = 5 boxes (equal to 2.9 and noted with a dashed line in both figures).

### 4.3    Community Based Filtering

From this study, we can conclude that we can use the community structure as a filter for those clauses that have been already selected using an LBD threshold.

Practically, we propose the following strategy: sharing all the clauses with an LBD $\leq$ 3 (without any community limit) as well as those with an LBD $\leq$ 4

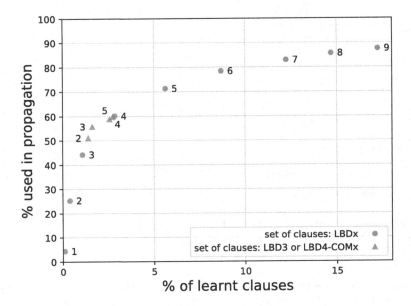

**Fig. 4.** Propagation usage of learnt clause considering LBD (Color figure online)

and a COM $\leq$ 3. Indeed, the former set of clauses is small while being very useful. Whereas, to select the clauses with COM = 3 among the set of clauses with LBD = 4 should allow a higher usage ratio while keeping the sharing at a reasonable ratio.

To verify this assertion, and validating the effectiveness of the chosen filter threshold (COM $\leq$ 3), we extend the study presented in Sect. 3.1. Thus, Fig. 4 and 5 take up the same points of Fig. 1, while separating propagation and conflict analysis results (to increase readability). To those points, the new figure shows, by triangles, the usage ratio for different filter values.

As expected, the point COM = 3 (green triangle) is at the inflection of the curves. Consequently, selecting this set of clauses allows us to reach a better usage ratio, close to the unfiltered LBD $\leq$ 4, while preserving a comparable number of clauses with LBD $\leq$ 3. This convinces us that this metrics should lead to increased performance in parallel SAT solving. The following section verifies these measures in practice.

## 5 Derived Parallel Strategy and Experimental Results

This section first describes parallel SAT solvers we have designed to evaluate our strategy, as well as the associated experimental protocol. It then presents and discusses our experimental results.

## 5.1  Solvers and Evaluation Protocol

As in Sect. 3.2, we use the solver P-MCOMSPS as a reference to validate our proposal. The only difference between the original solver and the newly developed resides in their sharing strategies. These are as follows:

– P-MCOMSPS: the same strategy, based on incremental values for LBD.
– P-MCOMSPS-L4C3: only learnt clauses with an LBD $\leq$ 3 or LBD = 4 and a COM value $\leq$ 3 are shared.

In P-MCOMSPS-L4C3, a special component (called $sp$) is dedicated to compute the community structure (using the Louvain algorithm). Meanwhile, the remaining components execute the CDCL algorithm to solve the formula, and share clauses with an LBD $\leq$ 3. As noted in Sect. 4.2, sharing all these clauses should not alter performances. Once communities have been computed, $sp$ starts to operate the CDCL algorithm (as others), and the initial sharing strategy is augmented by clauses characterized by an LBD = 4 and a COM $\leq$ 3. Preliminary experiments showed that $sp$ does not need more than a minute to finish Louvain for almost all instances of all benchmarks. Therefore, the augmented filter is activated early in the resolution of a formula.

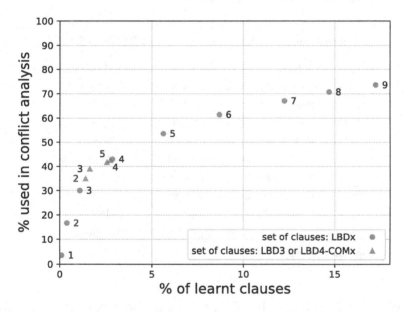

**Fig. 5.** Conflict analysis usage of learnt clause considering LBD (Color figure online)

For the evaluation, we used the main benchmark from SAT competitions 2016[5], 2017[6], 2018[7], and 2019[8]. All solvers were launched on the same machines and with the same configuration than Sect. 3.2. The results we observed are discussed in the next section.

## 5.2   Results and Discussion

Table 2 presents the experimental results on the aforementioned benchmarks. When considering the number of solved instances, we clearly observe that the new sharing strategy outperforms the one used in P-MCOMSPS, on all the SAT competition benchmarks. We add that P-MCOMSPS-L4C3 solves 29 new instances compared to P-MCOMSPS and fails to solve 16 instances that the latter solves.

Coming to the PAR-2 metrics, things seem to be more mitigated. We study the sharing strategy of P-MCOMSPS to find an explanation: P-MCOMSPS starts the resolution with a low LBD threshold and increments this threshold if it deems that the shared clauses throughput is not sufficient. This incremental strategy can help the solver to learn relevant information leading to the resolution of some edge cases. On the contrary, our restrictive sharing strategy can miss those relevant information for these particular cases.

**Table 2.** Evaluation of the performance of P-MCOMSPS-L4C3

|          | Solvers         | PAR-2     | # solved instances |
|----------|-----------------|-----------|--------------------|
| 2019     | P-MCOMSPS-L4C3  | 386h14    | **274**            |
|          | P-MCOMSPS       | **380h03**| 273                |
| 2018     | P-MCOMSPS-L4C3  | **408h01**| **268**            |
|          | P-MCOMSPS       | 430h13    | 258                |
| 2017     | P-MCOMSPS-L4C3  | **352h31**| **238**            |
|          | P-MCOMSPS       | 355h42    | 237                |
| 2016     | P-MCOMSPS-L4C3  | 355h08    | **183**            |
|          | P-MCOMSPS       | **354h39**| 182                |
| **All**  | P-MCOMSPS-L4C3  | 1 501h54  | **963**            |
| **together** | P-MCOMSPS   | 1 520h37  | 950                |

Finally, as our strategy is based on the study made on a sequential solver, we want to verify the evolution of the usage ratio in the parallel context. Let us conduct a new evaluation: using the same protocol as the one developed in the

---

[5] https://baldur.iti.kit.edu/sat-competition-2016/downloads/app16.zip.

[6] https://baldur.iti.kit.edu/sat-competition-2017/benchmarks/Main.zip.

[7] http://sat2018.forsyte.tuwien.ac.at/benchmarks/Main.zip.

[8] http://satcompetition.org/sr2019benchmarks.zip.

sequential setting of Sect. 4.2, we compute the clause usage ratio in propagation and conflict analysis of our parallel solver P-MCOMSPS-L4C3. The resulting number is the sum of all underlying sequential CDCL engines. These logs concern 100 instances (randomly taken) from the benchmark of the SAT competition 2018.

The collected data are presented in the box-plots of Fig. 6 (the left pair shows propagation and the right pair displays conflict analysis). Box-plots noted S-L4C3 represent the usage ratio of the corresponding set of clauses in Maple-COMSPS (the used sequential solver), while those noted P-L4C3 do the same for P-MCOMSPS-L4C3. Note that a clause is imported with its original LBD value. This is the LBD value reported in these figures.

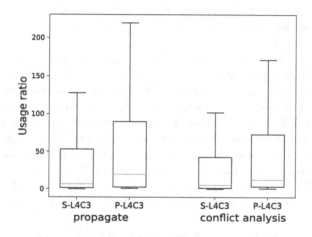

**Fig. 6.** Efficiency of our proposed sharing strategy.

The shared clauses in P-MCOMSPS-L4C3, clearly have a positive impact on the intrinsic behaviour of the underlying sequential engines. They bring new useful information for both unit propagation and conflict analysis procedures. For example, comparing the usage ratio in unit propagation of clauses in S-L4C3 and P-L4C3, we see that the ratio of these clauses goes beyond 50 in only $\approx 25\%$ of the problems for S-L4C3, reaching an upper bound of 130. In P-L4C3, this ratio goes beyond 90 in $\approx 25\%$ of the problems and reaches an upper bound slightly greater than 200. The same observation holds for the conflict analysis procedure. Besides, the medians for box-plots of the parallel approach are all higher than the corresponding medians of the sequential ones.

## 6 Related Works

The notable community structure of industrial SAT formulas has been identified in [1]. Newsham et al. think that such a structure is one main reason for the noticeable performances of SAT solvers on industrial problems [29].

Thus, several works exploit communities to improve solver performances. It has been used to split formulas to divide the work. Ansótegui et al. developed a pre-processor that solves community-based sub-formulas [2]. The aim is to collect useful information used to solve the whole formula afterwards. Martins et al. show that splitting the formula using communities helps for solving Max-SAT problem in parallel [25]. Community structure has also been used to diversify the decision order of workers in the context of parallel SAT portfolios [31].

It is also worth noting that another concept from the graph theory has been successfully used within the SAT context, namely, the centrality. Katsirelos et al. exhibited that variables selected for branching based on VSIDS are likely to have high eigenvector centrality [18]. The *betweenness centrality* has been incorporated successfully to CDCL: for branching, by using special bonus factors while bumping VSIDS of highly central variables [16]; and for cleaning learnt clauses database, by giving more chance to central clauses (the one with more central variables) [17].

Besides, multiple works present metrics to improve clause sharing for parallel SAT solving. `Penelope` [3] implements the progress saving based quality measure (*psm*). The *psm* of a clause is the size of the intersection between the clause and the phase saving of the solvers. The greater is the psm the more likely the clause will be satisfied. While receiving learnt clauses, a worker can decide to keep them or not. The drawback is that clauses are exchanged and then filtered which can induce some overhead, and an a priori criteria such as LBD is often used as a balance. In `Syrup` parallel solver [6], when a worker learns a clause, it waits for the clause to be used at least once before sending it to the others. The idea behind this is to send only clauses that seem to be useful because already used locally.

While most of the approaches focus on limiting the number of exchanged clauses, Lazaar et al. proposed to select workers allowed to communicate together [20]. This selection is formalized as a multi-armed bandit problem and several metrics are explored as gain functions: size, LBD, activity.

## 7    Conclusion and Future Works

Most of parallel SAT solvers use local quality metrics (the most relevant being LBD) to select learnt clauses that should be shared. In this paper, we proposed to combine this metric with a more global quality measure (COM) based on the community structure of the input SAT formulas. The guiding principle is to use the community criterion as a filter for set of clauses selected by LBD, in other to increase the usage ratio of shared clauses.

We have designed a tool to track and report learnt clauses' characteristics in a sequential context. As a result of this analysis, we derived a learnt clause sharing policy, which combines LBD and COM, in a parallel context. We attested this strategy by implementing it in `P-MCOMSPS`, which outperforms the competing solver on benchmarks from the SAT competition in 2016, 2017, 2018, and 2019.

We have in mind different ways to improve this work. First, we can look for a more dynamic approach in our filtering method. This would allow to address the problem we mentioned in the analysis of the result in Sect. 5.2. We noted that some instances benefit from an "unlimited" sharing of clauses: we believe that we could use a delta around some threshold value for the LBD metrics, while maintaining our threshold for the COM value. We could study the shared clauses throughput required for different types of instances as well as the throughput allowed by different types of hardware to increase or decrease the LBD accordingly. It is worth noting that Hamadi et al. developed a similar idea but using the size as a metric [14].

Second, we would like to study the effect of using communities as a metric for garbage collection. This latter is one of the main components of a SAT solver, as the solver can learn millions of clauses while exploring solutions: keeping all these clauses slows down the propagation and leads to memory problems. This is amplified in the parallel context where a sequential solver has its own set of learnt clauses enriched by other solvers too. In P-MCOMSPS, clauses are deleted depending on their LBD values. It makes sense to use a local metrics for the logic of a local component. However, the encouraging results shown in this paper let us think that extending the use of communities to garbage collection would improve furthermore the sequential engines.

# References

1. Ansótegui, C., Giráldez-Cru, J., Levy, J.: The community structure of SAT formulas. In: Cimatti, A., Sebastiani, R. (eds.) SAT 2012. LNCS, vol. 7317, pp. 410–423. Springer, Heidelberg (2012). https://doi.org/10.1007/978-3-642-31612-8_31
2. Ansótegui, C., Giráldez-Cru, J., Levy, J., Simon, L.: Using community structure to detect relevant learnt clauses. In: Heule, M., Weaver, S. (eds.) SAT 2015. LNCS, vol. 9340, pp. 238–254. Springer, Cham (2015). https://doi.org/10.1007/978-3-319-24318-4_18
3. Audemard, G., Hoessen, B., Jabbour, S., Lagniez, J.-M., Piette, C.: Revisiting clause exchange in parallel SAT solving. In: Cimatti, A., Sebastiani, R. (eds.) SAT 2012. LNCS, vol. 7317, pp. 200–213. Springer, Heidelberg (2012). https://doi.org/10.1007/978-3-642-31612-8_16
4. Audemard, G., Simon, L.: Predicting learnt clauses quality in modern SAT solvers. In: Proceedings of the 21st International Joint Conferences on Artifical Intelligence (IJCAI), pp. 399–404. AAAI Press (2009)
5. Audemard, G., Simon, L.: Glucose in the SAT 2014 competition. In: Proceedings of SAT Competition 2014: Solver and Benchmark Descriptions, p. 31. Department of Computer Science, University of Helsinki, Finland (2014)
6. Audemard, G., Simon, L.: Lazy clause exchange policy for parallel SAT solvers. In: Sinz, C., Egly, U. (eds.) SAT 2014. LNCS, vol. 8561, pp. 197–205. Springer, Cham (2014). https://doi.org/10.1007/978-3-319-09284-3_15
7. Balyo, T., Sinz, C.: Parallel satisfiability. Handbook of Parallel Constraint Reasoning, pp. 3–29. Springer, Cham (2018). https://doi.org/10.1007/978-3-319-63516-3_1
8. Biere, A.: Splatz, lingeling, plingeling, treengeling, yalsat entering the SAT competition 2016. In: Proceedings of SAT Competition 2016: Solver and Benchmark Descriptions, p. 44. Department of Computer Science, University of Helsinki, Finland (2016)

9. Biere, A., Heule, M., van Maaren, H.: Handbook of Satisfiability, vol. 185. IOS press (2009)
10. Blondel, V.D., Guillaume, J.L., Lambiotte, R., Lefebvre, E.: Fast unfolding of communities in large networks. J. Stat. Mech: Theory Exp. **2008**(10), P10008 (2008)
11. Brandes, U., et al.: On modularity clustering. IEEE Trans. Knowl. Data Eng. **20**(2), 172–188 (2007)
12. Cook, S.A.: The complexity of theorem-proving procedures. In: Proceedings of the 3rd ACM Symposium on Theory of Computing (STOC), pp. 151–158. ACM (1971)
13. Eén, N., Biere, A.: Effective preprocessing in SAT through variable and clause elimination. In: Bacchus, F., Walsh, T. (eds.) SAT 2005. LNCS, vol. 3569, pp. 61–75. Springer, Heidelberg (2005). https://doi.org/10.1007/11499107_5
14. Hamadi, Y., Jabbour, S., Sais, J.: Control-based clause sharing in parallel SAT solving. In: Hamadi, Y., Monfroy, E., Saubion, F. (eds.) Autonomous Search, pp. 245–267. Springer, Heidelberg (2011). https://doi.org/10.1007/978-3-642-21434-9_10
15. Hamadi, Y., Jabbour, S., Sais, L.: ManySAT: a parallel SAT solver. J. Satisf. Boolean Model. Comput. **6**(4), 245–262 (2009)
16. Jamali, S., Mitchell, D.: Centrality-based improvements to CDCL heuristics. In: Beyersdorff, O., Wintersteiger, C.M. (eds.) SAT 2018. LNCS, vol. 10929, pp. 122–131. Springer, Cham (2018). https://doi.org/10.1007/978-3-319-94144-8_8
17. Jamali, S., Mitchell, D.: Simplifying CDCL clause database reduction. In: Janota, M., Lynce, I. (eds.) SAT 2019. LNCS, vol. 11628, pp. 183–192. Springer, Cham (2019). https://doi.org/10.1007/978-3-030-24258-9_12
18. Katsirelos, G., Simon, L.: Eigenvector centrality in industrial SAT instances. In: Milano, M. (ed.) CP 2012. LNCS, pp. 348–356. Springer, Heidelberg (2012). https://doi.org/10.1007/978-3-642-33558-7_27
19. Kautz, H.A., Selman, B., et al.: Planning as satisfiability. In: Proceedings of the 10th European Conference on Artificial Intelligence (ECAI), vol. 92, pp. 359–363 (1992)
20. Lazaar, N., Hamadi, Y., Jabbour, S., Sebag, M.: Cooperation control in parallel SAT solving: a multi-armed bandit approach. Technical Report RR-8070, INRIA (2012)
21. Le Frioux, L., Baarir, S., Sopena, J., Kordon, F.: PaInleSS: a framework for parallel SAT solving. In: Gaspers, S., Walsh, T. (eds.) SAT 2017. LNCS, vol. 10491, pp. 233–250. Springer, Cham (2017). https://doi.org/10.1007/978-3-319-66263-3_15
22. Le Frioux, L., Metin, H., Baarir, S., Colange, M., Sopena, J., Kordon, F.: Painless-mcomsps and painless-mcomsps-sym. In: Proceedings of SAT Competition 2018: Solver and Benchmark Descriptions, pp. 33–34. Department of Computer Science, University of Helsinki, Finland (2018)
23. Liang, J.H., Oh, C., Ganesh, V., Czarnecki, K., Poupart, P.: MapleCOMSPS, mapleCOMSPS LRB, mapleCOMSPS CHB. In: Proceedings of SAT Competition 2016: Solver and Benchmark Descriptions, p. 52. Department of Computer Science, University of Helsinki, Finland (2016)
24. Marques-Silva, J.P., Sakallah, K.: GRASP: a search algorithm for propositional satisfiability. IEEE Trans. Comput. **48**(5), 506–521 (1999)
25. Martins, R., Manquinho, V., Lynce, I.: Community-based partitioning for MaxSAT solving. In: Järvisalo, M., Van Gelder, A. (eds.) SAT 2013. LNCS, vol. 7962, pp. 182–191. Springer, Heidelberg (2013). https://doi.org/10.1007/978-3-642-39071-5_14

26. Massacci, F., Marraro, L.: Logical cryptanalysis as a SAT problem. J. Automated Reasoning **24**(1), 165–203 (2000)
27. Moskewicz, M.W., Madigan, C.F., Zhao, Y., Zhang, L., Malik, S.: Chaff: engineering an efficient SAT solver. In: Proceedings of the 38th Design Automation Conference (DAC), pp. 530–535. ACM (2001)
28. Newman, M.E., Girvan, M.: Finding and evaluating community structure in networks. Phys. Rev. E **69**(2), 026113 (2004)
29. Newsham, Z., Ganesh, V., Fischmeister, S., Audemard, G., Simon, L.: Impact of community structure on SAT solver performance. In: Sinz, C., Egly, U. (eds.) SAT 2014. LNCS, vol. 8561, pp. 252–268. Springer, Cham (2014). https://doi.org/10.1007/978-3-319-09284-3_20
30. Pipatsrisawat, K., Darwiche, A.: A lightweight component caching scheme for satisfiability solvers. In: Marques-Silva, J., Sakallah, K.A. (eds.) SAT 2007. LNCS, vol. 4501, pp. 294–299. Springer, Heidelberg (2007). https://doi.org/10.1007/978-3-540-72788-0_28
31. Sonobe, T., Kondoh, S., Inaba, M.: Community branching for parallel portfolio SAT solvers. In: Sinz, C., Egly, U. (eds.) SAT 2014. LNCS, vol. 8561, pp. 188–196. Springer, Cham (2014). https://doi.org/10.1007/978-3-319-09284-3_14
32. Zhang, H., Bonacina, M.P., Hsiang, J.: PSATO: a distributed propositional prover and its application to quasigroup problems. J. Symb. Comput. **21**(4), 543–560 (1996)

# Clause Size Reduction with all-UIP Learning

Nick Feng[✉] and Fahiem Bacchus[✉]

Department of Computer Science, University of Toronto, Toronto, Canada
{fengnick,fbacchus}@cs.toronto.edu

**Abstract.** Almost all CDCL SAT solvers use the 1-UIP clause learning scheme for learning new clauses from conflicts, and our current understanding of SAT solving provides good reasons for using that scheme. In particular, the 1-UIP scheme yields asserting clauses, and these asserting clauses have minimum LBD among all possible asserting clauses. As a result of these advantages, other clause learning schemes, like $i$-UIP and all-UIP, that were proposed in early work are not used in modern solvers. In this paper, we propose a new technique for exploiting the all-UIP clause learning scheme. Our technique is to employ all-UIP learning under the constraint that the learnt clause's LBD does not increase (over the minimum established by the 1-UIP clause). Our method can learn clauses that are significantly smaller than the 1-UIP clause while preserving the minimum LBD. Unlike previous clause minimization methods, our technique is not limited to learning a sub-clause of the 1-UIP clause. We show empirically that our method can improve the performance of state of the art solvers.

## 1  Introduction

Clause learning is an essential technique in SAT solvers. There is good evidence to indicate that it is, in fact, the most important technique used in modern SAT solvers [6]. In early SAT research a number of different clause learning techniques were proposed [5,19,20,25]. However, following the revolutionary performance improvements achieved by the Chaff SAT solver, the field has converged on using the 1-UIP (first Unique Implication Point) scheme [25] employed in Chaff [13] (as well as other techniques pioneered in the Chaff solver).[1] Since then almost all SAT solvers have employed the 1-UIP clause learning scheme, along with clause minimization [21], as their primary method for learning new clauses.

However, other clause learning schemes can be used in SAT solvers without changes to the main data structures. Furthermore, advances in our understanding allow us to better understand the potential advantages and disadvantages of these alternate schemes. In this paper we reexamine these previously proposed schemes with a focus on the schemes described in [25]. Improved understanding

---

[1] The idea of UIP clauses was first mentioned in [19], and 1-UIP clauses along with other UIP clauses were learnt and used in the earlier GRASP SAT solver.

© Springer Nature Switzerland AG 2020
L. Pulina and M. Seidl (Eds.): SAT 2020, LNCS 12178, pp. 28–45, 2020.
https://doi.org/10.1007/978-3-030-51825-7_3

of SAT solvers, obtained from the last decade of research, allows us to see that in their original form these alternative clause learning schemes suffer significant disadvantages over 1-UIP clause learning.

One of the previously proposed schemes was the all-UIP scheme [25]. In this paper we propose a new way to exploit the main ideas of this scheme that avoids its main disadvantage which is that it can learn clauses with higher LBD scores. In particular, we propose to use a all-UIP like clause learning scheme to generate smaller learnt clauses which retain the good properties of standard 1-UIP clauses. Our method is related to, but not the same as, various clause minimization methods that try to remove redundant literals from the 1-UIP clause yielding a clause that is a subset of the 1-UIP clause, e.g., [10,21,24]. Our method is orthogonal to clause minimization. In particular, our approach can learn a clause that is not a subset of the 1-UIP clause but which still serves all of the same purposes as the 1-UIP clause. Clause minimization techniques can be applied on top of our method to remove redundant literals.

We present various versions of our method and show that these variants are often capable of learning shorter clauses than the 1-UIP scheme, and that this can lead to useful performance gains in state of the art SAT solvers.

## 2   Clause Learning Framework

We first provide some background and a framework for understanding clause learning as typically used in CDCL SAT solvers. A propositional formula $F$ expressed in Conjunctive Normal Form (CNF) contains a set of variables $V$. A literal is a variable $v \in V$ or its negation $\neg v$. For a literal $\ell$ we let var$(\ell)$ denote its underlying variable. A CNF consists of a conjunction of clauses, each of which is a disjunction of literals. We often view a clause as being a set of literals and employ set notation, e.g., $\ell \in C$ and $C' \subset C$.

Two clauses $C_1$ and $C_2$ can be *resolved* when they contain *conflicting* literals $\ell \in C_1$ and $\neg \ell \in C_2$. Their resolvent $C_1 \bowtie C_2$ is the new clause $(C_1 \cup C_2) - \{\ell, \neg \ell\}$. The resolvent will be a tautology (i.e., a clause containing a literal $x$ and its negation $\neg x$) if $C_1$ and $C_2$ contain more than one pair of conflicting literals.

We assume the reader is familiar with the operations of CDCL SAT solvers, and the main data structures used in such solvers. A good source for this background is [18].

*The Trail.* CDCL SAT solvers maintain a **trail**, $\mathcal{T}$, which is a *non-contradictory, non-redundant sequence of literals* that have been assigned TRUE by the solver; i.e. $\ell \in \mathcal{T} \rightarrow \neg \ell \notin \mathcal{T}$, and $\mathcal{T}$ contains no duplicates. Newly assigned literals are added to the end of the trail, and on backtrack literals are removed from the end of the trail and unassigned. If literal $\ell$ is on the trail let $\iota(\ell)$ denote its index on the trail, i.e, $\mathcal{T}[\iota(\ell)] = \ell$. For convenience, we also let $\iota(\ell) = \iota(\neg \ell) = \iota(\text{var}(\ell))$ even though neither $\neg \ell$ nor var$(\ell)$ are actually on $\mathcal{T}$. If $x$ and $y$ are both on the trail and $\iota(x) < \iota(y)$ we say that $x$ *appears before $y$ on the trail.*

Two types of true literals appear on the trail: *decision literals* that have been assumed to be true by the solver, and *unit propagated literals* that are *forced to* be *true* because they are the sole remaining unfalsified literal of a clause. Each literal $\ell \in \mathcal{T}$ has a decision level $decLvl(\ell)$. Let $k$ be the number of decision literals appearing before $\ell$ on the trail. When $\ell$ is a unit propagated literal $decLvl(\ell) = k$, and when $\ell$ is a decision literal $decLvl(\ell) = k + 1$. For example, $decLvl(d) = 1$ for the first decision literal $d \in \mathcal{T}$, and $decLvl(\ell) = 0$ for all literals $\ell$ appearing before $d$ on the trail. The set of literals on $\mathcal{T}$ that have the same decision level forms a contiguous subsequence of $\mathcal{T}$ that starts with a decision literal $d_i$ and ends just before the next decision literal $d_{i+1}$. If $decLvl(d_i) = i$ we call this subsequence of $\mathcal{T}$ the *i-th decision level*.

Each literal $\ell \in \mathcal{T}$ also has a clausal reason $reason(\ell)$. If $\ell$ is a unit propagated literal, $reason(\ell)$ is a clause of the formula such that $\ell \in reason(\ell)$ and $\forall x \in reason(\ell). x \neq \ell \rightarrow (\neg x \in \mathcal{T} \wedge \iota(\neg x) < \iota(\ell))$. That is, $reason(\ell)$ is a clause that has become unit implying $\ell$ due to the literals on the trail above $\ell$. If $\ell$ is a decision literal then $reason(\ell) = \varnothing$.

In most SAT solvers, clause learning is initiated as soon as a clause is falsified by $\mathcal{T}$. In this paper we will be concerned with the subsequent clause learning process which uses $\mathcal{T}$ to derive a new clause. We will try to make as few assumptions about how $\mathcal{T}$ is managed by the SAT solver as possible. One assumption we will make is that $\mathcal{T}$ *remains intact during clause learning* and is only changed after the new clause is learnt (by backtracking).

Say that $\mathcal{T}$ falsifies a clause $C_I$, and that the last decision literal $d_k$ in $\mathcal{T}$ has decision level $k$. Consider $\mathcal{T}_{k-1}$ the prefix of $\mathcal{T}$ above the last decision level, i.e., the sequence of literals $\mathcal{T}[0]$—$\mathcal{T}[\iota(d_k) - 1]$. We will assume that $\mathcal{T}_{k-1}$ is **unit propagation complete**, although the full trail $\mathcal{T}$ might not be. This means that (a) no clause was falsified by $\mathcal{T}_{k-1}$. And (b) if $C_u$ is a clause containing the literal $x$ and all literals in $C_u$ except for $x$ are falsified by $\mathcal{T}_{k-1}$, then $x \in \mathcal{T}_{k-1}$ and $decLvl(x) \leq \max\{decLvl(y) | y \in C_u \wedge y \neq x\}$. This means that if $x$ appears in a clause made unit it must have been added to the trail, and added at or before decision level the clause became unit. Note that more than one clause might be made unit by $\mathcal{T}$ forcing $x$, or $x$ might be set as a decision before being forced. This condition ensures that $x$ appears in $\mathcal{T}$ at or before the first decision level it is forced by any clause.

Any clause falsified by $\mathcal{T}$ is called a **conflict**. When a conflict is found, the final level of the trail, $k$, need not be unit propagation complete as the solver typically stops propagation as soon as it finds a conflict. This means that (a) other clauses might be falsified by $\mathcal{T}$ besides the conflict found, and (b) other literals might be unit implied by $\mathcal{T}$ but not added to $\mathcal{T}$.

**Definition 1 (Trail Resolvent).** *A trail resolvent is a clause arising from resolving a conflict against the reason clause of some literal $\ell \in \mathcal{T}$. Every trail resolvent is also a conflict.*

The following things can be noted about trail resolvents: (1) trail resolvents are never tautological, as the polarity of all literals in $reason(\ell)$ other than $\ell$ must agree with the polarity of all literals in the conflict (they are all falsified by $\mathcal{T}$); (2) one polarity of the variable var$(\ell)$ resolved on must be a unit propagated literal whose negation appears in the conflict; and (3) any variable in the conflict that is unit propagated in $\mathcal{T}$ can be resolved upon (the variable must appear in different polarities in the conflict and in $\mathcal{T}$).

**Definition 2 (Trail Resolution).** *A trail resolution is a sequence of trail resolvents applied to an initial conflict $C_I$ yielding a new conflict $C_L$. A trail resolution is* **ordered** *if the sequence of variables $v_1, \ldots, v_m$ resolved have strictly decreasing trail indices: $\iota(v_{i+1}) < \iota(v_i)$ $(1 \le i < m)$. (Note that this implies that no variable is resolved on more than once).*

Ordered trail resolutions resolve unit propagated literals from the end of the trail to the beginning. W.l.o.g we can require that all trail resolutions be ordered.

**Observation 1.** *If the unordered trail resolution $U$ yields the conflict clause $C_L$ from an initial conflict $C_I$, then there exists an ordered trail resolution $O$ that yields a conflict clause $C'_L$ such that $C'_L \subseteq C_L$.*

**Proof.** Let $U$ be the sequence of clauses $C_I = C_0,\ C_1,\ \ldots,\ C_m = C_L$ obtained by resolving on the sequence of variables $v_1,\ \ldots,\ v_m$ whose corresponding literals on $\mathcal{T}$ are $l_1,\ \ldots,\ l_m$. Reordering these resolution steps so that the variables are resolved in order of decreasing trail index and removing duplicates yields an ordered trail resolution $O$ with the desired properties. Since no reason clause contains literals with higher trail indices, $O$ must be a valid trail resolution if $U$ was, and furthermore $O$ yields the clause $C'_L = \bigcup_{i=1}^{m} reason(l_i) - \{l_1, \neg l_1, \ldots, l_m, \neg l_m\}$. Since $U$ resolves on the same variables (in a different order) using the same reason clauses we must have $C'_L \subseteq C_L$. It can, however, be the case that $C'_L$ is proper subset of $C_L$: if $l_i$ is resolved away it might be reintroduced when resolving on $l_{i+1}$ if $\iota(l_{i+1}) > \iota(l_i)$. $\qquad\square$

The relevance of trail resolutions is that all proposed clause learning schemes we are aware of use trail resolutions to produce learnt clauses. Furthermore, the commonly used technique for clause minimization [21] is also equivalent to a trail resolution that yields the minimized clause from the un-minimized clause. Interestingly, it is standard in SAT solver implementations to perform resolution going backwards along the trail. That is, these implementations are typically using ordered trail resolutions. Observation 1 shows that this is correct.

Ordered trail resolutions are a special case of *trivial resolutions* [2]. Trail resolutions are specific to the trail data structure typically used in SAT solvers. If $\mathcal{T}$ falsifies a clause at its last decision level, then its associated implication graph [20] contains a conflict node. Cuts in the implication graph that separate the conflict from the rest of the graph correspond to conflict clauses [2]. It is not difficult to see that the proof Proposition 4 of [2] applies also to trail resolutions. This means that *any conflict clause in the trail's implication graph can be derived using a trail resolution.*

### 2.1   Some Alternate Clause Learning Schemes

A number of different clause learning schemes for generating a new learnt clause from the initial conflict have been presented in prior work, e.g., [5, 19, 20, 25].

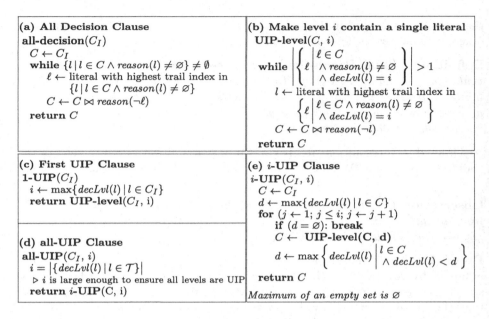

**Fig. 1.** Some different clause learning schemes. All use the current trail $\mathcal{T}$ and take as input an initial clause $C_I$ falsified by $\mathcal{T}$ at its deepest level.

Figure 1 gives a specification of some of these methods: (a) the all-decision scheme which resolves away all implied literals leaving a learnt clause over only decision literals; (c) the 1-UIP scheme which resolves away literals from the deepest decision level leaving a learnt clause with a single literal at the deepest level; (d) the all-UIP scheme which resolves away literals from each decision level leaving a learnt clause with a single literal at each decision level; and (e) the $i$-UIP scheme which resolves away literals from the $i$ deepest decision levels leaving a learnt clause with a single literal at its $i$ deepest decision levels. It should be noted that when resolving away literals at decision level $i$, new literals at decision levels less than $i$ might be introduced into the clause. Hence, it is important in the $i$-UIP and all-UIP schemes to use ordered trail resolutions.

Both the all-decision and all-UIP schemes yield a clause with only one literal at each decision level, and the all-UIP clause will be no larger that the all-decision clause. Furthermore, it is known [20] that once we reduce the number of literals at a decision level $d$ to one, we could continue performing resolutions and later achieve a different single literal at the level $d$. In particular, a decision level might contain more than one unique implication point, and in some contexts the term all-UIP is used to refer to all the unique implication points that exist in a particular decision level [17] rather than the all-UIP clause learning scheme as is used here. The algorithms given in Fig. 1 stop at the first UIP of a level, except for the all-decision schemes with stops at the last UIP of each level.

## 2.2   Asserting Clauses and LBD—Reasons to Prefer 1-UIP Clauses

An **asserting clause** [15] is a conflict clause $C_L$ that has exactly one literal $\ell$ at its deepest level, i.e., $\forall x \in C_L.decLvl(x) \leq decLvl(\ell) \wedge (decLvl(x) = decLvl(\ell) \rightarrow x = \ell)$. All of the clause learning schemes in Fig. 1 produced learnt clauses that are asserting.

The main advantage of asserting clauses is that they are 1-Empowering [15], i.e., they allow unit propagation to derive a new forced literal. Hence, asserting clauses can be used to guide backtracking—the solver can backtrack from the current deepest level to the point the learnt clause first becomes unit, and then use the learnt clause to add a new unit implicant to the trail. Since all but the deepest level was unit propagation complete, this means that the asserting clause must be a brand new clause; otherwise that unit implication would already have been made. On the other hand, if the learnt clause $C_L$ is not asserting then it could be that it is a duplicate of another clause already in the formula.

*Example 1.* Suppose that $a$ is a unit propagated literal and $d$ is a decision literal with $decLvl(d) > decLvl(a)$. Let the sequence of clauses watched by $\neg d$ be $(\neg d, x, \neg a)$, $(\neg d, y, \neg x, \neg a)$, $(\neg d, \neg y, \neg x, \neg a)$, $(\neg d, \neg x, \neg a)$. When $d$ is unit propagated the clauses on $\neg d$'s watch list will be checked in this order.

Hence, unit propagation of $d$ will extend the trail by first adding the unit propagated literal $x$ (with $reason(x) = (x, \neg d, \neg a)$) and then the unit propagated literal $y$ (with $reason(y) = (y, \neg x, \neg a, \neg d)$). Now the third clause on $\neg d$'s watch list, $(\neg d, \neg y, \neg x, \neg a)$ is detected to be a conflict.

Clause learning can now be initiated from conflict $C_I = (\neg d, \neg y, \neg x, \neg a)$. This clause has 3 literals at level $decLvl(d) = 10$. If we stop clause learning before reaching an asserting clause, then it is possible to simply resolve $C_I$ with $reason(y)$ to obtain the learnt clause $C_L = (\neg d, \neg x, \neg a)$. However, this non-asserting learnt clause is a duplicate of the fourth clause on $\neg d$'s watch list which is already in the formula.[2] This issue can arise whenever $C_L$ contains two or more literals at the deepest level (i.e., whenever $C_L$ is not asserting). In such cases $C_L$ might be a clause already in the formula with its two watches not yet fully unit propagated (and thus $C_L$ is not detected by the SAT solver to be a conflict) since propagation is typically stopped as soon as a conflict is detected.

The LBD of the learnt clause $C_L$ is the number of different decision levels in it: $\text{LBD}(C_L) = \left| \{ decLvl(l) \mid l \in C_L \} \right|$ [1]. Empirically LBD is a successful predictor of clause usefulness: clauses with lower LBD tend to be more useful. As noted in [1], from the initial falsified clause $C_I$ the 1-UIP scheme will produce a clause $C_L$ whose LBD is minimum among all asserting clauses that can be learnt from $C_I$. If $C'$ is a trail resolvent of $C$ and a reason clause $reason(l)$, then $\text{LBD}(C') \geq \text{LBD}(C)$ since $reason(l)$ must contain at least one other literal

---

[2] In this example, the fourth clause on $\neg d$'s watch list subsumes the third clause. But it is not difficult to construct more elaborate examples where there are no subsumed clauses and we still obtain learnt clauses that are duplicates of clauses already in the formula.

with the same decision level as $l$ and might contain literals with decision levels not in $C$. That is, the each trail resolution step might increase the LBD of the learnt clause and can never decrease the LBD. Hence, the 1-UIP scheme yields an asserting clause with minimum LBD as it performs the minimum number of trail resolutions required to generate an asserting clause.

The other schemes must perform more trail resolutions. In fact, all of these schemes (all-decision, all-UIP, i-UIP) use trail resolutions in which the 1-UIP clause appears. That is, they all must first generate the 1-UIP clause and then continue with further trail resolution steps. These extra resolution steps can introduce many addition decision levels into the final clause. Hence, these schemes learn clauses with LBD at least as large as the 1-UIP clauses.

Putting these two observations together we see that the 1-UIP scheme produces asserting clauses with lowest possible LBD. This is a compelling reasons for using this scheme. Hence, it is not surprising that modern SAT solvers almost exclusively use 1-UIP clause learning.[3]

## 3    Using all-UIP Clause Learning

Although learning clauses with low LBD has been shown empirically to be more important in SAT solving than learning short clauses [1], clause size is still important. Smaller clauses consume less memory and help to decrease the size of future learnt clauses. They are also semantically stronger than longer clauses.

The all-UIP scheme will tend to produce small clauses since the clauses contain at most one literal per decision level. However, the all-UIP clause can have much higher LBD. Since LBD is more important than size, our approach is to use all-UIP learning when, and only when, it succeeds in reducing the size of the clause *without increasing its LBD*. The all-UIP scheme first computes the 1-UIP clause when it reduces the deepest level to a single UIP literal. It then proceeds to reduce the shallower levels (see all-UIP's for loop in Fig. 1). So our approach will start with the 1-UIP clause and then try to apply all-UIP learning to reduce other levels to single literals. As noted above, clause minimization is orthogonal to our approach, so we also first apply standard clause minimization [21] to the 1-UIP clause. That is, our algorithm *stable-alluip* (Algorithm 1), starts with the clause that most SAT solvers learn from a conflict, a minimized 1-UIP clause.

Algorithm 1 tries to compute a clause shorter than the inputted 1-UIP clause $C_1$. If a clause shorter than $C_1$ cannot be computed the routine returns $C_1$ unchanged. Line 2 uses the parameter $t_{gap}$ to predict if Algorithm 1 will be successful in producing a shorter clause. This predication is described below. If the prediction is negative $C_1$ is immediately returned and Algorithm 1 is not attempted. Otherwise, a copy of $C_1$ is made in $C_i$ and $n_{tries}$, which counts the number of times Algorithm 1 is attempted, is incremented.

---

[3] Knuth in his sat13 CDCL solver [7] uses an all-decision clause when the 1-UIP clause is too large. In this context an all-UIP clause could also be used as it would be no larger than the all decision clause.

---

**Algorithm 1.** *stable-alluip*

---

**Require:** $C_1$ is minimized 1-UIP clause
**Require:** *config* a set of configuration parameters to give different versions *stable-alluip*.
**Require:** $t_{gap} \geq 0$ is a global parameter, $n_{tries}$ and $n_{succ}$ are used to dynamically adjust $t_{gap}$

```
 1:  stable-alluip(C_1, T)
 2:      if (|C_1| - LBD(C_1) < t_gap) return C_1
 3:      n_tries++
 4:      C_i ← C_1
 5:      decLvls ← decision levels in C_1 in descending order        ▷ These never change
 6:      for (i = 1; i < |decLvls|; i++)              ▷ skip the deepest level decLvls[0]
 7:          C_i ← try-uip-level (C_i, decLvls[i])       ▷ Try to reduce this level to UIP
 8:          if |{ℓ | ℓ ∈ C_i ∧ decLvl(ℓ) ≥ decLvls[i]}| + (|decLvls| - (i + 1)) ≥ |C_1|
 9:              return C_1                              ▷ can't generate smaller clause
10:      if pure-alluip ∈ config
11:          C_i ← minimize(C_i)
12:      if (|C_i| < |C_1| ∧ alluip-active ∈ config → (AvgVarAct(C_i) > AvgVarAct(C_1)))
13:          n_succ++, return C_i            ▷ C_i is smaller than the input clause
14:      else
15:          return C_1

16:  try-uip-level(C_i, i)                    ▷ Do not add new decision levels
17:      C_try = C_i
18:      L_i = {ℓ | ℓ ∈ C_try ∧ decLvl(l) = i}
19:      while |L_i| > 1
20:          p ← remove lit with the highest trail index from L_i
21:          if (∃q ∈ reason(¬p). decLvl(q) ∉ decLvls)  ▷ Would add new decision levels
22:              if (pure-alluip ∈ config)
23:                  return C_i                          ▷ Abort, can't UIP this level
24:              else if (min-alluip ∈ config)
25:                  continue                    ▷ Don't try to resolve away p
26:          else
27:              C_try ← C_try ⋈ reason(¬p)
28:              L_i = L_i ∪ {ℓ | ℓ ∈ reason(¬p) ∧ ℓ ≠ ¬p ∧ decLvl(ℓ) = i}
29:      return C_try
```

---

Then the decision levels of $C_1$ are computed and stored in *decLvls* in order from largest to lowest. The for loop of lines 6–9 is then executed for each decision level $decLvls[i]$. In the loop the subroutine *try-uip-level* tries to reduce the set of literals at $decLvls[i]$ down to a single UIP literal using a sequence of trail resolutions. Since $C_1$ is a 1-UIP clause $decLvls[0]$ (the deepest level) already contains only one literal, so we can start at $i = 1$.

After the call to *try-uip-level* a check (line 8) is made to see if we can abort further processing. At this point the algorithm has finished processing levels $decLvls[0]$–$decLvls[i]$ so the literals at those levels will not change. Furthermore, we know that the best that can be done from this point on is to reduce the remaining $|decLvls| - (i + 1)$ levels down to a single literal each. Hence, adding these two numbers gives a lower bound on the size of the final computed clause.

If that lower bound is as large as the size of the initial 1-UIP clause we can terminate and return the initial 1-UIP clause.

After processing all decision levels, if *try-uip-level* is using the *pure-alluip* configuration, additional reduction in the clause size might be achieved by an another round of clause minimization (line 11). Finally, if the newly computed clause $C_i$ is smaller that the input clause $C_1$ it is returned. Otherwise the original clause $C_1$ is returned. Additionally, if the configuration *alluip-active*, described in Sect. 3.1, is being used, then we also require that the average activity level of the new clause $C_i$ be larger than $C_1$ before we can return the new clause $C_i$.

*try-uip-level* ($C_i$, $i$) attempts to resolve away the literals at decision level $i$ in the clause $C_i$, i.e., those in the set $L_i$ (line 18), in order of decreasing trail index, until only one literal at level $i$ remains. If the resolution step will not introduce any new decision levels (line 26), it is performed updating $C_{try}$. In addition, all new literals added to $C_{try}$ at level $i$ are added to $L_i$.

On the other hand, if the resolution step would introduce new decision levels (line 21) then there are two options. The first option we call *pure-alluip*. With *pure-alluip* we abort our attempt to UIP this level and return the clause with level $i$ unchanged. In the second option, called *min-alluip*, we continue without performing the resolution, keeping the current literal $p$ in $C_{try}$. *min-alluip* then continues to try to resolve away the other literals in $L_i$ (note that $p$ is no longer in $L_i$) until $L_i$ is reduced to a single literal. Hence, *min-alluip* can leave multiple literals at level $i$—all of those with reasons containing new levels along with one other.[4] Observe that the number of literals at level $i$ can not be increased after processing it with *pure-alluip*. *min-alluip* can, however, potentially increase the number of literals at level $i$. In resolving away a literal $l$ at level $i$, more literals might be introduced into level $i$, and some of these might not be removable by *min-alluip* if their reasons contain new levels. However, both *pure-alluip* and *min-alluip* can increase the number of literals at levels less that $i$ as new literals can be introduced into those levels when the literals at level $i$ are resolved away. These added literals at the lower levels might not be removable from the clause, and thus both methods can yield a longer clause than the input 1-UIP clause.

After trying to UIP each level the clause $C_i$ is obtained. If we were using *pure-alluip* we can once again apply recursive clause minimization (line 11) [21], but this would be useless when using *min-alluip* as all but one literal of each level introduces a new level and thus cannot be recursively removed.[5]

$t_{gap}$: *stable-alluip* can produce significantly smaller clauses. However, when it does not yield a smaller clause, the cost of the additional resolution steps can hurt the solver's performance. Since resolution cannot reduce a clause's LBD, the maximum size reduction obtainable from *stable-alluip* is the difference between the 1-UIP clause's size and its LBD: $\text{gap}(C_1) = |C_1| - \text{LBD}(C_1)$. When $\text{gap}(C_1)$

---

[4] Since the sole remaining literal $u \in L_i$ is at a lower trail index than all of the other literals there is no point in trying to resolve away $u$—either it will be the decision literal for level $i$ having no reason, or its reason will contain at least one other literal at level $i$.

[5] Other more powerful minimization techniques could still be applied.

is small, applying *stable-alluip* is unlikely to be cost effective. Our approach is to dynamically set a threshold on gap($C_1$), $t_{gap}$, such that when gap($C_1$) < $t_{gap}$ we do not attempt to reduce the clause (line 2). Initially, $t_{gap} = 0$, and we count the number of times *stable-alluip* is attempted ($n_{tries}$) and the number of times it successfully yields a shorter clause ($n_{succ}$) (line 3 and 13). On every restart if the success rate since the last restart is greater than 80% (less than 80%), we decrease (increase) $t_{gap}$ by one not allowing it to become negative.

*Example 2.* Consider the trail $\mathcal{T} = \ldots, \ell_1, a_2, b_2, c_2, d_2, \ldots, \ldots, e_5, f_5, g_5, h_6,$ $i_6, j_6, k_6, \ldots, m_{10}, \ldots$ where the subscript indicates the decision level of each literal and the literals are in order of increasing trail index.

| $C_a = \varnothing$ | $C_b = (b_2, \neg \ell_3, \neg a_2)$ | $C_c = (c_2, \neg a_2, \neg b_2)$ |
|---|---|---|
| $C_d = (d_2, \neg b_2, \neg c_2)$ | $C_\ell = \varnothing$ | $C_e = \varnothing$ |
| $C_f = (f_5, \neg e_5, \neg \ell_1)$ | $C_g = (g_5, \neg a_2, \neg f_5)$ | $C_h = \varnothing$ |
| $C_i = (i_6, \neg e_5, \neg h_6)$ | $C_j = (j_6, \neg f_5, \neg i_6)$ | $C_k = (k_6, \neg f_5, \neg j_6)$ |

Let the clauses $C_x$, show above, denote the reason clause for literal $x_i$. Suppose 1-UIP learning yields the clause $C_1 = (\neg m_{10}, \neg k_6, \neg j_6, \neg i_6, \neg h_6, \neg g_5, \neg d_2, \neg c_2)$ where $\neg m_{10}$ is the UIP from the conflicting level. *stable-alluip* first tries to find the UIP for level 6 by resolving $C_1$ with $C_k$, $C_j$ and then $C_i$ producing the clause $C^* = (\neg m_{10}, \neg h_6, \neg g_5, \neg f_5, \neg e_5, \neg d_2, \neg c_2)$ where $\neg h_6$ is the UIP for level 6.

*stable-alluip* then attempts to find the UIP for level 5 by resolving $C^*$ with $C_g$ and then $C_f$. However, resolving with $C_f$ would introduce $\ell_1$ and a new decision level into $C^*$. *pure-alluip* thus leaves level 5 unchanged. *min-alluip*, on the other hand, skips the resolution with $C_f$ leaving $f_5$ in $C^*$. Besides $f_5$ only one other literal at level 5 remains in the clause, $e_5$, so *min-alluip* does not do any further resolutions at this level. Hence, *pure-alluip* yields $C^*$ unchanged, while *min-alluip* yields $C^*_{min} = (\neg m_{10}, \neg h_6, \neg f_5, \neg e_5, \neg d_2, \neg c_2, \neg a_2)$.

Finally, *stable-alluip* processes level 2. Resolving away $d_2$ and then $c_2$ will lead to an attempt to resolve away $b_2$. But again this would introduce a new decision level with the literal $\ell_1$. So *pure-alluip* will leave level 2 unchanged and *min-alluip* will leave $b_2$ unresolved. The final clauses produced by *pure-alluip* would be $(\neg m_{10}, \neg h_6, \neg f_5, \neg e_5, \neg d_2, \neg c_2, \neg a_2)$, a reduction of 1 over the 1-UIP clause, and by *min-alluip* would be $(\neg m_{10}, \neg h_6, \neg f_5, \neg e_5, \neg b_2, \neg a_2)$, a reduction of 2 over the 1-UIP clause.  □

## 3.1   Variants of *stable-alluip*

We also developed and experimented with a few variants of the *stable-alluip* algorithm which we describe below.

***alluip-active*: Clauses with Active Variables.** *stable-alluip* learning might introduce literals with low variable activity into $C_i$. Low activity variables are variables that have had low recent participation in clause learning. Hence, clauses with variables of low activity might not be as currently useful to the solver. Our variant *alluip-active* (line 12) in Algorithm 1) computes the average variable activity of the newly produced all-UIP clause $C_i$ and the original 1-UIP clause

$C_1$. The new clause $C_i$ will be returned only if it is both smaller and has higher average variable activity than the original 1-UIP clause. There are, of course, generalizations of this approach where one has a weighted trade-off between these factors that allows preferring the new clause when it has large gains in one metric even though it has small losses in the other. We did not, however, experiment with such generalizations.

**Adjust Variable Activity.** An alternative to filtering clauses with low average variable (*alluip-active*) is to alter the way variable activities are updated to account for our new clause learning method. The popular branching heuristics VSIDS [13] and LBR [8] bump the variable activity for all literals appearing in the learnt clause $C_L$ and all literals resolved away during the conflict analysis that yielded $C_L$ from the initially detected conflict $C_I$ (all literals on the conflict side).

We did not apply this approach to the *stable-alluip* clause, as we did not want to bump the activity of the literals above the deepest decision level that *stable-alluip* resolves away. Intuitively, these literals did not directly contribute to generating the conflict. Instead, we tried two modifications to the variable activity bumping schemes.

Let $C_1$ be the 1-UIP learnt clause and $C_i$ be the *stable-alluip* learnt clause. First, we kept all of the variable activity bumps normally done by 1-UIP learning.[6] Then, when the *stable-alluip* scheme was successful, i.e., $C_i$ was to be used as the new learnt clause, we perform further updates to the variable activities. In the *alluip-inclusive* approach all variables variables appearing in $C_i$ that are not in $C_1$ have their activities bumped. Intuitively, since the clause $C_i$ is being added to the clause database we want to increase the activity of all of its variables. On the other hand, in the *alluip-exclusive* approach in addition to bumping the activity of the new variables in $C_i$ we also remove the activity bumps of those variables in $C_1$ that are no longer in $C_i$.

In sum, the two modified variable activity update schemes we experimented with were (1) ***alluip-inclusive*** $\equiv \forall l \in C_i - C_1 . bumpActivity(l)$ and (2) ***alluip-exclusive*** $\equiv \forall l \in C_i - C_1 . bumpActivity(l) \land (\forall l \in C_1 - C_i . unbumpActivity(l))$.

**Chronological Backtracking.** We tested our new clause learning schemes on solvers that utilized Chronological Backtracking [12,14]. When chronological backtracking is used, the literals on the trail might no longer be sorted by decision level. So resolving literals in the conflict by highest trail index first no longer works. However, we can define a new ordering on the literals to replace the trail index ordering. Let $l_1$ and $l_2$ be two literals on the trail $\mathcal{T}$. We say that $l_1 >_{chron} l_2$ if $decLvl(l_1) > decLvl(l_2) \lor (decLvl(l_1) = decLvl(l_2) \land \iota(l_1) > \iota(l_2))$. That is, literals with higher decision level come first, and if that is equal then the literal with higher trail index comes first.

---

[6] So extra techniques used by the underlying solver, like reason side rate and locality [8], were kept intact.

Exploiting the analysis of [12], it can be observed that all clause learning schemes continue to work as long as literals are resolved away from the initial conflict in decreasing $>_{chron}$ order. In our implementation we used a heap (priority queue) to achieve this ordering of the literal resolutions in order to add our new schemes to those solvers using chronological backtracking.

## 4    Implementation and Experiments

We implemented *stable-alluip* learning schemes on *MapleCOMSPS-LRB* [9], the winner of SAT Race 2016 application track. We then evaluated these schemes and compare against the 1-UIP baseline on the full set of benchmarks from SAT RACE 2019 main track which contains 400 instances. We ran our experiments on 2.70 GHz XeonE5-2680 CPUs, allowing 5000 seconds per instance and a maximum of 12 GB memory.

| Solver | # solved (SAT, UNSAT) | PAR-2 | avg. clause Size |
|---|---|---|---|
| 1-UIP | 221 (132, 89) | 5018.89 | 62.6 |
| *pure-alluip* | **228 (135, 93) +7** | 4867.37 | 49.88 |
| *min-alluip* | 226 (135, 91) +5 | 4890.67 | 45.2 |
| *alluip-active* | 226 (135, 91) +5 | **4866.94** | 47.7 |
| *alluip-inclusive* | 225 **(138, 87) +4** | 4958.49 | 52.12 |
| *alluip-exclusive* | 223 (134, 89) +2 | 5015.23 | **43.2** |

**Fig. 2.** Results of *MapleCOMSPS-LRB* with 1-UIP, *pure-alluip*, *min-alluip*, *alluip-active*, *alluip-inclusive*, and *alluip-exclusive* on SAT2019 race main track.

Figure 2 shows each learning scheme's solved instances count, PAR-2 score, and average learnt clause size. We found that the *stable-alluip* learning schemes improved solved instances, PAR-2 scores, and learnt clause size over 1-UIP. More specifically, *pure-alluip* solved the most instances (+7 over 1-UIP) and the most UNSAT instances (+4); *alluip-inclusive* solved the most SAT instances (+6); and *alluip-active* yields the best PAR-2 score (−151 than 1-UIP). In all cases the *stable-alluip* schemes learnt significantly smaller clauses on average.

**Clause Reduction with *stable-alluip*.** To precisely measure *stable-alluip*'s clause reduction power, we compare each instance's learnt clause size from *min-alluip* and *pure-alluip* against 1-UIP. Figure 3 shows the probability density distribution (PDF) of the relative clause size of the *stable-alluip* learning schemes (*min-alluip* in green and *pure-alluip* in red) for each instance. *min-alluip* (*pure-alluip* resp.) produces shorter clauses for 88.5% (77.7%) of instances, and the average relative reduction ratio over 1-UIP is 18.5% (9.6%). Figure 4 compares the average learnt clause size of *min-alluip*, *pure-alluip* and 1-UIP per instance. Both *stable-alluip* schemes generally yield smaller clauses, and the size reduction is more significant for instances with larger 1-UIP clauses.

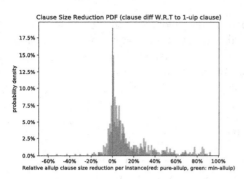

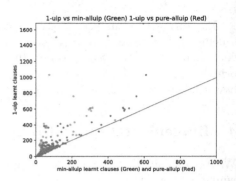

**Fig. 3.** Relative clause size reduction distribution. The $X$ axis indicates the relative size of difference between all-UIP and 1-UIP clauses (calculated as $\dfrac{|C_1| - |C_i|}{|C_1|}$ ) for each instance, and the $Y$ axis shows the probability density. (Color figure online)

**Fig. 4.** Average clause size comparison plot. Each point in the plot represents an instance. The $X$ and $Y$ axes shows the clause length from *stable-alluip* and 1-UIP, respectively. Each green (red) dot represents an compared instance between *MapleCOMSPS-LRB* and Maple-*min-alluip* (*pure-alluip*). (Color figure online)

The results in Fig. 2, 3 and 4 indicate *min-alluip* often achieves higher clause reduction than *pure-alluip*. We also observed that *min-alluip* attempted algorithm 1 more frequently than *pure-alluip* (28.8% vs 16.1%), and is more likely to succeed (59.3% vs 43.4%). This observation agrees with our experiment results.

**Reduced Proof Sizes with *stable-alluip*.** A learning scheme that yields smaller clauses (lemmas) might also construct smaller causal proofs. For 88 UNSAT instances solved mutually by *pure-alluip*, *min-alluip* and 1-UIP schemes, we additionally compared the size of the optimized DRAT proof from the three learning schemes. We used the DRAT-trim tool [23] with a 30000 second timeout to check and optimize every DRAT proof once[7].

The average optimized DRAT proof from *min-alluip* and *pure-alluip* are 556.6MB and 698.5MB, respectively. Both sizes are significantly smaller than the average optimized proof size from 1-UIP, 824.9MB. The average proof size reduction per instance for *min-alluip* and *pure-alluip* is 16.5% and 3.6% against 1-UIP, which roughly correlate with our clause size observation in Fig. 3.

***stable-alluip* in Modern SAT Solvers.** To validate *stable-alluip* in modern SAT solvers, we implemented *stable-alluip* in the winners of 2017, 2018 and 2019 SAT Race [10,16,22] and in the *expMaple-CM-GCBumpOnlyLRB* [11] (*expMaple*) and *CaDiCaL* [4] solvers. *expMaple* is a top ten solver from 2019 SAT race which uses random walk simulation to help branching. We chose *expMaple* because the random walk simulation branching heuristic is different from local branching heuristic (VSIDS and LRB) that we have considered in

---

[7] Applying DRAT-trim multiple times can further reduce the proof size until a fixpoint. However, the full optimization is too time consuming for our experiments.

| Solver | #solved (SAT, UNSAT) $\Delta$ | PAR-2 | avg. clause Size |
|---|---|---|---|
| SAT 2017 Winner *MapleLCMDist* | 232 (135, 97) | 4755.96 | 61.9 |
| *MapleLCMDist*-all-pure | **244 (146, 98)** +12 | **4504.18** | 43.76 |
| *MapleLCMDist*-all-min | 240 (144, 96) +8 | 4601.25 | **36.97** |
| *MapleLCMDist*-all-act | 237 (140, 97) +5 | 4678.434 | 43.62 |
| *MapleLCMDist*-all-inclusive | 234 (137, 97) +2 | 4718.03 | 37.96 |
| SAT 2018 Winner *MapleCB* | 236 (138, 98) | 4671.81 | 61.69 |
| *MapleCB*-all-pure | **241 (142, 99)** +5 | **4598.18** | 44.19 |
| *MapleCB*-all-min | 236 (141, 95) +0 | 4683.92 | 38.05 |
| *MapleCB*-all-act | 240 (141, **99**) +4 | 4626.99 | 41.16 |
| *MapleCB*-all-inclusive | 240 (**142**, 98) +4 | 4602.13 | **37.52** |

**Fig. 5.** Benchmark results of 1-UIP, *pure-alluip. min-alluip, alluip-active* and *alluip-inclusive* on SAT2019 race main track instances.

| SAT 2019 Winner *MapleCB-DL* | 238 (140, **98**) | 4531.24 | 60.91 |
|---|---|---|---|
| *MapleCB-DL*-all-pure | 240 (142, **98**) +2 | 4519.08 | 43.32 |
| *MapleCB-DL*-all-min | 244 (**148**, 96) +6 | **4419.84** | **36.88** |
| *MapleCB-DL*-all-act | 243 (146, 97) + 5 | 4476.73 | 40.65 |
| *MapleCB-DL*-all-inclusive | **243 (148**, 95) +5 | 4455.76 | 37.02 |
| SAT 2019 Competitor *expMaple* | 237 (137, 100) | 4628.96 | 63.19 |
| *expMaple*-all-pure | 235 (136, 99) −2 | 4668.96 | 48.26 |
| *expMaple*-all-min | 241 (143, 98) +4 | 4524.28 | 46.29 |
| *expMaple*-all-act | 244 (143, **101**) +7 | **4460.92** | 47.25 |
| *expMaple*-all-inclusive | **245 (146**, 99) +8 | 4475.76 | **45.33** |
| *CaDiCaL*version 1.2.1 *CaDiCaL-default* | 249 (150, 99) | **4311.76** | 101.96 |
| *CaDiCaL-default*-all-pure | 248 (151, 97) −1 | 4373.38 | 82.44 |
| *CaDiCaL-default*-all-min | 248 (149, 99) −1 | 4398.31 | 43.93 |
| *CaDiCaL-default*-all-act | **252 (152, 100)** +3 | 4331.56 | 47.34 |
| *CaDiCaL-default*-all-inclusive | 251 (**153**, 98) +2 | 4335.88 | **42.61** |

**Fig. 6.** Benchmark results of 1-UIP, *pure-alluip. min-alluip, alluip-active* and *alluip-inclusive* on SAT2019 race main track instances.

*alluip-active, alluip-inclusive,* and *alluip-exclusive.* We chose *CaDiCaL* because its default configuration (*CaDiCaL-default*) solved the most instances in the 2019 SAT Race (244). For this experiment, we used the latest available version of *CaDiCaL-default* instead of the 2019 SAT Race version [3]. We compared these solvers' base 1-UIP learning scheme with *pure-alluip, min-alluip* and the top two *stable-alluip* variants, *alluip-active* and *alluip-inclusive,* on the SAT Race 2019 main track benchmarks. We report solved instances, PAR-2 score and the average clause size.

Figures 5 and 6 show the results of the *stable-alluip* configurations in our suite of modern solvers. Overall, we observed similar performance gain on all modern solvers as we have seen on *MapleCOMSPS-LRB* in Fig. 2. More specifically, almost all configurations improved on solved instance (+3.9 instances in average) and PAR-2 score (−57.7 in average). The average clause size reduction is consistent across all solvers. Each configuration also exhibits different strengths: *pure-alluip* solved the most instances with the best PAR-2 score on two solvers, *min-alluip* yields small clauses, *alluip-inclusive* solved the most SAT instances, and *alluip-active* has stable performance.

On the SAT 2017 race winner *MapleLCMDist*, all four configurations of *stable-alluip* solved more instances than 1-UIP learning. *pure-alluip* solved more UNSAT and SAT instances while the other configurations improved on solving SAT instances. The clause size reduction of *stable-alluip* is more significant on this solver than on *MapleCOMSPS-LRB*. The SAT 2018 race winner *MapleCB* uses chronological backtracking (CB); three out of four configurations outperformed 1-UIP. On the SAT 2019 race winner *MapleCB-DL*, all four *stable-alluip* configurations solved more instances than 1-UIP. *MapleCB-DL* prioritizes clauses that are learned multiple times. We observed that *stable-alluip* clauses are less likely to be duplicated. As an example, *min-alluip* on average, added 12% less duplicated clauses into the core clause database than 1-UIP. This observation is surprising, and the cause is unclear.

On *expMaple*, three out of four *stable-alluip* configurations solved more instances than 1-UIP learning. We noticed that both *alluip-active* and *alluip-inclusive* show better performance than *min-alluip* and *pure-alluip* on this solver. The random walk simulation branching heuristic, however, didn't impact the performance of *stable-alluip* schemes significantly.

*CaDiCaL-default* with 1-UIP solved 249 instances. Applying *alluip-active* and *alluip-inclusive* helped the solver solve 3 and 2 more instances, respectively. The 1-UIP clauses in *CaDiCaL-default* were much larger than other solvers on average (101 vs 60) but the *stable-alluip* configurations yielded similar clause sizes.

## 5   Conclusion

In this paper we introduced a new clause learning scheme, *stable-alluip*, that preserves the strengths 1-UIP learning while learning shorter clauses. We provided empirical evidence that using *stable-alluip* and its variants in modern CDCL solvers achieves significant clause reduction and yields useful performance gains.

Our scheme extends 1-UIP learning by performing further resolution beyond the deepest decision level in an attempt to find the UIP at each level in the learnt clause. Since resolutions may increase the clause's LBD by introducing literals from new decision levels, we presented two methods to block such literals from entering the clause. Although our learning scheme is conceptually simple, and we presented optimizations to reduce and balance the learning cost. We additionally presented variants of our schemes to account for features used in state of the art solvers, e.g., local branching heuristics and chronological backtracking.

Although the field of SAT solving has converged on using the 1-UIP learning scheme, we have shown the possibility of developing an effective alternative through understanding the strengths and weaknesses of 1-UIP and clause learning schemes. Our learning scheme can be generalized and further improved by exploring more fine-grained trade-offs between different clause quality metrics beyond clause size and LBD. We also plan to study the interaction between clause learning and variable branching. Since most of the branching heuristics are tailored for 1-UIP scheme, their interactions with other learning schemes requires further study.

# References

1. Audemard, G., Simon, L.: Predicting learnt clauses quality in modern SAT solvers. In: Boutilier, C. (ed.) IJCAI 2009, Proceedings of the 21st International Joint Conference on Artificial Intelligence, Pasadena, California, USA, 11–17 July 2009, pp. 399–404 (2009). http://ijcai.org/Proceedings/09/Papers/074.pdf
2. Beame, P., Kautz, H.A., Sabharwal, A.: Towards understanding and harnessing the potential of clause learning. J. Artif. Intell. Res. **22**, 319–351 (2004). https://doi.org/10.1613/jair.1410
3. Biere, A.: CADICAL at the SAT race 2019. In: Heule, M.J.H., Järvisalo, M., Suda, M. (eds.) Proceedings of SAT Competition 2019 Solver and Benchmark Descriptions. University of Helsinki (2019). http://hdl.handle.net/10138/306988
4. Biere, A.: Cadical SAT solver (2019). https://github.com/arminbiere/cadical
5. Bayardo Jr, R.J., Schrag, R.: Using CSP look-back techniques to solve real-world SAT instances. In: Kuipers, B., Webber, B.L. (eds.) Proceedings of the Fourteenth National Conference on Artificial Intelligence and Ninth Innovative Applications of Artificial Intelligence Conference, AAAI 97, IAAI 97, 27–31 July 1997, Providence, Rhode Island, USA, pp. 203–208. AAAI Press/The MIT Press (1997). http://www.aaai.org/Library/AAAI/1997/aaai97-032.php
6. Katebi, H., Sakallah, K.A., Marques-Silva, J.P.: Empirical study of the anatomy of modern SAT solvers. In: Sakallah, K.A., Simon, L. (eds.) SAT 2011. LNCS, vol. 6695, pp. 343–356. Springer, Heidelberg (2011). https://doi.org/10.1007/978-3-642-21581-0_27
7. Knuth, D.E.: Implementation of algorithm 7.2.2.2c (conflict-driven clause learning SAT solver). https://www-cs-faculty.stanford.edu/~knuth/programs/sat13.w
8. Liang, J.H., Ganesh, V., Poupart, P., Czarnecki, K.: Learning rate based branching heuristic for SAT solvers. In: Creignou, N., Le Berre, D. (eds.) SAT 2016. LNCS, vol. 9710, pp. 123–140. Springer, Cham (2016). https://doi.org/10.1007/978-3-319-40970-2_9
9. Liang, J.H., Oh, C., Ganesh, V., Czarnecki, K., Poupart, P.: Maple-COMSPS, MapleCOMSPS_LRB, MapleCOMSPS_CHB. In: Balyo, T., Heule, M.J.H., Järvisalo, M.J. (eds.) Proceedings of SAT Competition 2016 Solver and Benchmark Descriptions. University of Helsinki (2016). http://hdl.handle.net/10138/164630
10. Luo, M., Li, C., Xiao, F., Manyà, F., Lü, Z.: An effective learnt clause minimization approach for CDCL SAT solvers. In: Sierra, C. (ed.) Proceedings of the Twenty-Sixth International Joint Conference on Artificial Intelligence, IJCAI 2017, Melbourne, Australia, 19–25 August 2017, pp. 703–711 (2017). ijcai.org, https://doi.org/10.24963/ijcai.2017/98

11. Chowdhury, M.S., Müller, M., You, J.H.: Four CDCL SAT solvers based on exploration and glue variable bumping. In: Heule, M.J.H., Järvisalo, M., Suda, M. (eds.) Proceedings of SAT Competition 2019 Solver and Benchmark Descriptions. University of Helsinki (2019). http://hdl.handle.net/10138/306988

12. Möhle, S., Biere, A.: Backing backtracking. In: Janota, M., Lynce, I. (eds.) SAT 2019. LNCS, vol. 11628, pp. 250–266. Springer, Cham (2019). https://doi.org/10.1007/978-3-030-24258-9_18

13. Moskewicz, M.W., Madigan, C.F., Zhao, Y., Zhang, L., Malik, S.: Chaff: engineering an efficient SAT solver. In: Proceedings of the 38th Design Automation Conference, DAC 2001, Las Vegas, NV, USA, 18–22 June 2001, pp. 530–535. ACM (2001). https://doi.org/10.1145/378239.379017

14. Nadel, A., Ryvchin, V.: Chronological backtracking. In: Beyersdorff, O., Wintersteiger, C.M. (eds.) SAT 2018. LNCS, vol. 10929, pp. 111–121. Springer, Cham (2018). https://doi.org/10.1007/978-3-319-94144-8_7

15. Pipatsrisawat, K., Darwiche, A.: On the power of clause-learning SAT solvers as resolution engines. Artif. Intell. **175**(2), 512–525 (2011). https://doi.org/10.1016/j.artint.2010.10.002

16. Ryvchin, V., Nadel, A.: Maple_LCM_Dist ChronoBT: featuring chronological backtracking. In: Heule, M.J.H., Järvisalo, M., Suda, M. (eds.) Proceedings of SAT Competition 2018 Solver and Benchmark Descriptions. University of Helsinki (2018). https://hdl.handle.net/10138/237063

17. Sabharwal, A., Samulowitz, H., Sellmann, M.: Learning back-clauses in SAT. In: Cimatti, A., Sebastiani, R. (eds.) SAT 2012. LNCS, vol. 7317, pp. 498–499. Springer, Heidelberg (2012). https://doi.org/10.1007/978-3-642-31612-8_53

18. Silva, J.P.M., Lynce, I., Malik, S.: Conflict-driven clause learning SAT solvers. In: Handbook of Satisfiability, pp. 131–153. IOS Press (2009). https://doi.org/10.3233/978-1-58603-929-5-131

19. Silva, J.P.M., Sakallah, K.A.: GRASP - a new search algorithm for satisfiability. In: Rutenbar, R.A., Otten, R.H.J.M. (eds.) Proceedings of the 1996 IEEE/ACM International Conference on Computer-Aided Design, ICCAD 1996, San Jose, CA, USA, 10–14 November 1996, pp. 220–227. IEEE Computer Society/ACM (1996). https://doi.org/10.1109/ICCAD.1996.569607

20. Silva, J.P.M., Sakallah, K.A.: GRASP: a search algorithm for propositional satisfiability. IEEE Trans. Comput. **48**(5), 506–521 (1999). https://doi.org/10.1109/12.769433

21. Sörensson, N., Biere, A.: Minimizing learned clauses. In: Kullmann, O. (ed.) SAT 2009. LNCS, vol. 5584, pp. 237–243. Springer, Heidelberg (2009). https://doi.org/10.1007/978-3-642-02777-2_23

22. Kochemazov, S., Zaikin, O., Kondratiev, V., Semenov, A.: MapleLCMDistchronoBT-DL, duplicate learnts heuristic-aided solvers at the SAT race 2019. In: Heule, M.J.H., Järvisalo, M., Suda, M. (eds.) Proceedings of SAT Competition 2019 Solver and Benchmark Descriptions. University of Helsinki (2019). http://hdl.handle.net/10138/306988

23. Wetzler, N., Heule, M.J.H., Hunt, W.A.: DRAT-trim: efficient checking and trimming using expressive clausal proofs. In: Sinz, C., Egly, U. (eds.) SAT 2014. LNCS, vol. 8561, pp. 422–429. Springer, Cham (2014). https://doi.org/10.1007/978-3-319-09284-3_31
24. Wieringa, S., Heljanko, K.: Concurrent clause strengthening. In: Järvisalo, M., Van Gelder, A. (eds.) SAT 2013. LNCS, vol. 7962, pp. 116–132. Springer, Heidelberg (2013). https://doi.org/10.1007/978-3-642-39071-5_10
25. Zhang, L., Madigan, C.F., Moskewicz, M.W., Malik, S.: Efficient conflict driven learning in Boolean satisfiability solver. In: Ernst, R. (ed.) Proceedings of the 2001 IEEE/ACM International Conference on Computer-Aided Design, ICCAD 2001, San Jose, CA, USA, 4–8 November 2001, pp. 279–285. IEEE Computer Society (2001). https://doi.org/10.1109/ICCAD.2001.968634

# Trail Saving on Backtrack

Randy Hickey[(✉)] and Fahiem Bacchus[(✉)]

Department of Computer Science, University of Toronto, Toronto, Canada
{rhickey,fbacchus}@cs.toronto.edu

**Abstract.** A CDCL SAT solver can backtrack a large distance when it learns a new clause, e.g, when the new learnt clause is a unit clause the solver has to backtrack to level zero. When the length of the backtrack is large, the solver can end up reproducing many of the same decisions and propagations when it redescends the search tree. Different techniques have been proposed to reduce this potential redundancy, e.g., partial/chronological backtracking and trail saving on restarts. In this paper we present a new trail saving technique that is not restricted to restarts, unlike prior trail saving methods. Our technique makes a copy of the part of the trail that is backtracked over. This saved copy can then be used to improve the efficiency of the solver's subsequent redescent. Furthermore, the saved trail also provides the solver with the ability to look ahead along the previous trail which can be exploited to improve its efficiency. Our new trail saving technique offers different tradeoffs in comparison with chronological backtracking and often yields superior performance. We also show that our technique is able to improve the performance of state-of-the-art solvers.

## 1 Introduction

The vast majority of modern SAT solvers that are used to solve real-world problems are based on the conflict-driven clause learning (CDCL) algorithm. In a CDCL SAT solver, backtracking occurs after every conflict, where all literals from one or more decision levels become unassigned before the solver resumes making decisions and performing unit propagations. Traditionally, CDCL solvers would backtrack to the conflict level, which is the second highest decision level remaining in the conflict clause after conflict analysis has resolved away all but one literal from the current decision level [9]. Recently, however, it has been shown that partial backtracking [6] or chronological backtracking, C-bt, (i.e., backtracking only to the previous level after conflict analysis) [8,11] can be effective on many instances. Partial backtracking has been used in the solvers that won the last two SAT competitions. Although chronological backtracking breaks some of the conventional invariants of CDCL solvers, it has been formalized and proven correct [8] (also see related formalizations [10,12]).

The motivation for using C-bt is the observation that when a solver backtracks across many levels, many of the literals that are unassigned during the backtrack might be re-assigned again in roughly the same order when the solver

© Springer Nature Switzerland AG 2020
L. Pulina and M. Seidl (Eds.): SAT 2020, LNCS 12178, pp. 46–61, 2020.
https://doi.org/10.1007/978-3-030-51825-7_4

redescends. This observation was first made in the context of restarts by van der Tak et al. [14]. Their technique backtracks to the minimum change level, i.e., the first level at which the solver's trail can change on redescent. However, their technique cannot be used when backtracking from a conflict: the solver's trail is going to be changed at the backtrack level so the minimum change level is the same as the backtrack level.

Chronological backtracking or partial backtracking instead allows a reduction in the length of the backtrack by placing literals on the trail out of decision level order. By reducing the length of the backtrack the solver can keep more of its assignment trail intact. This can save it from the work involved in reconstructing a lot of its trail. Using C-bt is not a panacea however. Its application must be limited for peak effectiveness. This indicates that it is sometimes beneficial for the solver to backtrack fully and redo its trail, even if this takes more work. We will expand on why this might be the case below.

In this paper we present a new trail saving method whereby we save the backtracked part of the solver's trail and attempt to use that information to make the solver's redescent more efficient. Unlike C-bt, our trail saving method preserves the traditional invariants of the SAT solver and its basic version is very simple to implement. It allows the search to retain complete control over the order of decisions, but helps make propagation faster. We develop some enhancements to make the idea more effective, and demonstrate experimentally that it performs as well as and often better than chronological backtracking. We also show that with our enhancements we are able to improve the performance of state-of-the-art solvers.

## 2   Background

SAT solvers determine the satisfiability of a propositional formula $\mathcal{F}$ expressed in Conjunctive Normal Form (CNF). $\mathcal{F}$ contains a set of variables $V$. A literal is a variable $v \in V$ or its negation $\neg v$, and for a literal $l$ we let $\text{var}(l)$ denote its underlying variable. A CNF consists of a conjunction of clauses, each of which is a disjunction of literals. We often view a clause as being a set of literals and employ set notation, e.g., $\ell \in C$ and $C' \subset C$. We will assume that the reader is familiar with the basic operations of CDCL SAT solvers. A good source for this background is [13].

*Trails.* CDCL SAT solvers maintain a trail which is the sequence of literals that have currently been assigned TRUE by the solver. During its operation a SAT solver will add newly assigned literals to the end of the trail, and on backtrack remove literals from the end of the trail. For convenience, we will regard *literals as having been assigned* TRUE *if and only if they are on the trail.* So removing/adding a literal to the trail is equivalent to unassigning/assigning the literal TRUE.

A SAT solver's trail satisfies a number of conditions. However, in this work we will need some additional flexibility in our definitions, as we will sometimes be working with trails that would never be constructed by a SAT solver. Hence, we define a *trail* to be a sequence of literals each of which is either a *decision* literal or an *implied* literal, and each of which has a *reason*. These two types of literals are distinguished by their *reasons*. Decision literals $d$ have a null reason, $reason(d) = \varnothing$. Implied literals $l$ have as a reason a clause of the formula $\mathcal{F}$, $reason(l) = C \in \mathcal{F}$. (The clause $reason(l)$ can be a learnt clause that has been added to $\mathcal{F}$).

If literal $\ell$ is on the trail $\mathcal{T}$ let $\iota_{\mathcal{T}}(\ell)$ denote its index on the trail, i.e, $\mathcal{T}[\iota_{\mathcal{T}}(\ell)] = \ell$. If $x$ and $y$ are both on the trail and $\iota_{\mathcal{T}}(x) < \iota_{\mathcal{T}}(y)$ we say that $x$ *appears before* $y$ on the trail. For convenience, when the trail being discussed is clear from context we simply write $\iota$ instead of $\iota_{\mathcal{T}}$.

Each literal $\ell \in \mathcal{T}$ has a decision level $decLvl(\ell)$ which is equal to the number of decision literals appearing on the trail up to and including $\ell$ ; hence, $decLvl(d) = 1$ for the first decision literal $d \in \mathcal{T}$. The set of literals on $\mathcal{T}$ that have the same decision level forms a *contiguous* subsequence[1] that starts with a decision literal $d_i$ and ends just before the next decision literal $d_{i+1}$. We will often need to refer to different decision level subsequences of $\mathcal{T}$. Hence, we let $\mathcal{T}[[i]]$ denote the subsequence of literals at decision level $i$; and let $\mathcal{T}[[i \ldots j]]$ denote the subsequence of literals at decision levels $k$ for $i \leq k \leq j$.

**Definition 1.** *A clause $C$ has **been made unit by $\mathcal{T}$ implying** $l$ when $l \in C \wedge (\forall x \in C.x \neq l \rightarrow \neg x \in \mathcal{T})$. That is, all literals in $C$ except $l$ must have been falsified by $\mathcal{T}$*

Now we define the following properties that a trail $\mathcal{T}$ can have.

**non-contradictory:** A variable cannot appear in both polarities in the trail: $l \in \mathcal{T} \rightarrow \neg l \notin \mathcal{T}$.

**non-redundant:** A literal can only appear once on $\mathcal{T}$.

**reason-sound:** For each implied literal $l \in \mathcal{T}$ we have that its reason clause $reason(l) = C$ has been made unit by $\mathcal{T}$ implying $l$, and for each $x \in C$ with $x \neq l$ we have that $\neg x$ appears before $l$ on $\mathcal{T}$: $\forall l \in \mathcal{T}. reason(l) \neq \varnothing \rightarrow l \in reason(l) \wedge (\forall x \in reason(l).x \neq l \rightarrow \neg x \in \mathcal{T} \wedge \iota(\neg x) < \iota(l))$.

**propagation-complete:** Unit propagation has been run to completion at all decisions levels of $\mathcal{T}$. This means that literals appear on $\mathcal{T}$ at the first decision level they were unit implied. Formally, this can be captured by the condition: $\forall i \in \{decLvl(l) \mid l \in \mathcal{T}\}.(\exists C \in \mathcal{F}.C$ is made unit by $\mathcal{T}[[0 \ldots i]]$ implying $l) \rightarrow l \in \mathcal{T}[[0 \ldots i]]$. Note that propagation completeness implies that $reason(l) \neq \varnothing$ must contain at least one other literal $y \neq l$ with $decLvl(y) = decLvl(l)$.

**conflict-free:** No clause of $F$ is falsified by $\mathcal{T}$. Clauses $C \in F$ falsified by $\mathcal{T}$ are typically called *conflicts*.

---

[1] Our approach uses standard trails in which the decision levels are contiguous. Chronological backtracking [6,8,11] generates trails with non-contiguous decision levels.

In CDCL solvers using standard conflict directed backtracking all properties hold of the prefix of the solver's trail consisting of all decisions levels but the deepest. The full trail might, however, contain a conflict at its deepest level so is not necessarily conflict-free. The full trail might also not be propagation-complete, as unit propagation at the deepest level is typically terminated early if a conflict is found. It can further be noted that the first four properties imply that if a clause $C$ is falsified at decision level $k$, then $C$ must contain at least two literals at level $k$ (otherwise $C$ would have become unit at a prior level and then satisfied by making its last unfalsified literal TRUE).

*Standard Backtracking.* In CDCL SAT solving the solver extends its trail by adding new decision literals followed by finding and adding all unit implied literals arising from that new decision. This continues until it reaches a decision level $L_{deep}$ where a conflict $C$ is found.

In standard backtracking, the solver then constructs a new 1-UIP clause by resolving away all but one literal at level $L_{deep}$ from the conflict $C$ using the reason clauses of these literals. (As noted above $C$ must contain at least two literals at level $L_{deep}$). Hence, the new clause $C_{1\text{-}UIP}$ will contain one literal $\ell_{deep}$ at level $L_{deep}$ and have all of its other literals a levels less than $L_{deep}$. The solver then backtracks to $L_{back}$ the second deepest level in $C_{1\text{-}UIP}$. This involves changing $T$ to its prefix $T[[0 \ldots L_{back}]]$ (by our convention all literals removed from $T$ are now unassigned). The new clause $C_{1\text{-}UIP}$ is made unit by $T[[0 \ldots L_{back}]]$ implying $\ell_{deep}$, so the solver then adds $\ell_{deep}$ to the trail and executes another round of unit propagation at level $L_{back}$, after which it continues by once again growing the trail with new decisions and unit implied literals until a new conflict or a satisfying assignment is found.

In standard backtracking, the difference between the backtrack level, $L_{back}$ and the current deepest level $L_{deep}$ can be very large. During its new descent from $L_{back}$ the solver can reproduce a large number of the same decisions and unit propagations, essentially wasting work. This potential inefficiency has been noted in prior work [6,8,11,14].

In [14] a technique for reducing the length of the backtrack during restarts was presented. In restarts, the solver backtracks to level 0, and this technique involves computing a new deeper backtrack level $M > 0$ for which it is known that on redescent the first $M + 1$ levels of the trail will be unchanged (except perhaps for the ordering of the literals). This technique removes the redundant work of reproducing the first $M$ trail levels. When backtracking from a conflict, however, the trail will be changed at level $L_{back}$ ($\ell_{deep}$ will be newly inserted at this level). Hence this technique cannot reduce the length of the backtrack. In this paper we will show that although we have to backtrack to $L_{back}$ we can make the subsequent redescent much more efficient.

*Chronological Backtracking.* Chronological backtracking (C-bt) and partial backtracking in the context of clause learning solvers are alternatives to standard backtracking which allow the solver to execute a shorter backtrack. That is, with these techniques the solver can avoid having to go all the way back to the second deepest level in the learnt clause, as in standard backtracking.

Formalisms for partial backtracking in clause learning solvers have been presented in [10,12]. In [6] practical issues of implementation were addressed, and experiments shown with a CDCL solver using partial backtracking. In [11] improved and more efficient implementation techniques were developed which allowed C-bt to make improvements to state-of-the-art SAT solvers, and [8] presented additional implementation ideas and details along with correctness results for these methods.

The aim of partial backtracking is to reduce the redundant work that might be done by the SAT solver on its redescent from the backtrack level $L_{back}$. The technique allows the solver to backtrack to any level $j$ in the range $L_{back} \leq j \leq L_{deep}-1$ (where $L_{deep}$ is the level the conflict was discovered). Nadel and Ryvchin [11] proposed to always backtrack chronologically to $L_{deep}-1$ while Möhle and Biere [8] returned to the proposal of [6] of flexibly backtracking to any level in the allowed range. Note that the new learnt 1-UIP clause $C_{1\text{-}UIP}$ is made unit at every level in this range. So after backtracking to level $j$ the newly implied literal $\ell_{deep}$ is added to the trail with $reason(\ell_{deep}) = C_{1\text{-}UIP}$, and $decLvl(\ell_{deep})$ is set to $L_{back}$ (the second deepest level in $C_{1\text{-}UIP}$).

This means that the decision levels on the trail are no longer contiguous, as $\ell_{deep}$ has a different level than the other literals at level $j$ (if $j \neq L_{back}$). This change has a number of consequences for the SAT solver's operation, all of which were described in [6]. Möhle and Biere [8] showed that despite these consequences partial backtracking can be made to preserve the soundness of a CDCL solver.

## 3    Chronological Backtracking Effects on Search

In this paper we present a new technique that allows the SAT solver to use standard backtracking, but also allows saving some redundant work on its redescent. Our method has more overhead than C-bt so the first question that must be addressed is why not just use chronological backtracking.

Although C-bt is able to avoid a lot of redundant work it also has other effects on the SAT solver search. These effects are sometimes detrimental to the solver's performance and so it is not always beneficial to use C-bt. In fact, in both [11] and [8] it was found that fairly limited application of C-bt performed best. In [11] C-bt was applied only when the length of the standard backtrack, $L_{back} - L_{deep}$ was greater than a given threshold $T$. In their experiments they found that $T = 100$ was the best value, i.e., C-bt is done only on longer backtracks. In practice, this meant that C-bt was relatively infrequent; in our measurements with their solver only about 3% of the solver backtracks were C-bt backtracks. In [8] the value $T = 100$ was also applied. However, they introduced an additional technique to add some applications of C-bt when the length of the backtrack is less than $T$. This allowed [8] to utilize C-bt in about 15% of the backtracks.

Although it is difficult to know precisely why C-bt is not always beneficial, we can identify some different ways in which C-bt can affect the SAT solver's search. With standard backtracking the literal $\ell_{deep}$ is placed on the trail at the end of $L_{back}$ and then unit propagated. This could impact the trail in at least the following ways. First, some literals might become unit at earlier levels. This could include decision literals becoming forced which might compress some decision levels together. Second, different decisions might be made due to changes in the variable scores arising from the newly learnt clause. And third, literals might be unit implied with different reasons. C-bt can change all of these things, each of which could have an impact on the future learnt clauses, and thus on the solver's overall efficiency.

The second impact, changing variable scores, is partially addressed in [8] who utilize the ideas of [14] to backtrack to a level where the decisions would be unchanged. However, if the length of the backtrack is greater than 100 there could still be a divergence between the variable decisions generated in standard backtracking and C-bt. An argument is also given in [14] that the third impact, changing literal reasons, is not significant. However, the experiments in [14] were run before good notions of clause quality were known [1]. Our empirical results indicate that once clause quality is accounted for, changing the literal reasons can have a significant impact.

The first impact is worth discussing since it was mentioned in [6] but not in the subsequent works. This is the issue of changing the decision levels of literals on the trail. C-bt computes the decision level of each implied literal based on the decision levels of the literals in its reason, but it does not go backwards to change the decisions levels of literals earlier on the trail.

*Example 1.* For example, suppose that $(x, \neg y) \in \mathcal{F}$, the literal $x$ is a decision literal on the trail with $decLvl(x) = 2$, and that the solver is currently at level 150 where it encounters a conflict. If this conflict yields the unit clause $(y)$, standard backtracking would backtrack to level 0, where $x$ would be implied. On redescent, $x$ would no longer form a new decision level and it would not appear in any new clauses (as it is entailed by $\mathcal{F}$). C-bt, on the other hand, would backtrack to level 149. On its trail $x$ would still be at level 2. Until a backtrack past level 2 occurs, learnt clauses might contain $\neg x$, and thus have level 2 added to their set of levels (potentially changing their LBD score). Only when backtrack past level 2 occurs would $x$ be restored to its correct level 0, and it would require inprocessing simplifications to remove $x$ from the learnt clauses.

In sum, although these impacts of C-bt on the SAT solver's search might or might not be harmful to the SAT solver, they do exist. In fact, there are two pieces of evidence that these impacts can sometimes be harmful. First, as mentioned above, previous work found that it is best to only apply C-bt on large backtracks where it has the potential to save the most work. If there were no harmful effects it would always be effective to apply C-bt. And second, in our empirical results below we show that our new trail saving technique, which always uses standard backtracking, can often outperform C-bt. Although our technique reduces the solver's work on redescent it does not completely eliminate it like C-bt does. Hence its superior performance can only occur if C-bt is sometimes harmful.

It is possible to combine C-bt with our trail saving technique to reduce the amount of work required whenever the solver performs non-chronological backtracking. However, C-bt greatly reduces the potential savings that could be achieved by our method since most of its non-chronological backtracks are relatively short (less than threshold $T$ levels). In our preliminary experiments this combination did not seem promising.

Nevertheless, there is good evidence that C-bt can improve SAT solver performance.[2] Hence, it should be that it is better to perform C-bt in some branches. Hence, an interesting direction for future work would be to develop better heuristics about when to use C-bt in a branch and when to use standard backtracking augmented by our trail saving method.

## 4   Trail Saving

Our approach is to save the trail $T$ on backtrack, and to use the saved trail $T_{save}$ when the solver redescends to improve the efficiency of propagations without affecting the decisions the solver wants to make. The saved trail $T_{save}$ also provides a secondary "lookahead mechanism" that the SAT solver can exploit as it redescends.

Suppose that the solver is at $L_{deep}$ where it has encountered a conflict. From the 1-UIP clause it learns, $C_{1\text{-}UIP}$, it now has to backtrack to $L_{back}$. This is accomplished by calling BACKTRACK($L_{back}$), shown in Fig. 1, which saves the backtracked portion of the trail.

Note that BACKTRACK does not save the deepest level of $T$. The full $T$ contains a conflict (at its deepest level). Hence the solver will never reproduce all the same levels, and it would be useless to save all of them. Note also that in addition to saving the literals in $T_{save}$ we also save the clause reason of the unit implied literals in a separate $reason_{save}$ vector. Finally, we see that after backtrack the first literal on $T_{save}$ is a decision literal: it is the first literal of $T$ at decision level $L_{back}+1$. Literals will be removed from $T_{save}$ during its use, but always in units of complete decision levels. So $T_{save}[0]$ will always be a (previous) decision literal.

After backtrack the solver will add $\ell_{deep}$ to the end of the updated $T$ with $reason(\ell_{deep}) = C_{1\text{-}UIP}$ and then invoke unit propagation. $T_{save}$ is exploited during propagation by the version of PROPAGATE shown in Fig. 1, which will initially be invoked with the argument $\iota(\ell_{deep})$ (i.e., the trail index of the newly added implicant). The saved trail will be continually consulted during the solver's descent whenever unit propagation is performed. When backtrack occurs $T_{save}$ will be overwritten to store the new backtracked portion of $T$.

$T_{save}$ is consulted in the procedure USESAVEDTRAIL (Fig. 1). This procedure tries to add saved implied literals and their reasons to the solver's trail, when

---

[2] C-bt can also be extremely useful in contexts where each descent can be very expensive, e.g., when doing theory propagation in SMT solving, or component analysis in #SAT solving. In these cases, C-bt, by avoiding backtracking and subsequent redescent, has considerable potential for improving solver performance.

```
 1:  BACKTRACK(L_back)
 2:     ∀ℓ ∈ T[[L_back+1 ... L_deep−1]] reason_save(ℓ) = reason(ℓ)
 3:     T_save = T[[L_back+1 ... L_deep−1]]
 4:     T = T[0 ... L_back]]
```

```
 1:  PROPAGATE(idx)
 2:     while idx < T.size()
 3:         c ← USESAVEDTRAIL ()
 4:         if (c ≠ ∅) return c                    ▷ Found conflict from T_save
 5:         ℓ ← T[idx]
 6:         for each clause c ∈ watchlist(¬ℓ)
 7:             if (c is unit implying x)
 8:                 T.addToEnd(x);  reason(x) ← c
 9:             else if (c is falsified) return c   ▷ Found conflict from unit prop.
10:             else Update c's watches.
11:         idx++
```

```
12:  USESAVEDTRAIL()
13:     idx ← 0; c ← ∅
14:     for ( ; idx < T_save.size(); idx++)
15:         l_save ← T_save[idx]
16:         if (reason_save(l_save) = ∅)                 ▷ Decision on T_save
17:             if (l_save ∈ T) continue     ▷ TRUE in solver, we can use its implied lits
18:             else break            ▷ Only solver can set decisions, so we stop here
19:         else                                 ▷ Implied Literal on T_save
20:             if (l_save ∈ T) continue                   ▷ ignore redundant lits
21:             else if (¬l_save ∈ T)      ▷ contradiction, return conflict and reset idx
22:                 c ← reason_save(l_save); idx ← 0; break
23:             else                          ▷ unset in solver, add to solver's trail
24:                 T.addToEnd(l_save)
25:                 reason(l_save) ← reason_save(l_save)
26:     for (i ← 0; i < idx; i++)              ▷ T_save is unchanged when idx = 0
27:         T_save.removeFront()
28:     return c
```

**Fig. 1.** Using $T_{save}$ in unit propagation and conflict detection

these implications are valid. We will show below that those implications that are added are in fact valid. We do not interfere with the solver's variable decisions. Instead we opportunistically test to see if literals implied on $T_{save}$ are valid implications for the solver given the solver's current decisions.

$T_{save}[0]$ is always a (previous) decision literal $d$ with $reason_{save}(d) = ∅$. Note that, since new literals (e.g., $ℓ_{deep}$) have been added to $T$, $d$ might now be an implied literal on $T$ (i.e., $reason(d) ≠ ∅$) even though before the backtrack it was previously a decision (i.e., $reason_{save}(d) = ∅$). If $d$ has not been assigned TRUE by the solver (i.e., $¬d ∈ T$), we cannot add any implied literals below it on $T_{save}$ to $T$ as these implied literals depend on $d$ being assigned TRUE. In this case we stop looking for more literals to add to $T$ (line 18).

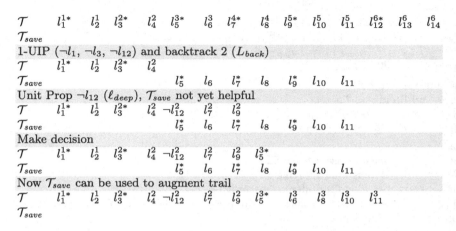

**Fig. 2.** Use of $\mathcal{T}_{save}$ from Example 2. The literal's decision level is indicated in its superscript, and a * superscript indicates that the literal is a decision.

On the other hand if $d$ has been made TRUE by the solver we can continue to add all of the implied literals below it (up to but not including the next decision literal on $\mathcal{T}_{save}$) to $\mathcal{T}$ (line 24), reusing their saved reasons. Any literals that have already been made true by the solver can be skipped (line 20). Finally, if we encounter a literal that has already been falsified by the solver, then its saved reason clause must be falsified by the solver and we can return it as a conflict (line 21). If a conflict is encountered we leave $\mathcal{T}_{save}$ unchanged by resetting $idx$ to zero. Otherwise, $idx$ will be the number of literals at the front of $\mathcal{T}_{save}$ that have been moved to $\mathcal{T}$ (or skipped over since they are already on $\mathcal{T}$). We then remove the first $idx$ literals from $\mathcal{T}_{save}$ (line 27), and return the conflict (equal to $\varnothing$ if no conflict was found).

*Example 2.* Figure 2 provides an example of how $\mathcal{T}_{save}$ is used. Initially the literals $l_1$ to $l_{14}$ are on the solver's $\mathcal{T}$, and $\mathcal{T}_{save}$ is empty. This is shown in the first two lines of the figure. In the figure the superscript on the literals indicates their decision level, and a superscripted * indicates that the literal is a decision. Hence $l_1^{1*}$ indicates that $decLvl(l_1) = 1$ and that $l_1$ is a decision.

Then a conflict is found at level 6 and the 1-UIP clause $(\neg l_1, \neg l_3, \neg l_{12})$ is learnt. Thus the solver will backtrack to level 2, where it will add $\neg l_{12}$ as a unit implicant. The next two lines show $\mathcal{T}$ and $\mathcal{T}_{save}$ right after the backtrack to level 2: the backtracked levels have been copied into $\mathcal{T}_{save}$ omitting the conflict level 6.

The new unit $\neg l_{12}$ is now added to $\mathcal{T}$ and unit propagation performed adding $l_7$ and $l_9$ to level 2. Since the first literal on $\mathcal{T}_{save}$, $l_5$, has $reason_{save}(l_5) = \varnothing$ ($l_5$ was a decision on $\mathcal{T}$ at the time backtrack occurred) and is not yet TRUE, $\mathcal{T}_{save}$ is not helpful at this stage. The status of $\mathcal{T}$ and $\mathcal{T}_{save}$ at this point is shown in the figure.

After unit propagation is finished the solver makes a new decision, which happens to be (but is not forced to be) $l_5$. Now $\mathcal{T}_{save}$ can be used: $l_5$ is TRUE so it is removed, $l_6$ is unassigned so it is added to $\mathcal{T}$, $l_7$ is TRUE and so removed,

$l_8$ is unassigned so it is added to $\mathcal{T}$, $l_9$ is TRUE and removed, and finally $l_{10}$ and $l_{11}$ are unassigned and so are added to $\mathcal{T}$. In this example, $\mathcal{T}_{save}$ is emptied, and cannot contribute more to $\mathcal{T}$.

All of these units are added to $\mathcal{T}$ before the solver starts to unit propagate $l_5$. Since, new literals have been added to $\mathcal{T}$ before $l_5$ the solver must propagate $l_5$ and all of the literals that follow it before making its next decision.

As noted in the previous example unit propagation has to be rerun on all saved literals added to $\mathcal{T}$ from $\mathcal{T}_{save}$. Thus our technique, unlike C-bt, does not completely remove the overhead of reproducing the trail on the solver's redescent. Nevertheless, trail saving improves the efficiency of this redescent in three different ways. First, by adding more forced literals to the trail before continuing propagating the next literal, propagation can potentially gain a quadratic speedup [2,5]. Second, propagation does not need to examine the reason clause of the added literals. If these literals were not added by USESAVEDTRAIL, propagation would have to traverse each of these reason clauses to determine that they have in fact become unit. Third, when a conflict is returned by USESAVEDTRAIL all further propagations can be halted. The added literals and their reasons will be sufficient to perform clause learning from the conflict returned by USESAVED-TRAIL. Since trail saving can sometimes save hundreds or thousands of literals at a time these savings can in sum be significant.

## 4.1   Correctness

Now we will prove that our use of $\mathcal{T}_{save}$ preserves the SAT solver's soundness. In particular, $\mathcal{T}_{save}$ is only used in the procedure USESAVEDTRAIL, in which it either adds new literals to the solver's trail, or returns conflict clauses to the solver. Hence, we only need to show that these new literals are in fact unit implied and the conflicts are in fact falsified by the solver's trail. Since both $\mathcal{T}$ and $\mathcal{T}_{save}$ are sequences of literals (with associated reasons) we can consider their concatenation denoted as $\mathcal{T} + \mathcal{T}_{save}$.

**Theorem 1.** *If $\mathcal{T} + \mathcal{T}_{save}$ is **reason sound** (Sect. 2) then the following holds. If the first $i$ literals on $\mathcal{T}_{save}$ are all in $\mathcal{T}$ ($\forall j. 0 \leq j < i. \mathcal{T}_{save}[j] \in \mathcal{T}$) and $\mathcal{T}_{save}[i] = l$ is an implied literal with $reason_{save}(l) = C$, then $C$ has been made unit by $\mathcal{T}$ implying $l$.*

*Proof:* Since $\mathcal{T} + \mathcal{T}_{save}$ is reason sound, every literal in $C$ other than $l$ appears negated before $l$ in the sequence $\mathcal{T} + \mathcal{T}_{save}$. Thus for $x \in C$ we have $\neg x \in \mathcal{T}$ or $\neg x \in \mathcal{T}_{save}[0] \ldots \mathcal{T}_{save}[i-1]$. But in the later case we also have $\neg x \in \mathcal{T}$.     □

This theorem shows that USESAVEDTRAIL's processing is sound. In this procedure, an implied literal from $\mathcal{T}_{save}$ is added to $\mathcal{T}$ (line 24) only when all prior literals on $\mathcal{T}_{save}$ are already on $\mathcal{T}$ (i.e., previously on $\mathcal{T}$ or already added to $\mathcal{T}$). Thus each new addition is sound given the inductive soundness of the previous additions, with the base case covered by Theorem 1. If $l$ is to be added, the theorem shows that every other literal in $reason_{save}(l)$ has been falsified by $\mathcal{T}$.

Hence if $l$ is also falsified by $\mathcal{T}$ then $reason_{save}(l)$ is a clause that is falsified by $\mathcal{T}$, thus it is a sound conflict for the solver.

Now we only have to show that $\mathcal{T} + \mathcal{T}_{save}$ is always **reason sound** during the operation of the solver.

**Proposition 1.** *If $\mathcal{T} + \mathcal{T}_{save}$ is reason sound then $\mathcal{T}' + \mathcal{T}'_{save}$ is reason sound in all of the following cases.*

1. $\mathcal{T}_{save}[0] \in \mathcal{T}$, $\mathcal{T}' = \mathcal{T}$, and $\mathcal{T}'_{save} = \mathcal{T}_{save}.\text{removeFront}()$.
2. $\mathcal{T}' = \mathcal{T} + \mathcal{T}_{save}[0]$ and $\mathcal{T}'_{save} = \mathcal{T}_{save}.\text{removeFront}()$.
3. $\mathcal{T}' = \mathcal{T} + \mathcal{T}_{new}$ and $\mathcal{T}'_{save} = \mathcal{T}'_{save}$ and $\mathcal{T}'$ is reason sound.
4. *We also have that $\mathcal{T}$ is reason sound if $\mathcal{T}$ was generated by the solver.*

*Proof:* (1) $\mathcal{T}_{save}[0]$ already appears earlier in the $\mathcal{T}$ so it can be removed without affecting the soundness of any reason following it. (2) is obvious as the sequence is unchanged. (3) the reasons in $\mathcal{T} + \mathcal{T}_{new}$ are sound by assumption. Those in $\mathcal{T}_{save}$ remain sound as they depend only on the literals in $\mathcal{T}$ and prior literals on $\mathcal{T}_{save}$, both of which are unchanged. (4) is obvious from the operation of unit propagation in the solver. □

**Theorem 2.** *$\mathcal{T} + \mathcal{T}_{save}$ is always reason sound during the operation of the solver.*

*Proof:* $\mathcal{T}_{save}$ starts off being empty, so $\mathcal{T} + \mathcal{T}_{save} = \mathcal{T}$ is reason sound as it was generated by the solver (4). In procedure BACKTRACK $\mathcal{T} + \mathcal{T}_{save}$ is set to a trail that was previously generated by the solver (4). The solver can add to $\mathcal{T}$ by decisions and propagations without using $\mathcal{T}_{save}$. In this case $\mathcal{T}' = \mathcal{T} + \mathcal{T}_{new}$, and $\mathcal{T}'$ is reason sound by (4), thus the new $\mathcal{T}' + \mathcal{T}_{save}$ is reason sound by (3). Finally, in procedure USESAVEDTRAIL either (a) literals at the front of $\mathcal{T}_{save}$ are discarded since they already appear on $\mathcal{T}$, or (b) literals are moved from $\mathcal{T}_{save}$ to $\mathcal{T}$. Under both of these changes $\mathcal{T} + \mathcal{T}_{save}$ remains reason sound by (1) and (2). □

## 4.2   Enhancements

We developed three enhancements of the base trail saving method described above. In this section we present these enhancements.

*Saving the Trail over Multiple Backtracks.* It can often be the case that when the solver finds a conflict and backtracks to $L_{back}$ it might immediately find a another conflict at $L_{back}$ causing a further backtrack. In the procedure BACKTRACK every backtrack causes $\mathcal{T}_{save}$ to be overwritten. Hence, in these cases most of the trail will not be saved—only the portion from the last backtrack. Our first extension addresses this potential issue and also provides more general trail saving in other contexts as well.

This extension is simply to add the latest backtrack to the front of $\mathcal{T}_{save}$ leaving all of the previous contents of $\mathcal{T}_{save}$ intact. Specifically, we replace line 3 of BACKTRACK by the new line:

3.     $\mathcal{T}_{save} = \mathcal{T}[[L_{back}+1 \ldots L_{deep}-1]] + \mathcal{T}_{save}$

It is not difficult to show that this change preserves soundness. Only Theorem 2 is potentially affected. However, we know that $\mathcal{T}_{save}$ is unchanged at the level at which a conflict occurs: either the conflict is detected without consulting $\mathcal{T}_{save}$ or if the conflict comes from $\mathcal{T}_{save}$ then USESAVEDTRAIL leaves $\mathcal{T}_{save}$ unchanged (line 21). Hence, at the level before the conflict occurred we have inductively that $\mathcal{T}[[0 \ldots L_{deep}-1]] + \mathcal{T}_{save}$ was reason sound, and hence so is $\mathcal{T}' + \mathcal{T}'_{save}$ with $\mathcal{T}' = \mathcal{T}[[0 \ldots L_{back}]]$ and $\mathcal{T}'_{save} = \mathcal{T}[[L_{back}+1 \ldots L_{deep}-1]] + \mathcal{T}_{save}$.

When adding to the front of $\mathcal{T}_{save}$ in this manner $\mathcal{T}_{save}$ can grow indefinitely. So we prune $\mathcal{T}_{save}$ when it gets too large by (a) removing $\mathcal{T}_{save}[i]$ if $\mathcal{T}_{save}[i] = \mathcal{T}_{save}[j]$ for some $j < i$ ($\mathcal{T}_{save}[i]$ is redundant), and (b) removing the suffix of $\mathcal{T}_{save}$ starting at $\mathcal{T}_{save}[i]$ when $\mathcal{T}_{save}[j] = \neg\mathcal{T}_{save}[i]$ for some $j < i$ ($\mathcal{T}_{save}[i]$ will never be useful as its negation, $\mathcal{T}_{save}[j]$, would have to be added to $\mathcal{T}$ first). In this way $\mathcal{T}_{save}$ need never become larger than the number of variables in $\mathcal{F}$.

*Lookahead for Conflicts.* In USESAVEDTRAIL we stop adding literals from $\mathcal{T}_{save}$ to $\mathcal{T}$ once we reach a decision literal $d$ on $\mathcal{T}_{save}$ that is not yet on $\mathcal{T}$ (line 18 of USESAVEDTRAIL). This is done so that the solver has full control over variable decisions without interference from the trail saving mechanism (unlike the case with C-bt). However, another option would be to force the solver to use $d$ as its next decision literal, which would then allow us to further add all of $d$'s implied literals on $\mathcal{T}_{save}$ onto $\mathcal{T}$. This can be done for the first $k$ decisions on $\mathcal{T}_{save}$ for any $k$. But in general, we do not want to remove the solver's autonomy by forcing it to make potentially different decisions than it might have wanted to.

However, if there is a literal $l \in \mathcal{T}_{save}$ for which $\neg l \in \mathcal{T}$, we can observe that forcing the solver to make all of the decisions of $\mathcal{T}_{save}$ that lie above $l$ will immediately generate a conflict in the solver: $reason_{save}(l)$ will be falsified. In fact, in this situation we would not even need to perform unit propagation over the literals added from $\mathcal{T}_{save}$; the literals and their reasons obtained from $\mathcal{T}_{save}$ would be sufficient to perform 1-UIP learning from $reason_{save}(l)$.

We experimented with this "lookahead for conflicts" idea using various values of $k$. We found that $k = 2$, i.e., forcing up to two decisions from $\mathcal{T}_{save}$ to be made by the solver if this yields a conflict, often enhanced the solver's performance. Limiting the lookahead to only one decision level of $\mathcal{T}_{save}$ was not as good, and looking ahead more than 2 decisions of $\mathcal{T}_{save}$ also degraded performance. This provides some evidence that taking too much control away from the solver and forcing it to make too many decisions from $\mathcal{T}_{save}$ can lead to conflicts that are not as useful to the solver.

*Reason Quality.* The saved trail can be thought of as remembering the solver's recent trajectory. Sometimes we want to follow the past trajectory, but perhaps sometimes we do not. In particular, when adding literals from $\mathcal{T}_{save}$ to $\mathcal{T}$ we can examine the quality of the saved reasons to see if they are worth using. Once we encounter a literal with a low quality saved reason we stop adding literals from $\mathcal{T}_{save}$ to the solver's trail. In particular, we can change lines 24–25 of USESAVED-TRAIL to the following:

23.5      **if** (lowQuality($reason_{save}(l_{save})$)) **break**
24.       $\mathcal{T}$.addToEnd($l_{save}$)
25.       $reason(l_{save}) \leftarrow reason_{save}(l_{save})$

Note that the solver will still set the un-added literals as they are unit implied by $\mathcal{T}$, but it might be able to find better reasons for these implicants. There is of course no guarantee that better reasons will be found, but our empirical results show that sometimes this does happen. We experimented with two quality metrics, clause size and clause LBD, obtaining positive results with both. These results also provides evidence against the argument given in [14] that changing literal reasons is not impactful. With an appropriate clause quality metric the changing of literal reasons can have an impact.

## 5   Experiments and Results

We implemented our techniques in two different SAT solvers, MapleSAT and Cadical,[3] both of which have finished at or near the top of SAT competitions for the past several years [3,4]. We then ran each solver on the 800 total benchmark instances used in the main tracks of the 2018 SAT Competition and 2019 SAT Race. The experiments were executed on a cluster of 2.7 GHz Intel cores with 5000 s CPU time and 7 GB memory limits for each instance. We chose not to output or verify the proofs generated by any of the solvers. The Par-2 scores obtained and total instances solved by each solver are reported in Figs. 3, 5, and 6. We also show the cactus plot of the new version of cadical in Fig. 4.

In Fig. 3 we used the newest version of cadical (downloaded as of January 1, 2020) as the baseline solver, in Fig. 5 we used the version of cadical published in [8] as the baseline, and in Fig. 6 we used MapleLCMDist [7,15] as the baseline. Each of the baselines were run with standard non-chronological backtracking. We then refer to versions of each solver with additional features implemented on top by adding suffixes. "-chrono" refers to the solver with C-bt enabled (using the solver's default settings), "-trail" refers to the baseline with plain trail saving added (as described in Fig. 1), "-trail-multipleBT" refers to the baseline with trail saving plus the first enhancement of saving over multiple backtracks, "-trail-multipleBT-lookahead" also adds the enhancement of lookahead for conflicts by 2 decision levels, and "-trail-multipleBT-lookahead-reason" also adds the final enhancement to cease trail saving once a reason of "low quality" is reached. For more details on the enhancements, please see Sect. 4.2.

Interestingly, C-bt made the newest version of cadical perform worse than the baseline (Fig. 3). This demonstrates that C-bt is not always beneficial. Trail saving alone did not impact the performance of this solver significantly, but adding all of the enhancements on top of trail saving resulted in solving six more instances and yielding a better Par-2 score than the baseline. The key enhancement for this solver seemed to be the last one where we stop using the saved trail once we detect a reason of "low quality". We tried both clause size

---

[3] Our implementation is available at https://github.com/rgh000/cadical-trail.

and lbd as the clause quality metric, and both yielded a positive gain, with clause size being slightly more effective.

|  | Total | SAT | UNSAT | Avg. Par-2 |
|---|---|---|---|---|
| cadical | 532 | 314 | 218 | 4025 |
| cadical-chrono | 525 | 308 | 217 | 4086 |
| cadical-trail | 529 | 310 | **219** | 4060 |
| cadical-trail-multipleBT | 530 | 313 | 217 | 3930 |
| cadical-trail-multipleBT-lookahead | 531 | 313 | 218 | 4020 |
| cadical-trail-multipleBT-lookahead-reason | **538** | **319** | **219** | **3854** |

**Fig. 3.** Table of results for cadical, version pulled from github as of January 1, 2020.

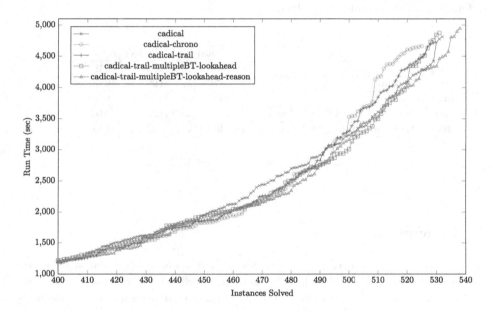

**Fig. 4.** Cactus plot for the newest version of cadical comparing standard non-chronological backtracking to C-bt and various configurations of trail saving. The first 400 problems were solved in less than 1200 s, so that part of the plot is truncated.

The version of cadical used in Fig. 5 did show benefits from C-bt in agreement with previously published results [8]. Trail saving alone did not significantly impact this solver, but adding all of the enhancements on top of trail saving resulted in solving the same number of instances as the solver with C-bt did, albeit with a slight increase in the Par-2 score.

MapleLCMDist (in Fig. 6) is another solver that benefited from C-bt. In this solver trail saving alone solved two more instances than the solver with C-bt

| | Total | SAT | UNSAT | Avg. Par-2 |
|---|---|---|---|---|
| cadical | 492 | 289 | 203 | 4378 |
| cadical-chrono | **498** | **292** | **206** | **4300** |
| cadical-trail | 493 | 290 | 203 | 4403 |
| cadical-trail-multipleBT-lookahead | 496 | **292** | 204 | 4356 |
| cadical-trail-multipleBT-lookahead-reason | **498** | **292** | **206** | 4351 |

**Fig. 5.** Table of results for cadical or "chrono", version published in [8].

did. Adding the first two enhancements on top of trail saving resulted in solving only one more instance but yielded a better Par-2 score than the solver with C-bt. Adding the last enhancement of ceasing trail saving on a "low quality" reason made the performance worse, whether clause size or lbd was used as the clause quality metric. This suggests that the enhancements to trail saving have different impacts on different solvers.

| | Total | SAT | UNSAT | Avg. Par-2 |
|---|---|---|---|---|
| maple | 458 | 259 | 199 | 4746 |
| maple-chrono | 470 | 271 | 199 | 4613 |
| maple-trail | **472** | 271 | **201** | 4618 |
| maple-trail-multipleBT-lookahead | 471 | **272** | 199 | **4597** |
| maple-trail-multipleBT-lookahead-reason | 469 | 271 | 198 | 4633 |

**Fig. 6.** Table of results for MapleLCMDist.

## 6   Conclusion

We have shown that our trail saving technique can speed up two state-of-the-art SAT solvers, cadical and MapleSAT, as or more effectively than chronological backtracking can. We also introduced three enhancements one can implement when using a saved trail and demonstrated experimentally that these enhancements can sometimes improve a solver's performance by a significant amount. We have shown that trail saving and all enhancements we proposed are sound.

There are many avenues that can be pursued in future work, such as using the saved trail to help make inprocessing techniques faster or using the saved trail to learn multiple clauses from a single conflict. It is also possible to combine trail saving with chronological backtracking, but it would require further work to determine whether or not this would be useful and how to best approach it.

# References

1. Audemard, G., Simon, L.: Predicting learnt clauses quality in modern SAT solvers. In: Boutilier, C. (ed.) IJCAI 2009, Proceedings of the 21st International Joint Conference on Artificial Intelligence, Pasadena, California, USA, 11–17 July 2009, pp. 399–404 (2009). http://ijcai.org/Proceedings/09/Papers/074.pdf
2. Gent, I.P.: Optimal implementation of watched literals and more general techniques. J. Artif. Intell. Res. **48**, 231–251 (2013). https://doi.org/10.1613/jair.4016
3. Heule, M., Järvisalo, M., Suda, M. (eds.): Proceedings of SAT Competition 2018: Solver and Benchmark Descriptions. University of Helsinki (2018). http://hdl.handle.net/10138/237063
4. Heule, M., Järvisalo, M., Suda, M. (eds.): Proceedings of SAT Race 2019: Solver and Benchmark Descriptions. University of Helsinki (2019). http://hdl.handle.net/10138/306988
5. Hickey, R., Bacchus, F.: Speeding up assumption-based SAT. In: Janota, M., Lynce, I. (eds.) SAT 2019. LNCS, vol. 11628, pp. 164–182. Springer, Cham (2019). https://doi.org/10.1007/978-3-030-24258-9_11
6. Jiang, C., Zhang, T.: Partial backtracking in CDCL solvers. In: McMillan, K., Middeldorp, A., Voronkov, A. (eds.) LPAR 2013. LNCS, vol. 8312, pp. 490–502. Springer, Heidelberg (2013). https://doi.org/10.1007/978-3-642-45221-5_33
7. Luo, M., Li, C., Xiao, F., Manyà, F., Lü, Z.: An effective learnt clause minimization approach for CDCL SAT solvers. In: Sierra, C. (ed.) Proceedings of the Twenty-Sixth International Joint Conference on Artificial Intelligence, IJCAI 2017, Melbourne, Australia, 19–25 August 2017, pp. 703–711. ijcai.org (2017). https://doi.org/10.24963/ijcai.2017/98
8. Möhle, S., Biere, A.: Backing backtracking. In: Janota, M., Lynce, I. (eds.) SAT 2019. LNCS, vol. 11628, pp. 250–266. Springer, Cham (2019). https://doi.org/10.1007/978-3-030-24258-9_18
9. Moskewicz, M.W., Madigan, C.F., Zhao, Y., Zhang, L., Malik, S.: Chaff: engineering an efficient SAT solver. In: Proceedings of the 38th Design Automation Conference, DAC 2001, Las Vegas, NV, USA, 18–22 June 2001, pp. 530–535. ACM (2001). https://doi.org/10.1145/378239.379017
10. Nadel, A.: Understanding and improving a modern SAT solver. Ph.D. thesis, Tel Aviv University (2009)
11. Nadel, A., Ryvchin, V.: Chronological backtracking. In: Beyersdorff, O., Wintersteiger, C.M. (eds.) SAT 2018. LNCS, vol. 10929, pp. 111–121. Springer, Cham (2018). https://doi.org/10.1007/978-3-319-94144-8_7
12. Nieuwenhuis, R., Oliveras, A., Tinelli, C.: Solving SAT and SAT modulo theories: from an abstract Davis-Putnam-Logemann-Loveland procedure to DPLL(T). J. ACM **53**(6), 937–977 (2006). https://doi.org/10.1145/1217856.1217859
13. Silva, J.P.M., Lynce, I., Malik, S.: Conflict-driven clause learning SAT solvers. In: Handbook of Satisfiability, pp. 131–153. IOS Press (2009). https://doi.org/10.3233/978-1-58603-929-5-131
14. van der Tak, P., Ramos, A., Heule, M.: Reusing the assignment trail in CDCL solvers. JSAT **7**(4), 133–138 (2011). https://satassociation.org/jsat/index.php/jsat/article/view/89
15. Xiao, F., Luo, M., Li, C.M., Manya, F., Lü, Z.: MapleLRB LCM, Maple LCM, Maple LCM dist, MapleLRB LCMoccRestart and glucose-3.0+ width in SAT competition 2017. In: Balyo, T., Heule, M.J.H., Järvisalo, M.J. (eds.) Proceedings of SAT Competition 2017: Solver and Benchmark Descrptions, pp. 22–23. University of Helsinki (2017). http://hdl.handle.net/10138/224324

# Four Flavors of Entailment

Sibylle Möhle[1]([✉])[iD], Roberto Sebastiani[2][iD], and Armin Biere[1][iD]

[1] Johannes Kepler University Linz, Linz, Austria
`sibylle.moehle-rotondi@jku.at`
[2] DISI, University of Trento, Trento, Italy

**Abstract.** We present a novel approach for enumerating partial models of a propositional formula, inspired by how theory solvers and the SAT solver interact in lazy SMT. Using various forms of dual reasoning allows our CDCL-based algorithm to enumerate partial models with no need for exploring and shrinking full models. Our focus is on model enumeration without repetition, with potential applications in weighted model counting and weighted model integration for probabilistic inference over Boolean and hybrid domains. Chronological backtracking renders the use of blocking clauses obsolete. We provide a formalization and examples. We further discuss important design choices for a future implementation related to the strength of dual reasoning, including unit propagation, using SAT or QBF oracles.

## 1 Introduction

Model enumeration is a key task in various activities, such as lazy Satisfiability Modulo Theories [29], predicate abstraction [13], software product line engineering [7], model checking [2,18,31], and preimage computation [14,30].

Whereas in some applications enumerating models multiple times causes no harm, in others avoiding repetitions is crucial. Examples are weighted model counting (WMC) for probabilistic reasoning in Boolean domains and weighted model integration (WMI), which generalizes WMC for hybrid domains [22,23]. There, the addends are *partial* satisfying assignments, i.e., some variables remain unassigned. Each of these assignments represents a set of *total* assignments, and consequently, the number of the addends is reduced. A formula might be represented in a concise manner by the disjunction of its pairwise contradicting partial models, which is of interest in digital circuit synthesis [1]. Partial models are relevant also in predicate abstraction [13], preimage computation [14,30], and existential quantification [4]. They can be obtained by shrinking total models [32]. Alternatively, dual reasoning, where the formula is considered together with its negation, allows for pruning the search space early and detecting partial models. It is also applied in the context of model counting [3,19].

If only a subset $X$ of the variables is significant, the models are *projected* onto these *relevant* variables. We say that we *existentially quantify* the formula over the *irrelevant* variables $Y$ and write $\exists Y \, [\, F(X,Y) \,]$, where $F(X,Y)$ is a formula over variables $X$ and $Y$ such that $X \cap Y = \emptyset$. Projected model enumeration

© Springer Nature Switzerland AG 2020
L. Pulina and M. Seidl (Eds.): SAT 2020, LNCS 12178, pp. 62–71, 2020.
https://doi.org/10.1007/978-3-030-51825-7_5

occurs in automotive configuration [34], existential quantifier elimination [4], image computation [9,10], predicate abstraction [13], and bounded model checking [31].

To avoid finding models multiple times, blocking clauses might be added to the formula under consideration [11,18]. This method suffers from a potentially exponential blowup of the formula and consequent slowdown of unit propagation. Toda and Soh [33] address this issue by a variant of conflict analysis, which is motivated by Gebser et al. [8] and is exempt from blocking clauses. Chronological backtracking in Grumberg et al. [9] and our previous work [21] ensures that the search space is traversed in a systematic manner, similarly to DPLL [5], and the use of blocking clauses is avoided. Whenever a model is found, the last (relevant) decision literal is flipped. No clause asserting this flipped decision is added, which might cause problems during later conflict analysis. This problem is addressed by modifying the implication graph [9] or by an alternative first UIP scheme [33].

*Our contribution.* We lift the way how theory and SAT solver interact in SMT to propositional projected model enumeration without repetition. Based on the notion of logical entailment, combined with dual reasoning, our algorithm detects partial models in a forward manner, rendering model shrinking superfluous. The test for entailment is crucial in our algorithm. Anticipating a future implementation, we present it in four flavors with different strengths together with examples. The main enumeration engine uses chronological CDCL [25], is exempt from blocking clauses, and thus does not suffer from a formula blowup. Its projection capabilities make it suitable also for applications requiring model enumeration with projection. We conclude our presentation by a formalization of our algorithm and a discussion of the presented approach. Our work is motivated by projected model counting and weighted model integration. We therefore focus on (projected) model enumeration without repetition. Contrarily to Oztok and Darwiche [26], we use an oracle and build a Disjoint Sum-of-Products (DSOP) [1]. The work by Lagniez and Marquis [12] is orthogonal to ours. It is led by a disjunctive decomposition of the formula under consideration after a full model is found and also decomposes it into disjoint connected components.

## 2    Preliminaries

A *literal* $\ell$ is a variable $v$ or its negation $\neg v$. We denote by $V(\ell)$ the variable of $\ell$ and extend this notation to sets and sequences of literals. We write $\bar{\ell}$ for the complement of $\ell$, i.e., $\bar{\ell} = \neg \ell$, defining $\neg\neg\ell = \ell$. A formula in *conjunctive normal form (CNF)* over variables $V$ is defined as a conjunction of *clauses*, which are disjunctions of literals with variable in $V$, whereas a formula in *disjunctive normal form (DNF)* is a disjunction of *cubes*, which are conjunctions of literals. We might interpret formulae, clauses, and cubes also as sets of clauses or cubes, and literals and write $C \in F$ for referring to a clause or cube $C$ in a formula $F$ and $\ell \in C$ where $\ell$ is a literal in $C$. The empty CNF formula and the empty cube are denoted by 1, the empty DNF formula and the empty clause by 0.

A *total assignment* is a mapping from the set of variables $V$ to the truth values 1 (true) and 0 (false). A *trail* $I = \ell_1 \ldots \ell_n$ is a non-contradictory sequence of literals, which might also be interpreted as a *(possibly partial) assignment*, where $I(\ell) = 1$ if $\ell \in I$ and $I(\ell) = 0$ if $\neg\ell \in I$. We denote the empty trail by $\varepsilon$ and the set of variables of the literals on $I$ by $V(I)$. Trails and literals might be concatenated, written $I = JK$ and $I = J\ell$, provided $V(J) \cap V(K) = \emptyset$ and $V(J) \cap V(\ell) = \emptyset$. We interpret $I$ also as a set of literals and write $\ell \in I$ to denote a literal $\ell$ on $I$. The *residual* of a formula $F$ under a trail $I$, written $F|_I$, is obtained by replacing the literals $\ell$ in $F$, where $V(\ell) \in V(I)$, by their truth value, and by recursively propagating truth values through Boolean connectives. In particular, for a CNF formula this consists in removing satisfied clauses as well as falsified literals. By "=" in $F|_I = 1$ and $F|_I = 0$, notably by omitting quantifiers, we explicitly mean syntactical equality and consider the (possibly partial) assignment represented by $I$, i.e., only the literals on $I$. The notion of residual is extended similarly to clauses and literals. We denote by $X - I$ the unassigned variables in $X$. By $\pi(I, X)$ we refer to the projection of $I$ onto $X$ and extend this notation to sets of literals.

The *decision level function* $\delta \colon V \mapsto \mathbb{N} \cup \{\infty\}$ returns the decision level of a variable $v$. If $v$ is unassigned, we have $\delta(v) = \infty$, and $\delta$ is updated whenever $v$ is assigned or unassigned. We define $\delta(\ell) = \delta(V(\ell))$ for a literal $\ell$, $\delta(C) = \max\{\delta(\ell) \mid \ell \in C\}$ for a clause $C \neq 0$, and $\delta(I) = \max\{\delta(\ell) \mid \ell \in I\}$ for a sequence of literals $I \neq \varepsilon$. Further, $\delta(L) = \max\{\delta(\ell) \mid \ell \in L\}$ for a set of literals $L \neq \emptyset$. We define $\delta(0) = \delta(\varepsilon) = \delta(\emptyset) = 0$. The updated function $\delta$, in which $V(\ell)$ is assigned to decision level $d$, is denoted by $\delta[\ell \mapsto d]$. If all literals in $V$ are unassigned, we write $\delta[V \mapsto \infty]$ or $\delta \equiv \infty$. The function $\delta$ is left-associative, i.e., $\delta[I \mapsto \infty][\ell \mapsto d]$ first unassigns all literals on $I$ and then assigns literal $\ell$ to decision level $d$. We mark the decision literals on $I$ by a superscript, i.e., $\ell^d$, and denote the set consisting of the decision literals on $I$ by $\mathsf{decs}(I) = \{\ell \mid \ell^d \in I\}$. Similarly, we denote the set of unit literals in $F$ or its residual under $I$ by $\mathsf{units}(F)$ or $\mathsf{units}(F|_I)$. Trails are partitioned into *decision levels*, and $I_{\leqslant n}$ is the subsequence of $I$ consisting of all literals $\ell$ where $\delta(\ell) \leqslant n$.

Following Sebastiani [28], we say that a (partial) assignment $I$ *entails* a formula $F$, if all total extensions of $I$ satisfy $F$. In this work it was noticed that, if $I$ entails $F$, we can not conclude that $F|_I = 1$, but only that $F|_I$ is valid. Consider as an example $F = (x \wedge y) \vee (x \wedge \neg y)$ over variables $X = \{x\}$ and $Y = \{y\}$ and the trail $I = x$ ranging over $X \cup Y$. The possible extensions of $I$ are $I' = xy$ and $I'' = x\neg y$. We have $F|_{I'} = F|_{I''} = 1$, therefore $I$ entails $F$. Notice that $F|_I = y \vee \neg y$ is valid but it syntactically differs from 1.

## 3   Early Pruning for Projected Model Enumeration

Our approach is inspired by how theory solvers and the SAT solver interact in lazy SMT. A general schema is described in Fig. 1. Let $F(X, Y)$ be a formula over relevant variables $X$ and irrelevant variables $Y$ such that $X \cap Y = \emptyset$. A SAT solver executes enumeration, either DPLL-based [5,6] or CDCL-based [17,24],

**Input:**    formula $F(X, Y)$ over variables $X \cup Y$ such that $X \cap Y = \emptyset$,
              trail $I$, decision level function $\delta$

**Output:**   DNF $M$ consisting of models of $F$ projected onto $X$

  Enumerate ( $F$ )

```
 1   I := ε                              // empty trail
 2   δ := ∞                              // unassign all variables
 3   M := 0                              // empty DNF
 4   forever do
 5       C := PropagateUnits ( F, I, δ )
 6       if C ≠ 0 then                   // conflict
 7           c := δ(C)                    // conflict level
 8           if c = 0 then
 9               return M
10           AnalyzeConflict ( F, I, C, c )
11       else if all variables in X ∪ Y are assigned then    // I is total model
12           if V(decs(I)) ∩ X = ∅ then  // no relevant decision left
13               return M ∨ π(I, X)       // record I projected onto X
14           M := M ∨ π(I, X)
15           b := δ(decs(π(I, X)))        // highest relevant decision level
16           Backtrack ( I, b − 1 )       // flip last relevant decision
17       else if Entails ( I, F ) then   // I is partial model
18           if V(decs(I)) ∩ X = ∅ then  // no relevant decision left
19               return M ∨ π(I, X)       // record I projected onto X
20           M := M ∨ π(I, X)
21           b := δ(decs(π(I, X)))        // highest relevant decision level
22           Backtrack ( I, b − 1 )       // flip last relevant decision
23       else
24           Decide ( I, δ )
```

**Fig. 1.** Early pruning for projected model enumeration. Lines 1–16 and 23–24 list CDCL-based model enumeration with chronological backtracking. If after unit propagation no conflict occurs and not all variables are assigned, an oracle might be called to check whether $I$ entails $F$ (line 17). If Entails returns 1, the relevant decision literal with highest decision level might be flipped. Otherwise, a decision is taken (line 24). Notice that lines 12–16 and lines 18–22 are identical.

on $F$, maintaining a trail $I$ over variables $X \cup Y$. In lines 1–16 and 23–24, we consider the CDCL-based enumeration engine with chronological backtracking of our framework [21]. Now assume unit propagation has been carried out until completion, no conflict occurred and there are still unassigned variables (line 17). The trail $I$ already might entail $F$, although $F|_I \neq 1$. We can check whether $I$ entails $F$ by an incremental call to an "oracle" [16] Entails on $I$ and $F$. If Entails returns 1, then the procedure does not need to test any total extension of $I$, since all of them are models of $F$. It can proceed and flip the relevant decision literal with highest decision level (line 21–22). If Entails returns 0, a decision needs to be taken (line 24). Notice that lines 12–16 and lines 18–22 are identical. Our method is based on chronological backtracking and follows the scheme in our framework [21], the functions PropagateUnits() and AnalyzeConflict() are taken from our previous work [20]. Entails plays the role of an "early pruning call" to a theory solver in SMT, and $F$ plays the role of the theory [29]. Redundant work is saved by applying unit propagation until completion before calling Entails.

*Quantified Entailment Condition.* We use quantifiers with QBF semantics, and quantified formulae are always closed. A closed QBF formula evaluates to either 1 or 0. Consider $\varphi = \forall X \forall Y [F|_I]$, where $F$ is a formula over variables $X \cup Y$ and the trail $I$ ranges over $X \cup Y$. In $\varphi$, the remaining variables $(X \cup Y) - I$ are quantified. Accordingly, by $\forall X \forall Y [F|_I] = 1$, we express that all possible total extensions of $I$ satisfy $F$, in contrast to $F|_I = 1$, expressing syntactic equality according to Sect. 2. The latter fact implies the former, but not vice versa.

*Entailment Under Projection.* If Entails implements the notion of entailment described in Sect. 2, then by calling it on $I$ and $F$, we check whether $F|_J = 1$ for all total extensions $J$ of $I$, i.e., whether $\forall X \forall Y [F|_I] = 1$. However, since we are interested in the models of $F$ projected onto $X$, it suffices to check that for each possible assignment $J_X$ to the unassigned variables in $X$, there exists *one* assignment $J_Y$ to the unassigned variables in $Y$ such that $F|_{I'} = 1$ where $I' = I \cup J_X \cup J_Y$. In essence, we need to determine the truth of the QBF formula $\forall X \exists Y [F|_I]$, which, in general, might be expensive, computationally. In some cases, however, a computationally cheaper (but weaker) test might be sufficient. Entails in line 17 of Enumerate can be seen as a black box pooling four entailment tests of different strengths, which we discuss in the next section.

## 4   Testing Entailment

Consider the original entailment condition, $\forall X \forall Y [F|_I] = 1$. Now we have that $\forall X \forall Y [F|_I] = 1 \iff \exists X \exists Y [\neg F|_I] = 0$. Therefore, to check whether $I$ entails $F$, a SAT solver might be called to check whether $\neg F \wedge I$ is unsatisfiable. The SAT solver returns "unsat", if and only if $I$ entails $F$. This observation motivates the use of dual reasoning for testing entailment in cases where cheaper tests fail. We present four flavors of the entailment test and provide examples.

1) $F|_I = 1$ *(syntactic check).* If $F|_I = 1$, also $\forall X \forall Y [F|_I] = 1$, and $I$ entails $F$.
2) $F|_I \approx 1$ *(incomplete check in* **P***).* Alternatively, if $F|_I$ differs from 1, an incomplete algorithm might be used, to check whether $\neg F \wedge I$ is unsatisfiable, by for instance executing only unit propagation or aborting after a predefined number of decision levels.
3) $F|_I \equiv 1$ *(semantic check in* **coNP***).* A SAT oracle runs on $\neg F \wedge I$ until termination. Basically, it checks the unsatisfiability of $\neg F \wedge I$, i.e., whether it holds that $\exists X \exists Y [\neg F|_I] = 0$. If it answers "unsat", then $I$ entails $F$.
4) $\forall X \exists Y [F|_I] = 1$ *(check in* $\mathbf{\Pi_2^P}$*).* A QBF oracle is called to check whether the 2QBF formula $\forall X \exists Y [F|_I]$ is 1.

Modern SAT solvers mostly work on CNFs. Thus, following our dualization approach [19], we may convert $F(X, Y)$ and $\neg F(X, Y)$ into CNF formulae $P(X, Y, S)$ and $N(X, Y, T)$, where $S$ and $T$ denote the variables introduced by the CNF encoding. Notice that $I \wedge \neg F$ is unsatisfiable iff $I \wedge N$ is unsatisfiable.

**Table 1.** Examples of formulae $F$ over relevant variables $X$ and irrelevant variables $Y$. For a concise representation of formulae, we represent conjunction by juxtaposition and negation by overline. In all examples, $I$ entails $F$ projected onto $X$. The entailment tests are listed from left to right in ascending order by their strength. Here, "✓" denotes the fact that $I$ passes the test in the column, if applied to the formula in the row.

| $F$ | $X$ | $Y$ | $I$ | $= 1$ | $\approx 1$ | $\equiv 1$ | 2QBF |
|---|---|---|---|---|---|---|---|
| $(x_1 \vee y \vee x_2)$ | $\{x_1, x_2\}$ | $\{y\}$ | $x_1$ | ✓ | ✓ | ✓ | ✓ |
| $x_1 y \vee \bar{y} x_2$ | $\{x_1, x_2\}$ | $\{y\}$ | $x_1 x_2$ | | ✓ | ✓ | ✓ |
| $x_1(\overline{x_2}\,\bar{y} \vee \overline{x_2} y \vee x_2 \bar{y} \vee x_2 y)$ | $\{x_1, x_2\}$ | $\{y\}$ | $x_1$ | | | ✓ | ✓ |
| $x_1(x_2 \leftrightarrow y)$ | $\{x_1, x_2\}$ | $\{y\}$ | $x_1$ | | | | ✓ |

Table 1 lists four examples, which differ in the strength of the required entailment test. The first column lists the formula $F$, the second and third column show the definitions of $X$ and $Y$. For a concise representation of formulae, we represent conjunction by juxtaposition and negation by overline. The fourth column contains the current trail $I$. The fifth to eighth column denote the tests, in ascending order by their strength: $F|_I = 1$, $F|_I \approx 1$, $F|_I \equiv 1$, $\forall X \exists Y [F|_I] = 1$. In all examples, $I$ entails $F$, and "✓" denotes the fact that $I$ passes the test in the column, if applied to the formula in the row.

Consider the first example, $F = (x_1 \vee y \vee x_2)$ and $I = x_1$. We have $F|_I = 1$, and $I$ entails $F$, which is detected by the syntactic check. For the second example, $F = x_1 y \vee \bar{y} x_2$, we have $F|_I = y \vee \bar{y}$, which is valid, but it syntactically differs from 1. The SAT solver therefore calls Entails on $\neg F \wedge I$. For $\neg F = (\overline{x_1} \vee \bar{y})(y \vee \overline{x_2})$, we find $\neg F|_I = (\bar{y})(y)$. After propagating $\bar{y}$, a conflict at decision level zero occurs, hence Entails returns 1, and an incomplete test is sufficient. In this example, $\neg F$ is already in CNF. The key idea conveyed by it can easily be lifted to the case where additional variables are introduced by the CNF transformation of $\neg F$. For the third example, $F = x_1(\overline{x_2}\,\bar{y} \vee \overline{x_2} y \vee x_2 \bar{y} \vee x_2 y)$, both $P|_I$ and $N|_I$ are undefined and contain no units. However, $N|_I$ is unsatisfiable, the SAT oracle call on $N \wedge I$ terminates with "unsat", and Entails returns 1. Hence, this example requires at least a SAT oracle. For the last example, $F = x_1(x_2 \leftrightarrow y)$, we define

$$P = (x_1)(s_1 \vee s_2)(\overline{s_1} \vee x_2)(\overline{s_1} \vee y)(\overline{s_2} \vee \overline{x_2})(\overline{s_2} \vee \bar{y}) \quad \text{with } S = \{s_1, s_2\} \text{ and}$$
$$N = (\overline{x_1} \vee t_1 \vee t_2)(\overline{t_1} \vee x_2)(\overline{t_1} \vee \bar{y})(\overline{t_2} \vee \overline{x_2})(\overline{t_2} \vee y) \quad \text{with } T = \{t_1, t_2\}$$

We have $P|_I \neq 1$. Neither $P|_I$ nor $N|_I$ contains a unit literal, hence the incomplete test is too weak. Assume a SAT solver is called to check unsatisfiability of $N \wedge I$, and $x_2$ is decided first. After propagating $\overline{t_2}$, $t_1$ and $\bar{y}$, a total model of $N$ is found. The SAT solver answers "sat", and Entails returns 0. A QBF solver checking $\varphi = \forall X \exists Y [x_2 y \vee \overline{x_2}\,\bar{y}]$ returns 1. In fact, $\varphi$ is true for $I = x_2 y$ and $I = \overline{x_2}\,\bar{y}$, and Entails answers 1. Thus, at least a QBF oracle is needed.

EndTrue:    $(F, I, M, \delta) \rightsquigarrow_{\mathsf{EndTrue}} M \vee m$   if   $V(\mathrm{decs}(I)) \cap X = \emptyset$ and
         $m \overset{\mathrm{def}}{=} \pi(I, X)$   and   $\forall X \exists Y \, [\, F|_I \,] = 1$

EndFalse:   $(F, I, M, \delta) \rightsquigarrow_{\mathsf{EndFalse}} M$   if   exists $C \in F$   and   $C|_I = 0$   and
         $\delta(C) = 0$

---

Unit:       $(F, I, M, \delta) \rightsquigarrow_{\mathsf{Unit}} (F, I\ell, M, \delta[\ell \mapsto a])$   if   $F|_I \neq 0$   and
         exists $C \in F$   with   $\{\ell\} = C|_I$   and   $a \overset{\mathrm{def}}{=} \delta(C \setminus \{\ell\})$

---

BackTrue:   $(F, I, M, \delta) \rightsquigarrow_{\mathsf{BackTrue}} (F, UK\ell, M \vee m, \delta[L \mapsto \infty][\ell \mapsto b])$   if
         $UV \overset{\mathrm{def}}{=} I$   and   $D \overset{\mathrm{def}}{=} \overline{\pi(\mathrm{decs}(I), X)}$   and   $b + 1 \overset{\mathrm{def}}{=} \delta(D) \leqslant \delta(I)$   and
         $\ell \in D$   and   $b = \delta(D \setminus \{\ell\}) = \delta(U)$   and   $m \overset{\mathrm{def}}{=} \pi(I, X)$   and
         $K \overset{\mathrm{def}}{=} V_{\leqslant b}$   and   $L \overset{\mathrm{def}}{=} V_{>b}$   and   $\forall X \exists Y \, [\, F|_I \,] = 1$

BackFalse: $(F, I, M, \delta) \rightsquigarrow_{\mathsf{BackFalse}} (F, UK\ell, M, \delta[L \mapsto \infty][\ell \mapsto j])$   if
         exists $C \in F$   and   exists $D$   with   $UV \overset{\mathrm{def}}{=} I$   and   $C|_I = 0$   and
         $c \overset{\mathrm{def}}{=} \delta(C) = \delta(D) > 0$   such that   $\ell \in D$   and   $\bar{\ell} \in \mathrm{decs}(I)$   and
         $\bar{\ell}|_V = 0$   and   $F \wedge \overline{M} \models D$   and   $j \overset{\mathrm{def}}{=} \delta(D \setminus \{\ell\})$   and
         $b \overset{\mathrm{def}}{=} \delta(U) = c - 1$   and   $K \overset{\mathrm{def}}{=} V_{\leqslant b}$   and   $L \overset{\mathrm{def}}{=} V_{>b}$

---

DecideX:   $(F, I, M, \delta) \rightsquigarrow_{\mathsf{DecideX}} (F, I\ell^d, M, \delta[\ell \mapsto d])$   if   $F|_I \neq 0$   and
         $\mathrm{units}(F|_I) = \emptyset$   and   $\delta(\ell) = \infty$   and   $d \overset{\mathrm{def}}{=} \delta(I) + 1$   and   $V(\ell) \in X$

DecideY:   $(F, I, M, \delta) \rightsquigarrow_{\mathsf{DecideY}} (F, I\ell^d, M, \delta[\ell \mapsto d])$   if   $F|_I \neq 0$   and
         $\mathrm{units}(F|_I) = \emptyset$   and   $\delta(\ell) = \infty$   and   $d \overset{\mathrm{def}}{=} \delta(I) + 1$   and   $V(\ell) \in Y$   and
         $X - I = \emptyset$

**Fig. 2.** Rules for Enumerate.

## 5   Formalization

The algorithm listed in Fig. 1 can be expressed by means of a formal calculus (Fig. 2). It extends our previous calculus [21] by projection and by a generalized precondition modeling an incremental call to an oracle for checking entailment (lines 17–22 in function Enumerate). Notably, in our work [21], only total models are found, while entailment in our actual work enables the detection of partial models. The variables in $Y$ and $S$ (from the CNF encoding) are treated equally with respect to unit propagation and decision. We therefore merge those two variable sets into $Y$ to simplify the formalization. This does not affect the outcome of the entailment test. In favor of a concise description of the rules, we emphasize the differences to our previous framework [21] and refer to this work for more details.

     The procedure terminates as soon as either a conflict at decision level zero occurs (rule EndFalse) or a possibly partial model is found and $I$ contains no relevant decision literal (rule EndTrue). Requiring that no relevant decision is left on the trail prevents the recording of redundant models. The projection of $I$ onto $X$ is recorded. Rule Unit remains unchanged except for the missing precondition $F|_I \neq 1$. If $I$ entails $F$ and contains relevant decision literals, the one at the highest decision level is flipped, and the projection of $I$ onto $X$ is recorded (rule BackTrue). Requiring that the last relevant decision literal is flipped prevents the recording of redundant models. Rule BackFalse remains unchanged. A decision is taken whenever $F|_I \neq 0$ and $F|_I$ contains no unit. Relevant variables are prioritized (rule DecideX) over irrelevant ones (rule DecideY).

Although not mandatory for correctness, the applicability of rule Unit might be restricted to the case where $F|_I \neq 1$. Similarly, a decision might be taken only if $I$ does not entail $F$. Notice that in rules Unit, DecideX, and DecideY, the precondition $F|_I \neq 0$ can also be omitted.

# 6 Conclusion

In many applications (projected) partial models play a central role. For this purpose, we have presented an algorithm and its formalization inspired by how theory solvers and the SAT solver interact in SMT. The basic idea was to detect partial assignments entailing the formula on-the-fly. We presented entailment tests of different strength and computational cost and discussed examples.

The syntactic check "$F|_I = 1$" is cheapest, using clause watches or counters for keeping track of the number of satisfied clauses or alternatively the number of assigned variables (line 11 in Fig. 1). It is also weakest, since $F|_I$ must syntactically coincide with 1. The incomplete check, denoted by "$F|_I \approx 1$", is slightly more involved. It calls a SAT solver on the negation of the formula, restricted, e.g., to unit propagation or a limited number of decision levels, and also might return "unknown". The SAT oracle executes an unsatisfiability check of $\neg F \wedge I$, given a (partial) assignment $I$, which might be too restrictive. The QBF oracle is the most powerful test, but also the most expensive one. It captures entailment under projection in a precise manner expressed by $\forall X \exists Y [F|_I] = 1$. Combining dual reasoning with oracle calls allows to avoid shrinking of total models. Finally, chronological CDCL renders the use of blocking clauses superfluous.

We claim that this is the first method combining dual reasoning and chronological CDCL for partial model detection. It is anticipated that applications with short partial models benefit most, since oracle calls might be expensive. We plan to implement our method and validate its competitiveness on applications from weighted model integration and model counting with or without projection. We also plan to investigate methods concerning the implementation of QBF oracles required by flavour 4), e.g., dependency schemes introduced by Samer and Szeider [27] or incremental QBF solving proposed by Lonsing and Egly [15].

**Acknowledgments.** The work was supported by the LIT Secure and Correct Systems Lab funded by the State of Upper Austria, by the QuaSI project funded by D-Wave Systems Inc., and by the Italian Assocation for Artificial Intelligence (AI*IA). We thank the anonymous reviewers and Mathias Fleury for suggesting many textual improvements.

# References

1. Bernasconi, A., Ciriani, V., Luccio, F., Pagli, L.: Compact DSOP and partial DSOP forms. Theory Comput. Syst. **53**(4), 583–608 (2013). https://doi.org/10.1007/s00224-013-9447-2
2. Biere, A., Cimatti, A., Clarke, E., Zhu, Y.: Symbolic model checking without BDDs. In: Cleaveland, W.R. (ed.) TACAS 1999. LNCS, vol. 1579, pp. 193–207. Springer, Heidelberg (1999). https://doi.org/10.1007/3-540-49059-0_14
3. Biere, A., Hölldobler, S., Möhle, S.: An abstract dual propositional model counter. In: YSIP, CEUR Workshop Proceedings, vol. 1837, pp. 17–26. CEUR-WS.org (2017)
4. Brauer, J., King, A., Kriener, J.: Existential quantification as incremental SAT. In: Gopalakrishnan, G., Qadeer, S. (eds.) CAV 2011. LNCS, vol. 6806, pp. 191–207. Springer, Heidelberg (2011). https://doi.org/10.1007/978-3-642-22110-1_17
5. Davis, M., Logemann, G., Loveland, D.W.: A machine program for theorem-proving. Commun. ACM **5**(7), 394–397 (1962)
6. Davis, M., Putnam, H.: A computing procedure for quantification theory. J. ACM **7**(3), 201–215 (1960)
7. Galindo, J.A., Acher, M., Tirado, J.M., Vidal, C., Baudry, B., Benavides, D.: Exploiting the enumeration of all feature model configurations: a new perspective with distributed computing. In: SPLC, pp. 74–78. ACM (2016)
8. Gebser, M., Kaufmann, B., Neumann, A., Schaub, T.: *clasp*: a conflict-driven answer set solver. In: Baral, C., Brewka, G., Schlipf, J. (eds.) LPNMR 2007. LNCS (LNAI), vol. 4483, pp. 260–265. Springer, Heidelberg (2007). https://doi.org/10.1007/978-3-540-72200-7_23
9. Grumberg, O., Schuster, A., Yadgar, A.: Memory efficient all-solutions SAT solver and its application for reachability analysis. In: Hu, A.J., Martin, A.K. (eds.) FMCAD 2004. LNCS, vol. 3312, pp. 275–289. Springer, Heidelberg (2004). https://doi.org/10.1007/978-3-540-30494-4_20
10. Gupta, A., Yang, Z., Ashar, P., Gupta, A.: SAT-based image computation with application in reachability analysis. In: Hunt, W.A., Johnson, S.D. (eds.) FMCAD 2000. LNCS, vol. 1954, pp. 391–408. Springer, Heidelberg (2000). https://doi.org/10.1007/3-540-40922-X_22
11. Klebanov, V., Manthey, N., Muise, C.: SAT-based analysis and quantification of information flow in programs. In: Joshi, K., Siegle, M., Stoelinga, M., D' Argenio, P.R. (eds.) QEST 2013. LNCS, vol. 8054, pp. 177–192. Springer, Heidelberg (2013). https://doi.org/10.1007/978-3-642-40196-1_16
12. Lagniez, J., Marquis, P.: A recursive algorithm for projected model counting. In: AAAI. pp. 1536–1543. AAAI Press (2019)
13. Lahiri, S.K., Nieuwenhuis, R., Oliveras, A.: SMT techniques for fast predicate abstraction. In: Ball, T., Jones, R.B. (eds.) CAV 2006. LNCS, vol. 4144, pp. 424–437. Springer, Heidelberg (2006). https://doi.org/10.1007/11817963_39
14. Li, B., Hsiao, M.S., Sheng, S.: A novel SAT all-solutions solver for efficient preimage computation. In: DATE, pp. 272–279. IEEE Computer Society (2004)
15. Lonsing, F., Egly, U.: Incremental QBF solving. In: O' Sullivan, Barry (ed.) CP 2014. LNCS, vol. 8656, pp. 514–530. Springer, Cham (2014). https://doi.org/10.1007/978-3-319-10428-7_38
16. Marques-Silva, J.: Computing with SAT oracles: past, present and future. In: Manea, F., Miller, R.G., Nowotka, D. (eds.) CiE 2018. LNCS, vol. 10936, pp. 264–276. Springer, Cham (2018). https://doi.org/10.1007/978-3-319-94418-0_27

17. Marques-Silva, J.P., Sakallah, K.A.: GRASP: a search algorithm for propositional satisfiability. IEEE Trans. Comput. **48**(5), 506–521 (1999)
18. McMillan, K.L.: Applying SAT methods in unbounded symbolic model checking. In: Brinksma, E., Larsen, K.G. (eds.) CAV 2002. LNCS, vol. 2404, pp. 250–264. Springer, Heidelberg (2002). https://doi.org/10.1007/3-540-45657-0_19
19. Möhle, S., Biere, A.: Dualizing projected model counting. In: Tsoukalas, L.H., Grégoire, É., Alamaniotis, M. (eds.) ICTAI 2018, pp. 702–709. IEEE (2018)
20. Möhle, S., Biere, A.: Backing backtracking. In: Janota, M., Lynce, I. (eds.) SAT 2019. LNCS, vol. 11628, pp. 250–266. Springer, Cham (2019). https://doi.org/10.1007/978-3-030-24258-9_18
21. Möhle, S., Biere, A.: Combining conflict-driven clause learning and chronological backtracking for propositional model counting. In: GCAI, EPiC Series in Computing, vol. 65, pp. 113–126. EasyChair (2019)
22. Morettin, P., Passerini, A., Sebastiani, R.: Efficient weighted model integration via SMT-based predicate abstraction. In: IJCAI, pp. 720–728. ijcai.org (2017)
23. Morettin, P., Passerini, A., Sebastiani, R.: Advanced SMT techniques for weighted model integration. Artif. Intell. **275**, 1–27 (2019)
24. Moskewicz, M.W., Madigan, C.F., Zhao, Y., Zhang, L., Malik, S.: Chaff: engineering an efficient SAT solver. In: DAC, pp. 530–535. ACM (2001)
25. Nadel, A., Ryvchin, V.: Chronological backtracking. In: Beyersdorff, O., Wintersteiger, C.M. (eds.) SAT 2018. LNCS, vol. 10929, pp. 111–121. Springer, Cham (2018). https://doi.org/10.1007/978-3-319-94144-8_7
26. Oztok, U., Darwiche, A.: An exhaustive DPLL algorithm for model counting. J. Artif. Intell. Res. **62**, 1–32 (2018)
27. Samer, M., Szeider, S.: Backdoor sets of quantified Boolean formulas. J. Autom. Reasoning **42**(1), 77–97 (2009)
28. Sebastiani, R.: Are you satisfied by this partial assignment? https://arxiv.org/abs/2003.04225
29. Sebastiani, R.: Lazy satisfiability modulo theories. JSAT **3**(3–4), 141–224 (2007)
30. Sheng, S., Hsiao, M.S.: Efficient preimage computation using a novel success-driven ATPG. In: DATE, pp. 10822–10827. IEEE Computer Society (2003)
31. Shtrichman, O.: Tuning SAT checkers for bounded model checking. In: Emerson, E.A., Sistla, A.P. (eds.) CAV 2000. LNCS, vol. 1855, pp. 480–494. Springer, Heidelberg (2000). https://doi.org/10.1007/10722167_36
32. Tibebu, A.T., Fey, G.: Augmenting all solution SAT solving for circuits with structural information. In: DDECS, pp. 117–122. IEEE (2018)
33. Toda, T., Soh, T.: Implementing efficient all solutions SAT solvers, ACM J. Exp. Algorithmics, 21(1), 1.12:1–1.12:44 (2016)
34. Zengler, C., Küchlin, W.: Boolean quantifier elimination for automotive configuration – a case study. In: Pecheur, C., Dierkes, M. (eds.) FMICS 2013. LNCS, vol. 8187, pp. 48–62. Springer, Heidelberg (2013). https://doi.org/10.1007/978-3-642-41010-9_4

# Designing New Phase Selection Heuristics

Arijit Shaw$^{(\boxtimes)}$ and Kuldeep S. Meel$^{(\boxtimes)}$

School of Computing, National University of Singapore,
Singapore, Singapore
if.arijit@gmail.com, meel@comp.nus.edu.sg

**Abstract.** CDCL-based SAT solvers have transformed the field of auto-mated reasoning owing to their demonstrated efficiency at handling prob-lems arising from diverse domains. The success of CDCL solvers is owed to the design of clever heuristics that enable the tight coupling of differ-ent components. One of the core components is phase selection, wherein the solver, during branching, decides the polarity of the branch to be explored for a given variable. Most of the state-of-the-art CDCL SAT solvers employ *phase-saving* as a phase selection heuristic, which was pro-posed to address the potential inefficiencies arising from *far-backtracking*. In light of the emergence of chronological backtracking in CDCL solvers, we re-examine the efficiency of phase saving. Our empirical evaluation leads to a surprising conclusion: The usage of saved phase and random selection of polarity for decisions following a chronological backtracking leads to an indistinguishable runtime performance in terms of instances solved and PAR-2 score.

We introduce Decaying Polarity Score (DPS) to capture the *trend* of the polarities attained by the variable, and upon observing lack of perfor-mance improvement due to DPS, we turn to a more sophisticated heuris-tic seeking to capture the activity of literals and the trend of polarities: Literal State Independent Decaying Sum (LSIDS). We find the 2019 win-ning SAT solver, Maple_LCM_Dist_ChronoBTv3, augmented with LSIDS solves 6 more instances while achieving a reduction of over 125 seconds in PAR-2 score, a significant improvement in the context of the SAT competition.

## 1 Introduction

Given a Boolean formula $F$, the problem of Boolean Satisfiability (SAT) asks whether there exists an assignment $\sigma$ such that $\sigma$ satisfies $F$. SAT is a funda-mental problem in computer science with wide-ranging applications including bioinformatics [24], AI planning [18], hardware and system verification [7,9], spectrum allocation, and the like. The seminal work of Cook [10] showed that SAT is NP-complete and the earliest algorithmic methods, mainly based on local search and the DPLL paradigm [11], suffered from scalability in practice. The arrival of Conflict Driven Clause Learning (CDCL) in the early '90s [35] ushered

---

The full version of the paper is available at http://arxiv.org/abs/2005.04850.

© Springer Nature Switzerland AG 2020
L. Pulina and M. Seidl (Eds.): SAT 2020, LNCS 12178, pp. 72–88, 2020.
https://doi.org/10.1007/978-3-030-51825-7_6

in an era of sustained interest from theoreticians and practitioners leading to a medley of efficient heuristics that have allowed SAT solvers to scale to instances involving millions of variables [25], a phenomenon often referred to as *SAT revolution* [2, 6, 12, 22, 26–28, 35].

The progress in modern CDCL SAT solving over the past two decades owes to the design and the tight integration of the core components: *branching* [19, 34], *phase selection* [31], *clause learning* [3, 23], *restarts* [4, 14, 16, 21], and *memory management* [5, 30]. The progress has often been driven by the improvement of the state of the art heuristics for the core components. The annual SAT competition [17] is witness to the pattern where development of the heuristics for one core component necessitates and encourages the design of new heuristics for other components to ensure a tight integration.

The past two years have witnessed the (re-)emergence of chronological backtracking, a regular feature of DPLL techniques, after almost a quarter-century since the introduction of non-chronological backtracking (NCB), thanks to Nadel and Ryvchin [28]. The impact of chronological backtracking (CB) heuristics is evident from its quick adoption by the community, and the CB-based solver, Maple_LCM_Dist _ChronoBT [32], winning the SAT Competition in 2018 and a subsequent version, Maple_LCM_Dist _ChronoBTv3, the SAT Race 2019 [15] winner. The 2nd best solver at the SAT Race 2019, CaDiCaL, also implements chronological backtracking. The emergence of chronological backtracking necessitates re-evaluation of the heuristics for the other components of SAT solving.

We turn to one of the core heuristics whose origin traces to the efforts to address the inefficiency arising due to loss of information caused by non-chronological backtracking: the *phase saving* [31] heuristic in the phase selection component. When the solver decides to branch on a variable $v$, the phase selection component seeks to identify the polarity of the branch to be explored by the solver. The idea of phase-saving traces back to the field of constraint satisfaction search [13] and SAT checkers [33], and was introduced in CDCL by Pipatsrisawat and Darwiche [31] in 2007. For a given variable $v$, phase saving returns the polarity of $v$ corresponding to the last time $v$ was assigned (via decision or propagation). The origin of phase saving traces to the observation by Pipatsrisawat and Darwiche that for several problems, the solver may forget a valid assignment to a subset of variables due to non-chronological backtracking and be forced to *re-discover* the earlier assignment. In this paper, we focus on the question: *is phase saving helpful for solvers that employ chronological backtracking? If not, can we design a new phase selection heuristic?*

The primary contribution of this work is a rigorous evaluation process to understand the efficacy of phase saving for all decisions following a chronological backtracking and subsequent design of improved phase selection heuristic. In particular,

1. We observe that in the context of 2019's winning SAT solver Maple_LCM_Dist _ChronoBTv3 (referred to as mldc henceforth)[1], phase saving heuristic for

---

[1] Acronyms in sans serif font denote solvers or solvers with some specific configurations. mldc is used as abbreviation for Maple_LCM_Dist_ChronoBTv3.

decisions following a chronological backtracking performs no better than the random heuristic which assigns positive or negative polarity randomly with probability 0.5.

2. To address the inefficacy of phase saving for decisions following a chronological backtracking, we introduce a new metric, *decaying polarity score* (DPS), and DPS-based phase selection heuristic. DPS seeks to capture the *trend* of polarities assigned to variables with higher priority given to recent assignments. We observe that augmenting mldc with DPS leads to almost the same performance as the default mldc, which employs phase saving as the phase selection heuristic.

3. To meet the dearth of performance gain by DPS, we introduce a sophisticated variant of DPS called *Literal State Independent Decaying Sum* (LSIDS), which performs additive bumping and multiplicative decay. While LSIDS is inspired by VSIDS, there are crucial differences in computation of the corresponding activity of literals that contribute significantly to the performance. Based on empirical evaluation on SAT 2019 instances, mldc augmented with LSIDS, called mldc-lsids-phase solves 6 more instances and achieves the PAR-2 score of 4475 in comparison to 4607 seconds achieved by the default mldc.

4. To determine the generality of performance improvement of mldc-lsids-phase over mldc; we perform an extensive case study on the benchmarks arising from *preimage attack* on SHA-1 cryptographic hash function, a class of benchmarks that achieves significant interest from the security community.

The rest of the paper is organized as follows. We discuss background about the core components of the modern SAT solvers in Sect. 2. Section 3 presents an empirical study to understand the efficacy of phase saving for decisions following a chronological backtracking. We then present DPS-based phase selection heuristic and the corresponding empirical study in Sect. 4. Section 5 presents the LSIDS-based phase selection heuristic. We finally conclude in Sect. 6.

## 2   Background

A literal is a propositional variable $v$ or its negation $\neg v$. A Boolean formula $F$ over the set of variables $V$ is in Conjunctive Normal Form (CNF) if $F$ is expressed as conjunction of clauses wherein each clause is a disjunction over a subset of literals. A truth assignment $\sigma : V \mapsto \{0, 1\}$ maps every variable to 0 (FALSE) or 1 (TRUE). An assignment $\sigma$ is called satisfying assignment or solution of $F$ if $F(\sigma) = 1$. The problem of Boolean Satisfiability (SAT) seeks to ask whether there exists a satisfying assignment of $F$. Given $F$, a SAT solver is expected to return a satisfying assignment of $F$ if there exists one, or a proof of unsatisfiability [36].

### 2.1   CDCL Solver

The principal driving force behind the so-called *SAT revolution* has been the advent of the Conflict Driven Clause Learning (CDCL) paradigm introduced by Marques-Silva and Sakallah [35], which shares syntactic similarities with the DPLL paradigm [11] but is known to be exponentially more powerful in theory. The power of CDCL over DPLL is not just restricted to theory, and its practical impact is evident from the observation that all the winning SAT solvers in the main track have been CDCL since the inception of SAT competition [17].

On a high-level, a CDCL-based solver proceeds with an empty set of assignments and at every time step maintains a *partial assignment*. The solver iteratively assigns a subset of variables until the current partial assignment is determined not to satisfy the current formula, and the solver then backtracks while learning the reason for the unsatisfiability expressed as a conflict clause. The modern solvers perform frequent restarts wherein the partial assignment is set to empty, but information from the run so far is often stored in form of different statistical metrics. We now provide a brief overview of the core components of a modern CDCL solver.

1. **Decision.** The decision component selects a variable $v$, called the *decision variable* from the set of unassigned variables and assigns a truth value, called the *decision phase* to it. Accordingly, a Decision heuristic is generally a combination of two different heuristics – a *branching heuristic* decides the decision variable and a *phase selection heuristic* selects the decision phase. A *decision level* is associated with each of the decision variables while it gets assigned. The count for decision level starts at 1 and keeps on incrementing with every decision.

2. **Propagation.** The propagation procedure computes the direct implication of the current partial assignment. For example, some clauses become *unit* (all but one of the literals are FALSE) with the decisions recently made by the solver. The remaining unassigned literal of that clause is asserted and added to the partial assignment by the propagation procedure. All variables that get assigned as a consequence of the variable $v$ get the same decision level as $v$.

3. **Conflict Analysis.** Propagation may also reveal that the formula is not satisfiable with the current partial assignment. The situation is called a *conflict*. The solver employs a *conflict analysis* subroutine to deduce the reason for unsatisfiability, expressed as a subset of the current partial assignment. Accordingly, the *conflict analysis* subroutine returns the negation of the literals from the subset as a clause $c$, called a *learnt clause* or *conflict clause* which is added to the list of the existing clauses. The clauses in the given CNF formula essentially imply the learnt clause.

4. **Backtrack.** In addition to leading a learnt clause, the solver then seeks to undo a subset of the current partial assignment. To this end, the conflict analysis subroutine computes the backtrack decision level $l$, and then the solver deletes assignment to all the variables with decision level greater than $l$. As the backtrack intimates removing assignment of last decision level only, backtracking for more than one level is also called *non-chronological backtracking* or, *backjumping*.

The solver keeps on repeating the procedures as mentioned above until it finds a satisfying assignment or finds a conflict without any assumption. The ability of modern CDCL SAT solvers to solve real-world problems with millions of variables depends on its highly sophisticated heuristics employed in different components of the solver. Now we discuss some terms related to CDCL SAT solving that we use extensively in the paper.

- *Variable state independent decaying sum* (VSIDS) introduced in Chaff [27] refers to a branching heuristic, where a score called *activity* is maintained for every variable. The variable with the highest *activity* is returned as the decision variable. Among different variations of VSIDS introduced later, the most effective is *Exponential VSIDS* or, EVSIDS [8,20] appeared in MiniSat [12]. The EVSIDS score for variable $v$, $activity[v]$, gets incremented additively by a factor $f$ every time $v$ appears in a learnt clause. The factor $f$ itself also gets incremented multiplicatively after each conflict. A constant factor $g = 1/f$ periodically decrements the *activity* of all the variables. The act of increment is called *bump*, and the decrement is called *decay*. The heuristic is called *state independent* because the *activity* of a variable is not dependent of the current state (e.g., current assumptions) of the solver.
- *Phase saving* [31] is a phase selection heuristic used by almost all solver modern solvers, with few exceptions such as the CaDiCaL solver in SAT Race 19 [15]. Every time the solver backtracks and erases the current truth assignment, phase saving stores the erased assignment. For any variable, only the last erased assignment is stored, and the assignment replaces the older stored assignment. Whenever the branching heuristic chooses a variable $v$ as the decision variable and asks phase saving for the decision phase, phase saving returns the saved assignment.
- *Chronological backtracking.* When a non-chronological solver faces a *conflict*, it *backtracks* for multiple levels. Nadel et al. [28] suggested non-chronological backtracking (NCB) might not always be helpful, and advocated backtracking to the previous decision level. The modified heuristic is called *chronological backtracking* (CB). We distinguish a decision based on whether the last backtrack was chronological or not. If the last backtrack is chronological, we say the solver is in *CB-state*, otherwise the solver is in *NCB-state*.

## 2.2    Experimental Setup

In this work, our methodology for the design of heuristics has focused on the implementation of heuristics on a base solver and conduction of an experimental evaluation on a high-performance cluster for SAT 2019 benchmarks. We now describe our experimental setup in detail. All the empirical evaluations in this paper used this setup, unless mentioned otherwise.

1. **Base Solver:** We implemented the proposed heuristics on top of the solver Maple_LCM_Dist_ChronoBTv3 (mldc), which is the winning solver for SAT Race 2019. Maple_LCM_Dist_ChronoBTv3 is an modification of

Maple_LCM_Dist_ChronoBT (2018), which implements chronological back-tracking on top of Maple_LCM_Dist (2017). Maple_LCM_Dist, in turn, evolved from MiniSat (2006) through Glucose (2009) and MapleSAT (2016). The years in parenthesis represent the year when the corresponding solver was published.

2. **Code Blocks:** The writing style of this paper is heavily influenced from the presentation of MiniSat by Eén and Sörensson [12]. Following Eén and Sörensson, we seek to present implementation details as code blocks that are intuitive yet detailed enough to allow the reader to implement our heuristics in their own solver. Furthermore, we seek to present not only the final heuristic that performed the best, but we also attempt to present closely related alternatives and understand their performance.

3. **Benchmarks:** Our benchmark suite consisted of the entire suite, totaling 400 instances, from SAT Race '19.

4. **Experiments:** We conducted all our experiments on a high-performance computer cluster, with each node consists of E5-2690 v3 CPU with 24 cores and 96 GB of RAM. We used 24 cores per node with memory limit set to 4 GB per core, and all individual instances for each solver were executed on a single core. Following the timeout used in SAT competitions, we put a timeout of 5000 seconds for all experiments, if not otherwise mentioned. In contrast to SAT competition, the significant difference in specifications of the system lies in the size of RAM: our setup allows 4 GB of RAM in comparison to 128 GB of RAM allowed in SAT race '19.

We computed the number of SAT and UNSAT instances the solver can solve with each of the heuristics. We also calculated the PAR-2 score. The PAR-2 score, an acronym for penalized average runtime, used in SAT competitions as a parameter to decide winners, assigns a runtime of two times the time limit (instead of a "not solved" status) for each benchmark not solved by the solver.[2]

# 3 Motivation

The impressive scalability of CDCL SAT solvers owes to the tight coupling among different components of the SAT solvers wherein the design of heuristic is influenced by its impact on other components. Consequently, the introduction of a new heuristic for one particular component requires one to analyze the efficacy of the existing heuristics in other components. To this end, we seek to examine the efficacy of phase saving in the context of recently introduced heuristic, Chronological Backtracking (CB). As mentioned in Sect. 1, the leading SAT solvers have incorporated CB and therefore, we seek to revisit the efficacy of other heuristics in light of CB. As a first step, we focus on the evaluation of phase selection heuristic.

---

[2] All experimental data are available at https://doi.org/10.5281/zenodo.3817476.

Phase saving was introduced to tackle the loss of precious work due to *far-backtracking* [31]. Interestingly, CB was introduced as an alternative to *far-bactracking*, i.e., when the conflict analysis recommends that the solver should backtrack to a level $\hat{l}$ such that $|l - \hat{l}|$ is greater than a chosen threshold (say, *thresh*), CB instead leads the solver to backtrack to the previous level. It is worth noting that if the conflict analysis returns $\hat{l}$ such that $l - \hat{l} < thresh$, then the solver does backtrack to $\hat{l}$. Returning to CB, since the solver in CB-state does not perform far-backtracking, it is not clear if phase saving in CB-state is advantageous. To analyze empirically, we conducted preliminary experiments with mldc, varying the phase-selection heuristics while the solver is performing CB. We fix the phase selection to phase saving whenever the solver performs NCB and vary the different heuristics while the solver performs CB:

1. *Phase-saving:* Choose the saved phase as polarity, default in mldc.
2. *Opposite of saved phase:* Choose the negation of the saved phase for the variable as polarity.
3. *Always false:* The phase is always set to FALSE, a strategy that was originally employed in MiniSat 1.0.
4. *Random:* Randomly choose between FALSE and TRUE.

Our choice of *Random* among the four heuristics was driven by our perception that a phase selection strategy should be expected to perform better than *Random*. Furthermore, to put the empirical results in a broader context, we also experimented with the random strategy for both NCB and CB. The performance of different configurations is presented in Table 1, which shows a comparison in terms of the number of SAT, UNSAT instances solved, and PAR-2 score.

**Table 1.** Performance of mldc on 400 SAT19 benchmarks while aided with different phase selection heuristics. SAT, UNSAT, and total columns indicate the number of SAT, UNSAT, and SAT+UNSAT instances solved by the solver when using the heuristic. A lower PAR2 score indicates a lower average runtime, therefore better performance of the solver.

| Phase selection heuristic used | | SAT | UNSAT | Total | PAR-2 |
|---|---|---|---|---|---|
| In NCB-state | In CB-state | | | | |
| *Random* | *Random* | 133 | 89 | 222 | 5040.59 |
| *Phase-saving* | *Phase-saving* | 140 | 97 | 237 | 4607.61 |
| *Phase-saving* | *Random* | 139 | 100 | 239 | 4537.65 |
| *Phase-saving* | *Always false* | 139 | 98 | 237 | 4597.06 |
| *Phase-saving* | *Opp. of saved phase* | 137 | 98 | 235 | 4649.13 |

We first observe that the mldc solves 237 instances and achieves a PAR-2 score of 4607 – a statistic that will be the baseline throughout the rest of the paper. Next, we observe that usage of random both in CB-state and NCB-state leads to significant degradation of performance: 15 fewer instances solved with an increase of 440 s for PAR-2. Surprisingly, we observe that random phase selection in CB-state while employing phase saving in NCB-state performs as good as phase-saving for CB-state. Even more surprisingly, we do not notice a significant performance decrease even when using *Always false* or *Opposite of saved phase*. These results strongly indicate that phase saving is not efficient when the solver is in CB-state, and motivate the need for a better heuristic. In the rest of the paper, we undertake the task of the searching for a better phase selection heuristic.

## 4   Decaying Polarity Score for Phase Selection

To address the ineffectiveness of phase saving in CB-state, we seek to design a new phase selection heuristic while the solver is in CB-state. As a first step, we view phase saving as remembering only the last assigned polarity and we intend to explore heuristic design based on the recent history of polarities assigned to the variable of interest. Informally, we would like to capture the weighted *trend* of the polarities assigned to the variable with higher weight to the recently assigned polarity. To this end, we maintain a score, represented as a floating-point number, for every variable and referred to as *decaying polarity score* (DPS). Each time the solver backtracks to level $l$, the assignments of all the variables with decision level higher than $l$ are removed from the partial assignment. We update the respective *decaying polarity score* of all these variables, whose assignment gets canceled, using the following formula:

$$dps[v] = pol(v) + dec \times dps[v] \qquad (1)$$

where,

- $dps[v]$ represent the *decaying polarity score* of the variable $v$.
- $pol(v)$ is +1 if polarity was set to TRUE at the last assignment of the variable, $-1$ otherwise.
- The decay factor $dec$ is chosen from $(0, 1)$. The greater the value of $dec$ is, the more preference we put on polarities selected in older conflicts.

Whenever the branching heuristic picks a variable $v$ to branch on, the DPS-based phase selection heuristic returns positive polarity if $dps[v]$ is positive; otherwise, negative polarity is returned.

### 4.1   Experimental Results

To test the efficiency of DPS-based phase selection heuristic, we augmented[3] our base solver, mldc, with DPS-based phase selection heuristic in CB-state. We set the value of $dec = 0.7$. As discussed in Subsect. 2.2, we conducted our empirical evaluation on SAT-19 benchmarks. Table 2 presents the comparison of the number of instances solved and PAR-2 score. We first note that the usage

---

[3] https://github.com/meelgroup/duriansat/tree/decay-pol.

**Table 2.** Performance comparison of decaying polarity score with phase saving on SAT19 instances. mldc-dec-phase represent the solver using DPS.

| System | SAT | UNSAT | Total | PAR-2 |
|--------|-----|-------|-------|-------|
| mldc | 140 | 97 | 237 | 4607.61 |
| mldc-dec-phase | 141 | 96 | 237 | 4604.66 |

of DPS did not result in a statistically significant change in the performance of mldc. It is worth noting that there are significantly many instances where mldc attains more than 20% improvement over mldc-dec-phase and vice-versa. The interesting behavior demonstrated by heuristic indicates, while DPS-based phase selection heuristic fails to attain such an objective, it is possible to design heuristics that can accomplish performance improvement over phase saving. In the next section, we design a more sophisticated scheme that seeks to achieve the above goal.

## 5    LSIDS: A VSIDS Like Heuristic for Phase Selection

We now shift to a more sophisticated heuristic that attempts to not only remember the trend of activity but also aims to capture the *activity* of the corresponding literal. To this end, we introduce a scoring scheme, called Literal State Independent Decay Sum (LSIDS), that performs additive bumping and multiplicative decay, à la VSIDS and EVISDS style. The primary contribution lies in the construction of policies regarding literal bumping. We maintain activity for every literal, and the activity is updated as follows:

1. **Literal bumping** refers to incrementing *activity* for a literal. With every *bump*, the activity for a literal is incremented (additive) by $inc * mult$, where $mult$ is a fixed constant while at every conflict, $inc$ gets multiplied by some factor $g > 1$. *Literal bumping* takes place in the following two different phases of the solver.
   **Reason-based bumping**
      When a clause $c$ is learnt, for all the literals $l_i$ appearing in $c$, the *activity* for $l_i$ is bumped. For example, if we learn the clause that consists of literals $v_5$, $\neg v_6$ and $v_3$, then we bump the *activity* of literals $v_5$, $\neg v_6$ and $v_3$.
   **Assignment-based bumping**
      While an assignment for a variable $v$ gets canceled during backtrack; if the assignment was TRUE, then the solver bumps *activity* for $v$, otherwise the *activity* for $\neg v$ is bumped.
2. **Literal decaying** denotes the incident of multiplying the parameter $inc$ by a factor $>1$ at every conflict. The multiplication of $inc$ implies the next bumps will be done by a higher $inc$. Therefore, the older bumps to *activity* will be relative smaller than the newer bumps. The name *decaying* underscores the

```
void litBumpActivity(Lit l, double mult)
   activity[l] += inc * mult
   if (activity[l] > 1e100)
      litRescore()

void litDecayActivity()
   inc *= 1/decay

void litRescore()
   for (int i = 0; i < nVars(); i++)
      activity[i] *= 1e−100
      activity[¬i] *= 1e−100
      inc *= 1e−100
```

**Fig. 1.** Bump, Decay and Rescore procedures for LSIDS activity.

fact that the effect of increasing *inc* is equivalent to decreasing (or, *decay*-ing) the *activity* of all literals.

3. **Literal rescoring:** As the *activity* gets incremented by a larger and larger factor every time, the value for *activity* reaches the limit of a floating-point number at some time. At this point the *activity* of all the literals are scaled down.

When the branching component returns a variable $v$, the LSIDS-based phase selection return positive if $activity[v] > activity[\neg v]$, and negative otherwise. One can view the proposed scheme as an attempt to capture both the participation of literals in learnt clause generation, in spirit similar to VSIDS, and storing the information about trend, à la phase saving/decay polarity score.

### 5.1   Implementation Details

Figure 1 shows the methods to bump and decay the LSIDS scores. Figure 2 shows blocks of code from MiniSat, where the activity of literals is bumped. Figure 3 showcases the subroutine to pick the branching literal based on LSIDS. Of particular note is the setting of *mult* to 2 for assignment-based bumping while setting *mult* to 0.5 for Reason-based bumping. In order to maintain consistency with constants in EVSIDS, the constants in *litRescore* are the same as that of EVSIDS employed in the context of branching in mldc.

### 5.2   Experimental Results

To test the efficiency of LSIDS as a phase selection heuristic, we implemented[4] the heuristic on mldc, replacing the existing phase saving heuristic. We call the mldc augmented with LSIDS phase selection heuristic as mldc-lsids-phase. Similar

---

[4] https://github.com/meelgroup/duriansat/tree/lsids.

```
BUMP LITERAL SCORES FOR LITERALS IN LEARNT CLAUSE
void Solver.analyze(Constr confl)
    c : conflict clause
    litDecayActivity()
    for (int j = 0; j < c.size(); j++)
        Lit q = c[j];
        litBumpActivity(¬q, .5);


BUMP LITERAL SCORES WHEN DELETING ASSIGNMENT
void Solver.cancelUntil(int bLevel)              ▷ bLevel : backtrack level
    - for all elements which are getting cancelled
    for (int c = trail.size()−1; c >= trailLim[bLevel]; c−−)
        Var x = var(trail[c])
        Lit l = mkLit(x, polarity[x])
        litBumpActivity(¬l, 2);
```

**Fig. 2.** Sections in MiniSat like code, where LSIDS score is bumped and decayed.

```
Lit pickBranchLit()
    next : variable returned by branching heuristic
    CBT : denotes whether the last backtrack was chronological

    if (CBT)
        bool pol = pickLsidsBasedPhase(next)
        return mkLit(next, pol)
    else
        return mkLit(next, polarity[next])

bool pickLsidsBasedPhase(Var v)
    if ( activity[posL] > activity[negL] )
        return true
    else
        return false
```

**Fig. 3.** Method to choose branching literal

to the previous section; we tested the implementations on SAT19 benchmarks using the setup mentioned in Subsect. 2.2.

*Solved Instances and PAR-2 Score Comparison.* Table 3 compares numbers of instances solved by the solver mldc and mldc-lsids-phase. First, observe that mldc-lsids-phase solves 243 instances in comparison to 237 instances solved by mldc, which amounts to the improvement of 6 in the number of solved instances. On a closer inspection, we discovered that mldc performs CB for at least 1% of backtracks only in 103 instances out of 400 instances. Since mldc-lsids-phase is identical to mldc for the cases when the solver does not perform chronologi-

**Table 3.** Performance comparison of LSIDS based phase selection with phase saving on 400 SAT19 instances.

| System | SAT | UNSAT | Total | PAR-2 |
|---|---|---|---|---|
| mldc | 140 | 97 | 237 | 4607.61 |
| mldc-lsids-phase | 147 | 96 | 243 | 4475.22 |

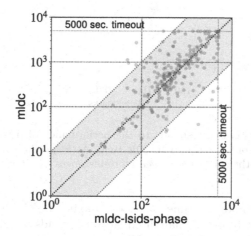

**Fig. 4.** Performance comparison of mldc-lsids-phase vis-a-vis mldc

cal backtracking, the improvement of 6 instances is out of the set of roughly 100 instances. It perhaps fits the often quoted paraphrase by Audemard and Simon [4]: *solving 10 or more instances on a fixed set (of size nearly 400) of instances from a competition by using a new technique, generally shows a critical feature.* In this context, we would like to believe that the ability of LSIDS-based phase selection to achieve improvement of 6 instances out of roughly 100 instances qualifies LSIDS-base phase saving to warrant serious attention by the community.

Table 3 also exhibits enhancement in PAR-2 score due to LSIDS-based phase selection. In particular, we observe mldc-lsids-phase achieved reduction 2.87% in PAR-2 score over mldc, which is significant as per SAT competitions standards. In particular, the difference among the winning solver and runner-up solver for the main track in 2019 and 2018 was 1.27% and 0.81%, respectively. In Fig. 4, we show the scatter plot comparing instance-wise runtime performance of mldc vis-a-vis mldc-lsids-phase. While the plot shows that there are more instances for which mldc-lsids-phase achieves speedup over mldc than vice-versa, the plot also highlights the existence of several instances that time out due to the usage of mldc-lsids-phase but could be solved by mldc.

*Solving Time Comparison.* Figure 5 shows a cactus plot comparing performance of mldc and mldc-lsids-phase on SAT19 benchmarks. We present number of

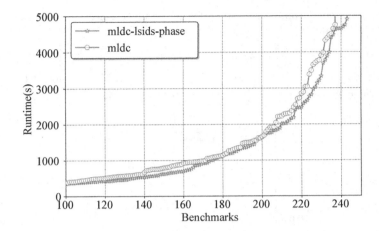

**Fig. 5.** Each of the curve corresponds to the performance of a solver, by means of number of instances solved within a specific time. At a specific runtime $t$, a curve to further right denotes the solver has solved more instances by time $t$.

instances on $x$-axis and the time taken on $y$-axis. A point $(x, y)$ in the plot denotes that a solver took less than or equal to $y$ seconds to solve $y$ benchmarks out of the 400 benchmarks in SAT19. The curves for mldc and mldc-lsids-phase indicate, for every given timeout, the number of instances solved by mldc-lsids-phase is greater than or equal to mldc.

*Percentage of Usage and Difference of Selected Phase.* Among the instances solved by mldc-lsids-phase, percentage of decisions taken with LSIDS phase selections is on average 3.17% over the entire data set. Among the decisions taken with LSIDS phase selection, the average fraction of decisions where the selected phase differs from that of phase saving is 4.67%; It is worth remarking that maximum achieved is 88% while the minimum is 0%. Therefore, there are benchmarks where LSIDS and phase selection are entirely the same while there are benchmarks where they agree for only 12% of the cases. The numbers thereby demonstrate that the LSIDS-based phase selection can not be simply simulated by random or choosing phase opposite of phase selection.

*Applicability of LSIDS in NCB-State.* The performance improvements owing the usage of LSIDS during CB-state raise the question of whether LSIDS is beneficial in NCB-state as well. To this end, we augmented mldc-lsids-phase to employ LSIDS-based phase selection during both NCB-state as well as CB-state. Interestingly, the augmented mldc-lsids-phase solved 228 instances, nine less compared to mldc, thereby providing evidence in support of our choice of usage of LSIDS during CB-state only.

*Deciding the Best Combination of CB and NCB.* Nadel and Ryvchin [28] inferred that SAT solvers benefit from an appropriate combination of CB and NCB rather

**Table 4.** Performance comparison of LSIDS based phase selection with phase saving on 400 SAT19 instances with different $T$ and $C$.

| | | T = 100 | | | | C = 4000 | | | |
|---|---|---|---|---|---|---|---|---|---|
| | | C = 2000 | 3000 | 4000 | 5000 | T = 25 | 90 | 150 | 200 |
| SAT | mldc | 137 | **141** | 140 | 137 | 139 | 137 | 134 | 138 |
| | mldc-lsids-phase | **143** | 139 | **147** | **139** | **142** | **141** | **139** | **142** |
| UNSAT | mldc | 98 | 96 | 97 | 97 | 98 | **97** | 95 | 97 |
| | mldc-lsids-phase | **99** | **101** | 96 | **100** | **99** | 96 | **99** | 97 |
| Total | mldc | 235 | 237 | 237 | 234 | 237 | 233 | 229 | 235 |
| | mldc-lsids-phase | **242** | **240** | **243** | **239** | **241** | **238** | **238** | **239** |
| PAR-2 | mldc | 4663 | 4588 | 4607 | 4674 | 4609 | 4706 | 4773 | 4641 |
| | mldc-lsids-phase | **4506** | **4558** | **4475** | **4575** | **4555** | **4556** | **4622** | **4583** |

than solely reliance on CB or NCB. To this end, they varied two parameters, $T$ and $C$ according to the following rules to heuristically decide the best combination.

- If the difference between the current decision level and backtrack level returned by conflict analysis procedure is more than $T$, then perform CB.
- For the first $C$ conflicts, perform NCB. This rule supersedes the above one.

Following the process, we experimented with different sets of $T$ and $C$ to determine the best combination of $T$ and $C$ for mldc-lsids-phase. For each configuration ($T$ and $C$), we computed the performance of mldc too. The results are summerized in Table 4. It turns out that $T = 100, C = 4000$ performs best in mldc-lsids-phase. Interestingly, for most of the configurations, mldc-lsids-phase performed better than mldc.

## 5.3  Case Study on Cryptographic Benchmarks

Following SAT-community traditions, we have concentrated on SAT-19 benchmarks. But the complicated process of selection of benchmarks leads us to be cautious about confidence in runtime performance improvement achieved by LSIDS-based phase selection. Therefore, in a bid to further improve our confidence in the proposed heuristic, we performed a case study on benchmarks arising from security analysis of *SHA-1 cryptographic hash functions*, a class of benchmarks of special interest to our industrial collaborators and to the security community at large. For a message $\mathcal{M}$, a cryptographic hash function $F$ creates a hash $\mathcal{H} = F(\mathcal{M})$. In a preimage attack, given a hash $\mathcal{H}$ of a message $\mathcal{M}$, we are interested to compute the original message $\mathcal{M}$. In the benchmark set generated, we considered SHA-1 with 80 rounds, 160 bits for hash are fixed, and $k$ bits out of 512 message bits are fixed, $485 < k < 500$. The solution to the preimage attack problem is to give the remaining $(512 - k)$ bits. Therefore, the brute complexity of these problems will range from $O(2^{12})$ to $O(2^{27})$. The

**Table 5.** Performance comparison of LSIDS based phase selection with phase saving on 512 cryptographic instances. Name of systems are same as Table 3.

| System | Total solved | PAR-2 |
|---|---|---|
| mldc | 291 | 9939.91 |
| mldc-lsids-phase | 299 | 9710.42 |

CNF encoding of these problems was created using the SAT instance generator for SHA-1 [29]. Note that by design, each of the instances is satisfiable. In our empirical evaluation, we focused on a suite comprising of 512 instances[5] and every experiment consisted of running a given solver with 3 h of timeout on a particular instance.

Table 5 presents the runtime performance comparison of mldc vis-a-vis mldc-lsids-phase for our benchmark suite. First, we observe that mldc-lsids-phase solves 299 instances in comparison to 291 instances solved by mldc, demonstrating an increase of 8 instances due to LSIDS-based phase selection. Furthermore, we observe a decrease of 229 in PAR-2 score, corresponding to a relative improvement of 2.30%, which is in the same ballpark as the improvement in PAR-2 score observed in the context of SAT-19 instances.

# 6   Conclusion

In this paper, we evaluated the efficacy of phase saving in the context of the recently emerged usage of chronological backtracking in CDCL solving. Upon observing indistinguishability in the performance of phase saving vis-a-vis random polarity selection, we propose a new score: Literal State Independent Decay Sum (LSIDS) that seeks to capture both the activity of a literal arising during clause learning and also the history of polarities assigned to the variable. We observed that incorporating LSIDS to Maple_LCM_Dist_ChronoBTv3 leads to 6 more solved benchmarks while attaining a decrease of 132 seconds in PAR-2 score. The design of a new phase selection heuristic due to the presence of CB leads us to believe that the community needs to analyze the efficiency of heuristics for other components in the presence of chronological backtracking.

**Acknowledgment.** We are grateful to the anonymous reviewers for constructive comments that significantly improved the final version of the paper. We are grateful to Mate Soos for identifying a critical error in the early drafts of the paper. This work was supported in part by National Research Foundation Singapore under its NRF Fellowship Programme[NRF-NRFFAI1-2019-0004] and AI Singapore Programme [AISG-RP-2018-005], and NUS ODPRT Grant [R-252-000-685-13]. The computational work for this article was performed on resources of the National Supercomputing Centre, Singapore [1].

---

[5] Benchmarks are available at https://doi.org/10.5281/zenodo.3817476.

# References

1. ASTAR, NTU, NUS, SUTD: National Supercomputing Centre (NSCC) Singapore (2018). https://www.nscc.sg/about-nscc/overview/
2. Audemard, G., Simon, L.: Glucose: a solver that predicts learnt clauses quality. SAT Competition, pp. 7–8 (2009)
3. Audemard, G., Simon, L.: Predicting learnt clauses quality in modern SAT solvers. In: Twenty-First International Joint Conference on Artificial Intelligence (2009)
4. Audemard, G., Simon, L.: Refining restarts strategies for SAT and UNSAT. In: Milano, M. (ed.) CP 2012. LNCS, pp. 118–126. Springer, Heidelberg (2012). https://doi.org/10.1007/978-3-642-33558-7_11
5. Audemard, G., Simon, L.: On the glucose SAT solver. Int. J. Artif. Intell. Tools **27**(01), 1840001 (2018)
6. Biere, A.: Lingeling, plingeling and treengeling entering the SAT competition 2013. In: Proceedings of SAT Competition 2013, p. 1 (2013)
7. Biere, A., Cimatti, A., Clarke, E., Zhu, Y.: Symbolic model checking without BDDs. In: Cleaveland, W.R. (ed.) TACAS 1999. LNCS, vol. 1579, pp. 193–207. Springer, Heidelberg (1999). https://doi.org/10.1007/3-540-49059-0_14
8. Biere, A., Fröhlich, A.: Evaluating CDCL variable scoring schemes. In: Heule, M., Weaver, S. (eds.) SAT 2015. LNCS, vol. 9340, pp. 405–422. Springer, Cham (2015). https://doi.org/10.1007/978-3-319-24318-4_29
9. Clarke, E., Kroening, D., Lerda, F.: A tool for checking ANSI-C programs. In: Jensen, K., Podelski, A. (eds.) TACAS 2004. LNCS, vol. 2988, pp. 168–176. Springer, Heidelberg (2004). https://doi.org/10.1007/978-3-540-24730-2_15
10. Cook, S.A.: The complexity of theorem-proving procedures. In: Proceedings of the Third Annual ACM Symposium on Theory of Computing, pp. 151–158 (1971)
11. Davis, M., Logemann, G., Loveland, D.: A machine program for theorem-proving. Commun. ACM **5**(7), 394–397 (1962)
12. Eén, N., Sörensson, N.: An extensible SAT-solver. In: Giunchiglia, E., Tacchella, A. (eds.) SAT 2003. LNCS, vol. 2919, pp. 502–518. Springer, Heidelberg (2004). https://doi.org/10.1007/978-3-540-24605-3_37
13. Frost, D., Dechter, R.: In search of the best constraint satisfaction search. AAAI **94**, 301–306 (1994)
14. Gomes, C.P., Selman, B., Kautz, H., et al.: Boosting combinatorial search through randomization. AAAI/IAAI **98**, 431–437 (1998)
15. Heule, M.J., Järvisalo, M., Suda, M.: Proceedings of SAT race 2019: solver and benchmark descriptions (2019)
16. Huang, J., et al.: The effect of restarts on the efficiency of clause learning. IJCAI **7**, 2318–2323 (2007)
17. Järvisalo, M., Le Berre, D., Roussel, O., Simon, L.: The international SAT solver competitions. Ai Mag. **33**(1), 89–92 (2012)
18. Kautz, H.A., Selman, B., et al.: Planning as satisfiability. In: ECAI, vol. 92, pp. 359–363. Citeseer (1992)
19. Liang, J.H., Ganesh, V., Poupart, P., Czarnecki, K.: Learning rate based branching heuristic for SAT solvers. In: Creignou, N., Le Berre, D. (eds.) SAT 2016. LNCS, vol. 9710, pp. 123–140. Springer, Cham (2016). https://doi.org/10.1007/978-3-319-40970-2_9
20. Liang, J.H., Ganesh, V., Zulkoski, E., Zaman, A., Czarnecki, K.: Understanding VSIDS branching heuristics in conflict-driven clause-learning SAT solvers. In: Piterman, N. (ed.) HVC 2015. LNCS, vol. 9434, pp. 225–241. Springer, Cham (2015). https://doi.org/10.1007/978-3-319-26287-1_14

21. Liang, J.H., Oh, C., Mathew, M., Thomas, C., Li, C., Ganesh, V.: Machine learning-based restart policy for CDCL SAT solvers. In: Beyersdorff, O., Wintersteiger, C.M. (eds.) SAT 2018. LNCS, vol. 10929, pp. 94–110. Springer, Cham (2018). https://doi.org/10.1007/978-3-319-94144-8_6

22. Liang, J.H., Poupart, P., Czarnecki, K., Ganesh, V.: An empirical study of branching heuristics through the lens of global learning rate. In: Gaspers, S., Walsh, T. (eds.) SAT 2017. LNCS, vol. 10491, pp. 119–135. Springer, Cham (2017). https://doi.org/10.1007/978-3-319-66263-3_8

23. Luo, M., Li, C.M., Xiao, F., Manya, F., Lü, Z.: An effective learnt clause minimization approach for CDCL SAT solvers. In: Proceedings of the 26th International Joint Conference on Artificial Intelligence, pp. 703–711 (2017)

24. Lynce, I., Marques-Silva, J.: SAT in bioinformatics: making the case with haplotype inference. In: Biere, A., Gomes, C.P. (eds.) SAT 2006. LNCS, vol. 4121, pp. 136–141. Springer, Heidelberg (2006). https://doi.org/10.1007/11814948_16

25. Marques-Silva, J., Lynce, I., Malik, S.: Conflict-driven clause learning SAT solvers. In: Handbook of Satisfiability, pp. 131–153. IOS Press (2009)

26. Marques-Silva, J.P., Sakallah, K.A.: GRASP: a search algorithm for propositional satisfiability. IEEE Trans. Comput. **48**(5), 506–521 (1999)

27. Moskewicz, M.W., Madigan, C.F., Zhao, Y., Zhang, L., Malik, S.: Chaff: engineering an efficient SAT solver. In: Proceedings of the 38th Annual Design Automation Conference, pp. 530–535. ACM (2001)

28. Nadel, A., Ryvchin, V.: Chronological backtracking. In: Beyersdorff, O., Wintersteiger, C.M. (eds.) SAT 2018. LNCS, vol. 10929, pp. 111–121. Springer, Cham (2018). https://doi.org/10.1007/978-3-319-94144-8_7

29. Nossum, V.: Instance generator for encoding preimage, second-preimage, and collision attacks on SHA-1. In: Proceedings of the SAT Competition, pp. 119–120 (2013)

30. Oh, C.: Improving SAT solvers by exploiting empirical characteristics of CDCL. Ph.D. thesis, New York University (2016)

31. Pipatsrisawat, K., Darwiche, A.: A lightweight component caching scheme for satisfiability solvers. In: Marques-Silva, J., Sakallah, K.A. (eds.) SAT 2007. LNCS, vol. 4501, pp. 294–299. Springer, Heidelberg (2007). https://doi.org/10.1007/978-3-540-72788-0_28

32. Ryvchin, V., Nadel, A.: Maple LCM dist ChronoBT: featuring chronological backtracking. In: Proceedings of SAT Competition, p. 29 (2018)

33. Shtrichman, O.: Tuning SAT checkers for bounded model checking. In: Emerson, E.A., Sistla, A.P. (eds.) CAV 2000. LNCS, vol. 1855, pp. 480–494. Springer, Heidelberg (2000). https://doi.org/10.1007/10722167_36

34. Silva, J.P.M., Sakallah, K.A.: Grasp—a new search algorithm for satisfiability. In: Kuehlmann, A. (ed.) The Best of ICCAD, pp. 73–89. Springer, Heidelberg (2003). https://doi.org/10.1007/978-1-4615-0292-0_7

35. Silva, J.M., Sakallah, K.A.: Conflict analysis in search algorithms for satisfiability. In: Proceedings Eighth IEEE International Conference on Tools with Artificial Intelligence, pp. 467–469. IEEE (1996)

36. Wetzler, N., Heule, M.J.H., Hunt, W.A.: Mechanical verification of SAT refutations with extended resolution. In: Blazy, S., Paulin-Mohring, C., Pichardie, D. (eds.) ITP 2013. LNCS, vol. 7998, pp. 229–244. Springer, Heidelberg (2013). https://doi.org/10.1007/978-3-642-39634-2_18

# On the Effect of Learned Clauses on Stochastic Local Search

Jan-Hendrik Lorenz[✉] and Florian Wörz[✉]

Institute of Theoretical Computer Science, Ulm University, 89069 Ulm, Germany
{jan-hendrik.lorenz,florian.woerz}@uni-ulm.de

**Abstract.** There are two competing paradigms in successful SAT solvers: Conflict-driven clause learning (CDCL) and stochastic local search (SLS). CDCL uses systematic exploration of the search space and has the ability to learn new clauses. SLS examines the neighborhood of the current complete assignment. Unlike CDCL, it lacks the ability to learn from its mistakes. This work revolves around the question whether it is beneficial for SLS to add new clauses to the original formula. We experimentally demonstrate that clauses with a large number of correct literals w. r. t. a fixed solution are beneficial to the runtime of SLS. We call such clauses *high-quality* clauses.

Empirical evaluations show that short clauses learned by CDCL possess the high-quality attribute. We study several domains of randomly generated instances and deduce the most beneficial strategies to add high-quality clauses as a preprocessing step. The strategies are implemented in an SLS solver, and it is shown that this considerably improves the state-of-the-art on randomly generated instances. The results are statistically significant.

**Keywords:** Stochastic Local Search · Conflict-Driven Clause Learning · Learned clauses

## 1 Introduction

The *satisfiability problem* (*SAT*) asks to determine if a given propositional formula $F$ has a satisfying assignment or not. Since Cook's NP-completeness proof of the problem [21], SAT is believed to be computationally intractable in the worst case. However, in the field of applied SAT solving, there were enormous improvements in the performance of SAT solvers in the last 20 years. Motivated by these significant improvements, SAT solvers have been applied to an increasing number of areas, including bounded model checking [15,19], cryptology [23], or even bioinformatics [35], to name just a few. Two algorithmic paradigms turned out to be especially promising to construct solvers. The first to mention

The authors acknowledge support by the state of Baden-Württemberg through bwHPC.

F. Wörz—Supported by the Deutsche Forschungsgemeinschaft (DFG).

L. Pulina and M. Seidl (Eds.): SAT 2020, LNCS 12178, pp. 89–106, 2020.
https://doi.org/10.1007/978-3-030-51825-7_7

is *conflict-driven clause learning* (*CDCL*) [13,36,39]. The CDCL procedure systematically explores the search space of all possible assignments for the formula $F$ in question, by constructing partial assignments in an exhaustive branching and backtracking search. Whenever a conflict occurs, a reason (a conflict clause) is learned and added to the original clause set [16,42]. The other successful paradigm is *stochastic local search* (see [16, Chapter 6] for an overview). Starting from a complete initial assignment for the formula, the neighborhood of this assignment is explored by moving from assignment to assignment in the space of solution candidates while trying to minimize the number of unsatisfied clauses by the assignment (or some other criterion). This movement is usually performed by *flipping* one variable of such a complete assignment. Both paradigms are described in Sect. 2 in more detail.

Besides the difference in the completeness of assignments considered during a run of those algorithms, another major difference between both paradigms is the completeness and incompleteness of the solvers (i. e., being able to certify both satisfiability and unsatisfiability, or not) [42]. CDCL solvers are complete, while SLS algorithms are incomplete. More interestingly for practitioners, perhaps, is the complimentary behavior these paradigms exhibit in terms of performance: CDCL seems well suited for application instances, whereas SLS excels on random formulas[1]. An interesting question thus is, if it is possible to combine the strength of both solvers or to eliminate the weaknesses of one paradigm by an oracle-assistance of the other. This challenge was posed by Selman et al. in [43, Challenge 7] (and again later in [30]):

> "Demonstrate the successful combination of stochastic search and systematic search techniques, by the creation of a new algorithm that outperforms the best previous examples of both approaches."

This easy to state question turns out to be surprisingly challenging. There have been some advances towards this goal, as we survey below. However, the performance of most algorithms that try to combine the strength of both paradigms, so-called *hybrid solvers*, is far from those of CDCL solvers (or other non-hybrids), especially on application instances [2] or even any wider range of benchmark problems [30]. In [27], the DPLL algorithm SATZ [34] was used to derive implication dependencies and equivalencies between literals in WALKSAT [37].

The effect of a restricted form of resolution to clause weighting solvers was investigated in [1]. A similar approach was previously studied in [18], where new resolvent clauses based on unsatisfied clauses at local minima and randomly selected neighboring clauses are added.

Local Search over partial assignments instead of complete assignments extended with constraint propagation was studied in [29]. In case of a conflict, a conflict clause is learned, and local search is used to repair this conflict. A similar approach to construct a hybrid solver using SLS as the main solver and a

---

[1] It is, however, noteworthy that the winning solver in the random track of the SAT Competition 2018 was SPARROW2RISS, a CDCL solver.

complete CDCL solver as a sub-solver was also studied in [4, 7], where the performance of the solvers HYBRIDGM, HYBRIDGP, and HYBRIDPP was empirically analyzed. The idea of these solvers is to build a partial assignment around one complete assignment from the search trajectory of the SLS solver. This partial assignment can then be applied to the formula resulting in a simpler one, which is solved by a complete CDCL solver.

A shared memory approach for multi-core processor architectures was proposed in [32]. In this case, DPLL can provide guidance for an SLS solver, being run simultaneously on a different core.

The solver HBISAT, introduced in [24], uses the partial assignments calculated by CDCL to initialize the SLS solver. The SLS solver sends unsatisfied clauses to the CDCL solver to either identify an unsatisfiable subformula of those clauses or satisfy them. This approach was later significantly improved by Letombe and Marques-Silva in [33].

Audemard et al. [2] introduced SATHYS, where both components cooperate by alternating between them. E. g., when CDCL chooses a variable to branch, its polarity is extracted from the best complete assignment found by the SLS solver. On the other hand, CDCL helps SLS out of local minima (i. e., CDCL is invoked conditionally in this solver).

**Our Contribution.** Our hybrid solver GAPSAT differs from the approaches described above in the sense that CDCL is used as a preprocessor for PROB-SAT [11] (an SLS solver), terraforming the landscape in advance. This approach eliminates, in many cases, the possibility for PROBSAT to get stuck in local minima, eliminating the necessity of further, more complicated interactions between both paradigms.

We examine the question, whether it is beneficial for SLS to add new clauses to the original formula in a *preprocessing step*, by invoking a complete CDCL solver. As it turns out, not all additional clauses are created equal. Experimentally, we demonstrate that adding clauses that contain a larger number of correct literals w. r. t. a fixed solution, drastically improves the performance of SLS solvers. However, clauses that only contain few correct literals w. r. t. the fixed solution can be deceptive for SLS. This effect can exponentially increase the runtime as measured in the number of flips of PROBSAT, a very simple SLS solver. In practice, one has to resort to known complete algorithms or proof systems to generate helpful clauses. We, in particular, investigate the effect of new clauses learned by CDCL and depth-limited resolution (Sect. 3). With the help of experiments, we conclude that CDCL (or resolution limited to depth 2) produces distinctively more helpful clauses for PROBSAT than resolution limited to depth 1, as was studied in the past [1]. We, therefore, focus our effort on CDCL as a clause learning mechanism for PROBSAT.

Motivated by these insights, we study the quality CDCL-learned clauses have for SLS in more detail. In training experiments that are described in Sect. 4, we systematically deduce parameter settings by statistical analysis that increase this quality. For example, shorter clauses learned by CDCL are more beneficial. As it turns out, however, the specific width depends on the underlying formula.

Another interesting observation is that the amount of added clauses has to be carefully restricted. Again, the specific restriction to use depends on the underlying formula.

To finally test the concrete effect of these ideas, we compared the performance of a newly designed solver with the winner of the 2018 random track competition, SPARROW2RISS [10]. Our observations were implemented in GAPSAT, which forms a combination of GLUCOSE and PROBSAT. A comprehensive experimental evaluation on 255 instances provides statistical evidence that the performance of our proposed solver GAPSAT exceeds SPARROW2RISS' substantially. In particular, GAPSAT was able to solve more instances in just 30 s than SPARROW2RISS in 5000 s. We present a summary of our experimental evaluation in Sect. 5.

## 2 Preliminaries

We briefly reiterate the notions necessary for this work. For a thorough introduction to the field, we refer the reader to [42]. A *literal* over a Boolean variable $x$ is either $x$ itself or its negation $\overline{x}$. A *clause* $C = a_1 \vee \cdots \vee a_\ell$ is a (possibly empty) disjunction of literals $a_i$ over pairwise disjoint variables. A *CNF formula* $F = C_1 \wedge \cdots \wedge C_m$ is a conjunction of clauses. A CNF formula is a *k-CNF* if all clauses in it have at most $k$ variables. An *assignment* $\alpha$ for a CNF formula $F$ is a function that maps some subset of Vars($F$) to $\{0, 1\}$. Given a complete assignment $\alpha$, the act of changing the truth value of precisely one variable of $\alpha$ is called a *flip*. *Resolution* is the proof system with the single derivation rule $\frac{B \vee x \quad C \vee \overline{x}}{B \vee C}$, where $B$ and $C$ are clauses.

**CDCL.** CDCL solvers, introduced in [13,36,39], construct a partial assignment. When some clause is falsified by the constructed assignment, the CDCL solver adds a new clause to the original formula $F$. This clause is a logical consequence of $F$. A more detailed description of CDCL can be found in [16,40,42]. Modern SAT solvers are additionally equipped with incremental data structures, restart policies [26], and activity-based variable selection heuristics (VSIDS) [39]. In this work, we use the CDCL solver GLUCOSE [3] (based on MINISAT [22]).

**probSAT.** Contrary to CDCL-like algorithms, algorithms based on *stochastic local search (SLS)* operate on complete assignments for a formula $F$. These solvers are started with a randomly generated complete initial assignment $\alpha$. If $\alpha$ satisfies $F$, a solution is found. Otherwise, the SLS solver tries to find a solution by repeatedly flipping the assignment of variables according to some underlying heuristic. That is, they perform a random walk over the set of complete assignments for the underlying formula.

In [11], the PROBSAT class of solvers was introduced. Over the last few years, PROBSAT-based solvers performed excellently on random instances: PROBSAT won the random track of the SAT competition 2013, DIMETHEUS [9] in 2014 and 2016, YALSAT [14] won in 2017. Only recently, in 2018, other types of solvers significantly exceeded PROBSAT based algorithms. This performance is the reason for choosing PROBSAT in this study.

The idea behind the solver is that a function $f$ is used, which gives a high probability to a variable if flipping this variable is deemed advantageous. A description of PROBSAT is given in Algorithm 1. This class of solvers is related to Schöning's random walk algorithm introduced in [41].

---

**Input:** Formula $F$, *maxFlips*, function $f$
$\alpha :=$ randomly generated complete assignment for $F$
**for** $i = 1$ **to** *maxFlips* **do**
  **if** $\alpha$ *satisfies* $F$ **then return** "satisfiable"
  Choose a clause $C = (u_1 \vee u_2 \vee \cdots \vee u_\ell)$ that is falsified under $\alpha$
  Choose $j \in \{1, \ldots, \ell\}$ with probability $\frac{f(u_j, \alpha)}{\sum_{u \in C} f(u, \alpha)}$
  Flip the assignment of the chosen variable $u_j$ and update $\alpha$

**Algorithm 1.** probSAT without restarts.

---

In [11], the *break-only-poly-algorithm* with $f(x, \alpha) := \left(\varepsilon + \text{break}(x, \alpha)\right)^{-b}$ was considered for 3-SAT, where $\text{break}(x, \alpha)$ is the number of clauses that are satisfied under $\alpha$ but will be falsified when the assignment of $x$ is flipped. For $k \neq 3$, the *break-only-exp-algorithm* $f(x, \alpha) := b^{-\text{break}(x, \alpha)}$ was studied. Balint and Schöning [11] found good choices for the parameters of these two functions. In this work, we have adopted these parameter settings.

## 3   The Quality of Learned Clauses

In this section, we investigate the effect logically equivalent formulas have on the SLS solver PROBSAT. More precisely, we use a formula $F$ as a base and add a set of clauses $S = \{C_1, \ldots, C_t\}$ to $F$ to obtain a new formula $G := F \cup S$. In general, adding new clauses to a formula $F$ does not yield a logically equivalent new formula $G$.

Thus, we observe two artificial models related to the *backbone* (see, e. g., [31]) and consider the 3-CNF case in the following. The backbone $\mathcal{B}(F)$ are the literals appearing in all satisfying assignments of $F$. In the first model, each new clause consists of one backbone literal $x \in \mathcal{B}(F)$ and two literals $y, z$ such that their complements are backbone literals, i. e., $y$ and $z$ do not occur in any solution. We call this the *deceptive model*. In the second model, each new clause has one backbone literal and two randomly chosen literals. This is the *general model*.

Figure 1 displays the effect of both models on PROBSAT. On the left is the deceptive model. Generally, a large number of deceptive clauses have a harmful effect on the runtime of PROBSAT. That is, the average runtime of PROBSAT increases exponentially with the number of added clauses.

The right-hand side of Fig. 1 shows the general model. Here, we can observe a strong, positive effect on the behavior of PROBSAT. The average runtime of PROBSAT improved by two orders of magnitude by adding 200 new clauses

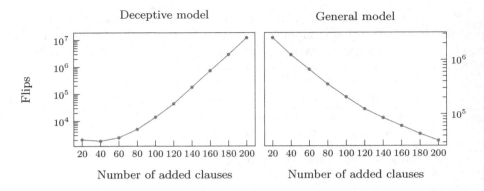

**Fig. 1.** On the left, the effect of *deceptive* clauses is displayed on an instance with 100 variables and 423 clauses. On the right, the effect of *general* clauses is displayed on an instance with 500 variables and 2100 clauses. The $x$-axes denote the number of additional clauses, and the $y$-axes denote the average runtime of 100 runs of PROBSAT as measured in the number of flips. Both $y$-axes are scaled logarithmically.

generated by the general model. Even though Fig. 1 depicts the data of only one instance, the general shape of the plot is similar on all tested instances.

Clauses generated by the deceptive model seem to give rise to new local minima which are far away from the solutions. Once PROBSAT is stuck in such a local minimum, the break-value makes it unlikely that PROBSAT escapes the region of the local minimum. On the other hand, the prevalence of correct literals in the general model seems to guide PROBSAT towards a solution. Due to this interpretation, we call clauses that have a high number of correct literals w. r. t. a fixed solution *high-quality clauses*. The view that clauses with few correct literals have a detrimental effect on local search solvers is also supported by the literature [28, 42].

From the considerations described above, it should be evident that it is crucial which clauses are added to the formula. Clearly, neither the deceptive nor the general model can be applied to real instances: The solution space would have to be known in advance to generate the clauses. In contrast, approaches like resolution and CDCL can be applied to real instances. All clauses which can be derived by resolution are already implied by the original formula. Accordingly, adding such a clause to the original formula yields a logically equivalent formula. Similarly, clauses learned by a CDCL algorithm can be added to obtain a logically equivalent formula.

In the following, we compare two models based on resolution and one model based on CDCL. In particular, let $F$ be a formula and let $B, C \in F$ be clauses such that there is a resolvent $R$. We call $R$ a *level 1* resolvent. Secondly, let $D, E$ be clauses such that there is a resolvent $S$ and let $D$ or $E$ (or both) be level 1 resolvents. We call $S$ a *level 2* resolvent. As a representative for CDCL solvers, we use GLUCOSE [3].

Let $F$ be a 3-CNF formula with $m$ clauses. New and logically equivalent formulas $F_1$, $F_2$, and $F_C$ are obtained in the following manner.

$F_1$ Randomly select at most $m/10$ level 1 resolvents of maximum width 4 and add them to $F$.

$F_2$ Randomly select at most $m/10$ level 2 resolvents of maximum width 4 and add them to $F$.

$F_C$ Randomly select at most $m/10$ learned clauses with maximum width 4 from GLUCOSE (with a time limit of 300 s) and add them to $F$.

The average behavior of PROBSAT over 1000 runs per instance on the instance types $F_1$, $F_2$, and $F_C$ is observed. We use a small testbed of 23 uniformly generated 3-CNF instances with 5000 to 11 600 variables and a clause-to-variable ratio of 4.267. The instances of type $F_1$ were the most challenging for PROBSAT; as a matter of fact, $F_1$ instances were considerably harder to solve than the original instances. On instances of type $F_2$, PROBSAT performed better, and on $F_C$, it was even more efficient. The t-test [44] confirms the observations: $F_2$ instances are easier on average than $F_1$ instances ($p < 0.01$), and $F_C$ instances are easier than $F_2$ instances ($p < 0.05$). Sections 4 and 5 present an in-depth examination of the effect clauses of type $F_C$ have on PROBSAT.

These results lead us to believe that level 1 clauses are of low quality while level 2 and CDCL clauses are generally of higher quality. It is impractical to confirm this suspicion on uniformly generated instances of the above-mentioned size. Hence, we use randomly generated models with hidden solutions [12] and judge the quality of learned clauses based on the hidden solution.

The SAT competition 2018 incorporated three types of models with hidden solutions. All three types are generated in a similar manner; they just differ in the choice of the parameters. Here, we compare the average quality of the new clauses on each of the three models. For each instance, the set of all level 1, level 2, and CDCL clauses is computed, and the quality is measured w. r. t. the hidden solution.

For the most part, the results confirm the observations from the uniformly generated instances: On all three models, level 2 clauses have a statistically significantly higher quality than level 1 clauses (t-test, all $p < 0.01$). On two of three domains, CDCL clauses have higher quality than level 2 clauses (t-test, both $p < 10^{-5}$), while level 2 clauses have higher quality on the remaining domain (t-test, $p < 10^{-8}$).

As a side note, CDCL is capable of learning unit and binary clauses. Nevertheless, this did not influence the quality of the clauses in any meaningful way: In the 120 test instances, only a single binary and no unit clause was learned.

In conclusion, we conjecture that level 2 and CDCL clauses have higher quality than level 1 clauses. On the uniform random testbed, CDCL performs better than level 2 clauses; also, CDCL clauses have higher quality than level 2 clauses on two of the three hidden solution domains.

## 4    Training Experiments

In the previous section, we argued that adding supplementary clauses to an instance can have a positive effect on the behavior of PROBSAT. The focus of this section lies on the question which clauses and how many should be added.

Especially for 3-SAT instances, an initial guess might be adding all clauses acquirable by so-called ternary resolution [17]. Informally speaking, ternary resolution is the restriction of the resolution rule to ternary clauses such that the resolvent is either a binary or ternary clause. Ternary resolution is performed until saturation. In [8], the effect of (amongst other techniques) ternary resolution on another SLS solver, SPARROW (see [6]), is observed. They empirically show that ternary resolution has a negative effect on the performance on satisfiable hard combinatorial instances. Anbulagan et al. [1] study the effect of ternary resolution on uniform random instances. They found that SLS solvers do not benefit from ternary resolution. They even conjecture that ternary resolution has a harmful impact on the runtime of SLS solvers on uniform instances. We performed some experiments on our own and can confirm this suspicion for PROBSAT. On medium-sized uniform instances, ternary resolution slowed PROBSAT down by 0.5% on average. As a consequence, we focus on methods to improve the runtime behavior of PROBSAT with clauses learned by GLUCOSE for the rest of this work.

The supplementary clauses are all learned by GLUCOSE within a 300 second time window; we only distinguish the learned clauses by their width. The number of supplementary clauses is measured in percent of the number of original clauses. To put it differently, we are interested in the maximal length of the new clauses and what percentage of the modified formula should be new clauses. The results of this section are used to configure GAPSAT.

### Description of Training Experiments

We split the experiments into two phases. In the preliminary phase, promising intervals for the *maximal width* and the *maximal percentage* of new clauses are obtained. In the subsequent phase, the most advantageous parameter combination is sought. Hereafter, we describe the setup of the experiments and their results.

**Training Data.** We used a set of training instances $\mathcal{C}$, which is assembled as follows: All instances of the SAT Competitions random tracks[2] 2014 to 2017 were gathered. We filtered these instances by proven satisfiability: An instance was added to the training set $\mathcal{C}$ if and only if at least one participating solver showed satisfiability. Since not enough uniform random 3-SAT instances of medium size were in $\mathcal{C}$, we added all instances of this kind from the SAT Competition 2013 as well. In total, $\mathcal{C}$ consists of 377 instances which can be divided into the following three domains:

---

[2] See http://www.satcompetition.org/. In 2015 there was no random track.

- 120 randomly generated instances with a *hidden* solution [12],
- 149 uniformly generated random 3, 5, and 7-SAT instances of *medium size*. The clause-to-variable ratio is close to the satisfiability threshold [38].
- 108 uniform random 3, 5, and 7-SAT instances of *huge size*, i. e., with over 50 000 variables. The clause-to-variable ratio of each instance is somewhat far from the satisfiability threshold.

**Training Setup.** The experiments were performed on the bwUniCluster and a local server. Sputnik [45] helped to parallelize the trials. The setup of the computer systems is heterogeneous. Therefore, the runtimes are not directly comparable to another. Consequently, we do not use the runtimes for these experiments. Instead, the number of variable flips performed by PROBSAT is used, which is a hardware-independent performance measure.

In this section, we use a timeout of $10^9$ flips for 3-SAT instances (5-SAT: $5 \cdot 10^8$; 7-SAT: $2.5 \cdot 10^8$). This timeout corresponds to roughly 10 min runtime on medium-sized instances on our hardware. Each instance from $\mathcal{C}$ is run 1000 times for each parameter combination. The primary performance indicator in this section is the number of timeouts per instance. Furthermore, the average *par2* value is sometimes used as a secondary performance indicator. The *par2* value is the *number of flips* if a solution was found or twice the timeout otherwise. For the rest of this work, PROBSAT refers to PROBSAT version SC13_v2 [5].

### Results of Training Experiments

We conducted a thorough statistical analysis of the data that was obtained in the training experiments described above. We describe our findings in condensed form below. The main part of the remainder of this section is concerned with uniform, medium-sized instances. The results for uniform, huge instances, and instances with a hidden solution are briefly discussed at the end of this section.

**3-SAT.** We found that adding all clauses up to width 4 that GLUCOSE could find within 300 s is the most beneficial configuration. Not limiting the number of added clauses is in stark contrast to the 5-SAT and 7-SAT cases. For 3-SAT, the relationship between the number of clauses and the performance is explored in Fig. 2. Each blue dot corresponds to one medium-sized 3-SAT instance from $\mathcal{C}$. We compare the average par2 value on the original instance with the average par2 value on the instance with all clauses up to width 4 added. Whenever the blue dot lies below the zero-baseline, then the performance of PROBSAT on the modified instance was better. The blue line is obtained by linear regression. By its slope, we can tell that, on average, adding more clauses is beneficial. The light blue area denotes the 95% confidence interval, which is calculated by bootstrapping [25]. The confidence interval shows that this relationship is unlikely to be due to chance. We conclude that the number of new clauses should not be limited for 3-SAT instances. On the other hand, the maximal width of the new clauses should be no more than four. Our experiments showed that adding longer clauses deteriorates the performance of probSAT.

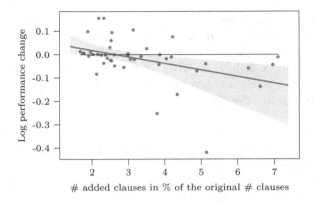

**Fig. 2.** In this plot, PROBSAT is compared on the original 3-SAT instances and the modified instances. The number of added clauses of the 3-SAT instances in % is on the $x$-axis. The $y$-axis is the logarithm of par2(modified)/par2(orig). The blue line denotes a linear regression fit, and the light blue area is the 95% confidence interval obtained by 1000 bootstrapping steps [25]. (Color figure online)

The left-hand side of Fig. 3 shows the overall performance of PROBSAT on instances with all clauses up to width 4. The $y$-axis denotes the difference of timeouts between the modified instance with all width 4 clauses and the original instance. Whenever the dot lies below the zero-baseline, the performance of PROBSAT was better on the modified instance. The color of the dots stands for the hardness as measured in the average par2 on the original instance. We can see that adding additional clauses has a positive effect, especially on hard instances with 4000 to 9000 variables. Nonetheless, the effect reverses for more than 9000 variables. Overall the results on 3-SAT instances are not statistically significant (t-test, $p = 0.0595$). However, we believe that the main reason for this is the bad performance of PROBSAT on the modified instances with more than 9000 variables. Furthermore, with a slightly larger sample size, the results might turn out to be statistically significant. We have used these observations in the configuration of GAPSAT, as depicted in Fig. 5.

**5-SAT.** In preliminary experiments, we found that the maximal width of the new clauses should be in the interval $\{7, 8, 9\}$, and the maximal number of new clauses should be at most 15% of the original clauses. The effect of adding more clauses is especially pronounced: Adding more than 15% of the clauses diminished the performance of PROBSAT dramatically, in contrast to the 3-SAT case where more clauses turned out to be beneficial.

In the detailed phase of the experiments, we found that the best configuration is adding clauses up to width 8 and using a limit of at most 5% of the original clauses. The results of this parameter configuration are shown on the right-hand side of Fig. 3. Again, the performance of PROBSAT was better on the modified instances if the dot lies below the zero-baseline. The color of the dots describes the hardness of the instance. Overall, the modification has a favorable impact

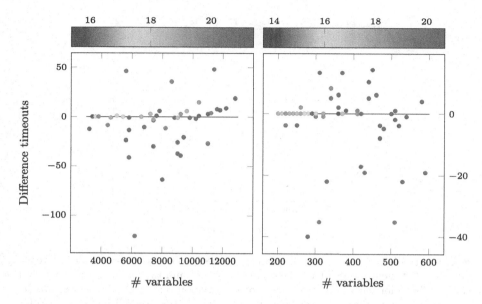

**Fig. 3.** Scatterplot of the number of variables of the 3-SAT (left) and 5-SAT (right) instance against difference in timeouts between PROBSAT on the original instance and the instance with the modified strategy. The color encodes the hardness of the original instance as measured in the logarithmic average par2 on 1000 runs of PROBSAT on the original instances. (Color figure online)

on the performance of PROBSAT over the full domain. The effect is statistically significant (t-test, $p = 0.0348$). Also, it appears to be increasing as the number of variables increase. However, we did not further investigate this relationship.

**7-SAT.** The preliminary experiments showed that the maximal width of the new clauses should be in the interval $\{9, 10, 11\}$. Moreover, similarly to the 5-SAT case, the number of new clauses should be limited. In the preliminary phase, we found that at most 3% of the original clauses should be added, otherwise the performance of PROBSAT decreases.

The detailed phase showed that clauses up to width 9 and a limit of at most 1% is the most advantageous combination. Figure 4 shows the results of this combination. The modified strategy was better on average if the corresponding dot lies below the zero-baseline. Again, we observe that the performance of PROB-SAT clearly benefits from the modified instances, especially on hard instances (red dots). This observation is also confirmed by the t-test ($p = 0.0062$). Additionally, similar to 5-SAT instances, the effect seems to increase as the number of variables increase.

**Hidden Solution.** In our training set $\mathcal{C}$, all instances with a hidden solution are 3-SAT instances with few variables (at most 540). The results are similar in nature to those discussed in the paragraph about uniform medium 3-SAT

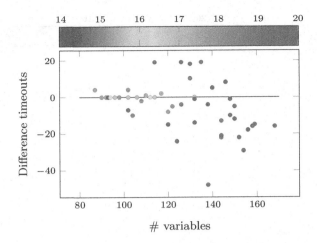

**Fig. 4.** Scatterplot of the number of variables of the 7-SAT instances against difference in timeouts between PROBSAT on the original instance and the instance with modified strategy. The color encodes the hardness of the original instance as measured in the logarithmic average par2 on 1000 runs of PROBSAT on the original instances. (Color figure online)

instances. That is, the addition of new clauses of maximal width 4 and no limit on the number of new clauses is generally beneficial.

**Huge Instances.** The huge instances in the training set $\mathcal{C}$ often have several million original clauses. In contrast, only a few new clauses are learned during the preprocessing time. Consequently, the effect of additional clauses is negligible. The preprocessing step should, therefore, be avoided on these instances.

### Description of GapSAT

The name GAPSAT stands for *Glucose assisted probSAT*, hinting towards the combination of PROBSAT as the core solver, that is being helped by a GLUCOSE preprocessing phase. The exact functioning principle of GAPSAT is depicted in the flowchart of Fig. 5.

As was noticeable in Fig. 3, if the 3-CNF formula contained more than approx. 9000 variables, the act of adding new clauses slows down PROBSAT. Furthermore, on huge instances, the preprocessing step yields no advantage. Thus, for over 9000 variables, the strategy of GAPSAT falls back to just PROBSAT on the original formula. Otherwise, in each case, a short run of PROBSAT is used to filter out very easy to solve instances. The runtime is limited by the number of flips. If the instance could not be solved, we employ the strategy (depending on the maximal clause width in the formula) that was deemed most promising in the evaluations described in the previous subsection. That is, we first let GLU-COSE extract clauses. The runtime of glucose was limited by 300 s in all cases. In the 5-SAT and 7-SAT case, GLUCOSE could finish earlier, if the restrictions on

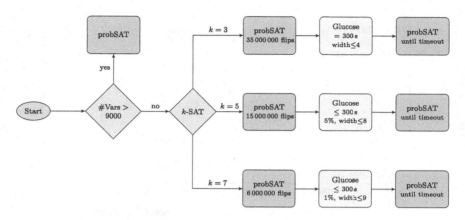

**Fig. 5.** Flowchart description of GapSAT.

the number of added clauses were met. We again emphasize that Fig. 2 explains the difference between the 3-SAT case when compared to the 5-SAT and 7-SAT case. Not restricting the number of learned clauses in the 3-SAT case turned out to be the superior strategy in our training experiments. One should further observe that GLUCOSE has the possibility to solve the instance during its runtime. If this was not successful, PROBSAT is restarted on the formula, that was modified by running GLUCOSE and adding the clauses corresponding to the strategy as developed in the previous subsection. It is noteworthy that GAPSAT does not use any additional preprocessing techniques. We refer to Sect. 6 for a further discussion of that point.

## 5 Experimental Evaluation

In the following, the performance of GAPSAT is evaluated. We compare GAPSAT with the winner of the random track at the SAT competition 2018, SPARROW2RISS, and with the original version of PROBSAT.

All experiments were executed on a computer with 32 Intel Xeon E5-2698 v3 CPUs running at 2.30 GHz. We set the time limit to 5000 s and used no memory limit. The benchmarks consist of all 255 instances of the random track at the SAT competition 2018. Unlike the experiments in Sect. 4, the performance of each solver is measured based on its par2 value w. r. t. the runtime *in seconds*. In the following, the *score* denotes the sum of the par2 values over all instances.

**Table 1.** GAPSAT, SPARROW2RISS, and PROBSAT are compared based on the number of solved instances and the corresponding score.

|  | # solved | score |
|---|---|---|
| PROBSAT | 133 | 1 234 986.01 |
| SPARROW2RISS | 189 | 672 335.89 |
| GAPSAT | **223** | **347 156.40** |

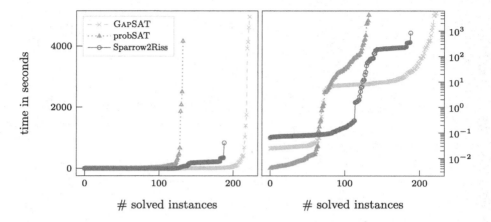

**Fig. 6.** Cactus plot comparing PROBSAT, SPARROW2RISS, and GAPSAT on the instances of the random track of the SAT competition 2018. On the left, the plot is linearly-scaled; on the right, it is logarithmically-scaled.

As can be observed in Table 1, GAPSAT solved substantially more instances than PROBSAT and SPARROW2RISS. The score of GAPSAT is nearly halved compared to the score of SPARROW2RISS. Figure 6 demonstrates that GAPSAT is especially efficient within the first few seconds. GAPSAT solved more instances within 30 s than SPARROW2RISS solved within the standard timeout of 5000 s is a case in point. This behavior can be observed in the logarithmically-scaled part of Fig. 6. Furthermore, there are no instances that could be solved with SPARROW2RISS, but not with GAPSAT within 5000 s.

We used statistical testing to evaluate the performance of GAPSAT compared to SPARROW2RISS and PROBSAT. The t-test [44] shows that the score of GAPSAT is better than both other solvers. We also used the Wilcoxon signed-rank test [46] to show that the median runtime of GAPSAT is superior to SPARROW2RISS and PROBSAT. All results are statistically significant, with $p$-values less than $10^{-9}$. Cohen's $d$ value [20] is 0.39 for the comparison with SPARROW2RISS and 0.73 for the comparison with PROBSAT.

The instances of the random track at the SAT competition 2018 can be split into three domains. Some instances are generated uniformly at random with a medium number of variables and a clause-to-variable ratio close to the satisfiability threshold [38]. Similarly, there are uniform random instances with a huge number of variables but with a clause-to-variable ratio not too close to the phase transition. Finally, there are randomly generated instances with a hidden solution [12]. Table 2 shows the performance of all three solvers on each domain. GAPSAT was the fastest solver on all three domains. It should be stated that the performances of GAPSAT and PROBSAT are interchangeable on huge, uniform instances since the differences are just due to random noise on this domain. Lastly, the average time needed to learn the new clauses was 103.17 s. That said, the actual time to perform the clause learning process is much shorter on most instances: The median time is just 7.89 s.

**Table 2.** GAPSAT, SPARROW2RISS, and PROBSAT are compared on three domains based on the score.

|  | Hidden | Medium | Huge |
|---|---|---|---|
| PROBSAT | 872 938.74 | 137 396.83 | 224 650.43 |
| SPARROW2RISS | 8 589.12 | 171 492.91 | 492 253.86 |
| GAPSAT | **851.36** | **127 982.19** | **218 322.85** |

We conclude that future generations of local search solvers should incorporate some kind of clause learning mechanisms, for example, as a prepossessing step as used by GAPSAT.

# 6  Conclusion and Outlook

In this work, a novel combination of CDCL as a preprocessing step and local search as the main solver is introduced. We empirically show on several domains that short clauses learned by CDCL have a high number of correct literals w. r. t. a fixed solution. Consequently, these new clauses guide local search solvers towards a solution. Using this knowledge, we design a new SAT solver GAP-SAT which uses the CDCL solver GLUCOSE in a preprocessing step to find new clauses. It then proceeds to use PROBSAT on the modified formula to find a solution. We show that GAPSAT improves the state-of-the-art on randomly generated instances.

The GAPSAT solver can be improved even further: Besides the techniques described in this paper, no preprocessing steps are performed. We believe that further, finely tuned preprocessing may help to increase the performance of GAP-SAT on instances where it struggled to find a solution. When tuning GAPSAT, we used the original settings of PROBSAT (i. e., we use the parameters from [5]). The only tuned parameters are the number of new clauses and their length. An interesting direction for further research is to obtain even better performance by simultaneously tuning these parameters together with the PROBSAT settings. Furthermore, we argued that the clauses which are added to the formula have a substantial effect on the performance of SLS algorithms. Even though clauses learned by GLUCOSE have good properties on average, it would be beneficial to devise a clause selection heuristic for local search algorithms. If clauses having a negative impact on local search can be avoided, then the overall performance of solvers like GAPSAT should improve significantly. Another general question that could be investigated is about the clauses from MINISAT, that is being used in GLUCOSE. Clauses learned by GLUCOSE are generated by conflict analysis but may depend on clauses generated by the MINISAT preprocessing. It may be the case that short clauses are missed because they are only considered inside the solver, but never by the learning mechanism (when generated in the preprocessing step).

**Supplementary Material.** The source code of GAPSAT and all evaluations are available under https://doi.org/10.5281/zenodo.3776052.

# References

1. Anbulagan, A., Pham, D.N., Slaney, J.K., Sattar, A.: Old resolution meets modern SLS. In: Proceedings of the 20th National Conference on Artificial Intelligence and the 17th Innovative Applications of Artificial Intelligence Conference (AAAI/IAAI 2005), pp. 354–359 (2005)
2. Audemard, G., Lagniez, J.-M., Mazure, B., Saïs, L.: Boosting local search thanks to CDCL. In: Fermüller, C.G., Voronkov, A. (eds.) LPAR 2010. LNCS, vol. 6397, pp. 474–488. Springer, Heidelberg (2010). https://doi.org/10.1007/978-3-642-16242-8_34
3. Audemard, G., Simon, L.: Predicting learnt clauses quality in modern SAT solvers. In: Proceedings of the 21st International Joint Conference on Artificial Intelligence (IJCAI 2009), pp. 399–404 (2009)
4. Balint, A.: Engineering stochastic local search for the satisfiability problem. Ph.D. thesis, University of Ulm (2014)
5. Balint, A.: Original implementation of probSAT (2015). https://github.com/adrianopolus/probSAT
6. Balint, A., Fröhlich, A.: Improving stochastic local search for SAT with a new probability distribution. In: Strichman, O., Szeider, S. (eds.) SAT 2010. LNCS, vol. 6175, pp. 10–15. Springer, Heidelberg (2010). https://doi.org/10.1007/978-3-642-14186-7_3
7. Balint, A., Henn, M., Gableske, O.: A novel approach to combine a SLS- and a DPLL-solver for the satisfiability problem. In: Kullmann, O. (ed.) SAT 2009. LNCS, vol. 5584, pp. 284–297. Springer, Heidelberg (2009). https://doi.org/10.1007/978-3-642-02777-2_28
8. Balint, A., Manthey, N.: Boosting the performance of SLS and CDCL solvers by preprocessor tuning. In: POS@ SAT, pp. 1–14 (2013)
9. Balint, A., Manthey, N.: Dimetheus. In: Proceedings of SAT Competition 2016: Solver and Benchmark Descriptions, vol. B-2016-1, pp. 37–38. Department of Computer Science Series of Publications B, University of Helsinki (2016)
10. Balint, A., Manthey, N.: SparrowToRiss 2018. In: Proceedings of SAT Competition 2018: Solver and Benchmark Descriptions, vol. B-2018-1, pp. 38–39. Department of Computer Science Series of Publications B, University of Helsinki (2018)
11. Balint, A., Schöning, U.: Choosing probability distributions for stochastic local search and the role of make versus break. In: Cimatti, A., Sebastiani, R. (eds.) SAT 2012. LNCS, vol. 7317, pp. 16–29. Springer, Heidelberg (2012). https://doi.org/10.1007/978-3-642-31612-8_3
12. Balyo, T., Chrpa, L.: Using algorithm configuration tools to generate hard SAT benchmarks. In: Proceedings of the 11th Annual Symposium on Combinatorial Search (SoCS 2018), pp. 133–137. AAAI Press (2018)
13. Bayardo Jr., R.J., Schrag, R.: Using CSP look-back techniques to solve real-world SAT instances. In: Proceedings of the 14th National Conference on Artificial Intelligence (AAAI 1997), pp. 203–208. AAAI Press (1997)

14. Biere, A.: Cadical, lingeling, plingeling, treengeling, YalSAT entering the SAT competition 2017. In: Proceedings of SAT Competition 2017 - Solver and Benchmark Descriptions B-2017-1, pp. 14–15 (2017)
15. Biere, A., Cimatti, A., Clarke, E.M., Strichman, O., Zhu, Y.: Bounded model checking. Adv. Comput. **58**, 117–148 (2003)
16. Biere, A., Heule, M., van Maaren, H., Walsh, T. (eds.): Handbook of Satisfiability. Frontiers in Artificial Intelligence and Applications, vol. 185. IOS Press (2009)
17. Billionnet, A., Sutter, A.: An efficient algorithm for the 3-satisfiability problem. Oper. Res. Lett. **12**(1), 29–36 (1992)
18. Cha, B., Iwama, K.: Adding new clauses for faster local search. In: Proceedings of the 13th National Conference on Artificial Intelligence and 8th Innovative Applications of Artificial Intelligence Conference (AAAI/IAAI 1996), pp. 332–337 (1996)
19. Clarke, E.M., Biere, A., Raimi, R., Zhu, Y.: Bounded model checking using satisfiability solving. Formal Methods Syst. Des. **19**(1), 7–34 (2001). https://doi.org/10.1023/A:1011276507260
20. Cohen, J.: Statistical Power Analysis for the Behavioral Sciences. Routledge (2013)
21. Cook, S.A.: The complexity of theorem-proving procedures. In: Proceedings of the 3rd Annual ACM Symposium on Theory of Computing (STOC 1971), pp. 151–158 (1971)
22. Eén, N., Sörensson, N.: An extensible SAT-solver. In: Giunchiglia, E., Tacchella, A. (eds.) SAT 2003. LNCS, vol. 2919, pp. 502–518. Springer, Heidelberg (2004). https://doi.org/10.1007/978-3-540-24605-3_37
23. Eibach, T., Pilz, E., Völkel, G.: Attacking bivium using SAT solvers. In: Kleine Büning, H., Zhao, X. (eds.) SAT 2008. LNCS, vol. 4996, pp. 63–76. Springer, Heidelberg (2008). https://doi.org/10.1007/978-3-540-79719-7_7
24. Fang, L., Hsiao, M.S.: A new hybrid solution to boost SAT solver performance. In: Proceedings of the Design, Automation and Test in Europe Conference and Exposition (DATE 2007), pp. 1307–1313 (2007)
25. Givens, G.H., Hoeting, J.A.: Computational Statistics, vol. 703. Wiley (2012)
26. Gomes, C.P., Selman, B., Kautz, H.A.: Boosting combinatorial search through randomization. In: Proceedings of the 15th National Conference on Artificial Intelligence and 10th Innovative Applications of Artificial Intelligence Conference (AAAI/IAAI 1998), pp. 431–437 (1998)
27. Habet, D., Li, C.M., Devendeville, L., Vasquez, M.: A hybrid approach for SAT. In: Van Hentenryck, P. (ed.) CP 2002. LNCS, vol. 2470, pp. 172–184. Springer, Heidelberg (2002). https://doi.org/10.1007/3-540-46135-3_12
28. Hirsch, E.A.: SAT local search algorithms: worst-case study. J. Autom. Reason. **24**(1–2), 127–143 (2000). https://doi.org/10.1023/A:1006318521185
29. Jussien, N., Lhomme, O.: Local search with constraint propagation and conflict-based heuristics. Artif. Intell. **139**(1), 21–45 (2002). (Preliminary version in IAAI 2000)
30. Kautz, H., Selman, B.: Ten challenges *Redux*: recent progress in propositional reasoning and search. In: Rossi, F. (ed.) CP 2003. LNCS, vol. 2833, pp. 1–18. Springer, Heidelberg (2003). https://doi.org/10.1007/978-3-540-45193-8_1
31. Kilby, P., Slaney, J., Thiébaux, S., Walsh, T., et al.: Backbones and backdoors in satisfiability. AAAI **5**, 1368–1373 (2005)
32. Kroc, L., Sabharwal, A., Gomes, C.P., Selman, B.: Integrating systematic and local search paradigms: a new strategy for MaxSAT. In: Proceedings of the 21st International Joint Conference on Artificial Intelligence (IJCAI 2009), pp. 544–551 (2009)

33. Letombe, F., Marques-Silva, J.: Improvements to hybrid incremental SAT algorithms. In: Kleine Büning, H., Zhao, X. (eds.) SAT 2008. LNCS, vol. 4996, pp. 168–181. Springer, Heidelberg (2008). https://doi.org/10.1007/978-3-540-79719-7_17

34. Li, C.M., Anbulagan, A.: Heuristics based on unit propagation for satisfiability problems. In: Proceedings of the 15th International Joint Conference on Artificial Intelligence (IJCAI 1997), pp. 366–371 (1997)

35. Lynce, I., Marques-Silva, J.: SAT in bioinformatics: making the case with haplotype inference. In: Biere, A., Gomes, C.P. (eds.) SAT 2006. LNCS, vol. 4121, pp. 136–141. Springer, Heidelberg (2006). https://doi.org/10.1007/11814948_16

36. Marques-Silva, J.P., Sakallah, K.A.: GRASP–a new search algorithm for satisfiability. In: Proceedings of the IEEE/ACM International Conference on Computer-Aided Design (ICCAD 1996), pp. 220–227 (1996)

37. McAllester, D.A., Selman, B., Kautz, H.A.: Evidence for invariants in local search. In: Proceedings of the 14th National Conference on Artificial Intelligence and 9th Innovative Applications of Artificial Intelligence Conference (AAAI/IAAI 1997), pp. 321–326 (1997)

38. Mertens, S., Mézard, M., Zecchina, R.: Threshold values of random $K$-SAT from the cavity method. Random Struct. Algorithms **28**(3), 340–373 (2006)

39. Moskewicz, M.W., Madigan, C.F., Zhao, Y., Zhang, L., Malik, S.: Chaff: engineering an efficient SAT solver. In: Proceedings of the 38th Design Automation Conference (DAC 2001), pp. 530–535 (2001)

40. Pipatsrisawat, K., Darwiche, A.: On the power of clause-learning SAT solvers with restarts. In: Gent, I.P. (ed.) CP 2009. LNCS, vol. 5732, pp. 654–668. Springer, Heidelberg (2009). https://doi.org/10.1007/978-3-642-04244-7_51

41. Schöning, U.: A probabilistic algorithm for $k$-SAT based on limited local search and restart. Algorithmica **32**(4), 615–623 (2002). https://doi.org/10.1007/s00453-001-0094-7. (Preliminary version in FOCS 1999)

42. Schöning, U., Torán, J.: The Satisfiability Problem: Algorithms and Analyses. Mathematics for Applications (Mathematik für Anwendungen), vol. 3. Lehmanns Media (2013)

43. Selman, B., Kautz, H.A., McAllester, D.A.: Ten challenges in propositional reasoning and search. In: Proceedings of the 15th International Joint Conference on Artificial Intelligence (IJCAI 1997), pp. 50–54 (1997)

44. Student: The probable error of a mean. Biometrika **6**(1), 1–25 (1908)

45. Völkel, G., Lausser, L., Schmid, F., Kraus, J.M., Kestler, H.A.: Sputnik: ad hoc distributed computation. Bioinformatics **31**(8), 1298–1301 (2015)

46. Wilcoxon, F.: Individual comparisons by ranking methods. Biometrics **1**(6), 80–83 (1945)

# SAT Heritage: A Community-Driven Effort for Archiving, Building and Running More Than Thousand SAT Solvers

Gilles Audemard[1]([✉]), Loïc Paulevé[2], and Laurent Simon[2]

[1] CRIL, Artois University, Lens, France
audemard@cril.fr
[2] Univ. Bordeaux, Bordeaux INP, CNRS, LaBRI, UMR5800, 33400 Talence, France
{loic.pauleve,lsimon}@labri.fr

**Abstract.** SAT research has a long history of source code and binary releases, thanks to competitions organized every year. However, since every cycle of competitions has its own set of rules and an adhoc way of publishing source code and binaries, compiling or even running any solver may be harder than what it seems. Moreover, there has been more than a thousand solvers published so far, some of them released in the early 90's. If the SAT community wants to archive and be able to keep track of all the solvers that made its history, it urgently needs to deploy an important effort.

We propose to initiate a community-driven effort to archive and to allow easy compilation and running of all SAT solvers that have been released so far. We rely on the best tools for archiving and building binaries (thanks to Docker, GitHub and Zenodo) and provide a consistent and easy way for this. Thanks to our tool, building (or running) a solver from its source (or from its binary) can be done in one line.

## 1    Introduction

As Donald Knuth wrote in [11], "The story of satisfiability is the tale of a triumph of software engineering". In this success story of computer science, the availability of SAT solvers source code have been crucial. Archiving and maintaining this important amount of knowledge may be as important as archiving the scientific papers that made this domain. The release of the source code of MiniSat [6] had, for instance, a dramatic impact on the field. However, nothing has yet been done to ensure that source code and recipes to build SAT solvers will be archived in the best possible way. This is a recent but important concern in the more broadly field of computer science. The Software Heritage [3] initiative is, for instance, a recent and strong initiative to handle this. In the domain of SAT solvers, however, collecting and archiving may not be sufficient: we must embed the recipe to build the code and to run it in the most efficient way. As input format for SAT solvers remains the same since more than 25 years [4],

© Springer Nature Switzerland AG 2020
L. Pulina and M. Seidl (Eds.): SAT 2020, LNCS 12178, pp. 107–113, 2020.
https://doi.org/10.1007/978-3-030-51825-7_8

it is always possible to compare the performances of all existing solvers, given a suitable way of compiling and running them. At that time, some code was using EGCS, a fork of GCC 2.8 including more features. Facebook and Google didn't exist and Linux machines were running with kernels 1.X. Solvers were distributed with source code to be compiled on Intel or SPARC computers. Fortunately enough, binaries for Intel 386 machines distributed at that time are still executable on recent computers, given the availability of compatible libraries.

Collecting and distributing SAT solvers source code is, luckily, not new. SAT competitions, organized since the beginning of the 21st century, have almost always forced the publication of the source code of submitted solvers. If source code was not distributed, binaries were often available. However, since the first competitions, the landscape of computer science has changed a lot. New technologies like `Docker` [5] are now available, changing the way tools are distributed.

We propose in this work to structure and bootstrap a collective effort to maintain a comprehensive and user-friendly library of all the solvers that shaped the SAT world. We build our tool, called `SAT Heritage`, on top of other recent tools, typically developed for archiving and distributing source code and applications, like `Docker` [5], `GitHub` [8], `Guix` [9], `Zenodo` [22]. The community is invited to contribute by archiving, from now on, all the solvers used in competitions (and papers). We also expect authors of previous solvers to contribute by adding informations about their solvers or special command lines not especially used during competitive events. Our tool allows, for instance, to add a DOI (thanks to `Zenodo`) to the exact version of any solver used in a paper, allowing simple but powerful references to be used.

In summary, the goals of our open-source tool are to:

- Collect and archive all SAT solvers, binaries and sources,
- Easily retrieve a `Docker` image with the binary of any solver, directly from the `Docker` Hub, or, when source code is available, by locally building the image from the source code of the solver,
- Allow to easily run any SAT solver that have ever been available (typically in the last 30 years), by a one line call (consistent over all solvers),
- Open an convenient solution for reproducibility (binaries, source code and receipt to build binaries are archived in a consistent way), thanks to strong connection with tools like `Guix` and `Zenodo`.

## 2    History of SAT Solvers Releases and Publications

The first SAT competitions happened in the 90's [1, 2]. Their goals were multiple: collect and compare SAT solvers performances in the fairest possible way, collect and distribute benchmarks, and also take a snapshot of the performances reached so far. Table 1 reports the number of SAT solvers that took part in the different competitions. We counted more than a thousand solvers, but even counting them was not an easy task: one source code can hide a number of subversions (with distinct parameters) and distinct tracks, and some information were only partially available.

**Table 1.** Number of solvers to the different competitions. Note that some solvers may be counted twice or more (some solvers did not change from year to the next or have been included in a competition as reference). (*) binaries and sources are available, but by navigating individually to each solver result. Different numbers indicate different organizers and different way of distributing results, source code (s) and binaries (b).

| Date | #Solvers | Collection | Type | Date | #Solvers | Collection | Type |
|------|----------|------------|------|------|----------|------------|------|
| ≤2000 | 24 | Satex | s/b | 2011 | 104 | Contest (2) | s/b |
| 2002 | 27 | Contest (1) | b | 2012 | 65 | Challenge | - |
| 2003 | 33 | Contest (1) | b | 2013 | 140 | Contest (3) | s(*)/b(*) |
| 2004 | 63 | Contest (1) | b | 2014 | 150 | Contest (3) | s(*)/b(*) |
| 2005 | 47 | Contest (1) | b | 2015 | 31 | Race (2) | - |
| 2006 | 16 | Race (1) | - | 2016 | 32 | Contest (4) | s/b |
| 2007 | 31 | Contest (2) | s/b | 2017 | 71 | Contest (4) | s/b |
| 2008 | 19 | Race (1) | - | 2018 | 66 | Contest (4) | s/b |
| 2009 | 64 | Contest (2) | s/b | 2019 | 55 | Race (3) | s/b |
| 2010 | 20 | Race (1) | - | Total | 1058 | | |

Following the ideas of these first competitions organized in the 90's, and thanks to the development of the web, the satex [17] website published solvers and benchmarks gathered by the website maintainer. satex was running SAT solvers on only one personal computer. Some solvers were modified to comply with the input/output of the satex framework (like a normalized exit code value). It was a personal initiative, made possible by the relatively few solvers available (all solvers of the initial satex are available in our tool).

During the first cycle of competitions (numbered 1 in Table 1) [16], submitters had to compile a static binary of their solver (to prevent library dependencies) via remote access to the same computer. To ensure the deployment of their solver, this computer had the exact same Linux version as the one deployed on the cluster used to run the contest. Some solvers were coming from industry, which explains why no open source code was mandatory: the priority was to draw the most accurate picture of solvers performances. However, it was quickly decided (competitions numbered 2 in the above table) that it was even more important to require submitters to open their code. Binaries were then allowed to enter the competition, but only in the demonstration category (no prizes). More recently, thanks to the starexec environment [19], compilation of solvers was somehow normalized (an image of a virtual Linux machine on which the code would be built and run was distributed). With each cycle of competition or race, came its own set of rules with an *ad hoc* way of publishing source code and binaries, with a non uniform way of providing details on which parameters to use. For example, since 2016, solvers must provide a certificate for unsatisfiable instances [10, 21]. One has thus to go through all the solvers to find the correct parameters for running them without proof logging.

Thus, despite the increasing importance of software archiving [3], the way SAT solvers are distributed had not really changed in the last 25 years. It is still mainly done via personal websites, or SAT competitions and races websites, each cycle of events defining its own rules for this. As a result, it is often unclear how to recover any SAT solver (same code, same arguments) used in many papers, old or recent. It is even more questionable whether, despite the importance of SAT solvers source code, we are able to correctly archive and maintain them.

## 3    SAT Heritage Docker Images

The SAT Heritage project provides a centralized repository of instructions to build and execute the SAT solvers involved in competitions since the early ages of SAT. To that aim, it relies on Docker images which are self-contained Linux-based environments to execute binaries. Docker allows to explicitly mention all the packages needed to compile the source code and to build a temporary image (the "builder") for compiling the solver. Then, the compiled solver is embedded in another, lighter, image which contains only the libraries required to execute it. So, each version of each collected solver is made available in a dedicated Docker image. Thanks to the layer structure of images, all solvers sharing the same environment will share the major part of the image content, thus substantially saving disk space. At the end, the Docker image will not be much heavier than the binary of the solver.

Docker images can be executed on usual operating systems. On Linux, Docker offers the same performance as native binaries: only filesystem and network operations have a slight overhead due to the isolation [7], which is not of concern for SAT solvers. On other systems, the images are executed within a virtual machine, adding a noticeable performance overhead, although considerably reduced on recent hardware [7].

### 3.1    Architecture

The instructions to build and run the collected solvers are hosted publicly on GitHub [13], on which the community is invited to contribute.

The solvers are typically grouped by year of competition. Images are then named as satex/<solver-name>:<year>.

The images are built by compiling solver sources whenever available. The compiling environment matches with a Linux distribution of the time of the competition. We selected the Debian GNU/Linux distribution which provides Docker images for each of its version since 2000. For instance, the solvers from the 2000 competition are built using the Debian "Potato" as it was back at that time. In principle, each solver can have its own recipe and environment for building and execution. Nevertheless, we managed to devise Docker recipes compatible with several generations of competitions. The architecture of the repository also allows custom sets of solvers. For example, the SAT Heritage collection includes the different Knuth's solvers or solvers with Java or Python.

The image building `Docker` recipes indicate where to download the sources or the binaries whenever the former are not available. At the time of the writing of this article, most recipes use URL from the website of the SAT competitions. In order to provide as most as persistent locations as possible, we are regularly moving more resources on `Zenodo` services to host sources and binaries in a near future [15] (currently, only the binaries of the original `satex` and the 2002's competition are hosted on it).

The images can be built locally from the git repository, and are also available for download from the main public `Docker` repository [14], that distributes "official" binaries of solvers. This allows to directly run any collected (or compiled) solver very quickly.

## 3.2   Running Solvers

We provide a Python script, called `satex`, which eases the execution and management of available `Docker` images, although images can be directly run without it. The script can be installed using `pip` utility: `pip3 install -U satex`.

The list of available solvers can be fetched using the command `satex list`.

We provide a generic wrapper in each image giving a unified mean to invoke the solver: a DIMACS file (possibly gzipped) as first argument, and optionally an output file for the proof:

```
# run a solver on a cnf file
satex run cadical:2019 file.cnf
# run and produce a proof
satex run glucose:2019 file.cnf proof
```

The `satex info` command gives, together with general information on the solver and the image environment, the specific options used for the run. Alternatively, custom options can be used with the `satex run-raw` command. If the image has not been built locally, it will attempt to fetch it from the online `Docker` repository. See the `satex -h` for other available commands, such as extracting binaries out of `Docker` images and invoking shells within a given image.

## 3.3   Building and Adding New Solvers

The building of images, which involve the compilation of the solvers when possible, also relies on `Docker` images, and thus only requires `Docker` and Python for the `satex` command. The following command, executed at the root of the `sat-heritage/docker-images` repository, will build the matching solvers with their adequate recipe:

```
satex build '*:2000' # build all 2000 solvers
```

Sets of solvers are added by specifying which `Docker` recipes to use for building the images and how to invoke the individual solvers. Managing sets of solvers allows sharing common configurations (such as linux distributions, compilers and so on) for docker images. A complete and up-to-date documentation can be found in the README file of the repository.

## 4    Ensuring Reproducibility

Reproducibility is a corner stone of science. In computer science, it recently appealed for significant efforts by researchers, institutions and companies to devise good practices and provide adequate infrastructures. Among the numerous initiatives, Software Heritage [3,18] and Zenodo [12,22] are probably the most important efforts for archiving source code, repositories, datasets, and binaries, for which they provide persistent storage, URLs, and references (DOI). Another example is the GitHub Archive Program, a repository on a 500-years lifespan storage preserved in the Artic World Archive [20]. Created more recently, the Guix [9] initiative aims at keeping the details of any Linux machine configuration, thanks to a declarative system configuration. External URL used for building any image are also archived. Our tool produces Docker images that can be easily frozen thanks to Guix, by building Guix images from the Dockerfile recipe. It is also worth mentioning that Guix has strong connections with Software Heritage and GitHub.

If we look at reproducibility of SAT solvers experiments on a longer time scale, we can expect that, some day, current binaries (for i386) will not genuinely run on computers any more. We can expect, however, that there will be i386 emulators. Once such an emulator is set up, we can also expect Docker to be available on it, and then all the images we built will be handled natively. If not, as Docker recipes are plain text, it will be easy to convert them to another framework.

Therefore, facilitating the accessibility of software in time now boils down to simple habits, such as using source versioning platforms, taking advantage of services like Zenodo or Software Heritage to freeze packages dependencies, source code, binaries, and benchmarks, and provide Docker images to give both environments and recipes to build and run your software.

## 5    Conclusion

We presented a tool for easily archiving and running all SAT solvers produced so far. Such a tool is needed because of (1) source code and experiments are crucial for the SAT community and (2) there are already too many SAT solvers produced so far, with many different ways of publishing sources.

In order to complete our tool we think at further improvements, like including Docker images for compiling SAT solvers for other architectures than i386 (ARM for instance), but also initiating another important effort for the community: including Docker images for benchmarks generations and maintenance. Many benchmarks are combinatoric ones, typically generated by short programs. These generators are generally not distributed by the different competitive events and may contain a lot of information on the structure of the generated problems. We also think that our tool could be very interesting for SAT solvers configurations and easy cloud-deployment in a portfolio way. We also expect our work to give the community the best possible habits for state of the art archiving and reproducibility practices.

# References

1. Buro, M., Buning, H.K.: Report on a SAT competition. Technical report (1992)
2. Crawford, J.: International competition and symposium on satisfiability testing (1996)
3. Di Cosmo, R., Zacchiroli, S.: Software heritage: why and how to preserve software source code. In: International Conference on Digital Preservation, pp. 1–10 (2017)
4. Second challenge on satisfiability testing organized by the center for discrete mathematics and computer science of Rutgers University (1993)
5. https://www.docker.com
6. Eén, N., Sörensson, N.: An extensible SAT-solver. In: Giunchiglia, E., Tacchella, A. (eds.) SAT 2003. LNCS, vol. 2919, pp. 502–518. Springer, Heidelberg (2004). https://doi.org/10.1007/978-3-540-24605-3_37
7. Felter, W., Ferreira, A., Rajamony, R., Rubio, J.: An updated performance comparison of virtual machines and Linux containers. In: 2015 IEEE International Symposium on Performance Analysis of Systems and Software (ISPASS), pp. 171–172 (2015)
8. https://www.github.com
9. https://guix.gnu.org
10. Heule, M., Hunt Jr., W.A., Wetzler, N.: Trimming while checking clausal proofs. In: Formal Methods in Computer-Aided Design, FMCAD, pp. 181–188 (2013)
11. Knuth, D.E.: The art of computer programming, vol. 4, p. iv, Fascicle 6 (2015)
12. Peters, I., Kraker, P., Lex, E., Gumpenberger, C., Gorraiz, J.I.: Zenodo in the spotlight of traditional and new metrics. Front. Res. Metrics Anal. **2**, 13 (2017)
13. https://github.com/sat-heritage/docker-images
14. https://hub.docker.com/u/satex
15. https://zenodo.org/communities/satex
16. Simon, L., Le Berre, D., Hirsch, E.A.: The SAT2002 competition report. Ann. Math. Artif. Intell. **43**(1), 207–342 (2005)
17. Simon, L., Chatalic, P.: SATEx: a web-based framework for SAT experimentation. Electron. Notes Discret. Math. **9**, 129–149 (2001)
18. https://www.softwareheritage.org
19. Stump, A., Sutcliffe, G., Tinelli, C.: StarExec: a cross-community infrastructure for logic solving. In: Demri, S., Kapur, D., Weidenbach, C. (eds.) IJCAR 2014. LNCS (LNAI), vol. 8562, pp. 367–373. Springer, Cham (2014). https://doi.org/10.1007/978-3-319-08587-6_28
20. Thorkildsen, M., Sjøvik, J.-F., Bryde, B.: Preserving irreplaceable national digital cultural heritage in the arctic world archive. In: Archiving Conference, vol. 2019, pp. 39–41. Society for Imaging Science and Technology (2019)
21. Wetzler, N., Heule, M.J.H., Hunt, W.A.: DRAT-trim: efficient checking and trimming using expressive clausal proofs. In: Sinz, C., Egly, U. (eds.) SAT 2014. LNCS, vol. 8561, pp. 422–429. Springer, Cham (2014). https://doi.org/10.1007/978-3-319-09284-3_31
22. https://zenodo.org

# Distributed Cube and Conquer with Paracooba

Maximilian Heisinger, Mathias Fleury[✉][ID], and Armin Biere[ID]

Johannes Kepler University Linz, Linz, Austria
{maximilian.heisinger,mathias.fleury,armin.biere}@jku.at

**Abstract.** Cube and conquer is currently the most effective approach to solve hard combinatorial problems in parallel. It organizes the search in two phases. First, a look-ahead solver splits the problem into many sub-problems, called cubes, which are then solved in parallel by incremental CDCL solvers. In this tool paper we present the first fully integrated and automatic distributed cube-and-conquer solver Paracooba targeting cluster and cloud computing. Previous work was limited to multi-core parallelism or relied on manual orchestration of the solving process. Our approach uses one master per problem to initialize the solving process and automatically discovers and releases compute nodes through elastic resource usage. Multiple problems can be solved in parallel on shared compute nodes, controlled by a custom peer-to-peer based load-balancing protocol. Experiments show the scalability of our approach.

## 1 Introduction

SAT solvers have been successfully applied in many practical domains, including cryptanalysis, hardware and software verification but also with increasing interest have been used to solve hard mathematical problems [17,21,26]. Sequential state-of-the-art SAT solving combines the well-known conflict-driven-clause-learning procedure (CDCL) [33,34] with sophisticated preprocessing techniques [10,23] and other efficient heuristics for variable selection [6,28,30], restarts [2,7,32], and clause database reduction [32]. While some authors argue that there was "no major performance breakthrough in close to two decades" [29], at the same time computers have become more and more powerful thanks to the ubiquitous availability of multi-core processors and the increasing usage of computers in the cloud. Thus improving the efficiency of parallel SAT solving remains an important topic. Accordingly, beside the traditional parallel track, the SAT Competition 2020 [19] features for the first time also a cloud track.

One approach to solve large problems in parallel consists in splitting the problem into smaller, more manageable instances, for example, using cube and conquer [16,20]. All these sub-problems are subsequently solved independently in parallel. This method was used by Heule to settle some long-standing mathematical conjectures [17,21]. Splitting the problems was done automatically by a tool, but then required to manually distribute instances for parallel solving.

© Springer Nature Switzerland AG 2020
L. Pulina and M. Seidl (Eds.): SAT 2020, LNCS 12178, pp. 114–122, 2020.
https://doi.org/10.1007/978-3-030-51825-7_9

In this paper, we present PARACOOBA [15]. After splitting a problem with the look-ahead solver MARCH, PARACOOBA transfers the sub-problems in an efficient way to many nodes (including over network). It detects when new instances become online and balances the work across all available nodes.

Other attempts for automatic and efficient distribution of problems exist, but use divide and conquer: Problems are dynamically split when nodes are underused (Sect. 2). In contrast, PARACOOBA assumes the problem is already split. Each node runs at least one instance of the SAT solver CADICAL [5]. The sub-problems are solved incrementally to reuse information from the previous solving. PARACOOBA relies on a custom protocol to automatically detect nodes that are underused and balances work across all nodes, including newly joining ones. It also supports disconnecting nodes by rebalancing the jobs (Sect. 3).

In the experiments, we focus on a single CNF `cruxmiter`, a miter for 32-bit adder trees [25], which is considered a challenge for resolution-based solvers and exemplary for the difficulties that arise in the verification of arithmetic circuits (see also [24]). Such benchmarks were also used in the SAT Race 2019. Already in the original work on cube and conquer similar multiplier equivalence checking problems were shown to benefit from the cube-and-conquer approach. Our results in Sect. 4 show that we get linear scaling with respect to the number of threads.

## 2   Preliminaries and Related Work

We use standard notations and refer the reader to the *Handbook of Satisfiability* for an introduction to SAT [8] as well as to the chapter on parallel SAT solving [4] in the *Handbook of Parallel Constraint Reasoning* [14].

One idea to improve solving of large instances is to distribute the work across different machines, via either a *diversification* of the search or *splitting* of the search space. In the first approach, several solvers are used as portfolio. By changing some parameters used by SAT solvers, they heuristically search on different parts of the search space and share some of the clauses they learned. ManySAT [13] pioneered the approach, which is now used in various tools like CRYPTOMINISAT [35], HORDESAT [3], PLINGELING [5], and SYRUP [1]. As soon as any instance derives SAT or UNSAT, then the problem is solved.

We use another approach that divides the search space explicitly as pioneered in [9,22,36] and refined in [16,20]. Solving the formula $\varphi$ is equivalent to splitting it into the two formulas $\varphi \wedge x$ and $\varphi \wedge \neg x$ and solving them. Unlike diversification, the overall problem is only considered to be UNSAT if all sub-problems are. Still, if any sub-problem is SAT, the overall problem is SAT, too. Splitting can be done dynamically during solving whenever a problem is deemed too hard. This is used for instance by PAINLESS [27] or MAPLEAMPHAROS [31]. These tools also share clauses to get some of the benefits of portfolio solvers.

Splitting can also be done upfront by look-ahead. By splitting the formula recursively, we obtain a formula of the form $\varphi \wedge c_1, ..., \varphi \wedge c_n$ where the conjunctions $c_i$ are called *cubes*. We use MARCH [18] to split the problem: It produces cubes, e.g., of the form $L_1 L_2 L_3, L_1 L_2 \neg L_3, ..., \neg L_1 \neg L_2 \neg L_3$. The cubes can be represented as a binary tree, the *cube tree*, where cubes are a path to a leaf: At each node, either the left (positive) or the right path (negative) is taken.

## 3   Architecture

PARACOOBA distinguishes between the masters that initiated work and workers that do the actual solving. Each node can either explore the cube tree deeper by sending work further (see Sect. 3.1) or solve the problem itself if a leaf node of the cube tree has been reached (Sect. 3.2). Nodes are also responsible for sending the result SAT or UNSAT back. PARACOOBA supports joining of new nodes dynamically, and the leaf nodes are able to wait for new tasks without consuming resources or shut down automatically, which is important if PARACOOBA is run in the cloud (Sect. 3.3). Figure 1 gives an overview of the solving process.

### 3.1   Static Organization

To combine fast local solving with automatic distribution to networked compute nodes, PARACOOBA sees tasks as *paths* in the cube tree. It distinguishes between *assigned tasks* (path to leafs) that are waiting for an available local worker and *unassigned tasks*. Only unassigned tasks are distributed further. A compute node is mapped to one PARACOOBA process which contains a fixed-size thread pool of local workers. Beside maintaining information on available nodes, every compute node has a unique 64-bit ID.

Connections between compute nodes are established at any time either by an integrated auto-discovery protocol or by providing a known peer at startup. Once connected, each compute node receives the full formula sent by the master. Then it announces that it is ready to receive tasks. Each compute node has a *solving context* for every master with the problem and the cubes to solve, a queue for unassigned tasks, and one for assigned tasks. Only paths in cube trees are exchanged during solving and similar assigned tasks are solved by the same solver. New contexts are created whenever a new master becomes online, and old ones are deleted if its master becomes offline. By using low-level socket functionality (UDP/TCP), PARACOOBA can be run without setting up a specialized environment (as needed for MPI [12]).

New (unassigned) tasks received by a compute node are inserted into the queue. When a compute node becomes idle, tasks with paths to leafs are instantiated into assigned tasks to be solved locally, whereas shorter paths are split

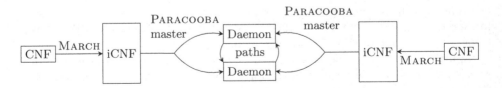

**Fig. 1.** Workflow of PARACOOBA with two different daemons

(by going deeper in the cube tree) into unassigned tasks that are distributed further. The overall strategy is to solve tasks with longer paths locally (as we are closer to leafs), while other tasks are distributed to further known compute nodes.

The SAT solver CADICAL [5] solves the assigned tasks incrementally [11]. It makes use of efficient preprocessing, including variable elimination and relying on efficiently restoring preprocessing steps if necessary [11]. This also provides a motivation for solving long paths locally: the cubes after a long shared path will be similar, making it possible to reuse more information compared to solving diverse cubes, where most of the preprocessing will have to be undone. If hard sub-problems are clustered on a single compute node, some can be offloaded.

## 3.2  Solving

We use the look-ahead solver MARCH [18] to generate cubes. PARACOOBA takes the output file containing the formula and the cubes as argument. This PARA-COOBA instance is the master node. All compute nodes parse both formula and cubes (reusing CADICAL's parser). After parsing, the initial task consisting of the empty path is created on the master compute node which will then branch on the first variable of the cube tree and create new unassigned tasks. These are either solved directly on the master or distributed to other compute nodes.

Paths in the cube tree are often transmitted across the network and should, therefore, have a compact representation. We represent them as 64-bit unsigned integers, where the first 58 bits describe the path in the binary tree and the last 6 bits specify the length of the path. This representation entails a maximum tree height of 58, which limits the number of different tasks to $2^{58}$. This constraint is not an issue, since it is 11 orders of magnitude larger than the one million cubes used for Heule's proof for Pythagorean Triples [21] that already created a 200TB proof. Communication between compute nodes is done using a custom protocol, which defines messages sent over UDP and TCP. The former is unreliable (packages can be dropped) and is used for non-critical messages, like auto-discovery, while the (reliable) latter is used for transmission of formulas, tasks, results, and status updates. Once a new compute node becomes known, all other nodes establish a TCP connection to it, which is used for all remaining transfers in order to circumvent UDP reliability issues in larger environments.

A sample interaction between a master and two daemon compute nodes is given in Fig. 2. First, the master starts with a problem to solve. It broadcasts an announcement request to all devices on the network. The daemons 1 and 2 answer the request and receive the formula in iCNF and a job initiator message. After that, solving starts and a path is sent from master to daemon 1. Work is rebalanced from daemon 1 to daemon 2. Once the problem is solved, the status is bubbled up to master and each node is responsible for collecting the results of offloaded jobs. Finally, master can conclude (UN)SAT.

Every daemon and every master sends a status message at every "tick", i.e., in configurable intervals with default 100 ms, to all compute nodes it knows.

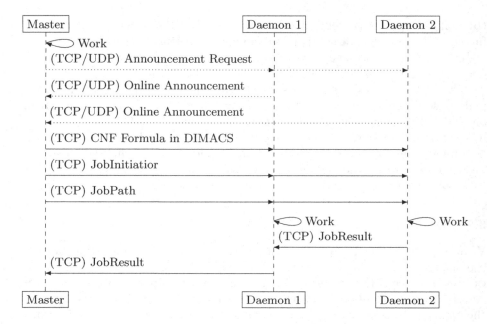

**Fig. 2.** Interaction between master and two daemons, without status messages

These messages describe the current queue sizes and are used by the distribution algorithm to decide whether and where tasks should be offloaded.

PARACOOBA allows an "$m$ to $n$" relation between masters and daemons, where daemons are used by different masters at the same time. Jobs are scheduled based on path length, not on the identity of the master.

When distributing tasks to other compute nodes, the ID of the original master, of the distribution target, and of the sender are always referenced, making all PARACOOBA instances aware of senders and receivers of each task. For the same reason, status messages of daemon compute nodes also contain a list of all current contexts to announce the formulas for which they can solve tasks.

### 3.3    System Management

By automatically discovering compute nodes in the same network, PARACOOBA can manage its overall resources automatically. Every daemon that is newly discovered by a master gets the formula and the cubes and, once ready, can receive tasks from all other connected compute nodes. Whenever a master node goes offline, it sends an offline announcement, which removes its solving context from all connected daemons, including all results and solver instances.

Compute nodes maintain a moving average of time between status messages for all other connected nodes. If a remote compute node does not send a status update early enough, it gets removed from the list of known nodes and all tasks sent to that node get re-added to the local unassigned-task queue (and can, for example, be offloaded again).

To save compute resources, an auto-shutdown timer can be enabled to measure the time a compute node has been idle without active tasks to shut down the compute node, if no new tasks are added before the timer runs out. Because tasks get distributed to inactive nodes quickly, the timeout can be set to low values (e.g., 3 s) to reduce cost, making PARACOOBA suited for cloud scenarios.

**Table 1.** Time to solve the `cruxmiter` depending on the number of threads.

| Threads $t$ | Nodes $n$ | Wall-Clock time $T_t^n$ | Speedup $T_t^n/T_1^1$ | Network speedup $T_t^n/T_{16}^1$ |
|---|---|---|---|---|
| 1 | 1 | 23 h 27 min 50 s | 1.00 | 0.05 |
| 2 | 1 | 9 h 19 min 40 s | 2.52 | 0.14 |
| 4 | 1 | 4 h 57 min 59 s | 4.72 | 0.25 |
| 8 | 1 | 2 h 33 min 47 s | 9.15 | 0.49 |
| 16 | 1 | 1 h 15 min 38 s | 18.61 | 1.00 |
| 32 | 2 | 31 min 51 s | 44.20 | 2.37 |
| 64 | 4 | 14 min 18 s | 98.45 | 5.29 |
| 128 | 8 | 7 min 58 s | 176.72 | 9.49 |
| 256 | 16 | 5 min 10 s | 272.48 | 14.64 |
| 512 | 32 | 3 min 22 s | 418.17 | 22.47 |

## 4 Experiments

As motivated in the introduction, we tested our tool PARACOOBA on a 32-bit `cruxmiter` problem [25], for which MARCH takes less than 10 s to split the initial problem into 52 520 cubes. We then run PARACOOBA on our compute cluster of 32 nodes connected through cheap commodity 1 Gbit ethernet cards. Each node contains two 8-core Intel Xeon E5-2620 v4 CPUs running at 2.10 GHz (turbo-mode disabled) and 128 GB main memory. Thus every node has 16 cores.

Table 1 shows the performance with respect to the number of threads. The run-time distribution for solving cubes is heavily skewed. Most tasks need only a few seconds, but some take more than a minute, limiting the performance improvement that can be achieved by using more threads, as, following Amdahl's law, the possible speedup is limited by the time required to solve the slowest cube. After 2 min, 5 instances of CADICAL are still running and it takes another minute to solve those. PARACOOBA outperforms static scheduling done by splitting the cubes upfront over 512 threads and solving each resulting iCNF for each group of cubes incrementally by CADICAL (4 min 17 s wall-clock time). We experimented with resplitting cubes, but could not improve solving time.

## 5 Conclusion

PARACOOBA is the first distributed cube-and-conquer solver. It relies on the state-of-the-art look-ahead solver MARCH to split the problem and then efficiently distributes the cubes over as many nodes as available. Our experiments reveal that the speedup is larger than the number of cores until saturation is reached.

As future work, it would be interesting to support proof generation in the nodes and store them in the master node.

**Acknowledgment.** This work is supported by the Austrian Science Fund (FWF), NFN S11408-N23 (RiSE), the LIT project LOGTECHEDU, and the LIT AI Lab funded by the State of Upper Austria. Daniela Kaufmann, Sibylle Möhle, and the reviewers suggested many textual improvements.

# References

1. Audemard, G., Lagniez, J.-M., Szczepanski, N., Tabary, S.: A distributed version of SYRUP. In: Gaspers, S., Walsh, T. (eds.) SAT 2017. LNCS, vol. 10491, pp. 215–232. Springer, Cham (2017). https://doi.org/10.1007/978-3-319-66263-3_14
2. Audemard, G., Simon, L.: Refining restarts strategies for SAT and UNSAT. In: Milano, M. (ed.) CP 2012. LNCS, pp. 118–126. Springer, Heidelberg (2012). https://doi.org/10.1007/978-3-642-33558-7_11
3. Balyo, T., Sanders, P., Sinz, C.: HordeSat: a massively parallel portfolio SAT solver. In: Heule, M., Weaver, S. (eds.) SAT 2015. LNCS, vol. 9340, pp. 156–172. Springer, Cham (2015). https://doi.org/10.1007/978-3-319-24318-4_12
4. Balyo, T., Sinz, C.: Parallel satisfiability. Handbook of Parallel Constraint Reasoning, pp. 3–29. Springer, Cham (2018). https://doi.org/10.1007/978-3-319-63516-3_1
5. Biere, A.: CaDiCaL, Lingeling, Plingeling, Treengeling and YalSAT entering the SAT Competition 2018. In: Heule, M., Järvisalo, M., Suda, M. (eds.) Proceedings of SAT Competition 2018 - Solver and Benchmark Descriptions. Department of Computer Science Series of Publications B, vol. B-2018-1, pp. 13–14. University of Helsinki (2018)
6. Biere, A., Fröhlich, A.: Evaluating CDCL variable scoring schemes. In: Heule, M., Weaver, S. (eds.) SAT 2015. LNCS, vol. 9340, pp. 405–422. Springer, Cham (2015). https://doi.org/10.1007/978-3-319-24318-4_29
7. Biere, A., Fröhlich, A.: Evaluating CDCL restart schemes. In: Berre, D.L., Järvisalo, M. (eds.) POS 2015/POS 2018. EPiC Series in Computing, vol. 59, pp. 1–17. EasyChair (2018). http://www.easychair.org/publications/paper/RdBL
8. Biere, A., Heule, M.J.H., van Maaren, H., Walsh, T. (eds.): Handbook of Satisfiability. Frontiers in Artificial Intelligence and Applications, vol. 185. IOS Press (2009)
9. Blochinger, W., Sinz, C., Küchlin, W.: Parallel propositional satisfiability checking with distributed dynamic learning. Parallel Comput. **29**(7), 969–994 (2003). https://doi.org/10.1016/S0167-8191(03)00068-1
10. Eén, N., Biere, A.: Effective preprocessing in SAT through variable and clause elimination. In: Bacchus, F., Walsh, T. (eds.) SAT 2005. LNCS, vol. 3569, pp. 61–75. Springer, Heidelberg (2005). https://doi.org/10.1007/11499107_5
11. Fazekas, K., Biere, A., Scholl, C.: Incremental inprocessing in SAT solving. In: Janota, M., Lynce, I. (eds.) SAT 2019. LNCS, vol. 11628, pp. 136–154. Springer, Cham (2019). https://doi.org/10.1007/978-3-030-24258-9_9
12. Graham, R.L.: The MPI 2.2 standard and the emerging MPI 3 standard. In: Ropo, M., Westerholm, J., Dongarra, J. (eds.) EuroPVM/MPI 2009. LNCS, vol. 5759, p. 2. Springer, Heidelberg (2009). https://doi.org/10.1007/978-3-642-03770-2_2

13. Hamadi, Y., Jabbour, S., Sais, L.: ManySAT: a parallel SAT solver. JSAT **6**(4), 245–262 (2009)
14. Hamadi, Y., Sais, L. (eds.): Handbook of Parallel Constraint Reasoning. Springer, Cham (2018). https://doi.org/10.1007/978-3-319-63516-3
15. Heisinger, M.: https://github.com/maximaximal/Paracooba.git. Accessed Feb 2020
16. Heule, M.J.H., Kullmann, O., Wieringa, S., Biere, A.: Cube and conquer: guiding CDCL SAT solvers by lookaheads. In: Eder, K., Lourenço, J., Shehory, O. (eds.) HVC 2011. LNCS, vol. 7261, pp. 50–65. Springer, Heidelberg (2012). https://doi.org/10.1007/978-3-642-34188-5_8
17. Heule, M.J.H.: Schur number five. In: McIlraith, S.A., Weinberger, K.Q. (eds.) AAAI 2018, pp. 6598–6606. AAAI Press (2018). https://www.aaai.org/ocs/index.php/AAAI/AAAI18/paper/view/16952
18. Heule, M., Dufour, M., van Zwieten, J., van Maaren, H.: March_eq: implementing additional reasoning into an efficient look-ahead SAT solver. In: Hoos, H.H., Mitchell, D.G. (eds.) SAT 2004. LNCS, vol. 3542, pp. 345–359. Springer, Heidelberg (2005). https://doi.org/10.1007/11527695_26
19. Heule, M.J.H., Järvisalo, M., Suda, M., Iser, M., Balyo, T.: https://satcompetition.github.io/2020/track_cloud.html. Accessed Feb 2020
20. Heule, M.J.H., Kullmann, O., Biere, A.: Cube-and-conquer for satisfiability. Handbook of Parallel Constraint Reasoning, pp. 31–59. Springer, Cham (2018). https://doi.org/10.1007/978-3-319-63516-3_2
21. Heule, M.J.H., Kullmann, O., Marek, V.W.: Solving very hard problems: cube-and-conquer, a hybrid SAT solving method. In: Creignou, N., Berre, D.L. (eds.) IJCAI 2017. LNCS, vol. 9710, pp. 228–245. IJCAI, August 2017. https://doi.org/10.24963/ijcai.2017/683
22. Hyvärinen, A.E.J., Junttila, T., Niemelä, I.: A distribution method for solving SAT in grids. In: Biere, A., Gomes, C.P. (eds.) SAT 2006. LNCS, vol. 4121, pp. 430–435. Springer, Heidelberg (2006). https://doi.org/10.1007/11814948_39
23. Järvisalo, M., Heule, M.J.H., Biere, A.: Inprocessing rules. In: Gramlich, B., Miller, D., Sattler, U. (eds.) IJCAR 2012. LNCS (LNAI), vol. 7364, pp. 355–370. Springer, Heidelberg (2012). https://doi.org/10.1007/978-3-642-31365-3_28
24. Kaufmann, D., Biere, A., Kauers, M.: Verifying large multipliers by combining SAT and computer algebra. In: Barrett, C.W., Yang, J. (eds.) FMCAD 2019, pp. 28–36. IEEE (2019). https://doi.org/10.23919/FMCAD.2019.8894250
25. Kaufmann, D., Kauers, M., Biere, A., Cok, D.: Arithmetic verification problems submitted to the SAT Race 2019. In: Heule, M., Järvisalo, M., Suda, M. (eds.) Proceedings of SAT Race 2019 - Solver and Benchmark Descriptions. Department of Computer Science Series of Publications B, vol. B-2019-1, p. 49. University of Helsinki (2019)
26. Konev, B., Lisitsa, A.: Computer-aided proof of Erdős discrepancy properties. Artif. Intell. **224**, 103–118 (2015). https://doi.org/10.1016/j.artint.2015.03.004
27. Le Frioux, L., Baarir, S., Sopena, J., Kordon, F.: Modular and efficient divide-and-conquer SAT solver on top of the painless framework. In: Vojnar, T., Zhang, L. (eds.) TACAS 2019. LNCS, vol. 11427, pp. 135–151. Springer, Cham (2019). https://doi.org/10.1007/978-3-030-17462-0_8
28. Liang, J.H., Ganesh, V., Poupart, P., Czarnecki, K.: Learning rate based branching heuristic for SAT solvers. In: Creignou, N., Le Berre, D. (eds.) SAT 2016. LNCS, vol. 9710, pp. 123–140. Springer, Cham (2016). https://doi.org/10.1007/978-3-319-40970-2_9

29. Marques-Silva, J.P.: SAT: Disruption, demise & resurgence (2019). pOS'2019. http://www.pragmaticsofsat.org/2019/disruption.pdf
30. Moskewicz, M.W., Madigan, C.F., Zhao, Y., Zhang, L., Malik, S.: Chaff: engineering an efficient SAT solver. In: DAC 2001, pp. 530–535. ACM (2001). https://doi.org/10.1145/378239.379017
31. Nejati, S., et al.: A propagation rate based splitting heuristic for divide-and-conquer solvers. In: Gaspers, S., Walsh, T. (eds.) SAT 2017. LNCS, vol. 10491, pp. 251–260. Springer, Cham (2017). https://doi.org/10.1007/978-3-319-66263-3_16
32. Oh, C.: Between SAT and UNSAT: the fundamental difference in CDCL SAT. In: Heule, M., Weaver, S. (eds.) SAT 2015. LNCS, vol. 9340, pp. 307–323. Springer, Cham (2015). https://doi.org/10.1007/978-3-319-24318-4_23
33. Silva, J.P.M., Lynce, I., Malik, S.: Conflict-driven clause learning SAT solvers. In: Biere, A., Heule, M., van Maaren, H., Walsh, T. (eds.) Handbook of Satisfiability. Frontiers in Artificial Intelligence and Applications, vol. 185, pp. 131–153. IOS Press (2009). https://doi.org/10.3233/978-1-58603-929-5-131
34. Silva, J.P.M., Sakallah, K.A.: GRASP - a new search algorithm for satisfiability. In: Rutenbar, R.A., Otten, R.H.J.M. (eds.) ICCAD 1996, pp. 220–227. IEEE Computer Society/ACM (1996). https://doi.org/10.1109/ICCAD.1996.569607
35. Soos, M., Nohl, K., Castelluccia, C.: Extending SAT solvers to cryptographic problems. In: Kullmann, O. (ed.) SAT 2009. LNCS, vol. 5584, pp. 244–257. Springer, Heidelberg (2009). https://doi.org/10.1007/978-3-642-02777-2_24
36. Zhang, H., Bonacina, M.P., Hsiang, J.: PSATO: a distributed propositional prover and its application to quasigroup problems. J. Symb. Comput. **21**(4), 543–560 (1996). https://doi.org/10.1006/jsco.1996.0030

# Reproducible Efficient Parallel SAT Solving

Hidetomo Nabeshima[1][✉] and Katsumi Inoue[2]

[1] University of Yamanashi, Kofu, Japan
nabesima@yamanashi.ac.jp
[2] National Institute of Informatics, Chiyoda, Japan
inoue@nii.ac.jp

**Abstract.** In this paper, we propose a new reproducible and efficient parallel SAT solving algorithm. Unlike sequential SAT solvers, most parallel solvers do not guarantee reproducible behavior due to maximizing the performance. The unstable and non-deterministic behavior of parallel SAT solvers hinders a wider adoption of parallel solvers to the practical applications. In order to achieve robust and efficient parallel SAT solving, we propose two techniques to significantly reduce idle time in deterministic parallel SAT solving: delayed clause exchange and accurate estimation of execution time of clause exchange interval between solvers. The experimental results show that our reproducible parallel SAT solver has comparable performance to non-deterministic parallel SAT solvers even in a many-core environment.

## 1 Introduction

Most modern computers have multiple cores, and the number of cores is increasing. To exploit the performance of multi-core systems, parallel processing software which efficiently utilizes each core is required. The same applies to SAT solvers, and parallel SAT solving is an active area of research. The parallel track of the SAT Competition is continuously held since 2011[1].

There are mainly two approaches of parallel SAT solving: portfolio and divide-and-conquer approaches. The former approach launches multiple SAT solvers with different search strategies in parallel, and each solver tries to solve the same SAT instance competitively [1,2,4]. The latter approach divide a given SAT instance in an attempt to distribute the total workload among computing units, and then solve them in parallel [5–7,10,12,13]. In both approaches, clause exchange techniques are combined into parallel systems in order to share the pruning information of the search space between solvers [1,2,4,11,12].

Most of parallel SAT solvers do not provide reproducible behavior in both runtime and found solutions due to maximizing the performance. Even for the same instance and computational environment, the execution time often varies for each run, and found models or unsatisfiability proofs may also differ. This is

---

[1] http://www.satcompetition.org/.

L. Pulina and M. Seidl (Eds.): SAT 2020, LNCS 12178, pp. 123–138, 2020.
https://doi.org/10.1007/978-3-030-51825-7_10

because there is no specific order in clause exchange between computing units. The timing of sending and receiving clauses can change due to system workload, cache misses and/or communication delays. The non-deterministic behavior of parallel SAT solvers causes various difficulties. In model checking, one may find different bugs (corresponding to satisfiable assignments) for each run. In the case of scheduling, even if a good solution is found, it may not be reproduced next time. If a bug occurs in software with an embedded non-deterministic SAT solver, the bug may not be reproduced. Researchers of parallel SAT solvers should have a number of experiments for stable evaluation of solvers. In contrast, most sequential SAT solvers guarantee reproducible behavior. The above-mentioned issues can be avoided if we use sequential SAT solvers. Reproducibility is thus an important property that directly affects the usability of SAT solvers as tools.

ManySAT 2.0 [3] is the first parallel SAT solver that supports reproducible behavior[2]. It is a portfolio parallel SAT solver for shared memory multi-core systems. To achieve deterministic behavior, it periodically synchronizes all threads, each of which executes a SAT solver, before and after the clause exchange. After the former synchronization, each solver exchanges clauses according to a specific order of threads until the latter synchronization. In ManySAT, all threads need to be synchronized periodically. Hence, waiting threads frequently occur as the number of CPU cores increases. As a result, there is a performance gap between deterministic and non-deterministic modes of ManySAT.

In this paper, we present two techniques to reduce the waiting time of threads: (1) delayed clause exchange and (2) refining the interval of clause exchange. The former suppresses the fluctuation of intervals between clause exchange, and the latter enables accurate prediction of exchange timing. We demonstrate that our approach significantly reduces the waiting time of threads and achieves the comparable performance with non-deterministic parallel SAT solvers even in a many-core environment[3].

The outline of this paper is as follows. The next section experimentally demonstrates the non-deterministic and unstable behavior of parallel SAT solvers. Section 3 describes the mechanism of ManySAT to realize the reproducibility and shows the experimental evaluation of the performance. In Sects. 4 and 5, we present two techniques in order to reduce the waiting time: delayed clause exchange and refining the interval of clause exchange, respectively. Experimental results are presented in Sect. 6. We conclude in Sect. 7.

## 2   Non-deterministic Behavior in Parallel SAT Solvers

In this section, we reexamine the unreproducible behavior of existing parallel SAT solvers. Here we consider ManySAT and Glucose-syrup as such parallel solvers developed for shared memory multi-core systems. ManySAT is the first portfolio parallel SAT solver [4] developed as a non-deterministic parallel solver,

---

[2] ManySAT 2.0 supports both deterministic and non-deterministic behavior.

[3] The solver source code and experimental results (including colored graphs in this paper) are available at http://www.kki.yamanashi.ac.jp/~nabesima/sat2020/.

**Table 1.** Solved instances on SAT Competition 2018 and SAT Race 2019 (800 instances in total). "X (Y + Z)" denotes the number of solved instances (X), solved satisfiable instances (Y) and solved unsatisfiable instances (Z), respectively. Non-deterministic solvers (ManySAT with non-det and Glucose-syrup) were run three times, and the first and last lines of the results denote the best and worst results, respectively.

| Solver | # of solved instances | |
|---|---|---|
| | 4 threads | 64 threads |
| ManySAT 2.0 with non-det | 434 (265 + 169) | 475 (292 + 183) |
| | 425 (265 + 160) | 475 (294 + 181) |
| | 420 (257 + 163) | 473 (288 + 185) |
| ManySAT 2.0 with det-static | 414 (251 + 163) | 457 (284 + 173) |
| ManySAT 2.0 with det-dynamic | 418 (258 + 160) | 448 (281 + 167) |
| Glucose-syrup 4.1 | 465 (263 + 202) | 524 (301 + 223) |
| | 462 (263 + 199) | 519 (295 + 224) |
| | 458 (255 + 203) | 515 (293 + 222) |

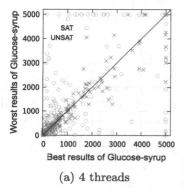
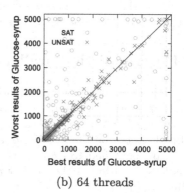

(a) 4 threads          (b) 64 threads

**Fig. 1.** Comparison of runtimes of the best and worst results of Glucose-syrup for each instance. Satisfiable instances are denoted with ∘, unsatisfiable instances with ×. Points on 5000 s mean that these instances are solved only by either the best or worst cases.

and ManySAT 2.0 supports both deterministic and non-deterministic strategies. Glucose-syrup [1] is one of the state-of-the-art parallel portfolio solvers.

We have run ManySAT 2.0 and Glucose-syrup 4.1 on instances from SAT Competition 2018 and SAT Race 2019, and show the experimental results as the numbers of solved instances in Table 1. In this work, we conducted all experiments on the following two computing environments: (1) a cluster equipped with 4-core Intel Core i5-6600 (3.3 GHz) machines using a memory limit of 8 GB, and (2) a cluster equipped with 68-core Intel Xeon Phi KNL (1.4 GHz) machines using a memory limit of 96 GB[4]. The time limit was set to 5000 s. We ran each

---

[4] We used the supercomputer of ACCMS, Kyoto University.

solver with 4 threads on the first environment and 64 threads on the second. "ManySAT 2.0 with non-det" denotes the non-deterministic mode of ManySAT, and "det-static" and "det-dynamic" mean the two deterministic modes described in the next section.

The results of non-deterministic solvers show that different runs yield different numbers of solved instances. For Glucose-syrup, the difference between the best and worst results is 7 and 9 instances on the 4 and 64 threads environments, respectively. Figure 1 gives scatter plots comparing the runtimes for each instance in which we compared the best and worst results of Glucose-syrup. The runtimes of satisfiable instances vary greatly with runs. Unsatisfiable instances have more stable results but there are some instances solved by either one. Such behavior is typically encountered when using parallel SAT solvers.

Clause exchange is a cooperative and fundamental mechanism in parallel SAT solvers in order to share the pruning information of the search space between computing units. Typically, the timing of sending and receiving clauses is affected by system workload, cache misses, and/or communication delays. However, most of parallel SAT solvers do not have synchronization mechanism of the timing in order to avoid the communication overhead and to maximize the performance. This is the cause of unreproducible behavior of parallel SAT solvers.

## 3    A Deterministic Parallel SAT Solver

In this section, we describe the algorithm called deterministic parallel DPLL ($(DP)^2LL$ in short) proposed by [3] which is implemented in the first deterministic parallel SAT solver ManySAT 2.0. The pseudo code is shown in Algorithm 1. Let $n$ be the number of solvers to be executed in parallel. Firstly, $n$ threads are launched to execute the function $search()$ (lines 2 and 4). Each thread is identified by an ID number $t \in \{1, \ldots, n\}$. After all threads have finished (line 5), the algorithm outputs the solution obtained by the thread with the lowest thread ID among all the threads which succeeded to decide the satisfiability of the instance (line 6). The reason for choosing the lowest ID is to avoid non-deterministic behavior if two or more threads find solutions at the same time.

The function $search()$ (lines 8–25) is the same as usual CDCL solvers, except for sending and receiving clauses. Each thread periodically receives clauses from the other threads. We call the reception interval a *period*. The function $endOfPeriod()$ decides whether the current period has ended (line 10). In ManySAT, it returns true when the number of conflicts in the period exceeds a certain threshold. In that case, all threads are synchronized *before* and *after* clause exchange by "$<$ barrier $>$" instruction[5] (lines 11 and 14). The former barrier is necessary for each thread to start importing clauses simultaneously. Suppose that a thread starts importing clauses at the end of period $x$. The latter barrier prevents the thread importing a clause which is exported from another thread at the next period $x+1$. In order to avoid deadlocks, when a thread finds

---

[5] The barrier is implemented by `#pragma omp barrier` directive in OpenMP.

---

**Algorithm 1:** Deterministic Parallel DPLL [3]

```
 1  Function solve(n)                            // n is the number of threads
 2      foreach t ∈ {1, ··· , n} do
 3          ansₜ ← unknown;
 4          launch thread t which executes ansₜ ← search(t);
 5      wait for all threads to finish;
 6      t_min ← min {t | ansₜ ≠ unknown};
 7      return ans_{t_min};

 8  Function search(t)
 9      loop
10          if endOfPeriod() = true then
11              nextPeriod: <barrier>;
12              if ∃i(ansᵢ ≠ unknown) then return ansₜ;
13              importExtraClauses(t);
14              <barrier>;
15          if propagate() = false then
16              if noDecision() = true then
17                  ansₜ = unsat;
18                  goto nextPeriod;
19              learnt ← analyze();
20              exportExtraClause(learnt);
21              backtrack();
22          else
23              if decide() = false then
24                  ansₜ = sat;
25                  goto nextPeriod;

26  Procedure importExtraClauses(t)
27      foreach i ∈ ⟨1, ··· , t − 1, t + 1, ··· , n⟩ do
28          import clauses from thread i;
```

---

a solution (lines 18 and 25), it needs to go to the first barrier on line 11 instead of exiting immediately. This is because other threads that have not found a solution are waiting there. After synchronization, each thread $t$ exits with its own status $ans_t$, if any thread finds a solution (line 12). The function $importExtraClauses()$ receives learnt clauses acquired by the other threads according to a fixed order of the threads (line 27), because different ordering of clauses will trigger off different ordering of unit propagations and consequently different behavior.

The rest of the search function follows CDCL algorithm. The $propagate()$ function (line 15) applies unit propagation (or Boolean constraint propagation) and returns $false$ if a conflict occurs, and $true$ otherwise. In the former case, if the conflict occurs without any decision (line 16), it means the unsatisfiability is proved. Otherwise, a cause of the conflict is analyzed (line 19) and a clause is learnt to prevent occurring the same conflict. If the learnt clause is eligible for export (for example, the length is short), it is marked to export. These exported clauses are periodically imported by the function $importExtraClauses()$. In the latter case, the function $decide()$ chooses an unassigned variable as the next

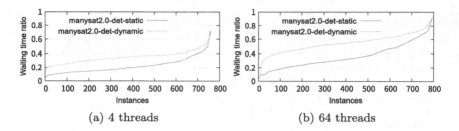

(a) 4 threads                          (b) 64 threads

**Fig. 2.** Waiting time ratio of ManySAT with static and dynamic periods. The ratio is defined for each instance as the total waiting time of all threads divided by the total solving time of all threads. Results are sorted by the ratio.

**Table 2.** Ratio of waiting time to the total solving time of all threads for all instances.

| Solver | 4 threads | 64 threads |
|---|---|---|
| ManySAT 2.0 with det-static | 23.0% | 40.6% |
| ManySAT 2.0 with det-dynamic | 34.5% | 56.6% |

decision and assigns it *true* or *false* (line 23). Otherwise it returns *false* as all the variable are assigned, that is, a model is found.

This $(DP)^2LL$ algorithm periodically requires synchronization with all threads. Runtime variation of periods on each thread causes idle time for each synchronization, that is, each thread should wait the slowest threads. A simple way to suppress the variation is to measure the execution time of threads and synchronize based on the elapsed time. However, the measurement of CPU or real time usually contains errors, so this approach is hard to hold reproducibility. In ManySAT, the length of a period is defined as the number of conflicts. There are two kinds of definitions of the period: *static* and *dynamic*. The *static period* is simply defined as a fixed number $c$ of conflicts ($c = 700$ in default). The *dynamic period* is intended to provide a better approximation of progression speed of each solver. Let $L_t^k$ denote the length of the $k$-th period of a thread $t$, defined as $L_t^k = c + c(1 - r_t^{k-1})$, where $r_t^{k-1}$ is the ratio of the number of learnt clauses held by the thread $t$ to the maximum number of learnt clauses among all threads at the $(k-1)$-th period. In this modeling, a thread with a large (or small) number of learnt clauses (the ratio tending to 1 (or 0)) is considered to be slow (or fast) and the length of period becomes shorter (or longer).

In Table 1, "det-static" and "det-dynamic" denote the results of the static and dynamic periods, respectively. There is a performance gap between deterministic and non-deterministic solvers. The cause is high waiting time ratio to the running time. Figure 2 shows the waiting time ratio of ManySAT for each instance, and the ratio on all instances is shown in Table 2. In our experiments, the waiting time ratio of the static period is lower than dynamic, but it reaches 23% for 4 threads and 40% for 64 threads environments. These results indicate that it is difficult to realize efficient solving by synchronizing all threads in a many-core environment.

## 4   Delayed Clause Exchange

In order to reduce the idle time in deterministic parallel SAT solvers, we propose a new clause exchange schema called *delayed clause exchange* (DCE). Figure 3 shows the runtime distribution of the static periods in ManySAT for two instances. These results indicate that the execution time of periods fluctuates very frequently, but in the long term it seems to be stable (roughly, 0.005 s for (a), 0.05 s for (b)). Other instances have a similar tendency. In order to take advantage of this property and absorb frequent fluctuations, we consider allowing clause reception to be delayed for a certain number of periods.

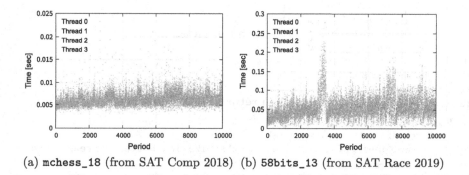

(a) `mchess_18` (from SAT Comp 2018)   (b) `58bits_13` (from SAT Race 2019)

**Fig. 3.** Distribution of period execution time (up to 10000 periods) on two instances by ManySAT 2.0 with det-static on the 4 threads environment. The results are sampled to half. The x-axis and y-axis show the number of periods and the execution time of the period, respectively. Each color represents different threads.

Let $n$ be the number of threads, $T = \{1, \ldots, n\}$ the set of thread IDs, $p_t$ the current period ID of a thread $t \in T$ ($p_t \geq 1$), $E_t^p$ a set of clauses exported by a thread $t$ at a period $p$ and $m$ an admissible delay, called *margin*, is denoted by the number of periods ($m \geq 0$). Algorithm 2 shows the pseudo code of $(DP)^2LL$ with DCE. There are two differences from Algorithm 1. The first point is clause reception. For each thread $t$, when the current period $p_t$ ends, then the thread increments the current period ID (line 14) and imports clauses from the other threads (line 15). If another thread $i$ has not yet finished the period $p_t - m$ (that is, $p_i < p_t - m$), then the thread $t$ should wait for it to complete (line 25). After that, the thread $t$ imports $E_i^{p_t - m}$. The second point concerns termination conditions. When multiple threads find solutions in DCE, to keep reproducibility, the algorithm select a thread that found at the earliest period. In case of a tie, the thread with the lowest ID is selected (line 7). Running threads that have

---

**Algorithm 2:** $(DP)^2LL$ with Delayed Clause Exchange

```
1  Function solve(n)                          // n is the number of threads
2  |   ans ← unknown;  pmin ← ∞;  tmin ← ∞;
3  |   foreach t ∈ {1, · · · , n} do
4  |   |   pt ← 1 ;                           // pt is the number of periods in thread t
5  |   |   launch thread t which executes the followings:
6  |   |   |   anst ← search(t);
7  |   |   |   if anst ≠ unknown and (pt < pmin or (pt = pmin and t < tmin))
   |   |   |      then ans ← anst;  pmin ← pt;  tmin ← t ;
8  |   wait for all threads to finish;
9  |   return ans;

10 Function search(t)
11 |   loop
12 |   |   if endOfPeriod() = true then
13 |   |   |   if ans ≠ unknown and pmin < pt then return unknown;
14 |   |   |   pt ← pt + 1;
15 |   |   |   importExtraClauses(t);
16 |   |   if propagate() = false then
17 |   |   |   if noDecision() = true then return unsat;
18 |   |   |   learnt ← analyze();
19 |   |   |   exportExtraClause(learnt);
20 |   |   |   backtrack();
21 |   |   else
22 |   |   |   if decide() = false then return sat;

23 Function importExtraClauses(t)
24 |   foreach i ∈ ⟨1, · · · , t − 1, t + 1, · · · , n⟩ do
25 |   |   wait until pi ≥ pt − m ;   // synchronization between thread t and i
26 |   |   import clauses from E_i^{pt−m};
```

---

not yet found a solution can be terminated if their periods exceed $p_{\min}$ (line 13). Note that when $m = 0$, $(DP)^2LL$ with DCE is same as $(DP)^2LL$.

DCE can reduce the total waiting time of threads. Firstly, we consider the best case of DCE. At some point, if for any two threads $i, j \in T$ $(i \neq j)$ the difference $p_i - p_j$ is less than or equal to $m$, then any thread can import clauses immediately without waiting for other threads at the point. This is because a set of clauses to be imported had already been exported by other threads. Secondly, the worst case is that only one thread is extremely slow and all other threads are ahead by $m$ periods. In this case, the fast $|T| - 1$ threads must wait the slowest thread until the difference less than or equal to $m$. In other cases, if there exists two threads $i, j \in T$ such that $p_i - p_j > m$, then the preceding thread $i$ should wait the postdating thread $j$ until the difference less than or equal to $m$. The execution time of periods fluctuates frequently, but if the total execution time of $m$ consecutive periods is almost the same for each thread, the DCE can be expected to reduce the waiting time. The disadvantage of DCE is that the clause reception is always delayed by $m$ periods, even if there is no difference in the period of each thread.

# 5    Refining Periods

In ManySAT, the length of a period is defined as the number of conflicts. The generation speed of conflicts is affected by the number and length of clauses. The number of clauses varies during search by learning and reduction of clauses, and the length of learnt clauses also changes sometimes significantly. As the result, runtime of a period fluctuates frequently as shown in Fig. 3. Accurate estimation of period execution time is important to reduce the waiting time. In this section, we introduce two new definitions of a period based on reproducible properties.

## 5.1    Refinement Based on Literal Accesses

Most of the memory used by SAT solvers is occupied by literals in clauses. Accessing literals in memory is a fundamental operation and occurs very frequently in unit propagation, conflict analysis, and so on. We consider defining the length of a period as the number of accesses to literals. The speed of accessing literals can be considered to be more stable than the generation speed of conflicts since it is almost independent of the number and length of clauses. With this definition, the function $endOfPeriod()$ (line 12 in Algorithm 2) returns true if the number of literal accesses in the period exceeds a certain threshold. In our implementation, we count the number of accesses to literals in unit propagation, conflict analysis and removal of clauses that are satisfied without any decision.

## 5.2    Refinement Based on Block Executions

In order to estimate the runtime of a period more accurately, we consider measuring not only the number of literal accesses, but also the number of executions of various operations performed by a SAT solver. It is similar to profiling a program which measures the number of calls and runtime of each function to detect performance bottlenecks. As a finer granularity than functions, we focus on compound statements called *blocks* (statements enclosed in curly braces in C++) and measure the number of the executions of each block during the search. For example, the runtime of one call of *propagate()* (line 16 in Algorithm 2) depends obviously on a given instance (proportional to the number of clauses). Whereas *propagate()* has a loop block that checks the value of each literal in a clause to determine whether the clause is unit or falsified. The time to execute the block once can be considered almost constant. We apply linear regression analysis to estimate the time required for one execution of each block.

Let $n$ be the number of blocks to be measured, $x_k^{i,j}$ the number of executions of a block $k$ of a thread $j$ in an instance $i$, $d_k$ the runtime to be required for one execution of a block $k$, and $y^{i,j}$ the execution time of a thread $j$ in an instance $i$ without waiting time. Each $d_k$ is non-negative. Hence, if a block $k$ has a nested block $l$, $d_k$ indicates the execution time of $k$ excluding $l$. Then, $y^{i,j}$ can be expressed as:

$$y^{i,j} = d_1 x_1^{i,j} + d_2 x_2^{i,j} + \cdots + d_n x_n^{i,j}. \tag{1}$$

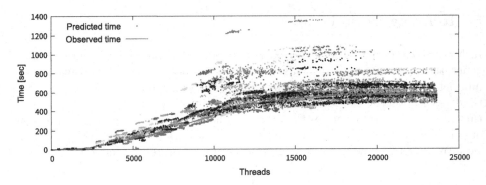

**Fig. 4.** Results of prediction of thread execution time on SAT Competition 2018. The solid line denotes the observed execution time for each thread and each instance (sorted by the execution time). Each point indicates the estimated execution time for each thread, and each color represents different instances.

With this definition, the function *endOfPeriod*() returns true if the value of (1) in a period exceeds a certain threshold.

Our developed deterministic parallel SAT solver (introduced in the next section) counts the number of executions for 71 blocks that almost cover the whole of operations performed by the solver. We used the application instances of SAT Competition 2016 and 2017 (300 and 350 instances, respectively) as training instances to estimate each $d_k$. The evaluation was performed on the 64 threads environment with a time limit of 600 s. To avoid overfitting, we manually selected 29 out of 71 blocks, which are mainly contained in unit propagation, conflict analysis and search loop (corresponding to *search*() function in Algorithm 2). Then, we estimated the regression coefficients $d_k$ from the results using the Elastic Nets method [14]. The coefficient of determination ($R^2$) was 0.94.

Figure 4 shows the results of prediction of thread execution time on SAT Competition 2018 as testing instances. Most of estimated results are close to the observed one. Some points are far from observations, but such points of the same color are often equidistant from observations. This means that the difference between the predicted and the observed time is approximately the same for each thread that solves the same instance, and in such cases, synchronization between threads can be expected to have less idle time.

## 6    Experimental Results

We have developed a new deterministic parallel SAT solver called ManyGlucose based on Glucose-syrup 4.1, which implements the delayed clause exchange and three types of periods (one is conflict based period used in ManySAT and the others are described in the previous section). In this work, we set the margin to a fixed value of 20 and adjust the length of the three types of periods. Suppose that $p_{conf}, p_{accs}$ and $p_{exec}$ denote the length of a period based on the number of

**Table 3.** Results of three types of periods. In this evaluation, we executed ManyGlucose with margin 20 for 400 instances used in SAT Competition 2018 for the 4 threads environment. The best result of each column in (a) is typeset in boldface.

(a) The numbers of solved instances

| $p_{conf}$ | # of solved | $p_{accs}$ | # of solved | $p_{exec}$ | # of solved |
|---|---|---|---|---|---|
| 50 | 231 (**128** + 103) | 1M | 231 (126 + **105**) | 0.2 | 230 (125 + **105**) |
| **100** | **232** (**128** + 104) | **2M** | **236** (**132** + 104) | 0.3 | 232 (127 + **105**) |
| 200 | 229 (126 + 103) | 4M | 231 (128 + 103) | 0.4 | 233 (129 + 104) |
| 300 | 226 (121 + **105**) | 6M | 225 (121 + 104) | **0.5** | **236** (**132** + 104) |
| 400 | 225 (122 + 103) | 8M | 230 (126 + 104) | 0.6 | 231 (126 + **105**) |

(b) The ratio of waiting time and average runtime per period

| $p_{conf}$ | Wait time ratio | Avg time/period | $p_{accs}$ | Wait time ratio | Avg time/period | $p_{exec}$ | Wait time ratio | Avg time/period |
|---|---|---|---|---|---|---|---|---|
| 50 | 16.9% | 0.044 | 1M | 9.5% | 0.029 | 0.2 | 6.0% | 0.043 |
| 100 | 15.6% | 0.087 | 2M | 8.8% | 0.056 | 0.3 | 5.6% | 0.065 |
| 200 | 13.9% | 0.172 | 4M | 8.0% | 0.112 | 0.4 | 5.3% | 0.087 |
| 300 | 13.1% | 0.263 | 6M | 7.5% | 0.167 | 0.5 | 5.1% | 0.109 |
| 400 | 12.6% | 0.350 | 8M | 7.2% | 0.219 | 0.6 | 4.8% | 0.131 |

conflicts, literal accesses and block executions (corresponding to the threshold in the function $endOfPeriod()$). We determine the appropriate length of each period by preliminary experiments.

Table 3 shows the results of three types of periods. As the length of period becomes longer, the waiting time is reduced since the number of clause exchanges is diminished, but the number of solved instances also tends to decreases. There is a trade-off between the number of clause exchanges and solved instances. From these results, we determined the appropriate length for each period type to be $p_{conf} = 100, p_{accs} = 2M$, and $p_{exec} = 0.5$. Table 3 (b) denotes that the period based on the block executions has the smallest waiting time ratio. The average runtime of periods when $p_{exec} = 0.5$ is 0.109 s. Hence, with this setting, time delay to receive learnt clauses acquired by other threads is about 2 s $(0.109 * 20)$.

Figure 5 shows the runtime distribution of periods based on these thresholds for some instances. These are results of unsolved instances by ManyGlucose without DCE within a 1000 s time limit (that is, the right end of x-axis corresponds to 1000 s). The execution time per period is normalized by the z-score to compare three period types. These graphs show that the period based on conflicts has large amplitude, and the block executions has small amplitude. For most instances, the block executions shows the best results, but Fig. 5 (c) is an example in which the literal access shows the best results.

We ran ManyGlucose configured with three types of periods and with and without DCE for the application instances used in SAT Competition 2018 and SAT Race 2019 in the 4 and 64 threads environments using the parameters obtained in the preliminary experiment (that is, $p_{conf} = 100, p_{accs} = 2M$, and $p_{exec} = 0.5$). ManyGlucose with DCE and block executions were run three times to show the robustness of our deterministic parallel SAT solver. Table 4 shows the number of solved instances and waiting time ratio for each solver and Fig. 6

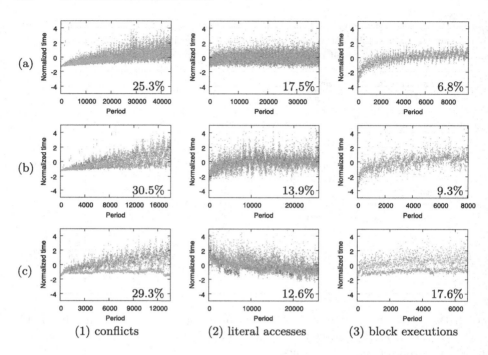

(1) conflicts          (2) literal accesses          (3) block executions

**Fig. 5.** Distribution of period execution time on (a) `mchess_18`, (b) `58bits_13` and (c) `eqspwtcl14bpwtcl14` (from SAT Race 2019). The results are sampled to 10%. The value at the bottom right of each graph shows the waiting time ratio.

and 7 are cactus plots of them. DCE can reduce the waiting time and increase the number of solved instances. The effect is remarkable in the 64 threads environment. ManyGlucose with DCE can solve 50, 36 and 34 more instances in conflict, literal access and block execution based periods than without DCE, respectively. For the ratio of waiting time, DCE reduces approximately 24%, 32% and 35% in conflict, literal access and block execution based periods than without DCE, respectively. The ratio of waiting time in 64 threads is greater than in 4 threads. When using block execution based period, it increases about 3.5 times without DCE (12.5% to 44.0%), but 1.5 times with DCE (5.7% to 8.8%). This means DCE is more effective in many-core systems. The regression coefficients of block execution based period are estimated in the 64 threads environment. Figure 7(a) shows that it is effective for reducing the waiting time even in the different environment. The difference of average runtime per period between 4 and 64 threads denotes the performance gap of sequential computation of each system. The 64 threads environment has a large number of CPU cores, although the sequential computing performance is not high. In the 64 threads environment, time delay to receive learnt clauses acquired by other threads is about 8 to 12 s.

Compared with Glucose-syrup, ManyGlucose shows the stable results due to its determinism. In the results of running ManyGlucose with DCE and block execution based period three times, the difference between the best and worst

**Table 4.** Results of (a) solved instances and (b) waiting time ratio on SAT Competition 2018 and SAT Race 2019. Results of Glucose-syrup are same as Table 1. The best result of each column in (a) is typeset in boldface. "confs", "lit accs" and "blk execs" mean the period type based on conflicts, literal accesses and block executions, respectively.

(a) The numbers of solved instances

| Solver | # of solved instances | |
|---|---|---|
| | 4 threads | 64 threads |
| Glucose-syrup 4.1 | **465 (263 + 202)** | **524 (301 + 223)** |
| | 462 (263 + 199) | 519 (295 + 224) |
| | 458 (255 + **203**) | 515 (293 + 222) |
| ManyGlucose + confs | 445 (250 + 195) | 444 (254 + 190) |
| ManyGlucose + DCE + confs | 456 (262 + 194) | 494 (283 + 211) |
| ManyGlucose + lit accs | 447 (252 + 195) | 476 (272 + 204) |
| ManyGlucose + DCE + lit accs | 462 (**265** + 197) | 512 (291 + 221) |
| ManyGlucose + blk execs | 456 (259 + 197) | 487 (275 + 212) |
| ManyGlucose + DCE + blk execs | 456 (258 + 198) | 521 (293 + **228**) |
| | 455 (258 + 197) | 521 (293 + **228**) |
| | 454 (258 + 196) | 521 (293 + **228**) |

(b) The ratio of waiting time and average runtime per period

| Solver | 4 threads | | 64 threads | |
|---|---|---|---|---|
| | Waiting time ratio | Avg time/period | Waiting time ratio | Avg time/period |
| ManyGlucose + confs | 29.7% | 0.099 | 58.3% | 0.528 |
| ManyGlucose + DCE + confs | 14.8% | 0.104 | 34.4% | 0.392 |
| ManyGlucose + lit accs | 17.9% | 0.058 | 51.8% | 0.398 |
| ManyGlucose + DCE + lit accs | 8.9% | 0.060 | 20.2% | 0.366 |
| ManyGlucose + blk execs | 12.5% | 0.122 | 44.0% | 0.621 |
| ManyGlucose + DCE + blk execs | 5.7% | 0.126 | 8.8% | 0.594 |
| | 5.6% | 0.125 | 8.8% | 0.593 |
| | 5.6% | 0.126 | 8.8% | 0.594 |

(a) 4 threads          (b) 64 threads

**Fig. 6.** Cactus plot comparing total instances solved within a given time bound for Glucose-syrup and ManyGlucose configured with three types of periods and with/without DCE. MG means ManyGlucose. The best and worst results of MG-DCE-blk-execs are almost overlapped.

results is 2 instances in the 4 threads and no difference in the 64 threads environment. We have confirmed that ManyGlucose can find the same model for each run for satisfiable instances. Our 4 threads environment is a cluster built on educational PCs and cannot be used exclusively, and the results fluctuate slightly. In contrast, the results of 64 threads are very stable due to the exclusive use of the system. Figure 8 shows comparisons of runtime of each instance in the best and worst results. The results for 4 threads vary slightly over time, while the results for 64 threads are almost completely distributed on the diagonal. In contrast to

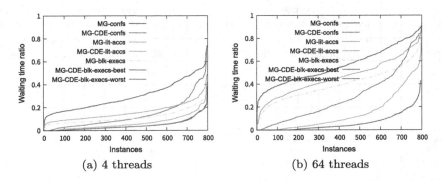

**Fig. 7.** Waiting time ratio of ManyGlucose. Results are sorted by the ratio.

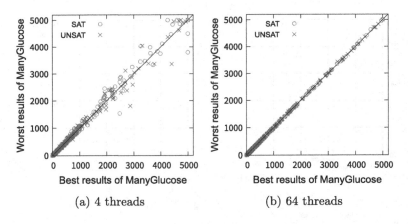

**Fig. 8.** Comparison of runtimes of the best and worst results of ManyGlucose for each instance.

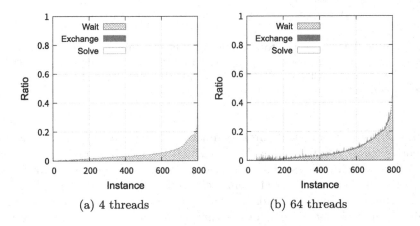

**Fig. 9.** Waiting time and clause exchanging time of ManyGlucose.

Fig. 1, this demonstrates the robustness of our deterministic parallel SAT solver. The number of solved instances in the best case of ManyGlucose exceeds the worst case of Glucose-syrup. This indicates that DCE and accurate period estimation can achieve performance comparable to non-deterministic solvers while holding deterministic behavior.

Figure 9 shows the time ratio required for clause exchange, which is very small ratio compared to the solving and waiting time. In DCE, for each thread $t$ and each period $p$, there is a database $E_t^p$ that stores clauses exported by the thread $t$ at the period $p$. If the current period of the thread $t$ is greater than $p$, then write access to $E_t^p$ no longer exists, so any thread can read it without mutual exclusive control. In contrast, shared clause databases in non-deterministic parallel SAT solvers usually have a mixture of write access to add clauses and read access to get clauses. Hence, the mutual exclusive control is required to access the clause database. One of the advantages of DCE is that it does not require the cost of mutual exclusive control to access clause databases.

## 7   Conclusion

The non-deterministic behavior of parallel SAT solvers is one of the obstacles to the promotion of application and research of parallel SAT solvers. In this paper, we have presented techniques to realize efficient and reproducible parallel SAT solving. The main technique is the delayed clause exchange (DCE), which absorbs fluctuations of intervals between clause exchanges. In order to enhance the effect of DCE, it is important to estimate exchange intervals accurately based on reproducible criterion. In this work, we presented two methods based on the number of literal accesses and block executions. The experimental results show that the combination of these techniques can achieve comparable performance to non-deterministic parallel SAT solvers even in many-core environments. Our approach can be applicable to deterministic parallel MaxSAT solving [9] which is based on the synchronization mechanism used in ManySAT. As future work it would be interesting to consider a general framework for building deterministic parallel SAT solvers (like PalnleSS [8] for non-deterministic parallel SAT solvers) in which state-of-the-art sequential solvers can easily participate.

**Acknowledgment.** This work was supported by JSPS KAKENHI Grant Number JP17K00300 and JP20K11934. In this research work we used the supercomputer of ACCMS, Kyoto University.

## References

1. Audemard, G., Simon, L.: Lazy clause exchange policy for parallel SAT solvers. In: Sinz, C., Egly, U. (eds.) SAT 2014. LNCS, vol. 8561, pp. 197–205. Springer, Cham (2014). https://doi.org/10.1007/978-3-319-09284-3_15
2. Biere, A.: Lingeling, Plingeling and Treengeling entering the SAT Competition 2013. In: SAT Competition 2013: Solver and Benchmark Descriptions, pp. 51–52 (2013)

3. Hamadi, Y., Jabbour, S., Piette, C., Sais, L.: Deterministic parallel DPLL. J. Satisf. Boolean Model. Comput. **7**(4), 127–132 (2011)
4. Hamadi, Y., Jabbour, S., Sais, L.: ManySAT: a parallel SAT solver. J. Satisf. Boolean Model. Comput. **6**(4), 245–262 (2009)
5. Heule, M.J.H., Kullmann, O., Wieringa, S., Biere, A.: Cube and conquer: guiding CDCL SAT solvers by lookaheads. In: Eder, K., Lourenço, J., Shehory, O. (eds.) HVC 2011. LNCS, vol. 7261, pp. 50–65. Springer, Heidelberg (2012). https://doi.org/10.1007/978-3-642-34188-5_8
6. Hyvärinen, A.E.J., Junttila, T., Niemelä, I.: A distribution method for solving SAT in grids. In: Biere, A., Gomes, C.P. (eds.) SAT 2006. LNCS, vol. 4121, pp. 430–435. Springer, Heidelberg (2006). https://doi.org/10.1007/11814948_39
7. Hyvärinen, A.E.J., Junttila, T., Niemelä, I.: Partitioning SAT instances for distributed solving. In: Fermüller, C.G., Voronkov, A. (eds.) LPAR 2010. LNCS, vol. 6397, pp. 372–386. Springer, Heidelberg (2010). https://doi.org/10.1007/978-3-642-16242-8_27
8. Le Frioux, L., Baarir, S., Sopena, J., Kordon, F.: PaInleSS: a framework for parallel SAT solving. In: Gaspers, S., Walsh, T. (eds.) SAT 2017. LNCS, vol. 10491, pp. 233–250. Springer, Cham (2017). https://doi.org/10.1007/978-3-319-66263-3_15
9. Martins, R., Manquinho, V.M., Lynce, I.: Deterministic parallel MaxSAT solving. Int. J. Artif. Intell. Tools **24**(3), 1550005:1–1550005:25 (2015)
10. Plaza, S., Markov, I., Bertacco, V.: Low-latency SAT solving on multicore processors with priority scheduling and XOR partitioning. In: Proceedings of the 17th International Workshop on Logic and Synthesis (2008)
11. Schubert, T., Lewis, M.D.T., Becker, B.: PaMira - A parallel SAT solver with knowledge sharing. In: Proceedings of Sixth International Workshop on Microprocessor Test and Verification (MTV 2005), Common Challenges and Solutions, pp. 29–36 (2005)
12. Sinz, C., Blochinger, W., Küchlin, W.: PaSAT- parallel SAT-checking with lemma exchange: implementation and applications. Electron. Notes Disc. Math. **9**, 205–216 (2001)
13. Zhang, H., Bonacina, M.P., Hsiang, J.: PSATO: a distributed propositional prover and its application to quasigroup problems. J. Symb. Comput. **21**(4), 543–560 (1996)
14. Zou, H., Hastie, T.: Regularization and variable selection via the elastic net. J. Roy. Stat. Soc. B **67**, 301–320 (2005)

# Improving Implementation of SAT Competitions 2017–2019 Winners

Stepan Kochemazov[(✉)] [iD]

Matrosov Institute for System Dynamics and Control Theory SB RAS,
Irkutsk, Russia
veinamond@gmail.com

**Abstract.** The results of annual SAT competitions are often viewed as
the milestones showcasing the progress in SAT solvers. However, their
competitive nature leads to the situation when the majority of this year's
solvers are based on previous year's winner. And since the main focus is
always on novelty, it means that there are times when some implementation details have a potential for improvement, but they are just inherited
from solver to solver for several years in a row. In this study we propose
small modifications of implementations of existing heuristics in several
related SAT solvers. These modifications mostly consist in employing
a deterministic strategy for switching between branching heuristics and
in augmentations of the treatment of *Tier2* and *Core* clauses. In our
experiments we show that the proposed changes have a positive effect
on solvers' performance both individually and in combination with each
other.

**Keywords:** SAT · SAT solvers · Heuristics

## 1 Introduction

The Conflict-Driven Clause Learning (CDCL) solvers [13] form the core of the
algorithms for solving the Boolean satisfiability problem (SAT) [4]. Every year
the community proposes new heuristics aimed at improving their performance.
To test them in close to real-world conditions, the SAT competitions are held.
They evaluate the prospective solvers' performance in the same computational
environment over the sets of test instances gathered from various areas where
the SAT solvers are applied.

To show that a new heuristic for SAT solving contributes to state of the art,
it is usually implemented on top of one of the well-known SAT-solvers such as
MiniSAT [5], glucose [1], CryptoMiniSat [20], Cadical [2] or, more often, the
most recent SAT competition(s) winner (e.g. [8,12,17]). The problem with the

The research was funded by Russian Science Foundation (project No. 16-11-10046)
with an additional support from Council for Grants of the President of the Russian
Federation (stipend no. SP-2017.2019.5).

L. Pulina and M. Seidl (Eds.): SAT 2020, LNCS 12178, pp. 139–148, 2020.
https://doi.org/10.1007/978-3-030-51825-7_11

latter is that the solver with a new heuristic must be compared with the original solver without modifications to show that the increase in performance is not due to fixing some aspects of original solver's implementation. This is not a bad thing as it is, but it can lead to solvers accumulating undesired traits. One of the examples of the latter is the non-deterministic switching between branching heuristics, first introduced in `MapleCOMSPS` [9] which won in the main track of SAT Competition 2016. This behavioral trait was inherited by the winners of SAT Competitions 2017 and 2018, and many participants of the SAT Race 2019.

In the present paper we propose several changes to the common implementation aspects of the solvers that won at the main tracks of SAT Competitions 2017 to 2019. The main contributions are as follows:

1. We analyze possible variants of deterministic switching between branching heuristics and show that some of them yield consistently better results compared to the non-deterministic switching at 2500 s.
2. We adjust the treatment of the so-called *Tier2* clauses and show that accumulating them in a slightly different manner results in better overall performance.
3. We show that contrary to the intuition provided in [18], it is better to sometimes purge some learnts from the *Core* tier in order to increase the propagation speed and the effectiveness of a solver on hard instances.

We use the `MapleLCMDistChronoBT` solver as the main object for analysis and experimentation. To evaluate improvements and modifications we use the set of SAT instances from the SAT Race 2019. After having figured out the modifications that serve our goals best, we implement them in the winners of SAT Competition 2017 and SAT Race 2019. Then we test the resulting six SAT solvers on a wide range of benchmarks.

## 2   Background

The presentation of the details on implementation of heuristics in SAT solvers implies that the reader is familiar with the architecture of CDCL SAT-solvers in general and with key advancements in the area during the recent years.

The SAT competitions are annual competitive events aimed at the development of SAT solving algorithms. In the course of the main track of the competition, the solvers are launched on each test instance with the time limit of 5000 s. The performance is evaluated using two criteria: Solution-Count Ranking (SCR) – the number of test instances successfully solved by an algorithm within the time limit, and Penalized Average Runtime (PAR-2) computed as the sum of solver runtimes on solved instances plus 2× the time limit for unsolved ones, divided by the total number of tests.

In the paper we study the solvers `MapleLCMDist`, `MapleLCMDistChronoBT` and `MapleLCMDistChronoBT-DL-v3`, which won at SAT Competitions 2017 to 2019. All three solvers are based on the `MapleCOMSPS` solver that won at SAT

Competition 2016 [8]. The latter uses the foundation of COMiniSatPS [18], which combined the better traits of the well-known MiniSAT [5] and Glucose [1] solvers.

One of the main novelties of COMiniSatPS was the special treatment of learnt clauses depending on their *literal block distance* value (lbd). The notion of lbd was first proposed in [1] and is equal to the number of distinct decision levels of the literals in a learnt clause. COMiniSatPS splits all learnt clauses into three tiers: *Core*, *Tier2* and *Local* and handles them differently. The second major novel feature of COMiniSatPS was to use two sets of activity values for the branching heuristic. In particular, the solver employed two sets of scores maintained via Variable State Independent Decaying Sum (VSIDS) [16]. They used different VSIDS decay values, but what's more important, their use was tied to periodic switching between restart strategies. For years after the introduction of Minisat [5], the SAT solvers mainly relied on *luby* restarts [11], but after the triumphal appearance of Glucose [1], the CDCL solvers mostly switched to the much faster *glucose*-restarts. COMiniSatPS combined both, and used one set of VSIDS scores with luby restarts and another set with glucose restarts.

In 2016 the *Learning Rate Branching* (LRB) heuristic [8] was proposed as an alternative to VSIDS. It was implemented in MapleCOMSPS as well as in several other solvers [9]. MapleCOMSPS is based on COMiniSatPS and uses LRB branching together with luby restarts and VSIDS-branching with glucose restarts. Unfortunately, the solver did not inherit the deterministic switching strategy employed by COMiniSatPS to combine different phases of solving. Instead, it relies on LRB + luby restarts for the first 2500 s of the search and VSIDS + glucose restarts for its remainder, which results in a non-deterministic behaviour. Having realized their oversight, the MapleCOMSPS's authors submitted to SAT competitions 2017–2019 only the deterministic variants of their solvers.

The winner of the main track of the SAT Competition 2017, MapleLCMDist introduced Learnt Clause Minimization [12] into MapleCOMSPS. It is an expensive inprocessing method [6] periodically applied to learnt clauses from *Tier2* and *Core* with the goal to remove irrelevant literals from them. The main track winner of SAT Competition 2018 called MapleLCMDistChronoBT had augmented MapleLCMDist with chronological backtracking [17], which was later studied in [15]. Finally, the winner of SAT Race 2019 was MapleLCMDistChronoBT-DL-v3, which used the duplicate learnts (DL) heuristic [7], that tracks duplicate learnts and sometimes adds them into *Tier2* or *Core*.

## 3    Improving Implementation of MapleLCMDistChronoBT

In this section we describe the three-phase experiment aimed at improving the implementation of MapleLCMDistChronoBT. The main question we strive to answer is if it is possible to retain its good overall performance but make the solver deterministic. In the first phase of experiments we implement several deterministic strategies for switching between branching heuristics and evaluate their performance. In the second phase we use the best-behaved switching strategy from the first phase to check whether it is possible to forego one of the branching heuristics and let the solver work with a single one. In the third phase we propose and implement a small change to handling *Tier2* and *Core* learnt clauses and see how it affects the general performance of a solver.

In all experiments we used a single node of the "Academician V.M. Matrosov" computing cluster of Irkutsk Supercomputer Center [14]. It is equipped with two 18-core Intel Xeon E5-2695 CPUs and 128 GB DDR4 RAM. The solvers were launched in 36 simultaneous threads.

### 3.1   On Switching Between Branching Heuristics

First we combed through the available deterministic implementations in the solvers from recent SAT competitions and found several variants for switching between branching heuristics that are summed up in Table 1. All of them assume that VSIDS is used with glucose restarts and LRB with luby restarts, thus we refer to them as to VSIDS and LRB phases, respectively. The `fcm` column corresponds to the implementation in `COMiniSatPS`. The variants `f1` and `f2` denote the implementations first appeared in `MapleCOMSPS_LRB` [9] and `MapleCOMSPS_LRB_VSIDS_2` [10], respectively. The `v3` variant employs the scheme used in the `MapleLCMDistChronoBT-DL-v3` solver [7], which uses propagations to measure the phases' sizes.

**Table 1.** The considered schemes for switching between branching heuristics

|  | `fcm` | `f1` | `f2` | `v3` |
|---|---|---|---|---|
| Initial `phase_allotment` | 100 conflicts | 100 conflicts | 10000 conflicts | 30 M props |
| VSIDS multiplier | 2 | 1 | 1 | 1 |
| `phase_allotment` multiplier | 1.1/1 | 1.1/1 | 2/1 | 1.1/1.1 |

All variants in Table 1 use `phase_allotment` variable to store the number of conflicts (or propagations) allocated for the next phase. They all first launch the LRB phase and then the VSIDS phase. Some of them allocate more resources to the VSIDS phase, in that case the value in the VSIDS multiplier row of Table 1 is different from 1. The value of `phase_allotment` is multiplied by a constant after each LRB/VSIDS phase and the cycle repeats anew. Actually, all strategies but `v3` only increase the value of `phase_allotment` at the end of the VSIDS phase.

We implemented the strategies from Table 1 in `MapleLCMDistChronoBT`. The performance of the resulting 4 solvers on the benchmarks from SAT Race 2019 is presented in form of a cactus plot in Fig. 1 with the corresponding detailed statistics in Table 1. They are accompanied by the original solver and 5 other solver variants, which will be described in detail below. Two of them (`-LRB` and `-VSIDS`) employ only a single branching heuristic and the other three use different combinations of augmented handling of *Tier2* clauses (`-t`) and the procedures for reducing *Core* database (`-rc`).

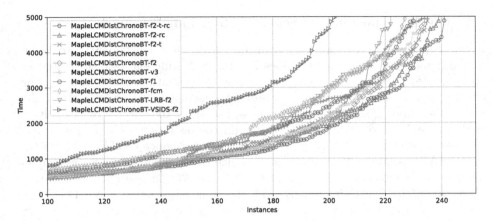

**Fig. 1.** Cactus plots depicting the performance of `MapleLCMDistChronoBT` with altered implementations of switching strategies between branching heuristics and of handling learnt clauses over the SAT Race 2019 benchmarks.

**Table 2.** The detailed statistics on `MapleLCMDistChronoBT` with different variants of switching between branching heuristics and heuristics for handling *Tier2* and *Core* learnts on SAT instances from SAT Race 2019

| Solver | SCR | SAT | UNSAT | PAR-2 |
|---|---|---|---|---|
| MapleLCMDistChronoBT | 236 | 138 | 98 | 4799 |
| MapleLCMDistChronoBT-f1 | 231 | 135 | 96 | 4882 |
| MapleLCMDistChronoBT-f2 | 235 | 140 | 95 | 4716 |
| MapleLCMDistChronoBT-v3 | 232 | 137 | 95 | 4787 |
| MapleLCMDistChronoBT-fcm | 228 | 133 | 95 | 4985 |
| MapleLCMDistChronoBT-LRB-f2 | 223 | 134 | 89 | 4996 |
| MapleLCMDistChronoBT-VSIDS-f2 | 203 | 115 | 88 | 5589 |
| MapleLCMDistChronoBT-f2-t | 236 | 139 | 97 | 4720 |
| MapleLCMDistChronoBT-f2-rc | 240 | 143 | 97 | 4631 |
| MapleLCMDistChronoBT-f2-t-rc | 242 | 143 | 99 | 4556 |

As it can be seen from both Table 2 and Fig. 1, the best deterministic variant in the considered conditions is `f2`. Its distinctive feature is the `phase_allotment` multiplier equal to 2. Thus, the solver alternates between long phases of rapidly increasing size in which it relies on a single heuristic. The similar behaviour is achieved by the `v3` scheme, where a large number of propagations is allocated to phases. Compared to `f2`, `v3` starts from longer phases, that grow in size significantly slower.

## 3.2   On the Importance of Separate Branching Strategies

It was first noted in [18] and later reaffirmed in [3] that it is beneficial to maintain different variable activity values for use with different restart strategies. Also, in [18] it was pointed out that a strong correlation exists between the effectiveness of glucose restarts for proving unsatisfiability and that of Luby restarts for finding satisfying assignments. Based on the success of the solvers that employ Luby restarts with LRB and glucose restarts with VSIDS it is easy to make a conclusion that LRB shows better performance on SAT and VSIDS on UNSAT instances. To evaluate at a glance the highlighted issues we implemented two versions of the `MapleLCMDistChronoBT-f2`: one that uses only a single set of LRB activity values, and one that relies only on VSIDS. They both still switch between luby restarts and glucose restarts deterministically after the `f2` fashion. We denote them as `MapleLCMDistChronoBT-LRB-f2` and `MapleLCMDistChronoBT-VSIDS-f2`, respectively.

The results of these solvers are included in Fig. 1 and Table 2. They point at the following two conclusions. First, that indeed two separate sets of activities for branching during different restart strategies work better than a single set. Second, that the relation between LRB, luby restarts and satisfiable instances, and VSIDS, glucose restarts and unsatisfiable instances is more complex than it might seem and warrants further investigation that goes out of the scope of the present paper.

## 3.3   Improving the Handling of *Core* and *Tier2* Learnts

Recall, that in [18,19], it was proposed to split learnt clauses into three tiers. *Core* tier accumulates and indefinitely holds the clauses with `lbd` not exceeding the value of the `core_lbd_cut` parameter, which is equal to 3 or 5 depending on a SAT instance. *Tier2* stores for a prolonged time period the clauses with slightly higher lbd than necessary to be included into *Core* (with `core_lbd_cut` $<$`lbd` $\leq 6$). The remaining clauses are put into the frequently purged *Local* tier.

We investigated how many learnts were pushed into these tiers by the considered solver over a wide range of benchmarks. It turned out that for about 80% instances from SAT Race 2019, `MapleLCMDistChronoBT-f2` accumulates more than 50 000 core learnts. The average sizes of *Tier2* also vary quite significantly since the distribution of `lbd` of conflict clauses is specific to each particular SAT instance. E.g. the capacity of *Tier2* can be as low as 200 learnts on average and as high as 12 000 learnts. In [19] it was specifically noted that the number of learnt clauses in *Tier2* may vary depending on the restart strategy and other factors and that it is perfectly normal. However, this observation did not take into account the Learnt Clause Minimization (LCM) [12] which was proposed in 2017 and is applied to *Core* and *Tier2* in `MapleLCMDistChronoBT`. Thus the instances with higher average `lbd` of learnt clauses do not benefit as much from LCM compared to the others due to the fact that such clauses often do not survive in *Tier2* long enough to be minimized.

These observations lead us to two small modifications in handling the *Tier2* and *Core* learnt clauses. The first one consists in periodically reducing the *Core* database. For this purpose as soon as the *Core* size exceeds a pre-specified limit (in our implementation it is equal to 50000 and is multiplied by 1.1 each time the procedure is invoked), we sort *Core* learnts in the ascending order based on their lbd and the size for equal lbd. Then we move all the clauses from the second half (with larger lbd and size), that did not participate in any of the most recent 100000 conflicts into *Tier2*.

The second improvement is to reorganize how the *Tier2* is purged. In the original implementation of `MapleLCMDistChronoBT` the *Tier2* is reduced every 10000 conflicts and as a result all clauses that have not participated in the most recent 30000 conflicts are moved to *Local*. We propose first to accumulate *Tier2* learnts until a pre-specified size limit (7000), and second during the purge to preserve only half of the clauses that participated in the most recent conflicts.

We implemented proposed modifications in `MapleLCMDistChronoBT-f2`. The inclusion of the procedure for reducing *Core* is labelled as `-rc` and the modification to handling of *Tier2* learnts is labelled as `-t`. The results of the corresponding variants are shown at Fig. 1 and in the lower part of Table 2. It is clear that the combination of proposed heuristics improves the solver's performance.

## 4    Experimental Evaluation

In the previous section we figured out the prospective deterministic configuration of the `MapleLCMDistChronoBT` solver. Since the proposed implementation changes do not depend on anything specific to chronological backtracking, we implemented the same changes into the winners of the SAT Competition 2017 and SAT Race 2019. We mark the improved variants of the solvers with `-f2-t-rc`.

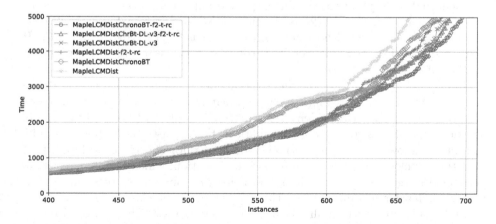

**Fig. 2.** Cactus plots showing the performance of improved implementations of SC2017-SR2019 winners on the combined benchmark set from SC2017, SC2018 and SR2019.

To evaluate the performance of constructed solvers we launched 3 original ones and 3 modifications on all benchmarks from SAT Competitions 2017 and 2018, and that from SAT Race 2019. The cactus plot of the resulting experiment is presented in Fig. 2. The detailed statistics is shown in Table 3. Here, MLCMD stands for MapleLCMDist, MLCMDCh for MapleLCMDistChronoBT and MLCMDCh-DL for MapleLCMDistChronoBT-DL-v3. An improved implementation of each solver is marked with a star.

**Table 3.** Detailed statistics of improved implementations of the winners of SC2017-SC2018-SR2019 on the instances from SC2017, SC2018 and SR2019. Each cell contains: SCR: SAT: UNSAT: PAR-2 (in seconds).

| | SC2017 | SC2018 | SR2019 | Total |
|---|---|---|---|---|
| MLCMD | 206:100:106:4706 | 229:129:100:4718 | 225:128:97:4999 | 660:357:303:4812 |
| MLCMD* | 204: 99:105:4693 | 237:135:102:4541 | 242:144:98:4597 | 683:378:305:4607 |
| MLCMDCh | 208: 97:111:4659 | 235:133:102:4573 | 236:138:98:4799 | 679:368:311:4678 |
| MLCMDCh* | 217:103:114:4373 | 239:136:103:4490 | 242:143:99:4556 | 698:382:316:4477 |
| MLCMDCh-DL | 208: 99:109:4586 | 243:142:101:4402 | 237:141:96:4675 | 688:382:306:4553 |
| MLCMDCh-DL* | 204: 92:112:4707 | 242:139:103:4423 | 243:146:97:4547 | 689:377:312:4553 |

Observe that the new implementation improved the performance of all three solvers. The least affected solver is the winner of SR2019: it already employs a viable deterministic strategy for switching between branching heuristics. Also the solver was tuned over benchmarks from SC2017 and SC2018. What is interesting is that the modified solver tackled 6 more unsatisfiable problems than the original, however at a cost of 5 satisfiable ones. This is the largest increase over three solvers on unsatisfiable instances. The largest overall increase in performance was achieved by MapleLCMDist. The improved implementation made it possible to solve 23 more instances (21 SAT and 2 UNSAT) and decrease the PAR-2 by about 4%. The overall best performing solver is MapleLCMDistChronoBT. It surpassed the closest competitor by 9 solved instances and solved the largest number of UNSAT instances.

Informally, from all the plots and experiments it follows that a proper deterministic strategy for switching between branching heuristics smooths a solver's performance. Observe the evident bump on the lines plotting the performance of non-deterministic solvers near 2500 s mark in Fig. 2. Also, the improved implementations show a drastic performance difference for shorter time limits that smooths out closer to 5000 s. The second modification, which mainly consists in periodical purges of the *Core* database, helps the solver to increase its performance for harder instances, when the amount of learnt clauses in *Core* starts to hinder the rate at which all CDCL procedures operate. The improved implementation shows the increase in both UNSAT and SAT instances solved. There are more SAT instances solved because the considered benchmarks contain more

SAT tests (solvable by at least one participating solver) than UNSAT. All implementations and logs of the experiments are available online[1].

**Acknowledgements.** The author thanks anonymous reviewers for their helpful comments and Oleg Zaikin and Alexander Semenov for many fruitful discussions.

# References

1. Audemard, G., Simon, L.: Predicting learnt clauses quality in modern SAT solvers. In: IJCAI, pp. 399–404 (2009)
2. Biere, A.: CaDiCaL, Lingeling, Plingeling, Treengeling, YalSAT entering the SAT competition 2017. In: Proceedings of SAT Competition 2017, vol. B-2017-1, pp. 14–15 (2017)
3. Biere, A.: CaDiCaL at the SAT Race 2019. In: Proceedings of SAT Race 2019, vol. B-2019-1, pp. 8–9 (2019)
4. Biere, A., Heule, M., van Maaren, H., Walsh, T. (eds.): Handbook of Satisfiability, Frontiers in Artificial Intelligence and Applications, vol. 185. IOS Press (2009)
5. Eén, N., Sörensson, N.: An extensible SAT-solver. In: Giunchiglia, E., Tacchella, A. (eds.) SAT 2003. LNCS, vol. 2919, pp. 502–518. Springer, Heidelberg (2004). https://doi.org/10.1007/978-3-540-24605-3_37
6. Järvisalo, M., Heule, M.J.H., Biere, A.: Inprocessing rules. In: Gramlich, B., Miller, D., Sattler, U. (eds.) IJCAR 2012. LNCS (LNAI), vol. 7364, pp. 355–370. Springer, Heidelberg (2012). https://doi.org/10.1007/978-3-642-31365-3_28
7. Kochemazov, S., Zaikin, O., Kondratiev, V., Semenov, A.: MapleLCMD istChronoBT-DL, duplicate learnts heuristic-aided solvers at the SAT Race 2019. In: Proceedings of SAT Race 2019, vol. B-2019-1, p. 24 (2019)
8. Liang, J.H., Ganesh, V., Poupart, P., Czarnecki, K.: Learning rate based branching heuristic for SAT solvers. In: Creignou, N., Le Berre, D. (eds.) SAT 2016. LNCS, vol. 9710, pp. 123–140. Springer, Cham (2016). https://doi.org/10.1007/978-3-319-40970-2_9
9. Liang, J.H., Oh, C., Ganesh, V., Czarnecki, K., Poupart, P.: MapleCOMSPS, MapleCOMSPS_LRB, MapleCOMSPS_CHB. In: Proceedings of SAT Competition 2016, vol. B-2016-1, pp. 52–53 (2016)
10. Liang, J.H., Oh, C., Ganesh, V., Czarnecki, K., Poupart, P.: MapleCOM SPS_LRB_VSIDS and MapleCOMSPS_CHB_VSIDS. In: Proceedings of SAT Competition 2017, vol. B-2017-1, pp. 20–21 (2017)
11. Luby, M., Sinclair, A., Zuckerman, D.: Optimal speedup of Las Vegas algorithms. Inf. Process. Lett. **47**(4), 173–180 (1993)
12. Luo, M., Li, C., Xiao, F., Manyà, F., Lü, Z.: An effective learnt clause minimization approach for CDCL SAT solvers. In: IJCAI, pp. 703–711 (2017)
13. Marques-Silva, J.P., Lynce, I., Malik, S.: Conflict-driven clause learning SAT solvers. In: Biere et al. [4], pp. 131–153
14. Irkutsk Supercomputer Center of SB RAS. http://hpc.icc.ru
15. Möhle, S., Biere, A.: Backing backtracking. In: Janota, M., Lynce, I. (eds.) SAT 2019. LNCS, vol. 11628, pp. 250–266. Springer, Cham (2019). https://doi.org/10.1007/978-3-030-24258-9_18

---

[1] https://github.com/veinamond/SAT2020.

16. Moskewicz, M.W., Madigan, C.F., Zhao, Y., Zhang, L., Malik, S.: Chaff: engineering an efficient SAT solver. In: Proceedings of the 38th Annual Design Automation Conference, DAC 2001, pp. 530–535 (2001)
17. Nadel, A., Ryvchin, V.: Chronological backtracking. In: Beyersdorff, O., Wintersteiger, C.M. (eds.) SAT 2018. LNCS, vol. 10929, pp. 111–121. Springer, Cham (2018). https://doi.org/10.1007/978-3-319-94144-8_7
18. Oh, C.: Between SAT and UNSAT: the fundamental difference in CDCL SAT. In: Heule, M., Weaver, S. (eds.) SAT 2015. LNCS, vol. 9340, pp. 307–323. Springer, Cham (2015). https://doi.org/10.1007/978-3-319-24318-4_23
19. Oh, C.: Improving SAT solvers by exploiting empirical characteristics of CDCL. Ph.D. thesis, New York University (2016)
20. Soos, M., Nohl, K., Castelluccia, C.: Extending SAT solvers to cryptographic problems. In: Kullmann, O. (ed.) SAT 2009. LNCS, vol. 5584, pp. 244–257. Springer, Heidelberg (2009). https://doi.org/10.1007/978-3-642-02777-2_24

# On CDCL-Based Proof Systems
# with the Ordered Decision Strategy

Nathan Mull[1($\boxtimes$)], Shuo Pang[2], and Alexander Razborov[1,2,3]

[1] Department of Computer Science, University of Chicago, Chicago, USA
{nmull,razborov}@cs.uchicago.edu
[2] Department of Mathematics, University of Chicago, Chicago, USA
{spang,razborov}@math.uchicago.edu
[3] Steklov Mathematical Institute, Moscow, Russia

**Abstract.** We prove that CDCL SAT-solvers with the ordered decision strategy and the DECISION learning scheme are equivalent to ordered resolution. We also prove that, by replacing this learning scheme with its opposite, which learns the first possible non-conflict clause, they become equivalent to general resolution. In both results, we allow nondeterminism in the solver's ability to perform unit propagation, conflict analysis, and restarts in a way that is similar to previous works in the literature. To aid the presentation of our results, and possibly future research, we define a model and language for CDCL-based proof systems – particularly those with nonstandard features – that allow for succinct and precise theorem statements.

## 1   Introduction

Since their conception, SAT-solvers have become significantly more efficient, but they have also become significantly more complex. Consequently, there has been increasing interest in understanding their theoretical limitations and strengths. Much of the recent literature has focused on the relationship between CDCL SAT-solvers[1] and the resolution proof system. Beame et al. [5] were the first to study this relationship and many followed suit (see [3,6,7,11,13–15,17–20] among others). In particular, Pipatsrisawat and Darwiche [18] show that, under a few assumptions, CDCL with the nondeterministic decision strategy (i.e., when the solver has to choose a variable to assign, it chooses both the variable and its assigned value nondeterministically) polynomially simulates resolution. An obvious question arises from this result: how much does the theoretical efficiency of CDCL depend on nondeterminism in the decision strategy? Along these lines, Atserias et al. [3] (concurrently with [18]) show that CDCL with the *random* decision strategy (i.e., both the variable and assigned value are chosen uniformly at random) simulates bounded-width resolution, under essentially

---

[1] In this paper, we focus solely on solvers which implement conflict-driven clause learning (CDCL). We refer to such solvers as CDCL SAT-solvers, CDCL solvers, or just CDCL for short.

© Springer Nature Switzerland AG 2020
L. Pulina and M. Seidl (Eds.): SAT 2020, LNCS 12178, pp. 149–165, 2020.
https://doi.org/10.1007/978-3-030-51825-7_12

the same assumptions as those in [18]. More recently, Vinyals [20] has shown that CDCL with the VSIDS decision strategy – among other common dynamic decision strategies – does not simulate general resolution. We attempt to make progress on this question by studying a simple decision strategy that we call the *ordered* decision strategy. This strategy is identical to the one studied by Beame et al. [4] in the context of DPLL without clause learning. It is defined naturally: when the solver has to choose a variable to assign, it chooses the smallest unassigned variable according to some fixed order and chooses its assigned value nondeterministically. If unit propagation is used, the solver may assign variables out of order; a unit clause does not necessarily correspond to the smallest unassigned variable. This possibility of "cutting the line" is precisely what makes the situation more subtle and nontrivial. Thus, our motivating question is the following:

*Is there a family of contradictory CNFs $\{\tau_n\}_{n=1}^{\infty}$ that possess polynomial size resolution refutations but require superpolynomial time for CDCL using the ordered decision strategy?*

We also note in passing that this question may be motivated as a way of understanding the strength of *static* decision strategies such as MINCE [1] and FORCE [2].

**Our Contributions.** A proof system that captures any class of CDCL solvers should be no stronger than general resolution, and if it captures solvers with the ordered decision strategy, it should be reasonably expected to be at least as strong as ordered resolution with respect to the same order. Our main results show that, depending on the learning scheme employed, both of these extremes are attained. More specifically, we prove

1. CDCL with the ordered decision strategy and a learning scheme we call DECISION-L is equivalent to ordered resolution (Theorem 1). In particular, it does not simulate general resolution.
2. CDCL with the ordered decision strategy and a learning scheme we call FIRST-L is equivalent to general resolution (Theorem 2).

*Remark 1.* As the name suggests, DECISION-L is the same as the so-called DECISION learning scheme.[2] FIRST-L is a learning scheme designed to directly simulate particular resolution steps in the presence of certain forms of nondeterminism, and is similar to FirstNewCut [5]. In the full version of this paper [16], we also prove linear width lower bounds which, combined with the second result, create a sharp contrast with the size-width relationship for general resolution proved by Ben-Sasson and Wigderson [8].

In these two results, the CDCL solver may arbitrarily choose the conflict/unit clause if there are several, may elect not to do conflict analysis/unit propagations

---

[2] The name DECISION-L better fits the naming conventions of our model.

at all, and may restart at any time. This substantial amount of nondeterminism allows us to identify two proof systems that are, more or less straightforwardly, equivalent to the corresponding CDCL variant. Determining the exact power of these systems constitutes our main technical contribution.

There are a couple points of interpretation to emphasize here. First, the implicit separation between CDCL solvers and general resolution in the first result applies to *actual* SAT-solver implementations, albeit with heuristics that are not usually used in practice, and could, in principle, be demonstrated by experiment. In contrast, the second result does not say anything substantial about actual SAT-solver implementations. But we also note that this is not unprecedented. The correspondence between proof systems and algorithms considered here is very similar to the correspondence between *regWRTI* and a variant of CDCL with similar features called DLL-LEARN, both introduced by Buss et al. [11]; nonstandard sources of nondeterminism manifest themselves naturally when translating CDCL into a proof system. Both lower and upper bounds on these systems are valuable; even if upper bounds do not apply directly to practice, they demonstrate, often nontrivially, what convenient features of simple proof systems must be dropped to potentially prove separations.

Finally, in order to aid the above work – and, perhaps, even facilitate further research in the area – we present a model and language for studying CDCL-based proof systems. This model is not meant to be novel, and is heavily influenced by previous work [3,13,17]. However, the primary goal of our model is to *highlight* possible nonstandard sources of nondeterminism in variants of CDCL, as opposed to creating a model completely faithful to applications. Our second result (Theorem 2) can be written in this language as:

*For any order $\pi$,* CDCL(FIRST-L, $\pi$-D) *is polynomially equivalent to general resolution.*

Due to space limitations, not all proofs are provided and there may be excluded details or remarks that, though not essential, are useful in understanding possible subtleties in the constructions and arguments. After presenting the preliminary material in Sect. 2, we give an nearly complete account of our first result mentioned above in Sect. 3, and reflect very briefly on our second result in Sect. 4. We refer the reader to the full version of this paper [16] for complete proofs and extended discussion.

## 2  Preliminaries

Throughout the paper, we assume that the set of propositional variables is fixed as $V \stackrel{\text{def}}{=} \{x_1, \ldots, x_n\}$. A *literal* is either a propositional variable or its negation. We will sometimes use the abbreviation $x^0$ for $\bar{x}$ and $x^1$ for $x$ (so that the Boolean assignment $x = a$ *satisfies* the literal $x^a$). A *clause* is a set of literals, thought of as their disjunction, in which no variable appears together with its negation. For a clause $C$, let Var($C$) denote the set of variables appearing in $C$. A *CNF* is a set of clauses thought of as their conjunction. For a CNF $\tau$, let Var($\tau$) denote the set of variables appearing in $\tau$, i.e., the union of $\mathrm{Var}(C)$ for all $C \in \tau$. We denote the empty clause by $0$. The *width* of a clause is the number of literals in it.

The *resolution proof system* is a Hilbert-style proof system whose lines are clauses and that has only one *resolution rule*

$$\frac{C \vee x_i^a \qquad D \vee x_i^{1-a}}{C \vee D}, \ a \in \{0,1\}. \tag{1}$$

We will sometimes denote the result of resolving $C \vee x_i^a$ and $D \vee x_i^{1-a}$ by $\mathrm{Res}(C \vee x_i^a, D \vee x_i^{1-a})$.

The *size* of a resolution proof $\Pi$, denoted as $|\Pi|$, is the number of lines in it. For a CNF $\tau$ and a clause $C$, let $S_R(\tau \vdash C)$ denote the minimal possible size of a resolution proof of the clause $C$ from clauses in $\tau$ ($\infty$ if $C$ is not implied by $\tau$). Likewise, let $w(\tau \vdash C)$ denote the minimal possible width of such a proof, defined as the maximal width of a clause in it. For a proof $\Pi$ that derives $C$ from $\tau$, the clauses in $\tau$ that appear in $\Pi$ are called *axioms*, and if $C = 0$ then $\Pi$ is called a *refutation*. Let $\mathrm{Var}(\Pi)$ denote the set of variables appearing in $\Pi$, i.e., the union of $\mathrm{Var}(C)$ for $C$ appearing in $\Pi$.

Note that the *weakening rule*

$$\frac{C}{C \vee D}$$

is *not* included by default. In the full system of resolution it is admissible in the sense that $S_R(\tau \vdash 0)$ does not change if we allow it. But this will not be the case for some of the CDCL-based fragments we will be considering below. Despite this, it is often convenient in analysis to consider intermediate systems that do allow the weakening rule. We make it clear when we do this by adding the annotation '+ weakening' to the system.

**Resolution Graphs.** Our results depend on the careful analysis of the structure of resolution proofs. It will, for example, be useful for us to maintain structural properties of the proof while changing the underlying clauses and derivations.

**Definition 1.** *For a resolution + weakening proof $\Pi$, its **resolution graph**, $G(\Pi)$, is an acyclic directed graph representing $\Pi$ in the natural way: each clause in $\Pi$ has a distinguished node, and for each node there are incoming edges from the nodes corresponding to the clauses from which it is derived. The set of nodes of $G(\Pi)$ is denoted by $V(\Pi)$, and the clause at $v \in V(\Pi)$ is denoted by $c_\Pi(v)$.*[3]

In the following collection of definitions, let $\Pi$ be an arbitrary resolution + weakening proof and let $S$ be an arbitrary subset of $V(\Pi)$. A vertex $u$ is *above* a vertex $v$ in $G(\Pi)$, written $u > v$, if there is a directed path from $v$ to $u$. We also say $v$ is *below* $u$. Moreover, $v$ is a *parent* of $u$ if $(v, u)$ is an edge in

---

[3] We do *not* assume that $c_\Pi$ is injective; we allow the same clause to appear in the proof several times.

$G(\Pi)$. $S$ is *independent* if any two of its nodes are incomparable. The *maximal* and *minimal* nodes of $S$ are $\max_\Pi S \stackrel{\text{def}}{=} \{v \in S \mid \forall u \in S \ (\neg(v < u))\}$ and $\min_\Pi S \stackrel{\text{def}}{=} \{v \in S \mid \forall u \in S \ (\neg(v > u))\}$, respectively. The *upward closure* and *downward closure* of $S$ in $G(\Pi)$ are $\text{ucl}_\Pi(S) \stackrel{\text{def}}{=} \{v \in V(\Pi) \mid \exists w \in S \ (v \geq w)\}$ and $\text{dcl}_\Pi(S) \stackrel{\text{def}}{=} \{v \in V(\Pi) \mid \exists w \in S \ (v \leq w)\}$, respectively. $S$ is *parent-complete* if it contains either both parents or neither parent of each of its nodes. $S$ is *path-complete* if it contains all nodes along any path in $G(\Pi)$ whose endpoints are in $S$. A resolution graph is *connected* if $|\max_\Pi V(\Pi)| = 1$, i.e., it has a unique sink. These definitions behave naturally, as demonstrated by the following useful proposition, which is easily verified.

**Proposition 1.** *Let $S \subseteq V(\Pi)$ be a nonempty set of nodes that is both parent-complete and path-complete. Then the induced subgraph of $G(\Pi)$ on $S$ is the graph of a subproof in $\Pi$ of $\max_\Pi S$ from $\min_\Pi S$.*

**Ordered Resolution.** Fix now an order $\pi \in S_n$. For any literal $x_k^a$, define $\pi(x_k^a) \stackrel{\text{def}}{=} \pi(k)$. For $k \in [n]$, let $\text{Var}_\pi^k$ denote the $k$ smallest variables according to $\pi$. A clause $C$ is *k-small* with respect to $\pi$ if $\text{Var}(C) \subseteq \text{Var}_\pi^k$.

The proof system $\pi$-*ordered resolution* is the subsystem of resolution defined by imposing the following restriction on the resolution rule (1):

$$\forall l \in C \vee D \ (\pi(l) < \pi(x_i)).$$

In the literature this system is usually defined differently, namely in a top-down manner (see, e.g., [10]). It is easy to see, however, that our version is equivalent.

**CDCL-Based Proof Systems.** Our approach to modeling CDCL is, in a sense, the opposite of what currently exists in the literature. Rather than attempting to model CDCL solver implementations as closely as possible and allowing non-determinism in various features, we rigorously describe a *basic* model that is very liberal and nondeterministic and intends to approximate the union of most conceivable features of CDCL solvers. Then models of actual interest will be defined by their *deviations* from the basic model. Due to space limitations, we present our model rather tersely (see the full version of this paper [16] for further details).

A few more definitions are in order before proceeding. A *unit clause* is a clause consisting of a single literal. An *assignment* is an expression of the form $x_i = a$ where $1 \leq i \leq n$ and $a \in \{0, 1\}$. A *restriction* $\rho$ is a set of assignments in which all variables are pairwise distinct. Let $\text{Var}(\rho)$ denote the set of all variables appearing in $\rho$. Restrictions naturally act on clauses, CNFs, and resolution proofs; we denote the result of this action by $C|_\rho$, $\tau|_\rho$, and $\Pi|_\rho$, respectively. An *annotated assignment* is an expression of the form $x_i \stackrel{*}{=} a$ where $1 \leq i \leq n$, $a \in \{0, 1\}$, and $* \in \{d, u\}$. See Definition 3 below for details about these annotations.

The underlying structure of our model is a labeled transition system whose states represent data maintained by a CDCL solver during runtime and whose labeled transitions are possible actions taken by a solver during runtime. We first define explicitly what constitutes a state.

**Definition 2.** *A **trail** is an ordered list of annotated assignments in which all variables are pairwise distinct. A trail acts on clauses, CNFs, and proofs just in the same way as does the restriction obtained from it by disregarding the order and the annotations on assignments. For a trail t and an annotated assignment $x_i \overset{*}{=} a$ such that $x_i$ does not appear in t, we denote by $[t, x_i \overset{*}{=} a]$ the trail obtained by appending $x_i \overset{*}{=} a$ to its end. $t[k]$ is the kth assignment of t. A **prefix** of a trail $t = [x_{i_1} \overset{*_1}{=} a_1, \ldots, x_{i_r} \overset{*_r}{=} a_r]$ is any trail of the form $[x_{i_1} \overset{*_1}{=} a_1, \ldots, x_{i_s} \overset{*_s}{=} a_s]$ where $0 \le s \le r$ and is denoted by $t[\le s]$. $\Lambda$ is the empty trail.*

*A **state** is a pair $(\tau, t)$, where $\tau$ is a CNF and t is a trail. The state $(\tau, t)$ is **terminal** if either $C|_t \equiv 1$ for all $C \in \tau$ or $\tau$ contains 0. All other states are nonterminal. We let $\mathbb{S}_n$ denote the set of all states (recall that n is reserved for the number of variables), and let $\mathbb{S}_n^o \subseteq \mathbb{S}_n$ be the set of all nonterminal states.*

We now describe the core of our (or, for that matter, any other) model, that is, transition rules between states.

**Definition 3.** *For a (nonterminal) state $S = (\tau, t) \in \mathbb{S}_n^o$, we define the finite set $Actions(S)$ and the function $Transition_S : Actions(S) \longrightarrow \mathbb{S}_n$; the fact $Transition_S(A) = S'$ will be usually abbreviated to $S \overset{A}{\Longrightarrow} S'$. Those are described as follows:*

$$Actions(S) \overset{\text{def}}{=} D(S) \,\dot\cup\, U(S) \,\dot\cup\, L(S),$$

*where the letters $D, U, L$ have the obvious meaning[4].*

- *$D(S)$ consists of all annotated assignments $x_i \overset{d}{=} a$ such that $x_i$ does not appear in t and $a \in \{0, 1\}$. We naturally let*

$$(\tau, t) \overset{x_i \overset{d}{=} a}{\Longrightarrow} (\tau, [t, x_i \overset{d}{=} a]). \tag{2}$$

- *$U(S)$ consists of all those assignments $x_i \overset{u}{=} a$ for which $\tau|_t$ contains the unit clause $x_i^a$; the transition function is given by the same formula (2) but with a different annotation:*

$$(\tau, t) \overset{x_i \overset{u}{=} a}{\Longrightarrow} (\tau, [t, x_i \overset{u}{=} a]). \tag{3}$$

- *As should be expected, $L(S)$ is the most sophisticated part of the definition (cf. [3, Section 2.3.3]). Let $t = [x_{i_1} \overset{*_1}{=} a_1, \ldots, x_{i_r} \overset{*_r}{=} a_r]$. By reverse induction on $k = r + 1, \ldots, 1$ we define the set $\mathbb{C}_k(S)$ that, intuitively, is the set of clauses that can be learned by backtracking up to the prefix $t[\le k]$. We let*

$$\mathbb{C}_{r+1}(S) \overset{\text{def}}{=} \{D \in \tau \mid D|_t = 0\}$$

---

[4] Restarts will be treated as a part of the learning scheme.

be the set of all **conflict clauses**.

For $1 \leq k \leq r$, we do the following: if the $k$th assignment of $t$ is of the form $x_{i_k} \overset{d}{=} a_k$, then $\mathbb{C}_k(S) \overset{\text{def}}{=} \mathbb{C}_{k+1}(S)$. Otherwise, it is of the form $x_{i_k} \overset{u}{=} a_k$, and we build up $\mathbb{C}_k(S)$ by processing every clause $D \in \mathbb{C}_{k+1}(S)$ as follows.

- If $D$ does not contain the literal $\overline{x_{i_k}^{a_k}}$ then we include $D$ into $\mathbb{C}_k(S)$ unchanged.
- If $D$ contains $\overline{x_{i_k}^{a_k}}$, then we resolve $D$ with all clauses $C \in \tau$ such that $C|_{t[\leq k-1]} = x_{i_k}^{a_k}$ and include all the results in $\mathbb{C}_k(S)$. The clause $D$ itself is not included.

To make sure that this definition is sound, we have to guarantee that $C$ and $D$ are actually resolvable (that is, they do not contain any other conflicting variables but $x_{i_k}$). For that we need the following observation, easily proved by reverse induction on $k$, simultaneously with the definition:

Claim. $D|_t = 0$ for every $D \in \mathbb{C}_k(S)$.

Finally, we let

$$\mathbb{C}(S) \overset{\text{def}}{=} \bigcup_{k=1}^{r} \mathbb{C}_k(S),$$

$$L(S) \overset{\text{def}}{=} \begin{cases} \{(0, \Lambda)\} & 0 \in \mathbb{C}(S) \\ \{(C, t^*) \mid C \in (\mathbb{C}(S) \setminus \tau) \text{ and} \\ \quad t^* \text{ is a prefix of } t \text{ such that } C|_{t^*} \neq 0\} & otherwise \end{cases} \quad (4)$$

and

$$(\tau, t) \overset{(C, t^*)}{\Longrightarrow} (\tau \cup \{C\}, t^*).$$

This completes the description of the basic model.

The *transition graph* $\Gamma_n$ is the directed graph on $\mathbb{S}_n$ defined by erasing the information about actions; thus $(S, S') \in E(\Gamma_n)$ if and only if $S' \in \text{im}(\text{Transition}_S)$. It is easy to see (by double induction on $(|\tau|, n - |t|)$) that $\Gamma_n$ is acyclic. Moreover, both the set $\{(S, A) \mid A \in \text{Actions}(S)\}$ and the function $(S, A) \mapsto \text{Transition}_S(A)$ are polynomial-time[5] computable. These observations motivate the following definition.

**Definition 4.** Given a CNF $\tau$, a **partial run** on $\tau$ from the state $S$ to the state $T$ is a sequence

$$S = S_0 \overset{A_0}{\Longrightarrow} S_1 \overset{A_1}{\Longrightarrow} \dots S_{L-1} \overset{A_{L-1}}{\Longrightarrow} S_L = T, \quad (5)$$

where $A_k \in \text{Actions}(S_k)$. In other words, a partial run is a labeled path in $\Gamma_n$. A **successful run** is a partial run from $(\tau, \Lambda)$ to a terminal state. A **CDCL**

---

[5] That is, polynomial in the size of the state $S$, not in $n$.

*solver* is a partial function[6] $\mu$ on $\mathbb{S}_n^o$ such that $\mu(S) \in Actions(S)$ whenever $\mu(S)$ is defined. The above remarks imply that when we repeatedly apply a CDCL solver $\mu$ starting at any initial state $(\tau, \Lambda)$, it will always result in a finite sequence like (5), with $T$ being a terminal state (successful run) or such that $\mu(T)$ is undefined (failure).

Theoretical analysis usually deals with *classes* (i.e., sets) of individual solvers rather than with individual implementations. We define such classes by prioritizing and restricting various actions.

**Definition 5.** *A **local** class of CDCL solvers is described by a collection of subsets AllowedActions(S) $\subseteq$ Actions(S) where $S \in \mathbb{S}_n^o$. It consists of all those solvers $\mu$ for which $\mu(S) \in AllowedActions(S)$, whenever $\mu(S)$ is defined.*

We will describe local classes of solvers in terms of *amendments* prescribing what actions should be *removed* from the set Actions(S) to form AllowedActions(S). The examples presented below illustrate how familiar restrictions look in this language. Throughout their description, we fix a nonterminal state $S = (\tau, t)$.

ALWAYS-C If $\tau|_t$ contains the empty clause, then $D(S)$ and $U(S)$ are removed from Actions(S). In other words, this amendment requires the solver to perform conflict analysis if it can do so.

ALWAYS-U If $\tau|_t$ contains a unit clause, then $D(S)$ is removed from Actions(S). This amendment insists on unit propagation, but leaves to nondeterminism the choice of the unit to propagate if there are several choices. Note that as defined, ALWAYS-U is a lower priority amendment than ALWAYS-C: if both a conflict and a unit clause are present, the solver must do conflict analysis.

DECISION-L In the definition (4), we shrink $\mathbb{C}(S) \setminus \tau$ to $\mathbb{C}_1(S) \setminus \tau$.

FIRST-L In the definition (4), we shrink $\mathbb{C}(S) \setminus \tau$ to those clauses that are obtained by resolving a conflict clause with one other clause in $\tau$. Such clauses are the first learnable clauses encountered in the process from Definition 3.

$\pi$-D, **where** $\pi \in S_n$ **is an order on the variables** We keep in $D(S)$ only the two assignments $x_i \overset{d}{=} 0$, $x_i \overset{d}{=} 1$, where $x_i$ is the *smallest* variable with respect to $\pi$ that does not appear in $t$. Note that this amendment does not have any effect on $U(S)$, and our main technical contributions can be phrased as answering under which circumstances this "loophole" can circumvent the severe restriction placed on the set $D(S)$.

NEVER-R In the definition (4), we require that $t^*$ is the *longest* prefix of $t$ satisfying $C|_{t^*} \neq 0$ (in which case $C|_{t^*}$ is necessarily a unit clause). As described, this amendment does not model nonchronological backtracking or require that the last assignment in the trail is a decision. However, this version is easier to state and it is not difficult to modify it to have the aforementioned properties.

WIDTH-$w$, **where** $w$ **is an integer** In the definition (4), we keep in $\mathbb{C}(S) \setminus \mathbb{C}$ only clauses of width $\leq w$. Note that this amendment still allows us to use wide clauses as intermediate results *within* a single clauses learning step.

---

[6] It is possible for Actions(S) to be empty.

Thus, our preferred way to specify local classes of solvers and the corresponding proof systems is by listing one or more amendments, with the convention that their effect is cumulative: an action is removed from Actions($S$) if and only if it should be removed according to at least one of the amendments present. More formally,

**Definition 6.** *For a finite set* $\mathcal{A}_1, \ldots, \mathcal{A}_r$ *of polynomial-time computable[7] amendments, we let* CDCL($\mathcal{A}_1, \ldots, \mathcal{A}_r$) *be the (possibly incomplete) proof system whose proofs are those successful runs in which none of the actions $A_i$ is affected by any of the amendments* $\mathcal{A}_1, \ldots, \mathcal{A}_r$.

Using this language, the main result from [18] can be very roughly summarized as

CDCL(DECISION-L, ALWAYS-C, ALWAYS-U) *is polynomially equivalent to general resolution.*

The open question asked in [3, Section 2.3.4] can be reasonably interpreted as whether CDCL(ALWAYS-C, ALWAYS-U, WIDTH-$w$) is as powerful as width-$w$ resolution, perhaps with some gap between the two width constraints. Our width lower bound mentioned in the introduction can be cast in this language as

*For any fixed order $\pi$ on the variables and every $\epsilon > 0$ there exist contradictory CNFs $\tau_n$ with $w(\tau_n \vdash 0) \leq O(1)$ not provable in* CDCL ($\pi$-D, WIDTH-$(1 - \epsilon)n$).

Finally, we would like to mention the currently open question about the exact strength of CDCL without restarts. This is one of the most interesting open questions in the area and has been considered heavily in the literature (see [6, 9, 11, 12, 15, 19] among others). It may be abstracted as

*Does* CDCL(ALWAYS-C, ALWAYS-U, NEVER-R) (*or at least* CDCL(NEVER-R)) *simulate general resolution?*

For both open questions mentioned above, we have taken the liberty of removing those amendments that do not appear immediately relevant.

At this point, since we discuss our main results in the introduction, we formulate them here more or less matter-of-factly.

**Theorem 1.** *For any fixed order $\pi$ on the variables, the system* CDCL (DECISION-L, $\pi$-D) *is polynomially equivalent to $\pi$-ordered resolution.*

**Theorem 2.** *For any fixed order $\pi$ on the variables, the system* CDCL (FIRST-L, $\pi$-D) *is polynomially equivalent to general resolution.*

---

[7] An amendment is polynomial-time computable if determining whether an action in *Action*($S$) is allowed by the amendment is polynomial-time checkable, given the state $S$.

# 3    CDCL(DECISION-L, $\pi$-D) $=_p$ $\pi$-Ordered Resolution

The proof of Theorem 1 is divided into two parts: we prove that each system is equivalent to an intermediate system we call $\pi$-*half-ordered resolution*.

Recall that, for the system $\pi$-ordered resolution, the variable that is resolved on must be $\pi$-maximal in each antecedent. In $\pi$-half-ordered resolution, this is required of *at least one* of the antecedents. That is, $\pi$-*half-ordered resolution* is the subsystem of resolution defined by imposing the restriction

$$(\forall l \in C \ (\pi(l) < \pi(x_i))) \vee (\forall l \in D \ (\pi(l) < \pi(x_i)))$$

on the resolution rule (1). Clearly, $\pi$-half-ordered resolution simulates $\pi$-ordered resolution but, somewhat surprisingly, it doesn't have any additional power over it.

**Theorem 3.** *For any fixed order $\pi$ on the variables, $\pi$-ordered resolution is polynomially equivalent to $\pi$-half-ordered resolution.*

We prove Theorem 3 by applying a sequence of transformations to a $\pi$-half-ordered refutation that, with the aid of the following definition, can be shown to make it incrementally closer to a $\pi$-ordered resolution refutation.

**Definition 7.** *A resolution refutation is $\pi$-**ordered up to** $k$ if it satisfies the property that if any two clauses are resolved on a variable $x_i \in Var_\pi^k$, then all resolution steps above it are on variables in $Var_\pi^{\pi(i)-1}$.*

The $\pi$-ordered refutations are then precisely those that are $\pi$-ordered up to $n - 1$. Now in order for these transformations not to blow up the size of the refutation, we need to carefully keep track of its structure throughout the process. As such, the proof of Theorem 3 depends heavily on resolution graphs and related definitions introduced in Sect. 2.

*Proof.* Let $\Pi$ be a $\pi$-half-ordered resolution refutation of $\tau$. Without loss of generality, assume $\pi = \mathrm{id}$; otherwise, rename variables. We will construct by induction on $k$ (satisfying $0 \leq k \leq n - 1$) a $\pi$-half-ordered resolution refutation $\Pi_k$ of $\tau$ which is ordered up to $k$. For the base case, let $\Pi_0 \stackrel{\mathrm{def}}{=} \Pi$. Suppose now $\Pi_k$ has been constructed. Without loss of generality, assume that $\Pi_k$ is connected; otherwise, take the subrefutation below any occurrence of 0.

Consider the set of nodes whose clauses are $k$-small. Note that this set is parent-complete. We claim that it is also upward-closed and, hence, path-complete. Indeed, let $u$ be a parent of $v$ and assume that $c_{\Pi_k}(u)$ is $k$-small. Then, since we disallow weakenings, $c_{\Pi_k}(v)$ is obtained by resolving on a variable $x_i \in Var_\pi^k$. Since $\Pi_k$ is ordered up to $k$, it follows that $Var(c_{\Pi_k}(v)) \subseteq Var_\pi^{i-1} \subseteq Var_\pi^k$; otherwise, some variable in $c_{\Pi_k}(v)$ would have remained unresolved on a path connecting $v$ to the sink (here we use the fact that $\Pi_k$ is connected). Hence $c_{\Pi_k}(v)$ is also $k$-small.

By Proposition 1, this set defines a subrefutation of the clauses labeling the independent set

$$L_k \stackrel{\text{def}}{=} \min_{\Pi_k} \{v \mid c_{\Pi_k}(v) \text{ is } k\text{ -small}\}. \tag{6}$$

Furthermore, $L_k$ splits $\Pi_k$ into two parts, i.e., $V(\Pi_k) = \text{ucl}_{\Pi_k}(L_k) \cup \text{dcl}_{\Pi_k}(L_k)$ and $L_k = \text{ucl}_{\Pi_k}(L_k) \cap \text{dcl}_{\Pi_k}(L_k)$. Let $D$ denote the subproof on $\text{dcl}_{\Pi_k}(L_k)$ and let $U$ denote the *subrefutation* on $\text{ucl}_{\Pi_k}(L_k)$. Note that $D$ is comprised of all nodes in $\Pi$ that are labeled by a clause that is not $k$-small or belong to $L_k$, and $U$ is comprised of all nodes labeled by a $k$-small clause. In particular, all axioms are in $D$, all resolutions in $U$ are on the variables in $\text{Var}_\pi^k$, and, since $\Pi_k$ is ordered up to $k$, all resolutions in $D$ are on the variables not in $\text{Var}_\pi^k$. Define

$$M \stackrel{\text{def}}{=} \min_D \{w \mid c_{\Pi_k}(w) \text{ is the result of resolving two clauses on } x_{k+1}\}. \tag{7}$$

If $M$ is empty, $\Pi_{k+1} \stackrel{\text{def}}{=} \Pi_k$. Otherwise, suppose $M = \{w_1, \ldots, w_s\}$ and define $A_i \stackrel{\text{def}}{=} \text{ucl}_D(\{w_i\})$. We will eliminate all resolutions on $x_{k+1}$ in $D$ by the following process, during which *the set of nodes stays the same*, while edges and clause-labeling function possibly change. More precisely, we update $D$ in $s$ rounds, defining a sequence of $\pi$-half-ordered resolution + *weakening* proofs $D_1, D_2, \ldots, D_s$. Initially $D_0 \stackrel{\text{def}}{=} D$. Fix now an index $i$. Let $c_{i-1}$ denote the clause-labeling $c_{D_{i-1}}$. To define the transformation of $D_{i-1}$ to $D_i$, we need the following structural properties of $D_{i-1}$, which are easily verified by induction simultaneously with the definition.

*Claim.* In the following properties, let $u$ and $v$ be arbitrary vertices in $V(D)$.

a. If $v$ is not above $u$ in $D$, then the same is true in $D_{i-1}$;
b. the clause $c_{i-1}(v)$ is equal to $c_D(v)$ or $c_D(v) \vee x_{k+1}$ or $c_D(v) \vee \overline{x_{k+1}}$;
c. if $v \notin \bigcup_{j=1}^{i-1} A_j$ then $c_{i-1}(v) = c_D(v)$ and, moreover, this clause is obtained in $D_{i-1}$ with the same resolution as in $D$;
d. $D_{i-1}$ is a $\pi$-half-ordered resolution + weakening proof.

Let us construct $D_i$ from $D_{i-1}$. By property (c) and the fact that $M$ is independent, the resolution step at $w_i$ is unchanged from $D$ to $D_{i-1}$. Let $w'$ and $w''$ denote the parents of $w_i$ in $D$ and let $c_D(w') = B \vee x_{k+1}$ and $c_D(w'') = C \vee \overline{x_{k+1}}$. Since $\Pi_k$ is $\pi$-half-ordered, either $B$ or $C$ is $k$-small. Assume without loss of generality that $B$ is $k$-small.

Recall that there is no resolution in $D$ on variables in $\text{Var}_\pi^k$. Thus, for all $v \in A_i$, it follows that $B$ is a subclause of $c_D(v)$, and by property (b), we have the following crucial property:

$$\text{For all } v \in A_i, B \text{ is a subclause of } c_{i-1}(v). \tag{8}$$

By property (a), $A_i$ remains upward closed in $D_{i-1}$. Accordingly, as the first step, for any $v \notin A_i$ we set $c_i(v):=c_{i-1}(v)$ and we leave its incoming edges unchanged.

Next, we update vertices $v \in A_i$ in an arbitrary $D$-topological order maintaining the property that $c_i(v) = c_{i-1}(v)$ or $c_i(v) = c_{i-1}(v) \vee \overline{x_{k+1}}$. In particular, $c_i(v) = c_{i-1}(v)$ whenever $c_{i-1}(v)$ contains the variable $x_{k+1}$.

First we set $c_i(w_i) := c_{i-1}(w_i) \vee \overline{x_{k+1}}$ and replace the incoming edges by a weakening edge from $w''$. This is possible since $c_{i-1}(w_i) = c_D(w_i)$ by property (c) and, hence, does not contain $x_{k+1}$ by virtue of being in $M$.

For $v \in A_i \setminus \{w_i\}$, we proceed as follows.

1. If $x_{k+1} \in c_{i-1}(v)$, keep the clause but replace incoming edges with a weakening edge $(w', v)$. This is well-defined by (8). Also, since $w' < w < v$ in $D$, we maintain property (a).
2. If $c_{i-1}(v) = \text{Res}(c_{i-1}(u), c_{i-1}(w))$ on $x_{k+1}$ where $\overline{x_{k+1}} \in c_{i-1}(u)$, set $c_i(v) := c_{i-1}(v) \vee \overline{x_{k+1}}$ – equivalently, $c_{i-1}(v) \vee c_i(u)$ – and replace incoming edges by a weakening edge $(u, v)$.
3. If $c_{i-1}(v)$ is weakened from $c_{i-1}(u)$ and $x_{k+1} \notin c_{i-1}(v)$, set $c_i(v) := c_{i-1}(v) \vee c_i(u)$. In other words, we append the literal $\overline{x_{k+1}}$ to $c_i(v)$ if and only if this was previously done for $c_i(u)$.
4. Otherwise, $x_{k+1} \notin c_{i-1}(v)$ and $c_{i-1}(v) = \text{Res}(c_{i-1}(u), c_{i-1}(w))$ on some $x_\ell$ where $\ell > k + 1$. In particular, $x_{k+1} \notin \text{Var}(c_{i-1}(u)) \cup \text{Var}(c_{i-1}(w))$. Set $c_i(v) := \text{Res}(c_i(u), c_i(w))$ that is, like in the previous item, we append $\overline{x_{k+1}}$ if and only if it was previously done for either $c_i(v)$ or $c_i(w)$. Since $\ell > k + 1$, this step remains $\pi$-half-ordered.

This completes our description of $D_i$. It is straightforward to verify that $D_s$ is a $\pi$-half-ordered resolution + weakening proof *without* resolutions on $x_{k+1}$.

To finally construct $\Pi_{k+1}$, we reconnect $D_s$ to $U$ along $L_k$ and then remove any weakenings introduced in $D_s$. This may require adding new nodes, as it may be the case that $c_s(v) \neq c_D(v)$ for some $v \in L_k$. But, in this case, it is straightforward to verify, using (8), that there is a vertex $w \in \text{dcl}_D(M) \setminus \{M\}$ such that $c_D(v) = \text{Res}(c_s(w), c_s(v))$ on $x_{k+1}$, and this resolution is half-ordered. In fact, $w$ can be taken to be a parent in $D$ of some nodes in $M$. Thus, when necessary we add to $D_s$ a new node $\tilde{v}$ labeled by $\text{Res}(c_s(w), c_s(v))$ and add the edges $(v, \tilde{v})$ and $(w, \tilde{v})$.

Denote by $\widetilde{\Pi}_{k+1}$ the result of connecting $D_s$ and $U$ along the vertices in $L_k$ and this newly added collection of vertices. Since neither $D_s$ nor $U$ contain resolutions on $x_{k+1}$ except for those in the derivations of the clauses just added to $D_s$, it follows that $\widetilde{\Pi}_{k+1}$ is a $\pi$-half-ordered resolution + weakening refutation that is *ordered up to $k + 1$*. Let $\Pi_{k+1}$ be obtained by contracting all weakenings.

It only remains to analyze its size (note that *a priori* it can be doubled at every step, which is unacceptable). Since

$$|\Pi_{k+1}| \leq |\Pi_k| + |L_k|, \tag{9}$$

we only have to control $|L_k|$. For that we will keep track of the invariant $|\text{dcl}_{\Pi_k}(L_k)|$; more precisely, we claim that

$$|\text{dcl}_{\Pi_{k+1}}(L_{k+1})| \leq |\text{dcl}_{\Pi_k}(L_k)|. \tag{10}$$

Let us prove this by constructing an injection from $\mathrm{dcl}_{\Pi_{k+1}}(L_{k+1})$ to $\mathrm{dcl}_{\Pi_k}(L_k)$; we will utilize the notation from above.

First note that the resolution + weakening refutation $\widetilde{\Pi}_{k+1}$ and its weakening-free contraction $\Pi_{k+1}$ can be related as follows. For every node $v \in V(\Pi_{k+1})$ there exists a node $v^* \in V(\widetilde{\Pi}_{k+1})$ with $c_{\widetilde{\Pi}_{k+1}}(v^*) \supseteq c_{\Pi_{k+1}}(v)$ which is *minimal* among those contracting to $v$. If $v$ is an axiom node of $\Pi_{k+1}$ then so is $v^*$ in $\widetilde{\Pi}_{k+1}$. Otherwise, if $u$ and $w$ are the two parents of $v$, and if $u'$ and $w'$ are the corresponding parents of $v^*$ ($v^*$ may not be obtained by weakening due to the minimality assumption), then $c_{\widetilde{\Pi}_{k+1}}(u^*)$ is a subclause of $c_{\widetilde{\Pi}_{k+1}}(u')$ and $c_{\widetilde{\Pi}_{k+1}}(w^*)$ is a subclause of $c_{\widetilde{\Pi}_{k+1}}(w')$. We claim that $(v \mapsto v^*)\mid_{\mathrm{dcl}_{\Pi_{k+1}}(L_{k+1})}$ (which is injective by definition) is the desired injection. We have to check that its image is contained in $\mathrm{dcl}_{\Pi_k}(L_k)$.

Fix $v \in \mathrm{dcl}_{\Pi_{k+1}}(L_{k+1})$. Then by definition of $L_k$, *either $v$ is an axiom or both its parents are not $(k+1)$-small*. By the above mentioned facts about the contraction $\widetilde{\Pi}_{k+1} \to \Pi_{k+1}$, this property is inherited by $v^*$. In particular, $v^* \notin \{\widetilde{w} \mid w \in L_k\}$ as all nodes in this set have at least one $(k+1)$-small parent due to half-orderedness. Finally, since the corresponding clauses in $D$ and $D_s$ differ only in the variable $x_{k+1}$, $v^*$ cannot be in $U$, for the same reason (recall that all axioms are in $D$). Hence $v^* \in V(D_s) = V(D) = \mathrm{dcl}_{\Pi_k}(L_k)$.

Having thus proved (10), we conclude by the obvious induction that $|L_k| \le |\mathrm{dcl}_{\Pi_k}(L_k)| \le |\mathrm{dcl}_{\Pi_0}(L_0)| \le |\Pi|$. Then (9) implies $|\Pi_{n-1}| \le n|\Pi|$, as desired.

At last, we must show the equivalence holds for the corresponding CDCL system. We provide a sketch of the proof.

**Theorem 4.** *For any fixed order $\pi$ on the variables, $\mathsf{CDCL}(\pi\text{-}\mathsf{D}, \mathsf{DECISION\text{-}L})$ is polynomially equivalent to $\pi$-half-ordered resolution.*

*Proof.* As above, assume $\pi = \mathrm{id}$. The fact that $\mathsf{CDCL}(\pi\text{-}\mathsf{D}, \mathsf{DECISION\text{-}L})$ polynomially simulates $\pi$-half ordered resolution is almost trivial. A $\pi$-half-ordered resolution step deriving $\mathrm{Res}(C \vee x_i, D \vee \overline{x_i})$ can be directly simulated by constructing a trail $t$ that falsifies $C \vee D$ and contains a single unit propagation on $x_i$. This is possible since $C$ or $D$ is $i$-small. Then $C \vee D$ can be easily learned using $t$.

The other direction is just slightly more involved. It suffices to show that for a learning step $(\tau, t) \overset{(D, t^*)}{\Longrightarrow} (\tau \cup \{D\}, t^*)$, there is a short $\pi$-half-ordered resolution proof of $D$ from $\tau$. Any learned clause can be thought of naturally as the result of a sequence of resolutions; there are clauses $C_1, \ldots, C_{k+1}$ in $\tau$ and variables $x_{i_1}, \ldots, x_{i_k}$ assigned by unit propagation in $t$ from which we can inductively define

$$C'_{k+1} \overset{\mathrm{def}}{=} C_{k+1} \qquad \text{and} \qquad C'_j \overset{\mathrm{def}}{=} \mathrm{Res}(C_j, C'_{j+1})$$

where $C_j$ and $C'_{j+1}$ are resolvable on $x_{i_j}$ and $D = C'_1$. These resolutions may not all be $\pi$-half-ordered, but they can be reordered and duplicated to derive the same clause while maintaining $\pi$-half-orderedness. Formally, we define by double induction a different collection of derivable clause: for $\gamma$ and $j$ in $[k+1]$ satisfying $j < \gamma$, $C_{\gamma,\gamma} \stackrel{\text{def}}{=} C_\gamma$ and

$$C_{\gamma,j} \stackrel{\text{def}}{=} \begin{cases} \text{Res}(C_{j,1}, C_{\gamma,j+1}) & C_{j,1} \text{ and } C_{\gamma,j+1} \text{ are resolvable on } x_{i_j} \\ C_{\gamma,j+1} & \text{otherwise.} \end{cases}$$

Because of $\pi$-D, the only literals appearing in $C_j$ that are potentially larger than $x_{i_j}$ (with respect to $\pi$) are the other variables assigned by unit propagation in $t$ and resolved on in the derivation of $C_1'$. One can show that the clause $C_{j,1}$ is the result of "washing out" these other literals so that $x_{i_j}$ appears maximally. Since all resolutions on are clauses of this form, they are all $\pi$-half-ordered. And for the learning scheme DECISION-L, the learned clause $C_1'$ contains *only* decision variables in $t$, so this reorganizing of resolutions does not affect the final derived clause; it can be verified that $C_{k+1,1} = C_1' = D$. In total, this derivation of $C_{k+1,1}$ (and, hence, $D$) is $\pi$-half ordered and has at most $n^2$ resolutions.

## 4    CDCL(FIRST-L, $\pi$-D) $=_p$ General Resolution

This result is by far the most technical. It would have been impossible to give a satisfying treatment in the space available, but in the interest of providing some idea of its formal aspects, we briefly discuss our approach. As in the previous section, the proof is divided into two parts: we prove that each system is equivalent to an intermediate system we call $\pi$-*trail resolution*.

**Definition 8.** *Fix an order $\pi$ on the variables. The proof system $\pi$-trail resolution is defined as follows. Its lines are either clauses or trails, where the empty trail is an axiom. It has the following rules of inference:*

$$\frac{t}{[t, x_i \stackrel{d}{=} a]}, \quad \textit{(Decision rule)}$$

*where $x_i$ is the $\pi$-smallest index such that $x_i$ does not appear in $t$ and $a \in \{0,1\}$ is arbitrary;*

$$\frac{t \qquad C}{[t, x_i \stackrel{u}{=} a]}, \quad \textit{(Unit propagation rule)}$$

*where $C|_t = x_i^a$;*

$$\frac{C \vee x_i^a \qquad D \vee x_i^{1-a} \qquad t}{C \vee D}, \quad \textit{(Learning rule)}$$

*where $(C \vee D)|_t = 0$, $(x_i \stackrel{*}{=} a) \in t$ and all other variables of $C$ appear before $x_i$ in $t$.*

Without the unit propagation rule, this is just $\pi$-half-ordered resolution, modulo additional traffic in trails. It follows almost directly from its definition that $\pi$-trail resolution is polynomially equivalent to CDCL(FIRST-L, $\pi$-D) (and even CDCL($\pi$-D)). Our main technical contribution is proving the following.

**Theorem 5.** *For any fixed order $\pi$ on the variables, $\pi$-trail resolution polynomially simulates general resolution.*

The key observation is that, due to the unit propagation rule, $\pi$-trail resolution becomes significantly more powerful when the underlying formula has many unit clauses. Thus, we design the simulation algorithm to output a derivation of *all* unit clauses appearing in the given refutation $\Pi$ and then recursively apply it to gain access to more unit clauses throughout the procedure. At first glance, it might seem reasonable to recursively apply the simulation algorithm to various restrictions of $\Pi$, but restriction as an operation has two flaws with regards to our objectives. First, the results of different restrictions on proofs often *overlap*; for example, when viewing restriction as an operation on resolution graphs, the graphs of $\Pi|_{x_i=0}$ and $\Pi|_{x_i=1}$ will likely share vertices from $G(\Pi)$. This leads to an exponential blow-up in the size of the output if one is not careful. Second, restrictions may *collapse* parts of the $\Pi$; for example, if $\rho$ falsifies an axiom of $\Pi$, then $\Pi|_\rho$ is the trivial refutation and it is impossible to extract anything from it by recursively applying our simulation algorithm.

To make this approach feasible, we introduce a new operator, which may be of independent interest, called *variable deletion*; it is an analogue of restriction for sets of variables as opposed to sets of variable assignments. This operator has the property that it always yields a nontrivial refutation (for proper subsets of variables), and its size and structure are highly regulated by the size and structure of the input refutation. This allows for a surgery-like process; we simulate small local pieces of the refutation and then stitch them together into a new global refutation. For complete details, see the full version of this paper [16].

# 5   Conclusion

Our work continues the line of research aimed at better understanding theoretical limitations of CDCL solvers. We have focused on the impact of decision strategies, and we have considered the simple strategy that always chooses the first available variable under a fixed ordering. We have shown that, somewhat surprisingly, the power of this model heavily depends on the learning scheme employed and may vary from ordered resolution to general resolution.

In practice, the fact that CDCL(DECISION-L, $\pi$-D, ALWAYS-C, ALWAYS-U) is not as powerful as resolution supports the observation that CDCL solvers with the ordered decision strategy are often less efficient than those with more powerful decision strategies like VSIDS. But, although DECISION-L is an asserting learning strategy, most solvers use more efficient asserting strategies like *1-UIP*. A natural open question then is what can be proved if DECISION-L is replaced with some other amendment modeling a different, possibly more practical asserting learning scheme? Furthermore, what is the exact strength of CDCL($\pi$-D, ALWAYS-C, ALWAYS-U)?

# References

1. Aloul, F.A., Markov, I.L., Sakallah, K.A.: MINCE: a static global variable-ordering for SAT and BDD. In: International Workshop on Logic and Synthesis, pp. 1167–1172 (2001)
2. Aloul, F.A., Markov, I.L., Sakallah, K.A.: FORCE: a fast & easy-to-implement variable-ordering heuristic. In: Proceedings of the 13th ACM Great Lakes Symposium on VLSI, pp. 116–119. ACM (2003)
3. Atserias, A., Fichte, J.K., Thurley, M.: Clause-learning algorithms with many restarts and bounded-width resolution. J. Artif. Intell. Res. **40**, 353–373 (2011)
4. Beame, P., Karp, R., Pitassi, T., Saks, M.: The efficiency of resolution and Davis-Putnam procedures. SIAM J. Comput. **31**(4), 1048–1075 (2002)
5. Beame, P., Kautz, H., Sabharwal, A.: Towards understanding and harnessing the potential of clause learning. J. Artif. Intell. Res. **22**, 319–351 (2004)
6. Beame, P., Sabharwal, A.: Non-restarting SAT solvers with simple preprocessing can efficiently simulate resolution. In: Proceedings of the Twenty-Eighth AAAI Conference on Artificial Intelligence, pp. 2608–2615 (2014)
7. Ben-Sasson, E., Johannsen, J.: Lower bounds for width-restricted clause learning on small width formulas. In: Strichman, O., Szeider, S. (eds.) SAT 2010. LNCS, vol. 6175, pp. 16–29. Springer, Heidelberg (2010). https://doi.org/10.1007/978-3-642-14186-7_4
8. Ben-Sasson, E., Wigderson, A.: Short proofs are narrow - resolution made simple. J. ACM **48**(2), 149–169 (2001)
9. Bonet, M.L., Buss, S., Johannsen, J.: Improved separations of regular resolution from clause learning proof systems. J. Artif. Intell. Res. **49**, 669–703 (2014)
10. Bonet, M.L., Esteban, J.L., Galesi, N., Johannsen, J.: On the relative complexity of resolution refinements and cutting planes proof systems. SIAM J. Comput. **30**(5), 1462–1484 (2000)
11. Buss, S.R., Hoffmann, J., Johannsen, J.: Resolution trees with lemmas: resolution refinements that characterize DLL-algorithms with clause learning. Logical Methods Comput. Sci. **4**(4), 1–28 (2008)
12. Buss, S.R., Kołodziejczyk, L.A.: Small stone in pool. Logical Methods Comput. Sci. **10**(2), 1–22 (2014). https://lmcs.episciences.org/852/pdf
13. Elffers, J., Johannsen, J., Lauria, M., Magnard, T., Nordström, J., Vinyals, M.: Trade-offs between time and memory in a tighter model of CDCL SAT solvers. In: Creignou, N., Le Berre, D. (eds.) SAT 2016. LNCS, vol. 9710, pp. 160–176. Springer, Cham (2016). https://doi.org/10.1007/978-3-319-40970-2_11
14. Hertel, P., Bacchus, F., Pitassi, T., Van Gelder, A.: Clause learning can effectively p-simulate general propositional resolution. In: Proceedings of the Twenty-Third AAAI Conference on Artificial Intelligence, pp. 283–290 (2008)
15. Li, C., Fleming, N., Vinyals, M., Pitassi, T., Ganesh, V.: Towards a complexity-theoretic understanding of restarts in SAT solvers. In: Pulina, L., Seidl, M. (eds.) Theory and Applications of Satisfiability Testing - SAT 2020. LNCS, vol. 12178, pp. 233–249. Springer, Cham (2020)
16. Mull, N., Pang, S., Razborov, A.: On CDCL-based proof systems with the ordered decision strategy. arXiv preprint arXiv:1909.04135 (2019)

17. Nieuwenhuis, R., Oliveras, A., Tinelli, C.: Solving SAT and SAT modulo theories: from an abstract Davis-Putnam-Logemann-Loveland procedure to DPLL($T$). J. ACM **53**(6), 937–977 (2006)
18. Pipatsrisawat, K., Darwiche, A.: On the power of clause-learning SAT solvers as resolution engines. Artif. Intell. **175**(2), 512–525 (2011)
19. Van Gelder, A.: Pool resolution and its relation to regular resolution and DPLL with clause learning. In: Sutcliffe, G., Voronkov, A. (eds.) LPAR 2005. LNCS (LNAI), vol. 3835, pp. 580–594. Springer, Heidelberg (2005). https://doi.org/10.1007/11591191_40
20. Vinyals, M.: Hard examples for common variable decision heuristics. In: Proceedings of the 34th AAAI Conference on Artificial Intelligence (AAAI 2020) (2020)

# Equivalence Between Systems Stronger Than Resolution

Maria Luisa Bonet[1]([⊠])(iD) and Jordi Levy[2]([⊠])(iD)

[1] Universitat Politècnica de Catalunya (UPC), Barcelona, Spain
bonet@cs.upc.edu
[2] Artificial Intelligence Research Institute, Spanish Research Council (IIIA-CSIC),
Barcelona, Spain
levy@iiia.csic.es

**Abstract.** In recent years there has been an increasing interest in studying proof systems stronger than Resolution, with the aim of building
more efficient SAT solvers based on them. In defining these proof systems, we try to find a balance between the power of the proof system
(the size of the proofs required to refute a formula) and the difficulty of
finding the proofs. Among those proof systems we can mention Circular
Resolution, MaxSAT Resolution with Extensions and MaxSAT Resolution with the Dual-Rail encoding.

In this paper we study the relative power of those proof systems from a
theoretical perspective. We prove that Circular Resolution and MaxSAT
Resolution with extension are polynomially equivalent proof systems.
This result is generalized to arbitrary sets of inference rules with proof
constructions based on circular graphs or based on weighted clauses. We
also prove that when we restrict the Split rule (that both systems use)
to bounded size clauses, these two restricted systems are also equivalent.
Finally, we show the relationship between these two restricted systems
and Dual-Rail MaxSAT Resolution.

## 1 Introduction

The Satisfiability (SAT) and Maximum Satisfiability (MaxSAT) problems are central in computer science. SAT is the problem of, given a CNF formula, deciding if it
has an assignment of 0/1 values that satisfy the formula. MaxSAT is the optimization version of SAT. Given a CNF formula, we want to know what is the maximum
number of clauses that can be satisfied by an assignment. SAT and the decision
version of MaxSAT are NP-Complete. Problems in many different areas like planning, computational biology, circuit design and verification, etc. can be solved by
encoding them into SAT or MaxSAT, and then using a SAT or MaxSAT solver.

Resolution based SAT solvers can handle huge industrial formulas successfully,
but on the other hand, seemingly easy tautologies like the Pigeonhole Principle

Research partially supported by the EU H2020 Research and Innovation Programme
under the LOGISTAR project (Grant Agreement No. 769142), the MINECO-FEDER
project TASSAT3 (TIN2016-76573-C2-2-P) and the MICINN project PROOFS
(PID2019-109137GB-C21).

L. Pulina and M. Seidl (Eds.): SAT 2020, LNCS 12178, pp. 166–181, 2020.
https://doi.org/10.1007/978-3-030-51825-7_13

require exponentially long Resolution refutations [8]. An important research direction is to implement SAT solvers based on stronger proof systems than Resolution. To be able to do that, the proof systems should not be too strong, given that the stronger a proof system is, the harder it is to find efficient algorithms to find refutations for the formulas. This is related to the notion of automatizability [2,5].

In the last few years, a number of proof systems somewhat stronger than Resolution have been defined. Among them are MaxSAT Resolution with Extension [13], Circular Resolution [1], Dual-Rail MaxSAT [9], Weighted Dual-Rail MaxSAT [3,14] and Sheraly-Adams proof system [7,15]. All these systems have polynomial size proofs of formulas like the Pigeonhole Principle. Atserias and Lauria [1] showed that Circular Resolution is equivalent to the Sheraly-Adams proof system. Larrosa and Rollón [13] showed that MaxSAT Resolution with Extension can simulate Dual-Rail MaxSAT. In this paper, we show that MaxSAT Resolution with Extension is equivalent to Circular Resolution. This equivalence is parametric on the set of inference rules used by both proof systems.

Both Circular Resolution and MaxSAT Resolution with Extension use a rule called SPLIT or EXTENSION, where from a clause $A$, we can obtain both $A \vee x$ and $A \vee \neg x$. We can add a restriction on this rule, and therefore on the proof system. If we bound the number of literals of $A$ to be used in the split rule by $k$, for $k \geq 0$, we can talk about *MaxSAT Resolution with k-Extension*, or about *k-Circular Resolution*. In the present article, we also prove the equivalence of both systems, k-Circular Resolution and MaxSAT Resolution with k-Extensions, and improve the result of [13], proving that these restricted proof systems can also simulate Dual-Rail MaxSAT and Weighted Dual-Rail MaxSAT.

This paper proceeds as follows. In the preliminary Sect. 2 we introduce Circular, Weighted and Dual-Rail proofs. In Sect. 3, we prove some basic facts about these proof systems. The equivalence of Circular Resolution and MaxSAT Resolution with Extension is proved in Sect. 4. In Sect. 5, we describe a restriction of these two proof systems, show that they are equivalent, and prove that they can simulate Weighted Dual-Rail MaxSAT.

## 2   Preliminaries

We consider a set $x_1, \ldots, x_n$ of variables, literals of the form $x_i$ or $\neg x_i$, clauses $A = l_1 \vee \cdots \vee l_r$ defined as sets of literals, and formulas defined as sets of clauses. Additionally, we also consider *weighted formulas*, defined as multisets of the form $\mathcal{F} = \{(A_1, u_1), \ldots, (A_r, u_r)\}$, where the $A_i$'s are clauses and the $u_i$'s are finite (positive or negative) integers. These integers $u_i$, when positive, describe the number of occurrences of the clause $A_i$. When they are negative, as we will see, they represent the obligation to prove these clauses in the future. Notice also that we do not require $A_i \neq A_j$, when $i \neq j$, thus we deal with multisets. We say that two weighted formulas are (fold-unfold) equivalent, noted $\mathcal{F}_1 \approx \mathcal{F}_2$, if for any clause $A$, we have $\sum_{(A,u) \in \mathcal{F}_1} u = \sum_{(A,v) \in \mathcal{F}_2} v$. Notice that, contrarily to traditional Partial MaxSAT formulas, we do not use clauses with infinite weight.

An inference rule is given by a multi-set of antecedent clauses and a multi-set of consequent clauses where any truth assignment that satisfies all the antecedents, also satisfies all the consequent clauses. Notice that the MAXSAT RESOLUTION rule was originally proposed to solve MaxSAT, and therefore, it satisfies additional properties. The rule preserves the number of unsatisfied clauses, for any assignment. Here, following the original idea of Ignatiev et al. [9], we use these MaxSAT techniques to solve SAT.

**Definition 1.** *These are some examples of inference rules defined in the literature (with possibly different names):*

$$\frac{x \vee A \qquad \neg x \vee B}{A \vee B} \text{ CUT} \qquad \frac{x \vee A \qquad \neg x \vee B}{\begin{array}{c} A \vee B \\ x \vee A \vee \overline{B} \\ \neg x \vee B \vee \overline{A} \end{array}} \text{ MAXSAT RESOLUTION}$$

$$\frac{x \vee A \qquad \neg x \vee A}{A} \text{ SYMMETRIC CUT} \qquad \frac{A}{x \vee A \qquad \neg x \vee A} \text{ SPLIT}$$

$$\frac{}{x \vee \neg x} \text{ AXIOM} \qquad \frac{\square}{x \qquad \neg x} \text{ 0-SPLIT}$$

*where, in the* MAXSAT RESOLUTION *rule, if* $A = x_1 \vee \cdots \vee x_r$, *then* $\overline{A}$ *denotes the set of* $|A|$ *clauses* $\{\neg x_1, x_1 \vee \neg x_2, x_1 \vee x_2 \vee \neg x_3, \ldots, x_1 \vee \cdots \vee x_{r-1} \vee \neg x_r\}$.

Notice that, in the previous definition, SYMMETRIC CUT is a special case of CUT where $A = B$. Similarly, it is also a special case of MAXSAT RESOLUTION where $A = B$, since $x \vee A \vee \overline{A}$ only contain tautologies that are removed. Notice that 0-SPLIT can be generalized to the K-SPLIT rule where SPLIT is applied only to clauses $A$ of length at most $k$. Below, we will also see that SPLIT and EXTENSION (defined in Sect. 3) are in essence the same inference rule.

Traditionally, fixed a set $\mathcal{R}$ of inference rules, a set of hypotheses $\mathcal{H}$ and a goal $C$, a *proof of* $\mathcal{H} \vdash C$ is a finite sequence of formulas that starts with $\mathcal{H}$, ends in $C$, and such that any other formula is one of the consequent of an inference rule in $\mathcal{R}$ whose antecedents are earlier in the sequence. These proofs can naturally be represented as bipartite DAGs, where nodes are either formulas or inference rules and edges denote the occurrence of a formula in the antecedents or the consequent of the rule.

In this paper we consider three distinct more complicated notions of *proof*, or proof systems: *Circular resolution* [1], *MaxSAT Resolution with Extension* [13], and *Dual-Rail MaxSAT* [9] or its generalization *Weighted Dual-Rail MaxSAT* [3].

All these proof systems will share the same inference rules, but they will use them in distinct ways (despite they have the same name). Thus, for instance, in the weighted context, the CUT rule will *replace* the premises $x \vee A$ and $\neg x \vee B$ by the conclusion $A \vee B$. Therefore, in the weighted context, after the application of the cut, these premises are no longer available for further cuts.

All these proof systems are able to prove the Pigeonhole Principle in polynomial size. In this paper we will study the relative power of these proof systems.

## 2.1 Circular Proofs

First, we introduce *Circular Proofs* as defined by Atserias and Lauria [1]:

**Definition 2 (Circular Proof).** *Fixed a set of inference rules $\mathcal{R}$, a set of hypotheses $\mathcal{H}$ and a goal $C$, a circular proof of $\mathcal{H} \vdash C$ is a bipartite directed graph $(I, J, E)$ where nodes are either inference rules ($R \in I$) or formulas[1] ($A \in J$), and edges $A \to R \in E$ denotes the occurrence of clause $A$ in the antecedents of rule $R$ and edges $R \to A \in E$ the occurrence of clause $A$ in the consequent of $R$.*

*Given a* flow assignment *$Flow : I \to \mathbb{N}^+$ to the rules, we define the* balance *of the clause as:*

$$Bal(A) = \sum_{R \in N^{in}(A)} Flow(R) - \sum_{R \in N^{out}(A)} Flow(R)$$

*where $N^{in}(A) = \{R \in I \mid R \to A\}$ and $N^{out}(A) = \{R \in I \mid A \to R\}$ are the sets of neighbours of a node.*

*In order to ensure soundness of a circular proof, it is required the existence of a flow assignment satisfying $Bal(A) \geq 0$, for any formula $A \in J \setminus \mathcal{H}$, and $Bal(C) > 0$, for the goal $C$.*

Atserias and Lauria [1] define *Circular Resolution* as the circular proof system where the set of inference rules is fixed to $\mathcal{R} = \{\text{AXIOM}, \text{SYMMETRIC CUT}, \text{SPLIT}\}$ and prove its soundness.

We will assume that the set of inference rules allows us to construct a constant size circular proof where formula $A$ is derivable from $A$ in one or more steps. The inference rule AXIOM is included in $\mathcal{R}$ for this purpose (in the third proof of Fig. 1, $x \lor \neg x$ is proved, which shows that AXIOM rule is indeed not necessary). If $A$ is the empty clause, we can use the SPLIT rule or even the 0-SPLIT rule and the CUT or the SYMMETRIC CUT rules. If $A$ is of the form $A = x \lor A'$, we have two possibilities, as shown in Fig. 1.

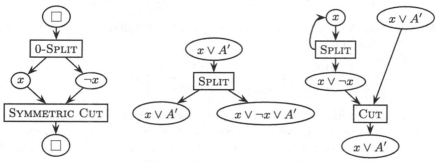

**Fig. 1.** Three ways to proof $A$ from $A$.

The *length* of a proof is defined as the number of nodes of the bipartite graph.

## 2.2   Weighted Proofs

Second, we also introduce *Weighted Proofs*, following the ideas of Larrosa and Heras [12] and Bonet et al. [4, 6] for positive weights and Larrosa and Rollón [13] for

---

[1] We keep the name *formula* for consistency with the original definition, although they are really *clauses*.

positive and negative weights. The main idea is that, when we apply an inference rule, we *replace* the antecedents by the consequent, instead of *adding* the consequent to the set of proved formulas. As a consequence, formulas cannot be *reused*. This is similar to the definition of *Read Once Resolution* [10]. We use weighted formulas, i.e. multi-sets of clauses instead of sets of clauses and, or more compactly, pairs $(A, u)$ where integer $u$ represents the number of occurrences of clause $A$. This makes sense since non-reusability of clauses implies that it is not the same having one or two copies of the same clause. Allowing the use of negative weights, we can represent clauses that are not proved yet and will be proved later. Notice that these proof systems were originally designed to solve MaxSAT. Here, we use them to construct proofs of SAT problems. Hence, the original formulas are unweighted, and we only use weighted formulas in the proofs.

**Definition 3 (Weighted Proof).** *Fixed a set of inference rules $\mathcal{R}$, a set of hypotheses $\mathcal{H}$ and a goal $C$, a weighted proof of $\mathcal{H} \vdash C$ is a sequence $\mathcal{F}_1 \vdash \cdots \vdash \mathcal{F}_n$ of weighted formulas such that:*

1. *$\mathcal{F}_1 = \{(A, u_A) \mid A \in \mathcal{H}\}$ for some arbitrary and positive weights $u_A \geq 0$.*
2. *$\mathcal{F}_n$ contains the goal $(C, u)$ with strictly positive weight $u > 0$, and possibly other clauses, all of them with positive weight.*
3. *For every proof step $\mathcal{F}_i \vdash \mathcal{F}_{i+1}$, either*
   (a) *(regular step) there exist an inference rule $A_1, \ldots, A_r \vdash B_1, \ldots, B_s \in \mathcal{R}$ and a positive weight $u > 0$ such that*

$$\mathcal{F}_{i+1} = \mathcal{F}_i \setminus \{(A_1, u), \ldots, (A_r, u)\} \cup \{(B_1, u), \ldots, (B_s, u)\}$$

   *or,*
   (b) *(fold-unfold step) $\mathcal{F}_{i+1} \approx \mathcal{F}_i$.*

Alternatively, fold-unfold steps may be defined as the application of the FOLD and UNFOLD rules defined as:

$$\frac{(C, u)(C, v)}{(C, u+v)} \text{ FOLD} \qquad \frac{(C, u+v)}{(C, u)(C, v)} \text{ UNFOLD}$$

Notice that, in regular steps, weights are positive integers. Instead, in the Fold and Unfold rules, $u$ and $v$ can be negative. Clauses with weight zero are freely removed from the formula, as well as tautological clauses $x \vee \neg x \vee A$. Notice that only one fold-unfold step is necessary between two regular steps, just to get $(A, u)$ for every antecedent $A$ of the next regular step, $u$ being the weight of this next step. Moreover, only a constant-bounded number of FOLD and UNFOLD rule applications is needed in this fold-unfold step. This motivates the definition of the *length of a proof* as its number of regular steps.

Larrosa and Heras [12] define *MaxSAT Resolution* as a method to solve MaxSAT using positive weighted proofs with the MAXSAT RESOLUTION rule. Bonet et al. [4,6] prove the completeness of this method for MaxSAT. The method is complete even if we restrict the hypotheses $\mathcal{H}$ to have weight one. Notice that the weighted proof system with the CUT rule is incomplete if we restrict hypotheses to have weight one. In other words, resolution is incomplete if we restrict hypotheses to be used only once like in Read Once Resolution [10].

Notice also that the MaxSAT resolution rule defined in [12] allows the weights of the antecedents to be different. Our version using equal weights for both antecedents is equivalent using the fold and unfold rules.

Recently, Larrosa and Rollón [13] define *MaxSAT Resolution with Extension* as the weighted proof system using $\mathcal{R} = \{\text{MaxSAT Resolution}, \text{Split}\}$ as inference rules. In fact, they use a rule called EXTENSION that, as we will see below, is equivalent to the SPLIT rule. They only explicitly mention the FOLD rule, but notice that the UNFOLD is a special case of the EXTENSION rule.

Traditionally, we say that formulas $F_1$ subsumes another formula $F_2$, noted $F_1 \subseteq F_2$, if for every clause $A_2 \in F_2$, there exists a clause $A_1 \in F_1$ such that $A_1 \subseteq A_2$. Instantiating a variable by true or false in a formula $F_2$ results in a formula $F_1 \subseteq F_2$ subsuming it. Moreover, for most proof systems, if $F_1 \subseteq F_2$ and we have a proof of $F_2$, we can easily construct a shorter proof of $F_1$. In the case of weighted proofs, we have to redefine these notions.

**Definition 4 (Subsumption).** *We say that a weighted formula $F_1$ subsumes another weighted formula $F_2$ if, either*

1. *$F_2 = \{(B_1, v_1), \ldots, (B_s, v_s)\}$ and there is a subset $\{(A_1, u_1), \ldots, (A_r, u_r)\} \subseteq F_1$ such that $\sum_{i=1}^{r} u_i \geq \sum_{j=1}^{s} v_j$ and, for all $i = 1, \ldots, r$ and $j = 1, \ldots, s$, $A_i \subseteq B_j$, or*
2. *We can decompose $F_1 \approx F_1' \cup F_1''$ and $F_2 \approx F_2' \cup F_2''$ such that $F_1'$ subsumes $F_2'$ and $F_1''$ subsumes $F_2''$.*

*We say that a set of inference rules $\mathcal{R}$ is* closed under subsumption *if whenever $F_2 \vdash_{\mathcal{R}} F_2'$ in one step and $F_1$ subsumes $F_2$, there exists a formula $F_1'$ such that $F_1 \vdash_{\mathcal{R}} F_1'$ in linear[2] number of steps and $F_1'$ subsumes $F_2'$.*

The definition of a proof system being closed under subsumption is a generalization of the definition of being closed under restrictions. If $F_1$ subsumes $F_2$, it is not necessarily true that $F_2$ under a restriction is equal to $F_1$. For instance, if $F_1 = \{a, \neg x \vee b\}$ and $F_2 = \{x \vee a, \neg x \vee b\}$, $F_1$ subsumes $F_2$, but $F_1$ is not the result of applying a restriction to $F_2$.

Notice that MaxSAT RESOLUTION is not closed under subsumption. For example, from $F_2 = \{(x \vee a, 1), (\neg x \vee b, 1)\}$ we derive $F_2' = \{(a \vee b, 1), (x \vee a \vee \neg b, 1), (\neg x \vee b \vee a, 1)\}$. However, from $F_1 = \{(a, 1), (\neg x \vee b, 1)\}$ that subsumes $F_2$ we cannot derive any formula subsuming $F_2'$. If in addition we also use SPLIT, from $F_1$, we can derive $F_1' = \{(a \vee b, 1), (a \vee \neg b, 1), (\neg x \vee b, 1)\}$ that subsumes $F_2'$.

**Lemma 1.** *Weighted proofs using $\mathcal{R} = \{\text{MaxSAT Resolution}, \text{Split}\}$, $\mathcal{R} = \{\text{Cut}\}$ or $\mathcal{R} = \{\text{Split}\}$ are all closed under subsumption.*

*The union of rule sets closed under subsumption is closed under subsumption.*

---

[2] Linear in the number of variables of $F_1$.

## 2.3   Weighted Dual-Rail Proofs

Third, we introduce the notion of *Weighted Dual-Rail Proofs* introduced by Bonet et al. [3] based on the notion of *Dual-Rail Proofs* introduced in [9]. Weighted Dual-Rail MaxSAT proofs may be seen as a special case of weighted proofs where all clause weights along the proof are positive.

The dual-rail encoding of the clauses $\mathcal{H}$ is defined as follows: Given a clause $A$ over the variables $\{x_1, \ldots, x_n\}$, $A^{dr}$ is the clause over the variables $\{p_1, \ldots, p_n, n_1, \ldots, n_n\}$ that results from replacing in $A$ the occurrences of $x_i$ by $\neg n_i$, and occurrences of $\neg x_i$ by $\neg p_i$. The semantics of $p_i$ is "variable $x_i$ is positive" and the semantics of $n_i$ is "the variable $x_i$ is negative".

**Definition 5 (Weighted Dual-Rail Proof).** *Fixed a set of hypotheses $\mathcal{H}$, a weighted dual-rail proof of $\mathcal{H} \vdash \square$ is a sequence $\mathcal{F}_1 \vdash \cdots \vdash \mathcal{F}_m$ of positively weighted formulas over the set of variables $\{p_1, \ldots, p_n, n_1, \ldots, n_n\}$, such that:*

1. *$\mathcal{F}_1 = \{(A^{dr}, u_A) \mid A \in \mathcal{H}\} \cup \{(\neg p_i \vee \neg n_i, u_i), (p_i, v_i), (n_i, v_i) \mid i = 1, \ldots, n\}$, for some arbitrary positive weights $u_A$, $u_i$ and $v_i$.*
2. *$(\square, 1 + \sum_{i=1}^{n} v_i) \in \mathcal{F}_m$.*
3. *For every step $\mathcal{F}_i \vdash \mathcal{F}_{i+1}$, we apply the MaxSAT Resolution rule (regular step) or the fold-unfold step like in weighted proofs.*

(Unweighted) Dual-Rail MaxSAT is the special case were weights $v_i$'s are equal to one. In the original definition [9], weights $u_A$'s and $u_i$ are equal to infinite. The use of infinite weights and negative weights together introduce some complications. Here, we prefer to use arbitrarily large, but finite weights, which result in an equivalent proof system.

Notice that, contrarily to generic weighted proofs, here weights are all positive. Notice also that, since weights $u_A$'s and $u_i$ are unrestricted, from clauses $\{(\neg p_i \vee \neg n_i, u_i), (p_i, v_i), (n_i, v_i)\}$ we can derive $\sum_{i=1}^{n} v_i$ copies of the empty clause plus $(p_i \vee n_i, v_i)$ for every $i$ using the MaxSAT Resolution rule. We have to derive at least one more empty clause to prove unsatisfiability.

## 3   Basic Facts

In this section we prove that MaxSAT Resolution with Extension [13] may be formulated in different but equivalent ways.

First, notice that the EXTENSION rule, defined in [13]:

$$\frac{}{(A, -u) \qquad (x \vee A, u) \qquad (\neg x \vee A, u)} \text{ EXTENSION}$$

does not fit to the weighted proof scheme (not all consequent formulas have the same weight in the inference rule). However, it is easy to prove that, in the construction of weighted proofs, this rule is equivalent to the SPLIT rule:

**Lemma 2.** *The SPLIT and EXTENSION rules are equivalent in weighted proof systems.*

*Proof.* We can simulate a step of SPLIT as:

$$\frac{\dfrac{(A, u)}{(A, u)}\text{EXTENSION}}{\dfrac{(A, -u)\ (x \vee A, u)\ (\neg x \vee A, u)}{(x \vee A, u)\ (\neg x \vee A, u)}\text{FOLD}}$$

Conversely, we can simulate a step of EXTENSION as:

$$\frac{\dfrac{}{(A, u)\ (A, -u)}\text{UNFOLD}}{(A, -u)\ (x \vee A, u)\ (\neg x \vee A, u)}\text{SPLIT}$$

∎

Second, we will prove that MaxSAT Resolution with Extension may be formulated using the SYMMETRIC CUT rule instead of the MAXSAT RESOLUTION rule. As we have already said, the SYMMETRIC CUT rule is a special case of the MAXSAT RESOLUTION rule, where $A = B$; i.e. all the clauses of the form $x \vee A \vee \neg B$ disappear (see comments after Definition 1). Interestingly, this limited form of MaxSAT Resolution is polynomially equivalent to the normal MaxSAT Resolution in the presence of SPLIT (or equivalently EXTENSION).

**Lemma 3.** *Weighted proofs using* $\mathcal{R} = \{\text{MAXSAT RESOLUTION, SPLIT}\}$ *are polynomially equivalent to weighted proofs using* $\mathcal{R} = \{\text{SYMMETRIC CUT, SPLIT}\}$.

*Proof.* SYMMETRIC CUT is a particular case of MAXSAT RESOLUTION, where $A = B$. Therefore, the equivalence is trivial in one direction.

In the opposite direction, we have to see how to simulate one step of MAXSAT RESOLUTION with a linear number of SYMMETRIC CUT and SPLIT steps. Let $A = a_1 \vee \cdots \vee a_r$ and $B = b_1 \vee \cdots \vee b_s$.

$$\frac{(x \vee A, u)\ (\neg x \vee B, u)}{(x \vee A \vee \neg b_1, u)\ (x \vee A \vee b_1, u)}\text{SPLIT}$$

$$\frac{(\neg x \vee B, u)}{(x \vee A \vee \neg b_1, u)\ (x \vee A \vee b_1 \vee \neg b_2, u)\ (x \vee A \vee b_1 \vee b_2, u)}\text{SPLIT}$$

$$\frac{(\neg x \vee B, u)}{(x \vee A \vee B, u)}(s - 2) \times \text{SPLIT}$$

$$\frac{(x \vee A \vee \neg b_1, u) \cdots (x \vee A \vee b_1 \vee \cdots \vee b_{s-1} \vee \neg b_s, u)}{(\neg x \vee B, u)} r \times \text{SPLIT}$$

$$\frac{(\neg x \vee B, u)}{(x \vee A \vee B, u)\ (\neg x \vee A \vee B, u)}$$

$$\frac{(x \vee A \vee \overline{B}, u)\ (\neg x \vee B \vee \overline{A}, u)}{(A \vee B, u)}\text{SYMMETRIC CUT}$$

$$(x \vee A \vee \overline{B}, u)\ (\neg x \vee B \vee \overline{A}, u)$$

In blue we mark the clauses that are added by the last inference step.    ∎

Notice that in the previous proof, the equivalence doesn't follow for the subsystem where the number of literals on the formula performing the SPLIT is bounded. So the previous argument does not show the equivalence between MaxSAT Resolution plus K-Split and Symmetric Cut plus K-Split.

Lemmas 2 and 3 allow us to conclude:

**Corollary 1.** *MaxSAT Resolution with Extension [13] is equivalent to weighted proofs using rules $\mathcal{R} = \{$Symmetric Cut, Split$\}$.*

The set $\mathcal{R}$ in Corollary 1 are precisely the rules used to define Circular Resolution [1] (except for the use of the Axiom rule that is added for a minor technical reason). This simplifies the proof of equivalence of both proof systems.

## 4    Equivalence of Circular Resolution and MaxSAT Resolution with Extension

In this Section we will prove a more general result: the equivalence between a proof system based on circular proofs with a set of inference rules $\mathcal{R}$ and a proof system based on weighted proofs using the same set of inference rules $\mathcal{R}$.

First we prove the more difficult direction, how we can simulate a circular proof with a weighted proof.

**Lemma 4.** *Weighted proofs polynomially simulate Circular proofs using the same set of inference rules.*

*Proof.* Assume we have a circular proof $(I, J, E)$ with formula nodes $J$, inference nodes $I$, edges $E$, hypotheses $\mathcal{H} \subset J$ and goal $C \in J$. Without loss of generality, we assume that the hypotheses formulas do not have incoming edges: for any $A \in \mathcal{H}$, we have $N^{in}(A) = \emptyset$. Notice that removing these incoming edges in a circular proof only decreases the balance of hypotheses formulas (that are already allowed to have negative balance) and increases the balance of the origin of these edges.

Now, assign a total ordering $\mu : I \cup J \rightarrow \{1, \ldots, |I| + |J|\}$ to each node with the following restrictions: 1) hypotheses nodes $\mathcal{H}$ go before any other node, and 2) for every inference node $R$, the formulas it generates are placed after $R$ in the ordering $\mu$. So, for any $R \in I$ and $A \in N^{out}(R)$ we have $\mu(R) < \mu(A)$. Notice that, if hypotheses nodes do not have incoming edges, there always exists such an ordering.

We construct the weighted proof $\mathcal{F}_0 \vdash \mathcal{F}_{|\mathcal{H}|} \vdash \cdots \vdash \mathcal{F}_{|I|+|J|}$ defined by:

$$
\begin{aligned}
\mathcal{F}_0 = \ & \{(A, -Bal(A)) \mid A \in \mathcal{H}\} \\
\mathcal{F}_m = \ & \{(A, Flow(R)) \mid (A \rightarrow R) \in E \wedge \mu(A) \leq m < \mu(R)\} \cup \\
& \{(A, -Flow(R)) \mid (A \rightarrow R) \in E \wedge \mu(R) \leq m < \mu(A)\} \cup \\
& \{(A, Flow(R)) \mid (R \rightarrow A) \in E \wedge \mu(R) \leq m < \mu(A)\} \cup \\
& \{(A, Bal(A)) \mid A \in J \setminus \mathcal{H} \wedge \mu(A) \leq m\}
\end{aligned}
$$

for any $m \in \{|\mathcal{H}|, \ldots, |I| + |J|\}$.

Notice that $\mathcal{F}_m$ only depends on the edges that connect a node smaller or equal to $m$ with a node bigger than $m$. Notice also that by definition of $\mu$ we never have the situation $(R \rightarrow A) \in E \wedge \mu(A) < \mu(R)$.

Now, we will prove that this is really a weighted proof.

Since all clauses from $\mathcal{H}$ only have outgoing edges, their balance is negative, and the weights in $\mathcal{F}_0$ are positive. Since, according to $\mu$, the smallest nodes are the hypotheses, and they do not have incoming edges, we have $\mathcal{F}_{|\mathcal{H}|} = \{(A, Flow(R)) \mid (A \in \mathcal{H} \wedge (A \to R) \in E\}$. Moreover, as $Bal(A) = -\sum_{R \in N^{out}(A)} Flow(R) \leq 0$, we can obtain $\mathcal{F}_{|\mathcal{H}|}$ from $\mathcal{F}_0$ by fold-unfold step.

For the rest of steps $\mathcal{F}_i \vdash \mathcal{F}_{i+1}$ with $i \geq |\mathcal{H}|$, we distinguish two cases according to the kind of node at position $i + 1$:

1. For formula nodes $A \in J$, with $\mu(A) = i + 1$.
   By definition of $\mu$, for any $R \in N^{in}(A)$, we have $\mu(R) < \mu(A)$. For the outgoing nodes, we can decompose them into $N^{out}(A) = I_1 \cup I_2$, where, for any $R \in I_1$, $\mu(R) > \mu(A)$, and for any $R \in I_2$, $\mu(R) < \mu(A)$. In $\mathcal{F}_i$, we have $(A, Flow(R))$, where $R \in N^{in}(A)$, and, for every $R \in I_2$, we also have $(A, -Flow(R))$. Applying the FOLD and UNFOLD rules, we derive the set of clauses $\{(A, Flow(R)) \mid R \in I_1\}$ plus $(A, m)$, where $m = \sum_{R \in N^{in}(A)} Flow(R) - \sum_{R \in N^{out}(A)} Flow(R) \geq 0$ is the balance of $A$.

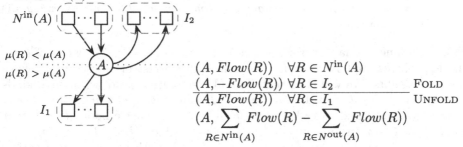

2. Inference nodes $R \in I$, with $\mu(R) = i + 1$.
   In this case, for all consequent $A \in N^{out}(R)$ of $R$, we have $\mu(A) > \mu(R)$. However, the antecedents can be decomposed into two subsets $J_1 = \{A \in N^{in}(R) \mid \mu(A) < \mu(R)\}$ and $J_2 = \{A \in N^{in}(R) \mid \mu(A) > \mu(R)\}$. In $\mathcal{F}_i$, we only have the clauses of $J_1$. In order to apply the same rule $R$ with weights, we have to also introduce the clauses of $J_2$. This can be done applying the UNFOLD rule that, from the empty set of antecedents, deduces $(A, Flow(R))$ and $(A, -Flow(R))$, for any $A \in J_2$. After that, we have no problem to apply the same rule $R$ with weight $Flow(R)$, obtaining $\mathcal{F}_{i+1}$.

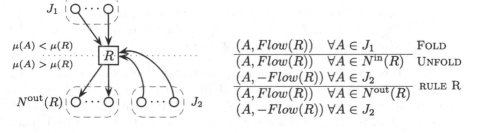

*Example 1.* Consider the third circular proof of Fig. 1. We construct the corresponding weighted proof, as described in the proof of Lemma 4, where nodes are ordered from top to bottom according to $\mu$, and all inference nodes have the same flow:

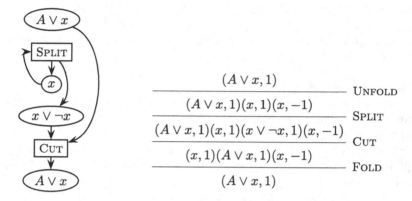

$$\frac{\displaystyle\frac{(A \vee x, 1)}{(A \vee x, 1)(x, 1)(x, -1)}}{\displaystyle\frac{(A \vee x, 1)(x, 1)(x \vee \neg x, 1)(x, -1)}{\displaystyle\frac{(x, 1)(A \vee x, 1)(x, -1)}{(A \vee x, 1)}}} \begin{array}{l} \text{UNFOLD} \\ \\ \text{SPLIT} \\ \\ \text{CUT} \\ \\ \text{FOLD} \end{array}$$

**Lemma 5.** *Circular proofs polynomially simulate weighted proofs using the same set of inference rules.*

*Proof.* Assume we have a weighted proof $\mathcal{F}_1 \vdash \mathcal{F}_2 \vdash \cdots \vdash \mathcal{F}_n$, where $\mathcal{F}_1$ are the hypotheses, and $\mathcal{F}_n$ contains the goal and the rest of clauses in $\mathcal{F}_n$ have positive weights. We will construct a circular resolution proof with three kinds of formula nodes: $J_1$ called *axiom nodes*, $J_2$ called *auxiliary nodes* (used only in the base case) and $J_3$ called *normal nodes* and inference nodes $I$ such that there exist a flow assignment $Flow : I \to \mathbb{N}$ and balance $Bal : J \to \mathbb{N}$ satisfying 1) the set of axiom nodes $A \in J_1$ is the set of hypotheses in $\mathcal{F}_1$ and satisfy $Bal(A) = -\sum_{(A,c)\in\mathcal{F}_1} c$ 2) the auxiliary nodes all have positive balance and 3) for every clause $(A, u)$ in $\mathcal{F}_n$, there exists a unique normal node $A \in J_3$ that satisfies $Bal(A) = \sum_{(A,c)\in\mathcal{F}_n} c$. The construction is by induction on $n$.

If $n = 1$, for any hypothesis $A$ of the weighted proof, let $u_A = \sum_{(A,u)\in\mathcal{F}_1} u$. We construct the constant-size circular proof that proves $A$ from $A$ with an axiom node $A$ with balance $-u_A$, a normal node $A$ with balance $u_A$, and the necessary auxiliary nodes. (Recall that we assume that the set of inference rules $\mathcal{R}$ allow us to infer $A$ with balance $u_A$, from $A$ with balance $-u_A$, for any clause $A$, using a constant-size circular proof).

Assume now, by induction hypothesis, that we have constructed a circular resolution corresponding to $\mathcal{F}_1 \vdash \cdots \vdash \mathcal{F}_i$. Depending on the MaxSAT resolution rule used in the step $\mathcal{F}_i \vdash \mathcal{F}_{i+1}$, we have two cases:

1. If the last MaxSAT inference is a FOLD or UNFOLD, the same circular resolution proof constructed for $\mathcal{F}_1 \vdash \cdots \vdash \mathcal{F}_i$ works for $\mathcal{F}_1 \vdash \cdots \vdash \mathcal{F}_{i+1}$. Only in the special case of unfolding $\emptyset \vdash (A, u), (A, -u)$, if there is no formula node corresponding to $A$, we add a lonely normal node $A$ (that will have balance zero) to ensure property 3).

2. If it corresponds to any other rule $R : A_1, \ldots, A_r \vdash B_1, \ldots, B_s$, we have
$\mathcal{F}_i = \mathcal{F}_{i-1} \setminus \{(A_1, u), \ldots, (A_r, u)\} \cup \{(B_1, u), \ldots, (B_s, u)\}$, for some weight $u$.
We add, to the already constructed circular resolution proof, a new inference
node $R$ with flow $Flow(R) = u$. We add edges from the formula nodes corresponding to $A_i$'s to the node $R$. If they do not exist, we add new normal
nodes $B_j$'s. Finally, we add an edge from $R$ to every $B_j$. The addition of these
nodes has the effect of reducing the balance of $A_i$'s by $u$, and creating nodes
$B_j$ with balance $u$, if they did not exist, or increasing the balance of $B_j$ in
$u$, if it existed. By induction hypothesis, this makes property 3) hold for the
new circular resolution proof and $\mathcal{F}_i$.

Notice that all clauses in $\mathcal{F}_1$ have strictly positive weight. Therefore, all axiom
formula nodes in $J_1$ have negative balance and all nodes in $J_2$ positive balance.
However, since clauses in $\mathcal{F}_i$, for $i \neq 1, n$ may have negative weight, balance of
normal nodes in $J_3$ can also be negative during the construction of the circular
resolution proof. Since all clauses in $\mathcal{F}_n$ have positive weight, at the end of the
construction process, all normal nodes will have positive balance. Therefore, at
the end of the process, the set of hypotheses $\mathcal{H}$ will be $J_1$.                        ∎

**Corollary 2.** *MaxSAT Resolution with Extension and Circular Resolution are
polynomially equivalent proof systems.*

*Proof.* From Lemmas 2, 3, 4 and 5.                                             ∎

## 5    Systems that Simulate Dual-Rail MaxSAT

In this Section, we analyze proof systems weaker than Circular Resolution and
MaxSAT Resolution with Extension. We replace the SPLIT rule by the 0-SPLIT
rule. Relaxing this rule forces us to use the non-symmetric version of the cut
rules, as the following example suggests.

*Example 2.* Weighted proofs and circular proofs using {SYMMETRIC CUT,
0-SPLIT} are not able to prove the unsatisfiability of the following formulas

$$F_1 = \{\neg x \vee y, \ \neg y \vee z, \ \neg z \vee \neg x, \ x \vee v, \ \neg v \vee w, \ \neg w \vee x\}$$

$$F_2 = \{\neg x, \ x \vee y, \ \neg y\}$$

The previous example suggests that we cannot base a complete proof system
in the SYMMETRIC CUT rule, when we restrict the power of the SPLIT rule. The
natural question is how to compare the power of the CUT and the MAXSAT
RESOLUTION rule, when we are in the context of weighted proofs, and the rules
replace the premises by the conclusions.

*Example 3.* Consider the formula $F_1 = \{x \vee y, \ x \vee \neg y, \ \neg x\}$.

Assigning weight one to all the hypotheses clauses we can deduce the empty clause with weight one using MAXSAT RESOLUTION:

$$(x \vee y, 1), \ (\neg x, 1), \ (x \vee \neg y, 1) \vdash (y, 1), \ (\neg x \vee \neg y, 1), \ (x \vee \neg y, 1)$$
$$\vdash (y, 1), \ (\neg y, 1) \vdash (\square, 1)$$

In the first step of this proof, since we are working with weighted proofs, after using $x \vee y$ and $\neg x$, these clauses disappear, and instead we obtain $y$ and $\neg x \vee \neg y$.

To simulate such a step with the CUT in the replacement form, we also use $x \vee y$ and $\neg x$, but only obtain $y$. In the following steps, we don't have $\neg x \vee \neg y$, but we can use $\neg x$, that subsumes it. However, we need to use clause $\neg x$ twice (or $\neg x$ with weight 2), one for the application of the first cut rule, and the other one to do the job of $\neg x \vee \neg y$. Repeating the same idea for the rest of the steps, we obtain the following proof with the CUT in the context of replacement rule with weights:

$$(x \vee y, 1), \ (\neg x, 2), \ (x \vee \neg y, 1) \vdash (y, 1), \ (\neg x, 1), \ (x \vee \neg y, 1)$$
$$\vdash (y, 1), \ (\neg y, 1) \vdash (\square, 1)$$

In this example, a deeper reorganization of the proof (cutting first $y$ and then $x$) allows us to derive the empty clause with weights one for all the premises:

$$(x \vee y, 1), \ (\neg x, 1), \ (x \vee \neg y, 1) \vdash (x, 1), \ (\neg x, 1) \vdash (\square, 1)$$

However, this is not always possible. For some unsatisfiable formulas, if we assign weight one to all the premises, we cannot obtain the empty clause using the CUT rule replacing premises by conclusions. This fact is deeply related to the incompleteness of the *Read Once Resolution* [10]. For instance, consider the unsatisfiable formula $\{x_1 \vee x_2, \ x_3 \vee x_4, \neg x_1 \vee \neg x_3, \ \neg x_1 \vee \neg x_4, \ \neg x_2 \vee \neg x_3, \ \neg x_2 \vee \neg x_4\}$ from [11]. In the context of weighted proofs, using the replacement CUT rule, we need to start with clauses with weight bigger than one in order to prove unsatisfiability. On the other hand, using MAXSAT RESOLUTION all hypotheses may have weight one, since Bonet et al. [4,6] prove that, for any unsatisfiable formula, we can derive the empty clause with the MAXSAT RESOLUTION rule and weight one for all the premises.

The previous example suggests us how we can simulate a weighted proof using MAXSAT RESOLUTION with a weighted proof using CUT, at the cost of increasing the weights of the initial clauses.

**Lemma 6.** *Let $\mathcal{R}$ be a set of inference rules closed under subsumption.*
*Weighted proofs using $\{\text{CUT}\} \cup \mathcal{R}$ are polynomially equivalent to weighted proofs using $\{\text{MAXSAT RESOLUTION}\} \cup \mathcal{R}$.*

*Proof.* In one direction the proof is trivial, since MAXSAT RESOLUTION has the same consequent as the CUT rule plus some additional clauses.

In the other direction, let $n$ be the number of variables of the formula. Let $\mathcal{F}_1 \vdash \cdots \vdash \mathcal{F}_m$ be a weighted proof using $\{\text{MAXSAT RESOLUTION}\} \cup \mathcal{R}$. We can find an equivalent proof $\mathcal{F}'_1 \vdash \cdots \vdash \mathcal{F}'_{m'}$ using $\{\text{CUT}\} \cup \mathcal{R}$, where $m' = m\,\mathcal{O}(n)$, such that 1) $\mathcal{F}'_1 = \{(A, v) \mid (A, u) \in \mathcal{F}_1 \wedge v \leq k^m\,u\}$, where $k = \mathcal{O}(n)$ and 2) $\mathcal{F}'_{m'}$ subsumes $\mathcal{F}_m$.

For the base case $m = 1$, it is trivial.

For the induction case $m > 1$, let $\mathcal{F}_1 \vdash \ldots F_{m+1}$ be the proof with MAXSAT RESOLUTION and let $\mathcal{F}'_1 \vdash \cdots \vdash \mathcal{F}'_{m'}$ be the proof with CUT given by the induction hypothesis for the first $m$ steps. There are three cases:

1. If the last inference step $\mathcal{F}_m \vdash \mathcal{F}_{m+1}$ is a FOLD, UNFOLD, then the same proof already works since $\mathcal{F}'_m$ also subsumes $\mathcal{F}_{m+1}$.
2. If this last step applies any rule from $\mathcal{R}$, by closure under subsumption of $\mathcal{R}$, applying a linear number of rules of $\mathcal{R}$ we can construct $\mathcal{F}'_{m'} \vdash \cdots \vdash \mathcal{F}_{m'+\mathcal{O}(n)}$, where $\mathcal{F}'_{m'+\mathcal{O}(n)}$ subsumes $\mathcal{F}_{m+1}$.
3. If the last inference step is an application of the MAXSAT RESOLUTION rule, let

$$\mathcal{F}_m = F \cup \{(x \vee A, u), (\neg x \vee B, u)\} \text{ and}$$
$$\mathcal{F}_{m+1} = F \cup \{(A \vee B, u), (x \vee A \vee \overline{B}, u), (\neg x \vee B \vee \overline{A}, u)\}.$$

Let $r = 1 + \max\{|A|, |B|\} = \mathcal{O}(n)$ and let $\mathcal{F}''_1 \vdash \cdots \vdash \mathcal{F}''_{m'}$ the same proof as $\mathcal{F}'_1 \vdash \cdots \vdash \mathcal{F}'_{m'}$, where every weight has been multiplied by $r$. By induction hypothesis, $\mathcal{F}'_{m'}$ subsumes $\mathcal{F}_m$. Hence, $\mathcal{F}''_{m'}$ contains a clause $(A', r\,u_1)$ corresponding to $(x \vee A, u)$, where $A' \subseteq x \vee A$ and $u' \geq u$, and a clause $(B', r\,u_2)$ corresponding to $(\neg x \vee B, u)$, where $B' \subseteq \neg x \vee B$ and $u_2 \geq u$. If $x \notin A'$ or $\neg x \notin B'$, applying the UNFOLD rule to $\mathcal{F}''_n$ we can split these clauses into clauses subsuming $\{(A \vee B, u), (x \vee A \vee \overline{B}, u), (\neg x \vee B \vee \overline{A}, u)\}$ with higher weights. Otherwise, we apply the UNFOLD rule to obtain $r$ copies of $(A', u)$ and $r$ copies of $(B', u)$, plus some useless clauses. The application of the CUT rule to one copy of $(A', u)$ and one of $(B', u)$ results in a clause that subsumes $(A \cup B, u)$. There are also at least $|B|$ more copies of $(A', u)$ that will subsume the clauses $(x \vee A \vee \overline{B}, u)$, and at least $|A|$ more copies of $(B', u)$ that will subsume the clauses $(\neg x \vee B \vee \overline{A}, u)$.

Notice that the length of the proof is multiplied by $\mathcal{O}(n)$. The weights are multiplied by $\mathcal{O}^m(n)$, hence its logarithmic representation is multiplied by $m\,\mathcal{O}(\log n)$. ∎

**Corollary 3.** *The circular proofs system using $\{\text{CUT}, \text{K-SPLIT}\}$ is polynomially equivalent to the weighted proofs system using $\{\text{MAXSAT RESOLUTION}, \text{K-SPLIT}\}$.*

*Proof.* From Lemma 6 and Lemmas 4 and 5.

In [13], it is proved that MaxSAT Resolution with Extension can simulate Dual-Rail MaxSAT. Next we prove that even using 0-SPLIT instead of SPLIT, it can simulate Weighted Dual-Rail, and as a consequence the weaker system Dual-Rail MaxSAT.

**Theorem 1.** *The weighted proof system using $\mathcal{R} = \{\text{MAXSAT RESOLUTION}, \text{0-SPLIT}\}$ polynomially simulates the Weighted Dual-Rail MaxSAT proof system.*

*Proof.* Let $\mathcal{F}_1 \vdash \cdots \vdash \mathcal{F}_m$ be a proof in weighted dual-rail MaxSAT. Let $A^{rdr}$ be the reverse of the dual-rail encoding, i.e. the substitution of variable $p_i$ by $x_i$ and of $n_i$ by $\neg x_i$. Applying this translation we get $\mathcal{F}_1^{rdr} = \{(A, u_A) \mid A \in \mathcal{H}\} \cup \{(x_i, v_i), (\neg x_i, v_i) \mid i = 1, \ldots, n\}$ and $\mathcal{F}_n^{rdr} = \{(\square, 1 + \sum_{i=1}^n v_i)\} \cup F$ where all clauses in $F$ have positive weight. Moreover, all steps in the proof satisfy $\mathcal{F}_i^{rdr} \vdash \mathcal{F}_{i+1}^{rdr}$, since all MaxSAT cuts between $p_i$ and $\neg p_i$, or between $n_i$ and $\neg n_i$ will become cuts of $x_i$ and $\neg x_i$. Notice that clauses $\neg p_i \vee \neg n_i$ are translated back as $x_i \vee \neg x_i$, hence tautologies and removed. Cuts in $\mathcal{F}_1 \vdash \cdots \vdash \mathcal{F}_m$ with $\neg p_i \vee \neg n_i$, when translated back, have not any effect (they replace $p_i$ by $\neg n_i$ or $n_i$ by $\neg p_i$, hence $x_i$ by $x_i$), thus they are removed. We can construct then the following weighted proof using $\mathcal{R}$:

$$
\begin{aligned}
&\{(A, u_A) \mid A \in \mathcal{H}\} \\
&\vdash \{(A, u_A) \mid A \in \mathcal{H}\} \cup \{(\square, -\sum_{i=1}^n v_i), (\square, v_i) \mid i = 1, \ldots, n\} \quad &&\textsc{Unfold} \\
&\vdash \{(A, u_A) \mid A \in \mathcal{H}\} \cup \{(\square, -\sum_{i=1}^n v_i), (x_i, v_i), (\neg x_i, v_i) \mid i = 1, \ldots, n\} \quad &&\text{0-}\textsc{Split} \\
&= \mathcal{F}_1^{rdr} \cup \{(\square, -\sum_{i=1}^n v_i)\} \\
&\cdots \\
&\vdash \mathcal{F}_n^{rdr} \cup \{(\square, -\sum_{i=1}^n v_i)\} \\
&= \{(\square, 1 + \sum_{i=1}^n v_i)\} \cup F \cup \{(\square, -\sum_{i=1}^n v_i)\} \\
&\vdash \{(\square, 1)\} \cup F \quad &&\textsc{Fold}
\end{aligned}
$$

that is a valid weighted proof for $\mathcal{H} \vdash \square$. ∎

**Corollary 4.** *The circular proof system using $\mathcal{R} = \{\textsc{Cut}, \text{0-}\textsc{Split}\}$ polynomially simulates the Weighted Dual-Rail MaxSAT proof system.*

*Proof.* Weighted Dual-Rail MaxSAT is polynomially simulated by weighted proofs using $\{\textsc{MaxSAT Resolution}, \text{0-}\textsc{Split}\}$, by Theorem 1. This is simulated by weighted proofs using $\{\textsc{Cut}, \text{0-}\textsc{Split}\}$, by Lemma 6. And this is simulated by circular proofs using $\{\textsc{Cut}, \text{0-}\textsc{Split}\}$, by Lemma 4.

## 6   Conclusions

We have shown how circular proofs and weighted proofs (with positive and negative weights), both parametric in the set of inference rules, are equivalent proof systems. We have also shown that if Split is one of such inference rules, then it does not matter if the other rule is Cut, MaxSAT Resolution or Symmetric Cut. In all the cases, we get polynomially equivalent proof systems. This proves the equivalence of Circular Resolution [1] and MaxSAT Resolution with extensions [13].

In these formalisms, if we restrict the Split rule to clauses of length zero (as $\square \vdash x, \neg x$), together with the Cut rule, we still get a strong enough proof system enable to simulate Weighted Dual-Rail MaxSAT [3,9] and to get polynomial proofs of the pigeonhole principle.

# References

1. Atserias, A., Lauria, M.: Circular (yet sound) proofs. In: Janota, M., Lynce, I. (eds.) SAT 2019. LNCS, vol. 11628, pp. 1–18. Springer, Cham (2019). https://doi. org/10.1007/978-3-030-24258-9_1
2. Atserias, A., Müller, M.: Automating resolution is NP-hard. In: Proceedings of the 60th IEEE Annual Symposium on Foundations of Computer Science, FOCS 2019, pp. 498–509 (2019). https://doi.org/10.1109/FOCS.2019.00038
3. Bonet, M.L., Buss, S., Ignatiev, A., Marques-Silva, J., Morgado, A.: MaxSAT resolution with the dual rail encoding. In: Proceedings of the 32nd AAAI Conference on Artificial Intelligence, AAAI 2018, pp. 6565–6572 (2018)
4. Bonet, M.L., Levy, J., Manyà, F.: A complete calculus for Max-SAT. In: Biere, A., Gomes, C.P. (eds.) SAT 2006. LNCS, vol. 4121, pp. 240–251. Springer, Heidelberg (2006). https://doi.org/10.1007/11814948_24
5. Bonet, M.L., Pitassi, T., Raz, R.: On interpolation and automatization for Frege systems. SIAM J. Comput. **29**(6), 1939–1967 (2000). https://doi.org/10.1137/ S0097539798353230
6. Bonet, M.L., Levy, J., Manyà, F.: Resolution for Max-SAT. Artif. Intell. **171**(8), 606–618 (2007). https://doi.org/10.1016/j.artint.2007.03.001
7. Dantchev, S., Martin, B., Rhodes, M.: Tight rank lower bounds for the Sherali-Adams proof system. Theor. Comput. Sci. **410**(21), 2054–2063 (2009). https://doi. org/10.1016/j.tcs.2009.01.002
8. Haken, A.: The intractability of resolution. Theor. Comput. Sci. **39**, 297–308 (1985). https://doi.org/10.1016/0304-3975(85)90144-6
9. Ignatiev, A., Morgado, A., Marques-Silva, J.: On tackling the limits of resolution in SAT solving. In: Gaspers, S., Walsh, T. (eds.) SAT 2017. LNCS, vol. 10491, pp. 164–183. Springer, Cham (2017). https://doi.org/10.1007/978-3-319-66263-3_11
10. Iwama, K., Miyano, E.: Intractability of read-once resolution. In: Proceedings of 10th Annual IEEE Conference on Structure in Complexity Theory, pp. 29–36 (1995)
11. Kleine Büning, H., Wojciechowski, P., Subramani, K.: On the computational complexity of read once resolution decidability in 2CNF formulas. In: Gopal, T.V., Jäger, G., Steila, S. (eds.) TAMC 2017. LNCS, vol. 10185, pp. 362–372. Springer, Cham (2017). https://doi.org/10.1007/978-3-319-55911-7_26
12. Larrosa, J., Heras, F.: Resolution in Max-SAT and its relation to local consistency in weighted CSPs. In: Proceedings of the 19th International Joint Conference on Artificial Intelligence, IJCAI 2005, pp. 193–198 (2005)
13. Larrosa, J., Rollón, E.: Augmenting the power of (partial) MaxSAT resolution with extension. In: Proceedings of the 34th National Conference on Artificial Intelligence, AAAI 2020 (2020)
14. Morgado, A., Ignatiev, A., Bonet, M.L., Marques-Silva, J., Buss, S.: DRMaxSAT with MaxHS: first contact. In: Janota, M., Lynce, I. (eds.) SAT 2019. LNCS, vol. 11628, pp. 239–249. Springer, Cham (2019). https://doi.org/10.1007/978-3-030-24258-9_17
15. Sherali, H.D., Adams, W.P.: A hierarchy of relaxations and convex hull characterizations for mixed-integer zero-one programming problems. Discret. Appl. Math. **52**(1), 83–106 (1994). https://doi.org/10.1016/0166-218X(92)00190-W

# Simplified and Improved Separations Between Regular and General Resolution by Lifting

Marc Vinyals[1]([⊠]), Jan Elffers[2,4], Jan Johannsen[3], and Jakob Nordström[2,4]

[1] Technion, Haifa, Israel
marcviny@cs.technion.ac.il
[2] Lund University, Lund, Sweden
jan.elffers@cs.lth.se
[3] Ludwig-Maximilians-Universität München, Munich, Germany
jan.johannsen@ifi.lmu.de
[4] University of Copenhagen, Copenhagen, Denmark
jn@di.ku.dk

**Abstract.** We give a significantly simplified proof of the exponential separation between regular and general resolution of Alekhnovich et al. (2007) as a consequence of a general theorem lifting proof depth to regular proof length in resolution. This simpler proof then allows us to strengthen the separation further, and to construct families of theoretically very easy benchmarks that are surprisingly hard for SAT solvers in practice.

## 1 Introduction

In the *resolution proof system* [17] the unsatisfiability of a formula in conjunctive normal form (CNF) is shown by iteratively deriving new disjunctive clauses until contradiction is reached (in the form of the empty clause). A resolution proof is said to be *regular* [59] if along the path of derivation steps from any input clause to contradiction every variable is eliminated, or *resolved*, at most once. This condition appears quite natural, since it essentially means that intermediate results should not be proven in a form stronger than what will later be used in the derivation, and indeed DPLL-style algorithms [26,27] can be seen to search for regular proofs. In view of this, it is natural to ask whether regularity can be assumed without loss of proof power, but this was ruled out in [40]. General resolution was shown to be superpolynomially stronger than regular resolution in [31], a separation that was improved to exponential in [2,61]. Regular resolution is in turn known to be exponentially stronger than *tree-like* resolution [11,19], where no intermediate clause can be used for further derivations more than once.

There is an interesting connection here to the quest for a better understanding of state-of-the-art SAT solvers based on *conflict-driven clause learning (CDCL)* [47,48].[1] Tree-like resolution corresponds to solvers without any clause

---

[1] A similar idea in the context of CSPs was independently developed in [5].

© Springer Nature Switzerland AG 2020
L. Pulina and M. Seidl (Eds.): SAT 2020, LNCS 12178, pp. 182–200, 2020.
https://doi.org/10.1007/978-3-030-51825-7_14

learning, whereas CDCL solvers have the potential to be as strong as general resolution [3,51]. The proofs of the latter result crucially use, among other assumptions, that solvers make frequent restarts, but it has remained open whether this is strictly needed, or whether "smarter" CDCL solvers without restarts could be equally powerful. To model CDCL without restarts, proof systems such as *pool resolution* [62] and different variants of *resolution trees with lemmas (RTL)* [20] have been introduced, which sit between regular and general resolution. Therefore, if one wants to prove that restarts increase the reasoning power of CDCL solvers, then formulas that could show this would, in particular, have to separate regular from general resolution. However, all known formulas witnessing this separation [2,61] have also been shown to have short pool resolution proofs [18,21]. It is therefore interesting to develop methods to find new formula families separating regular and general resolution. This brings us to our next topic of *lifting*.

In one sentence, a *lifting theorem* takes a weak complexity lower bound and amplifies it to a much stronger lower bound by simple syntactic manipulations. Focusing for concreteness on Boolean functions, one can take some moderately hard function $f : \{0,1\}^n \rightarrow \{0,1\}$ and compose it with a *gadget* $g : \{0,1\}^m \rightarrow \{0,1\}$ to obtain the new *lifted function* $f \circ g^n : \{0,1\}^{mn} \rightarrow \{0,1\}$ defined as $f(g(\boldsymbol{y}_1), g(\boldsymbol{y}_2), \ldots, g(\boldsymbol{y}_n))$, where $\boldsymbol{y}_j \in \{0,1\}^m$ for $j \in [n]$. If the gadget $g$ is carefully chosen, one can show that there is essentially no better way of evaluating $f \circ g^n$ than first computing $g(\boldsymbol{y}_j)$ for all $j \in [n]$ and then applying $f$ to the outputs. From this it follows that $f \circ g^n$ is a much harder function than $f$ or $g$ in isolation.

A seminal early paper implementing this paradigm is [54], and the rediscovery and strengthening of this work has led to dramatic progress on many long-standing open problems in communication complexity [33–35,37,38]. Other successful examples of the lifting paradigm include lower bounds in monotone complexity [52,53,58], extension complexity [32,43,45], and data structures [24]. Lifting has also been a very productive approach in proof complexity. Interestingly, many of the relevant papers [6,8,9,12,13,19,41,49,50] predate the "lifting revolution" and were not thought of as lifting papers at the time, but in later works such as [29,36,57] the connection is more explicit.

As described above, in the lifting construction different copies of the gadget $g$ are evaluated on disjoint sets of variables. In [55] it was instead proposed to let the variable domains for different gadgets overlap as specified by well-connected so-called *expander graphs*. This idea of recycling variables between gadgets has turned out to be very powerful, and an ingredient in a number of strong trade-off results between different complexity measures [15,16,56].

**Our Contributions.** The starting point of our work is the simple but crucial observation that the *stone formulas* in [2] can be viewed as lifted versions of *pebbling formulas* [14] with maximal overlap, namely as specified by complete bipartite graphs. This raises the question whether there is a lifting theorem waiting to be discovered here, and indeed we prove that the separation in [2] can be proven more cleanly as the statement that strong enough lower bounds on proof *depth* can be lifted to exponential lower bounds on proof *length* in regular resolution. This in turn implies that if one can find formulas that have short resolution proofs with only small clauses, but that require large depth, then lifting with overlap yields formulas that separate regular and general resolution.

This simpler, more modular proof of [2] is the main conceptual contribution of our paper, but this simplicity also opens up a path to further improvements. Originally, lifting with overlap was defined in [55] for low-degree expander graphs, and we show that our new lifting theorem can be extended to this setting also. Intuitively, this yields "sparse" versions of stone formulas that are essentially as hard as the original ones but much smaller. We use this finding for two purposes.

Firstly, we slightly improve the separation between regular and general resolution. It was known that there are formulas having general resolution proofs of length $L$ that require regular proofs of length $\exp\big(\Omega\big(L/((\log L)^7 \log \log L)\big)\big)$ [61]. We improve the lower bound to $\exp\big(\Omega\big(L/((\log L)^3 (\log \log L)^5)\big)\big)$.

Secondly, and perhaps more interestingly from an applied perspective, sparse stone formulas provide the first benchmarks separating regular and general resolution that are sufficiently small to allow meaningful experiments with CDCL solvers. Original stone formulas have the problem that they grow very big very fast. The so-called *guarded* formulas in [2,61] do not suffer from this problem, but the guarding literals ensuring the hardness in regular resolution are immediately removed during standard preprocessing, making these formulas very easy in practice. In contrast, sparse stone formulas exhibit quite interesting phenomena. Depending on the exact parameter settings they are either very dependent on frequent restarts, or very hard even with frequent restarts. This is so even though short proofs without restarts exist, which also seem to be possible to find algorithmically if the decision heuristic of the solver is carefully hand-coded.

**Outline of This Paper.** After reviewing some preliminaries in Sect. 2, we present our proof of [2] as a lifting result in Sect. 3. We extend the lower bound to sparse stone formulas in Sect. 4. We conclude with brief discussions of some experimental results in Sect. 5 and directions for future research in Sect. 6.

## 2    Preliminaries

**Resolution.** Throughout this paper 0 denotes false and 1 denotes true. A literal $a$ is either a variable $x$ or its negation $\overline{x}$. A clause $C$ is a disjunction $a_1 \vee \cdots \vee a_k$ of literals; the *width* of $C$ is $k$. A CNF formula $F = C_1 \wedge \cdots \wedge C_m$ is a conjunction of clauses, the *size* (or *length*) of which is $m$. We view clauses and CNF formulas as sets, so order is irrelevant and there are no repetitions.

A *resolution proof* for (the unsatisfiability of) $F$, also referred to as a *resolution refutation* of $F$, is a sequence of clauses, ending with the empty clause $\bot$ containing no literals, such that each clause either belongs to $F$ or is obtained by applying the resolution rule $C \vee x, \, D \vee \overline{x} \vdash C \vee D$ to two previous clauses. If we lay out the proof as a graph the result is a directed acyclic graph (DAG) where each node is labelled with a clause, where without loss of generality there is a single source labelled $\bot$, where each sink is a clause in $F$, and each intermediate node can be written on the form $C \vee D$ with edges to the children $C \vee x$ and

$D \vee \bar{x}$. The length of a refutation is the number of clauses, its width is the maximal width of a clause in it, and its depth is the longest path in the refutation DAG. The resolution length, width and depth of a formula are the minimum over all resolution refutations of it.

A restriction $\rho$ is a partial assignment of truth values to variables. We write $\rho(x) = *$ to denote that variable $x$ is unassigned. We obtain the restricted clause $C\!\restriction_\rho$ from $C$ by removing literals falsified by $\rho$, and the restricted formula $F\!\restriction_\rho$ from $F$ by removing clauses satisfied by $\rho$ and replacing other clauses $C$ by $C\!\restriction_\rho$.

If a formula $F$ has a resolution refutation $\pi$, then for every restriction $\rho$ the restricted formula $F\!\restriction_\rho$ has a refutation $\pi'$—denoted by $\pi\!\restriction_\rho$—the length, width and depth of which are bounded by the length, width and depth of $\pi$, respectively. If $\pi$ is regular, then so is $\pi\!\restriction_\rho$. We will need the following straightforward property of resolution depth.

**Lemma 1 ([60]).** *If $F$ requires resolution depth $D$, then for every variable $x$ in $F$ it holds for some $b \in \{0,1\}$ that $F\!\restriction_{\{x:=b\}}$ requires resolution depth $D - 1$.*

**Branching Programs.** In the *falsified clause search problem* for an unsatisfiable CNF formula $F$, the input is some (total) assignment $\alpha$ and a valid output is any clause of $F$ that $\alpha$ falsifies.

From a resolution refutation of $F$ we can build a *branching program* for the falsified clause search problem with the same underlying graph, where every non-source node queries a variable $x$ and has outgoing edges 0 and 1, and where any assignment $\alpha$ leads to a sink labelled by a clause that is a valid solution to the search problem for $F$. We maintain the invariant that an assignment $\alpha$ can reach a node labelled by $C$ if and only if $\alpha$ falsifies $C$—in what follows, we will be slightly sloppy and identify a node and the clause labelling it. In order to maintain the invariant, if a node $C \vee D$ has children $C \vee x$ and $D \vee \bar{x}$, we query variable $x$ at that node, move to the child with the new literal falsified by the assignment to $x$, and forget the value of any variable not in this child. A proof is regular if and only if it yields a *read-once* branching program, where any variable is queried at most once along any path, and it is tree-like if it yields a search tree.

**Pebbling Formulas.** Given a DAG $H$ of indegree 2 with a single sink, the pebbling formula over $H$ [14], denoted $Peb_H$, has one variable per vertex, a clause $u$ for each source $u$, a clause $\bar{u} \vee \bar{v} \vee w$ for each non-source $w$ with predecessors $u$ and $v$, and a clause $\bar{z}$ for the sink $z$.

Pebbling formulas over $n$-vertex DAGs $H$ have short, small-width refutations, of length $O(n)$ and width 3, but may require large depth. More precisely, the required depth coincides with the so-called *reversible pebbling number* of $H$ [22], and there exist graphs with pebbling number $\Theta(n/\log n)$ [30]. We will also need that so-called *pyramid graphs* have pebbling number $\Theta(\sqrt{n})$ [23,25].

**Lifting.** We proceed to define *lifting with overlap* inspired by [55]. Let $F$ be a formula with $n$ original variables $x_i$. We have $m$ new main variables $r_j$, which we often refer to as stone variables. Let $G$ be a bipartite graph of left degree $d$ and right degree $d'$ with original variables on the left side and main variables on the right side. We have $dn$ new selector variables $s_{i,j}$, one for each edge $(i, j)$ in $G$.

For convenience, let us write $y^1 = y$ and $y^0 = \bar{y}$ for the positive and negative literals over a variable $y$. Then the lifting of $x_i^b$ for $b \in \{0, 1\}$ is the conjunction of $d$ clauses $\mathcal{L}^G(x_i^b) = \bigwedge_{j \in N(i)} \bar{s}_{i,j} \vee r_j^b$. The lifting of a clause $C$ of width $w$ is the expression $\mathcal{L}^G(C) = \bigvee_{x_i^b \in C} \mathcal{L}^G(x_i^b)$, expanded into a CNF formula of width $2w$ and size $d^w$. The lifting of a CNF formula $F$ is the formula $\mathcal{L}^G(F) = \bigwedge_{C \in F} \mathcal{L}^G(C) \wedge \bigwedge_{i \in [n]} \bigvee_{j \in N(i)} s_{i,j}$ of size at most $d^w|F| + n$. We will omit the graph $G$ from the notation when it is clear from context.

If $G$ is a disjoint union of stars, then we obtain the usual lifting defined in [7], and if $G$ is a complete bipartite graph with $m \geq 2n$ and $F$ is a pebbling formula, then we obtain a stone formula [2]. We will need the fact, implicit in [13], that formulas with short, small-width refutations remain easy after lifting.

**Lemma 2.** *Let $\pi$ be a resolution refutation of $F$ of length $L$ and width $w$, and let $G$ be a bipartite graph of left degree $d$. Then there is a resolution refutation of $\mathcal{L}^G(F)$ of length $\mathrm{O}(d^{w+1}L)$.*

For the particular case of pebbling formulas, where there is a refutation where each derived clause is of width at most 2 even if some axioms are of width 3, the upper bound can be improved to $\mathrm{O}(d^3 L)$.

**Graphs.** In Sect. 3, we use complete bipartite graphs to reprove the known lower bounds on stone formulas. In Sect. 4, we consider bipartite random graphs sampled from the $\mathcal{G}(n, m, d)$ distribution, where the left and right sides $U$ and $V$ have $n$ and $m$ vertices respectively, and $d$ right neighbours are chosen at random for each left vertex.

A bipartite graph is an $(r, \kappa)$-*expander* if every left subset of vertices $U' \subseteq U$ of size $|U'| \leq r$ has at least $\kappa |U'|$ neighbours. It is well-known (see for instance [39]) that random graphs are good expanders.

**Lemma 3.** *With high probability a graph $G \sim \mathcal{G}(n, m, d)$ with $d = \Theta(\log(n/m))$ is an $(r, \kappa)$-expander with $\kappa = \Theta(d)$, $r = \Theta(m/\kappa)$, and right degree $d' \leq 2dn/m$.*

The following lemma, as well as its proof, is essentially the same as Lemmas 5 and 6 in [1] but adapted to vertex expansion.

**Lemma 4.** *If $G$ is an $(r, \kappa)$-expander, then for every set $V' \subseteq V$ of size at most $\kappa r/4$ there exists a set $U' \subseteq U$ of size at most $r/2$ such that the graph $G \setminus (U' \cup N(U') \cup V')$ obtained from $G$ by removing $U'$, $N(U')$, and $V'$ is an $(r/2, \kappa/2)$-expander.*

**Matchings and the Matching Game.** A matching $\mu$ in a bipartite graph is a set of vertex-disjoint edges. We write $\mu(u) = v$ if the edge $(u, v)$ is in $\mu$. The *matching game* [10] on a bipartite graph is played between two players Prover and Disprover, with $r$ fingers each numbered $1, \ldots, r$. In each round:

- either Prover places an unused finger $i$ on a free vertex $u \in U$, in which case Disprover has to place his $i$-th finger on a vertex $v \in N(u)$ not currently occupied by other fingers;
- or Prover removes one finger $i$ from a vertex, in which case Disprover removes his $i$-th finger.

Prover wins if at some point Disprover cannot answer one of his moves, and Disprover wins if the game can continue forever.

**Theorem 5 ([10, Theorem 4.2]).** *If a graph is an $(r, 1+\delta)$-bipartite expander, then Prover needs at least $\delta r/(2 + \delta)$ fingers to win the matching game.*

## 3    Lower Bound for Stone Formulas as a Lifting Theorem

We reprove the result in [2] by reinterpreting it as a lifting theorem.

**Theorem 6.** *If $F$ has resolution depth $D$ and $m \geq 2D$, then $\mathcal{L}^G(F)$ for $G$ the complete bipartite graph $K_{n,m}$ has regular resolution length $\exp(\Omega(D^2/n))$.*

When we choose as $F$ the pebbling formula of a graph of pebbling number $\Omega(n/\log n)$ [30] we reprove the result in [2], slightly improving the lower bound from $\exp(\Omega(n/\log^3 n))$ to $\exp(\Omega(n/\log^2 n))$.

**Corollary 7.** *There are formulas that have general resolution refutations of length $O(n^4)$ but require regular resolution length $\exp(\Omega(n/\log^2 n))$.*

We start with an overview and a few definitions common to this and the next section. The proof at a high level follows a common pattern in proof complexity: given some complexity measure on clauses, we apply a restriction to the resolution refutation that removes all complex clauses from a short enough proof. In a separate argument, we show that the restricted formula always requires complex clauses, contradicting our assumption of a short refutation.

To build a restriction we use the following concepts. Let $\mu\colon I \to J$ be a partial matching from original to stone variables. A matching $\mu$ induces an assignment $\rho$ to selector variables as follows.

$$\rho(s_{i,j}) = \begin{cases} 1 & \text{if } \mu(i) = j, \\ 0 & \text{if } i \in \mathrm{dom}(\mu) \text{ or } j \in \mathrm{im}(\mu) \text{ but } \mu(i) \neq j, \\ * & \text{otherwise.} \end{cases}$$

We say that an assignment $\rho$ whose restriction to selector variables is of this form *respects the lifting* because $\mathcal{L}^G(F)\!\restriction_\rho = \mathcal{L}^{G'}(F\!\restriction_\sigma)$, where $G'$ is the induced subgraph $G[(I \setminus \mathrm{dom}\,\mu) \cup (J \setminus \mathrm{im}\,\mu)]$ and $\sigma$ is the induced assignment to original variables $\sigma(x_i) = \rho(r_{\mu(i)})$ if $i \in \mathrm{dom}(\mu)$, and $\sigma(x_i) = *$ otherwise. An assignment that respects the lifting is *uninformative* if it induces an empty assignment to original variables, that is $\rho(r_j) = *$ whenever $j \in \mathrm{im}\,\mu$. Given an uninformative assignment $\rho$ and an assignment to original variables $\sigma$, we can extend the former to agree with the latter as $\rho(r_j) = \sigma(x_{\mu^{-1}(j)})$ if $j \in \mathrm{im}\,\mu$ and $\rho(r_j) = *$ otherwise.

The size of an assignment is the maximum of the size of the matching and the number of assigned stone variables.

A helpful complexity measure is the width of a clause; we use a complexity measure from [2] that enforces an additional structure with respect to the lifting.

**Definition 8.** *A clause $C$ is $(c, z)$-complex if either*

1. *$C$ contains at least $c$ stone variables,*
2. *there is a matching $\mu$ of size $c$ such that $C$ contains the literal $\overline{s}_{i,j}$ for each $(i, j) \in \mu$, or*
3. *there is a set $W$ of size $c$ where $C$ contains at least $z$ literals $s_{i,j}$ for each $i \in W$.*

In this section we only use $(c, c)$-complex clauses, which we refer to as $c$-complex. Note that $c$ can range from 1 to $m$. We also need the following lemma, which can be established by a straightforward calculation.

**Lemma 9.** *Consider a set of $s$ clauses $\mathcal{C}$ and a set of $n$ possibly dependent literals $\mathcal{L}$ such that after setting $\ln(s) \, n/p$ literals in $\mathcal{L}$ (plus any dependencies), for each clause $C \in \mathcal{C}$ there is a subset $\mathcal{L}_C \subseteq \mathcal{L}$ of at least $p$ literals, each of which satisfies $C$. Then there is a set of $\ln(s) \, n/p$ literals that satisfies $\mathcal{C}$.*

From now on we assume that $G$ is the complete bipartite graph $K_{n,m}$. The first step is to show that we can remove all complex clauses from a short proof.

**Lemma 10.** *There exists $\epsilon > 0$ such that if $\pi$ is a resolution refutation of $\mathcal{L}(F)$ of size $s = \exp(\epsilon c^3/mn)$, then there exists an uninformative restriction $\rho$ of size $c/2$ such that $\pi{\restriction}_\rho$ has no $c$-complex clauses.*

*Proof.* We build a restriction greedily. First we choose a matching $\mu$ so that after setting the corresponding selector variables with the restriction $\rho$ induced by $\mu$ we satisfy all $c$-complex clauses of type 2 and 3 in Definition 8. There are $mn$ positive selector literals $s_{ij}$. A clause of type 2 is satisfied if we set one of $c$ variables $s_{i,j} = 0$, and that happens if we assign a literal $s_{ij'} = 1$ with $j' \neq j$, for a total of $c(m - 1) \geq c^2$ choices. A clause of type 3 is satisfied if we set one of $c^2$ literals $s_{i,j} = 1$. After picking $k$ pairs to be matched there are still at least $(c - k)(m - k - 1) \geq (c - k)^2$ literals available to satisfy clauses of type 2, and $(c - k)^2$ literals available to satisfy clauses of type 3, so we can apply Lemma 9 and obtain that setting $q \leq \ln(s) \, mn/(c^2/4)$ literals is enough to satisfy all such clauses. Note that we used that $k \leq \ln(s) \, mn/(c^2/4) \leq c/2$.

Next we extend $\rho$ to $\rho'$ by setting some stone variables that are untouched by $\mu$ so that we satisfy all clauses of type 1. There are $m - q$ such variables, hence at most $2m$ literals, and a clause is satisfied when one of $c$ variables is picked with the appropriate polarity. After picking $k$ literals there are at least $c - q - k \geq c/2 - k$ choices left for each clause, so we can apply Lemma 9 and get that setting $q' = \log s \, m/(c/8)$ variables is enough to satisfy all clauses of type 1. Note that we used that $k \leq \ln(s) \, m/(c/8) \leq c/4$, which follows from $\ln(s) \leq c^2/16m \leq c^3/16mn$.

The size of the restriction $\rho'$ is then at most $c/2$.                    $\square$

Next we show that *regular* resolution proofs always contain a complex clause.

**Lemma 11.** *If $F$ requires depth $D$, then any regular resolution refutation of $\mathcal{L}(F)$ with $m < D$ has an $m/4$-complex clause.*

*Proof.* We build a path through the read-once branching program corresponding to the proof, using a decision tree $T$ for $F$ of depth $D$ to give the answers to some queries. We also keep a matching $\mu$, with the invariant that there is an edge $(i, j)$ in the matching if and only if $s_{ij} = 1$ or there are $m/4$ stones $j' \neq j$ such that $s_{ij'} = 0$. We can do so using the following strategy as long as at most $m/4$ stones are assigned and at most $m/4$ stones are matched.

- If the adversary queries $s_{ij}$ then if neither $i$ nor $j$ are matched we answer 1 and add $(i, j)$ to the matching, if $\mu(i) = j$ we answer 1, and otherwise we answer 0. If more than $m/4$ variables $s_{ij'}$ are 0 (for $i$ fixed and $j' \in [m]$) we choose one of the $m/4$ stones $j''$ that are not assigned, nor matched, nor have $s_{ij''} = 0$ and add $(i, j'')$ to the matching.
- If the adversary queries $r_j$ and $j$ is matched to $i$, we answer $b$ so that the depth of $T$ only shrinks by 1 when original variable $x_i$ is set to $b$, as given by Lemma 1. Otherwise we answer arbitrarily.
- If the adversary forgets a variable and there is an edge in the matching that does not respect the invariant, we remove it.

Assume for the sake of contradiction that we never reach an $m/4$-complex clause. Then we can maintain the invariant until we reach a leaf of the branching program, and that leaf never falsifies a clause of the form $\bigvee_{j \in [m]} s_{i,j}$. It follows that the path ends at a clause from $\mathcal{L}(D)$, at which point the depth of $T$ reduced to 0. Observe that the depth of $T$ only decreases by 1 when a stone variable is queried and that, since the branching program is read-once, these queries must be to $D$ different stones, but only $m < D$ stones are available. $\qquad\square$

We use these lemmas to complete the plan outlined at the beginning of this section and prove our lifting theorem.

*Proof (of Theorem 6).* Assume for the sake of contradiction that $\pi$ is a refutation of $\mathcal{L}(F)$ of length less than $\exp(\delta D^2/n)$, where $\delta = \epsilon/1024$ for the $\epsilon$ of Lemma 10.

We invoke Lemma 10 with $c = D/8$ to obtain that there is an uninformative restriction $\rho$ of size $D/16$ such that $\pi\restriction_\rho$ has no $D/8$-complex clauses. By Lemma 1 we can assign values to the matched stones in a way that the induced assignment to original variables $\sigma$ yields a formula of depth $15D/16$. We additionally assign all but the first $15D/16 - 1$ stones arbitrarily and set all selector variables that point to an assigned stone to 0. Let $\rho'$ be the new restriction.

The formula $F' = \mathcal{L}(F)\restriction_{\rho'}$ is the lifted version of a formula $F\restriction_\sigma$ of depth $D' = 15D/16$ with $m' = D' - 1$ stones, hence by Lemma 11 any refutation of $F'$ has an $m'/4$-complex clause. But since $m'/4 \geq 15D/64 - 1 > D/8$, this contradicts the fact that the refutation $\pi\restriction_{\rho'}$ has no $D/8$-complex clauses. $\qquad\square$

## 4    Lower Bound for Sparsely Lifted Formulas

We now generalize the lifting to sparse graphs. The first step is again to show that we can remove all complex clauses from a short proof, but this becomes a harder task so let us begin with an informal overview. Say that we start with a lifted formula whose selector variable graph is an expander and, as in Lemma 10, we want to make a few stones be assigned and a few stones be matched. After we remove these stone vertices from the graph, it will likely stop being an expander (e.g. because we will likely remove all the neighbours of some vertex).

Fortunately by Lemma 4 given a subset $V'$ of right vertices to remove there is a subset $U'$ of left vertices such that removing $V'$, $U'$, and $N(U')$ from the graph yields an expander, but this is still not enough because removing $U'$ forces us to a matching that may interfere with our plans. Maybe there is some vertex $i \in U'$ corresponding to an original variable that we want to assign to 0 but all of its neighbours are assigned to 1, or maybe there is some original variable $i \in U'$ all of whose neighbours are already matched to other original variables.

Our solution is to add one backup vertex for each stone vertex $j$, so that we can delay the expansion restoring step. Of course we cannot decide beforehand which vertices are primary and which are backup, otherwise it might be that all complex clauses would talk only about backup vertices and our assignment would not affect them, so we have to treat primary and backup vertices equally. But still we make sure that if a vertex $j$ is assigned 1, then its backup is assigned 0 and viceversa, taking care of the first problem; and that if a stone vertex $j$ is matched to some original variable $i$ then its backup is still free and viceversa, taking care of the second problem.

To make the concept of backup vertices formal, we say that a bipartite graph $G$ of the form $U \cup (V_0 \cup V_1)$ is a mirror if the subgraphs $G_0(U \cup V_0)$ and $G_1(U \cup V_1)$ are isomorphic, which we also refer to as the two halves of $G$.

We can state our sparse lifting theorem using the concept of mirror graphs.

**Theorem 12.** *If $F$ has resolution depth $D$, and $G$ is a mirror graph with $G_0 \sim \mathcal{G}(n, D/2, d)$, where $d = \Theta(\log(n/D))$, then with high probability $\mathcal{L}^G(F)$ has regular resolution length $\exp(\Omega(D^3/d^2 n^2))$.*

As before, if we choose for $F$ the pebbling formula of a graph of pebbling number $\Theta(n/\log n)$, then we get the following improved separation of regular and general resolution.

**Corollary 13.** *There are formulas that have general resolution refutations of length $O(n \log \log^3 n)$ but require regular resolution length $\exp(\Omega(n/\log^3 n \log \log^2 n))$.*

Let us establish some notation. After fixing an isomorphism $\Psi \colon G_0 \to G_1$ we name the vertices in pairs $j0$ and $j1$ so that $j1 = \Psi(j0)$. If $jb \in V_b$ and $a \in \{0, 1\}$, we let $jb + a$ denote the vertex $j(a + b \bmod 2) \in V_{a+b \bmod 2}$. Let $m = |V_0|$ so there are $2m$ right vertices in $G$. In this section $c$-complex stands for $(c, 1)$-complex and we assume that $d = \Theta(\log(n/m))$.

**Lemma 14.** *If $G$ is a mirror $(r, \kappa)$-expander with $\kappa > 2$, where $\kappa r = \Theta(m)$, and $\pi$ is a resolution proof of $\mathcal{L}^G(F)$ of size $s = \exp(O(c^2 m/d^2 n^2))$, where $c = \Theta(m)$, then there is a restriction $\rho$ such that $\pi{\restriction}_\rho$ is a proof of $\mathcal{L}^{G'}(F')$ that has no $c$-complex clauses, where $F'$ has resolution depth at least $D - r/2 - \kappa r/8$ and $G'$ is an $(r/2, \kappa/2)$-expander.*

*Proof.* We show that such a restriction exists using a hybrid between a random and a greedy restriction. We randomly partition the stone vertices in $V_0$ into free, assigned, and matched stones, and mirror the partition in $V_1$. Of the assigned stones, a set $A_0^-$ of $\kappa r/16$ stones are set to 0, and a set $A_0^+$ of $\kappa r/16$ stones are set to 1, while the stones in the corresponding sets $A_1^- = \psi(A_0^-)$ and $A_1^+ = \psi(A_0^+)$ are set to 1 and 0 respectively. We plan to use the sets $M_0$ and $M_1 = \psi(M_0)$ of $\kappa r/8$ matched stones each to greedily build a matching. The remaining $2(m - \kappa r/4)$ stone vertices are tentatively left untouched.

First we claim that, with high probability, all clauses of type 1 are satisfied. To show this we note that a clause $C$ of type 1 contains at least $c/4$ literals of the same polarity and referring to the same half of the graph. Assume without loss of generality that $C$ contains $c/4$ positive literals referring to stones in $V_0$ and let $C_0^+ = \{j0 \in V_0 : r_{j0} \in C\}$ be these stones.

The probability that no positive stone literal is satisfied is

$$\Pr[C_0^+ \cap A_0^+ = \emptyset] \le \frac{\binom{|V_0 \setminus C_0^+|}{|A_0^+|}}{\binom{|V_0|}{|A_0^+|}} \le (1 - c/4m)^{\kappa r/16} = \exp(-\Omega(\kappa r)) = \exp(-\Omega(m))$$

and since $\ln s = O(c^2 m/d^2 n^2) = O(m(c^2/d^2 n^2)) = o(m)$ the claim follows from a union bound over all clauses of type 1.

Next we greedily build a matching $\mu$ with the goal of satisfying all clauses of types 2 and 3. We ensure that overlaying both halves of the matching would also result in a matching; in other words if a vertex $jb$ is matched then we ensure that $jb + 1$ is not. For each edge $(i, jb)$ in the matching we set $s_{i,jb} = 1$, we set $s_{i',jb} = 0$ and $s_{i,j'b'} = 0$ for all $i' \ne i$, $j' \ne j$, and $b' \in \{0,1\}$, and we leave $s_{i,jb+1}$ tentatively unset for all $i$. Before we actually build the matching we need to prove that, with high probability, each of these clauses can be satisfied by choosing one of $c\kappa r/32m$ edges $(i, jb)$ with $j \in M_b$ to be in the matching.

For a clause $C$ of type 3 we assume without loss of generality that $c/2$ literals refer to stones in $V_0$. We can express the number of edges that satisfy $C$ as the random variable $x_C = \sum_{j0 \in V_0} x_{C,j0}$ where $x_{C,j0}$ takes the value $t_{C,j0} = |\{(i, j0) \in E : s_{i,j0} \in C\}|$ if $j0 \in M_0$ and 0 otherwise. We have that

$$E_C = E[x_C] = \sum_{j0 \in V_0} E[x_{C,j0}] = \sum_{j0 \in V_0} t_{C,j0} \cdot \Pr[j0 \in M_0]$$

$$= \frac{|M_0|}{m} \sum_{j0 \in V_0} t_{C,j0} \ge \frac{\kappa r}{8m} \cdot \frac{c}{2} = \frac{c\kappa r}{16m} = \Theta(c)$$

and each of $x_{C,j}$ is bounded by the right degree $d' \leq 2dn/m$, therefore by Hoeffding's inequality for sampling without replacement we obtain that

$$\Pr[x_C < \mathrm{E}_C/2] \leq \exp\left(-2\frac{(\mathrm{E}_C - \mathrm{E}_C/2)^2}{\sum_{j0 \in V_0} t_{C,j0}^2}\right) = \exp(-\Omega(c^2/d'c)) = \exp(-\Omega(cm/dn))$$

and the claim follows from a union bound over all clauses of type 3.

For clauses of type 2, for each literal $\overline{s_{i,j0}} \in C$ it is enough to choose as an edge one of the $(d-1)$ edges $(i, j'0)$ with $j' \neq j$. Hence the number of available choices is the random variable $x_C$ defined as before except that $t_{C,j0} = |\{(i, j0) \in E_0 : \exists j' \in V_0 \setminus \{j\}, \overline{s_{i,j'}} \in C\}|$. We have $\mathrm{E}_C = \mathrm{E}[x_C] \geq (d-1)c\kappa r/16m$ therefore $\Pr[x_C < \mathrm{E}_C] = \exp(-\Omega(cm/n))$ and the claim follows from a union bound.

Let us finish this step of the proof by building the matching. Observe that choosing an edge makes up to $d + d'$ incident edges ineligible, as well as up to $d + d'$ edges in the other half, for a total of $2(d + d') \leq 5d'$, hence after making $\ell$ choices there are still $e(\ell) = c\kappa r/32 - 5\ell d'$ choices available for each clause. By averaging, there is an edge that satisfies at least an $e(\ell)/dn$ fraction of the clauses of types 2 and 3. Hence after picking

$$k = e^{-1}(c\kappa r/64m) = \frac{c\kappa r/64m}{5d'} \leq \frac{c\kappa r}{320dn}$$

edges the remaining fraction of clauses is at most

$$\prod_{\ell=1}^{k}\left(1 - \frac{e(\ell)}{dn}\right) \leq \left(1 - \frac{e(k)}{dn}\right)^k = \left(1 - \frac{c\kappa r}{64mdn}\right)^{\frac{c\kappa r}{320dn}} = \exp\left(-\Omega\left(\frac{c^2 m}{d^2 n^2}\right)\right).$$

The last step is to ensure that after removing $V_0' = A_0 \cup M_0$ from $G_0$ we still have a good expander. By Lemma 4 there is a set $U'$ of size $r/2$ such that $G_0 \setminus U' \cup N(U') \cup V_0'$ is an $(r/2, \kappa/2)$-expander. Let $U'' = U' \setminus \mathrm{dom}\,\mu$. Let $\nu : U'' \to V_0$ be an injective mapping from indices to stones, which exists by Hall's theorem, and let $\sigma : U'' \to \{0,1\}$ be an assignment to $U''$ such that the depth of $F\restriction_\sigma$ reduces by at most $|\sigma|$.

We match each vertex $i \in U''$ to a stone as follows. If $\nu(i) \in A_0^-$ then $r_{\nu(i)+\sigma(i)} = \sigma(i)$ so we set $s_{i,\nu(i)+\sigma(i)} = 1$, while if $\nu(i) \in A_0^+$ then $r_{\nu(i)+\sigma(i)} = 1 - \sigma(i)$ so we set $s_{i,\nu(i)+\sigma(i)+1} = 1$. If $\nu(i) \in M_0$ then note that by construction of the matching $\mu$ at least one of $\nu(i)$ and $\nu(i) + 1$ is not matched; we let $jb$ be that stone and set $s_{i,jb} = 1$ and $r_{jb} = \sigma(i)$. Otherwise we add $\nu(i)$ to $M_0$ and $\nu(i) + 1$ to $M_1$, and do as in the previous case.

We also assign values to matched stones. Let $\mathrm{dom}\,\mu$ be the matched original variables and let $\tau : \mathrm{dom}\,\mu \to \{0,1\}$ be an assignment to $\mathrm{dom}\,\mu$ such that the depth of $F\restriction_{\sigma \cup \tau}$ reduces by at most $|\tau|$. For each vertex $i \in \mathrm{dom}\,\mu$ we set $r_{\mu(i)} = \tau(i)$. To obtain our final graph we set to 0 any variable $s_{i,j}$ with $i \in U' \cup \mathrm{dom}\,\mu$ or $j \in V_0' \cup N(U') \cup V_1$ that remains unassigned.

Let us recap and show that $\mathcal{L}^G(F)\restriction_\rho = \mathcal{L}^{G'}(F')$ where $G'$ is an expander and $F'$ has large depth as we claimed. $G'$ is the subgraph of $G$ induced by $U \setminus (U' \cup \mathrm{dom}\,\mu)$ and $V_0 \setminus (V_0' \cup N(U'))$, since we did not assign any selector variable corresponding to an edge between these two sets, but we did assign every other selector variable. The graph induced by $U \setminus U'$ and $V_0 \setminus (V_0' \cup N(U'))$ is an $(r/2, \kappa/2)$-expander by Lemma 4, and since removing left vertices does not affect expansion, so is $G'$. Regarding $F'$, for every variable $s_{i,j} = 1$ we have that $r_j = (\sigma \cup \tau)(i)$, so $F' = F\restriction_{\sigma \cup \tau}$ which has depth at least $D - r/2 - \kappa r/8$.    □

To prove an equivalent of Lemma 11 we use the *extended matching game*, where we allow the following additional move:

- Prover places an unused finger $i$ on a free vertex $v \in V$, in which case Disprover places his $i$-th finger on $v$ and optionally moves Prover's finger to a free vertex $u \in N(v)$.

**Lemma 15.** *If Prover needs $p$ fingers to win the matching game on a graph of right degree $d'$, then it needs $p - d'$ fingers to win the extended matching game.*

The proof can be found in the forthcoming full version.

Finally we are ready to prove our last lemma and complete the proof.

**Lemma 16.** *If $F$ has resolution depth $D$, and $G$ is a bipartite graph whose right hand side is of size $m < D$, duch that $G$ requires $r$ fingers in the extended matching game, then any regular resolution refutation of $\mathcal{L}^G(F)$ has an $r/3$-complex clause.*

*Proof.* At a high level we proceed as in the proof of Lemma 11, except that now keeping a matching is a more delicate task, and hence we use the extended matching game for it. We want to match any index $i$ for which we have information about, this is the value of a variable $s_{i,j}$ is remembered.

- If the adversary queries $r_j$ and $\mu(i) = j$ for some $i$, then we answer so that the depth of the decision tree only shrinks by 1.
- If the adversary queries $r_j$ where $j$ is not in the matching, then we play $j$ in the matching game. If we receive an answer $i$ we add $(i,j)$ to the matching and answer so that the depth of the decision tree only shrinks by 1. If instead we receive the answer $j$, we answer arbitrarily.
- If the adversary queries $s_{i,j}$ where either $i$ or $j$ are in the matching then we answer 1 if $(i,j)$ is in the matching and 0 otherwise.
- If the adversary queries $s_{i,j}$ where neither $i$ nor $j$ are in the matching then we play $i$ in the matching game and receive an answer $j'$. We add $(i,j')$ to the matching and answer 1 if $j = j'$ and 0 otherwise.
- If after the adversary forgets a variable there is an index $i$ such that $\mu(i) = j$ but none of $s_{i,j'}$ and $r_j$ are assigned, we forget $i$ in the matching game.

Assume for the sake of contradiction that Prover does not win the matching game. It follows that the branching program ends at a clause in $\mathcal{L}(D)$ for $D \in F$, at which point the depth of $T$ reduced to 0. Observe that the depth of $T$ only decreases by 1 when a stone variable is queried and that, since the branching program is read-once, these queries must be to $D$ different stones. However, only $m < D$ stones are available.

It follows that Prover eventually uses $r$ fingers in the matching game, at which point we claim that we are at an $r/3$-complex clause. Let us see why. For each finger $i$ in the matching game we remember either a selector literal $s_{i,j} = 1$, a selector literal $s_{i,j} = 0$, or a stone variable $r_j$, hence we remember at least $r/3$ variables of either type. In the first case we are at a clause of type 2, in the second at a clause of type 3, and in the third at a clause of type 1.    □

*Proof (of Theorem 12).* By Lemma 3, with high probability $G_0$ is an $(r, \kappa)$-expander for $r = \Theta(m/d)$ and $\kappa = \Theta(d)$, and has right degree at most $2dn/m$. Assume for the sake of contradiction that $\pi$ is a refutation of $\mathcal{L}^G(F)$ of length less than $\exp(\epsilon D^3/d^2 n^2)$.

Let $\rho$ be the restriction given by Lemma 14 so that $\pi\!\restriction_\rho$ is a regular resolution proof with no $c$-complex clauses with $c = \kappa r/75 = \Theta(m)$. The formula $\mathcal{L}(F)\!\restriction_\rho$ is the lifted version $\mathcal{L}^{G'}(F')$ of a formula $F'$ of depth at least $D - r/2 - \kappa r/8$, and the graph $G'$ is an $(r/2, \kappa/2)$-expander with $m' \leq m - \kappa r/8 \leq D - r/2 - \kappa r/8$. Since for each set $U$ of size at most $\kappa r/8$ and subset $U' \subseteq U$ of size $|U| \cdot 4/\kappa \leq r/2$ it holds that $|N(U)| \geq |N(U')| \geq \kappa/2|U'| = 2|U|$, $G'$ is also a $(\kappa r/8, 2)$-expander, hence by Theorem 5 and Lemma 15 $G'$ requires $\kappa r/24 - d' \geq \kappa r/25$ fingers in the extended matching game. By Lemma 11 any regular resolution proof of $\mathcal{L}^{G'}(F')$ has a $\kappa r/75$-complex clause. But this contradicts that the proof $\pi\!\restriction_\rho$ has no $\kappa r/75$-complex clauses.    □

It would also be interesting to prove a lower bound with plain random graphs, not relying on the additional mirror structure. Unfortunately, without backup vertices, the expansion restoring step would make $r/2$ right vertices ineligible to be matched, and that can prevent us from satisfying clauses of type 3 of complexity up to $d'r/2 \gg m$.

# 5   Experiments

We have run some experiments to investigate how hard sparse stone formulas are in practice and how restarts influence solvers running on this particular family.

As base formulas we use pebbling formulas over Gilbert–Tarjan graphs with butterflies [30,42], which require depth $\Theta(n/\log^2 n)$, and over pyramid graphs, which require depth $\Theta(\sqrt{n})$. Note that lifting the first type of formulas yields benchmarks that are provably hard for regular resolution, whereas for the second type of formulas we are not able to give any theoretical guarantees. Our experimental results are very similar, however, and so below we only discuss formulas obtained from pyramids, for which more benchmarks can be generated.

We used an instrumented version [28] of the solver Glucose [4] to make it possible to experiment with different heuristics. The results reported here are for the settings that worked best, namely VSIDS decision heuristic and preprocessing switched on. To vary the restart frequency we used Luby restarts with factors 1, 10, 100, and 1000 plus a setting with no restarts. The time-out limit was 24 h. For the record, we also ran some preliminary experiments for standard Glucose (with adaptive restarts) and Lingeling [46], but since the results were similar to those for Luby restarts with a factor 100 we did not run full-scale experiments with these configurations.

We illustrate our findings in Fig. 1 by plotting results from experiments using the pebbling formula over a pyramid graph of height 12 as the base formula and varying the number of stones. We used random graphs of left degree 6 as selector variable graphs. Note that once the pebbling DAG for the base formula has been fixed, changing the number of stones does not change the size of the formula too much. For the particular pebbling DAG in Fig. 1, the number of variables is in the interval from 550 to 650.

Empirically, the formulas are hardest when the number of stones is close to the proof depth for the base formula, which is also the scenario where the calculations in Sect. 4 yield the strongest bound. We expect the hardness to increase as the number of stones approaches from below the proof depth of the base formula, but as the number of stones grow further the formulas should get easier again. This is so since the fact that the selector graph left degree is kept constant means that the overlap decreases and ultimately vanishes, and pebbling formulas lifted without overlap are easy for regular resolution.

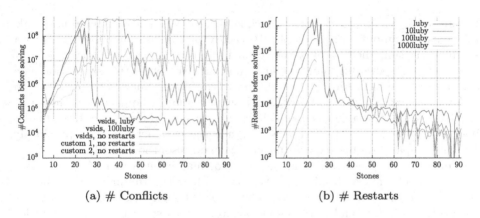

(a) # Conflicts                    (b) # Restarts

**Fig. 1.** Solving stone formulas over a pyramid of height 12.

Interestingly, the solver behaviour is very different on either side of this hardness peak. As we can see on the left in Fig. 1a, in the beginning the number of conflicts (and hence the running time) grows exponentially in the number of stones, independently of the number of restarts. With more stones, however, restarts become critical. The number of restarts used to solve a particular instance remains similar among all solver configurations, as shown on the right in Fig. 1b. Therefore, if the solver restarts more frequently it reaches this number of restarts faster and solves the formula faster, as shown by the conflict counts on the right in Fig. 1a.

To make CDCL solvers run as fast as possible, we crafted a custom decision order tailored to stone formulas over pyramids. With this decision order, no restarts, and very limited clause erasures, the solver decided dense stone formulas over pyramids of height $h$ with $h$ stones in a number of conflicts proportional to $h^{7.28}$ (where we note that these formulas have $O(h^3)$ variables and $O(h^5)$ clauses). For sparse stone formulas, we found one decision order (custom 1 in Fig. 1a) that worked reasonably well for small pyramids but failed for larger ones. A second attempt (custom 2) performed well for all pyramid sizes as long as the number of stones was below the hardness peak, but failed for more stones (when the formulas become easy for VSIDS with frequent restarts).

Summing up, even though stone formulas always possess short resolution refutations, and even though CDCL solvers can sometimes be guided to decide the formulas quickly even without restarts, these formulas can be surprisingly hard in practice for state-of-the-art solvers with default heuristics. The frequency of restarts seems to play a crucial role—which is an interesting empirical parallel of the theoretical results in [3,51]—but for some settings of stone formula parameters even frequent restarts cannot help the solver to perform well.

## 6   Concluding Remarks

In this work we employ lifting, a technique that has led to numerous breakthroughs in computational complexity theory in the last few years, to give a significantly simplified proof of the result in [2] that general resolution is exponentially more powerful than regular resolution. We obtain this separation as a corollary of a generic lifting theorem amplifying lower bounds on proof depth to lower bounds on regular proof length in resolution. Thanks to this new perspective we are also able to extend the result further, so that we obtain smaller benchmark formulas that slightly strengthen the parameters of the previously strongest separation between regular and general resolution in [61].

Furthermore, these new formulas are also small enough to make it possible to run experiments with CDCL solvers to see how the running time scales as the formula size grows. Our results show that although these formulas are theoretically very easy, and have resolution proofs that seem possible to find for CDCL solvers without restarts if they are given guidance about which variable decisions to make, in practice the performance depends heavily on settings such as frequent restarts, and is sometimes very poor even for very frequent restarts.

Our main result implies that if we can find CNF formulas that have resolution proofs in small width but require sufficiently large depth, then lifted versions of such formulas separate regular and general resolution. (This is so since proof width can only increase by a constant factor after lifting, and small-width proofs have to be short in general resolution by a simple counting argument.) Unfortunately, the only such formulas that are currently known are pebbling formulas. It would be very interesting to find other formulas with the same property.

Also, it would be desirable to improve the parameters of our lifting theorem. A popular family of pebbling graphs are pyramids, but the proof depth for pebbling formulas based on such graphs is right below the threshold where the lower bound amplification kicks in. Could the analysis in the proof of the lifting theorem be tightened to work also for, e.g., pebbling formulas over pyramids?

On the applied side, it is intriguing that sparse stone formulas can be so hard in practice. One natural question is whether one could find some tailor-made decision heuristic that always makes CDCL solvers run fast on such formulas, with or even without restarts. An even more relevant question is whether some improvement in standard CDCL heuristics could make state-of-the-art solvers run fast on these formulas (while maintaining performance on other formulas).

**Acknowledgements.** We are most grateful to Robert Robere for the interesting discussions that served as the starting point for this project. We also acknowledge the important role played by the Dagstuhl seminar 18051 "Proof Complexity," where some of this work was performed. Our computational experiments were run on resources provided by the Swedish National Infrastructure for Computing (SNIC). Our benchmarks were generated using the tool CNFgen [44].

The first author was supported by the Prof. R Narasimhan post-doctoral award. The second and fourth authors were funded by the Swedish Research Council (VR) grant 2016-00782. The fourth author was also supported by the Independent Research Fund Denmark (DFF) grant 9040-00389B.

# References

1. Alekhnovich, M., Hirsch, E.A., Itsykson, D.: Exponential lower bounds for the running time of DPLL algorithms on satisfiable formulas. J. Autom. Reason. **35**(1–3), 51–72 (2005). Preliminary version in ICALP 2004
2. Alekhnovich, M., Johannsen, J., Pitassi, T., Urquhart, A.: An exponential separation between regular and general resolution. Theory Comput. **3**(5), 81–102 (2007). Preliminary version in STOC 2002
3. Atserias, A., Fichte, J.K., Thurley, M.: Clause-learning algorithms with many restarts and bounded-width resolution. J. Artif. Intell. Res. **40**, 353–373 (2011). Preliminary version in SAT 2009
4. Audemard, G., Simon, L.: Predicting learnt clauses quality in modern SAT solvers. In: Proceedings of the 21st International Joint Conference on Artificial Intelligence (IJCAI 2009), pp. 399–404, July 2009
5. Bayardo Jr., R.J., Schrag, R.: Using CSP look-back techniques to solve real-world SAT instances. In: Proceedings of the 14th National Conference on Artificial Intelligence (AAAI 1997), pp. 203–208, July 1997
6. Beame, P., Beck, C., Impagliazzo, R.: Time-space tradeoffs in resolution: super-polynomial lower bounds for superlinear space. SIAM J. Comput. **45**(4), 1612–1645 (2016). Preliminary version in STOC 2012
7. Beame, P., Huynh, T., Pitassi, T.: Hardness amplification in proof complexity. In: Proceedings of the 42nd Annual ACM Symposium on Theory of Computing (STOC 2010), pp. 87–96, June 2010
8. Beame, P., Pitassi, T., Segerlind, N.: Lower bounds for Lovász-Schrijver systems and beyond follow from multiparty communication complexity. SIAM J. Comput. **37**(3), 845–869 (2007). Preliminary version in ICALP 2005

9. Beck, C., Nordström, J., Tang, B.: Some trade-off results for polynomial calculus. In: Proceedings of the 45th Annual ACM Symposium on Theory of Computing (STOC 2013), pp. 813–822, May 2013

10. Ben-Sasson, E., Galesi, N.: Space complexity of random formulae in resolution. Random Struct. Algorithms **23**(1), 92–109 (2003). Preliminary version in CCC 2001

11. Ben-Sasson, E., Impagliazzo, R., Wigderson, A.: Near optimal separation of tree-like and general resolution. Combinatorica **24**(4), 585–603 (2004)

12. Ben-Sasson, E., Nordström, J.: Short proofs may be spacious: an optimal separation of space and length in resolution. In: Proceedings of the 49th Annual IEEE Symposium on Foundations of Computer Science (FOCS 2008), pp. 709–718, October 2008

13. Ben-Sasson, E., Nordström, J.: Understanding space in proof complexity: separations and trade-offs via substitutions. In: Proceedings of the 2nd Symposium on Innovations in Computer Science (ICS 2011), pp. 401–416, January 2011

14. Ben-Sasson, E., Wigderson, A.: Short proofs are narrow-resolution made simple. J. ACM **48**(2), 149–169 (2001). Preliminary version in STOC 1999

15. Berkholz, C., Nordström, J.: Near-optimal lower bounds on quantifier depth and Weisfeiler-Leman refinement steps. In: Proceedings of the 31st Annual ACM/IEEE Symposium on Logic in Computer Science (LICS 2016), pp. 267–276, July 2016

16. Berkholz, C., Nordström, J.: Supercritical space-width trade-offs for resolution. SIAM J. Comput. **49**(1), 98–118 (2020). Preliminary version in ICALP 2016

17. Blake, A.: Canonical Expressions in Boolean Algebra. Ph.D. thesis, University of Chicago (1937)

18. Bonet, M.L., Buss, S., Johannsen, J.: Improved separations of regular resolution from clause learning proof systems. J. Artif. Intell. Res. **49**, 669–703 (2014)

19. Bonet, M.L., Esteban, J.L., Galesi, N., Johannsen, J.: On the relative complexity of resolution refinements and cutting planes proof systems. SIAM J. Comput. **30**(5), 1462–1484 (2000). Preliminary version in FOCS 1998

20. Buss, S.R., Hoffmann, J., Johannsen, J.: Resolution trees with lemmas: resolution refinements that characterize DLL-algorithms with clause learning. Logical Methods Comput. Sci. **4**(4:13) (2008)

21. Buss, S.R., Kołodziejczyk, L.: Small stone in pool. Logical Methods Comput. Sci. **10**(2), 16:1–16:22 (2014)

22. Chan, S.M.: Just a pebble game. In: Proceedings of the 28th Annual IEEE Conference on Computational Complexity (CCC 2013), pp. 133–143, June 2013

23. Chan, S.M., Lauria, M., Nordström, J., Vinyals, M.: Hardness of approximation in PSPACE and separation results for pebble games (Extended abstract). In: Proceedings of the 56th Annual IEEE Symposium on Foundations of Computer Science (FOCS 2015), pp. 466–485, October 2015

24. Chattopadhyay, A., Koucky, M., Loff, B., Mukhopadhyay, S.: Simulation beats richness: new data-structure lower bounds. In: Proceedings of the 50th Annual ACM Symposium on Theory of Computing (STOC 2018), pp. 1013–1020, June 2018

25. Cook, S.A.: An observation on time-storage trade off. J. Comput. Syst. Sci. **9**(3), 308–316 (1974). Preliminary version in STOC 1973

26. Davis, M., Logemann, G., Loveland, D.: A machine program for theorem proving. Commun. ACM **5**(7), 394–397 (1962)

27. Davis, M., Putnam, H.: A computing procedure for quantification theory. J. ACM **7**(3), 201–215 (1960)

28. Elffers, J., Giráldez-Cru, J., Gocht, S., Nordström, J., Simon, L.: Seeking practical CDCL insights from theoretical SAT benchmarks. In: Proceedings of the 27th International Joint Conference on Artificial Intelligence (IJCAI 2018), pp. 1300–1308, July 2018
29. Garg, A., Göös, M., Kamath, P., Sokolov, D.: Monotone circuit lower bounds from resolution. In: Proceedings of the 50th Annual ACM Symposium on Theory of Computing (STOC 2018), pp. 902–911, June 2018
30. Gilbert, J.R., Tarjan, R.E.: Variations of a pebble game on graphs. Technical Report STAN-CS-78-661, Stanford University (1978). http://infolab.stanford.edu/TR/CS-TR-78-661.html
31. Goerdt, A.: Regular resolution versus unrestricted resolution. SIAM J. Comput. **22**(4), 661–683 (1993)
32. Göös, M., Jain, R., Watson, T.: Extension complexity of independent set polytopes. SIAM J. Comput. **47**(1), 241–269 (2018)
33. Göös, M., Jayram, T.S., Pitassi, T., Watson, T.: Randomized communication vs. partition number. In: Proceedings of the 44th International Colloquium on Automata, Languages and Programming (ICALP 2017). Leibniz International Proceedings in Informatics (LIPIcs), vol. 80, pp. 52:1–52:15, July 2017
34. Göös, M., Kamath, P., Pitassi, T., Watson, T.: Query-to-communication lifting for PNP. In: Proceedings of the 32nd Annual Computational Complexity Conference (CCC 2017). Leibniz International Proceedings in Informatics (LIPIcs), vol. 79, pp. 12:1–12:16, July 2017
35. Göös, M., Lovett, S., Meka, R., Watson, T., Zuckerman, D.: Rectangles are non-negative juntas. In: Proceedings of the 47th Annual ACM Symposium on Theory of Computing (STOC 2015), pp. 257–266, June 2015
36. Göös, M., Pitassi, T.: Communication lower bounds via critical block sensitivity. SIAM J. Comput. **47**(5), 1778–1806 (2018). Preliminary version in STOC 2014
37. Göös, M., Pitassi, T., Watson, T.: Deterministic communication vs. partition number. In: Proceedings of the 56th Annual IEEE Symposium on Foundations of Computer Science (FOCS 2015), pp. 1077–1088, October 2015
38. Göös, M., Pitassi, T., Watson, T.: The landscape of communication complexity classes. Comput. Complex. **27**(2), 245–304 (2018). Preliminary version in ICALP 2016
39. Hoory, S., Linial, N., Wigderson, A.: Expander graphs and their applications. Bull. Am. Math. Soc. **43**(4), 439–561 (2006)
40. Huang, W., Yu, X.: A DNF without regular shortest consensus path. SIAM J. Comput. **16**(5), 836–840 (1987)
41. Huynh, T., Nordström, J.: On the virtue of succinct proofs: amplifying communication complexity hardness to time-space trade-offs in proof complexity (Extended abstract). In: Proceedings of the 44th Annual ACM Symposium on Theory of Computing (STOC 2012), pp. 233–248, May 2012
42. Järvisalo, M., Matsliah, A., Nordström, J., Živný, S.: Relating proof complexity measures and practical hardness of SAT. In: Milano, M. (ed.) CP 2012. LNCS, pp. 316–331. Springer, Heidelberg (2012). https://doi.org/10.1007/978-3-642-33558-7_25
43. Kothari, P.K., Meka, R., Raghavendra, P.: Approximating rectangles by juntas and weakly-exponential lower bounds for LP relaxations of CSPs. In: Proceedings of the 49th Annual ACM Symposium on Theory of Computing (STOC 2017), pp. 590–603, June 2017

44. Lauria, M., Elffers, J., Nordström, J., Vinyals, M.: CNFgen: a generator of crafted benchmarks. In: Gaspers, S., Walsh, T. (eds.) SAT 2017. LNCS, vol. 10491, pp. 464–473. Springer, Cham (2017). https://doi.org/10.1007/978-3-319-66263-3_30
45. Lee, J.R., Raghavendra, P., Steurer, D.: Lower bounds on the size of semidefinite programming relaxations. In: Proceedings of the 47th Annual ACM Symposium on Theory of Computing (STOC 2015), pp. 567–576, June 2015
46. Lingeling, Plingeling and Treengeling. http://fmv.jku.at/lingeling/
47. Marques-Silva, J.P., Sakallah, K.A.: GRASP: a search algorithm for propositional satisfiability. IEEE Trans. Comput. **48**(5), 506–521 (1999). Preliminary version in ICCAD 1996
48. Moskewicz, M.W., Madigan, C.F., Zhao, Y., Zhang, L., Malik, S.: Chaff: engineering an efficient SAT solver. In: Proceedings of the 38th Design Automation Conference (DAC 2001), pp. 530–535, June 2001
49. Nordström, J.: Narrow proofs may be spacious: separating space and width in resolution. SIAM J. Comput. **39**(1), 59–121 (2009). Preliminary version in STOC 2006
50. Nordström, J., Håstad, J.: Towards an optimal separation of space and length in resolution. Theory Comput. **9**, 471–557 (2013). Preliminary version in STOC 2008
51. Pipatsrisawat, K., Darwiche, A.: On the power of clause-learning SAT solvers as resolution engines. Artif. Intell. **175**(2), 512–525 (2011). Preliminary version in CP 2009
52. Pitassi, T., Robere, R.: Strongly exponential lower bounds for monotone computation. In: Proceedings of the 49th Annual ACM Symposium on Theory of Computing (STOC 2017), pp. 1246–1255, June 2017
53. Pitassi, T., Robere, R.: Lifting Nullstellensatz to monotone span programs over any field. In: Proceedings of the 50th Annual ACM Symposium on Theory of Computing (STOC 2018), pp. 1207–1219, June 2018
54. Raz, R., McKenzie, P.: Separation of the monotone NC hierarchy. Combinatorica **19**(3), 403–435 (1999). Preliminary version in FOCS 1997
55. Razborov, A.A.: A new kind of tradeoffs in propositional proof complexity. J. ACM **63**(2), 16:1–16:14 (2016)
56. Razborov, A.A.: On space and depth in resolution. Comput. Complex. **27**(3), 511–559 (2018)
57. de Rezende, S.F., Nordström, J., Vinyals, M.: How limited interaction hinders real communication (and what it means for proof and circuit complexity). In: Proceedings of the 57th Annual IEEE Symposium on Foundations of Computer Science (FOCS 2016), pp. 295–304, October 2016
58. Robere, R., Pitassi, T., Rossman, B., Cook, S.A.: Exponential lower bounds for monotone span programs. In: Proceedings of the 57th Annual IEEE Symposium on Foundations of Computer Science (FOCS 2016), pp. 406–415, October 2016
59. Tseitin, G.: On the complexity of derivation in propositional calculus. In: Silenko, A.O. (ed.) Structures in Constructive Mathematics and Mathematical Logic, Part II, pp. 115–125. Consultants Bureau, New York-London (1968)
60. Urquhart, A.: The depth of resolution proofs. Stud. Logica. **99**(1–3), 349–364 (2011)
61. Urquhart, A.: A near-optimal separation of regular and general resolution. SIAM J. Comput. **40**(1), 107–121 (2011). Preliminary version in SAT 2008
62. Van Gelder, A.: Pool resolution and its relation to regular resolution and DPLL with clause learning. In: Sutcliffe, G., Voronkov, A. (eds.) LPAR 2005. LNCS (LNAI), vol. 3835, pp. 580–594. Springer, Heidelberg (2005). https://doi.org/10.1007/11591191_40

# Mycielski Graphs and PR Proofs

Emre Yolcu[✉], Xinyu Wu, and Marijn J. H. Heule

Carnegie Mellon University, Pittsburgh, PA 15213, USA
{emreyolcu,xinyuwu,marijn}@cmu.edu

**Abstract.** Mycielski graphs are a family of triangle-free graphs $M_k$ with arbitrarily high chromatic number. $M_k$ has chromatic number $k$ and there is a short informal proof of this fact, yet finding proofs of it via automated reasoning techniques has proved to be a challenging task. In this paper, we study the complexity of clausal proofs of the uncolorability of $M_k$ with $k-1$ colors. In particular, we consider variants of the PR (propagation redundancy) proof system that are without new variables, and with or without deletion. These proof systems are of interest due to their potential uses for proof search. As our main result, we present a sublinear-length and constant-width PR proof without new variables or deletion. We also implement a proof generator and verify the correctness of our proof. Furthermore, we consider formulas extended with clauses from the proof until a short resolution proof exists, and investigate the performance of CDCL in finding the short proof. This turns out to be difficult for CDCL with the standard heuristics. Finally, we describe an approach inspired by SAT sweeping to find proofs of these extended formulas.

## 1 Introduction

Proof complexity investigates the relative strengths of Cook–Reckhow proof systems [7], defined in terms of the length of the shortest proof of a tautology as a function of the length of the tautology. Proof systems are separated with respect to their strengths by establishing lower and upper bounds on the lengths of the proofs of certain "difficult" tautologies in each system. Finding short proofs of such tautologies in a proof system is a method for proving small upper bounds, which provide evidence for the strength of a proof system. Similarly, the existence of a large lower bound implies that a proof system is relatively weak. The related field of SAT solving involves the study of search algorithms that have corresponding proof systems, and concerns itself with not only the existence of short proofs, but also the prospect of finding them automatically when they exist. As a result, the two areas interact. The long-term agenda of proof complexity is to prove lower bounds on proof systems of increasing strength towards concluding NP $\neq$ co-NP, whereas SAT solving benefits from strong proof systems with properties that make them suitable for automation. A recently proposed such system is PR (propagation redundancy) [14] and some of its variants SPR (subset PR), PR⁻ (without new variables), DPR⁻ (allowing deletion).

© Springer Nature Switzerland AG 2020
L. Pulina and M. Seidl (Eds.): SAT 2020, LNCS 12178, pp. 201–217, 2020.
https://doi.org/10.1007/978-3-030-51825-7_15

For several difficult tautologies, PR has been shown to admit proofs that are short (at most polynomial length), narrow (small clause width), and without extension (disallowing new variables) [5, 12, 13, 14]. From the perspective of proof search, these are favorable qualities for a proof system:

- Polynomial length is essentially a necessity.
- Small width implies that we may limit the search to narrow proofs.
- Eliminating extension drastically shrinks the search space.

Compared to strong proof systems with extension, a proof system with the above properties may admit a proof search algorithm that is effective in practice.

Mycielski graphs are a family of triangle-free graphs $M_k$ with arbitrarily high chromatic number. In particular, $M_k$ has chromatic number $k$. Despite having a simple informal proof, this has been a difficult fact to prove via automated reasoning techniques, and the state-of-the-art tools can only handle instances up to $M_6$ or $M_7$ [6, 9, 18, 19, 20, 21, 23]. Symmetry breaking [8], a crucial automated reasoning technique for hard graph coloring instances, is hardly effective on these graphs as the largest clique has size 2. Most short PR proofs for hard problems are based on symmetry arguments. Donald Knuth challenged us in personal communication[1] to explore whether short PR proofs exist for Mycielski graph formulas.

In this paper, we provide short proofs in PR$^-$ and DPR$^-$ for the colorability of Mycielski graphs [17]. Our proofs are of length quasilinear (with deletion and low discrepancy) and sublinear (without deletion but high discrepancy) in the length of the original formula, and include clauses that are at most ternary. With deletion allowed, the PR inferences have short witnesses, which allows us to additionally establish the existence of quasilinear-length DSPR$^-$ proofs. We also implement a proof generator and verify the generated proofs with dpr-trim[2]. Furthermore, we experiment with adding various combinations of the clauses in the proofs to the formulas and observe their effect on conflict-driven clause learning (CDCL) solver [3, 16] performance. It turns out that the resulting formulas are still difficult for state-of-the-art CDCL solvers despite the existence of short resolution proofs, reinforcing a recent result by Vinyals [22]. We then demonstrate an approach inspired by SAT sweeping [24] to solve these difficult formulas automatically.

## 2   Preliminaries

In this work we focus on propositional formulas in conjunctive normal form (CNF), which consist of the following: $n$ Boolean *variables*, at most $2n$ *literals* $p_i$ and $\overline{p_i}$ referring to different polarities of variables, and $m$ *clauses* $C_1, \ldots, C_m$ where each clause is a disjunction of literals. The CNF *formula* is the conjunction of all clauses. Formulas in CNF can be treated as sets of clauses, and clauses as

---

[1] Email correspondence on May 25, 2019
[2] https://github.com/marijnheule/dpr-trim

sets of literals. For two clauses $C, D$ such that $p \in C$, $\bar{p} \in D$, their *resolvent* on $p$ is the clause $(C \setminus \{p\}) \cup (D \setminus \{\bar{p}\})$. A clause is called a *tautology* if it includes both $p$ and $\bar{p}$. We denote the empty clause by $\bot$.

An assignment $\alpha$ is a partial mapping of variables in a formula to truth values in $\{0, 1\}$. We denote assignments by a conjunction of the literals they satisfy. As an example, the assignment $x \mapsto 1$, $y \mapsto 0$, $z \mapsto 1$ is denoted by $x \wedge \bar{y} \wedge z$. The set of variables assigned by $\alpha$ is denoted by $\mathrm{dom}(\alpha)$. We denote by $F|_\alpha$ the *restriction* of a formula $F$ under an assignment $\alpha$, the formula obtained by removing satisfied clauses and falsified literals from $F$. A clause $C$ is said to *block* the assignment $\alpha = \bigwedge_{p \in C} \bar{p}$, which we denote by $\overline{C}$.

A clause is called *unit* if it contains a single literal. *Unit propagation* refers to the iterative procedure where we assign the variables in a formula $F$ to satisfy the unit clauses, restrict the formula under the assignment, and repeat until no unit clauses remain. If this procedure yields the empty clause $\bot$, we say that unit propagation *derives a conflict* on $F$.

Assume for the rest of the paper that $F, H$ are formulas in CNF, $C$ is a clause, and $\alpha$ is the assignment blocked by $C$. Formulas $F, H$ are *equisatisfiable* if either they are both satisfiable or both unsatisfiable. $C$ is *redundant* with respect to $F$ if $F$ and $F \wedge C$ are equisatisfiable. $C$ is *blocked* with respect to $F$ if there exists a literal $p \in C$ such that for each clause $D \in F$ that includes $\bar{p}$, the resolvent of $C$ and $D$ on $p$ is a tautology [15]. $C$ is a *reverse unit propagation* (RUP) inference from $F$ if unit propagation derives a conflict on $F \wedge \alpha$ [11]. $F$ *implies $H$ by unit propagation*, denoted $F \vdash H$, if each clause $C \in H$ is a RUP inference from $F$. Let us state a lemma about implication by unit propagation for later use.

**Lemma 1 ([5]).** *Let $C, D$ be clauses such that $C \vee D$ is not a tautology and let $\alpha$ be the assignment blocked by $C$. Then*

$$F|_\alpha \vdash D \setminus C \quad \Longleftrightarrow \quad F|_\alpha \vdash D \quad \Longleftrightarrow \quad F \vdash C \vee D.$$

Letting $x_i$ be either a unit clause or a conjunction of unit clauses, we will use the notation $F \vdash x_1 \vdash x_2 \vdash \ldots \vdash x_N$ to mean that for each $i \in \{1, \ldots, N\}$ we have $F \wedge \bigwedge_{j=1}^{i-1} x_j \vdash x_i$. This serves as a compact way of writing a sequence of unit clauses that become true on the way to deriving $x_N$ from $F$ via unit propagation.

## 3   PR proof system

Redundancy is the basis for clausal proof systems. In a clausal proof of a contradiction, we start with the formula and introduce redundant clauses until we can finally introduce the empty clause. Since satisfiability is preserved at each step due to redundancy, introduction of the empty clause implies that the formula is unsatisfiable. The sequence of redundant clauses constitutes a proof of the formula. Also note that since only unsatisfiable formulas are of interest, we use "proof" and "refutation" interchangeably.

**Definition 1.** *For a formula $F$, a valid clausal proof of it is a sequence of clause–witness pairs $(C_1, \omega_1), \ldots, (C_N, \omega_N)$ where, defining $F_i := F \wedge \bigwedge_{j=1}^i C_j$, we have*

- *each clause $C_i$ is redundant with respect to the conjunction of the formula with the preceding clauses in the proof, that is, $F_{i-1}$ and $F_i = F_{i-1} \wedge C_i$ are equisatisfiable,*
- *there exists a predicate $r(F_{i-1}, C_i, \omega_i)$ computable in polynomial time that indicates whether $C_i$ is redundant with respect to $F_{i-1}$,*
- $C_N = \bot$.

For a clausal proof $P$ of length $N$, we call $\max_{i \in \{1,\ldots,N\}} |C_i|$ its *width*.

**Definition 2.** *$C$ is* propagation redundant *with respect to $F$ if there exists an assignment $\omega$ satisfying $C$ such that $F|_\alpha \vdash_1 F|_\omega$ where $\alpha$ is the assignment blocked by $C$.*

Note that propagation redundancy can be decided in polynomial time given a witness $\omega$ due to the existence of efficient unit propagation algorithms. Unit propagation is a core primitive in SAT solvers, and despite the prevalence of large collections of heuristics implemented in solvers, in practice the majority of the runtime of a SAT solver is spent performing unit propagation inferences.

**Theorem 1 ([14]).** *If $C$ is propagation redundant with respect to $F$, then it is redundant with respect to $F$.*

Theorem 1 allows us to define a specific clausal proof system:

**Definition 3.** *A* PR *proof is a clausal proof where the predicate $r(F_{i-1}, C_i, \omega_i)$ in Definition 1 computes the relation $F_{i-1}|_{\alpha_i} \vdash_1 F_{i-1}|_{\omega_i}$ where $\alpha_i$ is the assignment blocked by $C_i$.*

Resolvents, blocked clauses, and RUP inferences are propagation redundant. Hence they are valid steps in a PR proof.

Let us also mention a few notable variants of the PR proof system:

- SPR: For each clause–witness pair $(C_i, \omega_i)$ in the proof and $\alpha_i$ the assignment blocked by $C_i$, require that $\mathrm{dom}(\omega_i) = \mathrm{dom}(\alpha_i)$.
- PR$^-$: No clause $C$ in the proof can include a variable that does not occur in the formula $F$ being proven.
- DPR: In addition to introducing redundant clauses, allow deletion of a previous clause in the proof (or the original formula), that is, allow $F_i = F_{i-1} \setminus \{C\}$ for some $C \in F_{i-1}$.

Following the notation of Buss and Thapen [5], the prefix "D" denotes a variant of a proof system with deletion allowed, and the superscript "$-$" denotes a variant disallowing new variables.

### 3.1   Expressiveness of PR

**Intuition** PR allows us to introduce clauses that intuitively support the following reasoning:

If there exists a satisfying assignment, then there exists a satisfying assignment with a certain property $X$, described by the witness $\omega$. This is because we can take any assignment that does not have $X$, apply a transformation to it that does not violate any original constraints of the formula, and obtain a new satisfying assignment with property $X$. The validation of such a transformation in general is NP-hard. Transformations are limited such that they can be validated using unit propagation.

Hence, if our goal is to find some (not all) of the satisfying assignments to a formula or to refute it, then we can extend the formula by introducing useful assumptions without harming our goal since satisfiability is preserved with each assumption. The redundancy of each assumption is efficiently checkable using the blocked assignment $\alpha$ and the witness $\omega$ which together describe the transformation that we apply to a solution without property $X$ to obtain another with $X$. Having this kind of understanding and mentally executing unit propagation allows us to look for PR proofs while continuing to reason at a relatively intuitive level. This proves useful when working towards upper bounds.

**Upper bounds** For several difficult tautologies (pigeonhole principle, bit pigeonhole principle, parity principle, clique-coloring principle, Tseitin tautologies) short SPR$^-$ proofs exist [5,14]. Still, there are several problems mentioned by Buss and Thapen [5] for which there are no known PR$^-$ proofs of polynomial length. Furthermore, we do not know whether there are short SPR$^-$ proofs of the Mycielski graph formulas. Buss and Thapen [5] have a partial simulation result between SPR$^-$ and PR$^-$ depending on a notion called "discrepancy", defined as follows.

**Definition 4.** *For a PR inference, its discrepancy is* $|\mathrm{dom}(\omega) \setminus \mathrm{dom}(\alpha)|$.

**Theorem 2.** *Let $F$ be a formula with a PR refutation of length $N$ such that* $\max_{i \in \{1,\dots,N\}} |\mathrm{dom}(\omega_i) \setminus \mathrm{dom}(\alpha_i)| \leq \delta$. *Then, $F$ has an SPR refutation of length* $O(2^\delta N)$ *without using variables not in the PR refutation.*

As a result, a PR proof of length $N$ with maximum discrepancy at most $\log N$ directly gives an SPR proof of length $O(N^2)$. In our case, the maximum discrepancy of the PR$^-$ proof is $\Omega(N/(\log N)^2)$, hence we cannot utilize Theorem 2 to obtain a polynomial-length SPR$^-$ proof. For our DPR$^-$ proof, the maximum discrepancy is 2, and by Theorem 2 there do exist quasilinear-length DSPR$^-$ proofs of the Mycielski graph formulas.

# 4 Proofs of Mycielski graph formulas

## 4.1 Mycielski graphs

Let $G = (V, E)$ be a graph. Its *Mycielski graph* $\mu(G)$ is constructed as follows:

1. Include $G$ in $\mu(G)$ as a subgraph.

2. For each vertex $v_i \in V$, add a new vertex $u_i$ that is connected to all the neighbors of $v_i$ in $G$.
3. Add a vertex $w$ that is connected to each $u_i$.

Unless $G$ has a triangle $\mu(G)$ does not have a triangle, and $\mu(G)$ has chromatic number one higher than $G$. We denote the chromatic number of $G$ by $\chi(G)$.

Starting with $M_2 = K_2$ (the complete graph on 2 vertices) and applying $M_k = \mu(M_{k-1})$ repeatedly, we obtain triangle-free graphs with arbitrarily large chromatic number. We call $M_k$ the $k$th *Mycielski graph*. Since $\chi(M_2) = 2$ and $\mu$ increases the chromatic number by one, we have $\chi(M_k) = k$. The graph $M_k$ has $3 \cdot 2^{k-2} - 1 = \Theta(2^k)$ vertices and $\frac{1}{2}(7 \cdot 3^{k-2} + 1) - 3 \cdot 2^{k-2} = \Theta(3^k)$ edges [1].

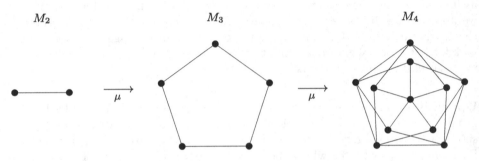

**Fig. 1.** The first few graphs in the sequence of Mycielski graphs.

Let us denote by $\mathrm{MYC}_k$ the contradiction that $M_k$ is colorable with $k-1$ colors. We will present short $\mathrm{PR}^-$ and $\mathrm{DPR}^-$ proofs of $\mathrm{MYC}_k$ in Section 4.2. Before doing so, let us present the short informal argument to prove that applying $\mu$ increases the chromatic number, which implies that $\chi(M_k) > k-1$.

**Proposition 1.** $\chi(\mu(G)) > \chi(G)$.

*Proof.* Assume we partition the vertices of $\mu(G)$ as $V \cup U \cup \{w\}$ where $V$ is the set of vertices of $G$ which is included as a subgraph, $U$ is the set of newly added vertices corresponding to each vertex in $V$, and $w$ is the vertex that is connected to all of $U$.

Let $k = \chi(G)$, and denote $[k] = \{1, 2, \ldots, k\}$. Denote the set of neighbors of a vertex $v$ by $N(v)$. Consider a proper $k$-coloring $\phi : V \cup U \to [k]$ of $\mu(G) \setminus \{w\}$. Assume that in this coloring $U$ uses only the first $k-1$ colors. Then we can define a $(k-1)$-coloring $\phi'$ of $G$ by setting $\phi'(v_i) = \phi(u_i)$ for $v_i$ with $\phi(v_i) = k$ and copying $\phi$ for the remaining vertices. The coloring $\phi'$ is proper, because for any two $v_i, v_j$,

- if $\phi(v_i) = \phi(v_j) = k$, then no edges exist between them;
- if $\phi(v_i), \phi(v_j) < k$, then their colors are not modified;
- if $\phi(v_i) \neq \phi(v_j) = k$, then $\phi'(v_i) = \phi(v_i) \neq \phi(u_j) = \phi'(v_j)$ since for all $v \in N(v_j)$ we have $\phi(v) \neq \phi(u_j)$.

As a result, we can obtain a proper $(k-1)$-coloring of $G$, contradiction. Hence, $U$ must use at least $k$ colors in a proper coloring of $\mu(G)$, and since $w$ then has to have a color greater than $k$ we have $\chi(\mu(G)) > k = \chi(G)$. $\qquad\square$

**Theorem 3.** $M_k$ *is not colorable with* $k-1$ *colors.*

*Proof.* Follows from the fact that $\chi(M_2) = 2$ and Proposition 1 via induction. $\square$

## 4.2    PR proofs

To obtain PR$^-$ and DPR$^-$ proofs, we follow a different kind of reasoning than that of the informal proof in the previous section. Let $k \geq 3$. Denote by $v_i$, $E_{k-1}$ the vertices and the edge set of the $(k-1)$th Mycielski graph, respectively. Assume we partition the vertices of $M_k$ as in the proof of Proposition 1 into $V \cup U \cup \{w\}$. Let $n_k = |V| = |U| = 3 \cdot 2^{k-3} - 1$.

In propositional logic, MYC$_k$ is defined on the variables $v_{i,c}$, $u_{i,c}$, $w_c$ for $i \in [n_k]$, $c \in [k-1]$. The variable $v_{i,c}$ indicates that the vertex $v_i \in V$ is assigned color $c$, and $u_{i,c}$, $w_c$ have similar meanings. MYC$_k$ consists of the clauses

$$\bigvee_{c \in [k-1]} v_{i,c} \quad \text{for each } i \in [n_k]$$

$$\bigvee_{c \in [k-1]} u_{i,c} \quad \text{for each } i \in [n_k]$$

$$\bigvee_{c \in [k-1]} w_c$$

$$\overline{v_{i,c}} \vee \overline{v_{j,c}} \quad \text{for each } v_i v_j \in E_{k-1},\ c \in [k-1]$$

$$\overline{u_{i,c}} \vee \overline{v_{j,c}} \quad \text{for each } i, j \text{ such that } v_i v_j \in E_{k-1},\ c \in [k-1]$$

$$\overline{u_{i,c}} \vee \overline{w_c} \quad \text{for each } i \in [n_k],\ c \in [k-1].$$

For both the PR$^-$ and the DPR$^-$ proofs, the high-level strategy is to introduce clauses that effectively insert edges between any $u_i, u_j$ for which $v_i v_j \in E_{k-1}$. In other words, if there is an edge $v_i v_j$, we introduce clauses that imply the existence of the edge $u_i u_j$, resulting in the modified graph $M_k'$ that has an induced subgraph $M_k'[U]$ isomorphic to $M_{k-1}$, and has all of its vertices connected to $w$. As an example, Figure 3a shows the result of this step on $M_4$. Then we partition the vertices of $M_k'[U]$ into new $V \cup U \cup \{w\}$ similar to the way we did for $M_k$. Such a partition exists as $M_k'[U]$ is isomorphic to $M_{k-1}$ which by construction has this partition. Then we inductively repeat the whole process. Figure 3c displays the result of repeating it once. Finally, the added edges result in a $k$-clique in $M_k$, as illustrated in Figure 3d. The vertices that participate in the clique are the two $u_i$'s of the subgraph we obtain at the last step that is isomorphic to $M_3$ and the $w$'s of all the intermediate graphs isomorphic to $M_{k'}$ for $k' \in [k] \setminus \{1,2\}$. Since we have $k-1$ colors available, the problem then reduces to the pigeonhole principle with $k$ pigeons and $k-1$ holes (denoted PHP$_{k-1}$), for which we know there exists a polynomial-length $PR^-$ proof due to Heule et al. [14]. At the end we simply concatenate the pigeonhole proof for the clique, which derives the empty clause as desired.

The primary difference between the versions of the proof with and without deletion is the discrepancy of the PR inferences. Deletion allows us to detach $M'_k[U]$ from $M'_k[V]$, as illustrated in Figure 3b, by removing each preceding clause that contains both a variable corresponding to some vertex in $U$ and another corresponding to some vertex in $V$. This makes it possible to introduce the PR clauses with discrepancy bounded by a constant. Without deletion, we instead introduce the PR inferences at each inductive step which imply that every $u_i \in U$ has the same color as its corresponding $v_i$, and this requires us to keep track of sets of equivalent vertices and assign them together in the witnesses. Figure 4 displays the effect of introducing these clauses on $M_4$.

For ease of presentation, we first describe the $DPR^-$ proof, followed by the $PR^-$ proof.

**Theorem 4.** $MYC_k$ *has quasilinear-length* $DPR^-$ *and* $DSPR^-$ *refutations.*

*Proof.* At each step below, let $F$ denote the conjunction of $MYC_k$ with the clauses introduced in the previous steps.

1. As the first step, we introduce $(2|U| + 1)\binom{k-1}{2}$ blocked clauses

$$
\begin{aligned}
\overline{v_{i,c}} \vee \overline{v_{i,c'}} \quad & \text{for each } i \in [n_k] \\
\overline{u_{i,c}} \vee \overline{u_{i,c'}} \quad & \text{for each } i \in [n_k] \\
\overline{w_c} \vee \overline{w_{c'}} \quad &
\end{aligned}
\tag{1}
$$

   for each $c, c' \in [k-1]$ such that $c < c'$. These clauses assert that each vertex in the graph can be assumed to have at most one color.

2. Then, we introduce $|U|(k-1)(k-2)$ PR clauses

$$
\begin{aligned}
\overline{v_{i,c}} \vee \overline{u_{i,c'}} \vee w_c \quad & \text{for each } i \in [n_k] \text{ and} \\
& \text{for each } c, c' \in [k-1], \ c \neq c'.
\end{aligned}
\tag{2}
$$

   Intuitively, these clauses introduce the assumption that if there exists a solution, then there exists a solution that does not simultaneously have $v_i$ colored $c$, $u_i$ colored $c'$, and $w$ not colored $c$. If $u_i$ has color $c'$, then we can switch its color to $c$ and still have a valid coloring. The validity of this new coloring is verifiable relying only on unit propagation inferences. It does not create any monochromatic edges between $u_i$ and $v_j \in N(v_i) \cap V$, as $v_j$ would already not have the color $c$. It also does not create a monochromatic edge between $u_i$ and $w$ since $w$ is already assumed not to have color $c$. Figure 2 shows this argument with a diagram. The corresponding witness for this transformation is $\omega = v_{i,c} \wedge \overline{u_{i,c'}} \wedge u_{i,c} \wedge \overline{w_c}$, leading to a discrepancy of 1.

3. Then, we introduce $|E_{k-1}|(k-1)(k-2)$ RUP inferences

$$
\begin{aligned}
\overline{u_{i,c}} \vee \overline{u_{j,c}} \vee \overline{v_{i,c'}} \quad & \text{for each } i, j \text{ such that } v_i v_j \in E_{k-1} \text{ and} \\
& \text{for each } c, c' \in [k-1], \ c \neq c'.
\end{aligned}
\tag{3}
$$

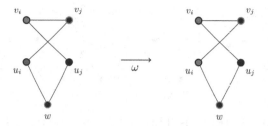

**Fig. 2.** Schematic form of the argument for the PR inference. With $c$ = Red and $c'$ = Blue, the above diagram shows the transformation we can apply to a solution to obtain another valid solution. A vertex colored black on the inside means that it does not have the outer color, i.e. $w$ has some color other than red. Unit propagation implies that $v_j$ is not colored red.

Let $C = \overline{u_{i,c}} \vee \overline{u_{j,c}} \vee \overline{v_{i,c'}}$ and $\alpha = \overline{C}$. Due to the previously introduced blocked and PR clauses (from (1) and (2)) we have

$$F|_\alpha \vdash w_{c'} \vdash \bigwedge_{\substack{1 \le d \le k-1 \\ d \ne c'}} \overline{w_d} \vdash \bigwedge_{\substack{1 \le d \le k-1 \\ d \ne c'}} \overline{v_{j,d}} \vdash v_{j,c'}$$

and also $F|_{v_{j,c'}} \vdash \overline{v_{i,c'}}$ due to the edge $v_i v_j$. These imply that $F|_\alpha \vdash \overline{v_{i,c'}}$. Then, since $\overline{v_{i,c'}} \in C$, we have $F|_\alpha \vdash \bot$ by Lemma 1.

4. Next, we introduce $|E_{k-1}|(k-1)$ RUP inferences

$$\overline{u_{i,c}} \vee \overline{u_{j,c}} \quad \text{for each } i, j \text{ such that } v_i v_j \in E_{k-1} \text{ and} \\ \text{for each } c \in [k-1]. \tag{4}$$

Let $D = \overline{u_{i,c}} \vee \overline{u_{j,c}}$ and $\beta = \overline{D}$. From the previous set of RUP inferences in (3) we have

$$F|_\beta \vdash \bigwedge_{\substack{1 \le d \le k-1 \\ d \ne c}} \overline{v_{i,d}} \vdash v_{i,c}.$$

Due to the edge $u_j v_i$ we also have $F|_{v_{i,c}} \vdash \overline{u_{j,c}}$ and consequently $F|_\beta \vdash \overline{u_{j,c}}$. Since $\overline{u_{j,c}} \in D$, we have $F|_\beta \vdash \bot$ by Lemma 1.

With the addition of this last set of assumptions, we have effectively copied the edges between $v_i$ to between $u_i$. Figure 3a visualizes the result of this step on $M_4$ with the red edges corresponding to the newly introduced assumptions.

5. After the addition of the new edges, we delete the clauses introduced in steps 2, 3, and the clauses corresponding to the edges between $U$ and $V$ of the current Mycielski graph. Figure 3b displays the graph after the deletions.

6. Then we inductively repeat steps 2–5, that is, we introduce clauses and delete the intermediate ones for each subgraph isomorphic to Mycielski graphs of descending order. Figure 3c shows the result of repeating the process on a subgraph isomorphic to $M_3$, with the blue edges corresponding to the latest assumptions.

(a) Introduction of edge assumptions to obtain a subgraph isomorphic to the Mycielski graph of the previous order.

(b) Deletions of the clauses introduced previously and the edges between $U, V$ to detach the subgraph.

(c) Repetition of the inductive step on the previously obtained subgraph isomorphic to $M_3$.

(d) Detached clique obtained after deleting the clauses corresponding to the edges leaving the clique.

**Fig. 3.** Illustrations of the proof steps in the case where $M_4$ is the initial graph, i.e. MYC$_4$ is the formula being refuted. The blue and the red edges correspond to the clauses introduced as RUP inferences, and the clauses corresponding to the faded edges are deleted.

7. After an edge is inserted between the two $u_i$ of the subgraph isomorphic to $M_3$, we obtain a $k$-clique on the two $u_i$ and all of the previous $w$'s. Then we delete all the clauses corresponding to the edges leaving the clique. This detaches the clique from the rest of the graph as illustrated for $M_4$ in Figure 3d. Since $(k-1)$-colorability of the $k$-clique is exactly the pigeonhole principle, we simply concatenate a PR$^-$ proof of the pigeonhole principle as described by Heule et al. [14], which has maximum discrepancy 2. This completes the DPR$^-$ proof that $M_k$ is not colorable with $k-1$ colors.

In total, the proof has length $O(3^k k^2)$ and the PR inferences have maximum discrepancy 2. Hence, by Theorem 2, there also exists a DSPR$^-$ proof of length $O(3^k k^2)$. Since MYC$_k$ has length $\Theta(3^k k)$, if we denote the length of the formula by $S$ then the proof is of quasilinear length $O(S \log S)$. □

**Theorem 5.** *MYC$_k$ has sublinear-length PR$^-$ refutations.*

*Proof.* At a high level, the proof is similar to the DPR$^-$ proof. However, in order to avoid deletion we introduce assumptions at each inductive step that imply the equivalence of every $u_i$ with its corresponding $v_i$. This eliminates the need to detach $M'_k[U]$ from $M'_k[V]$, but leads to sets of vertices forced to have the same color. As a result, the witnesses for the PR inferences after the first inductive step that refer to switching the color of a vertex $\nu$ need to also include all the previous vertices forced to have the same color as $\nu$.

1. We start by introducing the blocked clauses from (1).
2. Then we introduce the PR inferences from (2).

3. It becomes possible to infer the following $|U|(k-1)(k-2)$ clauses via PR.

$$\overline{u_{i,c}} \vee \overline{v_{i,c'}} \quad \text{for each } i \in [n_k] \text{ and} \\ \text{for each } c, c' \in [k-1], \ c \neq c'. \tag{5}$$

Let $\gamma = u_{i,c} \wedge v_{i,c'}$, and denote the conjunction of the formula and the clauses in (1) and (2) by $F$. In step 3 of the previous proof we showed that $F \vdash \overline{u_{i,c}} \vee \overline{u_{j,c}} \vee \overline{v_{i,c'}}$. Then, by Lemma 1, we have $F|_\gamma \vdash \overline{u_{j,c}}$. Hence, we can switch the color of $v_i$ from $c'$ to $c$. This does not result in any conflicts since $u_i$ having color $c$ implies that no $v_j \in N(v_i) \cap V$ has the color $c$, and $\overline{u_{j,c}}$ is implied by unit propagation. As a result, the clause $\overline{u_{i,c}} \vee \overline{v_{i,c'}}$ is PR with witness $\omega = u_{i,c} \wedge \overline{v_{i,c'}} \wedge v_{i,c}$. After the addition of these clauses, the equivalence $u_{i,c} \leftrightarrow v_{i,c}$ is implied via unit propagation. Due to the edge $v_i v_j$, the existence of the edge $u_i u_j$ is also implied via unit propagation. This step allows us to avoid deletion.

4. At this point, we inductively repeat steps 2–3 for each subgraph isomorphic to Mycielski graphs of descending order. However, due to the equivalences $u_{i,c} \leftrightarrow v_{i,c}$, any subsequent PR inference that argues by way of switching a vertex $\nu$'s color should include in its witness the same color switch for all the vertices that are transitively equivalent to $\nu$ from the previous steps. For instance, if a witness contains $\overline{\nu_{c'}} \wedge \nu_c$, then for each vertex $\eta$ that is equivalent to $\nu$ it also has to contain $\overline{\eta_{c'}} \wedge \eta_c$. The maximum number of such vertices for any $\nu$ occurring in the proof is $\Omega(2^k)$.

5. After the PR clauses are introduced for the subgraph isomorphic to $M_3$, the existence of a $k$-clique is implied via unit propagation. Figure 4 shows the equivalent vertices and the implied edges after the last inductive step when starting from $M_4$. At the end, we simply concatenate a proof of the pigeonhole principle as before, taking care to include in the witnesses all the equivalent vertices (as described in the previous step) to each vertex whose color is switched by a witness.

**Fig. 4.** Equivalent vertices and implied edges. Groups of equivalent vertices are highlighted. Dashed edges are implied by unit propagation.

The proof has length $O(2^k k^2)$, and $\text{MYC}_k$ has length $\Theta(3^k k)$. Letting $S$ denote the length of the formula, the proof has sublinear length $O(S^{\log_3 2}(\log S)^2)$. $\qquad \square$

In the $PR^-$ proof, the maximum discrepancy is $\Omega(2^k)$. Letting $N$ be the length of the proof, this becomes $\Omega(N/(\log N)^2)$. As a result, we cannot rely on Theorem 2, and the existence of a polynomial-length $SPR^-$ proof for Mycielski graph formulas remains open. While the existence of such a proof is plausible, we conjecture that it will not be of constant width as the ones we present.

## 5     Experimental results

All of the formulas, proofs, and the code for our experiments are available at https://github.com/emreyolcu/mycielski.

### 5.1     Proof verification

In order to verify the proofs we described in the previous section, we implemented two proof generators for $MYC_k$ and checked the $DPR^-$ and $PR^-$ proofs with dpr-trim for values of $k$ from 5 to 10. Figure 5 shows a plot of the lengths of the formulas and the proofs, and Table 1 shows their exact sizes.

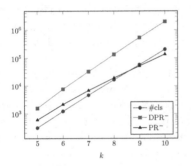

**Fig. 5.** Plot of the length of the formula and the lengths of the proofs versus $k$.

**Table 1.** Formula and proof sizes. For each formula $MYC_k$, this table shows the number of variables and clauses in the formula, and the lengths of the proofs.

| $k$ | #vars | #cls | $DPR^-$ | $PR^-$ |
|---|---|---|---|---|
| 5 | 92 | 307 | 1572 | 600 |
| 6 | 235 | 1227 | 7635 | 2165 |
| 7 | 570 | 4625 | 33178 | 6796 |
| 8 | 1337 | 16711 | 134855 | 19523 |
| 9 | 3064 | 58551 | 524456 | 52816 |
| 10 | 6903 | 200531 | 1976271 | 136905 |

### 5.2     Effect of redundant clauses on CDCL performance

Suppose we have a proof search algorithm for $DPR^-$ and that the redundant clauses we introduce in the $DPR^-$ proof are discovered automatically. Assuming they are found by some method, we look at their effect on the efficiency of CDCL at finding the rest of the proof automatically. In addition, we generate satisfiable instances of the coloring problem (denoted $MYC_k^+$ and stating that $M_k$ is colorable with $k$ colors) and compare how many of the satisfying assignments remain after the clauses are introduced. The reduction in the number of solutions suggests that the added clauses do a significant amount of work in breaking symmetries.

**Table 2.** Number of solutions left in $MYC_k^+$ after introducing redundant clauses. PR\BC is the version of the formula where we add the PR clauses but not the BC ones. For $k \geq 5$, it takes longer than 24 hours to count all solutions, so we only included the results for two small formulas here.

| $k$ | $MYC_k^+$ | BC | PR\BC | PR |
|---|---|---|---|---|
| 3 | 60 | 30 | 36 | 18 |
| 4 | 163680 | 12480 | 6576 | 792 |

**Table 3.** CDCL performance on formulas with additional clauses. Each cell shows the time (in seconds) it takes for CaDiCaL to prove unsatisfiability. The cells with dashes indicate that the solver ran out of time before finding a proof.

| $k$ | $MYC_k$ | BC | PR | R1 | R2 |
|---|---|---|---|---|---|
| 5 | 0.07 | 0.04 | 0.03 | 0.01 | 0.00 |
| 6 | 29.53 | 24.51 | 1.17 | 0.03 | 0.01 |
| 7 | — | — | 26.80 | 0.28 | 0.02 |
| 8 | — | — | 1503 | 1.33 | 0.19 |
| 9 | — | — | — | 22.99 | 0.88 |
| 10 | — | — | — | 196.18 | 12.88 |

Let us denote by

- BC: the blocked clauses that we add in step 1,
- PR: the PR clauses that we add inductively in step 2,
- R1: the RUP inferences that we add inductively in step 3,
- R2: the RUP inferences that we add inductively in step 4.

We consider extended versions of the formulas where we gradually include more of the redundant clauses. We cumulatively introduce the redundant clauses from each step, i.e. when we add the PR clauses we also add the BC clauses.

For the satisfiable formulas $MYC_k^+$, the remaining number of solutions are in Table 2. We used allsat[3] to count the exact number of solutions. Adding only the BC or PR clauses drastically reduces the number of solutions. Adding them both leaves a fraction of the solutions.

For the unsatisfiable formulas, we ran CaDiCaL[4] [3] with a timeout of 2000 seconds on the original formulas and the versions including the clauses introduced at each step. The results are in Table 3. These runtimes are somewhat unexpected as R1 and R2 can be derived from $MYC_k$+PR with relatively few resolution steps. One would therefore expect the performance on $MYC_k$+PR, $MYC_k$+R1, and $MYC_k$+R2 to be similar. We study this observation in the next subsection.

## 5.3    Difficult extended Mycielski graph formulas

The CDCL paradigm has been highly successful, because it has been able to find short refutations for problems arising from various applications. However, the above results show that there exist formulas for which CDCL cannot find the short refutations. In particular, the $MYC_k$+PR formulas have length $\Theta(3^k k)$ and there exist resolution refutations of length $O(3^k k^3)$: Each clause in R1 and R2, of

---

[3] https://github.com/marijnheule/allsat
[4] http://fmv.jku.at/cadical/

which there are $O(3^k k^2)$, can be derived in $O(k)$ steps of resolution. As for the clique, it is known that $\text{PHP}_{k-1}$ has resolution refutations of length $O(2^k k^3)$ [4].

This shows that, even if we devise an algorithm to discover the redundant PR clauses automatically, the Mycielski graph formulas still remain difficult for the standard tools. After the clauses in BC and PR become part of the formula, the difficulty lies in deriving the R2 clauses automatically. If we resort to incremental SAT solving [10] and provide the cubes $u_{i,c} \wedge u_{j,c}$ (negation of each clause in R2) as assumptions to the solver, the formulas become relatively easily solvable. For instance, $\text{MYC}_{10}+\text{PR}$ takes approximately 3 minutes on a single CPU. Although it is unlikely that a solver can run this efficiently without any explicit guidance, the small runtime provides evidence that the shortest resolution proof of $\text{MYC}_{10}+\text{PR}$ is of modest length.

In this section, we describe a method for discovering useful cubes automatically and using them to solve the $\text{MYC}_k+\text{PR}$ formulas. While inefficient, with this method it at least becomes possible to find proofs of these formulas in a matter of minutes, compared to CDCL which did not succeed even with a timeout of three days on $\text{MYC}_{10}+\text{PR}$. Given a formula $F$, the below procedure discovers binary clauses, inserts them to $F$, and attempts to solve $F$ via CDCL.

1. Iteratively remove the clause that has the largest number of resolution candidates until the formula becomes satisfiable. For $\text{MYC}_k+\text{PR}$, this corresponds to simply removing the clause $w_1 \vee \ldots \vee w_{k-1}$. Call the newly obtained formula, which is satisfiable, $F^-$.
2. Repeat:
   (a) Sample $M$ satisfying assignments for $F^-$ using a local search solver (we used YalSAT[5] [2]).
   (b) Find all pairs of literals $(l_i, l_j)$ that do not appear together in any of the solutions sampled so far. Form a list with the cubes $l_i \wedge l_j$, and shuffle it in order to avoid ordering the pairs with respect to variable indices. In the case of $\text{MYC}_k+\text{PR}$, the clause $\overline{u_{i,c}} \vee \overline{u_{j,c}}$ is implied by $F^-$, hence $(u_{i,c}, u_{j,c})$ must be among the pairs found.
   (c) If the number of pairs found did not decrease by more than 1 percent after the latest addition of satisfying assignments, break.
3. Repeat:
   (a) Partition the remaining cubes into $P$ pieces. Use $P$ workers in parallel to perform incremental solving with a limit of $L$ conflicts allowed on the instances of the formula $F$ using each separate piece as the set of assumptions. Aggregate a list of refuted cubes.
   (b) For each refuted cube $B$, append $\overline{B}$ to the formula $F$.
   (c) If the number of refuted cubes is less than half of the previous iteration, break.
4. Run CDCL on the final formula $F$ that includes negations of all the refuted cubes.

Table 4 displays the results for formulas with $k \in \{7, \ldots, 10\}$ and varying numbers of parallel workers $P$.

---

[5] http://fmv.jku.at/yalsat/

**Table 4.** Results on finding proofs for MYC$_k$+PR. From left to right, the columns correspond to the number of samples used for obtaining a list of cubes, the number of cubes obtained after filtering pairs of literals, time it takes to sample solutions using a local search solver with 20 workers and filter pairs of literals, maximum number $L$ of conflicts allowed to the incremental SAT solver, number of parallel workers $P$, total time it takes to refute cubes and prove unsatisfiability of the final formula $F$, percentage of time spent in the final CDCL run on $F$, number of iterations spent refuting cubes and adding them to the formula.

| $k$ | #samples | #cubes | time to cubes | $L$ | $P$ | time to solve | final% | #iter |
|---|---|---|---|---|---|---|---|---|
| 7 | 2000 | 9675 | 18.4s | 100 | 1 | 15.4s | 0.39% | 2 |
| | | | | | 12 | 5.7s | 0.87% | 3 |
| | | | | | 25 | 5.3s | 0.94% | 4 |
| | | | | | 50 | 6.3s | 0.80% | 4 |
| 8 | 2000 | 38255 | 2m 15s | 100 | 1 | 2m 50s | 0.12% | 2 |
| | | | | | 12 | 44.4s | 0.43% | 4 |
| | | | | | 25 | 30.6s | 0.65% | 4 |
| | | | | | 50 | 33.5s | 0.60% | 5 |
| 9 | 3000 | 148624 | 10m 37s | 100 | 1 | 38m 40s | 0.03% | 2 |
| | | | | | 12 | 7m 4s | 0.14% | 5 |
| | | | | | 25 | 5m 22s | 0.24% | 6 |
| | | | | | 50 | 3m 26s | 0.26% | 5 |
| 10 | 3000 | 568214 | 35m 18s | 100 | 1 | 11h 37m | 0.003% | 3 |
| | | | | | 12 | 1h 55m | 0.04% | 6 |
| | | | | | 25 | 1h 7m | 0.33% | 5 |
| | | | | | 50 | 42m 18s | 0.32% | 6 |

# 6  Conclusion

We showed that there exist short DPR$^-$, DSPR$^-$, and PR$^-$ proofs of the colorability of Mycielski graphs. Interesting questions about the proof complexity of PR variants remain. For instance, DPR$^-$ has not been shown to separate from ER or Frege, and even simpler questions regarding upper bounds for some difficult tautologies are open. It is also unknown, although plausible, whether there exists a polynomial-length SPR$^-$ proof of the Mycielski graph formulas.

Apart from our theoretical results, we encountered formulas with short resolution proofs for which CDCL requires substantial runtime. We developed an automated reasoning method to solve these formulas. In future work, we plan to study whether this method is also effective on other problems that are challenging for CDCL.

**Acknowledgements.** This work has been supported by the National Science Foundation (NSF) under grant CCF-1813993.

# References

1. The On-Line Encyclopedia of Integer Sequences. Published electronically at https://oeis.org/A122695
2. Biere, A.: CaDiCaL, Lingeling, Plingeling, Treengeling, YalSAT entering the SAT competition 2017. In: Proceedings of SAT Competition 2017 – Solver and Benchmark Descriptions. vol. B-2017-1, pp. 14–15 (2017)
3. Biere, A.: CaDiCaL at the SAT Race 2019. In: Proceedings of SAT Race 2019 – Solver and Benchmark Descriptions. vol. B-2019-1, pp. 8–9 (2019)
4. Buss, S., Pitassi, T.: Resolution and the weak pigeonhole principle. In: Computer Science Logic, pp. 149–156 (1998)
5. Buss, S., Thapen, N.: DRAT proofs, propagation redundancy, and extended resolution. In: Theory and Applications of Satisfiability Testing – SAT 2019. pp. 71–89 (2019)
6. Caramia, M., Dell'Olmo, P.: Coloring graphs by iterated local search traversing feasible and infeasible solutions. Discrete Applied Mathematics **156**(2), 201–217 (2008)
7. Cook, S.A., Reckhow, R.A.: The relative efficiency of propositional proof systems. The Journal of Symbolic Logic **44**(1), 36–50 (1979)
8. Crawford, J.M., Ginsberg, M.L., Luks, E.M., Roy, A.: Symmetry-breaking predicates for search problems. In: Proceedings of the Fifth International Conference on Principles of Knowledge Representation and Reasoning. pp. 148–159 (1996)
9. Desrosiers, C., Galinier, P., Hertz, A.: Efficient algorithms for finding critical subgraphs. Discrete Applied Mathematics **156**(2), 244–266 (2008)
10. Eén, N., Sörensson, N.: Temporal induction by incremental SAT solving. Electronic Notes in Theoretical Computer Science **89**(4), 543–560 (2003)
11. Goldberg, E., Novikov, Y.: Verification of proofs of unsatisfiability for CNF formulas. In: Proceedings of the Conference on Design, Automation and Test in Europe (DATE 2003). pp. 886–891 (2003)
12. Heule, M.J.H., Biere, A.: What a difference a variable makes. In: Tools and Algorithms for the Construction and Analysis of Systems. pp. 75–92 (2018)
13. Heule, M.J.H., Kiesl, B., Biere, A.: Clausal proofs of mutilated chessboards. In: NASA Formal Methods. pp. 204–210 (2019)
14. Heule, M.J.H., Kiesl, B., Biere, A.: Strong extension-free proof systems. Journal of Automated Reasoning **64**(3), 533–554 (2020)
15. Kullmann, O.: On a generalization of extended resolution. Discrete Applied Mathematics **96–97**, 149–176 (1999)
16. Marques-Silva, J.P., Sakallah, K.A.: GRASP—a new search algorithm for satisfiability. In: Proceedings of the 1996 IEEE/ACM International Conference on Computer-Aided Design. pp. 220–227 (1997)
17. Mycielski, J.: Sur le coloriage des graphs. Colloquium Mathematicae **3**(2), 161–162 (1955)
18. Ramani, A., Aloul, F.A., Markov, I.L., Sakallah, K.A.: Breaking instance-independent symmetries in exact graph coloring. In: Proceedings of the Conference on Design, Automation and Test in Europe (DATE 2004). pp. 324–329 (2004)
19. Schaafsma, B., Heule, M.J.H., van Maaren, H.: Dynamic symmetry breaking by simulating Zykov contraction. In: Theory and Applications of Satisfiability Testing – SAT 2009. pp. 223–236 (2009)
20. Trick, M.A., Yildiz, H.: A large neighborhood search heuristic for graph coloring. In: Integration of AI and OR Techniques in Constraint Programming for Combinatorial Optimization Problems. pp. 346–360 (2007)

21. Van Gelder, A.: Another look at graph coloring via propositional satisfiability. Discrete Applied Mathematics **156**(2), 230–243 (2008)
22. Vinyals, M.: Hard examples for common variable decision heuristics. In: Proceedings of the Thirty-Fourth AAAI Conference on Artificial Intelligence (2020)
23. Zhou, Z., Li, C.M., Huang, C., Xu, R.: An exact algorithm with learning for the graph coloring problem. Computers and Operations Research **51**, 282–301 (2014)
24. Zhu, Q., Kitchen, N., Kuehlmann, A., Sangiovanni-Vincentelli, A.: SAT sweeping with local observability don't-cares. In: Proceedings of the 43rd Annual Design Automation Conference. pp. 229–234 (2006)

# Towards a Better Understanding of (Partial Weighted) MaxSAT Proof Systems

Javier Larrosa[iD] and Emma Rollon[✉]

Universitat Politècnica de Catalunya, Barcelona Tech, Barcelona, Spain
{larrosa,erollon}@cs.upc.edu
http://www.cs.upc.edu

**Abstract.** MaxSAT is a very popular language for discrete optimization with many domains of application. While there has been a lot of progress in MaxSAT solvers during the last decade, the theoretical analysis of MaxSAT inference has not followed the pace. Aiming at compensating that lack of balance, in this paper we do a proof complexity approach to MaxSAT resolution-based proof systems. First, we give some basic definitions on completeness and show that refutational completeness makes compleness redundant, as it happens in SAT. Then we take three inference rules such that adding them sequentially allows us to navigate from the weakest to the strongest resolution-based MaxSAT system available (i.e., from standalone MaxSAT resolution to the recently proposed ResE), each rule making the system stronger. Finally, we show that the strongest system captures the recently proposed concept of Circular Proof while being conceptually simpler, since weights, which are intrinsic in MaxSAT, naturally guarantee the flow condition required for the SAT case.

**Keywords:** MaxSAT · Rule-based proof systems · Circular proofs

## 1 Introduction

*Proof Complexity* is the field aiming to understand the computational cost required to prove or refute statements. Different proof systems may provide different proofs for the same formula and some proof systems are provably more efficient than others. When that happens, proof complexity cares about which elements of the more powerful proof system really make the difference.

In propositional logic, proof systems that work with CNF formulas have attracted the interest of researchers for several decades. One of the reasons is that CNF is the working language of the extremely successful SAT solvers and the search spaces that they traverse can be understood and analyzed as proofs [5].

Projects TIN2015-69175-C4-3-R and RTI2018-094403-B-C33, funded by: FEDER/ Ministerio de Ciencia e Innovación − Agencia Estatal de Investigación, Spain.

© Springer Nature Switzerland AG 2020
L. Pulina and M. Seidl (Eds.): SAT 2020, LNCS 12178, pp. 218–232, 2020.
https://doi.org/10.1007/978-3-030-51825-7_16

(Partial Weighted) *MaxSAT* is the optimization version of SAT. Although discrete optimization problems can be modeled and solved with SAT solvers, many of these problems are more naturally treated as MaxSAT. For this reason the design of MaxSAT solvers has attracted the interest of researchers in the last decade. Interestingly, while some of the first efficient MaxSAT solvers were strongly influenced by MaxSAT inference [9], this influence has diminished along time. The currently most efficient algorithms solve MaxSAT by sophisticated sequences of calls to SAT solvers [1,4,11].

We think it is important to understand this scientific trend with a more formal approach and such understanding must go through an analysis of the possibilities and limitations of MaxSAT proof systems (how MaxSAT inference compares with obtaining the same result with a sequence of SAT inferences?). The purpose of this paper is to start contributing in that direction by improving the understanding of MaxSAT proof systems. With that aim we extend some classic proof complexity concepts (i.e, entailment, completeness, etc) to MaxSAT and analyze three proof systems of increasing complexity: from stand-alone MaxSAT resolution (**Res**) [9] to the recently proposed resolution with extension (**ResE**) [10]. For the sake of clarity, we break the extension rule of **ResE** into two atomic rules: *split* and *virtual*; and analyze their incremental power. Our results show that each add-on makes a provable stronger system. More precisely, we have observed that: **Res** is sound and refutationally complete. Adding the split rule (**ResS**) we get completeness and (unlike what happens in SAT) some exponential speed-up in certain refutations. Further adding the virtual rule (**ResSV**), which allows to keep negative weights during proofs, we get further speed-up by capturing the concept of *circular proofs* [3]. We also give the interesting and somehow unexpected result that in some cases rephrasing a MaxSAT refutation as a MaxSAT entailment may transform the problem from exponentially hard to polynomial when using **ResSV**.

The structure of the paper is as follows. In Sects. 2 and 3 we provide preliminaries on SAT and MaxSAT, respectively. In Sect. 4 we define some variations of the Pigeon Hole Problem that we need for the proofs of the theorems. In Sect. 5 we provide basic definition and properties on MaxSAT proof systems and introduce and analyze the different systems addressed in the paper. In Sect. 6 we show how the strongest proof system **ResSV** captures the notion of Circular Proof. Finally, in Sect. 7, we give some conclusions.

## 2 SAT Preliminaries

A *boolean variable* $x$ takes values on the set $\{0, 1\}$. A *literal* is a variable $x$ (positive literal) or its negation $\bar{x}$ (negative literal). A *clause* is a disjunction of literals. A clause $C$ is satisfied by a truth assignment $X$ if $X$ contains at least one of the literals in $C$. The *empty clause* is denoted $\square$ and cannot be satisfied. The negation of a clause $C = l_1 \vee l_2 \vee \ldots \vee l_p$ is satisfied if all its literals are falsified and this can be trivially expressed in CNF as the set of unit clause $\{\bar{l}_1, \bar{l}_2, \ldots, \bar{l}_p\}$.

A *CNF formula* $\mathcal{F}$ is a set of clauses (understood as a conjunction). A truth assignment satisfies a formula if it satisfies all its clauses. If such an assignment exists, we say that the assignment is a model and the formula is satisfiable, noted $SAT(\mathcal{F})$. Determining whether a formula is satisfiable constitutes the well-known SAT Problem.

We say that formula $\mathcal{F}$ *entails* formula $\mathcal{G}$, noted $\mathcal{F} \models \mathcal{G}$, if every model of $\mathcal{F}$ is also a model of $\mathcal{G}$. Two formulas $\mathcal{F}$ and $\mathcal{G}$ are *equivalent*, noted $\mathcal{F} \equiv \mathcal{G}$, if they entail each other.

An *inference rule* is given by a set of antecedent clauses and a set of consequent clauses. In SAT, the intended meaning of an inference rule is that if some clauses of a formula match the antecedents, the consequents can be *added*. The rule is *sound* if every truth assignment that satisfies the antecedents also satisfies the consequents. The process of applying an inference rule to a formula $\mathcal{F}$ is noted $\mathcal{F} \vdash \mathcal{F}'$.

Consider the following two rules [3,12],

$$\frac{x \vee A \qquad \overline{x} \vee B}{A \vee B} \qquad \qquad \frac{A}{A \vee x \qquad A \vee \overline{x}}$$

$$\text{(resolution)} \qquad\qquad\qquad \text{(split)}$$

where $A$ and $B$ are arbitrary (possibly empty) disjunctions of literals and $x$ is an arbitrary variable. In propositional logic it is customary to define rules with just one consequent because one rule with $s$ consequents can be obtained from $s$ one-consequent rules. As we will see, this is not the case in MaxSAT. For this reason, here we prefer to introduce the two-consequents split rule instead of the equivalent weakening rule [3] to keep the parallelism with MaxSAT more evident.

A *proof system* $\Delta$ is a set of inference rules. A *proof* of *length* $e$ under a proof system $\Delta$ is a finite sequence $\mathcal{F}_0 \vdash \mathcal{F}_1 \vdash \ldots \vdash \mathcal{F}_e$ where $\mathcal{F}_0$ is the original formula and each $\mathcal{F}_i$ is obtained by applying an inference rule from $\Delta$. We will use $\vdash^*$ to denote an arbitrary number of inference steps. A *short* proof is a proof whose length can be bounded by a polynomial on $|\mathcal{F}|$. A *refutation* is a proof such that $\square \in \mathcal{F}_e$. Refutations are important because they prove unsatisfiability.

A proof system is *sound* if all its rules are sound. All the SAT rules and proof systems considered in this paper are sound. A proof system is *complete* if for every $\mathcal{F}, \mathcal{G}$ such that $\mathcal{F} \models \mathcal{G}$, there is a proof $\mathcal{F} \vdash^* \mathcal{H}$ with $\mathcal{G} \subseteq \mathcal{H}$. Although completeness is a natural and elegant property, it has limited practical interest. For that reason a weaker version of completeness has been defined. A proof system is *refutationally complete* if for every unsatisfiable formula $\mathcal{F}$ there is a refutation starting in $\mathcal{F}$ (i.e, completeness is required only for refutations). It is usually believed that refutational completeness is enough for practical purposes. The reason is that $\mathcal{F} \models \mathcal{G}$ if and only if $\mathcal{F} \cup \overline{\mathcal{G}}$ is unsatisfiable, so any implicationally complete proof system can prove the entailment by deriving a refutation from a CNF formula equivalent to $\mathcal{F} \cup \overline{\mathcal{G}}$.

It is well-known that the proof system made exclusively of resolution is refutationally complete and adding the split rule makes the system complete. The following property says that adding the split rule does not give any advantage to resolution in terms of refutational power.

*Property 1* [(see Lemma 7 in [2]]. A proof system with resolution and split as inference rules cannot make shorter refutations than a proof system with only resolution.

## 3  MaxSAT Preliminaries

A *weight* $w$ is a positive number or $\infty$ (i.e, $w \in \mathbb{R}^+ \cup \{\infty\}$). We extend sum and substraction to weights defining $\infty + w = \infty$ and $\infty - w = \infty$ for all $w$. Note that $v - w$ is only defined when $w \leq v$.

A *weighted clause* is a pair $(C, w)$ where $C$ is a clause and $w$ is a weight associated to its falsification. If $w = \infty$ we say that the clause is *hard*, else it is *soft*. A weighted *MaxSAT CNF formula* is a set of weighted clauses $\mathcal{F} = \{(C_1, w_1), (C_2, w_2), \dots, (C_p, w_p)\}$. If all the clauses are hard, we say that the formula is hard. We say that $\mathcal{G} \subseteq \mathcal{F}$ if for all $(C, w) \in \mathcal{G}$ there is a $(C, w') \in \mathcal{F}$ with $w \leq w'$.

Given a formula $\mathcal{F}$, we define the cost of a truth assignment $X$, noted $\mathcal{F}(X)$, as the sum of weights over the clauses that are falsified by $X$. The MaxSAT problem is to find the minimum cost over the set of all truth assignments,

$$MaxSAT(\mathcal{F}) = \min_X \mathcal{F}(X)$$

This definition of MaxSAT including weights and hard clauses is sometimes referred to as Partial Weighted MaxSAT [11]. Note that any clause $(C, w)$ can be broken into two clauses $(C, u), (C, v)$ as long as $u + v = w$. In the following we will assume that clauses are separated and merged as needed.

We say that formula $\mathcal{F}$ *entails* formula $\mathcal{G}$, noted $\mathcal{F} \models \mathcal{G}$, if $\mathcal{G}(X)$ is a lower bound of $\mathcal{F}(X)$. That is, if for all $X$, $\mathcal{F}(X) \geq \mathcal{G}(X)$. We say that two formulas $\mathcal{F}$ and $\mathcal{G}$ are *equivalent*, noted $F \equiv G$, if they entail each other. That is, if forall $X$, $\mathcal{F}(X) = \mathcal{G}(X)$.

In the following Sections we will find useful to deal with negated clauses. Hence, the corresponding definitions and useful properties. Let $A$ and $B$ be arbitrary disjunctions of literals. Let $(A \vee \overline{B}, w)$ mean that falsifying $A \vee \overline{B}$ incurs a cost of $w$. Although $A \vee \overline{B}$ is not a clause, the following property shows that it can be efficiently transformed into a CNF equivalent,

*Property 2.* $\{(A \vee \overline{l_1 \vee l_2 \vee \dots \vee l_p}, w)\} \equiv \{(A \vee \overline{l_1}, w), (A \vee l_1 \vee \overline{l_2}, w), \dots, (A \vee l_1 \vee \dots \vee l_{p-1} \vee \overline{l_p}, w)\}$.

Observe that if we restrict the MaxSAT language to hard formulas we have standard SAT CNF formulas where $\infty$ corresponds to *false* and 0 corresponds to *true*. Note that all the previous definitions naturally instantiate to their SAT analogous.

## 4  Pigeon Hole Problem and Variations

We define the well-known Pigeon Hole Problem *PHP* and three MaxSAT versions $SPHP$, $SPHP^0$ and $SPHP^1$, that we will be using in the proof of our results.

In the *Pigeon Hole Problem PHP* the goal is to assign $m + 1$ pigeons to $m$ holes without any pair of pigeons sharing their hole. In the usual SAT encoding there is a boolean variable $x_{ij}$ (with $1 \leq i \leq m + 1$, and $1 \leq j \leq m$) which is true if pigeon $i$ is in hole $j$. There are two groups of clauses. For each pigeon $i$, we have the clause,

$$\mathcal{P}_i = \{x_{i1} \vee x_{i2} \vee \ldots \vee x_{im}\}$$

indicating that pigeon $i$ must be assigned to a hole. For each hole $j$ we have the set of clauses,

$$\mathcal{H}_j = \{\overline{x}_{ij} \vee \overline{x}_{i'j} \mid 1 \leq i < i' \leq m + 1\}$$

indicating that hole $j$ is occupied by at most one pigeon. Let $\mathcal{K}$ be the union of all these sets of clauses $\mathcal{K} = \cup_{1 \leq i \leq m+1} \mathcal{P}_i \cup_{1 \leq j \leq m} \mathcal{H}_j$. It is obvious that $\mathcal{K}$ is an unsatisfiable CNF formula. In MaxSAT notation the pigeon hole problem is,

$$PHP = \{(C, \infty) \mid C \in \mathcal{K}\}$$

and clearly $MaxSAT(PHP) = \infty$.

In the *soft Pigeon Hole Problem SPHP* the goal is to find the assignment that falsifies the minimum number of clauses. In MaxSAT language it is encoded as,

$$SPHP = \{(C, 1) \mid C \in \mathcal{K}\}$$

and it is obvious that $MaxSAT(SPHP) = 1$.

The $SPHP^0$ problem is like the soft pigeon hole problem but augmented with one more clause $(\Box, m^2 + m)$ where $m$ is the number of holes. Note that $MaxSAT(SPHP^0) = m^2 + m + 1$.

Finally, the $SPHP^1$ problem is like the soft pigeon hole problem but augmented with a set of unit clauses $\{(x_{ij}, 1), (\overline{x}_{ij}, 1) \mid 1 \leq i \leq m + 1, 1 \leq j \leq m\}$. Note that $MaxSAT(SPHP^1) = m^2 + m + 1$.

## 5   MaxSAT Proof Systems

A *MaxSAT inference rule* is given by a set of antecedent clauses and a set of consequent clause. In MaxSAT, the application of an inference rule is to *replace* the antecedents by the consequents. The process of applying an inference rule to a formula $\mathcal{F}$ is also noted $\mathcal{F} \vdash \mathcal{F}'$. The rule is *sound* if it preserves the equivalence of the formula.

As in the SAT case, given a proof system $\Delta$ (namely, a set of rules) a *proof* of length $e$ is a sequence $\mathcal{F}_0 \vdash \mathcal{F}_1 \vdash \ldots \vdash \mathcal{F}_e$ where $\mathcal{F}_0$ is the original formula and each $\mathcal{F}_i$ is obtained by applying an inference rule from $\Delta$. If $\mathcal{G} \subseteq \mathcal{F}_e$, we say that the proof is a proof of $\mathcal{G}$ from $\mathcal{F}$.

A proof system is *sound* if all its rules are sound. In this paper all MaxSAT rules and proof systems are sound. A proof system is *complete* if for every $\mathcal{F}$, $\mathcal{G}$

such that $\mathcal{F} \models \mathcal{G}$, there is a proof of $\mathcal{G}$ from $\mathcal{F}$. A *refutation* of $\mathcal{F}$ is a proof of $(\Box, k)$ from $\mathcal{F}$ with $k = MaxSAT(\mathcal{F})$. A proof system is *refutationally complete* if it can derive a refutation of every formula $\mathcal{F}$.

Next we show that, similarly to what happens in SAT, refutationally completeness is sufficient for practical purposes. The reason is that it can also be used to proof or disproof general entailment, making completeness somehow redundant. We need first to define the maximum soft cost of a formula as $\top(\mathcal{F}) = \sum_{(C,w)\in F|w\neq\infty} w$ and the negation of a MaxSAT formula as the negation of all its clauses $\overline{\mathcal{F}} = \{(\overline{C}, w)| (C, w) \in \mathcal{F}\}$. The following property tells the effect of negating a formula without hard clauses,

*Property 3.* If $\mathcal{F}$ is a CNF MaxSAT formula without hard clauses, then

$$\overline{\mathcal{F}}(X) = \top(\mathcal{F}) - \mathcal{F}(X)$$

*Proof.* Let $X$ be a truth assignment, $\mathcal{S}$ be the set of clauses satisfied by $X$ and $\mathcal{U}$ be the set of clauses falsified by $X$. It is clear that $\mathcal{F}(X) = \sum_{(C_i, w_i)\in\mathcal{U}} w_i$ while $\overline{\mathcal{F}}(X) = \sum_{(C_i, w_i)\in\mathcal{S}} w_i$. Since $\mathcal{S} \cap \mathcal{U} = \emptyset$ and $\mathcal{S} \cup \mathcal{U} = \mathcal{F}$, then $\sum_{(C_i, w_i)\in U} w_i + \sum_{(C_i, w_i)\in S} w_i = \top(\mathcal{F})$. Therefore, $\mathcal{F}(X) + \overline{\mathcal{F}}(X) = \top(\mathcal{F})$ and, as a consequence, $\overline{\mathcal{F}}(X) = \top(\mathcal{F}) - \mathcal{F}(X)$.

We can now show that an entailment $\mathcal{F} \models \mathcal{G}$ can be rephrased as a MaxSAT problem,

**Theorem 1.** *Let $\mathcal{F}$ and $\mathcal{G}$ be two MaxSAT formulas, possibly with hard clauses. Then,*

$$\mathcal{F} \models \mathcal{G} \text{ iff } MaxSAT(\mathcal{F} \cup \overline{\mathcal{G}}') \geq \top(\mathcal{G}')$$

*where $\mathcal{G}'$ is a softened version of $\mathcal{G}$ in which infinity weights are replaced by $\max\{\top(\mathcal{G}), \top(\mathcal{F})\} + 1$.*

*Proof.* Let us prove the if direction. $\mathcal{F} \models \mathcal{G}$ means that $\forall X, \mathcal{F}(X) \geq \mathcal{G}(X)$. Also, by construction $\forall X, \mathcal{G}(X) \geq \mathcal{G}'(X)$. Therefore, $\forall X, \mathcal{F}(X) \geq \mathcal{G}(X)$. Because $\mathcal{G}'$ does not contain hard clauses, $\mathcal{G}'(X) \neq \infty$, which means that, $\forall X, \mathcal{F}(X) - \mathcal{G}'(X) \geq 0$ Adding $\top(\mathcal{G}')$ to both sides of the disequality we get, $\forall X, \mathcal{F}(X) + \top(\mathcal{G}') - \mathcal{G}'(X) \geq \top(\mathcal{G}')$. By Property 3, we have, $\forall X, \mathcal{F}(X) + \overline{\mathcal{G}}'(X) \geq \top(\mathcal{G}')$ which clearly means that, $MaxSAT(\mathcal{F} \cup \overline{\mathcal{G}}') \geq \top(\mathcal{G}')$.

Let us proof the else if direction. $maxSAT(\mathcal{F} \cup \overline{\mathcal{G}}') \geq \top(\mathcal{G}')$ implies that $\forall X, \mathcal{F}(X) + \overline{\mathcal{G}}'(X) \geq \top(\mathcal{G}')$. Moreover, since $\overline{\mathcal{G}}'$ does not have hard clauses, from Property 3 we know that, $\forall X, \mathcal{F}(X) + \top(\mathcal{G}') - \mathcal{G}'(X) \geq \top(\mathcal{G}')$ so we have, $\forall X, \mathcal{F}(X) \geq \mathcal{G}'(X)$ and we need to have, $\forall X, \mathcal{F}(X) \geq \mathcal{G}(X)$. We reason on cases for truth assignment $X$:

1. If $\mathcal{G}'(X) <= \top(\mathcal{G})$, by definition of $\mathcal{G}'$, $\mathcal{G}'(X) = \mathcal{G}(X) \neq \infty$. Therefore, $\mathcal{F}(X) \geq \mathcal{G}'(X) = \mathcal{G}(X)$, which proofs this case.

2. If $\mathcal{G}'(X) > \top(\mathcal{G})$, by definition of $\mathcal{G}'$, $\mathcal{G}(X) = \infty$. We show that in this case, $\mathcal{F}(X) = \infty$.
   - if $\top(\mathcal{F}) \leq \top(\mathcal{G})$ then $\mathcal{F}(X) \geq \mathcal{G}'(X) > \top(\mathcal{G}) \geq \top(\mathcal{F})$. We show that $\mathcal{F}(X) > \top(\mathcal{F})$ implies that $\mathcal{F}(X) = \infty$. We proceed by contradiction. Let us suppose that $\mathcal{F}(X) > \top(\mathcal{F})$ and $\mathcal{F}(X) \neq \infty$. The latter means that $X$ satisfies all hard clauses. As a consequence, $\mathcal{F}(X) \leq \top(\mathcal{F})$, which contradicts the hypothesis.
   - if $\top(\mathcal{F}) > \top(\mathcal{G})$, then there are no $X$ such that $\top(\mathcal{G}) < \mathcal{G}'(X) \leq \top(\mathcal{F})$. By definition of $\mathcal{G}'$, forall $(C_i, \infty) \in G$, $(C_i, \top(\mathcal{F}) + 1) \in \mathcal{G}'$. Therefore, either $X$ satisfies all hard clauses in $\mathcal{G}$ and then $\mathcal{G}'(X) \leq \top(\mathcal{G})$ or $X$ falsifies at least one hard clause in $\mathcal{G}$ and then $\mathcal{G}'(X) > \top(\mathcal{F})$.

which proofs the theorem.

The application of the previous theorem to single clause entailment yields the following corollary.

**Corollary 1.** *Let $\mathcal{F}$ be a formula and $(C, w)$ be a weighted clause. Then,*

$$\mathcal{F} \models (C, w) \text{ iff } MaxSAT(\mathcal{F} \cup \{(\overline{C}, u)\} \geq u$$

*where* $u = \begin{cases} w, & \text{if } w \neq \infty \\ \top(\mathcal{F}) + 1, & \text{if } w = \infty \end{cases}$

A useful application of this corollary will be shown in Sect. 5.3.

In the rest of the section we introduce and analyze the incremental impact of the three inference rules.

## 5.1   Resolution

The MaxSAT *resolution* rule [8] is,

$$\frac{(x \vee A, v) \quad (\overline{x} \vee B, w)}{\begin{array}{c} (A \vee B, m) \\ (x \vee A, v - m) \quad (\overline{x} \vee B, w - m) \\ (x \vee A \vee \overline{B}, m) \quad (\overline{x} \vee B \vee \overline{A}, m) \end{array}}$$

where $A$ and $B$ are arbitrary (possibly empty) disjunctions of literals and $m = \min\{v, w\}$. When $A$ (resp. $B$) is empty, $\overline{A}$ (resp. $\overline{B}$) is constant true, so $x \vee \overline{A} \vee B$ (resp. $x \vee A \vee \overline{B}$) is tautological. Note that MaxSAT resolution, when applied to two hard clauses, corresponds to SAT resolution.

It is known that the proof system **Res** made exclusively of the resolution rule is refutationally complete,

**Theorem 2** [6,9]. ***Res** is refutationally complete.*

However, as we show next, it is not complete.

**Theorem 3.** *Res is not complete.*

*Proof.* Consider formula $\mathcal{F} = \{(x, 1), (y, 1)\}$. It is clear that $\mathcal{F} \models (x \vee y, 1)$ which cannot be derived with **Res**.

It is well-known that **Res** cannot compute short refutations for $PHP$ [12] or $SPHP$ [6]. However, it can efficiently refute $SPHP^1$. We write it as a property and sketch the proof (which is a direct adaptation of what was proved in [7] and [10]) because it will be instrumental in the proof of several results in the rest of this section,

*Property 4.* There is a short *Res* refutation of $SPHP^1$.

*Proof.* The proof is based on the fact that for each one of the $m + 1$ pigeons there is a short refutation

$$\{(x_{i1} \vee x_{i2} \vee \ldots \vee x_{im}, 1)\} \cup \{(\overline{x}_{ij}, 1) \mid 1 \leq j \leq m\} \vdash^* \mathcal{G} \cup (\Box, 1)$$

and for each one of the $m$ holes there is a short refutation

$$\{(\overline{x}_{ij} \vee \overline{x}_{i'j}, 1) \mid 1 \leq i < i' \leq m + 1\} \cup \{(x_{ij}, 1) \mid 1 \leq i \leq m + 1\} \vdash^* \mathcal{G} \cup \{(\Box, m)\}$$

Because each derivation is independent of the other we can concatenate them into,

$$SPHP^1 \vdash^* \mathcal{G} \cup \{(\Box, m^2 + m + 1)\}$$

which is a refutation of $SPHP^1$.

## 5.2  Split

The *split* rule,

$$\frac{(A, w)}{(A \vee x, w) \quad (A \vee \overline{x}, w)}$$

is the natural extension of its SAT counterpart. Consider the proof system **ResS**, made of resolution and split. We show that, as it happens in the SAT case, the split rule brings completeness,

**Theorem 4.** *ResS is complete.*

*Proof.* The proof is based on the following facts:

1. For every formula $\mathcal{F}$ there is a proof $\mathcal{F} \vdash^* \mathcal{F}^e$ where $\mathcal{F}^e$ is made exclusively of splits in which all the clauses of $F^e$ contain all the variables in the formula and there are no repeated clauses. Each clause $(C, w) \in \mathcal{F}$ can be expanded to a new variable not in $C$ using the split rule. This process can be repeated until all clauses in the current formula contain all the variables in the formula. Note that all clauses $(C', u)$, $(C', v)$ can be merged and, as a result, $\mathcal{F}^e$ does not contain repeated clauses.

2. If $\mathcal{F} \vdash^* \mathcal{F}^e$ then $\mathcal{F}^e \vdash^* \mathcal{F}$. Let $\mathcal{F} = \mathcal{F}_0 \vdash \mathcal{F}_1 \vdash \mathcal{F}_2 \vdash \ldots \vdash \mathcal{F}_p = \mathcal{F}^e$ be the proof from $\mathcal{F}$ to $\mathcal{F}^e$. Then, $\mathcal{F}^e = \mathcal{F}_p \vdash \ldots \vdash \mathcal{F}_2 \vdash \mathcal{F}_1 \vdash \mathcal{F}_0 = \mathcal{F}$ is done resolving the pairs of clauses in $\mathcal{F}_i$ that were splitted in the $\mathcal{F}_{i-1} \vdash \mathcal{F}_i$ step.

3. If $\mathcal{F}^e(X) = w$ then there exists a unique clause $(C, w) \in \mathcal{F}^e$ which is falsified by $X$

By fact (1), $\mathcal{F} \vdash^* \mathcal{F}^e$ and $\mathcal{G} \vdash^* \mathcal{G}^e$. Because of soundness, $\vdash$, $\forall X, \mathcal{F}(X) = \mathcal{F}^e(X)$ and $\forall X, \mathcal{G}(X) = \mathcal{G}^e(X)$. Since $\mathcal{F} \models \mathcal{G}$, $\forall X, F(X) \geq G(X)$. Therefore, $\forall X, F^e(X) \geq \mathcal{G}^e(X)$ which, by fact (3), means that for each $X$ there exists a unique $(C, F^e(X)) \in \mathcal{F}^e$ and $(C, G^e(X))$ which is falsified by $X$. Separating all $(C, \mathcal{F}^e(X))$ into $(C, \mathcal{G}^e(X))$, $(C, \mathcal{F}^e(X) - \mathcal{G}^e(X))$ we have $\mathcal{F}^e = \mathcal{G}^e \cup \mathcal{H}^e$. Therefore, $\mathcal{F} \vdash \mathcal{F}^e = \mathcal{G}^e \cup \mathcal{H}^e$. By fact (2), $\mathcal{G}^e \cup \mathcal{H}^e \vdash^* \mathcal{G} \cup \mathcal{H}^e$.

However, unlike what happens in the SAT case (see Property 1), **ResS** is stronger than **Res**,

**Theorem 5.** *ResS is stronger than **Res**.*

*Proof.* On the one hand, it is clear that **ResS** can simulate any proof of **Res** since it is a superset of **Res**. On the other hand, unlike **Res**, **ResS** can produce short refutations for $SPHP^0$, as shown below.

First, let us proof that **Res** cannot produce short refutations for $SPHP^0$. Since the resolution rule does not apply to the empty clause $(\square, w)$, if **Res** could refute $SPHP^0$ in polynomial time it would also refute $SPHP$ in polynomial time, which is impossible [6].

**ResS** can produce short refutations for $SPHP^0$ because it can transform $SPHP^0$ into $SPHP^1$ and then apply Property 4. The transformation is done by a sequence of splits,

$$\frac{(\square, 1)}{(x_{ij}, 1) \quad (\overline{x}_{ij}, 1)}$$

that move one unit of weight from the empty clause to every variable in the formula and its negation.

## 5.3    Virtual

In a recent paper [10] we proposed a proof system in which clauses with negative weights can appear during the proof. This is equivalent to adding to **ResS** the *virtual* rule,

$$\frac{}{(A, w) \quad (A, -w)}$$

which allows to introduce a fresh clause $(A, w)$ into the formula. To preserve soundness (i.e, cancel out the effect of the addition) it also adds $(A, -w)$.

Let **ResSV** be the proof system made of resolution, split and virtual (note that resolution and split are only defined for antecedents with positive weights).

It has been shown that if $\mathcal{F}_0 \vdash^* \mathcal{F}_e$ is a **ResSV** proof and $\mathcal{F}_e$ *does not contain any negative weight*, then for every $\mathcal{G} \subseteq \mathcal{F}_e$ we have that $\mathcal{F} \models \mathcal{G}$.

The following theorem shows that the virtual rule adds further strength to the proof system,

**Theorem 6.** *ResSV is stronger than ResS.*

*Proof.* On the one hand, it is clear that **ResSV** can simulate any proof of **ResS** since it is a superset of **ResS**. On the other hand, **ResSV** can produce a short refutation of $SPHP$ and **ResS** cannot.

The short refutation of **ResSV**, as shown in [10], is obtained by first virtually transforming $SPHP$ into $SPHP^1$. Then, it uses Property 4 to derive $(\Box, m^2 + m + 1)$. Finally, it splits one unit of the empty clause cost to each pair $x_{ij}, \overline{x}_{ij}$ to cancel out negative weights. At the end of the process all clauses have positive weight while still having $(\Box, 1)$.

It is clear that **ResS** cannot polynomially refute $SPHP$ because otherwise a SAT proof system with resolution and split rules would produce shorter refutations than a SAT proof system with only resolution, which contradicts Property 1.

We will finish this section showing that Theorem 1 has an unexpected application in the context of **ResSV**. Consider the problem of proving $PHP \models (\Box, \infty)$. This can be done with a refutation of $PHP$. Namely $PHP \vdash^* (\Box, \infty) \cup \mathcal{F}$ or using Corollary 1, which tells that $\mathcal{F} \models (\Box, \infty)$ if and only if $MaxSAT(\mathcal{F}) \geq 1$. The following two theorems shows that **ResSV** cannot do efficiently the first approach, but can do efficiently the second.

**Theorem 7.** *There is no short ResSV refutation of $PHP$.*

*Proof.* Virtual rule cannot introduce hard clauses and resolution and split rules only produce a hard consequence if they have hard antecedents. As a consequence, $(\Box, \infty)$ can only be obtained by resolving or splitting hard clauses in $PHP$. If **ResSV** produce a short refutation for $PHP$, **ResS** and, as a consequence **Res**, also produce the same short refutation for $PHP$, which contradicts Property 1.

**Theorem 8.** *There is a polynomial ResSV proof of $(\Box, 1)$ from $PHP$.*

*Proof.* We only need to apply the virtual rule,

$$\frac{}{(\Box, m^2 + m) \quad (\Box, -m^2 - m)}$$

and then split,

$$\frac{(\Box, 1)}{(x_{ij}, 1) \quad (\overline{x}_{ij}, 1)}$$

for each $i, j$. The resulting problem is similar to $SPHP^1$ but with hard clauses. At this point and adapting the proof of 4 we can derive $(\Box, m^2 + m + 1)$ cancel out the negative weight while still retaining $(\Box, 1)$.

# 6   MaxSAT Circular Proofs

In this section we study the relation between **ResSV** and the recently proposed concept of circular proofs [3]. Circular proofs allows the addition of an arbitrary set of clauses to the original formula. It can be seen that conclusions are sound as long as the added clauses are *re-derived* as many times as they are used. In the original paper this condition is characterized as the existence of a flow in a graphical representation of the proof. Here we show that the **ResSV** proof system naturally captures the same idea and extends it from SAT to MaxSAT with an arguably simpler notation. In particular, the virtual rule guarantees the existence of the flow.

## 6.1   SAT Circular Proofs

We start reviewing the SAT case, as defined in [3]. Given a CNF formula $\mathcal{F}$ a *circular pre-proof* of $C_r$ from $\mathcal{F}$ is a sequence,

$$\Pi = C_1, C_2, \ldots, C_p, C_{p+1}, C_{p+2}, \ldots, C_{p+q}, C_{p+q+1}, C_{p+q+2}, \ldots, C_r$$

such that $\mathcal{F} = \{C_1, C_2, \ldots, C_p\}$, $\mathcal{B} = \{C_{p+1}, C_{p+2}, \ldots, C_{p+q}\}$ is an arbitrary set of clauses and each $C_i$ $(i > p+q)$ is obtained from previous clauses by applying an inference rule in the proof system. Note that the same clause can be both derived and used several times during the proof.

A *circular pre-proof* $\Pi$ can be associated with a directed bi-partite graph $G(\Pi) = (I \cup J, E)$ such that there is one node in $J$ for each element of the sequence (called clause nodes) and one node in $I$ for each inference step (called inference nodes). There is an arc from $u \in J$ to $v \in I$ if $u$ is an antecedent clause in the inference step of $v$. There is an arc from $u \in I$ to $v \in J$ if $v$ is a consequent clause in the inference step of $u$. The graph is compacted by merging nodes whose associated clause is identical to one in $\mathcal{B}$. Note that before the compactation the graph is acyclic, but the compactation may introduce cycles. The set of in-neighbors and out-neighbors of node $C \in J$ are denoted $N^-(C)$ and $N^+(C)$, respectively.

A *flow assignment* for a circular pre-proof is an assignment $f : I \longrightarrow \mathbb{R}^+$ of positive reals to inference nodes. The *balance* of node $C \in J$ is the inflow minus the outflow,

$$b(C) = \sum_{R \in N^-(C)} f(R) - \sum_{R \in N^+(C)} f(R)$$

**Definition 1.** *A SAT circular proof of clause A from CNF formula $\mathcal{F}$ is a pre-proof $\Pi$ whose proof-graph $G(\Pi)$ admits a flow in which all clauses not in $\mathcal{F}$ have non-negative balance and $A \in J$ has a strictly positive balance.*

**Theorem 9 (soundness).** *Assuming a sound SAT proof system, if there is a SAT circular proof of A from formula $\mathcal{F}$ then $\mathcal{F} \models A$.*

*Property 5.* Using the proof system with the following two rules,

$$\frac{x \vee D \qquad \overline{x} \vee D}{D} \qquad\qquad \frac{D}{D \vee x \qquad D \vee \overline{x}}$$

(symmetric resolution)         (split)

there is a short circular refutation of *PHP*.

## 6.2   ResSV and MaxSAT Circular Proofs

Now we show that the MaxSAT **ResSV** proof system is a true extension of circular proofs from SAT to MaxSAT. The following two theorems show that, when restricted to hard formulas, **ResSV** and SAT circular proofs can simulate each other. Recall that specializing Corollary 1 to hard formulas, $\mathcal{F} \models (A, \infty)$ and $MaxSAT(\mathcal{F} \cup \{(\overline{A}, 1)\}) \geq 1$ is equivalent. Therefore, one can show $\mathcal{F} \models (A, \infty)$ with a proof $\mathcal{F} \cup \{(\overline{A}, 1)\} \vdash^* (\square, 1) \cup \mathcal{G}$.

**Theorem 10.** *Let $\Pi$ be a SAT circular proof of clause $A$ from formula $\mathcal{F} = \{C_1, \ldots, C_p\}$ using the proof system symmetric resolution and split. There is a MaxSAT **ResSV** proof of $(\square, 1)$ from $\mathcal{F}' = \{(C_1, \infty), \ldots, (C_p, \infty)\} \cup \{(\overline{A}, 1)\}$. The length of the proof is $O(|\Pi|)$.*

*Proof.* Let $G(\Pi) = (J \cup I, A)$ be the proof graph and and $f(\cdot)$ the flow of $\Pi$. By definition of SAT circular proof, $A \in J$ and $b(A) > 0$.

The ResSV proof starts with $\mathcal{F}' = \{(C_1, \infty), \ldots, (C_p, \infty)\} \cup \{(\overline{A}, 1)\}$ and consists in 3 phases. In the first phase, the virtual rule is applied for each $C \in J$ not in $\mathcal{F}$, introducing $\{(C, o), (C, -o)\}$ with $o = \sum_{R \in N^-(C)} f(R)$. In the second phase, there is an inference step for each $u \in I$. If $u$ is a SAT split, the inference step is a MaxSAT split generating two clauses with weight $f(u)$. If $u$ is a SAT symmetric resolution, the inference step is a MaxSAT resolution generating one clause with weight $f(u)$. Note that this phase never creates new clauses because all of them have been virtually added at the first phase. It only moves weights around the existing ones. Note as well that we guarantee by construction that at each step of the proof the antecedents are available no matter in which order the proof is done because the first phase has given enough weight to each added clause to guarantee it and original clauses are hard, so their weight never decreases. At the end of the second phase we have $\mathcal{F} \cup \{(\overline{A}, 1)\} \cup \mathcal{C}$ with $\mathcal{C} = \{(C, b(C) \mid C \in J, b(C) > 0\}$ with $b(C)$ being the balance of $C$. Therefore $(A, b(A))$ is in $\mathcal{C}$. The third phase is a final sequence of $q$ steps in which $(\square, 1)$ is derived from $\{(\overline{A}, 1), (A, b(A)\}$ which completes the proof. Note that the size of the proof is $O(|J + I|) = O(|\Pi|)$.

**Theorem 11.** *Consider a hard formula $\mathcal{F} = \{(C_1, \infty), \ldots, (C_p, \infty)\}$ and a **ResSV** proof $\mathcal{F} \cup \{(\overline{A}, 1)\} \vdash^* \mathcal{F} \cup \mathcal{G} \cup \{(\square, 1)\}$ with $e$ inference steps. There is a SAT circular proof $\Pi$ of $A$ from $\mathcal{F}' = \{C_1, \ldots, C_p\}$ with proof system symmetric resolution and split. Besides, $|\Pi| = O(e)$.*

*Proof.* We need to build a graph $G(\Pi) = (J \cup I, E)$ with $\mathcal{F}' \subset J$ and $A \in J$, and a flow $f(\cdot)$ that satisfies the balance conditions and with which $A$ has strictly positive balance.

Because the virtual rule does not have antecedents all its applications can be done at the beginning of the proof and all the cancellation of all the virtual clauses can be done at the end. Therefore, we can omit all those inference steps and assume without loss of generality that the proof is a ResS (that is, without virtual),

$$\mathcal{F} \cup \{(\overline{A}, 1)\} \cup \mathcal{B} \vdash \mathcal{F}_1 \vdash \mathcal{F}_2 \vdash \ldots \vdash \mathcal{F}_e = \mathcal{F} \cup \mathcal{G} \cup \{(\square, 1)\} \cup \mathcal{B}$$

where $\mathcal{B}$ is the set of virtually added clauses. Note as well that any application of MaxSAT resolution between $x \vee A$ and $\overline{x} \vee B$ can be simulated by a short sequence of splits to both clauses until their scope is the same and then one resolution step between $x \vee A \vee B$ and $\overline{x} \vee A \vee B$. So, again without loss of generality we can assume that the proof only contains splits and symmetric resolutions.

Our proof contains three phases. First, we are going to build an acyclic graph $G'(\Pi)$ which is an unfolded version of $G(\Pi)$ and a flow function $f'(\cdot)$ that may have $\infty$ flows. Second we will compute $f(\cdot)$ traversing the graph $G'(\Pi)$ bottom-up and replacing any infinite flow in $f'(\cdot)$ by a finite one that still guarantees the flow condition. In the third and final phase, we will compact the graph which will constitute the circular proof.

Phase 1:

We build $G'(\Pi) = (J' \cup I', E')$ by following the proof step by step. Let $G'_i = (J'_i \cup I'_i, E'_i)$ be the graph associated to proof step $i$. We define the *front* of $G_i$ as the set of clause nodes in $J'_i$ with strictly positive balance. By construction of $G'_i$ we will guarantee a connection between the current formula $\mathcal{F}_i$ and the front of the current graph $G'_i$

$$\mathcal{F}_i = \{(C, b(C)) \mid C \in J'_i, b(C) = \sum_{R \in N^-(C)} f'(R) - \sum_{R \in N^+(C)} f'(R)\}$$

where we define $\infty - \infty = \infty$.

$G'_1$ contains one clause node for each clause in $\mathcal{F}$, $\{(\overline{A}, 1)\}$ and $\mathcal{B}$, respectively. For each clause node there is one dummy inference node pointing to it. The flow $f'(\cdot)$ of the inference node is the weight of the clause it points to. This set of dummy inference nodes will be removed at step three. Then we proceed through the proof. At inference step $i$, we add a new inference node $i$ to $I$. Its in-neighbors will be nodes from the front (that must exist because of the invariant) and its out-neighbors will be newly added clause nodes. Its flow $f'(i)$ is the weight moved by the inference rule (which may be infinite). If the inference rule is split we add two clause nodes, one for each consequent and add the corresponding arcs. If the inference rule is a resolution we add one clause node for its consequent and add the corresponding arcs. Note that, the out-neighbors of node $i$ have a positive balance and in-neighbors of $i$ have their out-flow decreased, but cannot turn negative. Finally, we merge any pair of nodes in the front of $G'_i$ whose associated clause is the same (which preserves the property of balances being non-negative). Graph $G'$ is obtained after processing the last inference step. Note that the invariant guarantees that $\square$ is in $G'$ and its balance is 1.

Phase 2:

Now we traverse the inference nodes of $G'$ in the reverse order of how they were added transforming infinite flows into finite. When considering node $i$, because of the traversing order, we know that every $C \in N^+(i)$ has finite outgoing flow. We compute the flow value $f(i)$ as follows: if $f'(i)$ is finite, then $f(i) = f'(i)$, else $f(i)$ is the minimum value that guarantees that the balance of every $C \in N^+(i)$ is non-negative.

Phase 3:

We obtain $G$ by doing some final arrangements to $G'$. First, we remove dummy inference nodes pointing to clauses in $\mathcal{F}$, $(\overline{A}, 1)$ and $\mathcal{B}$ added in Phase 1. As a result, the balance of these nodes is negative. In particular, the balance of nodes representing $\overline{A}$ and $\mathcal{B}$ is its negative weight.

Since $\mathcal{B} \subseteq \mathcal{F}_e$, we know that all nodes representing $\mathcal{B}$ are included in the front of $G'$ with balance greater than or equal to its weight. We compact these nodes with the ones in $G_1'$ and, as a result, its balance is positive.

Finally, we add some split nodes with flow 1 from node $\square$ (recall that $b(\square) = 1$) in order to generate $A$ and $\overline{A}$, and we compact the latter ones with the ones in $G_1'$. As a result, the balance of $A$ is 1 and the balance of $\overline{A}$ nodes is positive.

# 7 Conclusions

This paper constitutes a first attempt towards MaxSAT resolution-based proof complexity analysis. We have provided some basic definitions and results emphasizing the similarities and differences with respect to SAT. In particular, we have shown that MaxSAT entailment can be rephrased as a MaxSAT refutation problem and, as a consequence, refutation completeness is sufficient for practical purposes. Interestingly, when such rephrasing is applied to hard formulas it transforms a SAT query into a MaxSAT one, and such transformation turns out to be relevant in our analysis of SAT circular proofs.

We have also provided three basic inference MaxSAT rules used in resolution-based proof systems (e.g. resolution, split and virtual) and have analysed their incremental effect in terms of refutation power. Finally, we have related **ResSV**, the strongest of the proof systems considered, with the recently proposed concept of circular proofs. We have shown that **ResSV** generalizes SAT circular proofs as defined in [3].

An additional contribution of the paper is to put together under a formal framework and common notation some ideas spread around in different recent papers such as [3,7,10].

# References

1. Ansótegui, C., Bonet, M.L., Levy, J.: SAT-based maxsat algorithms. Artif. Intell. **196**, 77–105 (2013). https://doi.org/10.1016/j.artint.2013.01.002
2. Atserias, A.: On sufficient conditions for unsatisfiability of random formulas. J. ACM **51**(2), 281–311 (2004). https://doi.org/10.1145/972639.972645
3. Atserias, A., Lauria, M.: Circular (yet sound) proofs. In: Janota, M., Lynce, I. (eds.) SAT 2019. LNCS, vol. 11628, pp. 1–18. Springer, Cham (2019). https://doi.org/10.1007/978-3-030-24258-9_1
4. Bacchus, F., Hyttinen, A., Järvisalo, M., Saikko, P.: Reduced cost fixing for maximum satisfiability. In: Lang, J. (ed.) Proceedings of the Twenty-Seventh International Joint Conference on Artificial Intelligence (IJCAI 2018), 13–19 July (2018), Stockholm, Sweden, pp. 5209–5213 (2018). ijcai.org. https://doi.org/10.24963/ijcai.2018/723
5. Biere, A., Heule, M., van Maaren, H., Walsh, T. (eds.): Handbook of Satisfiability, Frontiers in Artificial Intelligence and Applications, vol. 185. IOS Press, Amsterdam (2009)
6. Bonet, M.L., Levy, J., Manyà, F.: Resolution for Max-Sat. Artif. Intell. **171**(8–9), 606–618 (2007). https://doi.org/10.1016/j.artint.2007.03.001
7. Ignatiev, A., Morgado, A., Marques-Silva, J.: On tackling the limits of resolution in SAT solving. In: Gaspers, S., Walsh, T. (eds.) SAT 2017. LNCS, vol. 10491, pp. 164–183. Springer, Cham (2017). https://doi.org/10.1007/978-3-319-66263-3_11
8. Larrosa, J., Heras, F.: Resolution in Max-Sat and its relation to local consistency in weighted CSPs. In: Kaelbling, L.P., Saffiotti, A. (eds.) IJCAI-05, Proceedings of the Nineteenth International Joint Conference on Artificial Intelligence, Edinburgh, Scotland, UK, 30 July–5 August 2005, pp. 193–198. Professional Book Center (2005). http://ijcai.org/Proceedings/05/Papers/0360.pdf
9. Larrosa, J., Heras, F., de Givry, S.: A logical approach to efficient Max-SAT solving. Artif. Intell. **172**(2–3), 204–233 (2008). https://doi.org/10.1016/j.artint.2007.05.006
10. Larrosa, J., Rollon, E.: Augmenting the power of (partial) maxsat resolution with extension. In: Proceedings of the Thirty-third AAAI Conference on Artificial Intelligence (AAAI 2020), New York, New York, USA, 7–12 February 2020
11. Morgado, A., Heras, F., Liffiton, M.H., Planes, J., Marques-Silva, J.: Iterative and core-guided MaxSAT solving: a survey and assessment. Constraints **18**(4), 478–534 (2013). https://doi.org/10.1007/s10601-013-9146-2
12. Robinson, J.A.: A machine-oriented logic based on the resolution principle. J. ACM **12**(1), 23–41 (1965). https://doi.org/10.1145/321250.321253

# Towards a Complexity-Theoretic Understanding of Restarts in SAT Solvers

Chunxiao Li[1(✉)], Noah Fleming[2], Marc Vinyals[3], Toniann Pitassi[2], and Vijay Ganesh[1]

[1] University of Waterloo, Waterloo, Canada
chunxiao.ian.li@gmail.com
[2] University of Toronto, Toronto, Canada
[3] Technion, Haifa, Israel

**Abstract.** Restarts are a widely-used class of techniques integral to the efficiency of Conflict-Driven Clause Learning (CDCL) Boolean SAT solvers. While the utility of such policies has been well-established empirically, a theoretical understanding of whether restarts are indeed crucial to the power of CDCL solvers is missing.

In this paper, we prove a series of theoretical results that characterize the power of restarts for various models of SAT solvers. More precisely, we make the following contributions. First, we prove an exponential separation between a *drunk* randomized CDCL solver model with restarts and the same model without restarts using a family of satisfiable instances. Second, we show that the configuration of CDCL solver with VSIDS branching and restarts (with activities erased after restarts) is exponentially more powerful than the same configuration without restarts for a family of unsatisfiable instances. To the best of our knowledge, these are the first separation results involving restarts in the context of SAT solvers. Third, we show that restarts do not add any proof complexity-theoretic power vis-a-vis a number of models of CDCL and DPLL solvers with non-deterministic static variable and value selection.

## 1 Introduction

Over the last two decades, Conflict-Driven Clause Learning (CDCL) SAT solvers have had a revolutionary impact on many areas of software engineering, security and AI. This is primarily due to their ability to solve real-world instances containing millions of variables and clauses [2,6,15,16,18], despite the fact that the Boolean SAT problem is known to be an NP-complete problem and is believed to be intractable in the worst case.

This remarkable success has prompted complexity theorists to seek an explanation for the efficacy of CDCL solvers, with the aim of bridging the gap between theory and practice. Fortunately, a few results have already been established that lay the groundwork for a deeper understanding of SAT solvers viewed as proof systems [3,8,11]. Among them, the most important result is the one by Pipatsrisawat and Darwiche [18] and independently by Atserias et al. [2], that shows that

© Springer Nature Switzerland AG 2020
L. Pulina and M. Seidl (Eds.): SAT 2020, LNCS 12178, pp. 233–249, 2020.
https://doi.org/10.1007/978-3-030-51825-7_17

an idealized model of CDCL solvers with non-deterministic branching (variable selection and value selection), and restarts is *polynomially equivalent* to the general resolution proof system. However, an important question that remains open is whether this result holds even when restarts are disabled, i.e., whether configurations of CDCL solvers without restarts (when modeled as proof systems) are polynomial equivalent to the general resolution proof system. In practice there is significant evidence that restarts are crucial to solver performance.

This question of the "power of restarts" has prompted considerable theoretical work. For example, Bonet, Buss and Johannsen [7] showed that CDCL solvers with no restarts (but with non-deterministic variable and value selection) are strictly more powerful than regular resolution. Despite this progress, the central questions, such as whether restarts are integral to the efficient simulation of general resolution by CDCL solvers, remain open.

In addition to the aforementioned theoretical work, there have been many empirical attempts at understanding restarts given how important they are to solver performance. Many hypotheses have been proposed aimed at explaining the power of restarts. Examples include, the *heavy-tail* explanation [10], and the "restarts compact assignment trail and hence produce clauses with lower literal block distance (LBD)" perspective [14]. Having said that, the heavy-tailed distribution explanation of the power of restarts is not considered valid anymore in the CDCL setting [14].

### 1.1   Contributions

In this paper we make several contributions to the theoretical understanding of the power of restarts for several restricted models of CDCL solvers:

1. First, we show that CDCL solvers with backtracking, non-deterministic dynamic variable selection, randomized value selection, and restarts[1] are exponentially faster than the same model, but without restarts, with high probability (w.h.p)[2]. A notable feature of our proof is that we obtain this separation on a family of satisfiable instances. (See Sect. 4 for details.)
2. Second, we prove that CDCL solvers with VSIDS variable selection, phase saving value selection and restarts (where activities of variables are reset to zero after restarts) are exponentially faster (w.h.p) than the same solver configuration but without restarts for a class of unsatisfiable formulas. This result holds irrespective of whether the solver uses backtracking or backjumping. (See Sect. 5 for details.)
3. Finally, we prove several smaller separation and equivalence results for various configurations of CDCL and DPLL solvers with and without restarts. For example, we show that CDCL solvers with non-deterministic static variable

---

[1] In keeping with the terminology from [1], we refer any CDCL solver with randomized value selection as a *drunk* solver.
[2] We say that an event occurs with high probability (w.h.p.) if the probability of that event happening goes to 1 as $n \to \infty$.

selection, non-deterministic static value selection, and with restarts, are polynomially equivalent to the same model but without restarts. Another result we show is that for DPLL solvers, restarts do not add proof theoretic power as long as the solver configuration has non-deterministic dynamic variable selection. (See Sect. 6 for details.)

## 2  Definitions and Preliminaries

Below we provide relevant definitions and concepts used in this paper. We refer the reader to the Handbook of Satisfiability [6] for literature on CDCL and DPLL solvers and to [4,12] for literature on proof complexity.

We denote by $[c]$ the set of natural numbers $\{1, \ldots, c\}$. We treat CDCL solvers as proof systems. For proof systems $A$ and $B$, we use $A \sim_p B$ to denote that they are polynomially equivalent ($p$-equivalent). Throughout this paper it is convenient to think of the trail $\pi$ of the solver during its run on a formula $F$ as a restriction to that formula. We call a function $\pi \colon \{x_1, \ldots, x_n\} \to \{0, 1, *\}$ a restriction, where $*$ denotes that the variable is unassigned by $\pi$. Additionally, we assume that our Boolean Constraint Propagation (BCP) scheme is greedy, i.e., BCP is performed till "saturation".

**Restarts in SAT Solvers.** A restart policy is a method that erases part of the state of the solver at certain intervals during the run of a solver [10]. In most modern CDCL solvers, the restart policy erases the assignment trail upon invocation, but may choose not to erase the learnt clause database or variable activities. Throughout this paper, we assume that all restart policies are non-deterministic, i.e., the solver may (dynamically) non-deterministically choose its restart sequence. We refer the reader to a paper by Liang et al. [14] for a detailed discussion on modern restart policies.

## 3  Notation for Solver Configurations Considered

In this section, we precisely define the various heuristics used to define SAT solver configurations in this paper. By the term *solver configuration* we mean a solver parameterized with appropriate heuristic choices. For example, a CDCL solver with non-deterministic variable and value selection, as well as asserting learning scheme with restarts would be considered a solver configuration.

To keep track of these configurations, we denote solver configurations by the notation $M_{A,B}^{E,R}$, where $M$ indicates the underlying *solver model* (we use $C$ for CDCL and $D$ for DPLL solvers); the subscript $A$ denotes a *variable selection scheme*; the subscript $B$ is a *value selection scheme*; the superscript $E$ is a *backtracking scheme*, and finally the superscript $R$ indicates whether the solver configuration comes equipped with a restart policy. That is, the presence of the superscript $R$ indicates that the configuration has restarts, and its absence indicates that it does not. A $*$ in place of $A, B$ or $E$ denotes that the scheme is *arbitrary*, meaning that it works for any such scheme. See Table 1 for examples of solver configurations studied in this paper.

**Table 1.** Solver configurations in the order they appear in the paper. ND stands for non-deterministic dynamic.

| | Model | Variable selection | Value selection | Backtracking | Restarts |
|---|---|---|---|---|---|
| $C_{ND,RD}^{T,R}$ | CDCL | ND | Random dynamic | Backtracking | Yes |
| $C_{ND,RD}^{T}$ | CDCL | ND | Random dynamic | Backtracking | No |
| $C_{VS,PS}^{J,R}$ | CDCL | VSIDS | Phase saving | Backjumping | Yes |
| $C_{VS,PS}^{J}$ | CDCL | VSIDS | Phase saving | Backjumping | No |
| $C_{S,S}^{J,R}$ | CDCL | Static | Static | Backjumping | Yes |
| $C_{S,S}^{J}$ | CDCL | Static | Static | Backjumping | No |
| $D_{ND,*}^{T}$ | DPLL | ND | Arbitrary | Backtracking | No |
| $D_{ND,ND}^{T,R}$ | DPLL | ND | ND | Backtracking | Yes |
| $D_{ND,ND}^{T}$ | DPLL | ND | ND | Backtracking | No |
| $D_{ND,RD}^{T,R}$ | DPLL | ND | Random dynamic | Backtracking | Yes |
| $D_{ND,RD}^{T}$ | DPLL | ND | Random dynamic | Backtracking | No |
| $C_{ND,ND}^{J,R}$ | CDCL | ND | ND | Backjumping | Yes |
| $C_{ND,ND}^{J}$ | CDCL | ND | ND | Backjumping | No |

### 3.1 Variable Selection Schemes

**1. Static (S):** Upon invocation, the S variable selection heuristic returns the unassigned variable with the highest rank according to some predetermined, fixed, total ordering of the variables.

**2. Non-deterministic Dynamic (ND):** The ND variable selection scheme non-deterministically selects and returns an unassigned variable.

**3. VSIDS (VS)** [16]**:** Each variable has an associated number, called its *activity*, initially set to 0. Each time a solver learns a conflict, the activities of variables appearing on the conflict side of the implication graph receive a constant bump. The activities of all variables are decayed by a constant c, where $0 < c < 1$, at regular intervals. The VSIDS variable selection heuristic returns the unassigned variable with highest activity, with ties broken randomly.

### 3.2 Value Selection Schemes

**1. Static (S):** Before execution, a 1-1 mapping of variables to values is fixed. The S value selection heuristic takes as input a variable and returns the value assigned to that variable according to the predetermined mapping.

**2. Non-deterministic Dynamic (ND):** The ND value selection scheme non-deterministically selects and returns a truth assignment.

**3. Random Dynamic (RD):** A randomized algorithm that takes as input a variable and returns a uniformly random truth assignment.

**4. Phase Saving (PS):** A heuristic that takes as input an unassigned variable and returns the previous truth value that was assigned to the variable. Typically solver designers determine what value is returned when a variable has not been previously assigned. For simplicity, we use the phase saving heuristic that returns 0 if the variable has not been previously assigned.

### 3.3  Backtracking and Backjumping Schemes

To define different backtracking schemes we use the concept of *decision level* of a variable $x$, which is the number of decision variables on the trail prior to $x$. **Backtracking (T):** Upon deriving a conflict clause, the solver undoes the most recent decision variable on the assignment trail. **Backjumping (J):** Upon deriving a conflict clause, the solver undoes all decision variables with decision level higher than the variable with the second highest decision level in the conflict clause.

**Note on Solver Heuristics.** Most of our results hold irrespective of the choice of deterministic asserting clause learning schemes (except for Proposition 22). Additionally, it goes without saying that the questions we address in this paper make sense only when it is assumed that solver heuristics are polynomial time methods.

## 4  Separation for Drunk CDCL with and Without Restarts

Inspired by Alekhnovich et al. [1], where the authors proved exponential lower bound for drunk DPLL solvers over a class of satisfiable instances, we studied the behavior of restarts in a drunken model of CDCL solver. We introduce a class of satisfiable formulas, $Ladder_n$, and use them to prove the separation between $C_{ND,RD}^{T,R}$ and $C_{ND,RD}^{T}$. At the core of these formulas is a formula which is hard for general resolution even after any small restriction (corresponding to the current trail of the solver). For this, we use the well-known Tseitin formulas.

**Definition 1 (Tseitin Formulas).** *Let $G = (V, E)$ be a graph and $f: V \to \{0, 1\}$ a labelling of the vertices. The formula $Tseitin(G, f)$ has variables $x_e$ for $e \in E$ and constraints $\bigoplus_{uv \in E} x_{uv} = f(v)$ for each $v \in V$.*

For any graph $G$, $Tseitin(G, f)$ is unsatisfiable iff $\bigoplus_{v \in V} f(v) = 1$, in which case we call $f$ an *odd labelling*. The specifics of the labelling are irrelevant for our applications, any odd labelling will do. Therefore, we often omit defining $f$, and simply assume that it is odd.

The family of satisfiable $Ladder_n$ formulas are built around the Tseitin formulas, unless the variables of the formula are set to be consistent to one of two satisfying assignments, the formula will become unsatisfiable. Furthermore, the solver will only be able to backtrack out of the unsatisfiable sub-formula by first refuting Tseitin, which is a provably hard task for any CDCL solver [20].

The $Ladder_n$ formulas contain two sets of variables, $\ell_j^i$ for $0 \leq i \leq n - 2, j \in [\log n]$ and $c_m$ for $m \in [\log n]$, where $n$ is a power of two. We denote by $\ell^i$

the block of variables $\{\ell_1^i, \ldots, \ell_{\log n}^i\}$. These formulas are constructed using the following gadgets.

**Ladder gadgets:** $L^i := (\ell_1^i \vee \ldots \vee \ell_{\log n}^i) \wedge (\neg \ell_1^i \vee \ldots \vee \neg \ell_{\log n}^i)$.

Observe that $L^i$ is falsified only by the all-1 and all-0 assignments.

**Connecting gadgets:** $C^i := (c_1^{bin(i,1)} \wedge \ldots \wedge c_{\log n}^{bin(i,\log n)})$.

Here, $bin(i, m)$ returns the $m$th bit of the binary representation of $i$, and $c_m^1 := c_m$, while $c_m^0 := \neg c_m$. That is, $C^i$ is the conjunction that is satisfied only by the assignment encoding $i$ in binary.

**Equivalence gadget:** $EQ := \bigwedge_{i,j=0}^{n-2} \bigwedge_{m,k=1}^{\log n} (\ell_k^i \iff \ell_m^j)$.

These clauses enforce that every $\ell$-variable must take the same value.

**Definition 2 (Ladder formulas).** *For $G = (V, E)$ with $|E| = n - 1$ where $n$ is a power of two, let $Tseitin(G, f)$ be defined on the variables $\{\ell_1^0, \ldots, \ell_1^{n-2}\}$. $Ladder_n(G, f)$ is the conjunction of the clauses representing*

$$L^i \Rightarrow C^i, \qquad\qquad\qquad \forall 0 \le i \le n - 2$$

$$C^i \Rightarrow Tseitin(G, f), \qquad\quad \forall 0 \le i \le n - 2$$

$$C^{n-1} \Rightarrow EQ.$$

Observe that the $Ladder_n(G, f)$ formulas have polynomial size provided that the degree of $G$ is $O(\log n)$. As well, this formula is satisfiable only by the assignments that sets $c_m = 1$ and $\ell_j^i = \ell_q^p$ for every $m, j, q \in [\log n]$ and $0 \le i, p \le n - 2$.

These formulas are constructed so that after setting only a few variables, any drunk solver will enter an unsatisfiable subformula w.h.p. and thus be forced to refute the Tseitin formula. Both the ladder gadgets and equivalence gadget act as trapdoors for the Tseitin formula. Indeed, if any $c$-variable is set to 0 then we have already entered an unsatisfiable instance. Similarly, setting $\ell_j^i = 1$ and $\ell_q^p = 0$ for any $0 \le i, p \le n - 2$, $j, q \in [\log n]$ causes us to enter an unsatisfiable instance. This is because setting all $c$-variables to 1 together with this assignment would falsify a clause of the equivalence gadget. Thus, after the second decision of the solver, the probability that it is in an unsatisfiable instance is already at least $1/2$. With these formulas in hand, we prove the following theorem, separating backtracking $C_{ND,RD}^T$ solvers with and without restarts.

**Theorem 3.** *There exists a family of $O(\log n)$-degree graphs $G$ such that*

1. *$Ladder_n(G, f)$ can be decided in time $O(n^2)$ by $C_{ND,RD}^{T,R}$, except with exponentially small probability.*
2. *$C_{ND,RD}^T$ requires exponential time to decide $Ladder_n(G, f)$, except with probability $O(1/n)$.*

The proof of the preceding theorem occupies the remainder of this section.

### 4.1 Upper Bound on Ladder Formulas via Restarts

We present the proof for part (1) of Theorem 3. The proof relies on the following lemma, stating that given the all-1 restriction to the $c$-variables, $C_{ND,RD}^T$ will find a satisfying assignment.

**Lemma 4.** *For any graph $G$, $C_{ND,RD}^T$ will find a satisfying assignment to $Ladder_n(G, f)[c_1 = 1, \ldots, c_{\log n} = 1]$ in time $O(n \log n)$.*

*Proof.* When all $c$ variables are 1, we have $C^{n-1} = 1$. By the construction of the connecting gadget, $C^i = 0$ for all $0 \le i \le n - 2$. Under this assignment, the remaining clauses belong to $EQ$, along with $\neg L^i$ for $0 \le i \le n - 2$. It is easy to see that, as soon as the solver sets an $\ell$-variable, these clauses will propagate the remaining $\ell$-variables to the same value. □

Put differently, the set of $c$ variables forms a *weak backdoor* [23,24] for $Ladder_n$ formulas. Part (1) of Theorem 3 shows that, with probability at least $1/2$, $C_{ND,RD}^{T,R}$ can exploit this weak backdoor using only $O(n)$ number of restarts.

*Proof (of Theorem 3 Part (1)).* By Lemma 4, if $C_{ND,RD}^{T,R}$ is able to assign all $c$ variables to 1 before assigning any other variables, then the solver will find a satisfying assignment in time $O(n \log n)$ with probability 1. We show that the solver can exploit restarts in order to find this assignment. The strategy the solver adopts is as follows: query each of the $c$-variables; if at least one of the $c$-variables was assigned to 0, restart. We argue that if the solver repeats this procedure $k = n^2$ times then it will find the all-1 assignment to the $c$-variables, except with exponentially small probability. Because each variable is assigned 0 and 1 with equal probability, the probability that a single round of this procedure finds the all-1 assignment is $2^{-\log n}$. Therefore, the probability that the solver has not found the all-1 assignment after $k$ rounds is

$$(1 - 1/n)^k \le e^{-k/n} = e^{-n}.$$

□

## 4.2   Lower Bound on Ladder Formulas Without Restarts

We now prove part (2) of Theorem 3. The proof relies on the following three technical lemmas. The first claims that the solver is well-behaved (most importantly that it cannot learn any new clauses) while it has not made many decisions.

**Lemma 5.** *Let $G$ be any graph of degree at least $d$. Suppose that $C_{ND,RD}^T$ has made $\delta < \min(d - 1, \log n - 1)$ decisions since its invocation on $Ladder_n(G, f)$. Let $\pi_\delta$ be the current trail, then*

1. *The solver has yet to enter a conflict, and thus has not learned any clauses.*
2. *The trail $\pi_\delta$ contains variables from at most $\delta$ different blocks $\ell^i$.*

We defer the proof of this lemma to the arXiv version of the paper [13].

The following technical lemma states that if a solver with backtracking has caused the formula to become unsatisfiable, then it must *refute* that formula before it can backtrack out of it. For a restriction $\pi$ and a formula $F$, we say that the solver has *produced a refutation* of an unsatisfiable formula $F[\pi]$ if it has learned a clause $C$ such that $C$ is falsified under $\pi$. Note that because general resolution $p$-simulates CDCL, any refutation of a formula $F[\pi]$ implies a general resolution refutation of $F[\pi]$ of size at most polynomial in the time that the solver took to produce that refutation.

**Lemma 6.** *Let $F$ be any propositional formula, let $\pi$ be the current trail of the solver, and let $x$ be any literal in $\pi$. Then, $C^T_{ND,ND}$ backtracks $x$ only after it has produced a refutation of $F[\pi]$.*

*Proof.* In order to backtrack $x$, the solver must have learned a clause $C$ asserting the negation of some literal $z \in \pi$ that was set before $x$. Therefore, $C$ must only contain the negation of literals in $\pi$. Hence, $C[\pi] = \emptyset$.    □

The third lemma reduces proving a lower bound on the runtime of $C^T_{ND,ND}$ on the $Ladder_n$ formulas under any well-behaved restriction to proving a general resolution lower bound on an associated Tseitin formula.

**Definition 7.** *For any unsatisfiable formula $F$, denote by $Res(F \vdash \emptyset)$ the minimal size of any general resolution refutation of $F$.*

We say that a restriction (thought of as the current trail of the solver) $\pi$ to $Ladder_n(G, f)$ implies Tseitin if $\pi$ either sets some $c$-variable to 0 or $\pi[\ell^i_j] = 1$ and $\pi[\ell^p_q] = 0$ for some $0 \leq i, q \leq n - 2$, $j, q \in [\log n]$. Observe that in both of these cases the formula $Ladder_n(G, f)[\pi]$ is unsatisfiable.

**Lemma 8.** *Let $\pi$ be any restriction that implies Tseitin and such that each clause of $Ladder_n(G, f)[\pi]$ is either satisfied or contains at least two unassigned variables. Suppose that $\pi$ sets variables from at most $\delta$ blocks $\ell^i$. Then there is a restriction $\rho^*_\pi$ that sets at most $\delta$ variables of $Tseitin(G, f)$ such that*

$$Res(Ladder_n(G, f)[\pi] \vdash \emptyset) \geq Res(Tseitin(G, f)[\rho^*_\pi] \vdash \emptyset).$$

We defer the proof of this lemma to the arXiv version of the paper [13], and show how to use them to prove part (2) of Theorem 3. We prove this statement for any degree $O(\log n)$ graph $G$ with sufficient expansion.

**Definition 9.** *The expansion of a graph $G = (V, E)$ is*

$$e(G) := \min_{V' \subseteq V, |V'| \leq |V|/2} \frac{|E[V', V \setminus V']|}{|V'|},$$

*where $E[V', V \setminus V']$ is the set of edges in $E$ with one endpoint in $V'$ and the other in $V \setminus V'$.*

For every $d \geq 3$, *Ramanujan Graphs* provide an infinite family of $d$-regular expander graphs $G$ for which $e(G) \geq d/4$. The lower bound on solver runtime relies on the general resolution lower bounds for the Tseitin formulas [20]; we use the following lower bound criterion which follows immediately[3] from [5].

**Corollary 10** ([5]). *For any connected graph $G = (V, E)$ with maximum degree $d$ and odd weight function $f$,*

$$Res(Tseitin(G, f) \vdash \emptyset) = \exp\left(\Omega\left(\frac{(e(G)|V|/3 - d)^2}{|E|}\right)\right)$$

---

[3] In particular, this follows from Theorem 4.4 and Corollary 3.6 in [5], noting that the definition of expansion used in their paper is lower bounded by $3e(G)/|V|$ as they restrict to sets of vertices of size between $|V|/3$ and $2|V|/3$.

We are now ready to prove the theorem.

*Proof (of part (2) Theorem 3).* Fix $G = (V, E)$ to be any degree-$(8 \log n)$ graph on $|E| = n - 1$ edges such that $e(G) \geq 2 \log n$. Ramanujan graphs satisfy these conditions.

First, we argue that within $\delta < \log n - 1$ decisions from the solver's invocation, the trail $\pi_\delta$ will imply Tseitin, except with probability $1 - /2^{\delta-1}$. By Lemma 5, the solver has yet to backtrack or learn any clauses, and it has set variables from at most $\delta$ blocks $\ell^i$. Let $x$ be the variable queried during the $\delta$th decision. If $x$ is a $c$ variable, then with probability $1/2$ the solver sets $c_i = 0$. If $x$ is a variable $\ell^i_j$, then, unless this is the first time the solver sets an $\ell$-variable, the probability that it sets $\ell^i_j$ to a different value than the previously set $\ell$-variable is $1/2$.

Conditioning on the event that, within the first $\log n - 2$ decisions the trail of the solver implies Tseitin (which occurs with probability at least $(n - 8)/n$), we argue that the runtime of the solver is exponential in $n$. Let $\delta < \log n - 1$ be the first decision level such that the current trail $\pi_\delta$ implies Tseitin. By Lemma 6 the solver must have produced a refutation of $Ladder_n(G, f)[\pi_\delta]$ in order to backtrack out of the unsatisfying assignment. If the solver takes $t$ steps to refute $Ladder_n(G, f)[\pi_\delta]$ then this implies a general resolution refutation of size $\mathsf{poly}(t)$. Therefore, in order to lower bound the runtime of the solver, it is enough to lower bound the size of general resolution refutations of $Ladder_n(G, f)[\pi_\delta]$.

By Lemma 5, the solver has not learned any clauses, and has yet to enter into a conflict and therefore no clause in $Ladder_n(G, f)[\pi_\delta]$ is falsified. As well, $\pi_\delta$ sets variables from at most $\delta < \log n - 1$ blocks $\ell^i$. By Lemma 8 there exists a restriction $\rho^*_\pi$ such that $Res(Ladder_n(G, f)[\pi] \vdash \emptyset) \geq Res(Tseitin(G, f)[\rho^*_\pi] \vdash \emptyset)$. Furthermore, $\rho^*_\pi$ sets at most $\delta < \log n - 1$ variables and therefore cannot falsify any constraint of $Tseitin(G, f)$, as each clause depends on $8 \log n$ variables. Observe that if we set a variable $x_e$ of $Tseitin(G, f)$ then we obtain a new instance of $Tseitin(G_{\rho^*_\pi}, f')$ on a graph $G_{\rho^*_\pi} = (V, E \setminus \{e\})$. Therefore, we are able to apply Corollary 10 provided that we can show that $e(G_{\rho^*_\pi})$ is large enough.

**Claim 11.** *Let $G = (V, E)$ be a graph and let $G' = (V, E')$ be obtained from $G$ by removing at most $e(G)/2$ edges. Then $e(G') \geq e(G)/2$.*

*Proof.* Let $V' \subseteq V$ with $|V'| \leq |V|/2$. Then, $E'[V', V \setminus V'] \geq e(G)|V'| - e(G)/2 \geq (e(G)/2)|V'|$. □

It follows that $e(G_{\rho^*_\pi}) \geq \log n$. Note that $|V| = n/8 \log n$. By Corollary 10,

$$Res(Ladder_n(G, f)[\pi] \vdash \emptyset) = \exp(\Omega(((n - 1)/24 - 8 \log n)^2/n)) = \exp(\Omega(n)).$$

Therefore, the runtime of $C^T_{ND,ND}$ is $\exp(\Omega(n))$ on $Ladder_n(G, F)$ w.h.p. □

## 5   CDCL + VSIDS Solvers with and Without Restarts

In this section, we prove that CDCL solvers with VSIDS variable selection, phase saving value selection and restarts (where activities of variables are reset to zero after restarts) are exponentially more powerful than the same solver configuration but without restarts, w.h.p.

**Theorem 12.** *There is a family of unsatisfiable formulas that can be decided in polynomial time with $C_{VS,PS}^{J,R}$ but requires exponential time with $C_{VS,PS}^{J}$, except with exponentially small probability.*

We show this separation using pitfall formulas $\Phi(G_n, f, n, k)$, designed to be hard for solvers using the VSIDS heuristic [22]. We assume that $G_n$ is a constant-degree expander graph with $n$ vertices and $m$ edges, $f: V(G_n) \to \{0, 1\}$ is a function with odd support as with Tseitin formulas, we think of $k$ as a constant and let $n$ grow. We denote the indicator function of a Boolean expression $B$ with $[\![B]\!]$. These formulas have $k$ blocks of variables named $X_j$, $Y_j$, $Z_j$, $P_j$, and $A_j$, with $j \in [k]$, and the following clauses:

- $\left(\bigoplus_{e \ni v} x_{j,e} = f(v)\right) \vee \bigvee_{i=1}^{n} z_{j,i}$, expanded into CNF, for $v \in V(G_n)$ and $j \in [k]$;
- $y_{j,i_1} \vee y_{j,i_2} \vee \neg p_{j,i_3}$ for $i_1, i_2 \in [n]$, $i_1 < i_2$, $i_3 \in [m+n]$, and $j \in [k]$;
- $y_{j,i_1} \vee \bigvee_{i \in [m+n] \setminus \{i_2\}} p_{j,i} \vee \bigvee_{i=1}^{i_2 - 1} x_{j,i} \vee \neg x_{j,i_2}$ for $i_1 \in [n]$, $i_2 \in [m]$, and $j \in [k]$;
- $y_{j,i_1} \vee \bigvee_{i \in [m+n] \setminus \{m+i_2\}} p_{j,i} \vee \bigvee_{i=1}^{m} x_{j,i} \vee \bigvee_{i=1+[\![i_2=n]\!]}^{i_2 - 1} z_{j,i} \vee \neg z_{j,i_2}$ for $i_1, i_2 \in [n]$ and $j \in [k]$;
- $\neg a_{j,1} \vee a_{j,3} \vee \neg z_{j,i_1}$, $\neg a_{j,2} \vee \neg a_{j,3} \vee \neg z_{j,i_1}$, $a_{j,1} \vee \neg z_{j,i_1} \vee \neg y_{j,i_2}$, and $a_{j,2} \vee \neg z_{j,i_1} \vee \neg y_{j,i_2}$ for $i_1, i_2 \in [n]$ and $j \in [k]$; and
- $\bigvee_{j \in [k]} \neg y_{j,i} \vee \neg y_{j,i+1}$ for odd $i \in [n]$.

To give a brief overview, the first type of clauses are essentially a Tseitin formula and thus are hard to solve. The next four types form a pitfall gadget, which has the following easy-to-check property.

**Claim 13.** *Given any pair of variables $y_{j,i_1}$ and $y_{j,i_2}$ from the same block $Y_j$, assigning $y_{j,i_1} = 0$ and $y_{j,i_2} = 0$ yields a conflict.*

Furthermore, such a conflict involves all of the variables of a block $X_j$, which makes the solver prioritize these variables and it becomes stuck in a part of the search space where it must refute the first kind of clauses. Proving this formally requires a delicate argument, but we can use the end result as a black box.

**Theorem 14** ([22, Theorem 3.6]). *For $k$ fixed, $\Phi(G_n, f, n, k)$ requires time $\exp(\Omega(n))$ to decide with $C_{VS,PS}^{J}$, except with exponentially small probability.*

The last type of clauses, denoted by $\Gamma_i$, ensure that a short general resolution proof exists. Not only that, we can also prove that pitfall formulas have small backdoors [23,24], which is enough for a formula to be easy for $C_{VS,PS}^{J,R}$.

**Definition 15.** *A set of variables $V$ is a strong backdoor for unit-propagation if every assignment to all variables in $V$ leads to a conflict, after unit propagation.*

**Lemma 16.** *If $F$ has a strong backdoor for unit-propagation of size $c$, then $C_{VS,PS}^{J,R}$ can solve $F$ in time $n^{O(c)}$, except with exponentially small probability.*

*Proof.* We say that the solver learns a beneficial clause if it only contains variables in $V$. Since there are $2^c$ possible assignments to variables in $V$ and each beneficial clause forbids at least one assignment, it follows that learning $2^c$ beneficial clauses is enough to produce a conflict at level 0.

Therefore it is enough to prove that, after each restart, we learn a beneficial clause with large enough probability. Since all variables are tied, all decisions before the first conflict after a restart are random, and hence with probability at least $n^{-c}$ the first variables to be decided before reaching the first conflict are (a subset of) $V$. If this is the case then, since $V$ is a strong backdoor, no more decisions are needed to reach a conflict, and furthermore all decisions in the trail are variables in $V$, hence the learned clause is beneficial.

It follows that the probability of having a sequence of $n^{2c}$ restarts without learning a beneficial clause is at most

$$(1 - n^{-c})^{n^{2c}} \leq \exp(-n^{-c} \cdot n^{2c}) = \exp(-n^c) \tag{1}$$

hence by a union bound the probability of the algorithm needing more than $2^c \cdot n^{2c}$ restarts is at most $2^c \cdot \exp(-n^c)$.    □

We prove Theorem 12 by showing that $\Phi(G_n, f, n, k)$ contains a backdoor of size $2k(k+1)$.

*Proof (of Theorem 12).* We claim that the set of variables $V = \{y_{j,i} \mid (j, i) \in [k] \times [2k+2]\}$ is a strong backdoor for unit-propagation. Consider any assignment to $V$. Each of the $k+1$ clauses $\Gamma_1, \Gamma_3, \ldots, \Gamma_{2k+1}$ forces a different variable $y_{j,i}$ to 0, hence by the pigeonhole principle there is at least one block with two variables assigned to 0. But by Claim 13, this is enough to reach a conflict.

The upper bound follows from Lemma 16, while the lower bound follows from Theorem 14.    □

# 6    Minor Equivalences and Separations for CDCL/DPLL Solvers with and Without Restarts

In this section, we prove four smaller separation and equivalence results for various configurations of CDCL and DPLL solvers with and without restarts.

## 6.1    Equivalence Between CDCL Solvers with Static Configurations with and Without Restarts

First, we show that CDCL solvers with non-deterministic static variable and value selection without restarts $(C_{S,S}^J)$ is as powerful as the same configuration with restarts $(C_{S,S}^{J,R})$ for both satisfiable and unsatisfiable formulas. We assume that the BCP subroutine for the solver configurations under consideration is "fixed" in the following sense: if there is more than one unit clause under a partial assignment, the BCP subroutine propagates the clause that is added to the clause database first.

**Theorem 17.** $C_{S,S}^J \sim_p C_{S,S}^{J,R}$ *provided that they are given the same variable ordering and fixed mapping of variables to values for the variable selection and value selection schemes respectively.*

We prove this theorem by arguing for any run of $C_{S,S}^{J,R}$, that restarts can be removed without increasing the run-time.

*Proof.* Consider a run of $C_{S,S}^{J,R}$ on some formula $F$, and suppose that the solver has made $t$ restart calls. Consider the trail $\pi$ for $C_{S,S}^{J,R}$ up to the variable $l$ from the second highest decision from the last learnt clause before the first restart call. Now, observe that because the decision and variable selection orders are static, once $C_{S,S}^{J,R}$ restarts, it will force it to repeat the same decisions and unit propagations that brought it to the trail $\pi$. Suppose that this is not the case and consider the first literal on which the trails differ. This difference could not be caused by a unit propagation as the solver has not learned any new clauses since the restart. Thus, it must have been caused by a decision. However, because the clause databases are the same, this would contradict the static variable and value order. Therefore, this restart can be ignored, and we obtain a run of $C_{S,S}^{J,R}$ with $d-1$ restarts without increasing the run-time. The proof follows by induction. Once all restarts have been removed, the result is a valid run of $C_{S,S}^J$.     □

Note that in the proof of Theorem 17, not only we argue that $C_{S,S}^J$ is $p$-equivalent to $C_{S,S}^{J,R}$, we also show that the two configurations produce the same run. The crucial observation is that given any state of $C_{S,S}^{J,R}$, we can produce a run of $C_{S,S}^J$ which ends in the same state. In other words, our proof not only suggests that $C_{S,S}^{J,R}$ is equivalent to $C_{S,S}^J$ from a proof theoretic point of view, it also implies that the two configurations are equivalent for satisfiable formulas.

### 6.2   Equivalence Between DPLL Solvers with ND Variable Selection on UNSAT Formulas

We show that when considered as a proof system, a DPLL solver with non-deterministic dynamic variable selection, arbitrary value selection and no restarts ($D_{ND,*}^T$) is p-equivalent to DPLL solver with non-deterministic dynamic variable and value selection and restarts ($D_{ND,ND}^{T,R}$), and hence, transitively p-equivalent to tree-like resolution—the restriction of general resolution where each consequent can be an antecedent in only one later inference.

**Theorem 18.** $D_{ND,*}^T \sim_p D_{ND,ND}^T$.

*Proof.* To show that $D_{ND,ND}^T$ p-simulates $D_{ND,*}^T$, we argue that every proof of $D_{ND,ND}^T$ can be converted to a proof of same size in $D_{ND,*}^T$. Let $F$ be an unsatisfiable formula. Recall that a run of $D_{ND,ND}^T$ on $F$ begins with non-deterministically picking some variable $x$ to branch on, and a truth value to assign to $x$. W.l.o.g. suppose that the solver assigns $x$ to 1. Thus, the solver will first refute $F[x=1]$ before backtracking and refuting $F[x=0]$.

To simulate a run of $D^T_{ND,ND}$ with $D^T_{ND,*}$, since variable selection is non-deterministic, $D^T_{ND,*}$ also chooses the variable $x$ as the first variable to branch on. If the value selection returns $x = \alpha$ for $\alpha \in \{0, 1\}$, then the solver focus on the restricted formula $F[x = \alpha]$ first. Because there is no clause learning, whether $F[x = 1]$ or $F[x = 0]$ is searched first does not affect the size of the search space for the other. The proof follows by recursively calling $D^T_{ND,*}$ on $F[x = 1]$ and $F[x = 0]$. The converse direction follows since every run of $D^T_{ND,*}$ is a run of $D^T_{ND,ND}$.                                                                     □

**Corollary 19.** $D^T_{ND,*} \sim_p D^{T,R}_{ND,ND}$.

*Proof.* This follows from the fact that $D^{T,R}_{ND,ND} \sim_p D^T_{ND,ND}$. Indeed, with non-deterministic branching and without clause learning, restarts cannot help. If ever $D^{T,R}_{ND,ND}$ queries a variable $x = \alpha$ for $\alpha \in \{0, 1\}$ and then later restarts to assign it to $1 - \alpha$, then $D^T_{ND,ND}$ ignores the part of the computation when $x = \alpha$ and instead immediately non-deterministically chooses $x = 1 - \alpha$.                □

It is interesting to note that while the above result establishes a $p$-equivalence between DPLL solver models $D^T_{ND,*}$ and $D^{T,R}_{ND,ND}$, the following corollary implies that DPLL solvers with non-deterministic variable and randomized value selection are exponentially separable for satisfiable instances.

## 6.3   Separation Result for Drunk DPLL Solvers

We show that DPLL solvers with non-deterministic variable selection, randomized value selection and no restarts ($D^T_{ND,RD}$) is exponentially weaker than the same configuration with restarts ($D^{T,R}_{ND,RD}$).

**Corollary 20.** $D^T_{ND,RD}$ *runs exponentially slower on the class of satisfiable formulas* $Ladder_n(G, f)$ *than* $D^{T,R}_{ND,RD}$, *with high probability.*

The separation follows from the fact that our proof of the upper bound from Theorem 3 does not use the fact the solver has access to clause learning, which means the solver $D^{T,R}_{ND,RD}$ can also find a satisfying assignment for $Ladder_n(G, f)$ in time $O(n^2)$, except with exponentially small probability. On the other hand, the lower bound from Theorem 3 immediately implies an exponential lower bound for $D^T_{ND,RD}$, since $D^T_{ND,RD}$ is strictly weaker than $C^T_{ND,RD}$.

## 6.4   Separation Result for CDCL Solvers with WDLS

Finally, we state an observation of Robert Robere [19] on restarts in the context of the Weak Decision Learning Scheme (WDLS).

**Definition 21 (WDLS).** *Upon deriving a conflict, a CDCL solver with WDLS learns a conflict clause which is the disjunction of the negation of the decision variables on the current assignment trail.*

**Theorem 22.** $C^J_{ND,ND} + WDLS$ *is exponentially weaker than* $C^{J,R}_{ND,ND} + WDLS$.

*Proof.* The solver model $C^J_{ND,ND}$ with WDLS is only as powerful as $D^T_{ND,ND}$, since each learnt clause will only be used once for propagation after the solver backtracks immediately after learning the conlict clause, and remains satisfied for the rest of the solver run. This is exactly how $D^T_{ND,ND}$ behaves under the same circumstances. On the other hand, WDLS is an asserting learning scheme [17], and hence satisfies the conditions of the main theorem in [18], proving that CDCL with any asserting learning scheme and restarts p-simulates general resolution. Thus, we immediately have $C^{J,R}_{ND,ND}$ with WDLS is exponentially more powerful than the same solver but with no restarts (for unsatisfiable instances).    □

## 7    Related Work

**Previous Work on Theoretical Understanding of Restarts:** Buss et al. [8] and Van Gelder [21] proposed two proof systems, namely regWRTI and pool resolution respectively, with the aim of explaining the power of restarts in CDCL solvers. Buss et al. proved that regWRTI is able to capture exactly the power of CDCL solvers with *non-greedy BCP* and without restarts and Van Gelder proved that pool resolution can simulate certain configurations of DPLL solvers with clause learning. As both pool resolution and regWRTI are strictly more powerful than regular resolution, a natural question is whether formulas that exponentially separate regular and general resolution can be used to prove lower bounds for pool resolution and regWRTI, thus transitively proving lower bounds for CDCL solvers without restarts. However, since Bonet et al. [7] and Buss and Kołodziejczyk [9] proved that all such candidates have short proofs in pool resolution and regWRTI, the question of whether CDCL solvers without restarts are as powerful as general resolution still remains open.

**Previous Work on Empirical Understanding of Restarts:** The first paper to discuss restarts in the context of DPLL SAT solvers was by Gomes and Selman [10]. They proposed an explanation for the power of restarts popularly referred to as "heavy-tailed explanation of restarts". Their explanation relies on the observation that the runtime of randomized DPLL SAT solvers on satisfiable instances, when invoked with different random seeds, exhibits a heavy-tailed distribution. This means that the probability of the solver exhibiting a long runtime on a given input and random seed is non-negligible. However, because of the heavy-tailed distribution of solver runtimes, it is likely that the solver may run quickly on the given input for a different random seed. This observation was the motivation for the original proposal of the restart heuristic in DPLL SAT solvers by Gomes and Selman [10].

Unfortunately, the heavy-tailed explanation of the power of restarts does not lift to the context of CDCL SAT solvers. The key reason is that, unlike DPLL solvers, CDCL solvers save solver state (e.g., learnt clauses and variable activities) across restart boundaries. Additionally, the efficacy of restarts has been observed for both deterministic and randomized CDCL solvers, while the heavy-tailed explanation inherently relies on randomness. Hence, newer explanations have been proposed and validated empirically on SAT competition benchmarks. Chief among them is the idea that "restarts compact the assignment trail during its run and hence produce clauses with lower literal block distance (LBD), a key metric of quality of learnt clauses" [14].

**Comparison of Our Separation Results with Heavy-Tailed Explanation of Restarts:** A cursory glance at some of our separation results might lead one to believe that they are a complexity-theoretical analogue of the heavy-tailed explanation of the power of restarts, since our separation results are over randomized solver models. We argue this is not the case. First, notice that our main results are for drunk CDCL solvers that save solver state (e.g., learnt clauses) across restart boundaries, unlike the randomized DPLL solvers studied by Gomes et al. [10]. Second, we make no assumptions about independence (or lack thereof) of branching decisions across restarts boundaries. In point of fact, the variable selection in the CDCL model we use is non-deterministic. Only the value selection is randomized. More precisely, we have arrived at a separation result without relying on the assumptions made by the heavy-tailed distribution explanation, and interestingly we are able to prove that the "solver does get stuck in a bad part of the search space by making bad value selections". Note that in our model the solver is free to go back to "similar parts of the search space across restart boundaries". In fact, in our proof for CDCL with restarts, the solver chooses the same variable order across restart boundaries.

# 8    Conclusions

In this paper, we prove a series of results that establish the power of restarts (or lack thereof) for several models of CDCL and DPLL solvers. We first showed that CDCL solvers with backtracking, non-deterministic dynamic variable selection, randomized dynamic value selection, and restarts are exponentially faster than the same model without restarts for a class of satisfiable instances. Second, we showed CDCL solvers with VSIDS variable selection and phase saving without restarts are exponentially weaker than the same solver with restarts, for a family of unsatisfiable formulas. Finally, we proved four additional smaller separation and equivalence results for various configurations of DPLL and CDCL solvers.

By contrast to previous attempts at a "theoretical understanding the power of restarts" that typically assumed that variable and value selection heuristics in solvers are non-deterministic, we chose to study randomized or real-world models of solvers (e.g., VSIDS branching with phase saving value selection) that enabled us to more effectively isolate the power of restarts. This leads us to believe that the efficacy of restarts becomes apparent only when the solver models considered have weak heuristics (e.g., randomized or real-world deterministic) as opposed to models that assume that all solver heuristics are non-deterministic.

# References

1. Alekhnovich, M., Hirsch, E.A., Itsykson, D.: Exponential lower bounds for the running time of DPLL algorithms on satisfiable formulas. J. Autom. Reasoning **35**(1–3), 51–72 (2005)
2. Atserias, A., Fichte, J.K., Thurley, M.: Clause-learning algorithms with many restarts and bounded-width resolution. J. Artif. Intell. Res. **40**, 353–373 (2011)
3. Beame, P., Kautz, H.A., Sabharwal, A.: Towards understanding and harnessing the potential of clause learning. J. Artif. Intell. Res. **22**, 319–351 (2004)
4. Beame, P., Pitassi, T.: Propositional proof complexity: past, present, and future. In: Paun, G., Rozenberg, G., Salomaa, A. (eds.) Current Trends in Theoretical Computer Science, Entering the 21th Century, pp. 42–70. World Scientific, Singapore (2001)
5. Ben-Sasson, E., Wigderson, A.: Short proofs are narrow—resolution made simple. J. ACM **48**(2), 149–169 (2001)
6. Biere, A., Heule, M., van Maaren, H.: Handbook of Satisfiability, vol. 185. IOS press, Amsterdam (2009)
7. Bonet, M.L., Buss, S., Johannsen, J.: Improved separations of regular resolution from clause learning proof systems. J. Artif. Intell. Res. **49**, 669–703 (2014)
8. Buss, S.R., Hoffmann, J., Johannsen, J.: Resolution trees with lemmas: resolution refinements that characterize DLL algorithms with clauselearning. Log. Methods Comput. Sci. **4**(4) (2008)
9. Buss, S.R., Kołodziejczyk, L.: Small stone in pool. Log. Methods Comput. Sci. **10**(2), 16:1–16:22 (2014)
10. Gomes, C.P., Selman, B., Crato, N., Kautz, H.: Heavy-tailed phenomena in satisfiability and constraint satisfaction problems. J. Autom. Reasoning **24**(1–2), 67–100 (2000). https://doi.org/10.1023/A:1006314320276
11. Hertel, P., Bacchus, F., Pitassi, T., Gelder, A.V.: Clause learning can effectively P-simulate general propositional resolution. In: Fox, D., Gomes, C.P. (eds.) Proceedings of the Twenty-Third AAAI Conference on Artificial Intelligence (AAAI 2008), pp. 283–290. AAAI Press (2008)
12. Krajíček, J.: Proof Complexity, Encyclopedia of Mathematics and Its Applications, vol. 170. Cambridge University Press, Cambridge (2019)
13. Li, C., Fleming, N., Vinyals, M., Pitassi, T., Ganesh, V.: Towards a complexity-theoretic understanding of restarts in SAT solvers (2020)
14. Liang, J.H., Oh, C., Mathew, M., Thomas, C., Li, C., Ganesh, V.: Machine learning-based restart policy for CDCL SAT solvers. In: Beyersdorff, O., Wintersteiger, C.M. (eds.) SAT 2018. LNCS, vol. 10929, pp. 94–110. Springer, Cham (2018). https://doi.org/10.1007/978-3-319-94144-8_6
15. Marques-Silva, J.P., Sakallah, K.A.: GRASP: a search algorithm for propositional satisfiability. IEEE Trans. Comput. **48**(5), 506–521 (1999)
16. Moskewicz, M.W., Madigan, C.F., Zhao, Y., Zhang, L., Malik, S.: Chaff: engineering an efficient SAT solver. In: Proceedings of the 38th annual Design Automation Conference, pp. 530–535. ACM (2001)
17. Pipatsrisawat, K., Darwiche, A.: A new clause learning scheme for efficient unsatisfiability proofs. In: AAAI, pp. 1481–1484 (2008)
18. Pipatsrisawat, K., Darwiche, A.: On the power of clause-learning SAT solvers as resolution engines. Artif. Intell. **175**(2), 512–525 (2011)
19. Robere, R.: Personal communication (2018)
20. Urquhart, A.: Hard examples for resolution. J. ACM **34**(1), 209–219 (1987)

21. Gelder, A.: Pool resolution and its relation to regular resolution and DPLL with clause learning. In: Sutcliffe, G., Voronkov, A. (eds.) LPAR 2005. LNCS (LNAI), vol. 3835, pp. 580–594. Springer, Heidelberg (2005). https://doi.org/10.1007/11591191_40
22. Vinyals, M.: Hard examples for common variable decision heuristics. In: Proceedings of the 34th AAAI Conference on Artificial Intelligence (AAAI 2020), February 2020
23. Williams, R., Gomes, C., Selman, B.: On the connections between backdoors, restarts, and heavy-tailedness in combinatorial search. Structure **23**, 4 (2003)
24. Williams, R., Gomes, C.P., Selman, B.: Backdoors to typical case complexity. In: Proceedings of the Eighteenth International Joint Conference on Artificial Intelligence (IJCAI 2003), pp. 1173–1178 (2003)

# On the Sparsity of XORs in Approximate Model Counting

Durgesh Agrawal[1], Bhavishya[1], and Kuldeep S. Meel[2(✉)]

[1] Indian Institute of Technology Kanpur, Kanpur, India
[2] School of Computing, National University of Singapore, Singapore, Singapore
meel@comp.nus.edu.sg

**Abstract.** Given a Boolean formula $\varphi$, the problem of model counting, also referred to as #SAT, is to compute the number of solutions of $\varphi$. The hashing-based techniques for approximate counting have emerged as a dominant approach, promising achievement of both scalability and rigorous theoretical guarantees. The standard construction of strongly 2-universal hash functions employs *dense* XORs (i.e., involving half of the variables in expectation), which is widely known to cause degradation in the runtime performance of state of the art SAT solvers. Consequently, the past few years have witnessed an intense activity in the design of *sparse* XORs as hash functions. Such constructions have been proposed with beliefs to provide runtime performance improvement along with theoretical guarantees similar to that of dense XORs.

The primary contribution of this paper is a rigorous theoretical and empirical analysis to understand the effect of the sparsity of XORs. In contradiction to prior beliefs of applicability of analysis for sparse hash functions to all the hashing-based techniques, we prove a contradictory result. We show that the best-known bounds obtained for sparse XORs are still too weak to yield theoretical guarantees for a large class of hashing-based techniques, including the state of the art approach ApproxMC3. We then turn to a rigorous empirical analysis of the performance benefits of sparse hash functions. To this end, we first design, to the best of our knowledge, the most efficient algorithm called SparseCount2 using sparse hash functions, which achieves at least up to two orders of magnitude performance improvement over its predecessor. Contradicting the current beliefs, we observe that SparseCount2 still falls short of ApproxMC3 in runtime performance despite the usage of dense XORs in ApproxMC3. In conclusion, our work showcases that the question of whether it is possible to use short XORs to achieve scalability while providing strong theoretical guarantees is still wide open.

---

The author list has been sorted alphabetically by last name; this order should not be used to determine the extent of authors' contributions. Part of the work was carried out during the first two authors' internships at National University of Singapore.

© Springer Nature Switzerland AG 2020
L. Pulina and M. Seidl (Eds.): SAT 2020, LNCS 12178, pp. 250–266, 2020.
https://doi.org/10.1007/978-3-030-51825-7_18

# 1    Background and Introduction

Given a Boolean formula $\varphi$, the problem of model counting, also referred to as #SAT, is to compute the number of solutions of $\varphi$. Model counting is a fundamental problem in computer science with a wide range of applications ranging from quantified information flow, reliability of networks, probabilistic programming, Bayesian networks, and others [4,5,10,16,21–23].

Given the computational intractability of #SAT, attention has been focused on the approximation of #SAT [28,30]. In a breakthrough result, Stockmeyer provided a hashing-based randomized approximation scheme for counting that makes polynomially many invocations of an NP oracle [27]. The procedure, however, was computationally prohibitive in practice at that time, and no practical tools existed based on Stockmeyer's proposed algorithmic framework until the early 2000s [16]. Motivated by the success of SAT solvers, there has been a surge of interest in the design of hashing-based techniques for approximate model counting in the past decade [8,9,13,15,24,25].

The core idea of the hashing-based framework is to employ pairwise independent hash functions[1] to partition the solution space into *roughly equal-sized small* cells, wherein a cell is called *small* if it has solutions less than or equal to a pre-computed threshold, denoted by *thresh*. A SAT solver is employed to check if a cell is small by enumerating solutions one-by-one until either there are no more solutions or we have already enumerated *thresh* + 1 solutions. The current state of the art techniques can be broadly classified into two categories:

- The first category of techniques, henceforth called Cat1, consists of techniques that compute a constant factor approximation by setting *thresh* to a constant and use Stockmeyer's technique of constructing multiple copies of the input formula. [1,2,12,29,31]
- The second class of techniques, henceforth called Cat2, consists of techniques that directly compute an $(\varepsilon, \delta)$-estimate by setting *thresh* to $\mathcal{O}(\frac{1}{\varepsilon^2})$, and hence invoking the underlying NP oracle $\mathcal{O}(\frac{1}{\varepsilon^2})$ times [7–9,20,21,24,25].

The current state of the art technique, measured by runtime performance, is ApproxMC3, which falls into the class of Cat2 techniques [25]. The proofs of correctness for all the hashing-based techniques involve the use of concentration bounds due to pairwise independent hash functions.

The standard construction of pairwise independent hash functions employed in these techniques can be expressed as a conjunction of XOR constraints such that every variable is chosen with probability $f = 1/2$ for each XORs. As such, each XOR contains, on an average, $n/2$ variables. A SAT solver is invoked to enumerate solutions of the formula $\varphi$ in conjunction with these XOR constraints. The performance of SAT solvers, however, degrades with an increase in the size of XORs [15]. Therefore recent efforts have focused on the design of hash functions

---

[1] Pairwise independent hash functions were initially referred to as strongly 2-universal hash functions in [6]. The prior work on approximate counting often uses the term *2-universal hashing* to refer to strongly 2-universal hash functions.

where each variable is chosen with probability $f < 1/2$ [1,2,11,14,17]. We refer to the XOR constructed with $f = 1/2$ as dense XORs while those constructed with $f < 1/2$ as sparse XORs. In particular, given a hash function, $h$ and cell $\alpha$, the random variable of interest, denoted by $\mathsf{Cnt}_{\langle \varphi,h,\alpha \rangle}$ is the number of solutions of $\varphi$ that $h$ maps to cell $\alpha$. The pairwise independence of dense XORs is known to bound the variance of $\mathsf{Cnt}_{\langle \varphi,h,\alpha \rangle}$ by the expectation of $\mathsf{Cnt}_{\langle \varphi,h,\alpha \rangle}$, which is sufficient for their usage for both Cat1 and Cat2 techniques.

In a significant result, Asteris and Dimakis [3], and Zhao et al. [31] showed that $f = \mathcal{O}(\log n/n)$ asymptotically suffices for Cat1 techniques. It is worth pointing out that $f = \mathcal{O}(\log n/n)$ provides weaker guarantees on the variance of $\mathsf{Cnt}_{\langle \varphi,h,\alpha \rangle}$ as compared to the case when $f = 1/2$. However, Zhao et al. showed that the weaker guarantees are sufficient for Cat1 techniques with only polynomial overhead on the time complexity. Furthermore, Zhao et al. provided necessary and sufficient conditions on the required asymptotic value of $f$ and proposed a new algorithm SparseCount that uses the proposed family of hash functions. One would expect that the result of Zhao et al. would settle the quest for efficient hash functions. However, upon closer examination, few questions have been left unanswered in Zhao et al.'s work and subsequent follow-up studies [1,9,21].

1. Can the hash function constructed by Zhao et al. be used for Cat2 techniques, in particular for state of the art hashing-based techniques like ApproxMC3?
2. In practice, can the overhead due to the weakness of theoretical guarantees of sparse XORs proposed by Zhao et al. be compensated by the gain of performance due to sparse XORs in the runtime of SparseCount?
3. Is the runtime performance of SparseCount competitive to that of ApproxMC3? The reader may observe that Zhao et al.'s paper does not compare their proposed algorithm for $(\varepsilon, \delta)$-guarantees, called SparseCount, with state of the art algorithms at that time such as ApproxMC2, which is now in its third version, ApproxMC3 [25]. Therefore the question of whether the proposed sparse XORs are efficient in runtime was not settled. It is perhaps worth remarking that another line of work based on the construction of sparse XORs using low-density parity codes is known to introduce significant slowdown [1, 2] (See Section 9 of [1]).

The primary contribution of this paper is a rigorous theoretical and empirical analysis to understand the effect of sparse XORs for approximate model counters. In particular, we make the following key contributions:

1. We prove that the bounds obtained by Zhao et al., which are the strongest known bounds at this point, for the variance of $\mathsf{Cnt}_{\langle \varphi,h,\alpha \rangle}$, are still too weak for the analysis of ApproxMC3. To the best of our knowledge, this is the first time the need for stronger bounds in the context of Cat2 techniques has been identified.
2. Since the weakness of bounds prevents usage of sparse hash functions in ApproxMC3, we design the most efficient algorithm, to the best of our knowledge, using sparse hash functions. To this end, we propose an improvement of

SparseCount, called SparseCount2, that reduces the number of SAT calls from linear to logarithmic and significantly improves the runtime performance of SparseCount. The improvement from linear to logarithmic uses the idea of prefix-slicing introduced by Chakraborty, Meel, and Vardi [9] for ApproxMC2.

3. We next present a rigorous empirical study involving a benchmark suite totaling over 1800 instances of runtime performance of SparseCount2 vis-a-vis the state of the art approximate counting technique, ApproxMC3. Surprisingly and contrary to current beliefs, we discover that ApproxMC3, which uses dense XORs significantly outperforms SparseCount2 for every benchmark. It is worth remarking that both ApproxMC3 and SparseCount2 use identical SAT solver for underlying SAT calls and similar to other hashing-based techniques, over 99% for each of the algorithms is indeed consumed by the underlying SAT solver.

Given the surprising nature of our results, few words are in order. First of all, our work identifies the tradeoffs involved in the usage of sparse hash functions and demonstrates that the variance bounds offered by sparse hash functions are too weak to be employed in the state of the art techniques. Secondly, our work demonstrates that the weakness of variance bounds leads to such a large overhead that the algorithms using sparse hash functions scale much worse compared to the algorithms without sparse XORs. Thirdly and finally, we believe the negative results showcase that the question of the usage of sparse XORs to achieve scalability while providing strong theoretical guarantees is still wide open. In an upcoming work, Meel (r) Akshay[2] [20] define a new family of hash functions, called concentrated hashing, and provide a new construction of sparse hash functions belonging to concentrated hashing, and design a new algorithmic framework on top of ApproxMC, which is shown to achieve runtime improvements.

The rest of the paper is organized as follows. We discuss notations and preliminaries in Sect. 2. We then discuss the weakness of guarantees offered by sparse XORs in Sect. 3. In Sect. 4, we seek to design an efficient algorithm that utilizes all the advancements, to the best of our knowledge, in approximate model counting community. We present a rigorous empirical study comparing performance of SparseCount, SparseCount2, and ApproxMC3 in Sect. 5 and conclude in Sect. 6.

## 2    Preliminaries and Notations

Let $\varphi$ be a Boolean formula in conjunctive normal form (CNF), and let $\mathsf{Vars}(\varphi)$ be the set of variables appearing in $\varphi$. The set $\mathsf{Vars}(\varphi)$ is also called the *support* of $\varphi$. Unless otherwise stated, we will use $n$ to denote the number of variables in $\varphi$ i.e., $|\mathsf{Vars}(\varphi)|$. An assignment of truth values to the variables in $\mathsf{Vars}(\varphi)$ is called a *satisfying assignment* or *witness* of $\varphi$ if it makes $\varphi$ evaluate to true. We denote the set of all witnesses of $\varphi$ by $R_\varphi$.

---

[2] (r) is used to denote random author ordering, as suggested by the authors.

We write $\Pr[\mathcal{Z}]$ to denote the probability of outcome $\mathcal{Z}$. The expected value of $\mathcal{Z}$ is denoted $E[\mathcal{Z}]$ and its variance is denoted $\sigma^2[\mathcal{Z}]$.

The *propositional model counting problem* is to compute $|R_\varphi|$ for a given CNF formula $\varphi$. A *probably approximately correct* (PAC) counter is a probabilistic algorithm $\mathsf{ApproxCount}(\cdot, \cdot, \cdot)$ that takes as inputs a formula $F$, a tolerance $\varepsilon > 0$, and a confidence parameter $\delta \in (0, 1]$, and returns a count $c$ with $(\varepsilon, \delta)$-guarantees, i.e., $\Pr\left[|R_\varphi|/(1+\varepsilon) \le c \le (1+\varepsilon)|R_\varphi|\right] \ge 1 - \delta$.

In this work, we employ a family of universal hash functions. Let $H(n, m) \triangleq \{h : \{0,1\}^n \to \{0,1\}^m\}$ be a family of hash functions mapping $\{0,1\}^n$ to $\{0,1\}^m$. We use $h \xleftarrow{R} H$ to denote the probability space obtained by choosing a function $h$ uniformly at random from $H$.

In this work, we will use the concept of *prefix-slicing* introduced by Chakraborty et al. [9]. For $h \in H(n, m)$, formally, for every $j \in \{1, \ldots, m\}$, the $j^{th}$ prefix-slice of $h$, denoted $h^{(j)}$, is a map from $\{0,1\}^n$ to $\{0,1\}^j$, such that $h^{(j)}(y)[i] = h(y)[i]$, for all $y \in \{0,1\}^n$ and for all $i \in \{1, \ldots j\}$. Similarly, the $j^{th}$ prefix-slice of $\alpha$, denoted $\alpha^{(j)}$, is an element of $\{0,1\}^m$ such that $\alpha^{(j)}[i] = \alpha[i]$ for all $i \in \{1, \ldots j\}$. The randomness in the choices of $h$ and $\alpha$ induce randomness in the choices of $h^{(m)}$ and $\alpha^{(m)}$. However, the $(h^{(j)}, \alpha^{(j)})$ pairs chosen for different values of $j$ are no longer independent. Specifically, $h^{(k)}(y)[i] = h^{(\ell)}(y)[i]$ and $\alpha^{(k)}[i] = \alpha^{(\ell)}[i]$ for $1 \le k \le \ell \le m$ and for all $i \in \{1, \ldots k\}$.

For a formula $\varphi$, $h \in H(n, m)$, and $\alpha \in \{0,1\}^m$, we define $\mathsf{Cnt}_{\langle F, h^{(i)}, \alpha^{(i)} \rangle} := |\{y \in R_\varphi \mid h^{(i)}(y) = \alpha^{(i)}\}|$, i.e. the number of solutions of $\varphi$ mapped to $\alpha^{(i)}$ by $h^{(i)}$. For the sake of notational clarity, whenever $h^{(i)}$ and $\alpha^{(i)}$ are clear from the context, we will use $\mathsf{Cnt}_{\langle i \rangle}$ as a shorthand for $\mathsf{Cnt}_{\langle F, h^{(i)}, \alpha^{(i)} \rangle}$.

**Definition 1.** [6] *A family of hash functions $H(n, m)$ is pairwise independent (also known as strongly 2-universal) if $\forall \alpha_1, \alpha_2 \in \{0,1\}^m$, $\forall$ distinct $y_1, y_2 \in \{0,1\}^n$, $h \xleftarrow{R} H$, we have $\Pr[h(y_1) = \alpha_1 \wedge h(y_2) = \alpha_2] = \frac{1}{2^{2m}}$.*

**Definition 2.** *Let $A \in \{0,1\}^{m \times n}$ be a random matrix whose entries are Bernoulli i.i.d. random variables such that $f_i = \Pr[A[i, j] = 1]$ for all $j \in [n]$. Let $b \in \{0,1\}^m$ be chosen uniformly at random, independently from $A$. Let $h_{A,b}(y) = Ay + b$ and $H^{\{f_i\}}(n, m) = \{h_{A,b} : \{0,1\}^n \to \{0,1\}^m\}$, where $h_{A,b} \xleftarrow{R} H^{\{f_i\}}(n, m)$ is chosen randomly according to this process. Then, $H^{\{f_i\}}(n, m)$ is defined as hash family with $\{f_i\}$-sparsity.*

Since we can represent hash functions in $H^{\{f_i\}}(n, m)$ using a set of XORs; we will use *dense XORs* to refer to hash functions with $f_i = \frac{1}{2}$ for all $i$ while we use *sparse XORs* to refer to hash functions with $f_i < \frac{1}{2}$ for some $i$. Note that $H^{\{f_i = \frac{1}{2}\}}(n, m)$ is the standard pairwise independent hash family, also denoted as $H_{xor}(n, m)$ in earlier works [21].

**Definition 3.** [11] *Let $k \ge 0$ and $\delta > 2$. Let $Z$ be a random variable with $\mu = E[Z]$. Then $Z$ is strongly $(k, \delta)$-concentrated if $\Pr[|Z - \mu| \ge \sqrt{k}] \le 1/\delta$ and weakly $(k, \delta)$-concentrated if both $\Pr[Z \le \mu - \sqrt{k}] \le 1/\delta$ and $\Pr[Z \ge \mu + \sqrt{k}] \le 1/\delta$.*

## 2.1   Related Work

Gomes et al. [14] first identified the improvements in solving time due to the usage of sparse XORs in approximate model counting algorithms. The question of whether sparse XORs can provide the required theoretical guarantees was left open. A significant progress in this direction was achieved by Ermon et al. [11], who provided the first rigorous analysis of the usage of sparse XOR constraints. Building on Ermon et al., Zhao et al. [31] and Asteris and Dimakis [3] independently provided further improved analysis of Ermon et al. and showed that probability $f = \mathcal{O}(\frac{\log n}{n})$ suffices to provide constant factor approximation, which can be amplified to $(1 + \varepsilon)$ approximation.

While the above mentioned efforts focused on each entry of $A$ to be i.i.d., Achlioptas and Theodorpoulos [2], Achlioptas, Hammoudeh, and Theodorpoulos [1] investigated the design of hash functions where $A$ is a structured matrix by drawing on connections to the error correcting codes. While their techniques provide a construction of sparse constraints, the constants involved in asymptotics lead to impractical algorithms for $(\varepsilon, \delta)$ guarantees (See Sect. 9 of [1]). The work of Achlioptas et al. demonstrates the promise and limitations of structured random matrices in the design of hashing-based algorithms; however, there is no such study in the case when all the entries are i.i.d. In this paper, we theoretically improve the construction proposed by Asteris and Dimakis [3], and Zhao et al. [31] and perform a rigorous empirical study to understand the tradeoffs of sparsity.

## 3   Weakness of Guarantees Offered by Sparse XORs

In this section, we present the first contribution of this paper: demonstration of the weakness of theoretical guarantees obtained in prior work [3,11,31] for sparse XORs. To this end, we investigate whether the bounds offered by Zhao et al. on the variance of $\mathsf{Cnt}_{\langle i \rangle}$, which are the strongest bounds known on sparse XORs, can be employed in the analysis of Cat2 techniques. For clarity of exposition, we focus on the usage of sparse XOR bounds in ApproxMC3, but our conclusions extend to other Cat2 techniques, as pointed out below.

The analysis of ApproxMC3 employs the bounds on the variance of $\mathsf{Cnt}_{\langle i \rangle}$ using the following standard concentration bounds.

**Lemma 1.** *For every $\beta > 0, 0 < \varepsilon \leq 1$, $0 \leq i \leq n$, we have:*

*1. Chebyshev Inequality*

$$\Pr\left[\left|\mathsf{Cnt}_{\langle i \rangle} - \mathsf{E}[\mathsf{Cnt}_{\langle i \rangle}]\right| \geq \frac{\varepsilon}{1 + \varepsilon}\mathsf{E}[\mathsf{Cnt}_{\langle i \rangle}]\right] \leq \frac{(1 + \varepsilon)^2 \sigma^2[\mathsf{Cnt}_{\langle i \rangle}]}{\varepsilon^2 \mathsf{E}[\mathsf{Cnt}_{\langle i \rangle}]^2}$$

*2. Paley-Zygmund Inequality*

$$\Pr[\mathsf{Cnt}_{\langle i \rangle} \leq \beta\mathsf{E}[\mathsf{Cnt}_{\langle i \rangle}]] \leq \frac{1}{1 + \dfrac{(1 - \beta)^2 \mathsf{E}[\mathsf{Cnt}_{\langle i \rangle}]^2}{\sigma^2[\mathsf{Cnt}_{\langle i \rangle}]}}$$

The analysis of Cat2 techniques (and ApproxMC3 in particular) bounds the failure probability of the underlying algorithm by upper bounding the above expressions for appropriately chosen values of $i$. To obtain meaningful upper bounds, these techniques employ the inequality $\sigma^2[\mathsf{Cnt}_{\langle i \rangle}] \leq \mathsf{E}[\mathsf{Cnt}_{\langle i \rangle}]$ obtained via the usage of 2-universal hash functions[3].

Recall, that the core idea of the hashing-based framework is to employ 2-universal hash functions to partition the solution space into *roughly equal sized small* cells, wherein a cell is called *small* if it has solutions less than or equal to a pre-computed threshold, denoted by *thresh*, which is chosen as $\mathcal{O}(1/\varepsilon^2)$. To this end, the analysis lower bounds $\mathsf{E}[\mathsf{Cnt}_{\langle i \rangle}]$ by $\frac{thresh}{2}$, which allows the denominator to be lower bounded by a constant. Given that *thresh* can be set to $\mathcal{O}(\frac{1}{\varepsilon^2})^{1/c}$ for some $c > 0$, we can relax the requirement on the chosen hash family to ensuring $\sigma^2[\mathsf{Cnt}_{\langle i \rangle}] \leq \mathsf{E}[\mathsf{Cnt}_{\langle i \rangle}]^{2-c}$ for some $c > 0$. Note that pairwise independent hash functions based on dense XORs ensure $\sigma^2[\mathsf{Cnt}_{\langle i \rangle}] \leq \mathsf{E}[\mathsf{Cnt}_{\langle i \rangle}]$ (i.e., $c = 1$).

We now investigate the guarantees provided by sparse XORs. To this end, we first recall the following result, which follows from combining Theorem 1 and Theorem 3 of [11].

**Lemma 2.** $[11]^4$ *For* $2 \leq |R_F| \leq 2^n$, *let*

$$w^* = max\left\{ w \mid \sum_{j=1}^{w} \binom{n}{j} \leq |R_F| - 1 \right\}$$

$$q^* = |R_F| - 1 - \sum_{w=1}^{w^*} \binom{n}{w}$$

$$\eta = \frac{1}{|R_F| - 1} \left( q^* \left( \frac{1}{2} + \frac{1}{2}(1 - 2f)^{w^*+1} \right)^m + \sum_{w=1}^{w^*} \binom{n}{w} \left( \frac{1}{2} + \frac{1}{2}(1 - 2f)^w \right)^m \right)$$

*For* $h \xleftarrow{R} H^{\{f_j\}}(n, m)$, *we have:*

$$\sigma^2[\mathsf{Cnt}_{\langle i \rangle}] \leq \mathsf{E}[\mathsf{Cnt}_{\langle i \rangle}] + \eta\mathsf{E}[\mathsf{Cnt}_{\langle i \rangle}](|R_F| - 1) - \mathsf{E}[\mathsf{Cnt}_{\langle i \rangle}]^2.$$

Zhao et al. [31], building on Ermon et al. [11], obtain the following bound (see, Lemma 8 and Lemma 10 of [31]).

**Lemma 3.** *[31] Define* $k = 2^m \eta(1 - \frac{1}{|R_F|})$. *Then* $k \leq \gamma$ *for* $\gamma > 1$.

The bound on $\sigma^2[\mathsf{Cnt}_{\langle i \rangle}]$ from Zhao et al. can be stated as:

**Theorem 1.** $\sigma^2[\mathsf{Cnt}_{\langle i \rangle}] \leq \zeta$ *where* $\zeta \in \Omega(\mathsf{E}[\mathsf{Cnt}_{\langle i \rangle}]^2)$.

---

[3] While we are focusing on ApproxMC3, the requirement of $\sigma^2[\mathsf{Cnt}_{\langle i \rangle}] \leq \mathsf{E}[\mathsf{Cnt}_{\langle i \rangle}]$ holds for other Cat2 techniques.

[4] The expression stated in the Theorem can be found in the revised version at https://cs.stanford.edu/~ermon/papers/SparseHashing-revised.pdf (Accessed: May 10, 2020).

*Proof.*

$$\sigma^2[\mathsf{Cnt}_{\langle i \rangle}] \le \mathsf{E}[\mathsf{Cnt}_{\langle i \rangle}] + \eta \mathsf{E}[\mathsf{Cnt}_{\langle i \rangle}](|R_F| - 1) - \mathsf{E}[\mathsf{Cnt}_{\langle i \rangle}]^2.$$

(Substituting $|R_F| = \mathsf{E}[\mathsf{Cnt}_{\langle i \rangle}] \times 2^m$, we have)

$$\sigma^2[\mathsf{Cnt}_{\langle i \rangle}] \le \mathsf{E}[\mathsf{Cnt}_{\langle i \rangle}] + 2^m \eta \mathsf{E}[\mathsf{Cnt}_{\langle i \rangle}]^2(1 - 1/|R_F|) - \mathsf{E}[\mathsf{Cnt}_{\langle i \rangle}]^2$$

Substituting $k = 2^m \eta(1 - \frac{1}{|R_F|})$, we have:

$$\sigma^2[\mathsf{Cnt}_{\langle i \rangle}] \le \mathsf{E}[\mathsf{Cnt}_{\langle i \rangle}] + (k - 1)\mathsf{E}[\mathsf{Cnt}_{\langle i \rangle}]^2 = \zeta.$$

Using Corollary 3, we have $\zeta \in \Omega(\mathsf{E}[\mathsf{Cnt}_{\langle i \rangle}]^2)$.

Recall, the analysis of ApproxMC3 requires us to upper bound $\sigma^2[\mathsf{Cnt}_{\langle i \rangle}]$ by $\mathsf{E}[\mathsf{Cnt}_{\langle i \rangle}]^{2-c}$ for $c > 0$. Since the best-known bounds on $\sigma^2[\mathsf{Cnt}_{\langle i \rangle}]$ lower bound $\sigma^2[\mathsf{Cnt}_{\langle i \rangle}]$ by $\mathsf{E}[\mathsf{Cnt}_{\langle i \rangle}]^2$, these bounds are not sufficient to be used by ApproxMC3. At this point, one may wonder as to what is the key algorithmic difference between Cat1 and Cat2 that necessitates the use of stronger bounds: Cat1 techniques compute a constant factor approximation and then make use of Stockmeyer's argument to lift a constant factor approximation to $(1 + \varepsilon)$-approximation, whereas, Cat2 techniques directly compute a $(1 + \varepsilon)$-approximation, which necessitates the usage of stronger concentration bounds.

## 4    SparseCount2: An Efficient Algorithm for Sparse XORs

The inability of sparse XORs to provide good enough bounds on variance for usage in Cat2 techniques, in particular ApproxMC3, leads us to ask: how do we design the *most efficient* algorithm for approximate model counting making use of the existing gadgets in the model counting literature. We recall that Zhao et al. [31] provided matching necessary and sufficient conditions on the required asymptotic density of matrix $A$. Furthermore, they proposed a hashing-based algorithm, SparseCount, that utilizes sparser constraints.

As mentioned earlier, Chakraborty et al. [9] proposed the technique of using *prefix-slicing* of hash functions in the context of hashing-based techniques and their empirical evaluation demonstrated significant theoretical and empirical improvements owing to the usage of prefix hashing. In this work, we first show a dramatic reduction in the complexity of SparseCount by utilizing the concept of *prefix-slicing* and thereby improving the number of SAT calls from $\mathcal{O}(n \log n)$ to $\mathcal{O}((\log n)^2)$ for fixed $\varepsilon$ and $\delta$ The modified algorithm, called SparseCount2, significantly outperforms SparseCount, as demonstrated in Sect. 5.

Algorithm 1 shows the pseudo-code for SparseCount2. SparseCount2 assumes access to SAT oracle that takes in a formula $\varphi$ and returns YES if $\varphi$ is satisfiable, otherwise it returns NO. Furthermore, SparseCount2 assumes access to the subroutine MakeCopies that creates multiple copies of a given formula, a standard technique first proposed by Stockmeyer [27] to lift a constant factor approximation to that of $(1+\varepsilon)$-factor approximation for arbitrary $\varepsilon$. Similar to Algorithm

---

**Algorithm 1.** SparseCount2 $(\varphi, \varepsilon, \delta)$     ▷ Assume $\varphi$ is satisfiable

---
1: $\Delta \leftarrow 0.0042$
2: $\psi \leftarrow \mathsf{MakeCopies}(\varphi, \lceil \frac{1}{\log_4(1+\varepsilon)} \rceil)$
3: $m \leftarrow 0;\ iter \leftarrow 0;\ \mathcal{C} \leftarrow \mathsf{EmptyList};\ \hat{n} \leftarrow |\mathsf{Vars}(\psi)|$
4: $T \leftarrow \left\lceil \dfrac{\log(\hat{n}/\delta)}{\Delta} \right\rceil$
5: $\{f_j\} \leftarrow \mathsf{ComputeSparseDensities}(\hat{n})$
6: **repeat**
7:     $iter \leftarrow iter + 1$
8:     $m \leftarrow \mathsf{CoreSearch}(\psi, m, \{f_j\})$
9:     $\mathsf{AddToList}(\mathcal{C}, 2^m)$
10: **until** $iter < T$
11: $\hat{c} \leftarrow \mathsf{Median}(\mathcal{C})$
12: **return** $\hat{c}^{\lceil \log_4(1+\varepsilon) \rceil}$

---

**Algorithm 2.** CoreSearch$(\psi, \text{mPrev}, \{f_j\})$

---
1: Choose $h$ uniformly at random from $H^{\{f_j\}}(\hat{n}, \hat{n})$
2: Choose $\alpha$ uniformly at random from $\{0, 1\}^{\hat{n}}$
3: $Y \leftarrow \mathsf{SAT}(\psi \wedge h^{(\hat{n})}(\mathsf{Vars}(\psi)) = \alpha^{\hat{n}})$
4: **if** $Y$ is YES **then**
5:     **return** $\hat{n}$
6: $m \leftarrow \mathsf{LogSATSearch}(\psi, h, \alpha, \text{mPrev})$
7: **return** $m$

---

1 of [11], we choose $\{f_j\}$ in line 5, such that the resulting hash functions guarantee weak $(\mu_i^2, 9/4)$-concentration for the random variable $\mathsf{Cnt}_{\langle i \rangle}$ for all $i$, where $\mu_i = \mathsf{E}[\mathsf{Cnt}_{\langle i \rangle}]$. SparseCount2 shares similarity with SparseCount with the core difference in the replacement of linear search in SparseCount with the procedure CoreSearch. CoreSearch shares similarity with the procedure ApproxMC2Core of Chakraborty et al. [9]. The subroutine CoreSearch employs prefix search, which ensures that for all $i$, $\mathsf{Cnt}_{\langle i \rangle} \geq \mathsf{Cnt}_{\langle i+1 \rangle}$. The monotonicity of $\mathsf{Cnt}_{\langle i \rangle}$ allows us to perform a binary search to find the value of $i$ for which $\mathsf{Cnt}_{\langle i \rangle} \geq 1$ and $\mathsf{Cnt}_{\langle i+1 \rangle} = 0$. Consequently, we make $\mathcal{O}(\log n)$ calls to the underlying NP oracle during each invocation of CoreSearch instead of $\mathcal{O}(n)$ calls in case of SparseCount. Note that CoreSearch is invoked $T$ times, where $T = \left\lceil \frac{log(\hat{n}/\delta)}{\Delta} \right\rceil$ ($\hat{n}, \delta, \Delta$ as defined in the algorithm) and the returned value is added to the list $\mathcal{C}$. We then return the median of $\mathcal{C}$.

It is worth noting that SparseCount2 and ApproxMC3 differ only in the usage of *thresh*, which is set to 1 for SparseCount2 and a function of $\varepsilon$ for ApproxMC3, as observed in the discussion following Lemma 1. The usage of *thresh* dependent on $\varepsilon$ requires stronger bounds on variance, which can not be provided by sparse XORs as discussed in the previous section.

**Algorithm 3.** LogSATSearch($\psi, h, \alpha$, mPrev)

1: loIndex $\leftarrow 0$; hiIndex $\leftarrow \hat{n} - 1$; $m \leftarrow$ mPrev
2: BigCell[0] $\leftarrow 1$; BigCell[$\hat{n}$] $\leftarrow 0$
3: BigCell[$i$] $\leftarrow \perp \forall i \in [1, \hat{n} - 1]$
4: **while** true **do**
5:     $Y \leftarrow \mathsf{SAT}(\psi \wedge h^{(m)}(\mathsf{Vars}(\psi)) = \alpha^{(m)})$
6:     **if** $Y$ is YES **then**
7:         **if** BigCell[$m + 1$] $= 0$ **then**
8:             **return** $m + 1$
9:         BigCell[$i$] $\leftarrow 1 \ \forall i \in \{1, ...m\}$
10:        loIndex $\leftarrow m$
11:        **if** $|m - \text{mPrev}| < 3$ **then**
12:            $m \leftarrow m + 1$
13:        **else if** $2m < |\hat{n}|$ **then**
14:            $m \leftarrow 2m$
15:        **else** $m \leftarrow$ (hiIndex$+m$)$/2$
16:    **else**
17:        **if** BigCell[$m - 1$] $= 1$ **then**
18:            **return** $m$
19:        BigCell[$i$] $\leftarrow 0 \ \forall i \in \{m, ...\hat{n}\}$
20:        hiIndex $\leftarrow m$
21:        **if** $|m - \text{mPrev}| < 3$ **then** $m \leftarrow m - 1$
22:        **else** $m \leftarrow$ (loIndex$+m$)$/2$

## 4.1  Analysis of Correctness of SparseCount2

We now present the theoretical analysis of SparseCount2. It is worth asserting that the proof structure and technique for SparseCount2 and ApproxMC3 are significantly different, as is evident from the inability of ApproxMC3 to use sparse XORs. Therefore, while the algorithmic change might look minor, the proof of correctness requires a different analysis.

**Theorem 2.** *Let* SparseCount2 *employ* $H^{\{f_j\}_{j=0}^n}$ *hash families, where* $\{f_j\}_{j=0}^n$ *is chosen such that it guarantees weak* $(\mu_i^2, 9/4)$*-concentration for the random variable* $\mathsf{Cnt}_{\langle i \rangle}$ *for all* $i$, *then* SparseCount2 *returns count* $c$ *such that*

$$\Pr\left[\frac{|R_\varphi|}{1+\varepsilon} \le c \le (1+\varepsilon) \times |R_\varphi|\right] \ge 1 - \delta$$

*Proof.* Similar to [31], we assume that $|R_\varphi|$ is a power of 2; a relaxation of the assumption simply introduces a constant factor in the approximation. Let $|R_\psi| = 2^{i^*}$ and for we define the variable $\mathsf{Cnt}_{\langle i \rangle}^t$ to denote the value of $\mathsf{Cnt}_{\langle i \rangle}$ when iter $= t$. Let $\mu_i^t = \mathsf{E}[\mathsf{Cnt}_{\langle i \rangle}^t] = \frac{2^{i^*}}{2^i}$. Note that the choice of $f_i$ ensures that $\mathsf{Cnt}_{\langle i \rangle}^t$ is weakly $((\mu_i^t)^2, 9/4)$ concentrated.

Let $\mathcal{E}$ denote the event that $\hat{c} > 4 \times |R_\psi|$ or $\hat{c} < \frac{|R_\psi|}{4}$. We denote the event $\hat{c} > 4 \times |R_\psi|$ as $\mathcal{E}_H$ and the event $\hat{c} < \frac{|R_\psi|}{4}$ as $\mathcal{E}_L$. Note that $\Pr[\mathcal{E}] = \Pr[\mathcal{E}_L] + \Pr[\mathcal{E}_H]$. We now compute $\Pr[\mathcal{E}_L]$ and $\Pr[\mathcal{E}_H]$ as follows:

1. From Algorithm 1, we have $\hat{c} = \text{Median}(\mathcal{C})$. For $\hat{c} < \frac{|R_\psi|}{4}$, we have that for at least $\frac{T}{2}$ iterations of CoreSearch returns $m < i^* - 2$. For $t$-th invocation of CoreSearch (i.e., $iter = t$) to return $m - 1$, then it is necessarily the case that $\text{Cnt}^t_{\langle m \rangle} = 0$. Since $\{f_j\}^n_{j=0}$ is chosen such that the resulting hash function guarantees $((\mu^t_m)^2, 9/4)$-concentration for the random variable $\text{Cnt}^t_{\langle m \rangle}$, we have $\Pr[\text{Cnt}^t_{\langle m \rangle} \geq 1] \geq 5/9$ for $m \leq i^* - 2$.

   Let us denote, by $\mathcal{E}^i_L$, the event that at least for $\frac{T}{2}$ of $\{\text{Cnt}^t_{\langle i \rangle}\}^T_{t=0}$ we have $\text{Cnt}^t_{\langle i \rangle} = 0$. Therefore, by Chernoff bound we have $\Pr[\mathcal{E}^i_L] \leq e^{-\nu^{(1)}T}$ where $\nu^{(1)} = 2(4/9 - 1/2)^2$. By applying union bound, we have $\Pr[\mathcal{E}_L] \leq ne^{-\nu^{(1)}T}$

2. Again, from the Algorithm 1, we have $\hat{c} = \text{Median}(\mathcal{C})$. Therefore, for $\hat{c} > 4 \times |R_\psi|$, we have at least $\frac{T}{2}$ invocations of CoreSearch return $m > i^* + 2$. For $t$-th invocation of CoreSearch (i.e., $iter = t$) to return $m$, then it is necessarily the case that $\text{Cnt}^t_{\langle m-1 \rangle} \geq 1$.

   Noting, $\mathsf{E}[\text{Cnt}^t_{\langle m \rangle}] = 2^{i^* - m}$. For $m \geq i^* + 2$, we have for $m \geq i^* + 2$

$$\Pr[\text{Cnt}^t_{\langle m \rangle} \geq 1] \leq 1/4.$$

Let us denote by $\mathcal{E}^i_H$, the event that for at least $\frac{T}{2}$ of $\{\text{Cnt}^t_{\langle i \rangle}\}^T_{t=0}$ values, we have $\text{Cnt}^t_{\langle i \rangle} \geq 1$. By Chernoff bound for $m \geq i^* + 2$, we have $\Pr[\mathcal{E}^i_H] \leq e^{-\nu^{(2)}T}$ where $\nu^{(2)} = 2(1/4 - 1/2)^2$. By applying union bound, we have $\Pr[\mathcal{E}_H] \leq ne^{-\nu^{(2)}T}$.

Therefore, we have $\Pr[\mathcal{E}] = \Pr[\mathcal{E}_L] + \Pr[\mathcal{E}_H] \leq ne^{-\nu^{(1)}T} + ne^{-\nu^{(2)}T}$. Substituting $T$, we have

$$\Pr\left[\frac{|R_\psi|}{4} \leq \hat{c} \leq 4 \times |R_\psi|\right] \geq 1 - \delta.$$

Now notice that $|R_\psi| = |R_\varphi|^{\frac{1}{\log_4(1+\varepsilon)}}$; Therefore, $\frac{|R_\psi|}{4} \leq \hat{c} \leq 4 \times |R_\psi|$ ensures that we have $\frac{|R_\varphi|}{1+\varepsilon} \leq c \leq (1+\varepsilon) \times |R_\varphi|$. Therefore,

$$\Pr\left[\frac{|R_\varphi|}{1+\varepsilon} \leq c \leq (1+\varepsilon) \times |R_\varphi|\right] \geq 1 - \delta.$$

## 5    Empirical Studies

We focus on empirical study for comparison of runtime performance of SparseCount, SparseCount2, and ApproxMC3. All the three algorithms, SparseCount, SparseCount2, and ApproxMC3, are implemented in C++ and use the same underlying SAT solver, CryptoMiniSat [26] augmented with the BIRD framework introduced in [24,25]. CryptoMiniSat augmented with BIRD is state of the art SAT solver equipped to handle XOR constraints natively. *It is worth noting that for hashing-based techniques, over 99% of the runtime is consumed by the underlying SAT solver [25].* Therefore, the difference in the performance

**Table 1.** Table of comparison between SparseCount, SparseCount2, and ApproxMC3

| Benchmark (.cnf) | Vars | Clauses | Time (s) | | |
|---|---|---|---|---|---|
| | | | SparseCount | SparseCount2 | ApproxMC3 |
| blasted_case200 | 14 | 42 | 18.13 | 8.49 | 0.01 |
| blasted_case60 | 15 | 35 | 350.48 | 23.43 | 0.01 |
| s27_3_2 | 20 | 43 | 1581.63 | 30.28 | 0.01 |
| SetTest.sk_9_21 | 33744 | 148948 | 1679.62 | 171.02 | 0.81 |
| lss.sk_6_7 | 82362 | 259552 | 1959.39 | 405.61 | 1.63 |
| registerlesSwap.sk_3_10 | 372 | 1493 | 2498.02 | 60.23 | 0.03 |
| polynomial.sk_7_25 | 313 | 1027 | 2896.49 | 99.94 | 0.02 |
| 02A-3 | 5488 | 21477 | 3576.82 | 467.26 | 0.06 |
| blasted_case24 | 65 | 190 | TO | 125.25 | 0.05 |
| ConcreteActivityService.sk_13_28 | 2481 | 9011 | TO | 467.97 | 0.84 |
| GuidanceService2.sk_2_27 | 715 | 2181 | TO | 498.14 | 0.29 |
| ActivityService2.sk_10_27 | 1952 | 6867 | TO | 505.23 | 0.5 |
| UserServiceImpl.sk_8_32 | 1509 | 5009 | TO | 511.09 | 0.33 |
| or-100-10-4-UC-60 | 200 | 500 | TO | 608.86 | 0.05 |
| 02A-2 | 3857 | 15028 | TO | 1063.67 | 0.05 |
| LoginService2.sk_23_36 | 11511 | 41411 | TO | 1127.36 | 2.96 |
| 17.sk_3_45 | 10090 | 27056 | TO | 1299.15 | 1.69 |
| diagStencil.sk_35_36 | 319730 | 1774184 | TO | 2188.19 | 112.52 |
| tableBasedAddition.sk_240_1024 | 1026 | 961 | TO | TO | 2.17 |
| blasted_squaring9 | 1434 | 5028 | TO | TO | 5.04 |
| blasted_TR_b12_1_linear | 1914 | 6619 | TO | TO | 259.3 |

of the algorithms is primarily due to the number of SAT calls and the formulas over which the SAT solver is invoked. Furthermore, our empirical conclusions do not change even using the older versions of CryptoMiniSat.

We conducted experiments on a wide variety of publicly available benchmarks. Our benchmark suite consists of 1896 formulas arising from probabilistic inference in grid networks, synthetic grid structured random interaction Ising models, plan recognition, DQMR networks, bit-blasted versions of SMTLIB benchmarks, ISCAS89 combinational circuits, and program synthesis examples. Every experiment consisted of running a counting algorithm on a particular instance with a timeout of 4500 s. The experiments were conducted on a high-performance cluster, where each node consists of E5-2690 v3 CPU with 24 cores and 96GB of RAM. We set $\varepsilon = 0.8$ and $\delta = 0.2$ for all the tools.

The objective of our empirical study was to seek answers to the following questions:

1. How does SparseCount compare against SparseCount2 in terms of runtime performance?
2. How does SparseCount2 perform against ApproxMC3 in terms of runtime?

Overall, we observe that SparseCount2 significantly outperforms SparseCount. On the other hand, ApproxMC3 outperforms SparseCount2 with a mean speedup of 568.53×.

Our conclusions are surprising and stand in stark contrast to the widely held belief that the current construction of sparse XORs by Zhao et al. [31] and Ermon et al. [11] lead to runtime improvement [1,18,19].

Figure 1 shows the cactus plot for SparseCount, SparseCount2, and ApproxMC3. We present the number of benchmarks on $x$−axis and the time taken on $y$−axis. A point $(x, y)$ implies that $x$ benchmarks took less than or equal to $y$ seconds for the corresponding tool to execute. We present a runtime comparison of SparseCount2 vis-a-vis SparseCount and ApproxMC3 in Table 1. Column 1 of this table gives the benchmark name while column 2 and 3 list the number of variables and clauses, respectively. Column 4, 5, and 6 list the runtime (in seconds) of SparseCount, SparseCount2 and ApproxMC3, respectively. Note that "TO" stands for timeout. For lack of space, we present results only on a subset of benchmarks. The detailed logs along with list of benchmarks and the binaries employed to run the experiments are available at http://doi.org/10.5281/zenodo.3792748

We present relative comparisons separately for ease of exposition and clarity.

### 5.1    SparseCount vis-a-vis SparseCount2

As shown in Fig. 1, with a timeout of 4500 s, SparseCount could only finish execution on 90 benchmarks while SparseCount2 completed on 379 benchmarks. Note that SparseCount2 retains the same theoretical guarantees of SparseCount.

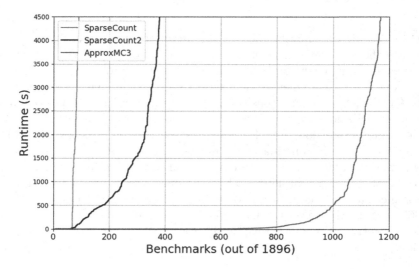

**Fig. 1.** Cactus plot of runtime performance (best viewed in color)

For a clear picture of performance gain achieved by SparseCount2 over SparseCount, we turn to Table 1. Table 1 clearly demonstrates that SparseCount2 outperforms SparseCount significantly. In particular, for all the benchmarks where both SparseCount and SparseCount2 did not timeout, the mean speedup is 10.94×.

**Explanation.** The stark difference in the performance of SparseCount and SparseCount2 is primarily due to a significant reduction in the number of SAT calls in SparseCount2. Recall, SparseCount invokes the underlying SAT solver $\mathcal{O}(n \log n)$ times while SparseCount invokes the underlying SAT solver only $\mathcal{O}(\log^2 n)$ times. As discussed above, such a difference was achieved due to the usage of *prefix-slices*.

### 5.2    ApproxMC3 vis-a-vis SparseCount2

With a timeout of 4500 s, SparseCount2 could only finish execution on 379 benchmarks while ApproxMC3 finishes execution on 1169 benchmarks. Furthermore, Table 1 clearly demonstrates that ApproxMC3 significantly outperforms SparseCount2. In particular, for all the formulas where both SparseCount2 and ApproxMC3 did not timeout, the mean speedup is 568.53×. In light of recent improvements in CryptoMiniSat, one may wonder if the observations reported in this paper are mere artifacts of how the SAT solvers have changed in the past few years and perhaps such a study on an earlier version of CryptoMiniSat may have led to a different conclusion. To account for this threat of validity, we conducted preliminary experiments using the old versions of CryptoMiniSat and again observed that similar observations hold. In particular, the latest improvements in CryptoMiniSat such as BIRD framework [24, 25] favor SparseCount and SparseCount2 relatively in comparison to ApproxMC3.

**Explanation.** The primary contributing factor for the difference in the runtime performance of SparseCount2 and ApproxMC3 is the fact that weaker guarantees for the variance of $Cnt_{(i)}$ necessitates the usage of Stockmeyer's trick of usage of the *amplification technique* wherein the underlying routines are invoked over $\psi$ instead of $\varphi$. Furthermore, the weak theoretical guarantees also lead to a larger value of $T$ as compared to its analogous parameter in ApproxMC3. It is worth noticing that prior work on the design of sparse hash function has claimed that the usage of sparse hash functions leads to runtime performance improvement of the underlying techniques. Such inference may perhaps be drawn based only on observing the time taken by a SAT solver on CNF formula with a fixed number of XORs and only varying the density of XORs. While such an observation does indeed highlight that sparse XORs are easy for SAT solvers, but it fails, as has been the case in prior work, to take into account the tradeoffs due to the weakness of theoretical guarantees of sparse hash functions. To emphasize this further, the best known theoretical guarantees offered by sparse XORs are so weak that one can not merely replace the dense XORs with sparse XORs. The state of the art counters such as ApproxMC3 require stronger guarantees than those known today.

## 6 Conclusion

Hashing-based techniques have emerged as a promising paradigm to attain scalability and rigorous guarantees in the context of approximate model counting. Since the performance of SAT solvers was observed to degrade with an increase in the size of XORs, efforts have focused on the design of sparse XORs. In this paper, we performed the first rigorous analysis to understand the theoretical and empirical effect of sparse XORs. Our conclusions are surprising and stand in stark contrast to the widely held belief that the current construction of sparse XORs by Zhao et al. [31] and Ermon et al. [11] lead to runtime improvement. We demonstrate that the theoretical guarantees offered by the construction as mentioned earlier are still too weak to be employed in the state of the art approximate counters such as ApproxMC3. Furthermore, the most efficient algorithm using sparse XORs, to the best of our knowledge, still falls significantly short of ApproxMC3 in runtime performance. While our analysis leads to negative results for the current state of the art sparse construction of hash functions, we hope our work would motivate other researchers in the community to investigate the construction of efficient hash functions rigorously. In this spirit, concurrent work of Meel Ⓡ Akshay [20] proposes a new family of hash functions called concentrated hash functions, and design a new family of sparse hash functions of the form $Ay + b$ wherein every entry of $A[i]$ is chosen with probability $p_i \in \mathcal{O}(\frac{\log n}{n})$. Meel Ⓡ Akshay propose an adaption of ApproxMC3 that can make use of the newly designed sparse hash functions, and in turn, obtain promising speedups on a subset of benchmarks.

**Acknowledgments.** The authors would like to sincerely thank the anonymous reviewers of AAAI-20 and SAT-20 (and in particular, Meta-Reviewer of AAAI-20) for providing detailed, constructive criticism that has significantly improved the quality of the paper. We are grateful to Mate Soos for many insightful discussions.

This work was supported in part by National Research Foundation Singapore under its NRF Fellowship Programme [NRF-NRFFAI1-2019-0004] and AI Singapore Programme [AISG-RP-2018-005], and NUS ODPRT Grant [R-252-000-685-13]. The computational work for this article was performed on resources of the National Supercomputing Centre, Singapore https://www.nscc.sg.

## References

1. Achlioptas, D., Hammoudeh, Z., Theodoropoulos, P.: Fast and flexible probabilistic model counting. In: Beyersdorff, O., Wintersteiger, C.M. (eds.) SAT 2018. LNCS, vol. 10929, pp. 148–164. Springer, Cham (2018). https://doi.org/10.1007/978-3-319-94144-8_10
2. Achlioptas, D., Theodoropoulos, P.: Probabilistic model counting with short XORs. In: Gaspers, S., Walsh, T. (eds.) SAT 2017. LNCS, vol. 10491, pp. 3–19. Springer, Cham (2017). https://doi.org/10.1007/978-3-319-66263-3_1

3. Asteris, M., Dimakis, A.G.: LDPC codes for discrete integration. Technical report, UT Austin (2016)
4. Baluta, T., Shen, S., Shinde, S., Meel, K.S., Saxena, P.: Quantitative verification of neural networks and its security applications. In: Proceedings of the 2019 ACM SIGSAC Conference on Computer and Communications Security, pp. 1249–1264 (2019)
5. Biondi, F., Enescu, M.A., Heuser, A., Legay, A., Meel, K.S., Quilbeuf, J.: Scalable approximation of quantitative information flow in programs. In: VMCAI 2018. LNCS, vol. 10747, pp. 71–93. Springer, Cham (2018). https://doi.org/10.1007/978-3-319-73721-8_4
6. Carter, J.L., Wegman, M.N.: Universal classes of hash functions. In: Proceedings of the Ninth Annual ACM Symposium on Theory of Computing, pp. 106–112. ACM (1977)
7. Chakraborty, S., Meel, K.S., Mistry, R., Vardi, M.Y.: Approximate probabilistic inference via word-level counting. In: Proceedings of AAAI (2016)
8. Chakraborty, S., Meel, K.S., Vardi, M.Y.: A scalable approximate model counter. In: Proceedings of CP, pp. 200–216 (2013)
9. Chakraborty, S., Meel, K.S., Vardi, M.Y.: Algorithmic improvements in approximate counting for probabilistic inference: from linear to logarithmic SAT calls. In: Proceedings of IJCAI (2016)
10. Duenas-Osorio, L., Meel, K.S., Paredes, R., Vardi, M.Y.: Counting-based reliability estimation for power-transmission grids. In: Proceedings of AAAI, February 2017
11. Ermon, S., Gomes, C.P., Sabharwal, A., Selman, B.: Low-density parity constraints for hashing-based discrete integration. In: Proceedings of ICML, pp. 271–279 (2014)
12. Ermon, S., Gomes, C.P., Sabharwal, A., Selman, B.: Optimization with parity constraints: from binary codes to discrete integration. In: Proceedings of UAI (2013)
13. Ermon, S., Gomes, C.P., Sabharwal, A., Selman, B.: Taming the curse of dimensionality: discrete integration by hashing and optimization. In: Proceedings of ICML, pp. 334–342 (2013)
14. Gomes, C.P., Hoffmann, J., Sabharwal, A., Selman, B.: Short XORs for model counting: from theory to practice. In: Marques-Silva, J., Sakallah, K.A. (eds.) SAT 2007. LNCS, vol. 4501, pp. 100–106. Springer, Heidelberg (2007). https://doi.org/10.1007/978-3-540-72788-0_13
15. Gomes, C.P., Sabharwal, A., Selman, B.: Model counting: a new strategy for obtaining good bounds. In: Proceedings of AAAI, vol. 21, pp. 54–61 (2006)
16. Gomes, C.P., Sabharwal, A., Selman, B.: Model counting. In: Biere, A., Heule, M., Maaren, H.V., Walsh, T. (eds.) Handbook of Satisfiability, Frontiers in Artificial Intelligence and Applications, vol. 185, pp. 633–654. IOS Press (2009)
17. Ivrii, A., Malik, S., Meel, K.S., Vardi, M.Y.: On computing minimal independent support and its applications to sampling and counting. Constraints, 1–18 (2015). https://doi.org/10.1007/s10601-015-9204-z
18. Kuck, J., Dao, T., Zhao, S., Bartan, B., Sabharwal, A., Ermon, S.: Adaptive hashing for model counting. In: Conference on Uncertainty in Artificial Intelligence (2019)
19. Kuck, J., Sabharwal, A., Ermon, S.: Approximate inference via weighted Rademacher complexity. In: Thirty-Second AAAI Conference on Artificial Intelligence (2018)
20. Meel, K.S., Akshay, S.: Sparse hashing for scalable approximate model counting: theory and practice. In: Proceedings of LICS (2020)
21. Meel, K.S., et al.: Constrained sampling and counting: universal hashing meets sat solving. In: Proceedings of Beyond NP Workshop (2016)

22. Roth, D.: On the hardness of approximate reasoning. Artif. Intell. **82**(1), 273–302 (1996). https://doi.org/10.1016/0004-3702(94)00092-1
23. Sang, T., Beame, P., Kautz, H.: Performing Bayesian inference by weighted model counting. In: Proceedings of AAAI, pp. 475–481 (2005)
24. Soos, M., Gocht, S., Meel, K.S.: Accelerating approximate techniques for counting and sampling models through refined CNF-XOR solving. In: Proceedings of International Conference on Computer-Aided Verification (CAV), July 2020
25. Soos, M., Meel, K.S.: Bird: engineering an efficient CNF-XOR Sat solver and its applications to approximate model counting. In: Proceedings of AAAI Conference on Artificial Intelligence (AAAI) (2019)
26. Soos, M., Nohl, K., Castelluccia, C.: Extending SAT solvers to cryptographic problems. In: Kullmann, O. (ed.) SAT 2009. LNCS, vol. 5584, pp. 244–257. Springer, Heidelberg (2009). https://doi.org/10.1007/978-3-642-02777-2_24
27. Stockmeyer, L.: The complexity of approximate counting. In: Proceedings of STOC, pp. 118–126 (1983)
28. Toda, S.: On the computational power of PP and (+)P. In: Proceedings of FOCS, pp. 514–519. IEEE (1989)
29. Trevisan, L.: Lecture notes on computational complexity. Notes written in Fall (2002). http://citeseerx.ist.psu.edu/viewdoc/download?doi=10.1.1.71.9877&rep=rep1&type=pdf
30. Valiant, L.: The complexity of enumeration and reliability problems. SIAM J. Comput. **8**(3), 410–421 (1979)
31. Zhao, S., Chaturapruek, S., Sabharwal, A., Ermon, S.: Closing the gap between short and long XORs for model counting. In: Proceedings of AAAI (2016)

# A Faster Algorithm for Propositional Model Counting Parameterized by Incidence Treewidth

Friedrich Slivovsky and Stefan Szeider[✉]

TU Wien, Vienna, Austria
{fs,sz}@ac.tuwien.ac.at

**Abstract.** The propositional model counting problem (#SAT) is known to be fixed-parameter-tractable (FPT) when parameterized by the width $k$ of a given tree decomposition of the incidence graph. The running time of the fastest known FPT algorithm contains the exponential factor of $4^k$. We improve this factor to $2^k$ by utilizing fast algorithms for computing the zeta transform and covering product of functions representing partial model counts, thereby achieving the same running time as FPT algorithms that are parameterized by the less general treewidth of the primal graph. Our new algorithm is asymptotically optimal unless the Strong Exponential Time Hypothesis (SETH) fails.

## 1 Introduction

Propositional model counting (#SAT) is the problem of determining the number of satisfying truth assignments of a given propositional formula. The problem arises in several areas of AI, in particular in the context of probabilistic reasoning [2,15]. #SAT is #P-complete [18], even for 2-CNF Horn formulas, and it is NP-hard to approximate the number of models of a formula with $n$ variables within $2^{n^{1-\varepsilon}}$, for any $\varepsilon > 0$ [15].

Since syntactic restrictions do not make the problem tractable, research generally focused on structural restrictions in terms of certain graphs associated with the input formula, which is often assumed to be in CNF. Popular graphical models are the primal graph (vertices are the variables, two variables are adjacent if they appear together in a clause), the dual graph (vertices are the clauses, two clauses are adjacent if they share a variable), and the incidence graph (vertices are variables and clauses, a variable and a clause are adjacent if the variable occurs in the clause). The structural complexity of a graph can be restricted in terms of the fundamental graph invariant *treewidth* [14] By taking the treewidth of the primal, dual, or incidence graph one obtains the *primal treewidth*, the *dual treewidth*, and the *incidence treewidth* of the formula, respectively. If we consider CNF formulas for which any of the three parameters is

Supported by the Austrian Science Fund (FWF) under grant P32441 and the Vienna Science and Technology Fund (WWTF) under grants ICT19-060 and ICT19-065.

© Springer Nature Switzerland AG 2020
L. Pulina and M. Seidl (Eds.): SAT 2020, LNCS 12178, pp. 267–276, 2020.
https://doi.org/10.1007/978-3-030-51825-7_19

bounded by a constant, the number of models can be computed in polynomial time. Indeed, the order of the polynomial is independent of the treewidth bound, and so #SAT is fixed-parameter tractable (FPT) when parameterized by primal, dual or incidence treewidth.

Incidence treewidth is considered the most general parameter among the three, as any formula of primal or dual treewidth $k$ has incidence treewidth at most $k + 1$. However, one can easily construct formulas of constant incidence treewidth and arbitrarily large primal and dual treewidth. Known model counting algorithms based on incidence treewidth have to pay for this generality with a significant larger running time: Whereas the number of models for formulas of primal or dual treewidth $k$ can be counted in time[1] $O^*(2^k)$, the best know algorithm for formulas of incidence treewidth $k$ takes time $O^*(4^k)$.[2] This discrepancy cannot be accounted for by a loose worst-case analysis, but is caused by the actual size of the dynamic programming tables constructed by the algorithms.

In this paper, we show that algebraic techniques can be used to bring down the running time to $O^*(2^k)$. Specifically, we prove that the most time-consuming steps can be expressed as *zeta transforms* and *covering products* of functions obtained from partial model counts. Since there are fast algorithms for computing these operations [3], we obtain the desired speedup.

## 2    Preliminaries

*Treewidth.* Let $G = (V(G), E(G))$ be a graph, $T = (V(T), E(T))$ be a tree, and $\chi$ be a labeling of the vertices of $T$ by sets of vertices of $G$. We refer to the vertices of $T$ as "nodes" to avoid confusion with the vertices of $G$. The tuple $(T, \chi)$ is a *tree decomposition* of $G$ if the following three conditions hold:

1. For every $v \in V(G)$ there exists a node $t \in V(T)$ such that $v \in \chi(t)$.
2. For every $vw \in E(G)$ there exists a node $t \in V(T)$ such that $v, w \in \chi(t)$.
3. For any three nodes $t_1, t_2, t_3 \in V(T)$, if $t_2$ lies on the path from $t_1$ to $t_3$, then $\chi(t_1) \cap \chi(t_3) \subseteq \chi(t_2)$.

The *width* of a tree decomposition $(T, \chi)$ is defined by $\max_{t \in V(T)} |\chi(t)| - 1$. The *treewidth* $tw(G)$ of a graph $G$ is the minimum width over all its tree decompositions. For constant $k$, there exists a linear-time algorithm that checks whether a given graph has treewidth at most $k$ and, if so, outputs a tree decomposition of minimum width [5]. However, the huge constant factor in the runtime of this algorithm makes it practically infeasible. For our purposes, it suffices to obtain tree decompositions of small but not necessarily minimal width. There

---

[1] The $O^*$ notation omits factors that are polynomial in the input size [19].

[2] Alternatively, one can convert a formula with incidence treewidth $k$ into a 3-CNF formula that has the same number of models and dual treewidth at most $3(k+1)$, or an equisatisfiable 3-CNF formula of primal treewidth at most $3(k+1)$ [17]. Applying one of these transformations followed by an algorithm for the corresponding width parameter results in an overall running time of $O^*(8^k)$.

exist several powerful tree decomposition heuristics that construct tree decompositions of small width for many cases that are relevant in practice [4,12], and the single-exponential FPT algorithm by Bodlander et al. [6] produces a factor-5 approximation of treewidth.

In this paper we also consider a particular type of tree decompositions. The triple $(T, \chi, r)$ is a *nice tree decomposition* of $G$ if $(T, \chi)$ is a tree decomposition, the tree $T$ is rooted at node $r$, and the following three conditions hold [11]:

1. Every node of $T$ has at most two children.
2. If a node $t$ of $T$ has two children $t_1$ and $t_2$, then $\chi(t) = \chi(t_1) = \chi(t_2)$; in that case we call $t$ a *join node*.
3. If a node $t$ of $T$ has exactly one child $t'$, then one of the following holds:
   (a) $|\chi(t)| = |\chi(t')| + 1$ and $\chi(t') \subset \chi(t)$; in that case we call $t$ an *introduce node*.
   (b) $|\chi(t)| = |\chi(t')| - 1$ and $\chi(t) \subset \chi(t')$; in that case we call $t$ a *forget node*.

It is known that one can transform efficiently any tree decomposition of width $k$ of a graph with $n$ vertices into a nice tree decomposition of width at most $k$ and at most $4n$ nodes [11, Lemma 13.1.3].

*Propositional Formulas.* We consider propositional formulas $F$ in conjunctive normal form (CNF) represented as set of clauses. Each clause in $F$ is a finite set of *literals*, and a literal is a negated or unnegated propositional *variable*. For a clause $C$ we denote by $var(C)$ the set of variables that occur (negated or unnegated) in $C$; for a formula $F$ we put $var(F) = \bigcup_{C \in F} var(C)$. The *size* of a clause is its cardinality. A *truth assignment* is a mapping $\tau : X \to \{0, 1\}$ defined on some set $X$ of variables. We extend $\tau$ to literals by setting $\tau(\neg x) = 1 - \tau(x)$ for $x \in X$. A truth assignment $\tau : X \to \{0, 1\}$ *satisfies* a clause $C$ if for some variable $x \in var(C) \cap X$ we have $x \in C$ and $\tau(x) = 1$, or $\neg x \in C$ and $\tau(x) = 0$. An assignment satisfies a set $F$ of clauses if it satisfies every clause in $F$. For a formula $F$, we call a truth assignment $\tau : var(F) \to \{0, 1\}$ a *model* of $F$ if $\tau$ satisfies $F$. We denote the number of models of $F$ by $\#(F)$. The *propositional model counting problem* #SAT is the problem of computing $\#(F)$ for a given propositional formula $F$ in CNF.

*Incidence Treewidth.* The *incidence graph* $G^*(F)$ of a CNF formula $F$ is the bipartite graph with vertex set $F \cup var(F)$; a variable $x$ and a clause $C$ are joined by an edge if and only if $x \in var(C)$. The *incidence treewidth* $tw^*(F)$ of a CNF formula $F$ is the treewidth of its incidence graph, that is $tw^*(F) = tw(G^*(F))$.

**Definition 1 (Zeta and Möbius Transforms).** *Let $V$ be a finite set and let $f : 2^V \to \mathbb{Z}$ be a function. The* zeta transform $\zeta f$ *of $f$ is defined as*

$$(\zeta f)(X) = \sum_{Y \subseteq X} f(Y), \tag{1}$$

*and the* Möbius transform $\mu f$ *of $f$ is given by*

$$(\mu f)(X) = \sum_{Y \subseteq X} (-1)^{|X \setminus Y|} f(Y). \tag{2}$$

**Theorem 1 (Kennes [10]).** *Let $V$ be an $k$-element set and let $f : 2^V \to \mathbb{Z}$ be a function. All values of $\zeta f$ and $\mu f$ can be computed using $O(2^k k)$ arithmetic operations.*

**Definition 2.** *The* covering product *of two functions $f, g : 2^V \to \mathbb{Z}$ is a function $(f *_c g) : 2^V \to \mathbb{Z}$ such that for every $Y \subseteq V$,*

$$(f *_c g)(Y) = \sum_{A \cup B = Y} f(A)g(B). \tag{3}$$

The covering product can be computed using zeta and Möbius transforms by applying the following two results (see Aigner [1]).

**Lemma 1.** *Given functions $f, g : 2^V \to \mathbb{Z}$, the zeta transform of the covering product of $f$ and $g$ is the pointwise product of the zeta-transformed arguments. That is, for each $X \subseteq V$*

$$\zeta(f *_c g)(X) = (\zeta f(X))(\zeta g(X)).$$

**Theorem 2 (Inversion formula).** *Let $f : 2^V \to \mathbb{Z}$. Then for every $X \subseteq V$*

$$f(X) = (\mu \zeta f)(X) = (\zeta \mu f)(X).$$

## 3    Faster Model Counting for Incidence Treewidth

Samer and Szeider presented an algorithm for #SAT with a running time of $O^*(4^k)$ [16], where $k$ is the width of a given tree decomposition of the incidence graph. In this section, we are going to show how to improve this to $O^*(2^k)$.

Their algorithm proceeds by bottom-up dynamic programming on a nice tree decomposition, maintaining tables that contain partial solution counts for each node. For the remainder of this section, let $F$ be an arbitrary but fixed CNF formula, and let $(T, \chi, r)$ be a nice tree-decomposition of the incidence graph $G^*(F)$ that has width $k$. For each node $t$ of $T$, let $T_t$ denote the subtree of $T$ rooted at $t$, and let $V_t = \bigcup_{t' \in V(T_t)} \chi(t')$ denote the set of vertices appearing in bags of $T_t$. Further, let $F_t$ denote the set of clauses in $V_t$, and let $X_t$ denote the set of all variables in $V_t$. We also use the shorthands $\chi_c(t) = \chi(t) \cap F$ and $\chi_v(t) = \chi(t) \cap var(F)$ for the set of clauses and the set of variables in $\chi(t)$, respectively. Let $t$ be a node of $T$. For each assignment $\alpha : \chi_v(t) \to \{0, 1\}$ and subset $A \subseteq \chi_c(t)$, we define $N(t, \alpha, A)$ as the set of assignments $\tau : X_t \to \{0, 1\}$ for which the following two conditions hold:

1. $\tau(x) = \alpha(x)$ for all variables $x \in \chi_v(t)$.
2. $A$ is exactly the set of clauses in $F_t$ that are not satisfied by $\tau$.

We represent the values of $n(t, \alpha, A) = |N(t, \alpha, A)|$ for all $\alpha : \chi_v(t) \to \{0, 1\}$ and $A \subseteq \chi_c(t)$ by a table $M_t$ with $|\chi(t)| + 1$ columns and $2^{|\chi(t)|}$ rows. The first $|\chi(t)|$ columns of $M_t$ contain Boolean values encoding $\alpha(x)$ for variables

$x \in \chi_v(t)$, and membership of $C$ in $A$ for clauses $C \in \chi_c(t)$. The last entry of each row contains the integer $n(t, \alpha, A)$.

Samer and Szeider showed that the entries of the table $M_t$ can be efficiently computed for each node $t$. More specifically, they showed how $M_t$ can be obtained for leaf nodes $t$, and how $M_t$ can be computed from the tables for the child nodes of introduce, forget, and join nodes $t$. Since the running time for join and variable introduce nodes are the bottleneck of the algorithm, we summarize the results concerning correctness and running time for the remaining node types as follows.

**Lemma 2 (Samer and Szeider** [16]**).** *If $t \in T$ is a leaf node, or a forget or clause introduce node with child $t'$ such that $M_{t'}$ has already been computed, the table $M_t$ can be obtained in time $2^k |\varphi|^{O(1)}$.*

The table entries for a join node can be computed as a sum of products from tables of its child nodes.

**Lemma 3 (Samer and Szeider** [16]**).** *Let $t \in T$ be a join node with children $t_1, t_2$. For each assignment $\alpha : \chi_v(t) \to \{0, 1\}$ and set $A \subseteq \chi_c(t)$ we have*

$$n(t, \alpha, A) = \sum_{\substack{A_1, A_2 \subseteq \chi_c(t), \\ A_1 \cap A_2 = A}} n(t_1, \alpha, A_1)\, n(t_2, \alpha, A_2). \tag{4}$$

A straight-forward algorithm for computing this sum requires an arithmetic operation for each pair $(A_1, A_2)$ where $A_1 \subseteq \chi_c(t_1)$, $A_2 \subseteq \chi_c(t_2)$, and thus $2^k 2^k = 4^k$ operations in the worst case. By using a fast algorithm for the covering product, we can significantly reduce the number of arithmetic operations and thus the running time. The key observation is that the sum of products in (4) can be readily expressed as a covering product (3).

**Lemma 4.** *Let $t$ be a join node of $T$ with children $t_1, t_2$ and let $\alpha : \chi_v(t) \to \{0, 1\}$ be a truth assignment. For $i \in \{1, 2\}$ let $f_i : 2^{\chi_c(t_i)} \to \mathbb{Z}$ be the function given by $f_i(A) := n(t_i, \alpha, \chi_c(t) \setminus A)$. Then $n(t, \alpha, \chi_c(t) \setminus A) = (f_1 *_c f_2)(A)$ for each subset $A \subseteq \chi_c(t)$.*

*Proof.* For $S \subseteq \chi_c(t)$, let $S^c = \chi_c(t) \setminus S$. We have

$$(f_1 *_c f_2)(A) = \sum_{\substack{A_1, A_2 \subseteq \chi_c(t) \\ A_1 \cup A_2 = A}} f_1(A_1) f_2(A_2)$$

$$= \sum_{\substack{A_1, A_2 \subseteq \chi_c(t) \\ A_1^c \cap A_2^c = A^c}} n(t_1, \alpha, A_1^c)\, n(t_2, \alpha, A_2^c) = n(t, \alpha, A^c).$$

$\square$

For variable introduce nodes the table entry for each assignment and subset of clauses can be computed as a sum over table entries of the child table.

**Lemma 5 (Samer and Szeider [16]).** *Let $t$ be an introduce node with child $t'$ such that $\chi(t) = \chi(t') \cup \{x\}$ for a variable $x$. Then, for each truth assignment $\alpha : \chi_v(t') \to \{0, 1\}$ and set $A \subseteq \chi_c(t)$, we have*

$$n(t, \alpha \cup \{(x, 0)\}, A) = \begin{cases} 0 & \text{if } \neg x \in C \text{ for some } C \in A \\ & \text{otherwise, where} \\ \sum_{B' \subseteq B} n(t', \alpha, A \cup B') & B = \{C \in \chi_c(t) \mid \neg x \in C\}; \end{cases} \quad (5)$$

$$n(t, \alpha \cup \{(x, 1)\}, A) = \begin{cases} 0 & \text{if } x \in C \text{ for some } C \in A; \\ & \text{otherwise, where} \\ \sum_{B' \subseteq B} n(t', \alpha, A \cup B') & B = \{C \in \chi_c(t) \mid x \in C\}. \end{cases} \quad (6)$$

A simple approach is to go through all $2^k$ assignments $\alpha$ and subsets $A$ and, if necessary, compute the sums in (5) and (6). Since there could be up to $2^k$ subsets to sum over, this again requires $4^k$ arithmetic operations in the worst case. The following lemma observes that we can instead use the zeta transform (1).

**Lemma 6.** *Let $t$ be an introduce node with child $t'$ such that $\chi(t) = \chi(t') \cup \{x\}$ for a variable $x$. Define $f(S) = n(t', \alpha, S)$ for $S \subseteq \chi_c(t')$. Then, for each truth assignment $\alpha : \chi_v(t') \to \{0, 1\}$ and set $A \subseteq \chi_c(t)$, we have*

$$n(t, \alpha \cup \{(x, 0)\}, A) = \begin{cases} 0 & \text{if } \neg x \in C \text{ for some } C \in A; \\ & \text{otherwise, where} \\ (\zeta f)(A \cup B) - (\zeta f)(A) & B = \{C \in \chi_c(t) \mid \neg x \in C\}; \end{cases} \quad (7)$$

$$n(t, \alpha \cup \{(x, 1)\}, A) = \begin{cases} 0 & \text{if } x \in C \text{ for some } C \in A; \\ & \text{otherwise, where} \\ (\zeta f)(A \cup B) - (\zeta f)(A) & B = \{C \in \chi_c(t) \mid x \in C\}. \end{cases} \quad (8)$$

*Proof.* We can rewrite the sums in (5) and (6) as

$$\sum_{A \subseteq S \subseteq A \cup B} f(S) = \sum_{S \subseteq A \cup B} f(S) - \sum_{S \subseteq A} f(S) = (\zeta f)(A \cup B) - (\zeta f)(A).$$

$\square$

By Theorem 1, the zeta and Möbius transform of a function $f : 2^V \to \mathbb{Z}$ can be computed using $O(2^k k)$ arithmetic operations, where $k = |V|$ is the size of the underlying set. In conjunction with Lemma 1 and Lemma 2, this lets us compute the covering product of two functions $f, g : 2^V \to \mathbb{Z}$ with $O(2^k k)$ operations.

How this translates into running time depends on the choice of computational model. Since the model count of a formula can be exponential in the number of variables, it is unrealistic to assume that arithmetic operations can be performed in constant time. Instead, we adopt a random access machine model where two $n$-bit integers can be added, subtracted, and compared in time $O(n)$, and multiplied in time $O(n \log n)$ [8]. For the purposes of proving a bound of $O^*(2^k)$, it is sufficient to show that the bit size of integers obtained as intermediate results while computing the zeta and Möbius transforms is polynomially bounded by the number of variables in the input formula. To verify that this is the case, we present the dynamic programming algorithms used to efficiently compute these transforms [3], following the presentation by Fomin and Kratsch [7].

**Theorem 3.** *Let $V$ be a $k$-element set and let $f : 2^V \to \mathbb{Z}$ be a function that can be evaluated in time $O(1)$ and whose range is contained in the interval $(-2^N, 2^N)$. All values of $\zeta f$ and $\mu f$ can be computed in time $2^k(k + N)^{O(1)}$.*

*Proof.* Let $V = \{1, \dots, k\}$. We compute intermediate values

$$\zeta_j(X) = \sum_{Y \subseteq X \cap \{1, \dots, j\}} f(Y \cup (X \cap \{j+1, \dots, k\})),$$

for $j = 0, \dots, k$. Note that $\zeta_k(X) = (\zeta f)(X)$. The values $\zeta_j$ can be computed as

$$\zeta_j(X) = \begin{cases} \zeta_{j-1}(X) & \text{when } j \notin X, \\ \zeta_{j-1}(X) + \zeta_{j-1}(X \setminus \{j\}) & \text{when } j \in X. \end{cases}$$

For the Möbius transform, we compute intermediate values

$$\mu_j(X) = \sum_{Y \subseteq X \cap \{1, \dots, j\}} (-1)^{|(X \cap \{1, \dots, j\}) \setminus Y|} f(Y \cup (X \cap \{j+1, \dots, k\})).$$

Again we have $\mu_k(X) = (\mu f)(X)$, and the values $\mu_j$ can be computed as

$$\mu_j(X) = \begin{cases} \mu_{j-1}(X) & \text{when } j \notin X, \\ \mu_{j-1}(X) - \mu_{j-1}(X \setminus \{j\}) & \text{when } j \in X. \end{cases}$$

In both cases, this requires $k$ arithmetic operations for each set $X \subseteq V$, and the intermediate values are contained in the interval $(-2^k 2^N, 2^k 2^N)$. $\qquad\square$

**Corollary 1.** *Let $V$ be an $k$-element set and let $f, g : 2^V \to \mathbb{Z}$ be functions that can be evaluated in time $O(1)$ and whose range is contained in the interval $(-2^N, 2^N)$. All values of $f *_c g$ can be computed in time $2^k(k + N)^{O(1)}$.*

Having obtained these bounds on the time required to compute the zeta transform and the covering product, we can now state improved time bounds for obtaining the entries of the tables $M_t$ of join and variable introduce nodes $t$.

**Lemma 7.** *The table $M_t$ for a join node $t \in T$ can be computed in time $O^*(2^k)$ given the tables $M_{t_1}$ and $M_{t_2}$ of its child nodes $t_1$ and $t_2$.*

*Proof.* Let $p = |\chi_v(t)|$ and $q = |\chi_c(t)|$. For each assignment $\alpha : \chi_v(t) \to \{0, 1\}$, we perform the following steps. For $i \in \{1, 2\}$, we first modify the table $M_{t_i}$ by flipping the values encoding membership of a clause $C$ in the set $A$ so as to obtain a table $M'_i$ containing the values $f_i(A) := n(t_i, \alpha, \chi_c(t) \setminus A)$ for each $A \subseteq \chi_c(t)$. Clearly, this can be done in time $O^*(2^p)$. By Lemma 4, the values $(f_1 *_c f_2)(A)$

correspond to $n(t, \alpha, \chi_c(t) \setminus A)$. Each entry of the tables $M_{t_i}$ represents a partial model count that cannot exceed $2^n$, so we can compute all values of the covering product $f_1 *_c f_2$ in time $2^q(q + n)^{O(1)}$ by Corollary 1, where $n$ is the number of variables. Since $q \leq n$ this is in $O^*(2^q)$. There are at most $2^p$ assignments $\alpha : \chi_v(t) \to \{0,1\}$, so the overall running time is in $O^*(2^k)$. $\qquad\square$

**Lemma 8.** *The table $M_t$ for a variable introduce node $t \in T$ can be computed in time $O^*(2^k)$ given the table $M_{t'}$ of its child node $t'$.*

*Proof.* As before, let $p = |\chi_v(t')|$ and $q = |\chi_c(t')|$. For each truth assignment $\alpha : \chi_v(t') \to \{0,1\}$, we proceed as follows. We compute the value of the zeta transform $(\zeta f)(A)$ for all subsets $A \subseteq \chi_c(t')$. Again, each entry of $M_{t'}$ represents a partial model count that is bounded by $2^n$, so we can do this in time $2^q(q + n)^{O(1)}$ by Theorem 1. Since $q \leq n$ this is in $O^*(2^q)$. Then, we iterate over all $A \subseteq \chi_c(t')$ and set the entries $M_t(\alpha, A)$ based on (7) and (8), using the values of $\zeta f$. This can again be done in time $O^*(2^q)$ and is correct by Lemma 6. The number of assignments $\alpha : \chi_v(t) \to \{0,1\}$ is $2^p$, so the overall running time is in $O^*(2^k)$. $\qquad\square$

**Theorem 4.** *Given a CNF formula $F$ and a nice tree decomposition of $G^*(F)$, we can compute $\#(F)$ in time $O^*(2^k)$, where $k$ is the width of the decomposition.*

*Proof.* Let $(T, \chi, r)$ be a nice tree decomposition of the incidence graph of $F$. We compute the tables $M_t$ for all nodes $t$ of $T$, starting from the leaf nodes of $T$. By Lemmas 2, 7, and 8, each table can be computed in time $O^*(2^k)$. We can compute $\#(F) = \sum_{\alpha : \chi_v(r) \to \{0,1\}} n(r, \alpha, \emptyset)$ at the root $r$. $\qquad\square$

## 4    Discussion

The space requirements of the algorithm remain unchanged by the proposed improvements. Each table $M_t$ has at most $2^{k+1}$ entries, and each entry requires up to $n$ bits. By keeping as few tables in memory as possible and discarding tables whenever they are no longer needed, no more than $\lfloor 1 + \log_2(N + 1) \rfloor$ tables need to be stored in memory at any point, where N is the number of nodes in the tree decomposition [16, Proposition 3].

Let $s_r := \inf\{ \delta \mid$ there exists an $O^*(2^{\delta n})$ algorithm that decides the satisfiability of $n$-variable $r$-CNF formulas with parameter $n \}$ and let $s_\infty := \lim_{r \to \infty} s_r$. Impagliazzo et al. [9] introduced the Strong Exponential Time Hypothesis (SETH), which states that $s_\infty = 1$. SETH has served as a very useful hypothesis for establishing tight bounds on the running time for NP-hard problems [13]. For instance, an immediate consequence of the SETH is that the satisfiability of an $n$-variable CNF formula cannot be solved in time $O^*((2 - \varepsilon))^n)$ for any $\varepsilon > 0$. However, for the incidence graph of an $n$-variable CNF formula $F = \{C_1, \ldots, C_m\}$ we can always give a tree decomposition $(T, \chi)$ of width $n$ (recall that the width of a tree decomposition is the size of its largest bag minus one) by taking as $T$ a star with center $t$ and leaves $t_1, \ldots, t_m$, and by putting

$\chi(t) = var(F)$ and $\chi(t_i) = var(F) \cup \{C_i\}$, for $1 \leq i \leq n$. Thus, if the bound in Theorem 4 could be improved from $O^*(2^k)$ to $O^*((2 - \varepsilon)^k)$, we would have an $O^*((2 - \varepsilon)^n)$ SAT-algorithm, and hence a contradiction to the SETH. We can, therefore, conclude that Theorem 4 is tight under the SETH.

**Acknowledgements.** We thank Andreas Björklund for the suggestion of using covering products to improve the running time of SAT algorithms for instances of bounded incidence treewidth.

# References

1. Aigner, M.: Combinatorial Theory. Springer, Heidelberg (2012). https://doi.org/10.1007/978-3-642-59101-3
2. Bacchus, F., Dalmao, S., Pitassi, T.: Algorithms and complexity results for #SAT and Bayesian inference. In: 44th Annual IEEE Symposium on Foundations of Computer Science (FOCS 2003), pp. 340–351 (2003)
3. Björklund, A., Husfeldt, T., Kaski, P., Koivisto, M.: Fourier meets Möbius: fast subset convolution. In: Johnson, D.S., Feige, U. (eds.) Proceedings of the 39th Annual ACM Symposium on Theory of Computing, San Diego, California, USA, 11–13 June 2007, pp. 67–74. Association for Computing Machinery, New York (2007)
4. Bodlaender, H.L.: Discovering treewidth. In: Vojtáš, P., Bieliková, M., Charron-Bost, B., Sýkora, O. (eds.) SOFSEM 2005. LNCS, vol. 3381, pp. 1–16. Springer, Heidelberg (2005). https://doi.org/10.1007/978-3-540-30577-4_1
5. Bodlaender, H.L.: A linear-time algorithm for finding tree-decompositions of small treewidth. SIAM J. Comput. **25**(6), 1305–1317 (1996)
6. Bodlaender, H.L., Drange, P.G., Dregi, M.S., Fomin, F.V., Lokshtanov, D., Pilipczuk, M.: A $c^k n$ 5-approximation algorithm for treewidth. SIAM J. Comput. **45**(2), 317–378 (2016)
7. Fomin, F.V., Kratsch, D.: Exact Exponential Algorithms. Springer, Heidelberg (2010). https://doi.org/10.1007/978-3-642-16533-7
8. Harvey, D., Van Der Hoeven, J.: Integer multiplication in time $O(n \log n)$. HAL archives ouvertes (hal-02070778) (2019)
9. Impagliazzo, R., Paturi, R., Zane, F.: Which problems have strongly exponential complexity? J. Comput. Syst. Sci. **63**(4), 512–530 (2001)
10. Kennes, R.: Computational aspects of the mobius transformation of graphs. IEEE Trans. Syst. Man Cybern. **22**(2), 201–223 (1992)
11. Kloks, T.: Treewidth: Computations and Approximations. Springer, Berlin (1994). https://doi.org/10.1007/BFb0045375
12. Koster, A.M.C.A., Bodlaender, H.L., van Hoesel, S.P.M.: Treewidth: Computational experiments. Electr. Notes Discrete Math. **8** (2001)
13. Lokshtanov, D., Marx, D., Saurabh, S.: Lower bounds based on the exponential time hypothesis. Bull. Eur. Assoc. Theor. Comput. Sci. **105**, 41–72 (2011)
14. Robertson, N., Seymour, P.D.: Graph minors. II. Algorithmic aspects of tree-width. J. Algorithms **7**(3), 309–322 (1986)
15. Roth, D.: On the hardness of approximate reasoning. Artif. Intell. **82**(1–2), 273–302 (1996)
16. Samer, M., Szeider, S.: Algorithms for propositional model counting. J. Discrete Algorithms **8**(1), 50–64 (2010)

17. Samer, M., Szeider, S.: Constraint satisfaction with bounded treewidth revisited. J. Comput. Syst. Sci. **76**(2), 103–114 (2010)
18. Valiant, L.G.: The complexity of computing the permanent. Theor. Comput. Sci. **8**(2), 189–201 (1979)
19. Woeginger, G.J.: Exact algorithms for NP-hard problems: a survey. In: Jünger, M., Reinelt, G., Rinaldi, G. (eds.) Combinatorial Optimization — Eureka, You Shrink!. LNCS, vol. 2570, pp. 185–207. Springer, Heidelberg (2003). https://doi.org/10.1007/3-540-36478-1_17

# Abstract Cores in Implicit Hitting Set MaxSat Solving

Jeremias Berg[1]([envelope])[iD], Fahiem Bacchus[2], and Alex Poole[2]

[1] HIIT, Department of Computer Science, University of Helsinki, Helsinki, Finland
jeremias.berg@helsinki.fi
[2] Department of Computer Science, University of Toronto, Toronto, Canada
fbacchus@cs.toronto.edu

**Abstract.** Maximum Satisfiability (MaxSat) solving is an active area of research motivated by numerous successful applications to solving NP-hard combinatorial optimization problems. One of the most successful approaches to solving MaxSat instances arising from real world applications is the Implicit Hitting Set (IHS) approach. IHS solvers are complete MaxSat solvers that harness the strengths of both Boolean Satisfiability (SAT) and Integer Linear Programming (IP) solvers by decoupling core-extraction and optimization. While such solvers show state-of-the-art performance on many instances, it is known that there exist MaxSat instances on which IHS solvers need to extract an exponential number of cores before terminating. Motivated by the structure of the simplest of these problematic instances, we propose a technique we call abstract cores that provides a compact representation for a potentially exponential number of regular cores. We demonstrate how to incorporate abstract core reasoning into the IHS algorithm and report on an empirical evaluation demonstrating that including abstract cores into a state-of-the-art IHS solver improves its performance enough to surpass the best performing solvers of the most recent 2019 MaxSat Evaluation.

**Keywords:** Combinatorial optimization · Maximum Satisfiability · MaxSat · Implicit Hitting Set · IHS

## 1 Introduction

Maximum Satisfiability (MaxSat), the optimisation extension of the Boolean Satisfiability (SAT) problem, has in recent years matured into a competitive and thriving constraint optimisation paradigm with several successful applications in a variety of domains [7,8,11,16,18,19,31]. As a consequence, the development of MaxSat solvers is an active area of research with the state-of-the-art solvers evaluated annually in the MaxSat Evaluations [4,5].

In this work, we focus on improving the Implicit Hitting Set (IHS) approach to complete MaxSat solving [4,14,29]. As witnessed by the results of the annual evaluations, IHS solvers are, together with core-guided [2,20,24–26] and

© Springer Nature Switzerland AG 2020
L. Pulina and M. Seidl (Eds.): SAT 2020, LNCS 12178, pp. 277–294, 2020.
https://doi.org/10.1007/978-3-030-51825-7_20

model improving [22] algorithms, one of the most successful approaches to solving MaxSat instances encountered in practical applications. IHS solvers decouple MaxSat solving into separate *core extraction* and *optimisation* steps. By using a Boolean Satisfiability (SAT) solver for core extraction and an Integer Linear Programming (IP) optimizer, the IHS approach is able to exploit the disparate strengths of these different technologies.

Through this separation IHS solvers avoid increasing the complexity of the underlying SAT instance by deferring all numerical reasoning to the optimizer [13]. One drawback of the approach, however, is that on some problems an exponential number of cores need to be extracted by the SAT solver and given to the optimizer. In this paper we identify a seemingly common pattern that appears in the simplest problems exhibiting this exponential worse case. We propose a technique, which we call *abstract cores*, for addressing problems with this pattern. Abstract cores provide a compact representation for a potentially exponential number of ordinary cores. Hence, by extracting abstract cores and giving them to the optimizer we can in principle achieve an exponential reduction in the number of constraints the SAT solver has to extract and supply to the optimizer. The net effect can be significant performance improvements.

In the rest of the paper we formalize the concept of abstract cores and explain how to incorporate them into the IHS algorithm both in theory and practice. Finally, we demonstrate empirically that adding abstract cores to a state-of-the-art IHS solver improves its performance enough to surpass the best performing solvers of the 2019 MaxSat evaluation.

## 2    Preliminaries

MaxSat problems are expressed as CNF formulas $\mathcal{F}$ with weight annotations. A CNF formula consists of a conjunction ($\wedge$) of clauses, each of which is a disjunction ($\vee$) of literals, a literal is either a variable $v$ of $\mathcal{F}$ (a positive literal) or its negation $\neg v$ (a negative literal). We will often regard $\mathcal{F}$ and clauses $C$ as being sets of clauses and literals respectively. For example $l \in C$, indicates that literal $l$ is in the clause $C$ using set notation, and $(x, \neg y, z)$ denotes the clause $(x \vee \neg y \vee z)$.

A truth assignment $\tau$ maps Boolean variables to 1 (TRUE) or 0 (FALSE). It is extended to assign 1 or 0 to literals, clauses and formulas in the following standard way: $\tau(\neg l) = 1 - \tau(l)$, $\tau(C) = \max\{\tau(l) \mid l \in C\}$, and $\tau(\mathcal{F}) = \min\{\tau(C) \mid C \in \mathcal{F}\}$, for literals $l$, clauses $C$, and CNF formulas $\mathcal{F}$, respectively. We say that $\tau$ satisfies a clause $C$ (formula $\mathcal{F}$) if $\tau(C) = 1$ ($\tau(\mathcal{F}) = 1$), and that the formula $\mathcal{F}$ is satisfiable if there exists a truth assignment $\tau$ such that $\tau(\mathcal{F}) = 1$.

A MaxSat instance $\mathcal{I} = (\mathcal{F}, wt)$ is a CNF formula $\mathcal{F}$ along with a weight function that maps every clause $C \in \mathcal{F}$ to a integer weight $wt(C) > 0$. Clauses $C$ whose weight is infinite $wt(C) = \infty$ are called *hard clauses* while those with a finite weight are called *soft clauses*. $\mathcal{I}$ is said to be unweighted if all soft clauses have weight 1. We denote the set of hard and soft clauses of $\mathcal{F}$ by $\mathcal{F}_H$ and $\mathcal{F}_S$, respectively.

An assignment $\tau$ is a *solution* to $\mathcal{I}$ if it satisfies $\mathcal{F}_H$ ($\tau(\mathcal{F}_H) = 1$). The cost of a solution $\tau$, $cost(\mathcal{I}, \tau)$, is the sum of the weights of the soft clauses it falsifies, i.e., $cost(\mathcal{I}, \tau) = \sum_{C \in \mathcal{F}_S} (1 - \tau(C)) \cdot wt(C)$. When the instance is clear from context we shorten notation to $cost(\tau)$. A solution $\tau$ is *optimal* if it has minimum cost among all solutions: i.e. if $cost(\tau) \leq cost(\tau')$ holds for all solutions $\tau'$. The task in MaxSat solving is to find an (any) optimal solution. We will assume that at least one solution exists, i.e., that $\mathcal{F}_H$ is satisfiable.

To simplify our notation it will be useful to transform all of the soft clauses in $\mathcal{F}$ so that they become unit clauses containing a single negative literal. If $C \in \mathcal{F}_S$ is not in the right form we replace it by the soft clause $(\neg b)$ and the hard clause $(C \vee b)$, where $b$ is a brand new variable and $wt((\neg b)) = wt(C)$. This transformation preserves the set of solutions and their costs. We call the variables in the resulting set of unit soft clauses *blocking variables* or *b-variables* for short. Note that assigning a b-variable $b$ the value TRUE is equivalent to falsifying its corresponding soft clause $(\neg b)$. We denote the set of b-variables of the transformed formula by $\mathcal{F}_B$, and write $wt(b)$ for a b-variable $b$ to denote the weight of its underlying soft clause $wt(\neg b)$. With this convention we can write the cost of a solution $\tau$ more simply as $cost(\tau) = \sum_{b \in \mathcal{F}_B} wt(b) \cdot \tau(b)$. For any set $B$ of b-variables we write $cost(B)$ to denote the sum of their weights.

In the MaxSat context a *core* $\kappa$ is defined to be a set of soft clauses $\kappa \subseteq \mathcal{F}_S$ that are unsatisfiable given the hard clauses, i.e.. $\kappa \cup \mathcal{F}_H$ is unsatisfiable. This means that every solution $\tau$, which by definition must satisfy $\mathcal{F}_H$, must falsify at least one soft clause in $\kappa$. Given that the soft clauses are of the form $(\neg b)$ for some b-variable $b$ we can express every core as a clause $\kappa = \bigvee_{b \in \kappa} b$ containing only positive b-variables: one of these variables must be true. This clause is entailed by $\mathcal{F}_H$. We can also express $\kappa$ as a linear inequality $\sum_{\{b \mid (\neg b) \in \kappa\}} b \geq 1$ that is also entailed by $\mathcal{F}_H$. A MaxSat *correction set* $hs$ is dually defined to be a set of soft clauses $hs \subseteq \mathcal{F}_S$ whose removal renders the remaining soft clauses satisfiable with the hard clauses, i.e., $(\mathcal{F}_S - hs) \cup \mathcal{F}_H$ is satisfiable.

# 3 Implicit Hitting Set Based MaxSat Solving

Algorithm 1 shows the implicit hitting set (IHS) approach to MaxSat solving. Our specification generalizes the original specification of [13]. In particular, we use upper and lower bounds, terminating when these bounds meet, rather than waiting until the optimizer returns a correction set as in [13]. We use this reformulation as it makes it easier to understand our extension to abstract cores.

Starting from a lower bound of zero, an upper bound of infinity, and an empty set of cores $\mathcal{C}$ (line 3), the algorithm computes a minimum cost hitting set of its current set of cores $\mathcal{C}$. This is accomplished by expressing each core in $\mathcal{C}$ as its equivalent linear inequality $\sum_{b \in \kappa} b \geq 1$ and using the optimizer to find a solution $hs$ with the smallest weight of true $b$ variables (Fig. 1a). This corresponds to computing the minimum weight of soft clauses that need to be falsified in order to satisfy the constraints imposed by cores found so far. Hence, $cost(hs)$ must be a lower bound on the cost of any optimal solution: every solution must satisfy these constraints. This allows us to update the lower bound (line 6) and exit the while loop if the lower bound now meets the upper bound. (Note that since new cores are continually added to the optimizer's model the lower bound will never decrease).

The optimizer's solution is then used to extract more cores that can be added to the optimizer's constrains for the next iteration. Core extraction is done by the ex-cores procedure shown in Algorithm 2. ex-cores extracts cores until it finds a solution $\tau$. If the solution has lower cost than any previous solution the upper bound $UB$ will be updated and this best solution stored in $\tau_{best}$. The set of cores $K$ extracted are returned and added to the optimizer's model potentially increasing the lower bound.

```
1  Basic-IHS (F, wt)
     Input: A MaxSat instance (F, wt)
     Output: An optimal solution τ
2    LB ← 0; UB ← ∞;
3    τ_best ← ∅; C ← ∅ ;
4    while (TRUE) do
5      hs ← Min-Hs(F_B, C);
6      LB = cost(hs);
7      if (LB = UB) break;
8      K ← ex-cores (hs, UB, τ_best);
9      if (LB = UB) break;
10     C ← C ∪ K
11   return τ_best
```
**Algorithm 1:** IHS MaxSat

Min-Hs $(\mathcal{F}_B, \mathcal{C})$:

minimize: $\sum_{b \in \mathcal{F}_B} wt(b) \cdot b$

subject to:

$\sum_{b \in \kappa} b \geq 1 \qquad \forall \kappa \in \mathcal{C}$

$b \in \{0, 1\} \qquad \forall b \in \mathcal{F}_B$

return:

$\{b \mid b \text{ set to 1 in opt. soln}\}$

(a) IP for optimizing with cores

The original IHS formulation [13] extracted only one core from each optimizer solution, but this was shown to be a significant detriment to performance [15] requiring too many calls to the optimizer. The procedure ex-cores gives one simple way of extracting more than one core from the optimizer's solution $hs$. It can be extended in a variety ways to allow extracting large numbers (hundreds) of cores from each optimizer solution [12, 15, 28]. In our implementation we used such techniques.

ex-cores (Algorithm 2) uses a SAT solver and its assumption mechanism to extract cores. It first initializes the assumptions to force the SAT solver to satisfy every soft clause not in $hs$. More specifically, for every soft clause $(\neg b)$ not in $hs$, $\neg b$ is assumed, forcing the solver to satisfy this soft clause. Then it invokes ex-cores-sub which iteratively calls the SAT solver to find a solution satisfying $\mathcal{F}_H$ along with the current set of assumptions. After each core is found its b-variables are removed from the assumptions (line 11) so that on each iteration we require the SAT solver to satisfy fewer soft clauses. Since $\mathcal{F}_H$ is satisfiable, eventually the SAT solver will be asked to satisfy so few soft clauses that it will find a solution $\tau$ terminating the loop.

In the original IHS specification [13] IHS terminates with an optimal solution when the optimal hitting set $hs$ is a correction set. This condition will also cause termination in our specification. In particular, before calling ex-cores the lower bound $LB$ is set to $cost(hs)$ (Algorithm 1, line 6). If $hs$ is a correction set, a

```
1  ex-cores (hs, UB, τ_best)
2  │    assumps = {¬b | b ∈ (F_B − hs)};
3  │    return ex-cores-sub(assumps, UB, τ_best)

4  ex-cores-sub (UB, τ_best)
5  │    K ← {};
6  │    while TRUE do
7  │    │    (sat?, κ, τ) ← sat-assume(F_H, assumps) ;
8  │    │    if (sat?) then
9  │    │    │    if (cost(τ) < UB) then τ_best ← τ; UB ← cost(τ);
10 │    │    │    return K;
11 │    │    else K ← K ∪ {κ}; assumps ← assumps − {¬b|b ∈ κ}
```

**Algorithm 2:** Extracting multiple cores from a single optimizer solution

solution $\tau$ will be found by the SAT solver in the first iteration of Algorithm 2, (line 7). That $\tau$ will have $cost(\tau) = cost(hs)$ as it cannot falsify any soft clause not in $hs$ and cannot have cost less than the lower bound. Hence, on ex-cores's return Algorithm 1 will terminate with $UB = LB$. As shown in [13] the optimizer's must eventually return a correction set. This means that the original proof that IHS terminates, returning an optimal solution given in [13] continues to apply our reformulated Algorithm 1.

Algorithm 1 can also terminate before the optimizer returns a correction set. In particular, $\tau_{best}$ can be set to an optimal solution (Algorithm 2, line 9) well before we can verify its optimality. In this case termination can occur as soon as the optimizer has been given a sufficient number of cores to drive its lower bound up to $cost(\tau_{best})$, even if the optimizer's solution is not a correction set. In fact, termination in the IHS approach always requires that the optimizer be given enough constraints to drive the cost of its optimal solution up to the cost of the MaxSat optimal solution.

*Example 1.* With $F_H = \{(b_1, b_2), (b_2, b_3), (b_3, b_4)\}$ and $F_S = \{(\neg b_1), (\neg b_2), (\neg b_3), (\neg b_4)\}$ all having weight 1, Algorithm 1 will first obtain $hs = \emptyset$ from Min-Hs as there initially are no cores to hit. ex-cores will then SAT solve $F_H$ under the assumptions $\neg b_1$, $\neg b_2$, $\neg b_3$, $\neg b_4$ trying to satisfy all softs not in $hs$. This is *unsat* and any of a number of different cores could be returned. Say that the core $(b_1, b_2)$ is returned. ex-cores then attempts another SAT solve, this time with the assumptions $\neg b_3$ and $\neg b_4$. Now the SAT solver returns the core $(b_3, b_4)$. Finally, the SAT solver will be called to solve $F_H$ under the empty set of assumptions. Say that the solver finds the satisfying assignment $\tau = \{\neg b_1, b_2, \neg b_3, b_4\}$ setting $UB$ to 2 and $\tau_{best}$ to $\tau$. After returning to the main IHS routine, Min-Hs will be asked to compute an optimal solution to the set of cores $\{(b_1, b_2), (b_3, b_4)\}$. It might return $hs = \{b_1, b_4\}$ and set $LB = cost(hs) = 2$. Now $LB$ is equal to $UB$ and $\tau_{best}$ can be returned since it is an optimal solution. Note that in this example $hs$, is not a correction set.

As mentioned above IHS cannot terminate until its optimizer has been given enough constraints to drive the cost of an optimal solution up to be equal to the cost of an optimal MaxSat solution. As shown in [12] in the worst case this can require giving the optimizer an exponential number of constraints.

*Example 2.* Let $n$ and $r$ be integers with $0 < r < n$. Consider the MaxSat instance $\mathcal{F}^{n,r}$ with $\mathcal{F}_H{}^{n,r} = \text{CNF}(\sum_{i=1}^{n} b_i \geq r)$ and $\mathcal{F}_S{}^{n,r} = \{(\neg b_i) \mid 1 \leq i \leq n\}$, where $\text{CNF}(\sum_{i=1}^{n} b_i \geq r)$ is a CNF encoding of the cardinality constraint stating that at least $r$ soft clauses must be falsified. The cost of every optimal solution is thus $r$; the maximum number of soft clauses that can be satisfied is $n - r$; and every subset of $n - r + 1$ soft clauses must be a core. Let $\mathcal{C}$ be the set of all of such cores. From the results of [12] we have that if the optimizer is given all cores in $\mathcal{C}$ it would yield solutions $hs$ with $cost(hs) = r$; furthermore, if even one core of $\mathcal{C}$ is missing from the optimizer the optimizer solutions $hs$ would have $cost(hs) < r$. This means that Algorithm 1 will have to extract $\binom{n}{n-r+1}$ cores for the optimizer before it can reach the cost of an optimal MaxSat solution and terminate. When $r$ is close to $n/2$ the number of cores required for termination is exponential in $n$.

The results of the 2019 MaxSat Evaluation [4,5] witness this drawback in practice. The drmx-atmostk set of instances in the evaluation contain 11 instances with the same underlying structure as Example 2. Out of these, the IHS solver MaxHS [13,14], failed to solve 8 out of 11 when given an hour for each instance, while the best performing solvers were able to solve all 11 instances in under 10s.

# 4    Abstract Cores

Example 2 shows that a significant bottleneck for the IHS approach on some instances is the large number of cores that have to be given to the optimizer. Thus, a natural question to ask is whether or not there exists a more compact representation of this large number of cores that can still be efficiently reasoned with by the IHS algorithm. In this section we propose *abstract cores* as one such representation. As we will demonstrate, each abstract core compactly represents a large number of regular cores. By extracting abstract cores with the SAT solver and then giving them to the optimizer, we can communicate constraints to the optimizer that would have otherwise potentially required an exponential number of ordinary core constraints.

The structure of the instances $\mathcal{F}^{n,r}$ discussed in Example 2 provides some intuition for abstract cores. In these instances the identity of the variables does not matter, all that matters is how many are set to TRUE and how many are set to FALSE. For example, in any core $\kappa$ of $\mathcal{F}^{n,r}$ we can exchange any soft clause $C \in \kappa$ for any other soft clause $C' \notin \kappa$ and the result will still be a core of $\mathcal{F}^{n,r}$. In other words, every soft clause is exchangeable with every other soft clause in these instances. While it seems unlikely that complete exchangeability would hold for other instances, it is plausible that many instances might contain subsets of soft clauses that are exchangeable or nearly exchangeable. In particular, in any MaxSat instance the cost of a solution depends only on the number of soft clauses of each weight that it falsifies. The identity of the falsified soft clauses does not matter except to the extent that $\mathcal{F}_H$ might place logical constraints on the set of soft clauses that can be satisfied together.[1]

*Abstraction Sets.* Suppose we have a set of b-variables all with the same weight and we want to exploit any exchangeability that might exist between their corresponding soft clauses. This can be accomplished by forming an *abstraction set*. An abstraction set, $ab$, is a set of b-variables that have been annotated by adding $|ab|$ new variables, called $ab$'s count variables, used to indicate the number of true b-variables in $ab$ (i.e. the number of corresponding falsified soft clauses). The count variables allow us to abstract away from the identity of the particular b-variables that have been made false. We let $ab.c$ denote the sequence of $ab$'s count variables, and let the individual count variables be denoted by $ab.c[1]$, ..., $ab.c[|ab|]$. Every count variable has a corresponding definition, with the $i$'th count variable being defined by the constraint $ab.c[i] \leftrightarrow \sum_{b \in ab} b \geq i$. Note that these definitions can be encoded into CNF and added to the SAT solver using various known encodings for cardinality constraints [3,6,27,30].

Let $\mathcal{AB}$ be a collection of abstraction sets. We require that (1) the sets in $\mathcal{AB}$ are disjoint (so no b-variable is part of two different abstraction sets) and (2) that all of the b-variables in a specific abstraction set $ab \in \mathcal{AB}$ have the same weight (variables in different abstraction sets can have different weights). Let $\mathcal{AB}.c = \bigcup_{ab \in \mathcal{AB}} ab.c$ be the set of all count variables.

**Definition 1.** *An* abstract core *is a clause $C$ such that (1) all literals $C$ are either positive b-variables or count variables, $\forall l \in C$ ($l \in \mathcal{F}_B \lor l \in \mathcal{AB}.c$); and (2) $C$ is entailed by $\mathcal{F}_H$ and the conjunction of the count variable definitions, i.e., $\mathcal{F}_H \land \left( \bigwedge_{ab.c[k] \in \mathcal{AB}.c} (ab.c[k] \leftrightarrow \sum_{b \in ab} b \geq k) \right) \models C$.*

As pointed out in Sect. 2 every ordinary core is equivalent to a clause containing only positive b-variables that is entailed by $\mathcal{F}_H$. Abstract cores, can be ordinary cores containing only b-variables but they can also contain positive count variables. Like ordinary cores they also must be entailed by $\mathcal{F}_H$ (and the count variable definitions that are required to give meaning to the count variables they contain).

*Example 3.* Consider an instance $\mathcal{F}^{n,r}$ defined in Example 2. Say we form an single abstraction set, $ab$, from the full set of blocking variables $\mathcal{F}_B{}^{n,r}$. Then $\mathcal{F}^{n,r}$ will have among its abstract cores the unit clause $(ab.c[r])$ asserting that $\sum_{b \in \mathcal{F}_B{}^{n,r}} b \geq r$. This single abstract core is equivalent to the conjunction of $\binom{n}{n-r+1}$ non-abstract cores. In particular, with $n$ b-variables, asserting that at least $r$ must be true entails that every set of $n - r + 1$ b-variables must contain at least one true b-variable. That is, $(ab.c[r])$ entails $\binom{n}{n-r+1}$ different clauses each of which is equivalent to a non-abstract core. It is not difficult to show that entailment in the other direction also holds giving equivalence.

---

[1] This notion of exchangeability is clearly related to symmetries and exploring this connection is a worthwhile direction for future work.

This example demonstrates the expressive power of abstract cores. More generally, let $C$ be an abstract core containing the count literals $\{ab^1.c[c_1], \ldots, ab^k.c[c_k]\}$. Then, each $ab^i.c[c_i]$ is equivalent to the conjunction of $\binom{|ab^i|}{|ab^i|-c_i+1}$ clauses. Hence, $C$ is equivalent to the conjunction of $\prod_{i=1}^k \binom{|ab^i|}{|ab^i|-c_i+1}$ non-abstract cores. In other words, abstract cores achieve the desideratum of providing a compact representation of a large number of cores. We address the second desideratum of being able to reason efficiently with abstract cores in the IHS algorithm in the next section. It can also be noted that core-guided solvers use cardinality constraints and thus are able to generate abstract cores, although they use these cores in a different way than our proposed approach.

```
1  Abstract-IHS (F, wt);
2  LB ← 0; UB ← ∞; τbest ← ∅;
3  C ← ∅; AB ← ∅
4  while true do
5      hs ← Min-Abs(FB, AB, C)
6      LB = cost(hs);
7      AB ← update-abs(AB, K)
8      if (LB = UB) break;
9      K ← ex-abs-cores;
10              (hs, AB, UB, τbest);
11     if (LB = UB) break;
12     C ← C ∪ K;
13 return τbest
```

**Algorithm 3:** IHS with abstract cores

Min-Abs $(\mathcal{F}_B, \mathcal{AB}, \mathcal{C})$

minimize: $\displaystyle\sum_{b \in \mathcal{F}_B} wt(b) \cdot b$

subject to:

$\sum_{x \in \kappa} x \geq 1$ $\quad\quad\quad \forall \kappa \in \mathcal{C}$

$\sum_{b \in ab} b - k \cdot ab.c[k] \geq 0$ $\quad \forall ab.c[k] \in \mathcal{AB}.c$

$\sum_{b \in ab} b - |ab| \cdot ab.c[k] < k \; \forall ab.c[k] \in \mathcal{AB}.c$

$b \in \{0, 1\}$ $\quad\quad\quad\quad \forall b \in \mathcal{F}_B$

$ab.c[k] \in \{0, 1\}$ $\quad\quad\quad \forall ab.c[k] \in \mathcal{AB}.c$

return: $\{b \mid b$ set to 1 in opt. soln$\}$

(a) IP for optimizing with abstract cores

## 5   Abstract Cores in IHS MaxSat Solving

Algorithm 3 shows the IHS algorithm extended with abstract cores. Its processing follows the same steps as used earlier in the non-abstract IHS algorithm (Algorithm 1). There are however, three changes: (1) the optimizer must now solve a slightly different problem, (2) the abstraction sets are used in **ex-abs-cores** when extracting new constraints for the optimizer and (3) a collection of abstraction sets $\mathcal{AB}$ is maintained and dynamically updated by **update-abs** (line 7). We will assume that **update-abs** is also responsible for updating $\mathcal{F}_H$ so that $\mathcal{F}_H$ always includes all of the count variable definitions, $\bigcup_{ab.c[k] \in \mathcal{AB}.c} \text{CNF}(ab.c[k] \leftrightarrow \sum_{b \in ab} b \geq k)$ as new abstraction sets are added. In this way the other routines given below need only access $\mathcal{F}_H$ assuming that it already includes the count variable definitions.

*New Optimization Problem:* The optimization problem shown in Fig. 2a is very similar to the previous minimum cost hitting set optimization (Fig. 1a). It continues to minimize the cost of the set of b-variables that have to be set to true in order to satisfy the constraints. Each abstract core $\kappa \in \mathcal{C}$ is a clause and thus is equivalent to the linear constraint $\sum_{x \in \kappa} x \geq 1$, just like the non-abstract cores. The abstract cores can, however, contain count variables $ab.c[k]$ each of which has a specific definition. These definitions need to be given to the optimizer as linear constraints. For each count variable $ab.c[i]$ the constraints added are (a) $\sum_{b \in ab} b - k \cdot ab.c[k] \geq 0$ and (b) $\sum_{b \in ab} b - |ab| \cdot ab.c[k] < k$. That is, when $ab.c[k]$ is 1 (TRUE) constraint (a) ensures that the sum of $ab$'s b-variables is $\geq k$ and constraint (b) becomes trivial; and when $ab.c[k]$ is 0 (FALSE) constraint (a) becomes trivial and constraint (b) ensures that the sum of $ab$'s b-variables is $< k$. These definitions ensure the intended interaction between abstract cores and count variables. For example, if the optimizer has the abstract core constraint $b_1 + ab.c[5] + b_2 \geq 1$ it must be able to reason that if it chooses to satisfy this constraint by setting $ab.c[5] = 1$ then it must also set 5 of the b-variables in $ab$ to 1. The definitions allow this inference.

*Extracting Abstract Cores:* As before the optimizer's solution is used to create a set of assumptions for the SAT solver. Cores arise from the conflicts the SAT solver finds when using these assumptions. For ordinary cores ex-cores (Algorithm 2) used a set of negated b-variables as assumptions (ensuring that the corresponding set of soft clauses must be satisfied). If the SAT solver finds a conflict over these assumptions, the conflict will be a clause containing only negated assumptions; i.e, a clause containing only positive b-variables. Such clauses are ordinary cores. Hence, if we wish to extract abstract cores, we must give the SAT solver assumptions that consist of negated b-variables and negated count variables. Any conflicts that arise will then contain positive b-variables and positive count variables and will thus be abstract cores.

In the non-abstract case, the optimizer's solution $hs$ specifies a set of b-variables that can be set to true to obtain an optimal solution to the current set of constraints. That is, $hs$ provides a set of clauses that, if falsified, will most cost effectively block the cores found so far. In the abstract case, the optimizer's solution is also a set of b-variables with the same properties. All that has changed is the type of constraints the optimizer has optimized over.

Consider an abstraction set in the current set of abstractions $ab \in \mathcal{AB}$. Say that $ab$ is the set of b-variables $\{b_1, b_2, b_3, b_4\}$. Further, suppose that the optimizer returns the set $hs = \{b_1, b_4, b_5\}$ as its solution, and that the full set of b-variables is $\mathcal{F}_B = \{b_1, b_2, b_3, b_4, b_5, b_6\}$. In the non-abstract case, the SAT solver will be allowed to make $b_1$, $b_4$ and $b_5$ TRUE, while being forced to make $b_2$, $b_3$, and $b_6$ FALSE. In particular, the SAT solver will be called with the set of assumptions $\neg b_2$, $\neg b_3$ and $\neg b_6$, i.e., the set $\{\neg b \mid b \in (\mathcal{F}_B - hs)\}$ (line 7, Algorithm 2). Notice, that the SAT solver is being allowed to make specific b-variables in $ab \cap hs$ TRUE (namely $b_1$ and $b_4$), while being forced to make specific b-variables in $ab - (ab \cap hs)$ FALSE (namely $b_2$ and $b_3$). Given that we believe the b-variables in $ab$ to exchangeable, we can achieve abstraction by removing these specific choices. In particular, instead of assuming that $b_2$ and $b_3$ are FALSE and forcing the SAT solver to satisfy these specific soft clauses, we can instead assume $\neg ab.c[3]$. This means allowing at most two b-variables in $ab$ to be true,

```
1  ex-abs-cores (hs, AB, UB, τ_best)
2  |   assumps ← {¬b | b ∈ F_B − hs};
3  |   foreach ab ∈ AB do
4  |   |   assumps ← assumps − {¬b | b ∈ ab};
5  |   |   if |ab ∩ hs| = |ab| then continue;
6  |   |   assumps ← assumps ∪ {¬ab.c[|ab ∩ hs| + 1]};
7  |   K ← ex-cores-sub(assumps, UB, τ_best);
8  |   optionally: K ← K ∪ ex-cores(hs, UB, τ_best);
9  |   return K
```

**Algorithm 4:** Extracting Abstract cores from the optimizer solution

forcing the remaining $|ab| - 2$ ( $= 2$) b-variables to be FALSE. Hence, the SAT solver must satisfy at least two soft clauses from the set $\{(\neg b_1), (\neg b_2), (\neg b_3), (\neg b_4)\}$ corresponding to $ab$, but it is no longer forced to try to satisfy the specific clauses $(\neg b_2)$ and $(\neg b_3)$. Hence, we can use $\{\neg ab.c[3], \neg b_6\}$ as the SAT solver's assumptions and thus be able to extract an abstract core. Note also that since the weight of every b-variable in $ab$ is the same, the SAT solver is still being asked to find a solution of cost equal to $cost(hs)$. Using this insight we can specify the procedure ex-abs-cores used to extract abstract cores.

Algorithm 4 shows the procedure ex-abs-cores. Once it has set up its assumptions this procedure operates exactly like ex-cores, calling the same subroutine ex-cores-sub to iteratively extract some number of cores. It first adds the negation of all b-variables not in $hs$: the optimizer wants to satisfy all of these soft clauses. Then it performs abstraction. It removes the b-variables of each abstraction set $ab$ from the assumptions, and adds instead a single count variable from $ab$. The optimizer's solution has made $k = |hs \cap ab|$ of $ab$'s b-variables TRUE. So we permit the SAT solver to make this number of $ab$'s b-variables TRUE, but no more. This is accomplished by giving it the assumption $\neg ab.c[k + 1]$. Note that $\neg ab.c[k + 1] \leftrightarrow \sum_{b \in ab} b \leq k$ by the definition of the count variables. Finally, if every b-variable of $ab$ is in $hs$ we need not add anything to the set of assumptions (line 5): the SAT solver can freely make all of $ab$'s b-variables true.

ex-abs-cores also has the option of additionally extracting a set of non-abstract cores by invoking its non-abstract version (line 8). Abstract and non-abstract cores can be freely mixed in Abstract-IHS. Due to the indeterminism in the conflicts the SAT solver returns, the non-abstract cores need not be subsumed by the abstract cores. Hence, in practice it is often beneficial to extract both.

The correctness of the IHS algorithm with abstract cores is easily proved.

**Theorem 1.** *Let $(\mathcal{F}, wt)$ be a MaxSat instance with $\mathcal{F}_H$ being satisfiable and assume that (a) the optimizer correctly returns optimal solutions to its set of constraints, and (b) every conflict $C$ over assumptions returned by the SAT solver is a clause that is entailed by the formula it is solving. Then Abstract-IHS when called on $(\mathcal{F}, wt)$ must eventually terminate returning an optimal solution.*

*Proof.* First observe that the extra clauses $E$ used to define the count variables in $\mathcal{AB}.c$ do not change the set of solutions (models of $\mathcal{F}_H$) nor their costs as they are definitions. In particular, any model $\tau$ of $\mathcal{F}_H$ can be extended to a model of $\mathcal{F}_H \cup E$ by appropriately setting the value of each count variable, and any model $\tau^E$ of $\mathcal{F}_H \cup E$ becomes a model of $\mathcal{F}_H$ once we remove its assignments to the count variables. In both cases the cost of the model is preserved. Therefore, we will prove that Abstract-IHS eventually terminates returning an optimal solution to $(\mathcal{F} \cup E, wt)$ (with every clause in $E$ being hard): this optimal solution provides us with an optimal solution to $(\mathcal{F}, wt)$.

From the definitions of the count variables in $E$ and the soundness of the abstract cores computed as assumption conflicts by the SAT solver, we see that every constraint in the optimizer's model is entailed by $\mathcal{F}_H \cup E$. That is, every solution of $\mathcal{F}_H \cup E$ is also a solution of the optimizer's constraints. Therefore, the cost of the optimizer's optimal solutions, $LB$, is always a lower bound on the cost of an optimal solution of $\mathcal{F}_H \cup E$. Furthermore, $\tau_{best}$ is always a solution of $\mathcal{F}_H \cup E$ as it is found by the SAT solver. Therefore, when $UB = cost(\tau_{best}) = LB$, $\tau_{best}$ must be an optimal solution. Hence we have that when Abstract-IHS returns a solution, that solution must be optimal.

Furthermore, when the optimizer returns a solution $hs$ to its model and $hs$ does not cause termination, then Abstract-IHS will compute a new abstract core $\kappa$ that $hs$ does not satisfy. This follows from the fact $\kappa$ is falsified by all solutions that make FALSE exactly the same set of un-abstracted b-variables and exactly the same count of b-variables from each abstraction set as $hs$. Hence, once we add $\kappa$ to the optimizer we block the solution $hs$. There are only a finite number of solutions to the optimizer's constraints since the variables all $0/1$ variables, and every optimal MaxSat solution of $\mathcal{F}_H \cup E$ always satisfies the optimizer's model. Therefore, as more constraints are added to the optimizer it must eventually return one of these optimal MaxSat solutions causing Abstract-IHS to terminate. □

*Example 4.* Consider running Abstract-IHS on the formula used in Example 1: $\mathcal{F}_H = \{(b_1, b_2), (b_2, b_3), (b_3, b_4)\}$, $\mathcal{F}_S = \{(\neg b_1), (\neg b_2), (\neg b_3), (\neg b_4)\}$, and all weights equal to 1. First Min-Abs is called on an empty set of constraints, and it returns $hs = \emptyset$. Say that update-abs creates a single abstraction set, $\mathcal{AB} = \{ab\}$, with $ab = \{b_2, b_3\}$ and that it is unchanged during the rest of the run.

Using $hs$, ex-abs-cores will initialize its assumptions to $\{\neg b_1, \neg ab.c[1], \neg b_4\}$ and call the SAT solver. These assumptions are *unsat*. Let the conflict found be the unit clause $(ab.c[1])$. In ex-cores-sub the next SAT call will be with the assumptions $\{\neg b_1, \neg b_4\}$. These assumptions are satisfiable and the solution $\tau = \{\neg b_1, b_2, b_3, \neg b_4\}$ is returned. The upper bound $UB$ will be set to $cost(\tau) = 2$ and $\tau_{best}$ will be set to $\tau$.

ex-abs-cores now returns and the optimizer is called with the set of abstract cores $\mathcal{C} = \{(ab.c[1])\}$. The optimizer can return two different optimal solutions $\{b_2\}$ or $\{b_3\}$, and say that it returns the first one $hs = \{b_2\}$. This will set the lower bound $LB = cost(hs) = 1$. Then ex-abs-cores will be called again and from $hs$ it will initialize its assumptions to $\{\neg b_1, \neg ab.c[2], \neg b_4\}$, which is *unsat* with the unique conflict $(b_1, ab.c[2], b_4)$. Hence, the next SAT call will be with an empty set of assumptions and a solution will be found. Suppose that this solution is the same as before, so that neither $UB$ nor $\tau_{best}$ is changed. ex-abs-cores will then return and the optimizer called with the accumulated cores $\{(ab.c[1]),$ $(b_1, ab.c[2], b_4)\}$. There are different choices for the optimal solution, but say that it returns $\{b_2, b_3\}$ as its optimal solution. This will reset the lower bound $LB$ to 2, the lower bound will meet the upper bound, and the MaxSat optimal solution $\tau_{best} = \{\neg b_1, b_2, b_3, \neg b_4\}$ will be found.

Abstract cores can decrease the worst-case number of cores the IHS algorithm needs to extract. Consider the instances $\mathcal{F}^{n,r}$ from Examples 2 and 3. As discussed in Example 2 when $r$ is close to $n/2$ these instances have an exponential number of non-abstract cores all of which must be extracted by the IHS algorithm. If on the other hand all b-variables are placed into a single abstraction set $ab$ as in Example 3, Abstract-IHS will generate the sequence of abstract cores $(ab.c[1]), \ldots, (ab.c[r])$ after which the optimizer will return a solution of cost $r$ that will be a correction sets allowing Abstract-IHS to terminate. More generally, this strategy can be applied to any unweighted MaxSat instance.

**Proposition 1.** *Let $(\mathcal{F}, wt)$ be an unweighted MaxSat instance, i.e. $w(C) = 1$ for all $C \in \mathcal{F}_S$ and construct an abstraction set $ab$ containing all $\mathcal{F}_B$. Then Algorithm 3 needs to extract at most $|\mathcal{F}_s|$ cores before terminating.*

The solving strategy of Proposition 1 in fact mimics the Linear UNSAT-SAT algorithm [9,17] where the SAT solver solves the sequence of queries "can a solution of cost 1 be found", "can a solution of cost 2 be found", etc. As interesting future work, we note that the framework of abstract cores presented here can be used to mimic the behaviour several of the recently proposed core-guided algorithms [23,25,26].

*Computing Abstraction Sets:* When computing abstraction sets, there is an inherent trade-off between the overhead and potential benefits from abstraction; too large sets can lead to large CNF encodings for the count variable definitions, making SAT solving very inefficient while with too small sets the algorithm reverts back to non-abstract IHS with hardly any gain from abstraction.

Although the notion of exchangeability has intuitive appeal, it seems likely be computationally hard to identify exchangeable b-variables that can be grouped into abstraction sets. In our implementation we used a heuristic approach to finding abstraction sets motivated by the $\mathcal{F}^{n,r}$ instances (Example 1). In those instances, each b-variable appears in many cores with each of the other b-variables. We decided to build abstraction sets from sets of b-variables that often appear in cores together. This technique worked in practice (see Sect. 6), but a deeper understanding of how best to construct abstraction sets remains as future work.

To find b-variables that appear in many cores with each other, we used the set of cores found to construct a graph $G$. The graph has b-variables as nodes and weighted edges between two b-variables representing the number of times these two b-variables appeared together in a core. We then applied the Louvain

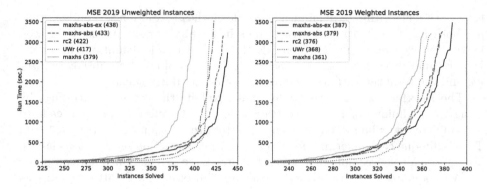

**Fig. 1.** Cactus plot of solver performance on the unweighted (left) and weighted (right) instances of MSEval 2019. The numbers in parenthesis are the number of instance solved within the time and memory limits

clustering algorithm [10] to $G$ obtain clusters of nodes such that the nodes in a cluster have a higher weight of edges between each other (i.e. appear in cores more often together) than with nodes in other clusters, these were then taken to define an abstraction set. We also monitored how effective the cores found were in increasing the lower bound generated by the optimizer. If the cores were failing to drive the optimizer's lower bound higher, we computed new abstraction sets by clustering the graph $G$, and updated $\mathcal{AB}$ with these new abstraction sets. If clustering had already been performed and the extracted cores were still not effective, the nodes of the b-variables in each abstraction set were merged into one new node and $G$ was reclustered. (The Louvain algorithm can compute hierarchical clusters). Any new clusters so generated will either be new abstraction sets or supersets of existing abstractions sets. New abstraction sets are formed from these new clusters and added to $\mathcal{AB}$. All subsets are removed from $\mathcal{AB}$ so that future abstractions will be generated using the larger abstraction sets.

We also found that abstraction was not cost effective on instances where the average core size was in the hundreds. The generated abstraction sets were so large that the CNF encoding of their count variables definitions slowed the SAT solver down too much. Finally, only we add the CNF encoding of the count variable definitions $ab.c[k] \leftrightarrow \sum_{b \in ab} b \geq k$ to the SAT solver when $\neg ab.c[k]$ first appears in the set of assumptions. Furthermore, we only add the encoding in the direction $ab.c[k] \leftarrow \sum_{b \in ab} b \geq k$.

## 6    Experimental Evaluation

We have implemented two versions of abstract cores on top of the MaxHS solver [12,14] using the version that had been submitted to the MaxSat 2019 evaluation (MSE 2019) [5]. The two new solvers are called **maxhs-abs** and **maxhs-abs-ex**. **maxhs-abs** implements the abstraction method described above, using the Louvain algorithm to dynamically decide on the abstraction sets and extracting both abstract and non-abstract cores in `ex-abs-cores`. We used the well known totalizer encoding [6] to encode the count variable definitions into CNF. In particular, each totalizer takes as input the b-variables of an abstraction set $ab$, and the totalizer outputs become the count variables $ab.c[k]$.

The **maxhs-abs-ex** solver additionally exploits the totalizer encodings by using the technique of core exhaustion [20]. This technique uses SAT calls to determine a lower bound on the number of totalizer outputs forced to be TRUE. This technique can sometimes force many of the abstraction set count variables. We impose a resource bound of 60s on the process so exhaustion is not complete.

We compare the new solvers to the base **maxhs** (MSE 2019 version) as well as to two other solvers: the MSE 2019 version of RC2 (**rc2**) [4,20], the best performing solver in both the weighted and unweighted track and a new solver in MSE 2019 called UWrMaxSat (**UWr**) [4,21]. Both implement the OLL algorithm [1,25] and differ mainly in how the cardinality constraints are encoded into CNF. As benchmarks, we used all 599 weighted and 586 unweighted instances from the complete track of the 2019 MaxSat Evaluation, drawn from a variety of different problem families. All experiments were run on a cluster of 2.4 GHz Intel machines using a per-instance time limit of 3600 s and memory limit of 5 GB.

Figure 1 show cactus plots comparing the solvers on the unweighted (left) and weighted (right) instances. Comparing **maxhs** and **maxhs-abs** we observe that abstract core reasoning is very effective, increasing the number of unweighted instances solved from 397 to 433 and weighted instances from 361 to 379 surpassing both **rc2** and **UWr** in both categories. **maxhs-abs-ex** improves even further with 438 unweighted and 387 weighted instances solved.

**Table 1.** The entry in cell $(X, Y)$ shows the number instances solved by solver $X$ that were not solved by solver $Y$ in the format #Unweighted/#Weighted.

| Solver | maxhs-abs-ex | maxhs-abs | rc2 | UWr | maxhs |
|---|---|---|---|---|---|
| maxhs-abs-ex | | 7/8 | 26/37 | 29/46 | 60/42 |
| maxhs-abs | 2/0 | | 27/33 | 31/43 | 55/35 |
| rc2 | 10/26 | 16/30 | | 12/29 | 61/50 |
| UWr | 8/27 | 15/32 | 7/21 | | 61/52 |
| maxhs | 19/16 | 19/17 | 36/35 | 41/45 | |

Table 1 gives a pair-wise solver comparison of the number of instances that could be solved by one solver but not by the other. We observe that even though the solvers can be ranked by number of instances solved, every solver was able to beat every other solver on some instances (except that **maxhs-abs** did not solve any weighted instances that **maxhs-abs-ex** could not). This speaks to the diversity of the instances, and indicates that truly robust solvers might have to employ a variety of different techniques.

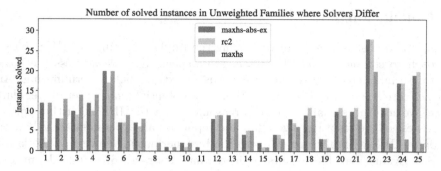

**Fig. 2.** Number of instances solved in the 25/48 families of unweighted instances on which **maxhs-abs-ex**, **rc2**, and **maxhs** solved different amounts of instances: 1. kbtree, 2. extension-enforcement, 3. optic, 4. logic-synthesis, 5. close-solutions, 6. min-fill, 7. atcoss, 8. set-covering, 9. maxcut, 10. aes, 11. gen-hyper-tw, 12. frb, 13. bcp, 14. HaplotypeAssembly, 15. scheduling, 16. CircuitTraceCompaction, 17. xai-mindset2, 18. MaxSATQueriesinInterpretableClassifiers, 19. reversi, 20. aes-key-recovery, 21. uaq, 22. MaximumCommonSub-GraphExtraction, 23. protein-ins, 24. drmx-atmostk, 25. fault-diagnosis

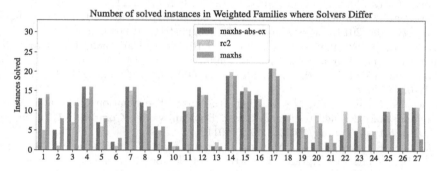

**Fig. 3.** Number of instances solved in the 27/39 families of unweighted instances on which **maxhs-abs-ex**, **rc2**, and **maxhs** solved different amounts of instances: 1. BTBNSL, 2. maxcut, 3. correlation-clustering, 4. auctions, 5. ParametricRBAC-Maintenance, 6. ramsey, 7. set-covering, 8. timetabling, 9. relational-inference, 10. hs-timetabling, 11. frb, 12. mpe, 13. railway-transport, 14. metro, 15. max-realizability, 16. MaxSATQueriesinInterpretableClassifiers, 17. haplotyping-pedigrees, 18. drmx-cryptogen, 19. spot5, 20. af-synthesis, 21. min-width, 22. css-refactoring, 23. shiftdesign, 24. lisbon-wedding, 25. tcp, 26. rna-alignment, 27. drmx-atmostk

Figures 2 and 3 show a breakdown by family for the three solvers **maxhs-abs-ex**, **rc2** and **maxhs**. The plots show only those families where the solvers exhibited different performance. We observe that **rc2** and **maxhs** achieve quite disparate performance with each one dominating the other on different families. **maxhs-abs-ex**, on the other hand, is often able to achieve the same performance as the better of the two other solvers on these different families.

# 7    Conclusion

We proposed abstract cores for improving the Implicit Hitting Set (IHS) based approach to complete MaxSat solving. More specifically, we address the large worst-case number of cores that IHS needs to extract before terminating. An abstract core is a compact representation of a (potentially large) set of (regular) cores. We incorporate abstract core reasoning into the IHS algorithm, prove correctness of the resulting algorithm and report on an experimental evaluation comparing IHS with abstract cores to the best performing solvers of the latest MaxSat Evaluation. The results indicate that abstract cores indeed improve the empirical performance of the IHS algorithm, resulting in state-of-the-art performance on the instances of the Evaluation.

# References

1. Andres, B., Kaufmann, B., Matheis, O., Schaub, T.: Unsatisfiability-based optimization in clasp. In: Dovier, A., Costa, V.S. (eds.) Technical Communications of ICLP. LIPIcs, vol. 17, pp. 211–221. Schloss Dagstuhl - Leibniz-Zentrum für Informatik (2012). https://doi.org/10.4230/LIPIcs.ICLP.2012.211
2. Ansótegui, C., Bonet, M.L., Levy, J.: SAT-based MaxSAT algorithms. Artif. Intell. **196**, 77–105 (2013). https://doi.org/10.1016/j.artint.2013.01.002
3. Asín, R., Nieuwenhuis, R., Oliveras, A., Rodríguez-Carbonell, E.: Cardinality networks and their applications. In: Kullmann, O. (ed.) SAT 2009. LNCS, vol. 5584, pp. 167–180. Springer, Heidelberg (2009). https://doi.org/10.1007/978-3-642-02777-2_18
4. Bacchus, F., Järvisalo, M., Martins, R.: MaxSAT evaluation 2018: new developments and detailed results. J. Satisf. Boolean Model. Comput. **11**(1), 99–131 (2019). https://doi.org/10.3233/SAT190119
5. Bacchus, F., Järvisalo, M., Martins, R. (eds.): MaxSAT Evaluation 2019: Solver and Benchmark Descriptions. Department of Computer Science Report Series B, Department of Computer Science, University of Helsinki, Finland (2019)
6. Bailleux, O., Boufkhad, Y.: Efficient CNF encoding of boolean cardinality constraints. In: Rossi, F. (ed.) CP 2003. LNCS, vol. 2833, pp. 108–122. Springer, Heidelberg (2003). https://doi.org/10.1007/978-3-540-45193-8_8
7. Berg, J., Hyttinen, A., Järvisalo, M.: Applications of MaxSAT in data analysis. In: Berre, D.L., Järvisalo, M. (eds.) Proc Pragmatics of SAT. EPiC Series in Computing, vol. 59, pp. 50–64. EasyChair (2018). http://www.easychair.org/publications/paper/6HpF
8. Berg, J., Järvisalo, M.: Cost-optimal constrained correlation clustering via weighted partial maximum satisfiability. Artif. Intell. **244**, 110–142 (2017). https://doi.org/10.1016/j.artint.2015.07.001
9. Berre, D.L., Parrain, A.: The SAT4J library, release 2.2. J. Satisf. Boolean Model. Comput. **7**(2–3), 59–60 (2010)
10. Blondel, V.D., Guillaume, J.L., Lambiotte, R., Lefebvre, E.: Fast unfolding of communities in large networks. J. Stat. Mech.: Theory Exp. **2008**(10), P10008 (2008). https://doi.org/10.1088%2F1742-5468%2F2008%2F10%2Fp10008
11. Chen, Y., Safarpour, S., Marques-Silva, J., Veneris, A.G.: Automated design debugging with maximum satisfiability. IEEE Trans. CAD Integr. Circuits Syst. **29**(11), 1804–1817 (2010). https://doi.org/10.1109/TCAD.2010.2061270

12. Davies, J.: Solving MAXSAT by decoupling optimization and satisfaction. Ph.D. thesis, University of Toronto (2013)
13. Davies, J., Bacchus, F.: Solving MAXSAT by solving a sequence of simpler SAT instances. In: Lee, J. (ed.) CP 2011. LNCS, vol. 6876, pp. 225–239. Springer, Heidelberg (2011). https://doi.org/10.1007/978-3-642-23786-7_19
14. Davies, J., Bacchus, F.: Exploiting the power of MIP solvers in MAXSAT. In: Järvisalo, M., Van Gelder, A. (eds.) SAT 2013. LNCS, vol. 7962, pp. 166–181. Springer, Heidelberg (2013). https://doi.org/10.1007/978-3-642-39071-5_13
15. Davies, J., Bacchus, F.: Postponing optimization to speed up MAXSAT solving. In: Schulte, C. (ed.) CP 2013. LNCS, vol. 8124, pp. 247–262. Springer, Heidelberg (2013). https://doi.org/10.1007/978-3-642-40627-0_21
16. Demirovic, E., Musliu, N., Winter, F.: Modeling and solving staff scheduling with partial weighted MaxSAT. Ann. OR **275**(1), 79–99 (2019). https://doi.org/10.1007/s10479-017-2693-y
17. Eén, N., Sörensson, N.: Translating pseudo-boolean constraints into SAT. J. Satisf. Boolean Model. Comput. **2**(1–4), 1–26 (2006). https://satassociation.org/jsat/index.php/jsat/article/view/18
18. Ghosh, B., Meel, K.S.: IMLI: an incremental framework for MaxSAT-based learning of interpretable classification rules. In: Conitzer, V., Hadfield, G.K., Vallor, S. (eds.) Proceedings of AIES, pp. 203–210. ACM (2019). https://doi.org/10.1145/3306618.3314283
19. Hosokawa, T., Yamazaki, H., Misawa, K., Yoshimura, M., Hirama, Y., Arai, M.: A low capture power oriented X-filling method using partial MaxSAT iteratively. In: Proc IEEE International Symposium on Defect and Fault Tolerance in VLSI and Nanotechnology Systems, DFT, pp. 1–6. IEEE (2019). https://doi.org/10.1109/DFT.2019.8875434
20. Ignatiev, A., Morgado, A., Marques-Silva, J.: RC2: an efficient MaxSAT solver. J. Satisf. Boolean Model. Comput. **11**(1), 53–64 (2019). https://doi.org/10.3233/SAT190116
21. Karpinski, M., Piotrów, M.: Encoding cardinality constraints using multiway merge selection networks. Constraints **24**(3–4), 234–251 (2019). https://doi.org/10.1007/s10601-019-09302-0
22. Koshimura, M., Zhang, T., Fujita, H., Hasegawa, R.: QMaxSAT: a partial MaxSAT solver. J. Satisf. Boolean Model. Comput. 8(1/2), 95–100 (2012). https://satassociation.org/jsat/index.php/jsat/article/view/98
23. Marques-Silva, J., Planes, J.: On using unsatisfiability for solving maximum satisfiability. CoRR abs/0712.1097 (2007)
24. Martins, R., Manquinho, V., Lynce, I.: Open-WBO: a modular MaxSAT solver'. In: Sinz, C., Egly, U. (eds.) SAT 2014. LNCS, vol. 8561, pp. 438–445. Springer, Cham (2014). https://doi.org/10.1007/978-3-319-09284-3_33
25. Morgado, A., Dodaro, C., Marques-Silva, J.: Core-Guided MaxSAT with Soft Cardinality Constraints. In: OSullivan, B. (ed.) CP 2014. LNCS, vol. 8656, pp. 564–573. Springer, Cham (2014). https://doi.org/10.1007/978-3-319-10428-7_41
26. Narodytska, N., Bacchus, F.: Maximum satisfiability using core-guided MaxSAT resolution. In: Brodley, C.E., Stone, P. (eds.) Proc AAAI, pp. 2717–2723. AAAI Press (2014). http://www.aaai.org/ocs/index.php/AAAI/AAAI14/paper/view/8513
27. Ogawa, T., Liu, Y., Hasegawa, R., Koshimura, M., Fujita, H.: Modulo based CNF encoding of cardinality constraints and its application to MaxSAT solvers. In: Proceedings of ICTAI, pp. 9–17. IEEE Computer Society (2013). https://doi.org/10.1109/ICTAI.2013.13

28. Saikko, P.: Re-implementing and extending a hybrid SAT-IP approach to maximum satisfiability. Master's thesis, University of Helsinki (2015). http://hdl.handle.net/10138/159186
29. Saikko, P., Berg, J., Järvisalo, M.: LMHS: a SAT-IP hybrid MaxSAT solver. In: Creignou, N., Le Berre, D. (eds.) SAT 2016. LNCS, vol. 9710, pp. 539–546. Springer, Cham (2016). https://doi.org/10.1007/978-3-319-40970-2_34
30. Sinz, C.: Towards an optimal CNF encoding of boolean cardinality constraints. In: van Beek, P. (ed.) CP 2005. LNCS, vol. 3709, pp. 827–831. Springer, Heidelberg (2005). https://doi.org/10.1007/11564751_73
31. Zhang, L., Bacchus, F.: MAXSAT heuristics for cost optimal planning. In: Hoffmann, J., Selman, B. (eds.) Proceedings of AAAI. AAAI Press (2012). http://www.aaai.org/ocs/index.php/AAAI/AAAI12/paper/view/5190

# MaxSAT Resolution and Subcube Sums

Yuval Filmus[1], Meena Mahajan[2(✉)], Gaurav Sood[2], and Marc Vinyals[1]

[1] Technion – Israel Institute of Technology, Haifa, Israel
{yuvalfi,marcviny}@cs.technion.ac.il
[2] The Institute of Mathematical Sciences (HBNI), Chennai, India
{meena,gauravs}@imsc.res.in

**Abstract.** We study the MaxRes rule in the context of certifying unsatisfiability. We show that it can be exponentially more powerful than tree-like resolution, and when augmented with weakening (the system MaxResW), $p$-simulates tree-like resolution. In devising a lower bound technique specific to MaxRes (and not merely inheriting lower bounds from Res), we define a new semialgebraic proof system called the Sub-CubeSums proof system. This system, which $p$-simulates MaxResW, is a special case of the Sherali–Adams proof system. In expressivity, it is the integral restriction of conical juntas studied in the contexts of communication complexity and extension complexity. We show that it is not simulated by Res. Using a proof technique qualitatively different from the lower bounds that MaxResW inherits from Res, we show that Tseitin contradictions on expander graphs are hard to refute in SubCubeSums. We also establish a lower bound technique via lifting: for formulas requiring large degree in SubCubeSums, their XOR-ification requires large size in SubCubeSums.

**Keywords:** Proof complexity · MaxSAT resolution · Subcube complexity · Sherali–Adams proofs · Conical juntas

## 1 Introduction

The most well-studied propositional proof system is Resolution (Res), [5,22]. It is a refutational line-based system that operates on clauses, successively inferring newer clauses until the empty clause is derived, indicating that the initial set of clauses is unsatisfiable. It has just one satisfiability-preserving rule: if clauses $A \vee x$ and $B \vee \neg x$ have been inferred, then the clause $A \vee B$ can be inferred. Sometimes it is convenient, though not necessary in terms of efficiency, to also allow a weakening rule: from clause $A$, a clause $A \vee x$ can be inferred. While there are several lower bounds known for this system, it is still very useful in practice and underlies many current SAT solvers.

While deciding satisfiability of a propositional formula is NP-complete, the MaxSAT question is an optimization question, and deciding whether its value is as given (i.e. deciding, given a formula and a number $k$, whether the maximum number of clauses simultaneously satisfiable is exactly $k$) is potentially harder

© Springer Nature Switzerland AG 2020
L. Pulina and M. Seidl (Eds.): SAT 2020, LNCS 12178, pp. 295–311, 2020.
https://doi.org/10.1007/978-3-030-51825-7_21

since it is hard for both NP and coNP. A proof system for MaxSAT was proposed in [7,14]. This system, denoted MaxSAT Resolution or more briefly MaxRes, operates on multi-sets of clauses. At each step, two clauses from the multi-set are resolved and removed. The resolvent, as well as certain "disjoint" weakenings of the two clauses, are added to the multiset. The invariant maintained is that for each assignment $\rho$, the number of clauses in the multi-set falsified by $\rho$ remains unchanged. The process stops when the multi-set has a satisfiable instance along with $k$ copies of the empty clause; $k$ is exactly the minimum number of clauses of the initial multi-set that must be falsified by every assignment.

Since MaxRes maintains multi-sets of clauses and replaces used clauses, this suggests a "read-once"-like constraint. However, this is not the case; read-once resolution is not even complete [13], whereas MaxRes is a complete system for certifying the MaxSAT value (and in particular, for certifying unsatisfiability). One could use the MaxRes system to certify unsatisfiability, by stopping the derivation as soon as one empty clause is produced. Such a proof of unsatisfiability, by the very definition of the system, can be $p$-simulated by Resolution. (The MaxRes proof is itself a proof with resolution and weakening, and weakening can be eliminated at no cost.) Thus, lower bounds for Resolution automatically apply to MaxRes and to MaxResW (the augmenting of MaxRes with an appropriate weakening rule) as well. However, since MaxRes needs to maintain a stronger invariant than merely satisfiability, it seems reasonable that for certifying unsatisfiability, MaxRes is weaker than Resolution. (This would explain why, in practice, MaxSAT solvers do not seem to use MaxRes – possibly with the exception of [20], but they instead directly call SAT solvers, which use standard resolution.) Proving this would require a lower bound technique specific to MaxRes.

Associating with each clause the subcube (conjunction of literals) of assignments that falsify it, each MaxRes step manipulates and rearranges multi-sets of subcubes. This naturally leads us to the formulation of a static semi-algebraic proof system that we call the SubCubeSums proof system. This system, by its very definition, $p$-simulates MaxResW and is a special case of the Sherali–Adams proof system. Given this position in the ecosystem of simple proof systems, understanding its capabilities and limitations seems an interesting question.

## Our Contributions and Techniques

1. We observe that for certifying unsatisfiability, the proof system MaxResW $p$-simulates the tree-like fragment of Res, TreeRes (Lemma 1). This simulation seems to make essential use of the weakening rule. On the other hand, we show that even MaxRes without weakening is not simulated by TreeRes (Theorem 1). We exhibit a formula, which is a variant of the pebbling contradiction [4] on a pyramid graph, with short refutations in MaxRes (Lemma 2), and show that it requires exponential size in TreeRes (Lemma 7).

2. We initiate a formal study of the newly-defined semialgebraic proof system SubCubeSums, which is a natural restriction of the Sherali–Adams proof system. We show that this system is not simulated by Res (Theorem 2).

3. We show that the Tseitin contradiction on an odd-charged expander graph is hard for SubCubeSums (Theorem 3) and hence also hard for MaxResW. While this already follows from the fact that these formulas are hard for Sherali–Adams [1], our lower-bound technique is qualitatively different; it crucially uses the fact that a stricter invariant is maintained in MaxResW and SubCubeSums refutations.

4. Abstracting the ideas from the lower bound for Tseitin contradictions, we devise a lower-bound technique for SubCubeSums based on lifting (Theorem 4). Namely, we show that if every SubCubeSums refutation of a formula $F$ must have at least one wide clause, then every SubCubeSums refutation of the formula $F \circ \oplus$ must have many cubes. We illustrate how the Tseitin contradiction lower bound can be recovered in this way.

The relations among these proof systems are summarized in the figure below, which also includes two proof systems discussed in Related Work.

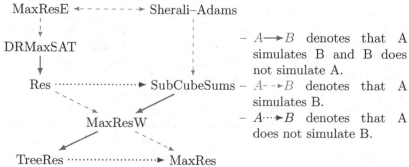

- $A \longrightarrow B$ denotes that A simulates B and B does not simulate A.
- $A \dashrightarrow B$ denotes that A simulates B.
- $A \cdots\!\!\rightarrow B$ denotes that A does not simulate B.

## Related Work

One reason why studying MaxRes is interesting is that it displays unexpected power after some preprocessing. As described in [12] (see also [18]), the PHP formulas that are hard for Resolution can be encoded into MaxHornSAT, and then polynomially many MaxRes steps suffice to expose the contradiction. The underlying proof system, DRMaxSAT, has been studied further in [6], where it is shown to p-simulate general Resolution. While DRMaxSAT gains power from the encoding, the basic steps are MaxRes steps. Thus, to understand how DRMaxSAT operates, a better understanding of MaxRes could be quite useful.

A very recent paper [15] studies a proof system MaxResE where MaxRes is augmented with an extension rule. The extension rule generalises a weighted version of MaxRes; as defined, it eliminates the non-negativity constraint inherent in MaxResW and SubCubeSums. This system happens to be equivalent to Circular Resolution [16], which in turn is equivalent to Sherali–Adams [2]. It is also worth mentioning that MaxResW appears in [16] as MaxRes with a split rule, or ResS.

In the setting of communication complexity and of extension complexity of polytopes, non-negative rank is an important and useful measure. As discussed in [11], the query-complexity analogue is *conical juntas*; these are non-negative combinations of subcubes. Our SubCubeSums refutations are a restriction of conical juntas to non-negative *integral* combinations. Not surprisingly, our lower bound for Tseitin contradictions is similar to the conical junta degree lower bound established in [10].

### Organisation of the Paper

We define the proof systems MaxRes, MaxResW, and SubCubeSums in Sect. 2. In Sect. 3 we relate them to TreeRes. In Sect. 4, we focus on the SubCubeSums proof system, showing the separation from Res (Sect. 4.1), the lower bound for SubCubeSums (Sect. 4.2), and the lifting technique (Sect. 4.3).

## 2   Defining the Proof Systems

For set $X$ of variables, let $\langle X \rangle$ denote the set of all total assignments to variables in $X$. For a (multi-) set of $F$ clauses, $\mathrm{viol}_F \colon \langle X \rangle \to \{0\} \cup \mathbb{N}$ is the function mapping $\alpha$ to the number of clauses in $F$ (counted with multiplicity) falsified by $\alpha$. A (sub)cube is the set of assignments falsifying a clause, or equivalently, the set of assignments satisfying a conjunction of literals.

The proof system Res has the resolution rule inferring $C \vee D$ from $C \vee x$ and $D \vee \overline{x}$, and optionally the weakening rule inferring $C \vee x$ from $C$ if $\overline{x} \notin C$. A refutation of a CNF formula $F$ is a sequence of clauses $C_1, \ldots, C_t$ where each $C_i$ is either in $F$ or is obtained from some $j, k < i$ using resolution or weakening, and where $C_t$ is the empty clause. The underlying graph of such a refutation has the clauses as nodes, and directed edge from $C$ to $D$ if $C$ is used in the step deriving $D$. The proof system TreeRes is the fragment of Res where only refutations in which the underlying graph is a tree are permitted. A proof system $P$ simulates ($p$-simulates) another proof system $P'$ if proofs in $P$ can be transformed into proofs in $P'$ with polynomial blow-up (in time polynomial in the size of the proof). See, for instance, [3], for more details.

### The MaxRes and MaxResW Proof Systems

The MaxRes proof system operates on sets of clauses, and uses the MaxSAT resolution rule [7], defined as follows:

$$\frac{\begin{array}{ll} x \vee a_1 \vee \ldots \vee a_s & (x \vee A) \\ \overline{x} \vee b_1 \vee \ldots \vee b_t & (\overline{x} \vee B) \end{array}}{a_1 \vee \ldots \vee a_s \vee b_1 \vee \ldots \vee b_t \quad \text{(the ``standard resolvent'')}}$$

$$\begin{array}{ll} \text{(weakenings of } x \vee A) & \text{(weakenings of } \overline{x} \vee B) \\ x \vee A \vee \overline{b}_1 & \overline{x} \vee B \vee \overline{a}_1 \\ x \vee A \vee b_1 \vee \overline{b}_2 & \overline{x} \vee B \vee a_1 \vee \overline{a}_2 \\ \vdots & \vdots \\ x \vee A \vee b_1 \vee \ldots \vee b_{t-1} \vee \overline{b}_t & \overline{x} \vee B \vee a_1 \vee \ldots \vee a_{s-1} \vee \overline{a}_s \end{array}$$

The weakening rule for MaxSAT resolution replaces a clause $A$ by the two clauses $A \lor x$ and $A \lor \overline{x}$. While applying either of these rules, the antecedents are removed from the multi-set and the non-tautologous consequents are added. If $F'$ is obtained from $F$ by applying these rules, then $\text{viol}_F$ and $\text{viol}_{F'}$ are the same function.

In the proof system MaxRes, a refutation of $F$ is a sequence $F = F_0, F_1, \ldots, F_s$ where each $F_i$ is a multi-set of clauses, each $F_i$ is obtained from $F_{i-1}$ by an application of the MaxSAT resolution rule, and $F_s$ contains the empty clause $\square$. In the proof system MaxResW, $F_i$ may also be obtained from $F_{i-1}$ by using the weakening rule. The size of the proof is the number of steps, $s$. In [7,14], MaxRes is shown to be complete for MaxSAT, hence also for unsatisfiability. Since the proof system MaxRes we consider here is a refutation system rather than a system for MaxSAT, we can stop as soon as a single $\square$ is derived.

### The SubCubeSums Proof System

The SubCubeSums proof system is a static proof system. For an unsatisfiable CNF formula $F$, a SubCubeSums proof is a multi-set $G$ of sub-cubes (or terms, or conjunctions of literals) satisfying $\text{viol}_F \equiv 1 + \text{viol}_G$.

We can view SubCubeSums as a subsystem of the semialgebraic Sherali–Adams proof system as follows. Let $F$ be a CNF formula with $m$ clauses in variables $x_1, \ldots, x_n$. Each clause $C_i$, $i \in [m]$, is translated into a polynomial equation $f_i = 0$; a Boolean assignment satisfies clause $C_i$ iff it satisfies equation $f_i = 0$. Boolean values are forced through the axioms $x_j^2 - x_j = 0$ for $j \in [n]$. A Sherali–Adams proof is a sequence of polynomials $g_i$, $i \in [m]$; $q_j$, $j \in [n]$; and a polynomial $p_0$ of the form

$$p_0 = \sum_{A,B \subseteq [n]} \alpha_{A,B} \prod_{j \in A} x_j \prod_{j \in B} (1 - x_j)$$

where each $\alpha_{A,B} \geq 0$, such that

$$\left( \sum_{i \in [m]} g_i f_i \right) + \left( \sum_{j \in [n]} q_j (x_j^2 - x_j) \right) + p_0 + 1 = 0$$

The degree or rank of the proof is the maximum degree of $g_i f_i, q_j(x_j^2 - x_j), p_0$.

The polynomials $f_i$ corresponding to the clauses, as well as the terms in $p_0$, are conjunctions of literals, thus special kinds of $d$-juntas (Boolean functions depending on at most $d$ variables). So $p_0$ is a non-negative linear combination of non-negative juntas, that is, in the nomenclature of [11], a *conical junta*.

The Sherali–Adams system is sound and complete, and verifiable in randomized polynomial time; see for instance [9].

Consider the following restriction of Sherali–Adams:

1. Each $g_i = -1$.
2. Each $\alpha_{A,B} \in \mathbb{Z}^{\geq 0}$, (non-negative integers), and $\alpha_{A,B} > 0 \implies A \cap B = \emptyset$.

This implies each $q_j$ must be 0, since the rest of the expression is multilinear. Hence, for some non-negative integral $\alpha_{A,B}$,

$$\sum_{A,B\subseteq[n]:A\cap B=\emptyset} \alpha_{A,B} \prod_{j\in A} x_j \prod_{j\in B}(1-x_j) + 1 = \sum_{i\in[m]} f_i$$

This is exactly the SubCubeSums proof system: the terms in $p_0$ are subcubes, and the right-hand-side is $\mathrm{viol}_F$. For each disjoint pair $A, B$, the SubCubeSums proof has $\alpha_{A,B}$ copies of the corresponding sub-cube. The degree of a SubCube-Sums proof is the maximum number of literals appearing in a conjunction. The size of a SubCubeSums proof is the number of subcubes, that is, $\sum_{A,B}\alpha_{A,B}$. The constraint $g_i = 1$ means that for bounded CNF formulas, the degree of a SubCubeSums proof is essentially the degree of $p_0$, i.e. the degree of the juntas.

The following proposition shows why this restriction remains complete.

**Proposition 1.** *SubCubeSums p-simulates MaxResW.*

*Proof.* If an unsatisfiable CNF formula $F$ with $m$ clauses and $n \geq 3$ variables has a MaxResW refutation with $s$ steps, then this derivation produces $\{\Box\}\cup G$ where the number of clauses in $G$ is at most $m + (n-2)s - 1$. (A weakening step increases the number of clauses by 1. A MaxRes step increases it by at most $n - 2$.) The subcubes falsifying the clauses in $G$ give a SubCubeSums proof. $\Box$

In fact, SubCubeSums is also implicationally complete in the following sense. We say that $f \geq g$ if for every Boolean $x$, $f(x) \geq g(x)$.

**Proposition 2.** *If $f$ and $g$ are polynomials with $f \geq g$, then there are subcubes $h_j$ and non-negative numbers $c_j$ such that on the Boolean hypercube, $f - g = \sum_j c_j h_j$. Further, if $f, g$ are integral on the Boolean hypercube, so are the $c_j$.*

# 3   MaxRes, MaxResW, and TreeRes

Since TreeRes allows reuse only of input clauses, while MaxRes does not allow any reuse of clauses but produces multiple clauses at each step, the relative power of these fragments of Res is intriguing. In this section, we show that MaxRes with the weakening rule, MaxResW, $p$-simulates TreeRes, is exponentially separated from it, and even MaxRes (without weakening) is not simulated by TreeRes.

**Lemma 1.** *For every unsatisfiable CNF $F$, size($F \vdash_{MaxResW} \Box$) $\leq 2size$ $(F \vdash_{TreeRes} \Box)$.*

*Proof.* Let $T$ be a tree-like derivation of $\Box$ from $F$ of size $s$. Without loss of generality, we may assume that $T$ is regular; no variable is used as pivot twice on the same path.

Since a MaxSAT resolution step always adds the standard resolvent, each step in a tree-like resolution proof can be performed in MaxResW as well, provided the antecedents are available. However, a tree-like proof may use an axiom (a clause in $F$) multiple times, whereas after it is used once in MaxResW it is no longer available, although some weakenings are available. So we need to work with weaker antecedents. We describe below how to obtain sufficient weakenings.

For each axiom $A \in F$, consider the subtree $T_A$ of $T$ defined by retaining only the paths from leaves labeled $A$ to the final empty clause. We will produce multiple disjoint weakenings of $A$, one for each leaf labelled $A$. Start with $A$ at the final node (where $T_A$ has the empty clause) and walk up the tree $T_A$ towards the leaves. If we reach a branching node $v$ with clause $A'$, and the pivot at $v$ is $x$, weaken $A'$ to $A' \vee x$ and $A' \vee \overline{x}$. Proceed along the edge contributing $x$ with $A' \vee x$, and along the other edge with $A' \vee \overline{x}$. Since $T$ is regular, no tautologies are created in this process, which ends with multiple "disjoint" weakenings of $A$.

After doing this for each axiom, we have as many clauses as leaves in $T$. Now we simply perform all the steps in $T$.

Since each weakening step increases the number of clauses by one, and since we finally produce at most $s$ clauses for the leaves, the number of weakening steps required is at most $s$.                                     □

We now show that even without weakening, MaxRes has short proofs of formulas exponentially hard for TreeRes. The formulas that exhibit the separation are *composed* formulas of the form $F \circ g$, where $F$ is a CNF formula, $g \colon \{0,1\}^\ell \to \{0,1\}$ is a Boolean function, there are $\ell$ new variables $x_1, \ldots, x_\ell$ for each original variable $x$ of $F$, and there is a block of clauses $C \circ g$, a CNF expansion of the expression $\bigvee_{x^b \in C} [\![ g(x_1, \ldots x_\ell) = b ]\!]$, for each original clause $C \in F$. We use the pebbling formulas on single-sink directed acyclic graphs: there is a variable for each node, variables at sources must be true, the variable at the sink must be false, and at each node $v$, if variables at origins of incoming edges are true, then the variable at $v$ must also be true.

We denote by PebHint($G$) the standard pebbling formula with additional hints $u \vee v$ for each pair of siblings $(u, v)$—that is, two incomparable vertices with a common predecessor—, and we prove the separation for PebHint($G$) composed with the OR function. More formally, if $G$ is a DAG with a single sink $z$, we define PebHint($G$) $\circ$ OR as follows. For each vertex $v \in G$ there are variables $v_1$ and $v_2$. The clauses are

- For each source $v$, the clause $v_1 \vee v_2$.
- For each internal vertex $w$ with predecessors $u, v$, the expression $((u_1 \vee u_2) \wedge (v_1 \vee v_2)) \to (w_1 \vee w_2)$, expanded into 4 clauses.
- The clauses $\overline{z_1}$ and $\overline{z_2}$ for the sink $z$.
- For each pair of siblings $(u, v)$, the clause $u_1 \vee u_2 \vee v_1 \vee v_2$.

Note that the first three types of clauses are also present in standard composed pebbling formulas, while the last type are the hints.

We prove a MaxRes upper bound for the particular case of pyramid graphs. Let $P_h$ be a pyramid graph of height $h$ and $n = \Theta(h^2)$ vertices.

**Lemma 2.** *The* PebHint($P_h$)$\circ$OR *formulas have* $\Theta(n)$ *size MaxRes refutations.*

*Proof.* We derive the clause $s_1 \vee s_2$ for each vertex $s \in P_n$ in layered order, and left-to-right within one layer. If $s$ is a source, then $s_1 \vee s_2$ is readily available as an axiom. Otherwise assume that for a vertex $s$ with predecessors $u$ and $v$ and siblings $r$ and $t$ – in this order – we have clauses $u_1 \vee u_2 \vee s_1 \vee s_2$ and $v_1 \vee v_2$, and let us see how to derive $s_1 \vee s_2$. (Except at the boundary, we don't have the clause $u_1 \vee u_2$ itself, since it has been used to obtain the sibling $r$ and doesn't exist anymore.) We also make sure that the clause $v_1 \vee v_2 \vee t_1 \vee t_2$ becomes available to be used in the next step.

In the following derivation we skip $\vee$ symbols, and we colour-code clauses so that green clauses are available by induction, axioms are blue, and red clauses, on the right side in steps with multiple consequents, are additional clauses that are obtained by the MaxRes rule but not with the usual resolution rule.

$$
\cfrac{
\cfrac{
\cfrac{\overline{u_1}\overline{v_1}s_1s_2 \quad u_1u_2s_1s_2}{u_2\overline{v_1}s_1s_2} \quad \cfrac{u_1u_2v_1s_1s_2 \quad \overline{u_1}\overline{v_2}s_1s_2}{u_2v_1\overline{v_2}s_1s_2}}{\cfrac{\overline{u_2}\overline{v_1}s_1s_2}{\overline{v_1}s_1s_2}} \quad \cfrac{\overline{u_2}\overline{v_2}s_1s_2}{v_1\overline{v_2}s_1s_2}
}{s_1s_2} \qquad \cfrac{v_1\overline{v_2}s_1s_2 \quad \cfrac{v_1v_2 \quad v_1v_2s_1\overline{s_2} \quad s_1s_2t_1t_2}{}}{v_1s_1s_2 \quad v_1v_2\overline{s_1} \quad v_1v_2s_1\overline{s_2} \quad s_1s_2t_1t_2}
$$

The case where some of the siblings are missing is similar: if $r$ is missing then we use the axiom $u_1 \vee u_2$ instead of the clause $u_1 \vee u_2 \vee s_1 \vee s_2$ that would be available by induction, and if $t$ is missing then we skip the steps that use $s_1 \vee s_2 \vee t_1 \vee t_2$ and lead to deriving $v_1 \vee v_2 \vee t_1 \vee t_2$.

Finally, once we derive the clause $z_1 \vee z_2$ for the sink, we resolve it with axiom clauses $\overline{z_1}$ and $\overline{z_2}$ to obtain a contradiction.

A constant number of steps suffice for each vertex, for a total of $\Theta(n)$.  □

We can prove a tree-like lower bound along the lines of [3], but with some extra care to respect the hints. As in [3] we use the *pebble game*, a game where the single player starts with a DAG and a set of pebbles, the allowed moves are to place a pebble on a vertex if all its predecessors have pebbles or to remove a pebble at any time, and the goal is to place a pebble on the sink using the minimum number of pebbles. Denote by bpeb$(P \to w)$ the cost of placing a pebble on a vertex $w$ assuming there are free pebbles on a set of vertices $P \subseteq V$ – in other words, the number of pebbles used outside of $P$ when the starting position has pebbles in $P$. For a DAG $G$ with a single sink $z$, bpeb$(G)$ denotes bpeb$(\emptyset \to z)$. For $U \subseteq V$ and $v \in V$, the subgraph of $v$ modulo $U$ is the set of vertices $u$ such that there exists a path from $u$ to $v$ avoiding $U$.

**Lemma 3** ([8]). bpeb$(P_h) = h + 1$.

**Lemma 4** ([3]). *For all $P, v, w$, we have*

$$\text{bpeb}(P \to v) \leq \max(\text{bpeb}(P \to w), \text{bpeb}(P \cup \{w\} \to v) + 1).$$

The canonical search problem of a formula $F$ is the relation Search($F$) where inputs are variable assignments $\alpha \in \{0,1\}^n$ and the valid outputs for $\alpha$ are the clauses $C \in F$ that $\alpha$ falsifies. Given a relation $f$, we denote by $\mathrm{DT}_1(f)$ the 1-query complexity of $f$ [17], that is the minimum over all decision trees computing $f$ of the maximum of 1-answers that the decision tree receives.

**Lemma 5.** *For all $G$ we have* $\mathrm{DT}_1(\mathrm{Search}(\mathrm{PebHint}(G))) \geq \mathrm{bpeb}(G) - 1$.

*Proof.* We give an adversarial strategy. Let $R_i$ be the set of variables that are assigned to 1 at round $i$. We initially set $w_0 = z$, and maintain the invariant that

1. there is a distinguished variable $w_i$ and a path $\pi_i$ from $w_i$ to the sink $z$ such that a queried variable $v$ is 0 iff $v \in \pi_i$; and
2. after each query the number of 1 answers so far is at least $\mathrm{bpeb}(G) - \mathrm{bpeb}(R_i \rightarrow w_i)$.

Assume that a variable $v$ is queried. If $v$ is not in the subgraph of $w_i$ modulo $R_i$ then we answer 0 if $v \in \pi_i$ and 1 otherwise. Otherwise we consider $p_0 = \mathrm{bpeb}(R_i \rightarrow v)$ and $p_1 = \mathrm{bpeb}(R_i \cup \{v\} \rightarrow w_i)$. By Lemma 4, $\mathrm{bpeb}(R_i \rightarrow w_i) \leq \max(p_0, p_1 + 1)$. If $p_0 \geq p_1$ then we answer 0, set $w_{i+1} = v$, and extend $\pi_i$ with a path from $w_{i+1}$ to $w_i$ that does not contain any 1 variables (which exists by definition of subgraph modulo $R_i$). This preserves item 1 of the invariant, and since $p_0 \geq \mathrm{bpeb}(R_i \rightarrow w_i)$, item 2 is also preserved. Otherwise we answer 1 and since $p_1 \geq \mathrm{bpeb}(R_i \rightarrow w_i) - 1$ the invariant is also preserved.

This strategy does not falsify any hint clause, because all 0 variables lie on a path, or the sink axiom, because the sink is assigned 0 if at all. Therefore the decision tree ends at a vertex $w_t$ that is set to 0 and all its predecessors are set to 1, hence $\mathrm{bpeb}(R_t \rightarrow w_t) = 1$. By item 2 of the invariant the number of 1 answers is at least $\mathrm{bpeb}(G) - 1$.                       $\square$

To complete the lower bound we use the Pudlák–Impagliazzo Prover–Delayer game [21] where Prover points to a variable, Delayer may answer 0, 1, or $*$, in which case Delayer obtains a point in exchange for letting Prover choose the answer, and the game ends when a clause is falsified.

**Lemma 6** ([21]). *If Delayer can win $p$ points, then all TreeRes proofs require size at least $2^p$.*

**Lemma 7.** *$F \circ \mathrm{OR}$ requires size* $\exp(\Omega(\mathrm{DT}_1(\mathrm{Search}(F))))$ *in tree-like resolution.*

*Proof.* We use a strategy for the 1-query game of Search($F$) to ensure that Delayer gets $\mathrm{DT}_1(F)$ points in the Prover–Delayer game. If Prover queries a variable $x_i$ then

- If $x$ is already queried we answer accordingly.
- Otherwise we query $x$. If the answer is 0 we answer 0, otherwise we answer $*$.

Our strategy ensures that if both $x_1$ and $x_2$ are assigned then $x_1 \vee x_2 = x$. Therefore the game only finishes at a leaf of the decision tree, at which point Delayer earns as many points as 1s are present in the path leading to the leaf. The lemma follows by Lemma 6.                                                          □

The formulas $\mathrm{PebHint}(P_n) \circ \mathrm{OR}$ are easy to refute in MaxRes (Lemma 2), but from Lemmas 3, 5, and 7, they are exponentially hard for TreeRes. Hence,

**Theorem 1.** *TreeRes does not simulate MaxResW and MaxRes.*

## 4    The SubCubeSums Proof System

In this section, we explore the power and limitations of the SubCubeSums proof system. On the one hand we show (Theorem 2) that it has short proofs of the subset cardinality formulas, known to be hard for resolution but easy for Sherali–Adams. On the other hand we show a lower bound for SubCubeSums for the Tseitin formulas on odd-charged expander graphs (Theorem 3). Finally, we establish a technique for obtaining lower bounds on SubCubeSums size: a degree lower bound in SubCubeSums for $F$ translates to a size lower bound in SubCubeSums for $F \circ \oplus$ (Theorem 4).

### 4.1    Res Does Not Simulate SubCubeSums

We now show that Res does not simulate SubCubeSums.

**Theorem 2.** *There are formulas that have SubCubeSums proofs of size $O(n)$ but require resolution length $\exp(\Omega(n))$.*

The separation is achieved using subset cardinality formulas [19,23,25]. These are defined as follows: we have a bipartite graph $G(U \cup V, E)$, with $|U| = |V| = n$. The degree of $G$ is 4, except for two vertices that have degree 5. There is one variable for each edge. For each left vertex $u \in U$ we have a constraint $\sum_{e \ni u} x_e \geq \lceil d(u)/2 \rceil$, while for each right vertex $v \in V$ we have a constraint $\sum_{e \ni v} x_e \leq \lfloor d(v)/2 \rfloor$, both expressed as a CNF. In other words, for each vertex $u \in U$ we have the clauses $\bigvee_{i \in I} x_i$ for $I \in \binom{E(u)}{\lfloor d(u)/2 \rfloor + 1}$, while for each vertex $v \in V$ we have the clauses $\bigvee_{i \in I} \overline{x_i}$ for $I \in \binom{E(v)}{\lfloor d(v)/2 \rfloor + 1}$.

The lower bound requires $G$ to be an expander, and is proven in [19, Theorem 6]. The upper bound is the following lemma.

**Lemma 8.** *Subset cardinality formulas have SubCubeSums proofs of size $O(n)$.*

*Proof.* Our plan is to reconstruct each constraint independently, so that for each vertex we obtain the original constraints $\sum_{e \ni u} x_e \geq \lceil d(u)/2 \rceil$ and $\sum_{e \ni v} \overline{x_e} \geq \lceil d(v)/2 \rceil$, and then add all of these constraints together.

Formally, if $F_u$ is the set of polynomials that encode the constraint corresponding to vertex $u$, we want to write the equations

$$\sum_{f \in F_u} f - \left( \lceil d(u)/2 \rceil - \sum_{e \ni u} x_e \right) = \sum_j c_{u,j} h_j \tag{1}$$

and

$$\sum_{f \in F_v} f - \left( \lceil d(v)/2 \rceil - \sum_{e \ni v} \overline{x_e} \right) = \sum_j c_{v,j} h_j \tag{2}$$

with $c_{u,j}, c_{v,j} \geq 0$ and $\sum_j c_{u,j} = O(1)$, so that with $c_j = \sum_{v \in U \cup V} c_{v,j}$, we get

$$\sum_{f \in F} f = \sum_{u \in U} \sum_{f \in F_u} f + \sum_{v \in V} \sum_{f \in F_v} f$$

$$= \sum_{u \in U} \left( \lceil d(u)/2 \rceil - \sum_{e \ni u} x_e + \sum_j c_{u,j} h_j \right)$$

$$+ \sum_{v \in V} \left( \lceil d(v)/2 \rceil - \sum_{e \ni v} \overline{x_e} + \sum_j c_{v,j} h_j \right)$$

$$= \sum_{u \in U} \lceil d(u)/2 \rceil + \sum_{v \in V} \lceil d(v)/2 \rceil - \sum_{e \in E} (x_e + \overline{x_e}) + \sum_j c_j h_j$$

$$= \left( 1 + \sum_{u \in U} 2 \right) + \left( 1 + \sum_{v \in V} 2 \right) - \sum_{e \in E} 1 + \sum_j c_j h_j$$

$$= (2n + 1) + (2n + 1) - (4n + 1) + \sum_j c_j h_j = 1 + \sum_j c_j h_j$$

Hence we can write $\sum_{f \in F} f - 1 = \sum_j c_j h_j$ with $\sum_j c_j = O(n)$ and each $c_j \geq 0$.

It remains to show how to derive Eqs. (1) and (2). The easiest way is to appeal to the implicational completeness of SubCubeSums, Proposition 2. We continue deriving Eq. (1), assuming for simplicity a vertex of degree $d$ and incident edges $[d]$. Let $\overline{x_I} = \prod_{i \in I} \overline{x_i}$, and let $\left\{ \overline{x_I} : I \in \binom{[d]}{d-k+1} \right\}$ represent a constraint $\sum_{i \in [d]} x_i \geq k$. Let $f = \sum_{I \in \binom{[d]}{d-k+1}} \overline{x_I}$ and $g = k - \sum_{i \in [d]} x_i$. For each point $x \in \{0, 1\}^d$ we have that either $x$ satisfies the constraint, in which case $f(x) \geq 0 \geq g(x)$, or it falsifies it, in which case we have on the one hand $g(x) = s > 0$, and on the other hand $f(x) = \binom{d-k+s}{d-k+1} = \frac{(d-k+s) \cdots s}{(d-k+1) \cdots 1} \geq s$.

We proved that $f \geq g$, therefore by Proposition 2 we can write $f - g$ as a sum of subcubes of size at most $2^d = O(1)$.

Equation (2) can be derived analogously, completing the proof.    □

## 4.2    A Lower Bound for SubCubeSums

Fix any graph $G$ with $n$ nodes and $m$ edges, and let $I$ be the node-edge incidence matrix. Assign a variable $x_e$ for each edge $e$. Let $b$ be a vector in $\{0, 1\}^n$ with

$\sum_i b_i \equiv 1 \bmod 2$. The Tseitin contradiction asserts that the system $IX = b$ has a solution over $\mathbb{F}_2$. The CNF formulation has, for each vertex $u$ in $G$, with degree $d_u$, a set $S_u$ of $2^{d_u-1}$ clauses expressing that the parity of the set of variables $\{x_e \mid e \text{ is incident on } u\}$ equals $b_u$.

These formulas are exponentially hard for Res [24], and hence are also hard for MaxResW. We now show that they are also hard for SubCubeSums. By Theorem 2, this lower bound cannot be inferred from hardness for Res.

We will use some standard facts: For connected graph $G$, over $\mathbb{F}_2$, if $\sum_i b_i \equiv 1 \bmod 2$, then the equations $IX = b$ have no solution, and if $\sum_i b_i \equiv 0 \bmod 2$, then $IX = b$ has exactly $2^{m-n+1}$ solutions. Furthermore, for any assignment $a$, and any vertex $u$, $a$ falsifies at most one clause in $S_u$.

A graph is a $c$-expander if for all $V' \subseteq V$ with $|V'| \leq |V|/2$, $|\delta(V')| \geq c|V'|$, where $\delta(V') = \{(u, v) \in E \mid u \in V', v \in V \setminus V'\}$.

**Theorem 3.** *Tseitin contradictions on odd-charged expanders require exponential size SubCubeSums refutations.*

*Proof.* Fix a graph $G$ that is a $d$-regular $c$-expander on $n$ vertices, where $n$ is odd; $m = dn/2$. Let $b$ be the all-1s vector. The Tseitin contradiction $F$ has $n2^{d-1}$ clauses. By the facts mentioned above, for all $a \in \{0,1\}^m$, $\operatorname{viol}_F(a)$ is odd. So $\operatorname{viol}_F$ partitions $\{0,1\}^m$ into $X_1, X_3, \ldots, X_{N-1}$, where $X_i = \operatorname{viol}_F^{-1}(i)$.

Let $\mathcal{C}$ be a SubCubeSums refutation of $F$, that is, $\operatorname{viol}_{\mathcal{C}} = \operatorname{viol}_F - 1 = g$, say. For a cube $C$, define $N_i(C) = |C \cap X_i|$. Then for all $C \in \mathcal{C}$, $N_1(C) = 0$, and so $C$ is partitioned by $X_i$, $i \geq 3$. Let $\mathcal{C}'$ be those cubes of $\mathcal{C}$ that have a non-empty part in $X_3$. We will show that $\mathcal{C}'$ is large. In fact, we will show that for a suitable $S$, the set $\mathcal{C}'' \subseteq \mathcal{C}'$ of cubes with $|C \cap X_5| \leq S|C \cap X_3|$ is large.

Defining the probability distribution $\mu$ on $\mathcal{C}'$ as

$$\mu(C) = \frac{|C \cap X_3|}{\sum_{D \in \mathcal{C}'} |D \cap X_3|} = \frac{N_3(C)}{\sum_{D \in \mathcal{C}'} N_3(D)},$$

$$|\mathcal{C}'| = \mathop{\mathbb{E}}_{C \sim \mu} \left[ \frac{1}{\mu(C)} \right] \geq \underbrace{\mathop{\mathbb{E}}_{C \sim \mu} \left[ \frac{1}{\mu(C)} \frac{|C \cap X_5|}{|C \cap X_3|} \leq S \right]}_{A} \cdot \underbrace{\Pr_{\mu} \left[ \frac{|C \cap X_5|}{|C \cap X_3|} \leq S \right]}_{B} \tag{3}$$

We want to choose a good value for $S$ so that $A$ is very large, and $B$ is $\Theta(1)$. To see what will be a good value for $S$, we estimate the expected value of $\frac{|C \cap X_5|}{|C \cap X_3|}$ and then use Markov's inequality. For this, we should understand the sets $X_3$, $X_5$ better. These set sizes are known precisely: for each odd $i$, $|X_i| = \binom{n}{i} 2^{m-n+1}$.

Now let us consider $C \cap X_3$ and $C \cap X_5$ for some $C \in \mathcal{C}'$. We rewrite the system $IX = b$ as $I'X' + I_C X_C = b$, where $X_C$ are the variables fixed in cube $C$ (to $a_C$, say). So $I'X' = b + I_C a_C$. An assignment $a$ is in $C \cap X_r$ iff it is of the form $a'a_C$, and $a'$ falsifies exactly $r$ equations in $I'X' = b'$ where $b' = b + I_C a_C$. This is a system for the subgraph $G_C$ where the edges in $X_C$ have been deleted. This subgraph may not be connected, so we cannot use our size expressions directly. Consider the vertex sets $V_1, V_2, \ldots$ of the components of $G_C$. The system $I'X' = b'$ can be broken up into independent systems; $I'(i)X'(i) = b'(i)$ for the *i*th connected component. Say a component is odd if $\sum_{j\in V_i} b'(i)_j \equiv 1 \mod 2$, even otherwise. Let $|V_i| = n_i$ and $|E_i| = m_i$. Any $a'$ falsifies an odd/even number of equations in an odd/even component.

For $a' \in C \cap X_3$, it must falsify three equations overall, so $G_C$ must have either one or three odd components. If it has only one odd component, then there is another assignment in $C$ falsifying just one equation (from this odd component), so $C \cap X_1 \neq \emptyset$, a contradiction. Hence $G_C$ has exactly three odd components, with vertex sets $V_1, V_2, V_3$, and overall $k \geq 3$ components. An $a \in C \cap X_3$ falsifies exactly one equation in $I(1), I(2), I(3)$. We thus arrive at the expression

$$|C \cap X_3| = \left( \prod_{i=1}^{3} n_i 2^{m_i - n_i + 1} \right) \left( \prod_{i \geq 4} 2^{m_i - n_i + 1} \right) = n_1 n_2 n_3 2^{m - w(C) - n + k}.$$

Similarly, an $a' \in C \cap X_5$ must falsify five equations overall. One each must be from $V_1, V_2, V_3$. The remaining 2 must be from the same component. Hence

$$|C \cap X_5| = n_1 n_2 n_3 \left( \sum_{i=1}^{3} \binom{n_i}{3} \frac{1}{n_i} + \sum_{i=4}^{k} \binom{n_i}{2} \right) 2^{m - w(C) - n + k}$$

$$\geq n_1 n_2 n_3 2^{m - w(C) - n + k} \left( \frac{1}{3} \sum_{i=1}^{k} \binom{n_i - 1}{2} \right)$$

Hence we have, for $C \in \mathcal{C}'$, $\dfrac{|C \cap X_5|}{|C \cap X_3|} \geq \dfrac{1}{3} \sum_{i=1}^{k} \binom{n_i - 1}{2}$.

We can deduce more by using the definition of $\mu$, and the following fact: Since $g = \text{viol}_F - 1$, an assignment in $X_3$ belongs to exactly two cubes in $\mathcal{C}$, and by definition these cubes are in $\mathcal{C}'$. Similarly, an assignment in $X_5$ belongs to exactly four cubes in $\mathcal{C}$, not all of which may be in $\mathcal{C}'$. Hence

$$\sum_{C \in \mathcal{C}'} |C \cap X_3| = 2|X_3| = 2\binom{n}{3} 2^{m-n+1}; \qquad \mu(C) = \frac{|C \cap X_3|}{2|X_3|}.$$

$$\sum_{C \in \mathcal{C}'} |C \cap X_5| \leq 4|X_5| = 4\binom{n}{5} 2^{m-n+1}.$$

Now we can estimate the average:

$$\mathbb{E}_\mu \left[ \frac{|C \cap X_5|}{|C \cap X_3|} \right] = \sum_{C \in \mathcal{C}'} \mu(C) \frac{|C \cap X_5|}{|C \cap X_3|} = \sum_{C \in \mathcal{C}'} \frac{|C \cap X_5|}{2|X_3|} \leq \frac{4|X_5|}{2|X_3|} \leq \frac{n^2}{10}$$

At $S = n^2/9$, by Markov's inequality, $B = \Pr_\mu \left[ \dfrac{|C \cap X_5|}{|C \cap X_3|} \leq S = \dfrac{n^2}{9} \right] \geq 1/10.$

Now we show that conditioned on $\frac{|C \cap X_5|}{|C \cap X_3|} \leq S$, the average value of $\frac{1}{\mu(C)}$ is large.

$$\frac{1}{\mu(C)} = \frac{2|X_3|}{|C \cap X_3|} = \frac{2\binom{n}{3} 2^{m-n+1}}{n_1 n_2 n_3 2^{m-w(C)-n+k}} = \frac{2\binom{n}{3} 2^{w(C)+1-k}}{n_1 n_2 n_3} \geq \frac{2^{w(C)+1-n}}{3}$$

So we must show that $w(C)$ must be large. Each literal in $C$ removes one edge from $G$ while constructing $G_C$. Counting the sizes of the cuts that isolate components of $G_C$, we count each deleted edge twice. So

$$2w(C) = \sum_{i=1}^{k} |\delta(V_i, V \setminus V_i)| = \underbrace{\sum_{i:n_i \leq n/2} |\delta(V_i, V \setminus V_i)|}_{Q1} + \underbrace{\sum_{i:n_i > n/2} |\delta(V_i, V \setminus V_i)|}_{Q2}$$

By the $c$-expansion property of $G$, $Q1 \geq cn_i$.
If $n_i > n/2$, it still cannot be too large because of the conditioning. Recall

$$S = \frac{n^2}{9} \geq \frac{|C \cap X_5|}{|C \cap X_3|} \geq \frac{1}{3} \sum_{i=1}^{k} \binom{n_i - 1}{2}$$

So each $n_i \leq 5n/6$. Thus even when $n_i > n/2$, we can conclude that $n_i/5 \leq n/6 \leq n - n_i < n/2$. By expansion of $V \setminus V_i$, we have $Q2 \geq c(n - n_i) \geq cn_i/5$.

$$2w(C) = \underbrace{\sum_{i:n_i \leq n/2} |\delta(V_i, V \setminus V_i)|}_{Q1} + \underbrace{\sum_{i:n_i > n/2} |\delta(V_i, V \setminus V_i)|}_{Q2} \geq \frac{cn}{5}$$

Choose $c$-expanders where $c$ ensures $w(C) + 1 - n = \Omega(n)$. (Any constant $c > 10$.) Going back to our calculation of $A$ from Eq. 3),

$$A = \mathop{\mathbb{E}}_{C \sim \mu}\left[\frac{1}{\mu(C)} \left\|\frac{|C \cap X_5|}{|C \cap X_3|} \leq S\right\|\right] \geq \mathop{\mathbb{E}}_{C \sim \mu}\left[\frac{2^{w(C)+1-n}}{3} \left\|\frac{|C \cap X_5|}{|C \cap X_3|} \leq S\right\|\right] = 2^{\Omega(n)}$$

for suitable $c > 10$. Thus $|\mathcal{C}| \geq |\mathcal{C}'| \geq A \cdot B \geq 2^{\Omega(n)} \cdot (1/10)$.    □

### 4.3    Lifting Degree Lower Bounds to Size

We describe a general technique to lift lower bounds on conical junta degree to size lower bounds for SubCubeSums.

**Theorem 4.** *Let $d$ be the minimum degree of a SubCubeSums refutation of an unsatisfiable CNF formula $F$. Then every SubCubeSums refutation of $F \circ \oplus$ has size $\exp(\Omega(d))$.*

Before proving this theorem, we establish two lemmas. For a function $h$: $\{0,1\}^n \rightarrow \mathbb{R}$, define the function $h \circ \oplus : \{0,1\}^{2n} \rightarrow \mathbb{R}$ as $(h \circ \oplus)(\alpha_1, \alpha_2) = h(\alpha_1 \oplus \alpha_2)$, where $\alpha_1, \alpha_2 \in \{0,1\}^n$ and the $\oplus$ in $\alpha_1 \oplus \alpha_2$ is taken bitwise.

**Lemma 9.** $\mathrm{viol}_F(\alpha_1 \oplus \alpha_2) = \mathrm{viol}_{F \circ \oplus}(\alpha_1, \alpha_2)$.

*Proof.* Fix assignments $\alpha_1$, $\alpha_2$ and let $\alpha = \alpha_1 \oplus \alpha_2$. We claim that for each clause $C \in F$ falsified by $\alpha$ there is exactly one clause $D \in F \circ \oplus$ that is falsified by $\alpha_1\alpha_2$. Indeed, by the definition of composed formula the assignment $\alpha_1\alpha_2$ falsifies $C \circ \oplus$, hence the assignment falsifies some clause $D \in C \circ \oplus$. However, the clauses in the CNF expansion of $C \circ \oplus$ have disjoint subcubes, hence $\alpha_1\alpha_2$ falsifies at most one clause from the same block. Observing that if $\alpha$ does not falsify $C$, then $\alpha_1\alpha_2$ does not falsify any clause in $C \circ \oplus$ completes the proof. □

**Corollary 1.** $\mathrm{viol}_{F \circ \oplus} - 1 = ((\mathrm{viol}_F) \circ \oplus) - 1 = (\mathrm{viol}_F - 1) \circ \oplus$.

**Lemma 10.** *If $f \circ \oplus_2$ has a (integral) conical junta of size $s$, then $f$ has a (integral) conical junta of degree $d = \mathrm{O}(\log s)$.*

*Proof.* Let $J$ be a conical junta of size $s$ that computes $f \circ \oplus_2$. Let $\rho$ be the following random restriction: for each variable $x$ of $f$, pick $i \in \{0, 1\}$ and $b \in \{0, 1\}$ uniformly and set $x_i = b$. Consider a term $C$ of $J$ of degree at least $d > \log_{4/3} s$. The probability that $C$ is not zeroed out by $\rho$ is at most $(3/4)^d < 1/s$, hence the probability that the junta $J{\upharpoonright}_\rho$ has degree larger than $d$ is at most $s \cdot (3/4)^d < 1$. Hence there is a restriction $\rho$ such that $J{\upharpoonright}_\rho$ is a junta of degree at most $d$, although not one that computes $f$. Since for each original variable $x$, $\rho$ sets exactly one of the variables $x_0, x_1$, flipping the appropriate surviving variables—those where $x_i$ is set to 1—gives a junta of degree at most $d$ for $f$. $\square$

*Proof* (of Theorem 4). We prove the contrapositive. Assume $F \circ \oplus$ has a Sub-CubeSums proof of size $s$. Let $H$ be the collection of $s$ cubes in this proof. So $\mathrm{viol}_{F \circ \oplus} - 1 = \mathrm{viol}_H$. By Corollary 1, there is an integral conical junta for $(\mathrm{viol}_F - 1) \circ \oplus$ of size $s$. By Lemma 10 there is an integral conical junta for $\mathrm{viol}_F - 1$ of degree $\mathrm{O}(\log s)$.                                           $\square$

*Recovering the Tseitin lower bound:* This theorem, along with the $\Omega(n)$ conical junta degree lower bound of [10], yields an exponential lower bound for the SubCubeSums and MaxResW refutation size for Tseitin contradictions.

*A candidate for separating Res from SubCubeSums:* We conjecture that the SubCubeSums degree of the pebbling contradiction on the pyramid graph, or on a minor modification of it (a stack of butterfly networks, say, at the base of a pyramid), is $n^{\Omega(1)}$. This, with Theorem 4, would imply that $F \circ \oplus$ is hard for Sub-CubeSums, thereby separating it from Res. We have not yet been able to prove the desired degree lower bound. We do know that SubCubeSums degree is not exactly the same as Res width – for small examples, a brute-force computation has shown SubCubeSums degree to be strictly larger than Res width.

## 5    Discussion

We placed MaxRes(W) in a propositional proof complexity frame and compared it to more standard proof systems, showing that MaxResW is between tree-like resolution (strictly) and resolution. With the goal of also separating MaxRes and resolution we devised a new lower bound technique, captured by SubCubeSums, and proved lower bounds for MaxRes without relying on Res lower bounds.

Perhaps the most conspicuous open problem is whether our conjecture that pebbling contradictions composed with XOR separate Res and SubCubeSums holds. It also remains open to show that MaxRes simulates TreeRes – or even MaxResW – or that they are incomparable instead.

**Acknowledgments.** Part of this work was done when the last author was at TIFR, Mumbai, India. Some of this work was done in the conducive academic environs of the Chennai Mathematical Institute (during the CAALM workshop of CNRS UMI ReLaX, 2019), Banff International Research Station BIRS (seminar 20w5144) and Schloss Dagstuhl Leibniz Centre for Informatics (seminar 20061). The authors thank Susanna de Rezende, Tuomas Hakoniemi, and Aaron Potechin for useful discussions.

# References

1. Atserias, A., Hakoniemi, T.: Size-degree trade-offs for sums-of-squares and Positivstellensatz proofs. In: Proceedings of the 34th Computational Complexity Conference (CCC 2019), pp. 24:1–24:20, July 2019
2. Atserias, A., Lauria, M.: Circular (yet sound) proofs. In: Janota, M., Lynce, I. (eds.) SAT 2019. LNCS, vol. 11628, pp. 1–18. Springer, Cham (2019). https://doi.org/10.1007/978-3-030-24258-9_1
3. Ben-Sasson, E., Impagliazzo, R., Wigderson, A.: Near optimal separation of tree-like and general resolution. Combinatorica **24**(4), 585–603 (2004)
4. Ben-Sasson, E., Wigderson, A.: Short proofs are narrow–resolution made simple. J. ACM **48**(2), 149–169 (2001). Preliminary version in STOC 1999
5. Blake, A.: Canonical expressions in boolean algebra. Ph.D. thesis, University of Chicago (1937)
6. Bonet, M.L., Buss, S., Ignatiev, A., Marques-Silva, J., Morgado, A.: MaxSAT resolution with the dual rail encoding. In: Proceedings of the 32nd AAAI Conference on Artificial Intelligence, (AAAI 2018), pp. 6565–6572 (2018)
7. Bonet, M.L., Levy, J., Manyà, F.: Resolution for Max-SAT. Artif. Intell. **171**(8), 606–618 (2007)
8. Cook, S.A.: An observation on time-storage trade off. J. Comput. Syst. Sci. **9**(3), 308–316 (1974). Preliminary version in STOC 1973
9. Fleming, N., Kothari, P., Pitassi, T.: Semialgebraic proofs and efficient algorithm design. Found. Trends Theor. Comput. Sci. **14**(1–2), 1–221 (2019)
10. Göös, M., Jain, R., Watson, T.: Extension complexity of independent set polytopes. SIAM J. Comput. **47**(1), 241–269 (2018)
11. Göös, M., Lovett, S., Meka, R., Watson, T., Zuckerman, D.: Rectangles are nonnegative juntas. SIAM J. Comput. **45**(5), 1835–1869 (2016). Preliminary version in STOC 2015
12. Ignatiev, A., Morgado, A., Marques-Silva, J.: On tackling the limits of resolution in SAT solving. In: Gaspers, S., Walsh, T. (eds.) SAT 2017. LNCS, vol. 10491, pp. 164–183. Springer, Cham (2017). https://doi.org/10.1007/978-3-319-66263-3_11
13. Iwama, K., Miyano, E.: Intractability of read-once resolution. In: Structure in Complexity Theory Conference, pp. 29–36. IEEE Computer Society (1995)
14. Larrosa, J., Heras, F., de Givry, S.: A logical approach to efficient Max-SAT solving. Artif. Intell. **172**(2–3), 204–233 (2008)
15. Larrosa, J., Rollon, E.: Augmenting the power of (partial) MaxSAT resolution with extension. In: Proceedings of the 34th AAAI Conference on Artificial Intelligence (2020)
16. Larrosa, J., Rollon, E.: Towards a better understanding of (partial weighted) MaxSAT proof systems. Technical report. 2003.02718. arXiv.org (2020)
17. Loff, B., Mukhopadhyay, S.: Lifting theorems for equality. In: Proceedings of the 36th Symposium on Theoretical Aspects of Computer Science (STACS 2019), pp. 50:1–50:19, March 2019

18. Marques-Silva, J., Ignatiev, A., Morgado, A.: Horn maximum satisfiability: reductions, algorithms and applications. In: Oliveira, E., Gama, J., Vale, Z., Lopes Cardoso, H. (eds.) EPIA 2017. LNCS (LNAI), vol. 10423, pp. 681–694. Springer, Cham (2017). https://doi.org/10.1007/978-3-319-65340-2_56

19. Mikša, M., Nordström, J.: Long proofs of (seemingly) simple formulas. In: Sinz, C., Egly, U. (eds.) SAT 2014. LNCS, vol. 8561, pp. 121–137. Springer, Cham (2014). https://doi.org/10.1007/978-3-319-09284-3_10

20. Narodytska, N., Bacchus, F.: Maximum satisfiability using core-guided MaxSAT resolution. In: Proceedings of the 28th AAAI Conference on Artificial Intelligence, pp. 2717–2723 (2014)

21. Pudlák, P., Impagliazzo, R.: A lower bound for DLL algorithms for $k$-SAT (preliminary version). In: Proceedings of the 11th Annual ACM-SIAM Symposium on Discrete Algorithms (SODA 2000), pp. 128–136, January 2000

22. Robinson, J.A.: A machine-oriented logic based on the resolution principle. J. ACM **12**, 23–41 (1965)

23. Spence, I.: sgen1: a generator of small but difficult satisfiability benchmarks. J. Exp. Algorithmics **15**, 1.2:1–1.2:15 (2010)

24. Urquhart, A.: Hard examples for resolution. J. ACM **34**(1), 209–219 (1987)

25. Van Gelder, A., Spence, I.: Zero-one designs produce small hard SAT instances. In: Strichman, O., Szeider, S. (eds.) SAT 2010. LNCS, vol. 6175, pp. 388–397. Springer, Heidelberg (2010). https://doi.org/10.1007/978-3-642-14186-7_37

# A Lower Bound on DNNF Encodings
# of Pseudo-Boolean Constraints

Alexis de Colnet$^{(\boxtimes)}$

CRIL, CNRS & Univ Artois, Lens, France
decolnet@cril.fr

**Abstract.** Two major considerations when encoding pseudo-Boolean (PB) constraints into SAT are the size of the encoding and its propagation strength, that is, the guarantee that it has a good behaviour under unit propagation. Several encodings with propagation strength guarantees rely upon prior compilation of the constraints into DNNF (decomposable negation normal form), BDD (binary decision diagram), or some other sub-variants. However it has been shown that there exist PB-constraints whose ordered BDD (OBDD) representations, and thus the inferred CNF encodings, all have exponential size. Since DNNFs are more succinct than OBDDs, preferring encodings via DNNF to avoid size explosion seems a legitimate choice. Yet in this paper, we prove the existence of PB-constraints whose DNNFs all require exponential size.

**Keywords:** PB constraints · Knowledge compilation · DNNF

## 1 Introduction

Pseudo-Boolean (PB) constraints are Boolean functions over $0/1$ Boolean variables $x_1, \ldots, x_n$ of the form $\sum_{i=1}^{n} w_i x_i$ 'op' $\theta$ where the $w_i$ are integer weights, $\theta$ is an integer threshold and 'op' is a comparison operator $<, \leq, >$ or $\geq$. PB-constraints have been studied extensively under different names (e.g. threshold functions [14], Knapsack constraints [13]) due to their omnipresence in many domains of AI and their wide range of practical applications [3,7,9,15,21].

One way to handle PB-constraints in a constraint satisfaction problem is to translate them into a CNF formula and feed it to a SAT solver. The general idea is to generate a CNF, possibly introducing auxiliary Boolean variables, whose restriction to variables of the constraint is equivalent to the constraint. Two major considerations here are the size of the CNF encoding and its propagation strength. One wants, on the one hand, to avoid the size of the encoding to explode, and on the other hand, to guarantee a good behaviour of the SAT instance under unit propagation – a technique at the very core of SAT solving. Desired propagation strength properties are, for instance, generalized arc consistency (GAC) [4] or propagation completeness (PC) [6]. Several encodings to CNF follow the same two-steps method: first, each constraint is represented in a compact form such as BDD (Binary Decision Diagram) or DNNF (Decomposable Negation Normal Form).

© Springer Nature Switzerland AG 2020
L. Pulina and M. Seidl (Eds.): SAT 2020, LNCS 12178, pp. 312–321, 2020.
https://doi.org/10.1007/978-3-030-51825-7_22

Second, the compact forms are turned into CNFs using Tseitin or other transformations. The SAT instance is the conjunction of all obtained CNFs. It is worth mentioning that there are GAC encodings of PB-constraints into polynomial size CNFs that do not follow this two-steps method [5]. However no similar result is known for PC encodings. PC encodings are more restrictive that GAC encodings and may be obtained via techniques requiring compilation to DNNF [17]. Thus the first step is a knowledge compilation task.

Knowledge compilation studies different representations for knowledge [10, 19] under the general idea that some representations are more suitable than others when solving specific reasoning problems. One observation that has been made is that the more reasoning tasks can be solved efficiently with particular representations, the larger these representations get in size. In the context of constraint encodings to SAT, the conversion of compiled forms to CNFs does not reduce the size of the SAT instance, therefore it is essential to control the size of the representations obtained by knowledge compilation.

Several representations have been studied with respect to different encoding techniques with the purpose of determining which properties of representations are sufficient to ensure propagation strength [1,2,11,12,16,17]. Popular representations in this context are DNNF and BDD and their many variants: deterministic DNNF, smooth DNNF, OBDD... As mentioned above, a problem occurring when compiling a constraint into such representations is that exponential space may be required. Most notably, it has been shown in [2,14] that some PB-constraints can only be represented by OBDDs whose size is exponential in $\sqrt{n}$, where $n$ is the number of variables. Our contribution is the proof of the following theorem where we lift the statement from OBDD to DNNF.

**Theorem 1.** *There is a class of PB-constraints $\mathcal{F}$ such that for any constraint $f \in \mathcal{F}$ on $n^2$ variables, any DNNF representation of $f$ has size $2^{\Omega(n)}$.*

Since DNNFs are exponentially more succinct than OBDDs [10], our result is a generalisation of the result in [2,14]. The class $\mathcal{F}$ is similar to that used in [2,14], actually the only difference is the choice of the threshold for the PB-constraints. Yet, adapting proofs given in [2,14] for OBDD to DNNF is not straightforward, thus our proof of Theorem 1 bears very little resemblance.

It has been shown in [18] that there exist sets of PB-constraints such that the whole *set* (so a conjunction of PB-constraints) requires exponential size DNNF to represent. Our result is a generalisation to *single* PB-constraints.

## 2   Preliminaries

**Conventions of Notation.** Boolean variables are seen as variables over $\{0, 1\}$, where 0 and 1 represent *false* and *true* respectively. Via this 0/1 representation, Boolean variables can be used in arithmetic expressions over $\mathbb{Z}$. For notational convenience, we keep the usual operators $\neg$, $\vee$ and $\wedge$ to denote, respectively, the negation, disjunction and conjunction of Boolean variables or functions. Given $X$ a set of $n$ Boolean variables, assignments to $X$ are seen as vectors in $\{0, 1\}^n$.

Single Boolean variables are written in plain text $(x)$ while assignments to several variables are written in bold $(\mathbf{x})$. We write $\mathbf{x} \leq \mathbf{y}$ when the vector $\mathbf{y}$ dominates $\mathbf{x}$ element-wise. We write $\mathbf{x} < \mathbf{y}$ when $\mathbf{x} \leq \mathbf{y}$ and $\mathbf{x} \neq \mathbf{y}$. In this framework, a Boolean function $f$ over $X$ is a mapping from $\{0,1\}^n$ to $\{0,1\}$. $f$ is said to *accept* an assignment $\mathbf{x}$ when $f(\mathbf{x}) = 1$, then $\mathbf{x}$ is called a *model* of $f$. The function is *monotone* if for any model $\mathbf{x}$ of $f$, all $\mathbf{y} \geq \mathbf{x}$ are models of $f$ as well. The set of models of $f$ is denoted $f^{-1}(1)$. Given $f$ and $g$ two Boolean functions over $X$, we write $f \leq g$ when $f^{-1}(1) \subseteq g^{-1}(1)$. We write $f < g$ when the inclusion is strict.

**Pseudo-Boolean Constraints.** *Pseudo-Boolean (PB) constraints* are inequalities the form $\sum_{i=1}^n w_i x_i$ 'op' $\theta$ where the $x_i$ are $0/1$ Boolean variables, the $w_i$ and $\theta$ are integers, and 'op' is one of the comparison operator $<, \leq, >$ or $\geq$. A PB-constraint is associated with a Boolean function whose models are exactly the assignments to $\{x_1, \ldots, x_n\}$ that satisfy the inequality. For simplicity we directly consider PB-constraints as Boolean functions – although the same function may represent different constraints – while keeping the term "constraints" when referring to them. In this paper, we restrict our attention to PB-constraints where 'op' is $\geq$ and all weights are positive integers. Note that such PB-constraints are monotone Boolean functions. Given a sequence of positive integer weights $W = (w_1, \ldots, w_n)$ and an integer threshold $\theta$, we define the function $w : \{0,1\}^n \to \mathbb{N}$ that maps any assignment to its weight by $w(\mathbf{x}) = \sum_{i=1}^n w_i x_i$. With these notations, a PB-constraint over $X$ for a given pair $(W, \theta)$ is a Boolean function whose models are exactly the $\mathbf{x}$ such that $w(\mathbf{x}) \geq \theta$.

*Example 1.* Let $n = 5$, $W = (1, 2, 3, 4, 5)$ and $\theta = 9$. The PB-constraint for $(W, \theta)$ is the Boolean function whose models are the assignments such that $\sum_{i=1}^5 i x_i \geq 9$. E.g. $\mathbf{x} = (0, 1, 1, 0, 1)$ is a model of weight $w(\mathbf{x}) = 10$.

For notational clarity, given any subset $Y \subseteq X$ and denoting $\mathbf{x}|_Y$ the restriction of $\mathbf{x}$ to variables of $Y$, we overload $w$ so that $w(\mathbf{x}|_Y)$ is the sum of weights activated by variables of $Y$ set to 1 in $\mathbf{x}$.

**Decomposable NNF.** A circuit in *negation normal form* (NNF) is a single output Boolean circuit whose inputs are Boolean variables and their complements, and whose gates are fanin-2 AND and OR gates. The *size* of the circuit is the number of its gates. We say that an NNF is *decomposable* (DNNF) if for any AND gate, the two sub-circuits rooted at that gate share no input variable, i.e., if $x$ or $\neg x$ is an input of the circuit rooted at the left input of the AND gate, then neither $x$ nor $\neg x$ is an input of the circuit rooted at the right input, and vice versa. A Boolean function $f$ is *encoded* by a DNNF $D$ if the assignments of variables for which the output of $D$ is 1 (*true*) are exactly the models of $f$.

**Rectangle Covers.** Let $X$ be a finite set of Boolean variables and let $\Pi = (X_1, X_2)$ be a partition of $X$ (i.e., $X_1 \cup X_2 = X$ and $X_1 \cap X_2 = \emptyset$). A *rectangle* $r$ with respect to $\Pi$ is a Boolean function over $X$ defined as the conjunction of two functions $\rho_1$ and $\rho_2$ over $X_1$ and $X_2$ respectively. $\Pi$ is called the *partition* of $r$. We say that the partition and the rectangle are *balanced* when $\frac{|X|}{3} \leq |X_1| \leq \frac{2|X|}{3}$ (thus the same holds for $X_2$). Whenever considering a partition $(X_1, X_2)$, we use for any assignment $\mathbf{x}$ to $X$ the notations $\mathbf{x}_1 := \mathbf{x}|_{X_1}$ and $\mathbf{x}_2 := \mathbf{x}|_{X_2}$. And for any two assignments $\mathbf{x}_1$ and $\mathbf{x}_2$ to $X_1$ and $X_2$, we note $(\mathbf{x}_1, \mathbf{x}_2)$ the assignment to $X$ whose restrictions to $X_1$ and $X_2$ are $\mathbf{x}_1$ and $\mathbf{x}_2$. Given $f$ a Boolean function over $X$, a *rectangle cover* of $f$ is a disjunction of rectangles over $X$, possibly with different partitions, equivalent to $f$. The *size* of a rectangle cover is the number of its rectangles. A cover is called *balanced* if all its rectangles are balanced.

*Example 2.* Going back to Example 1, consider the partition $X_1 := \{x_1, x_3, x_4\}$, $X_2 := \{x_2, x_5\}$ and define $\rho_1 := x_3 \wedge x_4$ and $\rho_2 := x_2 \vee x_5$. Then $r := \rho_1 \wedge \rho_2$ is a rectangle w.r.t. this partition that accepts only models of the PB-constraint from Example 1. Thus it can be part of a rectangle cover for this constraint.

Any function $f$ has at least one balanced rectangle cover as one can create a balanced rectangle accepting exactly one chosen model of $f$. We denote by $C(f)$ the size of the smallest balanced rectangle cover of $f$. The following result from [8] links $C(f)$ to the size of any DNNF encoding $f$.

**Theorem 2.** *Let $D$ be a DNNF encoding a Boolean function $f$. Then $f$ has a balanced rectangle cover of size at most the size of $D$.*

Theorem 2 reduces the problem of finding lower bounds on the size of DNNFs encoding $f$ to that of finding lower bounds on $C(f)$.

## 3   Restriction to Threshold Models of PB-Constraints

The strategy to prove Theorem 1 is to find a PB-constraint $f$ over $n$ variables such that $C(f)$ is exponential in $\sqrt{n}$ and then use Theorem 2. We first show that we can restrict our attention to covering particular models of $f$ with rectangles rather than the whole function. In this section $X$ is a set of $n$ Boolean variables and $f$ is a PB-constraint over $X$. Recall that we only consider constraints of the form $\sum_{i=1}^n w_i x_i \geq \theta$ where the $w_i$ and $\theta$ are positive integers.

**Definition 1.** *The* threshold models *of $f$ are the models $\mathbf{x}$ such that $w(\mathbf{x}) = \theta$.*

Threshold models should not be confused with minimal models (or minimals).

**Definition 2.** *A* minimal *of $f$ is a model $\mathbf{x}$ such that no $\mathbf{y} < \mathbf{x}$ is a model of $f$.*

For a monotone PB-constraint, a minimal model is such that its sum of weights drops below the threshold if we remove any element from it. Any threshold model is minimal, but not all minimals are threshold models. There even exist constraints with no threshold models (e.g. take even weights and an odd threshold) while there always are minimals for satisfiable constraints.

*Example 3.* The minimals of the PB-constraint from Example 1 are $(0, 0, 0, 1, 1)$, $(0, 1, 1, 1, 0)$, $(1, 0, 1, 0, 1)$ and $(0, 1, 1, 0, 1)$. The first three are threshold models.

Let $f^*$ be the Boolean function whose models are exactly the threshold models of $f$. In the next lemma, we prove that the smallest rectangle cover of $f^*$ has size at most $C(f)$. Thus, lower bounds on $C(f^*)$ are also lower bounds on $C(f)$.

**Lemma 1.** *Let $f^*$ be the Boolean function whose models are exactly the threshold models of $f$. Then $C(f) \geq C(f^*)$.*

*Proof.* Let $r := \rho_1 \wedge \rho_2$ be a balanced rectangle with $r \leq f$ and assume $r$ accepts some threshold models. Let $\Pi := (X_1, X_2)$ be the partition of $r$. We claim that there exist two integers $\theta_1$ and $\theta_2$ such that $\theta_1 + \theta_2 = \theta$ and, for any threshold model $\mathbf{x}$ accepted by $r$, there is $w(\mathbf{x_1}) = \theta_1$ and $w(\mathbf{x_2}) = \theta_2$. To see this, assume by contradiction that there exists another partition $\theta = \theta_1' + \theta_2'$ of $\theta$ such that some other threshold model $\mathbf{y}$ with $w(\mathbf{y_1}) = \theta_1'$ and $w(\mathbf{y_2}) = \theta_2'$ is accepted by $r$. Then either $w(\mathbf{x_1}) + w(\mathbf{y_2}) < \theta$ or $w(\mathbf{y_1}) + w(\mathbf{x_2}) < \theta$, but since $(\mathbf{x_1}, \mathbf{y_2})$ and $(\mathbf{y_1}, \mathbf{x_2})$ are also models of $r$, $r$ would accept a non-model of $f$, which is forbidden. Now let $\rho_1^*$ (resp. $\rho_2^*$) be the function whose models are exactly the models of $\rho_1$ (resp. $\rho_2$) of weight $\theta_1$ (resp. $\theta_2$). Then $r^* := \rho_1^* \wedge \rho_2^*$ is a balanced rectangle whose models are exactly the threshold models accepted by $r$.

Now consider a balanced rectangle cover of $f$ of size $C(f)$. For each rectangle $r$ of the cover, if $r$ accepts no threshold model then discard it, otherwise construct $r^*$. The disjunction of these new rectangles is a balanced rectangle cover of $f^*$ of size at most $C(f)$. Therefore $C(f) \geq C(f^*)$. □

## 4     Reduction to Covering Maximal Matchings of $K_{n,n}$

We define the class of hard PB-constraints for Theorem 1 in this section. Recall that for a hard constraint $f$, our aim is to find an exponential lower bound on $C(f)$. We will show, using Lemma 1, that the problem can be reduced to that of covering all maximal matchings of the complete $n \times n$ bipartite graph $K_{n,n}$ with rectangles. In this section, $X$ is a set of $n^2$ Boolean variables. For presentability reasons, assignments to $X$ are written as $n \times n$ matrices. Each variable $x_{i,j}$ has the weight $w_{i,j} := (2^i + 2^{j+n})/2$. Define the matrix of weights $W := (w_{i,j} : 1 \leq i, j \leq n)$ and the threshold $\theta := 2^{2n} - 1$. The PB-constraint $f$ for the pair $(W, \theta)$ is such that $f(\mathbf{x}) = 1$ if and only if $\mathbf{x}$ satisfies

$$\sum_{1 \leq i,j \leq n} \left( \frac{2^i + 2^{j+n}}{2} \right) x_{i,j} \geq 2^{2n} - 1 . \tag{1}$$

Constraints of this form constitute the class of hard constraints of Theorem 1. One may find it easier to picture $f$ writing the weights and threshold as binary numbers of $2n$ bits. Bits of indices 1 to $n$ form the *lower part* of the number and those of indices $n + 1$ to $2n$ form the *upper part*. The weight $w_{i,j}$ is the binary number where the only bits set to 1 are the $i$th bit of the lower part and the $j$th bit of the upper part. Thus when a variable $x_{i,j}$ is set to 1, exactly one bit of value 1 is added to each part of the binary number of the sum.

Assignments to $X$ uniquely encode subgraphs of $K_{n,n}$. We denote $U = \{u_1, \ldots, u_n\}$ the nodes of the left side and $V = \{v_1, \ldots, v_n\}$ those of the right side of $K_{n,n}$. The bipartite graph encoded by $\mathbf{x}$ is such that there is an edge between the $u_i$ and $v_j$ if and only if $x_{i,j}$ is set to 1 in $\mathbf{x}$.

*Example 4.* Take $n = 4$. The assignment $\mathbf{x} = \begin{pmatrix} 1 & 1 & 0 & 1 \\ 0 & 0 & 0 & 0 \\ 0 & 1 & 0 & 0 \\ 0 & 1 & 0 & 0 \end{pmatrix}$ encodes

$$
\begin{array}{cc}
u_1 \bullet & \bullet v_1 \\
u_2 \bullet & \bullet v_2 \\
u_3 \bullet & \bullet v_3 \\
u_4 \bullet & \bullet v_4
\end{array}
$$

**Definition 3.** *A* maximal matching assignment *(or* maximal matching model*) is an assignment* $\mathbf{x}$ *to* $X$ *such that*

- *for any* $i \in [n]$*, there is exactly one* $k$ *such that* $x_{i,k}$ *is set to 1 in* $\mathbf{x}$*,*
- *for any* $j \in [n]$*, there is exactly one* $k$ *such that* $x_{k,j}$ *is set to 1 in* $\mathbf{x}$*.*

As the name suggests, the maximal matching assignments are those encoding graphs whose edges form a maximal matching of $K_{n,n}$ (i.e., a maximum cardinality matching). One can also see them as encodings for permutations of $[n]$.

*Example 5.* The maximal matching model $\mathbf{x} = \begin{pmatrix} 0 & 0 & 1 & 0 \\ 1 & 0 & 0 & 0 \\ 0 & 0 & 0 & 1 \\ 0 & 1 & 0 & 0 \end{pmatrix}$ encodes

$$
\begin{array}{cc}
u_1 \bullet & \bullet v_1 \\
u_2 \bullet & \bullet v_2 \\
u_3 \bullet & \bullet v_3 \\
u_4 \bullet & \bullet v_4
\end{array}
$$

For a given $\mathbf{x}$, define $\mathrm{var}_k(\mathbf{x})$ by $\mathrm{var}_k(\mathbf{x}) := \{j \mid x_{k,j}$ is set to 1 in $\mathbf{x}\}$ when $1 \leq k \leq n$ and by $\mathrm{var}_k(\mathbf{x}) := \{i \mid x_{i,k-n}$ is set to 1 in $\mathbf{x}\}$ when $n + 1 \leq k \leq 2n$. $\mathrm{var}_k(\mathbf{x})$ stores the index of variables in $\mathbf{x}$ that directly add 1 to the $k$th bit of $w(\mathbf{x})$. Note that a maximal matching model is an assignment $\mathbf{x}$ such that $|\mathrm{var}_k(\mathbf{x})| = 1$ for all $k$. It is easy to see that maximal matching models are threshold models of $f$: seeing weights as binary numbers of $2n$ bits, for every bit of the sum the value 1 is added exactly once, so exactly the first $2n$ bits of the sum are set to 1, which gives us $\theta$. Note that not all threshold models of $f$ are maximal matching models, for instance the assignment from Example 4 does not encode a maximal matching but one can verify that it is a threshold model. Recall that $f^*$ is the function whose models are the threshold models of $f$. In the next lemmas, we prove that lower bounds on the size of rectangle covers of the maximal matching models are lower bounds on $C(f^*)$, and a fortiori on $C(f)$.

**Lemma 2.** *Let* $\Pi := (X_1, X_2)$ *be a partition of* $X$*. Let* $\mathbf{x} := (\mathbf{x}_1, \mathbf{x}_2)$ *and* $\mathbf{y} := (\mathbf{y}_1, \mathbf{y}_2)$ *be maximal matching assignments. If* $(\mathbf{x}_1, \mathbf{y}_2)$ *and* $(\mathbf{y}_1, \mathbf{x}_2)$ *both have weight* $\theta := 2^{2n} - 1$ *then both are maximal matching assignments.*

*Proof.* It is sufficient to show that $|\mathrm{var}_k(\mathbf{x}_1, \mathbf{y}_2)| = 1$ and $|\mathrm{var}_k(\mathbf{y}_1, \mathbf{x}_2)| = 1$ for all $1 \leq k \leq 2n$. We prove it for $(\mathbf{x}_1, \mathbf{y}_2)$ by induction on $k$. First observe that since $|\mathrm{var}_k(\mathbf{x})| = 1$ and $|\mathrm{var}_k(\mathbf{y})| = 1$ for all $1 \leq k \leq 2n$, the only possibilities for $|\mathrm{var}_k(\mathbf{x}_1, \mathbf{y}_2)|$ are 0, 1 or 2.

- For the base case $k = 1$, if $|\mathrm{var}_1(\mathbf{x}_1, \mathbf{y}_2)|$ is even then the first bit of $w(\mathbf{x}_1) + w(\mathbf{y}_2)$ is 0 and the weight of $(\mathbf{x}_1, \mathbf{y}_2)$ is not $\theta$. So $|\mathrm{var}_1(\mathbf{x}_1, \mathbf{y}_2)| = 1$.
- For the general case $1 < k \leq 2n$, assume that $|\mathrm{var}_1(\mathbf{x}_1, \mathbf{y}_2)| = \cdots = |\mathrm{var}_{k-1}(\mathbf{x}_1, \mathbf{y}_2)| = 1$. So the $k$th bit of $w(\mathbf{x}_1) + w(\mathbf{y}_2)$ depends only on the parity of $|\mathrm{var}_k(\mathbf{x}_1, \mathbf{y}_2)|$: the $k$th bit is 0 if $|\mathrm{var}_k(\mathbf{x}_1, \mathbf{y}_2)|$ is even and 1 otherwise. $(\mathbf{x}_1, \mathbf{y}_2)$ has weight $\theta$ so $|\mathrm{var}_k(\mathbf{x}_1, \mathbf{y}_2)| = 1$.

The argument applies to $(\mathbf{y}_1, \mathbf{x}_2)$ analogously. $\square$

**Lemma 3.** *Let $f$ be the PB-constraint (1) and let $\hat{f}$ be the function whose models are exactly the maximal matching assignments. Then $C(f) \geq C(\hat{f})$.*

*Proof.* By Lemma 1, it is sufficient to prove that $C(f^*) \geq C(\hat{f})$. We already know that $\hat{f} \leq f^*$. Let $r := \rho_1 \wedge \rho_2$ be a balanced rectangle of partition $\Pi := (X_1, X_2)$ with $r \leq f^*$, and assume $r$ accepts some maximal matching assignment. Let $\hat{\rho}_1$ (resp. $\hat{\rho}_2$) be the Boolean function over $X_1$ (resp. $X_2$) whose models are the $\mathbf{x}_1$ (resp. $\mathbf{x}_2$) such that there is a maximal matching assignment $(\mathbf{x}_1, \mathbf{x}_2)$ accepted by $r$. We claim that the balanced rectangle $\hat{r} := \hat{\rho}_1 \wedge \hat{\rho}_2$ accepts exactly the maximal matching models of $r$. On the one hand, it is clear that all maximal matching models of $r$ are models of $\hat{r}$. On the other hand, all models of $\hat{r}$ are threshold models of the form $(\mathbf{x}_1, \mathbf{y}_2)$, where $(\mathbf{x}_1, \mathbf{x}_2)$ and $(\mathbf{y}_1, \mathbf{y}_2)$ encode maximal matchings, so by Lemma 2, $\hat{r}$ accepts only maximal matching models of $r$.

Now consider a balanced rectangle cover of $f^*$ of size $C(f^*)$. For each rectangle $r$ of the cover, if $r$ accepts no maximal matching assignment then discard it, otherwise construct $\hat{r}$. The disjunction of these new rectangles is a balanced rectangle cover of $\hat{f}$ of size at most $C(f^*)$. Therefore $C(f^*) \geq C(\hat{f})$.     □

## 5     Proof of Theorem 1

**Theorem 1.** *There is a class of PB-constraints $\mathcal{F}$ such that for any constraint $f \in \mathcal{F}$ on $n^2$ variables, any DNNF encoding $f$ has size $2^{\Omega(n)}$.*

$\mathcal{F}$ is the class of constraints defined in (1). Thanks to Theorem 2 and Lemma 3, the proof boils down to finding exponential lower bounds on $C(\hat{f})$, where $\hat{f}$ is the Boolean function on $n^2$ variables whose models encode exactly the maximal matchings of $K_{n,n}$ (or equivalently, the permutations of $[n]$). $\hat{f}$ has $n!$ models. The idea is now to prove that rectangles covering $\hat{f}$ must be relatively small, so that covering the whole function requires many of them.

**Lemma 4.** *Let $\Pi = (X_1, X_2)$ be a balanced partition of $X$. Let $r$ be a rectangle with respect to $\Pi$ with $r \leq \hat{f}$. Then $|r^{-1}(1)| \leq n! / \binom{n}{n\sqrt{2/3}}$.*

The function $\hat{f}$ has already been studied extensively in the literature, often under the name $\mathrm{PERM}_n$ (for *permutations on* $[n]$), see for instance Chap. 4 of [22] or Sect. 6.2 of [20] where a statement similar to Lemma 4 is established. With Lemma 4 we can give the proof of Theorem 1.

*Proof (Theorem 1).* Let $\bigvee_{k=1}^{C(\hat{f})} r_k$ be a balanced rectangle cover of $\hat{f}$. We have $\sum_{k=1}^{C(\hat{f})} |r_k^{-1}(1)| \geq |\hat{f}^{-1}(1)| = n!$. Lemma 4 gives us $(C(\hat{f})n!)/\binom{n}{n\sqrt{2/3}} \geq n!$, thus

$$C(\hat{f}) \geq \binom{n}{n\sqrt{2/3}} \geq \left(\frac{n}{n\sqrt{2/3}}\right)^{n\sqrt{2/3}} = \left(\frac{3}{2}\right)^{n\frac{\sqrt{2/3}}{2}} \geq 2^{n\frac{\sqrt{2/3}}{4}} = 2^{\Omega(n)}$$

where we have used $\binom{a}{b} \geq (a/b)^b$ and $3/2 \geq \sqrt{2}$. Using Lemma 3 we get that $C(f) \geq C(\hat{f}) \geq 2^{\Omega(n)}$. Theorem 2 allows us to conclude.     □

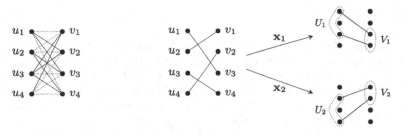

(a) Balanced partition $\Pi$ of $K_{4,4}$    (b) Partition of a maximal matching w.r.t. $\Pi$

**Fig. 1.** Partition of maximal matching

All that is left is to prove Lemma 4.

*Proof (Lemma 4).*

Let $r := \rho_1 \wedge \rho_2$ and $\Pi := (X_1, X_2)$. Recall that $U := \{u_1, \ldots, u_n\}$ and $V := \{v_1, \ldots, v_n\}$ are the nodes from the left and right part of $K_{n,n}$ respectively. Define $U_1 := \{u_i \mid \text{there exists } x_{i,l} \in X_1 \text{ such that a model of } \rho_1 \text{ has } x_{i,l} \text{ set to } 1\}$ and $V_1 := \{v_j \mid \text{there exists } x_{l,j} \in X_1 \text{ such that a model of } \rho_1 \text{ has } x_{l,j} \text{ set to } 1\}$. Define $U_2$ and $V_2$ analogously (this time using $X_2$ and $\rho_2$). Figure 1 illustrates the construction of these sets: Fig. 1a shows a partition $\Pi$ of the edges of $K_{4,4}$ (full edges in $X_1$, dotted edges in $X_2$) and Fig. 1b shows the contribution of a model of $r$ to $U_1$, $V_1$, $U_2$, and $V_2$ after partition according to $\Pi$.

Models of $\rho_1$ are clearly matchings of $K_{n,n}$. Actually they are matchings between $U_1$ and $V_1$ by construction of these sets. We claim that they are maximal. To verify this, observe that $U_1 \cap U_2 = \emptyset$ and $V_1 \cap V_2 = \emptyset$ since otherwise $r$ has a model that is not a matching. Thus if $\rho_1$ were to accept a non-maximal matching between $U_1$ and $V_1$ then $r$ would accept a non-maximal matching between $U$ and $V$. So $\rho_1$ accepts only maximal matchings between $U_1$ and $V_1$, consequently $|U_1| = |V_1|$. The argument applies symmetrically for $V_2$ and $U_2$. We note $k := |U_1|$. It stands that $U_1 \cup U_2 = U$ and $V_1 \cup V_2 = V$ as otherwise $r$ accepts matchings that are not maximal. So $|U_2| = |V_2| = n - k$. We now have $|\rho_1^{-1}(1)| \leq k!$ and $|\rho_2^{-1}(1)| \leq (n-k)!$, leading to $|r^{-1}(1)| \leq k!(n-k)! = n!/\binom{n}{k}$.

Up to $k^2$ edges may be used to build matchings between $U_1$ and $V_1$. Since $r$ is balanced we obtain $k^2 \leq 2n^2/3$. Applying the same argument to $U_2$ and $V_2$ gives us $(n-k)^2 \leq 2n^2/3$, so $n(1 - \sqrt{2/3}) \leq k \leq n\sqrt{2/3}$. Finally, the function $k \mapsto n!/\binom{n}{k}$, when restricted to some interval $[\![n(1-\alpha), \alpha n]\!]$, reaches its maximum at $k = \alpha n$, hence the upper bound $|r^{-1}(1)| \leq n!/\binom{n}{n\sqrt{2/3}}$. $\qquad\square$

**Acknowledgments.** This work has been partly supported by the PING/ACK project of the French National Agency for Research (ANR-18-CE40-0011).

# References

1. Abío, I., Gange, G., Mayer-Eichberger, V., Stuckey, P.J.: On CNF encodings of decision diagrams. In: Quimper, C.-G. (ed.) CPAIOR 2016. LNCS, vol. 9676, pp. 1–17. Springer, Cham (2016). https://doi.org/10.1007/978-3-319-33954-2_1

2. Abío, I., Nieuwenhuis, R., Oliveras, A., Rodríguez-Carbonell, E., Mayer-Eichberger, V.: A new look at BDDs for pseudo-boolean constraints. J. Artif. Intell. Res. **45**, 443–480 (2012). https://doi.org/10.1613/jair.3653

3. Aloul, F.A., Ramani, A., Markov, I.L., Sakallah, K.A.: Generic ILP versus specialized 0–1 ILP: an update. In: IEEE/ACM International Conference on Computer-aided Design, ICCAD, pp. 450–457 (2002). https://doi.org/10.1145/774572.774638

4. Bacchus, F.: GAC via unit propagation. In: Bessière, C. (ed.) CP 2007. LNCS, vol. 4741, pp. 133–147. Springer, Heidelberg (2007). https://doi.org/10.1007/978-3-540-74970-7_12

5. Bailleux, O., Boufkhad, Y., Roussel, O.: New encodings of pseudo-boolean constraints into CNF. In: Kullmann, O. (ed.) SAT 2009. LNCS, vol. 5584, pp. 181–194. Springer, Heidelberg (2009). https://doi.org/10.1007/978-3-642-02777-2_19

6. Bordeaux, L., Marques-Silva, J.: Knowledge compilation with empowerment. In: Bieliková, M., Friedrich, G., Gottlob, G., Katzenbeisser, S., Turán, G. (eds.) SOFSEM 2012. LNCS, vol. 7147, pp. 612–624. Springer, Heidelberg (2012). https://doi.org/10.1007/978-3-642-27660-6_50

7. Boros, E., Hammer, P.L., Minoux, M., Rader, D.J.: Optimal cell flipping to minimize channel density in VLSI design and pseudo-boolean optimization. Discrete Appl. Math. **90**(1–3), 69–88 (1999). https://doi.org/10.1016/S0166-218X(98)00114-0

8. Bova, S., Capelli, F., Mengel, S., Slivovsky, F.: Knowledge compilation meets communication complexity. In: International Joint Conference on Artificial Intelligence, IJCAI, pp. 1008–1014 (2016). http://www.ijcai.org/Abstract/16/147

9. Bryant, R.E., Lahiri, S.K., Seshia, S.A.: Deciding CLU logic formulas via boolean and pseudo-boolean encodings. In: International Workshop on Constraints in Formal Verification, CFV (2002)

10. Darwiche, A., Marquis, P.: A knowledge compilation map. J. Artif. Intell. Res. **17**, 229–264 (2002). https://doi.org/10.1613/jair.989

11. Eén, N., Sörensson, N.: Translating pseudo-boolean constraints into SAT. J. Satisf. Boolean Model. Comput. **2**(1–4), 1–26 (2006). https://satassociation.org/jsat/index.php/jsat/article/view/18

12. Gange, G., Stuckey, P.J.: Explaining propagators for s-DNNF circuits. In: Beldiceanu, N., Jussien, N., Pinson, É. (eds.) CPAIOR 2012. LNCS, vol. 7298, pp. 195–210. Springer, Heidelberg (2012). https://doi.org/10.1007/978-3-642-29828-8_13

13. Gopalan, P., Klivans, A.R., Meka, R., Štefankovič, D., Vempala, S.S., Vigoda, E.: An FPTAS for #Knapsack and related counting problems. In: IEEE Symposium on Foundations of Computer Science, FOCS, pp. 817–826 (2011). https://doi.org/10.1109/FOCS.2011.32

14. Hosaka, K., Takenaga, Y., Kaneda, T., Yajima, S.: Size of ordered binary decision diagrams representing threshold functions. Theor. Comput. Sci. **180**(1–2), 47–60 (1997). https://doi.org/10.1016/S0304-3975(97)83807-8

15. Ivănescu, P.L.: Some network flow problems solved with pseudo-boolean programming. Oper. Res. **13**(3), 388–399 (1965)

16. Jung, J.C., Barahona, P., Katsirelos, G., Walsh, T.: Two encodings of DNNF theories. In: ECAI Workshop on Inference Methods Based on Graphical Structures of Knowledge (2008)
17. Kučera, P., Savický, P.: Propagation complete encodings of smooth DNNF theories. CoRR abs/1909.06673 (2019). http://arxiv.org/abs/1909.06673
18. Le Berre, D., Marquis, P., Mengel, S., Wallon, R.: Pseudo-boolean constraints from a knowledge representation perspective. In: International Joint Conference on Artificial Intelligence, IJCAI, pp. 1891–1897 (2018). https://doi.org/10.24963/ijcai.2018/261
19. Marquis, P.: Compile! In: AAAI Conference on Artificial Intelligence, AAAI, pp. 4112–4118 (2015). http://www.aaai.org/ocs/index.php/AAAI/AAAI15/paper/view/9596
20. Mengel, S., Wallon, R.: Revisiting graph width measures for CNF-encodings. In: Janota, M., Lynce, I. (eds.) SAT 2019. LNCS, vol. 11628, pp. 222–238. Springer, Cham (2019). https://doi.org/10.1007/978-3-030-24258-9_16
21. Papaioannou, S.G.: Optimal test generation in combinational networks by pseudo-boolean programming. IEEE Trans. Comput. 26(6), 553–560 (1977). https://doi.org/10.1109/TC.1977.1674880
22. Wegener, I.: Branching programs and binary decision diagrams. SIAM (2000). http://ls2-www.cs.uni-dortmund.de/monographs/bdd/

# On Weakening Strategies for PB Solvers

Daniel Le Berre[1], Pierre Marquis[1,2], and Romain Wallon[1(✉)]

[1] CRIL, Univ Artois & CNRS, Lens, France
{leberre,marquis,wallon}@cril.fr
[2] Institut Universitaire de France, Paris, France

**Abstract.** Current pseudo-Boolean solvers implement different variants of the cutting planes proof system to infer new constraints during conflict analysis. One of these variants is *generalized resolution*, which allows to infer strong constraints, but suffers from the growth of coefficients it generates while combining pseudo-Boolean constraints. Another variant consists in using *weakening* and *division*, which is more efficient in practice but may infer weaker constraints. In both cases, weakening is mandatory to derive conflicting constraints. However, its impact on the performance of pseudo-Boolean solvers has not been assessed so far. In this paper, new application strategies for this rule are studied, aiming to infer strong constraints with small coefficients. We implemented them in *Sat4j* and observed that each of them improves the runtime of the solver. While none of them performs better than the others on all benchmarks, applying weakening on the *conflict* side has surprising good performance, whereas applying *partial* weakening and division on both the conflict and the reason sides provides the best results overall.

**Keywords:** PB constraint · Constraint learning · Cutting planes

## 1 Introduction

The last decades have seen many improvements in SAT solving that are at the root of the success of modern SAT solvers [5,13,15]. Despite their practical efficiency on many real-world instances, these solvers suffer from the weakness of the resolution proof system they use in their conflict analyses. Specifically, when proving the unsatisfiability of an input formula requires an exponential number of resolution steps – as for pigeonhole-principle formulae [9] – a SAT solver cannot find a refutation proof efficiently. This motivated the development of pseudo-Boolean (PB) solvers [17], which take as input conjunctions of *PB constraints* (linear inequations over Boolean variables) and apply *cutting planes based inference* to derive inconsistency [8,10,16]. This inference system is stronger than the resolution proof system, as it *p-simulates* the latter: any resolution proof can be translated into a cutting planes proof of polynomial size [2]. Using such a proof system may, in theory, make solvers more efficient: for instance, a pigeonhole principle formula may be refuted with a linear number of cutting planes steps.

L. Pulina and M. Seidl (Eds.): SAT 2020, LNCS 12178, pp. 322–331, 2020.
https://doi.org/10.1007/978-3-030-51825-7_23

However, in practice, PB solvers fail to keep the promises of the theory. In particular, they only implement *subsets* of the cutting planes proof system, which degenerate to resolution when given a CNF formula as input: they do not exploit the full power of the cutting planes proof system [20]. One of these subsets is *generalized resolution* [10], which is implemented in many PB solvers [1,4,11,19]. It consists in using the *cancellation* rule to combine constraints so as to resolve away literals during conflict analysis, as SAT solvers do with the resolution rule. Another approach has been introduced by *RoundingSat* [7], which relies on the *weakening* and *division* rules to infer constraints having smaller coefficients to be more efficient in practice. These proof systems are described in Sect. 2.

This paper follows the direction initiated by *RoundingSat* and investigates to what extent applying the *weakening* rule may have an impact on the performance of PB solvers. First, we show that applying a *partial* weakening instead of an aggressive weakening as proposed in [7] allows to infer stronger constraints while preserving the nice properties of *RoundingSat*. Second, we show that weakening operations can be extended to certain literals that are falsified by the current partial assignment to derive shorter constraints. Finally, we introduce a trade-off strategy, trying to get the best of both worlds. These new approaches are described in Sect. 3, and empirically evaluated in Sect. 4.

# 2   Pseudo-Boolean Solving

We consider a propositional setting defined on a finite set of classically interpreted propositional variables $V$. A *literal* $l$ is a variable $v \in V$ or its negation $\bar{v}$. Boolean values are represented by the integers 1 (true) and 0 (false), so that $\bar{v} = 1 - v$. A *PB constraint* is an integral linear equation or inequation over Boolean variables. Such constraints are supposed, w.l.o.g., to be in the normalized form $\sum_{i=1}^{n} \alpha_i l_i \geq \delta$, where $\alpha_i$ (the *coefficients* or *weights*) and $\delta$ (the *degree*) are positive integers and $l_i$ are literals. A *cardinality constraint* is a PB constraint with its weights equal to 1 and a *clause* is a cardinality constraint of degree 1.

Several approaches have been designed for solving PB problems. One of them consists in encoding the input into a CNF formula and let a SAT solver decide its satisfiability [6,14,18], while another one relies on lazily translating PB constraints into clauses during conflict analysis [21]. However, such solvers are based on the *resolution* proof system, which is somewhat *weak*: instances that are hard for resolution are out of reach of SAT solvers. In the following, we consider instead solvers based on the *cutting planes* proof system, the PB counterpart of the resolution proof system. Such solvers handle PB constraints natively, and are based on one of the two main subsets of cutting planes rules described below.

## 2.1   Generalized Resolution Based Solvers

Following the CDCL algorithm of SAT solvers, PB solvers based on *generalized resolution* [10] make *decisions* on variables, which force other literals to be satisfied. These *propagated* literals are detected using the *slack* of each constraint.

**Definition 1 (slack).** *Given a partial assignment $\rho$, the* slack *of a constraint* $\sum_{i=1}^{n} \alpha_i l_i \geq \delta$ *is the value* $\sum_{i=1, \rho(l_i) \neq 0}^{n} \alpha_i - \delta$.

*Observation 1.* Let $s$ be the slack of the constraint $\sum_{i=1}^{n} \alpha_i l_i \geq \delta$ under some partial assignment. If $s < 0$, the constraint is currently falsified. Otherwise, the constraint requires all unassigned literals having a weight $\alpha > s$ to be satisfied.

*Example 1.* Let $\rho$ be the partial assignement such that $\rho(a) = 1$, $\rho(c) = \rho(d) = \rho(e) = 0$ (all other variables are unassigned). Under $\rho$, the constraint $6\bar{b} + 6c + 4e + f + g + h \geq 7$ has slack 2. As $\bar{b}$ is unassigned and has weight $6 > 2$, this literal is propagated (the constraint is the *reason* for $\bar{b}$). This propagation falsifies the constraint $5a + 4b + c + d \geq 6$, which now has slack $-1$ (this is a *conflict*).

When a conflict occurs, the solver analyzes this conflict to derive an *assertive* constraint, i.e., a constraint propagating some of its literals. To do so, it applies successively the *cancellation* rule between the conflict and the reason for the propagation of one of its literals ("LCM" denotes the least common multiple):

$$\frac{\alpha l + \sum_{i=1}^{n} \alpha_i l_i \geq \delta \qquad \beta \bar{l} + \sum_{i=1}^{n} \beta_i l_i \geq \delta' \qquad \mu\alpha = \nu\beta = \mathrm{LCM}(\alpha, \beta)}{\sum_{i=1}^{n} (\mu\alpha_i + \nu\beta_i) l_i \geq \mu\delta + \nu\delta' - \mu\alpha} \text{ (canc.)}$$

To make sure that an assertive constraint will be eventually derived, the constraint produced by this operation has to be conflictual, which is not guaranteed by the cancellation rule. To preserve the conflict, one can take advantage of the fact that the slack is *subadditive*: the slack of the constraint obtained by applying the cancellation between two constraints is at most equal to the sum of the slacks of these constraints. Whenever the sum of both slacks is not negative, the constraint may not be conflictual, and the *weakening* and *saturation* rules are applied until the slack of the reason becomes low enough to ensure the conflict to be preserved (only literals that are not falsified may be weakened away).

$$\frac{\alpha l + \sum_{i=1}^{n} \alpha_i l_i \geq \delta}{\sum_{i=1}^{n} \alpha_i l_i \geq \delta - \alpha} \text{ (weakening)} \qquad \frac{\sum_{i=1}^{n} \alpha_i l_i \geq \delta}{\sum_{i=1}^{n} \min(\delta, \alpha_i) l_i \geq \delta} \text{ (saturation)}$$

*Example 2 (Example 1 cont'd).* As $5a + 4b + c + d \geq 6$ is conflicting and $\bar{b}$ was propagated by $6\bar{b} + 6c + 4e + f + g + h \geq 7$, the cancellation rule must be applied between these two constraints to eliminate $b$. To do so, the *conflict side* (i.e., the first constraint) has to be multiplied by 3 and the *reason side* (i.e., the second constraint) by 2, giving slack $-3$ and 4, respectively. As the sum of these values is equal to 1, the resulting constraint is not guaranteed to be conflicting. Thus, the reason is weakened on $g$ and $h$ and saturated to get $5\bar{b} + 5c + 4e + f \geq 5$, which has slack 1. To cancel $b$ out, this constraint is multiplied by 4 and the conflict by 5, giving $25a + 25c + 16e + 5d + 4f \geq 30$, which has slack $-1$.

This approach has several drawbacks. Observe in Example 2 the growth of the coefficients in just one derivation step. In practice, there are many such steps during conflict analysis, and the learned constraints will be reused later on, so that coefficients will continue to grow, requiring the use of arbitrary precision arithmetic. Moreover, after each weakening operation, the LCM of the coefficients must be recomputed to estimate the slack, and other literals to be weakened must be found. The cost of these operations motivated the development of alternative proof systems, such as those weakening the derived constraints to infer only cardinality constraints [1], or those based on the *division* rule.

## 2.2   Division Based Solvers

To limit the growth of the coefficients during conflict analysis, *RoundingSat* [7] introduced an aggressive use of the weakening and *division* rules.

$$\frac{\sum_{i=1}^n \alpha_i l_i \geq \delta \qquad r > 0}{\sum_{i=1}^n \lceil \frac{\alpha_i}{r} \rceil l_i \geq \lceil \frac{\delta}{r} \rceil} \text{ (division)}$$

When a conflict occurs, both the conflict and the reason are weakened so as to remove all literals not falsified by the current assignment and having a coefficient not divisible by the weight of the literal used as pivot for the cancellation, before being divided by this weight. This ensures that the pivot has a weight equal to 1, which guarantees that the result of the cancellation will be conflictual [3].

*Example 3 (Example 2 cont'd).* The weakening operation is applied on both the conflict $5a + 4b + c + d \geq 6$ and the reason $6\bar{b} + 6c + 4e + f + g + h \geq 7$, yielding $4b + c + d \geq 1$ and $6\bar{b} + 6c + 4e \geq 4$, respectively. Both constraints are then divided by the coefficient of the pivot (4 and 6, respectively), giving $b + c + d \geq 1$ and $\bar{b} + c + e \geq 1$. Applying the cancellation rule on these two constraints gives $2c + d + e \geq 1$, which is equivalent to the clause $c + d + e \geq 1$.

The *RoundingSat* approach succeeds in keeping coefficients small, and its aggressive weakening allows to find the literals to remove efficiently. However, some constraints inferred by this solver may be weaker than those inferred with generalized resolution (compare the constraints derived in Examples 2 and 3).

## 3   Weakening Strategies

As explained before, the weakening rule is mandatory in PB solvers to maintain the inferred constraints conflictual. In the following, we introduce different strategies for applying this rule in PB solvers, designed towards finding a tradeoff between the strength of the inferred constraints and their size.

### 3.1   Weakening Ineffective Literals for Shorter Constraints

Within CDCL solvers, one captures the reason for a conflict being encountered. A conflict occurs when a variable is propagated to both 0 and 1. Intuitively, understanding why such a conflict occurred amounts to understanding why these values have been propagated. In the PB case, a constraint may be conflicting (resp. propagate literals) even if it contains literals that are unassigned or already satisfied (see Example 1). However, conflicts (resp. propagations) depend only on *falsified* literals (the slack of a constraint changes only when one of its literals is falsified). Literals that are not falsified are thus *ineffective*: they do not play a role in the conflict (resp. propagation), and may thus be weakened away. We can go even further: when most literals are falsified, weakening some of them may still preserve the conflict (resp. propagation), as shown in the following example.

*Example 4.* Let $\rho$ be the partial assignment such that $\rho(a) = \rho(c) = \rho(f) = 0$ (all other variables are unassigned). Under $\rho$, $3\bar{a}+3\bar{b}+c+d+e \geq 6$ has slack 2, so that literal $\bar{b}$ is propagated to 1. This propagation still holds after weakening away $\bar{a}$, $d$ and $e$, giving after saturation $\bar{b} + c \geq 1$. Similarly, consider the conflicting constraint $2a+b+c+f \geq 2$. After the propagation of $\bar{b}$, weakening the constraint on $c$ and applying saturation produces $a + b + f \geq 1$, which is still conflicting. In both cases, the slack allows to detect whether a literal can be weakened.

Observe that the constraints obtained are shorter, but are always clauses. This guarantees that the resulting constraint will be conflictual, but, if this operation is performed on both sides, only clauses can be inferred, and the proof system boils down to resolution, as in *SATIRE* [21] or *Sat4j-Resolution* [11].

*Example 5 (Example 4 cont'd).* If a resolution step is performed between the weaker constraints $\bar{b} + c \geq 1$ and $a + b + f \geq 1$, the clause $a + c + f \geq 1$ is inferred. However, if only one side is weakened, for example the conflict side, the cancellation between $3\bar{a} + 3\bar{b} + c + d + e \geq 6$ and $a + b + f \geq 1$ produces the constraint $3f + c + d + e \geq 3$. Observe that, when the weakening operation is applied at the next step, the stronger clause $c+f \geq 1$ is inferred after saturation.

### 3.2   Partial Weakening for Stronger Constraints

To avoid the inference of constraints that are too weak to preserve the strength of the proof system, an interesting option is to use *partial weakening*. Indeed, the weakening rule, as described above, can be generalized as follows.

$$\frac{\alpha l + \sum_{i=1}^{n} \alpha_i l_i \geq \delta \qquad \varepsilon \in \mathbb{N} \qquad 0 < \varepsilon \leq \alpha}{(\alpha - \varepsilon)l + \sum_{i=1}^{n} \alpha_i l_i \geq \delta - \varepsilon} \text{ (partial weakening)}$$

This rule is rarely used in practice by PB solvers, and the weakening rule (i.e., the case when $\varepsilon = \alpha$) is often preferred. However, partial weakening gives more freedom when it comes to inferring new constraints, and allows to infer stronger constraints. We implemented a variant of *RoundingSat* which uses this rule as follows. Before cancelling a literal out during conflict analysis, all literals that are not currently falsified and have a coefficient not divisible by the weight of the pivot are *partially weakened* (instead of simply *weakened*). This operation is applied so that the resulting coefficient becomes a multiple of the weight of the pivot. This approach preserves the nice properties of *RoundingSat* (see [7, Proposition 3.1 and Corollary 3.2]), and in particular the fact that the produced constraint will be conflictual (the coefficient of the pivot will be equal to 1). It also allows to infer stronger constraints, as illustrated by the following example.

*Example 6.* Let $\rho$ be the partial assignment defined by $\rho(a) = 1$ and $\rho(b) = \rho(c) = \rho(d) = \rho(e) = 0$ (all other variables are unassigned). Consider the (conflicting) constraint $8a + 7b + 7c + 2d + 2e + f \geq 11$, where $b$ is the literal to be cancelled out. The above rule yields $7a + 7b + 7c + 2d + 2e \geq 9$ which, divided by 7, gives $a + b + c + d + e \geq 2$. This constraint is stronger than the clause $b + c + d + e \geq 1$ inferred by *RoundingSat*, which weakens away the literal $a$.

This variant has several advantages. First, its cost is comparable to that of *RoundingSat*: checking whether a coefficient is divisible by the weight of the pivot is computed with the remainder of the division of the former by the latter, which is the amount by which the literal must be partially weakened. Second, the constraints it infers may be stronger than that of *RoundingSat*. Yet, this strategy does not reduce the size of the constraints as much as the weakening of ineffective literals. To get the best of both worlds, we introduce tradeoff strategies.

## 3.3   Towards a Tradeoff

The previous sections showed that the weakening operation may help finding short explanations for conflicts, but may also infer weaker constraints. Several observations may guide us towards tradeoff applications of the weakening rule.

First, the key property motivating *RoundingSat* to round the coefficient of the pivot to 1 does not require it to be equal to 1 on *both* sides of the cancellation: actually, having a coefficient equal to 1 on only *one* side is enough to guarantee the resulting constraint to be conflicting [3]. Weakening only the reason or the conflict is thus enough to preserve this property, while maintaining coefficients low enough, as only one side of the cancellation may need to be multiplied.

Second, one may apply the weakening rule in a different manner to keep coefficients small so as to speed up arithmetic operations. A possible approach is the following, that we call *Multiply and Weaken*. Let $r$ be the coefficient of the pivot used in the cancellation appearing in the reason and $c$ that in the conflict. Find two values $\mu$ and $\nu$ such that $(\nu - 1) \cdot r < \mu \cdot c \leq \nu \cdot r$ (which can be done using Euclidean division). Then, multiply the reason by $\nu$, and apply weakening operations on this constraint so as to reduce the coefficient of the pivot to $\mu \cdot c$. Note that, in order to preserve the propagation, this coefficient cannot be weakened directly. Instead, ineffective literals (as described above) are successively weakened away so that the saturation rule produces the expected reduction on the coefficient. Since this operation does not guarantee to preserve the conflict, an additional weakening operation has to be performed, as for generalized resolution. Note that this approach may also derive clauses, even though this is not always the case, as shown in the following example.

*Example 7.* Let $\rho$ be the partial assignment such that $\rho(a) = \rho(d) = 0$ and $\rho(e) = 1$ (all other variables are unassigned). Under $\rho$ the constraint $5a + 5b + 3c + 2d + e \geq 6$ propagates $b$. The constraint $3\bar{b} + 2a + 2d + \bar{e} \geq 5$ becomes thus falsified. Instead of using the LCM of 3 and 5 (i.e., 15), the reason of $b$ is weakened on $e$ and partially on $c$ to get, after saturation, $3a + 3b + 2d + c \geq 3$. The cancellation produces then $5a + 4d + c + \bar{e} \geq 5$.

## 4     Experimental Results

This section presents an empirical evaluation of the various strategies introduced in this paper. To make sure that their comparison only takes care of the underlying proof systems, and not of implementation details, we integrated all of them in *Sat4j* [11] (including *RoundingSat* proof system). The source code is available on *Sat4j* repository[1].

All experiments presented in this section have been run on a cluster equiped with quadcore bi-processors Intel XEON E5-5637 v4 (3.5 GHz) and 128 GB of memory. The time limit was set to 1200 seconds and the memory limit to 32 GB. The whole set of decision benchmarks containing "small" integers used in the PB competitions since the first edition [12] was considered as input.

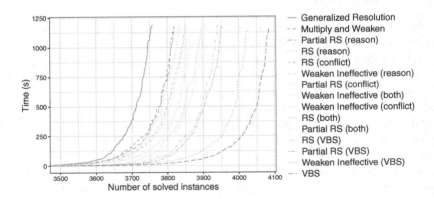

**Fig. 1.** Cactus plot of the different strategies implemented in Sat4j. For more readability, the first 3,500 easy instances are cut out.

As shown by Fig. 1, strategies applying heavily the weakening rule perform better than generalized resolution. Yet, among these strategies, none of them is strictly better than the others. In particular, the Virtual Best Solver (VBS),

---

[1] https://gitlab.ow2.org/sat4j/sat4j/tree/weakening-investigations.

obtained by choosing the best solver for each of the instances, performs clearly better than any individual strategy. Each of these individual strategies does not have an important contribution to the VBS, since the *both*, *conflict* and *reason* variants are very similar. However, if we consider the VBS of the different "main" strategies, and in particular that of *RS*, *Partial RS* and *Weaken Ineffective* (also represented on the cactus plot), their contributions become clearer: *Generalized Resolution* contributes to 6 instances, *RS* to 13 instances, *Partial RS* to 16 instances, and *Weaken Ineffective* to 83 instances. Even though *Multiply and Weaken* does not solve instances that are not solved by any other solver, it solves 13 instances more than 1 second faster than any other approach (5 among them are faster solved by more than 1 min). This suggests that choosing the *right* weakening strategy plays a key role in the performance of the solver.

The strategies showing the best and most robust performance are those applying the *RoundingSat* (*RS*) approach on both sides of the cancellation rule, as they widely take advantage of the inference power of the division rule. However, applying *partial* weakening instead of weakening gives better results, thanks to the stronger constraints it infers. In particular, *RS (both)* solves 3895 instances and *Partial RS (both)* solves 3903 instances (with 3845 common instances). The performance of *Partial RS (both)* is evidenced on the `tsp` family, especially on satisfiable instances: it solves 22 more such instances than *RS (both)*, i.e., 35 instances. For unsatisfiable instances, no common instances are solved: *Partial RS (both)* solves 7 instances, while *RS (both)* solves 5 distinct instances. In both cases, *Partial RS (both)* performs much more assignments per second than *RS (both)*, allowing it to go further in the search space within the time limit.

Surprisingly, another strategy exhibiting good performance consists in weakening ineffective literals on the *conflict* side at each cancellation (it contributes to 18 instances in the VBS). Similarly, *RoundingSat* strategies perform better when applied on the conflict side rather than the reason side. Since the early development of cutting planes based solvers, weakening has only been applied on the *reason* side (except for *RoundingSat* [7], which applies it on *both* sides). Our experiments show that it may be preferable to apply it only on the conflict side: literals introduced there when cancelling may still be weakened away later on.

The gain we observe between the different strategies has several plausible explanations. First, the solver does not explore the same search space from one strategy to another, and it may thus learn completely different constraints. In particular, they may be stronger or weaker, and this impacts the size of the proof built by the solver. Second, these constraints may be based on distinct literals, which may have side effects on the VSIDS heuristic [15]: different literals will be "bumped" during conflict analysis. Such side effects are hard to assess, due to the tight link between the heuristic and the other components of the solver.

# 5   Conclusion

In this paper, we introduced various strategies for applying the weakening rule in PB solvers. We showed that each of them may improve the runtime of the solver, but not on all benchmarks. Contrary to the approaches implemented in most PB solvers, the strategies consisting in applying an aggressive weakening only on the conflict side provide surprisingly good results. However, approaches based on *RoundingSat* perform better, but our experiments showed that partial weakening is preferable in this context. This suggests that weakening operations should be guided by the strength of the constraints to infer. To do so, a perspective for further research consists in searching for better tradeoffs in this direction.

**Acknowledgements.** The authors are grateful to the anonymous reviewers for their numerous comments, that greatly helped to improve the paper. Part of this work was supported by the French Ministry for Higher Education and Research and the Hauts-de-France Regional Council through the "Contrat de Plan État Région (CPER) DATA".

# References

1. Chai, D., Kuehlmann, A.: A fast pseudo-boolean constraint solver. IEEE Trans. CAD Integr. Circuits Syst. **24**, 305–317 (2005)
2. Cook, W., Coullard, C.R., Turán, G.: On the complexity of cutting-plane proofs. Discrete Appl. Math. **18**, 25–38 (1987)
3. Dixon, H.: Automating pseudo-boolean inference within a DPLL framework. Ph.D. thesis, University of Oregon (2004)
4. Dixon, H.E., Ginsberg, M.L.: Inference methods for a pseudo-boolean satisfiability solver. In: Proceedings of AAAI 2002, pp. 635–640 (2002)
5. Eén, N., Sörensson, N.: An extensible SAT-solver. In: Giunchiglia, E., Tacchella, A. (eds.) SAT 2003. LNCS, vol. 2919, pp. 502–518. Springer, Heidelberg (2004). https://doi.org/10.1007/978-3-540-24605-3_37
6. Een, N., Sörensson, N.: Translating pseudo-boolean constraints into SAT. JSAT **2**, 1–26 (2006)
7. Elffers, J., Nordström, J.: Divide and Conquer: towards faster pseudo-boolean solving. In: Proceedings of IJCAI 2018, pp. 1291–1299 (2018)
8. Gomory, R.E.: Outline of an algorithm for integer solutions to linear programs. Bull. Am. Math. Soc. **64**, 275–278 (1958)
9. Haken, A.: The intractability of resolution. Theor. Comput. Sci. **39**, 297–308 (1985)
10. Hooker, J.N.: Generalized resolution and cutting planes. Ann. Oper. Res. **12**, 217–239 (1988)
11. Le Berre, D., Parrain, A.: The SAT4J library, release 2.2, system description. JSAT **7**, 59–64 (2010)
12. Manquinho, V., Roussel, O.: The first evaluation of pseudo-boolean solvers (PB'05). JSAT **2**, 103–143 (2006)
13. Marques-Silva, J., Sakallah, K.A.: GRASP: a search algorithm for propositional satisfiability. IEEE Trans. Comput. **48**, 220–227 (1999)
14. Martins, R., Manquinho, V., Lynce, I.: Open-WBO: a modular MaxSAT solver. In: Sinz, C., Egly, U. (eds.) SAT 2014. LNCS, vol. 8561, pp. 438–445. Springer, Cham (2014). https://doi.org/10.1007/978-3-319-09284-3_33
15. Moskewicz, M.W., Madigan, C.F., Zhao, Y., Zhang, L., Malik, S.: Chaff: engineering an efficient SAT solver. In: Proceedings of DAC 2001, pp. 530–535 (2001)
16. Nordström, J.: On the interplay between proof complexity and SAT solving. ACM SIGLOG News **2**, 19–44 (2015)

17. Roussel, O., Manquinho, V.M.: Pseudo-boolean and cardinality constraints. In: Handbook of Satisfiability, chap. 22, pp. 695–733. IOS Press (2009)
18. Sakai, M., Nabeshima, H.: Construction of an ROBDD for a PB-constraint in band form and related techniques for PB-solvers. IEICE Trans. Inf. Syst. **98**, 1121–1127 (2015)
19. Sheini, H.M., Sakallah, K.A.: Pueblo: a hybrid pseudo-boolean SAT solver. JSAT **2**, 165–189 (2006)
20. Vinyals, M., Elffers, J., Giráldez-Cru, J., Gocht, S., Nordström, J.: In between resolution and cutting planes: a study of proof systems for pseudo-boolean SAT solving. In: Beyersdorff, O., Wintersteiger, C.M. (eds.) SAT 2018. LNCS, vol. 10929, pp. 292–310. Springer, Cham (2018). https://doi.org/10.1007/978-3-319-94144-8_18
21. Whittemore, J., Sakallah, J.K.K.A.: SATIRE: a new incremental satisfiability engine. In: Proceedings of DAC 2001, pp. 542–545 (2001)

# Reasoning About Strong Inconsistency in ASP

Carlos Mencía[1]([⊠]) and Joao Marques-Silva[2]

[1] University of Oviedo, Gijón, Spain
`menciacarlos@uniovi.es`
[2] ANITI, University of Toulouse, Toulouse, France
`joao.marques-silva@univ-toulouse.fr`

**Abstract.** The last decade has witnessed remarkable improvements in the analysis of inconsistent formulas, namely in the case of Boolean Satisfiability (SAT) formulas. However, these successes have been restricted to monotonic logics. Recent work proposed the notion of *strong inconsistency* for a number of non-monotonic logics, including Answer Set Programming (ASP). This paper shows how algorithms for reasoning about inconsistency in monotonic logics can be extended to the case of ASP programs, in the concrete case of strong inconsistency. Initial experimental results illustrate the potential of the proposed approach.

## 1 Introduction

The last decade and a half witnessed a remarkable evolution in algorithms for reasoning about inconsistency. This is the case with algorithms for the extraction and enumeration of minimal unsatisfiable subsets (MUSes) [4–6,8,9,33,34,37,42] and minimal correction subsets (MCSes) [3,6,21,26,27,36,39,40,45], but also algorithms for maximum satisfiability (MaxSAT) [1,2,18,35,41]. This work was motivated by earlier important advances [7,17,28,29,32,48]. Although most of this work was proposed in the context of propositional formulas it is also the case that most of the algorithms are amenable to generalization for different fragments of First-Order Logic (FOL). These algorithms specifically addressed monotonic logics, with propositional logic as a concrete example.

In the case of non-monotonic logics, minimal inconsistency is uninteresting [16], because of non-monotonicity. Recent work proposed the concept of *strong inconsistency* for non-monotonic logics [15,16], which enabled demonstrating that well-known properties of inconsistent sets in monotonic logics also apply in the case of strong inconsistency, with a reference example being the minimal hitting set relationship between minimal inconsistent subsets and minimal correction subsets [46]. Nevertheless, a limitation of this earlier work is

This research is supported by the Spanish Government under project TIN2016-79190-R, by the Principality of Asturias under grant IDI/2018/000176, and by ANITI, funded by the French program "Investing for the Future—PIA3" under Grant agreement n$^o$ ANR-19-PI3A-0004.

L. Pulina and M. Seidl (Eds.): SAT 2020, LNCS 12178, pp. 332–342, 2020.
https://doi.org/10.1007/978-3-030-51825-7_24

that the algorithms proposed aim at being illustrative, consisting of simple set enumeration approaches, known not to scale in practice [34].

This paper changes this state of affairs. Concretely, the paper proposes novel simple insights, which enable *any* algorithm for reasoning about inconsistency in the monotonic cases, to also be applicable to reasoning about strong inconsistency in the non-monotonic cases. The paper demonstrates the proposed ideas in the concrete setting of Answer Set Programming (ASP) [14,22], but these can be applied in other settings provided mild conditions hold. The significance of being able to reason efficiently about (strong) inconsistency in ASP should be highlighted. Whereas SAT reasoners represent a remarkable (and unique) problem solving technology, ASP blends efficient problem solving (by exploiting the technologies that are the hallmark of SAT solvers) with a well-established and widely used knowledge representation paradigm. The proposed algorithms enable new applications of ASP based on reasoning about (strong) inconsistency.

## 2    Preliminaries

*Boolean Satisfiability.* We consider definitions and notation standard in Boolean Satisfiability (SAT) [10]. Concretely, we consider propositional formulas in *conjunctive normal form* (CNF), defined as a conjunction, or set, of clauses $\mathcal{F} = \{c_1, ..., c_m\}$ over a set of variables $V(\mathcal{F}) = \{x_1, ..., x_n\}$ where a clause is a disjunction of literals, and a literal is a variable $x$ or its negation $\neg x$. An interpretation is a mapping $\mu: V(\mathcal{F}) \to \{0, 1\}$. If $\mu$ satisfies $\mathcal{F}$, it is referred to as a *model* of $\mathcal{F}$. $\mathcal{F} \vDash \mathcal{G}$ means that all the models of $\mathcal{F}$ are models of $\mathcal{G}$. A minimal (resp. maximal) model is such that the set of variables assigned value 1 (resp. 0) is irreducible. A formula is satisfiable ($\mathcal{F} \nvDash \bot$) if it has a model; and otherwise unsatisfiable ($\mathcal{F} \vDash \bot$). In the latter case, the following definitions apply:

**Definition 1 (MUS/MCS).** $\mathcal{M} \subseteq \mathcal{F}$ *is a* minimal unsatisfiable subset *(MUS) if and only if* $\mathcal{M} \vDash \bot$ *and for all* $\mathcal{M}' \subsetneq \mathcal{M}$, $\mathcal{M}' \nvDash \bot$. $\mathcal{C} \subseteq \mathcal{F}$ *is a* minimal correction subset *(MCS) if and only if* $(\mathcal{F} \backslash \mathcal{C}) \nvDash \bot$ *and for all* $\mathcal{C}' \subsetneq \mathcal{C}$, $\mathcal{F} \backslash \mathcal{C}' \vDash \bot$.

MUSes are minimal explanations of unsatisfiability, while MCSes are irreducible sets of clauses whose removal renders satisfiability. The complement of an MCS is a *maximal satisfiable subset* (MSS). MUSes and MCSes are hitting set duals: Every MCS is a minimal hitting set of all MUSes and vice versa [11,46].

*Example 1.* Let $F_{ex} = \{(\neg x_1), (x_1), (x_1 \vee x_2), (\neg x_2)\}$. $F_{ex}$ is unsatisfiable. It has two MUSes: $\mathcal{M}_1 = \{(x_1), (\neg x_1)\}$, $\mathcal{M}_2 = \{(\neg x_1), (x_1 \vee x_2), (\neg x_2)\}$; and three MCSes: $\mathcal{C}_1 = \{(\neg x_1)\}$, $\mathcal{C}_2 = \{(x_1), (x_1 \vee x_2)\}$, $\mathcal{C}_3 = \{(x_1), (\neg x_2)\}$.

*Minimal Sets over a Monotone Predicate.* Several problems in propositional logic can be reduced to computing a minimal set over a monotone predicate (MSMP) [37,38][1]. In this setting, a predicate $p: 2^\mathcal{R} \to \{0, 1\}$, defined over a

---

[1]  MSMP was proposed in [37,38], but it was inspired by earlier work [12,13].

reference set $\mathcal{R}$, is *monotone* if whenever $p(\mathcal{R}_0)$ holds, then $p(\mathcal{R}_1)$ also holds, with $\mathcal{R}_0 \subseteq \mathcal{R}_1 \subseteq \mathcal{R}$. $\mathcal{M} \subseteq \mathcal{R}$ is a minimal set over a predicate $p$ if $p(\mathcal{M})$ holds and, for all $\mathcal{M}' \subsetneq \mathcal{M}$, $p(\mathcal{M}')$ does not hold. As an example, given $\mathcal{F} \models \bot$, by setting $\mathcal{R} \triangleq \mathcal{F}$, the MUSes of $\mathcal{F}$ are the minimal sets over the monotone predicate $p(\mathcal{W}) \triangleq \neg\mathsf{SAT}(\mathcal{W})$, with $\mathcal{W} \subseteq \mathcal{R}$. The MCSes of $\mathcal{F}$ are the minimal sets over $p(\mathcal{W}) \triangleq \mathsf{SAT}(\mathcal{R}\backslash\mathcal{W})$, with $\mathcal{W} \subseteq \mathcal{R}$.

*Answer Set Programming & Strong Inconsistency.* We review basic concepts in ASP. A more detailed account can be found in [14,22].

A *(normal) logic program* $P = \{r_1, ..., r_n\}$ is a finite set of rules of the following form: $a \leftarrow b_1, ..., b_m, \mathsf{not}\ c_{m+1}, ..., \mathsf{not}\ c_n$, where $a$, $b_i$ and $c_i$ are atoms. A *literal* is an atom or its default negation $\mathsf{not}\ a$. Extended logic programs may include classical negation ($\neg$). For a rule $r$, $\mathsf{body}(r)$ denotes the literals $b_1, ..., b_m, \mathsf{not}\ c_{m+1}, ..., \mathsf{not}\ c_n$ and $\mathsf{head}(r)$ denotes the literal $a$. We write $B^+(r)$ for $b_1, ..., b_m$ and $B^-(r)$ for $c_{m+1}, ..., c_n$. A rule is a *fact* if it has an empty body. Further, we allow *choice rules* of the form $n \leq \{a_1, ..., a_k\}$, with $n \geq 0$. A program is *ground* if it does not contain any variables. A *ground instance* of a program $P$, denoted $\mathsf{grd}(P)$, is a ground program obtained by substituting the variables of $P$ by all constants from its Herbrand universe.

The semantics of ASP programs can be defined via a *reduct* [25]. A set $I$ of ground atoms is a *model* of a program $P$ if $\mathsf{head}(r) \in I$ whenever $B^+(r) \subseteq I$ and $B^-(r) \cap I = \emptyset$ for every $r \in \mathsf{grd}(P)$. The reduct of $P$ w.r.t. the set $I$, denoted $P^I$, is defined as $P^I = \{\mathsf{head}(r) \leftarrow B^+(r) \mid r \in \mathsf{grd}(P), I \cap B^-(r) = \emptyset\}$. The set $I$ is an *answer set* of $P$ if $I$ is a minimal model of $P^I$. The inclusion of choice rule $n \leq \{a_1, ..., a_k\}$ guarantees that any answer set contains at least $n$ atoms from $\{a_1, ..., a_k\}$. A program $P$ is *consistent* if it has at least one consistent answer set; otherwise, $P$ is *inconsistent*.

This paper focuses on the analysis of inconsistent ASP programs. Throughout, we will consider that programs are partitioned into two subsets: $P = B \cup S$, where $B$ denotes *background knowledge*, assumed to be consistent and which cannot be relaxed, and $S$ denotes the set of rules that can be dropped to achieve consistency. In contrast to propositional logic, logical entailment is not monotonic in ASP. Hence, supersets of an inconsistent program are not necessarily inconsistent, and a subset of a consistent program may be inconsistent. This way, MUSes and MCSes as defined for propositional logic do not capture their intended meaning and properties. To overcome this drawback, the notion of *strong inconsistency* [15,16][2] was recently proposed: Given an inconsistent program $P = B \cup S$, $P' = B \cup S'$, with $S' \subseteq S$, is *strongly P-inconsistent* if for all $S''$, with $S' \subseteq S'' \subseteq S$, $B \cup S''$ is inconsistent. In other words, strong inconsistency denotes that all supersets (up to $P$) of a given subprogram are inconsistent. Minimal explanations and corrections of inconsistent ASP programs can be defined in terms of strong inconsistency, as follows:

**Definition 2 (MSIS/MSICS).** *Given an inconsistent program $P = B \cup S$, the subset $M \subseteq S$ is a minimal strongly P-inconsistent subset (MSIS) iff $B \cup$*

---

[2] This notion was defined for arbitrary non-monotonic logics. We show it for ASP.

$M$ is strongly $P$-inconsistent and, for all $M' \subsetneq M$, $B \cup M'$ is not strongly $P$-inconsistent. $C \subseteq S$ is a minimal strong $P$-inconsistency correction subset (MSICS) iff $B \cup (S \backslash C)$ is not strongly $P$-inconsistent and, for all $C' \subsetneq C$, $B \cup (S \backslash C')$ is strongly $P$-inconsistent.

The complement of an MSICS is a *maximal consistent subset*. Besides, every MSIS is a minimal hitting set of the set of all MSICSes and vice versa [15,16].

*Example 2.* Consider the inconsistent program $P_{ex} = B_{ex} \cup S_{ex}$, with $B_{ex} = \emptyset$ and $S_{ex} = \{r_1 : a \leftarrow \text{not } a, \text{not } b., r_2 : b \leftarrow \text{not } a., r_3 : \neg b.\}$. There are two MSISes: $M_1 = \{r_1, r_3\}$, $M_2 = \{r_2, r_3\}$; and two MSICSes: $C_1 = \{r_1, r_2\}$, $C_2 = \{r_3\}$. Notice that although $\{r_1\}$ is inconsistent, it is not strongly $P_{ex}$-inconsistent, since $\{r_1, r_2\}$ is consistent (with the only answer set $\{b\}$).

*Related Work.* Debugging ASP programs has attracted a large body of research (see [20] for a survey). Systems as `spock` [24] or `Ouroboros` [43,44], based on meta-programming, enable pinpointing errors causing inconsistency, as unsupported atoms or unsatisfied rules. On the other hand, `DWASP` [19] allows for interactively debugging ASP programs by exploiting unsatisfiable cores. In contrast, our goal is computing MSISes and MSICSes, in the case of strong inconsistency. Our work is closely related to [30,31], which extended a number of algorithms for MSSes in SAT to maximal consistent subsets in ASP (and so MSICSes). Herein, we focus on computing MSISes as well, and on enumerating both kinds of sets. To our best knowledge, the only proposed approach for computing MSISes [15,16] relies on exhaustive set enumeration and was not evaluated empirically.

# 3    Reasoning About Strongly Inconsistent ASP Programs

## 3.1    Strong Inconsistency and MSMP

Strong inconsistency exhibits a monotonicity property, that all the supersets (up to $P$) of a strongly $P$-inconsistent program are strongly $P$-inconsistent too:

**Proposition 1.** *Let $P = B \cup S$, and $P_U = B \cup U$, with $U \subseteq S$, be strongly $P$-inconsistent. Then, for all $U \subseteq U' \subseteq S$, $P_{U'} = B \cup U'$ is strongly $P$-inconsistent.*

*Proof.* Since $P_U$ is strongly $P$-inconsistent, for all $U'$ with $U \subseteq U' \subseteq S$, $B \cup U'$ is inconsistent. Hence, for any superset $U'$ with $U \subseteq U' \subseteq S$, it holds that for all $U' \subseteq U'' \subseteq S$, $B \cup U''$ is inconsistent. So, $P_{U'}$ is strongly $P$-inconsistent.    □

Throughout, for a given program $P = B \cup S$, and $R \subseteq S$, $\mathsf{SAT}^+(B, S, R)$ indicates whether there is a superset of $R$ (up to $S$) that together with $B$ is consistent, i.e. it is true iff there exists $R'$, with $R \subseteq R' \subseteq S$, such that $P' = B \cup R'$ is consistent. Noticeably, $\mathsf{SAT}^+(B, S, R)$ is false iff $B \cup R$ is strongly $P$-inconsistent. We show that computing an MSIS is an instance of MSMP.

**Proposition 2.** *Computing an MSIS is an instance of the MSMP problem.*

---

**Algorithm 1:** Deletion-based minimal set computation

---
    **Input**: $p$: Monotone predicate, $\mathcal{R}$: Reference set
    **Output**: $\mathcal{M}$: Minimal set

1   $\mathcal{M} \leftarrow \mathcal{R}$;         `// M is over-approximation`

2   **foreach** $u \in \mathcal{M}$ **do**        `// Inv: p(M)`

3     **if** $p(\mathcal{M} \setminus \{u\})$ **then**      `// Do we need u?`

4       $\mathcal{M} \leftarrow \mathcal{M} \setminus \{u\}$;      `// If not, drop it`

5   **return** $\mathcal{M}$;       `// Final M is a minimal set`

---

*Proof.* Let $p(\mathcal{W}) \triangleq \neg\mathsf{SAT}^+(B, S, \mathcal{W})$ with $\mathcal{W} \subseteq \mathcal{R}$, and $\mathcal{R} \triangleq S$. We prove that $p$ is monotone and that any minimal set over $p$ is an MSIS of $P = B \cup S$.
*Monotonicity*: If $p(\mathcal{W})$ holds, $B \cup \mathcal{W}$ is strongly $P$-inconsistent. By Proposition 1, for all $\mathcal{W}'$, with $\mathcal{W} \subseteq \mathcal{W}' \subseteq S$, $B \cup \mathcal{W}'$ is strongly $P$-inconsistent, so $p(\mathcal{W}')$ holds.
*Correctness*: Let $\mathcal{M}$ be a minimal set for which $p(\mathcal{M})$ holds, i.e. $B \cup \mathcal{M}$ is strongly $P$-inconsistent. Since $\mathcal{M}$ is minimal, for any $\mathcal{M}' \subsetneq \mathcal{M}$, $p(\mathcal{M}')$ does not hold, i.e. $B \cup \mathcal{M}'$ is not strongly $P$-inconsistent. Thus, by Definition 2, $\mathcal{M}$ is an MSIS. $\square$

Computing an MSICS can also be reduced to MSMP. The proof is analogous, by defining $p(\mathcal{W}) \triangleq \mathsf{SAT}^+(B, S, S \setminus \mathcal{W})$ with $\mathcal{W} \subseteq \mathcal{R}$, and $\mathcal{R} \triangleq S$.

### 3.2 Computing Minimal Explanations and Corrections

The reductions above enable computing MSISes and MSICSes by using *any* algorithm for MSMP and an oracle implementing $\mathsf{SAT}^+(B, S, R)$.

*Extracting a Single Minimal Set.* Algorithms for computing a single minimal set in MSMP include Deletion [17], Progression [37] or QuickXplain [32], among others [8]. Herein we focus on the deletion-based approach, shown in Algorithm 1.

Given an inconsistent program $P = B \cup S$, by setting the predicate to $p(\mathcal{W}) \triangleq \neg\mathsf{SAT}^+(B, S, \mathcal{W})$ with $\mathcal{W} \subseteq \mathcal{R}$, and $\mathcal{R} \triangleq S$, Algorithm 1 proceeds as follows: Starting with $\mathcal{M} = \mathcal{R}$, the algorithm iteratively picks a rule $u \in \mathcal{M}$ and tests whether $B \cup (\mathcal{M} \setminus \{u\})$ is strongly $P$-inconsistent. If it is, $u$ is removed from $\mathcal{M}$; otherwise $u$ is kept in $\mathcal{M}$. After considering all the rules in $\mathcal{R}$, $\mathcal{M}$ is an MSIS.

An MSICS of $P$ can be computed using *basic linear search* (BLS) [6,36]: Starting with $\mathcal{S} = \emptyset$, iteratively pick a rule in $u \in S \setminus \mathcal{S}$ and test whether $\mathsf{SAT}^+(B, S, \mathcal{S} \cup \{u\})$ holds. If it does, $B \cup (\mathcal{S} \cup \{u\})$ is not strongly $P$-inconsistent, and $u$ is added to $\mathcal{S}$. On termination, the set of rules not added to $\mathcal{S}$ is an MSICS (and $\mathcal{S}$ is a maximal consistent subset). Besides, if the oracle for $\mathsf{SAT}^+(B, S, \mathcal{S} \cup \{u\})$ returns a witness after positive answers, all the elements in $S$ satisfied can be added to $\mathcal{S}$, saving some predicate tests. BLS is equivalent to Algorithm 1 using the predicate $p(\mathcal{W}) \triangleq \mathsf{SAT}^+(B, S, S \setminus \mathcal{W})$, with $\mathcal{W} \subseteq \mathcal{R}$, and $\mathcal{R} \triangleq S$.

---

**Algorithm 2:** Minimal set enumeration

---

    **Input:** $P = B \cup S$: Inconsistent ASP program
    **Output:** MSISes and MSICSes of $P$

1    $I \leftarrow \{p_i \mid r_i \in S\}$;

2    $\mathcal{H} \leftarrow \emptyset$;                  `// Block MSISes and MSICSes`

3    **while** true **do**

4      $(st, MxM) \leftarrow$ ComputeMaximalModel$(\mathcal{H})$;

5      **if not** $st$ **then return**

6      $R \leftarrow \{r_i \mid p_i \in MxM\}$;        `// Pick selected rules`

7      **if not** SAT$^+(B, S, R)$ **then**

8        $M \leftarrow$ ComputeMSIS$(B, S, R)$;    `// Extract MSIS from` $R$

9        ReportMSIS$(M)$;

10      $b \leftarrow \{\neg p_i \mid r_i \in M\}$;       `// Block the MSIS`

11      **else**

12        ReportMSICS$(S \setminus R)$;

13        $b \leftarrow \{p_i \mid p_i \in I \setminus MxM\}$;   `// Block the MSICS`

14      $\mathcal{H} \leftarrow \mathcal{H} \cup \{b\}$;

---

*Enumerating Minimal Sets.* MARCO [33] is a successful approach for enumerating MUSes and MCSes of CNF formulas. This algorithm exploits the hitting set duality between MUSes and MCSes. Since this relationship also holds between MSISes and MSICSes, MARCO can be adapted to ASP, as shown in Algorithm 2.

For a given inconsistent program $P = B \cup S$, the algorithm associates a propositional variable $p_i$ with each rule $r_i \in S$, and maintains a CNF formula $\mathcal{H}$ defined on these variables. The formula $\mathcal{H}$, initially empty, serves to subsequently avoid considering any superset (resp. subset) of previously found MSISes (resp. MSICSes). Iteratively, a maximal model $MxM$ of $\mathcal{H}$ is computed, which induces the set of rules $R$ whose associated variables are set to 1 in $MxM$. Then, if the program $B \cup R$ is strongly $P$-inconsistent (i.e. if SAT$^+(B, S, R)$ does not hold), an MSIS $M \subseteq R$ of $P$ is extracted (e.g. by using Algorithm 1, with $\mathcal{R} \triangleq R$), whose supersets are blocked by adding a negative clause on its associated variables to $\mathcal{H}$. Otherwise, $R$ is a maximal consistent subset, and so $S \setminus R$ is an MSICS of $P$, whose subsets are blocked by adding a positive clause on its associated variables to $\mathcal{H}$. The process is repeated until $\mathcal{H}$ becomes unsatisfiable, with the guarantee that all MSISes and MSICSes of $P$ have been computed.

Algorithm 2 is organized to give (heuristic) preference to finding MSISes quickly. We refer to it as eMax. A variant giving preference to finding MSICSes can be easily obtained, by computing minimal models of $\mathcal{H}$ (instead of maximal ones) and extracting an MSICS whenever SAT$^+(B, S, R)$ holds. This variant is referred to as eMin.

*Implementing* SAT$^+(B, S, R)$. It remains to discuss the way SAT$^+(B, S, R)$ can be implemented in ASP. We invoke an ASP solver on an modified program which includes *selector* atoms and choice rules. This approach was used in [30,31] to compute maximal consistent subsets. For a set of atoms $\mathcal{A}$, choice$(\mathcal{A})$ denotes the rule $0 \leq \{a_1, .., a_k\}$, with $a_i \in \mathcal{A}$. Modern ASP solvers allow choice rules, and their inclusion does not increase the complexity beyond NP [47].

For a given program $P = B \cup S$, we first build the program $P_s = B \cup S_s$, where $S_s$ is obtained from $S$ as follows: for each rule $r_i \in S$ we introduce a fresh atom $s_i$, and add the rule $\mathsf{head}(r_i) \leftarrow \mathsf{body}(r_i), s_i$ to $S_s$. Note that if the fact $s_i$ is added to $P_s$, the rule $r_i$ is *activated*, and relaxed otherwise. For a given subset $R \subseteq S$, we use $s(R)$ to denote the set of selector atoms for rules in $R$ in $S_s$, i.e. $s(R) = \{s_i \mid r_i \in R\}$. Then, the test $\mathsf{SAT}^+(B, S, R)$ is solved by invoking an ASP solver on the program $P' = P_s \cup \bigcup_{s \in s(R)} \{s\} \cup \mathsf{choice}(s(S \backslash R))$. Notice that each rule $r \in R$ is active in $P'$. Besides, the inclusion of the rule $\mathsf{choice}(s(S \backslash R))$ allows for activating any (or none) of the rules in $S \backslash R$ when looking for answer sets of $P'$. Hence, $P'$ is consistent iff the program $B \cup R$ is not strongly $P$-inconsistent.

*Example 3.* Let $P = B \cup S$ be the program in Example 2, and consider the test $\mathsf{SAT}^+(B, S, \{r_1\})$. We first build $P_s = \{a \leftarrow \mathsf{not}\, a, \mathsf{not}\, b, s_1., \quad b \leftarrow \mathsf{not}\, a, s_2., \neg b \leftarrow s_3.\}$. Then, we define $P' = P_s \cup \{s_1.\} \cup \mathsf{choice}(\{s_2, s_3\})$. $P'$ is consistent (with the unique answer set $\{b, s_1, s_2\}$), indicating that $\{r_1\}$ is not strongly $P$-inconsistent.

## 4    Preliminary Results

This section reports an initial experimental assessment of the proposed approaches. We implemented a prototype in Python 2.7, interfacing the ASP solver `clingo` [23] (v. 5.4.0), and ran a series of experiments on a Linux machine (2.26 GHz, 128 GB). Each process was limited to 3600 s and 4 GB. Below, `ComputeMSIS` (resp. `ComputeMSICS`) is Algorithm 1 using the predicate shown in Sect. 3 for computing an MSIS (resp. MSICS). Besides, witnesses are used in the extraction of MSICSes as an optimization, as described earlier. On the other hand, `eMax` corresponds to Algorithm 2, giving preference to finding MSISes quickly, and `eMin` is the variant that gives preference to MSICSes. In these cases, maximal and minimal models are computed using the tool mcsls [36][3].

Similarly to earlier work [31], we built a number of instances. We considered three problem domains (common in ASP competitions): *Graceful graphs, Knight tour with holes* and *Solitaire*. Each instance is an inconsistent ASP program $P = B \cup S$, where $B$ contains the rules encoding the problem domain (assumed correct) and $S$ contains the facts specific for each instance. Given the complexity of the tasks to solve, the instances are reasonably small. The benchmarks are as follows: *1) Graceful graphs*: Given a graph $(V, E)$ the goal is to label its vertices with distinct integers in the range $0..|E|$ so that each edge is labeled with the absolute difference between the labels of its vertices and all edge labels are distinct. $S$ contains the facts indicating the edges, so $|S| = |E|$. We considered values of $|V| \in \{10, 20\}$ and $|E| \in \{10, 20, 50\}$. *2) Knight tour with holes*: Given

---

[3] Computing a minimal/maximal model can be reduced to computing an MCS. For this purpose, several alternatives can be used (e.g. [3,36,39,40]).

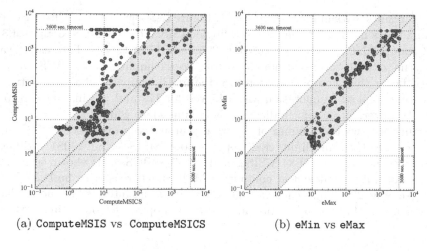

(a) ComputeMSIS vs ComputeMSICS    (b) eMin vs eMax

**Fig. 1.** Running times

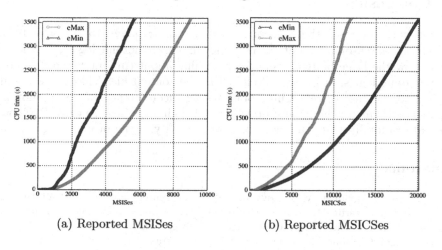

(a) Reported MSISes    (b) Reported MSICSes

**Fig. 2.** Number of reported sets (eMin vs eMax)

an $N \times N$ board with $H$ holes, the problem asks if a knight chess piece can visit all non-hole positions of the boards exactly once returning to the initial position. $S$ consists of facts with the positions of holes, so $|S| = H$. We considered values of $N \in \{7, 8\}$ and $H \in \{10, 20, 30\}$. *3) Solitaire*: Given a $7 \times 7$ board, with $2 \times 2$ corners removed (i.e. with 33 squares), an initial configuration is specified by facts $empty(L)$ and $full(L)$, indicating if each square $L$ is empty or contains a stone. A stone can be moved by two squares if it jumps over another stone, which is removed. The goal is to perform $T$ steps. $S$ contains the facts $empty(L)$ and $full(L)$, so $|S| = 33$. We considered values of $T \in \{8, 10, 12, 14, 16, 18\}$. For each configuration, we built 20 random instances, making 360 in all.

The results are summarized in Fig. 1. Figure 1a shows, for each instance, the running times needed for computing a single MSIS and an MSICS. `ComputeMSIS` and `ComputeMSICS` solved, respectively, 295 and 317 instances. The results vary across the set of instances, although in more cases computing an MSICS was performed faster than computing an MSIS. Figure 1b compares `eMax` and `eMin`. In this case, complete enumeration was achieved for 172 and 167 instances respectively. However, as the plot indicates, there is no clear winner.

Figure 2 shows the number of reported minimal sets over the whole benchmark set. By the time limit `eMax` reports 9008 MSISes and 12081 MSICSes, whereas `eMin` computes 5684 MSISes and 20057 MSICSes. As shown in Fig. 2a, `eMax` is much more efficient at computing MSISes, whereas `eMin` finds MSICSes faster (see Fig. 2b). Thus, each variant is effective at its intended purpose. These results suggest that a combination may be a good option for obtaining both sets quickly.

## 5   Conclusions

Recent work proposed the concept of *strong inconsistency* [15,16], which provides a way of reasoning about inconsistency in non-monotonic logics. This paper shows how the large body of work for reasoning about (minimal) inconsistency in monotonic logics, originally developed in the context of SAT, can be readily applied to the case of reasoning about strong inconsistency in non-monotonic logics. Furthermore, the paper applies these insights to the case of ASP. Experimental results illustrate the scope and applicability of the proposed approach.

## References

1. Ansótegui, C., Bonet, M.L., Levy, J.: Solving (weighted) partial maxSAT through satisfiability testing. In: Kullmann, O. (ed.) SAT 2009. LNCS, vol. 5584, pp. 427–440. Springer, Heidelberg (2009). https://doi.org/10.1007/978-3-642-02777-2_39
2. Ansótegui, C., Bonet, M.L., Levy, J.: SAT-based MaxSAT algorithms. Artif. Intell. **196**, 77–105 (2013)
3. Bacchus, F., Davies, J., Tsimpoukelli, M., Katsirelos, G.: Relaxation search: a simple way of managing optional clauses. In: AAAI, pp. 835–841 (2014)
4. Bacchus, F., Katsirelos, G.: Using minimal correction sets to more efficiently compute minimal unsatisfiable sets. In: Kroening, D., Pǎsǎreanu, C.S. (eds.) CAV 2015. LNCS, vol. 9207, pp. 70–86. Springer, Cham (2015). https://doi.org/10.1007/978-3-319-21668-3_5
5. Bacchus, F., Katsirelos, G.: Finding a collection of MUSes incrementally. In: Quimper, C.-G. (ed.) CPAIOR 2016. LNCS, vol. 9676, pp. 35–44. Springer, Cham (2016). https://doi.org/10.1007/978-3-319-33954-2_3
6. Bailey, J., Stuckey, P.J.: Discovery of minimal unsatisfiable subsets of constraints using hitting set dualization. In: Hermenegildo, M.V., Cabeza, D. (eds.) PADL 2005. LNCS, vol. 3350, pp. 174–186. Springer, Heidelberg (2005). https://doi.org/10.1007/978-3-540-30557-6_14
7. Bakker, R.R., Dikker, F., Tempelman, F., Wognum, P.M.: Diagnosing and solving over-determined constraint satisfaction problems. In: IJCAI, pp. 276–281 (1993)

8. Belov, A., Lynce, I., Marques-Silva, J.: Towards efficient MUS extraction. AI Commun. **25**(2), 97–116 (2012)
9. Bendík, J., Černá, I., Beneš, N.: Recursive online enumeration of all minimal unsatisfiable subsets. In: Lahiri, S.K., Wang, C. (eds.) ATVA 2018. LNCS, vol. 11138, pp. 143–159. Springer, Cham (2018). https://doi.org/10.1007/978-3-030-01090-4_9
10. Biere, A., Heule, M., van Maaren, H., Walsh, T. (eds.): Handbook of Satisfiability, Frontiers in Artificial Intelligence and Applications, vol. 185. IOS Press, Amsterdam (2009)
11. Birnbaum, E., Lozinskii, E.L.: Consistent subsets of inconsistent systems: structure and behaviour. J. Exp. Theor. Artif. Intell. **15**(1), 25–46 (2003)
12. Bradley, A.R., Manna, Z.: Checking safety by inductive generalization of counterexamples to induction. In: FMCAD, pp. 173–180 (2007)
13. Bradley, A.R., Manna, Z.: Property-directed incremental invariant generation. Formal Aspects of Comput. **20**(4–5), 379–405 (2008). https://doi.org/10.1007/s00165-008-0080-9
14. Brewka, G., Eiter, T., Truszczynski, M.: Answer set programming at a glance. Commun. ACM **54**(12), 92–103 (2011)
15. Brewka, G., Thimm, M., Ulbricht, M.: Strong inconsistency in nonmonotonic reasoning. In: IJCAI, pp. 901–907 (2017)
16. Brewka, G., Thimm, M., Ulbricht, M.: Strong inconsistency. Artif. Intell. **267**, 78–117 (2019)
17. Chinneck, J.W., Dravnieks, E.W.: Locating minimal infeasible constraint sets in linear programs. ORSA J. Comput. **3**(2), 157–168 (1991)
18. Davies, J., Bacchus, F.: Solving MAXSAT by solving a sequence of simpler SAT instances. In: Lee, J. (ed.) CP 2011. LNCS, vol. 6876, pp. 225–239. Springer, Heidelberg (2011). https://doi.org/10.1007/978-3-642-23786-7_19
19. Dodaro, C., Gasteiger, P., Musitsch, B., Ricca, F., Shchekotykhin, K.: Interactive debugging of non-ground ASP programs. In: Calimeri, F., Ianni, G., Truszczynski, M. (eds.) LPNMR 2015. LNCS (LNAI), vol. 9345, pp. 279–293. Springer, Cham (2015). https://doi.org/10.1007/978-3-319-23264-5_24
20. Fandinno, J., Schulz, C.: Answering the "why" in answer set programming - a survey of explanation approaches. Theory Pract. Log. Program. **19**(2), 114–203 (2019)
21. Felfernig, A., Schubert, M., Zehentner, C.: An efficient diagnosis algorithm for inconsistent constraint sets. AI EDAM **26**(1), 53–62 (2012)
22. Gebser, M., Kaminski, R., Kaufmann, B., Schaub, T.: Answer Set Solving in Practice. Synthesis Lectures on Artificial Intelligence and Machine Learning. Morgan & Claypool Publishers, San Rafael (2012)
23. Gebser, M., Kaminski, R., Kaufmann, B., Schaub, T.: Clingo = ASP + control: extended report. Technical report, University of Potsdam (2014)
24. Gebser, M., Pührer, J., Schaub, T., Tompits, H.: A meta-programming technique for debugging answer-set programs. In: AAAI, pp. 448–453 (2008)
25. Gelfond, M., Lifschitz, V.: The stable model semantics for logic programming. In: ICLP/SLP, pp. 1070–1080 (1988)
26. Grégoire, É., Izza, Y., Lagniez, J.: Boosting MCSes enumeration. In: IJCAI, pp. 1309–1315 (2018)
27. Grégoire, É., Lagniez, J., Mazure, B.: An experimentally efficient method for (MSS, CoMSS) partitioning. In: AAAI, pp. 2666–2673 (2014)
28. Grégoire, É., Mazure, B., Piette, C.: Extracting MUSes. In: ECAI, pp. 387–391 (2006)

29. Hemery, F., Lecoutre, C., Sais, L., Boussemart, F.: Extracting MUCs from constraint networks. In: ECAI, pp. 113–117 (2006)
30. Janota, M., Marques-Silva, J.: On minimal corrections in ASP. CoRR abs/1406.7838 (2014). http://arxiv.org/abs/1406.7838
31. Janota, M., Marques-Silva, J.: On minimal corrections in ASP. In: RCRA, pp. 45–54 (2017). http://ceur-ws.org/Vol-2011/paper5.pdf
32. Junker, U.: QUICKXPLAIN: preferred explanations and relaxations for over-constrained problems. In: AAAI, pp. 167–172 (2004)
33. Liffiton, M.H., Previti, A., Malik, A., Marques-Silva, J.: Fast, flexible MUS enumeration. Constraints 21(2), 223–250 (2015). https://doi.org/10.1007/s10601-015-9183-0
34. Liffiton, M.H., Sakallah, K.A.: Algorithms for computing minimal unsatisfiable subsets of constraints. J. Autom. Reasoning 40(1), 1–33 (2008). https://doi.org/10.1007/s10817-007-9084-z
35. Manquinho, V., Marques-Silva, J., Planes, J.: Algorithms for weighted Boolean optimization. In: Kullmann, O. (ed.) SAT 2009. LNCS, vol. 5584, pp. 495–508. Springer, Heidelberg (2009). https://doi.org/10.1007/978-3-642-02777-2_45
36. Marques-Silva, J., Heras, F., Janota, M., Previti, A., Belov, A.: On computing minimal correction subsets. In: IJCAI, pp. 615–622 (2013)
37. Marques-Silva, J., Janota, M., Belov, A.: Minimal sets over monotone predicates in Boolean formulae. In: Sharygina, N., Veith, H. (eds.) CAV 2013. LNCS, vol. 8044, pp. 592–607. Springer, Heidelberg (2013). https://doi.org/10.1007/978-3-642-39799-8_39
38. Marques-Silva, J., Janota, M., Mencía, C.: Minimal sets on propositional formulae. Problems and reductions. Artif. Intell. 252, 22–50 (2017)
39. Mencía, C., Ignatiev, A., Previti, A., Marques-Silva, J.: MCS extraction with sublinear oracle queries. In: Creignou, N., Le Berre, D. (eds.) SAT 2016. LNCS, vol. 9710, pp. 342–360. Springer, Cham (2016). https://doi.org/10.1007/978-3-319-40970-2_21
40. Mencía, C., Previti, A., Marques-Silva, J.: Literal-based MCS extraction. In: IJCAI, pp. 1973–1979 (2015)
41. Morgado, A., Heras, F., Liffiton, M.H., Planes, J., Marques-Silva, J.: Iterative and core-guided MaxSAT solving: a survey and assessment. Constraints 18(4), 478–534 (2013). https://doi.org/10.1007/s10601-013-9146-2
42. Narodytska, N., Bjørner, N., Marinescu, M.V., Sagiv, M.: Core-guided minimal correction set and core enumeration. In: IJCAI, pp. 1353–1361 (2018)
43. Oetsch, J., Pührer, J., Tompits, H.: Catching the ouroboros: on debugging non-ground answer-set programs. Theory Pract. Log. Program. 10(4–6), 513–529 (2010)
44. Polleres, A., Frühstück, M., Schenner, G., Friedrich, G.: Debugging non-ground ASP programs with choice rules, cardinality and weight constraints. In: Cabalar, P., Son, T.C. (eds.) LPNMR 2013. LNCS (LNAI), vol. 8148, pp. 452–464. Springer, Heidelberg (2013). https://doi.org/10.1007/978-3-642-40564-8_45
45. Previti, A., Mencía, C., Järvisalo, M., Marques-Silva, J.: Premise set caching for enumerating minimal correction subsets. In: AAAI, pp. 6633–6640 (2018)
46. Reiter, R.: A theory of diagnosis from first principles. Artif. Intell. 32(1), 57–95 (1987)
47. Simons, P., Niemelä, I., Soininen, T.: Extending and implementing the stable model semantics. Artif. Intell. 138(1–2), 181–234 (2002)
48. de Siqueira N., J.L., Puget, J.: Explanation-based generalisation of failures. In: ECAI, pp. 339–344 (1988)

# Taming High Treewidth with Abstraction, Nested Dynamic Programming, and Database Technology

Markus Hecher$^{(\boxtimes)}$, Patrick Thier$^{(\boxtimes)}$, and Stefan Woltran$^{(\boxtimes)}$

Institute of Logic and Computation, TU Wien, Vienna, Austria
{hecher,thier,woltran}@dbai.tuwien.ac.at

**Abstract.** Treewidth is one of the most prominent structural parameters. While numerous theoretical results establish tractability under the assumption of fixed treewidth, the practical success of exploiting this parameter is far behind what theoretical runtime bounds have promised. In particular, a naive application of dynamic programming (DP) on tree decompositions (TDs) suffers already from instances of medium width. In this paper, we present several measures to advance this paradigm towards general applicability in practice: We present nested DP, where different levels of abstractions are used to (recursively) compute TDs of a given instance. Further, we integrate the concept of hybrid solving, where subproblems hidden by the abstraction are solved by classical search-based solvers, which leads to an interleaving of parameterized and classical solving. Finally, we provide nested DP algorithms and implementations relying on database technology for variants and extensions of Boolean satisfiability. Experiments indicate that the advancements are promising.

## 1 Introduction

Treewidth [43] is a prominent structural parameter, originating from graph theory and is well-studied in the area of parameterized complexity [6,18,40]. For several problems hard for complexity class NP, there are results [12] showing so-called (fixed-parameter) tractability, which indicates a *fixed-parameter tractable (FPT)* algorithm running in polynomial time assuming that a given parameter (e.g., treewidth) is fixed. Practical implementations exploiting treewidth include generic frameworks [3,5,36], but also dedicated solvers that deal with problems ranging from (counting variants of) Boolean satisfiability (SAT) [25], over generalizations thereof [9,10] based on *Quantified Boolean Formulas (QBFs)*, to formalisms relevant to knowledge representation and reasoning [22]. For SAT, these solvers are of particular interest as there is a well-known correspondence between treewidth and resolution width [2]. QBFs extend Boolean logic by explicit universal and existential quantification over variables, which has applications in formal verification, synthesis, and AI problems such as planning [28]. Some of

© Springer Nature Switzerland AG 2020
L. Pulina and M. Seidl (Eds.): SAT 2020, LNCS 12178, pp. 343–360, 2020.
https://doi.org/10.1007/978-3-030-51825-7_25

these parameterized solvers are particularly efficient for certain fragments [37], and even successfully participated in problem-specific competitions [42].

Most of these systems are based on *dynamic programming (DP)*, where a tree decomposition (TD) is traversed in a post-order, i.e., from the leaves towards the root, and thereby for each TD node tables are computed. The size of these tables (and thus the computational efforts required) are bounded by a function in the treewidth of the instance. Although dedicated competitions [15] for treewidth advanced the state-of-the-art for efficiently computing treewidth and TDs [1,47], these DP approaches reach their limits when instances have higher treewidth; a situation which can even occur in structured real-world instances [38]. Nevertheless in the area of Boolean satisfiability, this approach proved to be successful for counting problems, such as, e.g., (weighted) model counting [24,25,44] and projected model counting [23].

To further increase the applicability of this paradigm, novel techniques are required which (1) rely on different levels of abstraction of the instance at hand; (2) treat subproblems originating in the abstraction by standard solvers whenever widths appear too high; and (3) use highly sophisticated data management in order to store and process tables obtained by dynamic programming.

**Contributions.** Above aspects are treated as follows.

1. To tame the beast of high treewidth, we propose *nested dynamic programming*, where only parts of an abstraction of a graph are decomposed. Then, each TD node also needs to solve a *subproblem* residing in the graph, but may involve vertices outside the abstraction. In turn, for solving such subproblems, the idea of nested DP is to subsequently repeat decomposing and solving more fine-grained graph abstractions in a nested fashion. This results not only in elegant DP algorithms, but also allows to deal with high treewidth. While candidates for obtaining abstractions often originate naturally from the problem, nested DP may require non-obvious sub-abstractions, for which we present a generic solution.
2. To further improve the capability of handling high treewidth, we show how to apply nested DP in the context of *hybrid solving*, where established, standard solvers (e.g., SAT solvers) and caching are incorporated in nested DP such that the best of two worlds are combined. Thereby, structured solving is applied to parts of the problem instance subject to counting or enumeration, while depending on results of subproblems. These subproblems (subject to search) reside in the abstraction only, and are solved via standard solvers.
3. We implemented a system based on a recently published tool called dpdb [24] for using database management systems (DBMS) to efficiently perform table manipulation operations needed during DP. Our system uses and significantly extends this tool in order to perform hybrid solving, thereby combining nested DP and standard solvers. As a result, we use DBMS for efficiently implementing the handling of tables needed by nested DP. Preliminary experiments indicate that nested DP with hybrid solving can be fruitful.

We exemplify these ideas on the problem of Projected Model Counting ($\#\exists$SAT) and discuss adaptions for other problems.

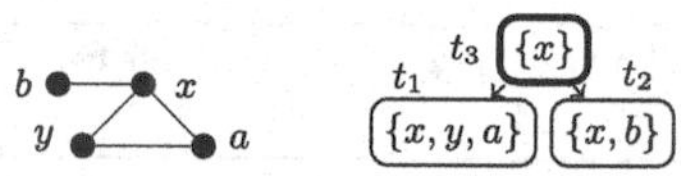

**Fig. 1.** Graph $G$ (left), a TD $\mathcal{T}$ of graph $G$ (right).

## 2 Background

**Projected Model Counting.** We define Boolean formulas in the usual way, cf., [28]. A literal is a Boolean variable $x$ or its negation $\neg x$. A *(CNF) formula* $\varphi$ is a set of *clauses* interpreted as conjunction. A clause is a set of literals interpreted as disjunction. For a formula or clause $X$, we abbreviate by $\text{var}(X)$ the variables that occur in $X$. An *assignment* of $\varphi$ is a mapping $I : \text{var}(\varphi) \to \{0, 1\}$. The formula $\varphi[I]$ *under assignment* $I$ is obtained by removing every clause $c$ from $\varphi$ that contains a literal set to 1 by $I$, and removing from every remaining clause of $\varphi$ all literals set to 0 by $I$. An assignment $I$ is *satisfying* if $\varphi[I] = \emptyset$. *Problem* #SAT asks to output the number of satisfying assignments of a formula. *Projected Model Counting* #∃SAT takes a formula $\varphi$ and a set $P \subseteq \text{var}(\varphi)$ of *projection variables*, and asks for #∃SAT$(\varphi, P) := |\{I^{-1}(1) \cap P \mid \varphi[I] = \emptyset\}|$. Consequently, SAT$(\varphi) := $ #∃SAT$(\varphi, \emptyset)$, and #SAT$(\varphi) := $ #∃SAT$(\varphi, \text{var}(\varphi))$. #∃SAT is #·NP-complete [19] and thus probably harder than #SAT (#P-complete).

**Tree Decomposition and Treewidth.** We assume familiarity with graph terminology, cf., [17]. A *tree decomposition (TD)* [43] of a given graph $G$ is a pair $\mathcal{T} = (T, \chi)$ where $T$ is a rooted tree and $\chi$ assigns to each node $t \in V(T)$ a set $\chi(t) \subseteq V(G)$, called *bag*, such that (i) $V(G) = \bigcup_{t \in V(T)} \chi(t)$, (ii) $E(G) \subseteq \{\, \{u, v\} \mid t \in V(T), \{u, v\} \subseteq \chi(t)\,\}$, and (iii) for each $r, s, t \in V(T)$, such that $s$ lies on the path from $r$ to $t$, we have $\chi(r) \cap \chi(t) \subseteq \chi(s)$. We let width$(\mathcal{T}) := \max_{t \in V(T)} |\chi(t)| - 1$. The *treewidth* tw$(G)$ of $G$ is the minimum width$(\mathcal{T})$ over all TDs $\mathcal{T}$ of $G$. For a node $t \in V(T)$, we say that type$(t)$ is *leaf* if $t$ has no children and $\chi(t) = \emptyset$; *join* if $t$ has children $t'$ and $t''$ with $t' \neq t''$ and $\chi(t) = \chi(t') = \chi(t'')$; *intr* ("introduce") if $t$ has a single child $t'$, $\chi(t') \subseteq \chi(t)$ and $|\chi(t)| = |\chi(t')| + 1$; *rem* ("removal") if $t$ has a single child $t'$, $\chi(t') \supseteq \chi(t)$ and $|\chi(t')| = |\chi(t)| + 1$. If for every node $t \in V(T)$, type$(t) \in \{leaf, join, intr, rem\}$, the TD is called *nice*. A nice TD can be computed from a given TD $\mathcal{T}$ in linear time without increasing the width [31], assuming the width of $\mathcal{T}$ is fixed.

*Example 1.* Figure 1 depicts a graph $G$ and a (non-nice) TD $\mathcal{T}$ of $G$ of width 2.

**Relational Algebra.** We formalize DP algorithms by means of relational algebra [11], similar to related work [24]. A *table* $\tau$ is a finite set of *rows* $r$ over a set att$(\tau)$ of *attributes*. Each *row* $r \in \tau$ is a set of pairs $(a, v)$ with $a \in \text{att}(\tau)$ and $v$ in *domain* dom$(a)$ of $a$, s.t. for each $a \in \text{att}(\tau)$ there is exactly one $(a, v) \in r$. Notably, apart from counters we use mainly binary domains in this paper.

*Selection* of rows in $\tau$ according to a Boolean formula $\varphi$ is defined by $\sigma_\varphi(\tau) := \{r \mid r \in \tau, \varphi[\text{ass}(r)] = \emptyset\}$, assuming that ass$(r)$ refers to the truth assignment over the attributes of binary domain of a given row $r \in \tau$. Given a relation $\tau'$

---

**Listing 1:** Table algorithm $\#\text{SAT}_t(\chi_t, \varphi_t, \langle \tau_1, \ldots, \tau_\ell \rangle)$ for solving $\#\text{SAT}$ on node $t$ of a nice tree decomposition, cf., [24].

---

**In:** Bag $\chi_t$, bag formula $\varphi_t$, child tables $\langle \tau_1, \ldots \tau_\ell \rangle$ of $t$.
**Out:** Table $\tau_t$.

1  **if** type$(t) = leaf$ **then** $\tau_t := \{\{(\text{cnt}, 1)\}\}$
2  **else if** type$(t) = intr$, and $a \in \chi_t$ is introduced **then**
3  $\quad | \quad \tau_t := \tau_1 \bowtie_{\varphi_t} \{\{(a, 0)\}, \{(a, 1)\}\}$
4  **else if** type$(t) = rem$, and $a \notin \chi_t$ is removed **then**
5  $\quad | \quad \tau_t := {}_{\chi_t}G_{\text{cnt}\leftarrow\text{SUM(cnt)}}(\Pi_{\text{att}(\tau_1)\setminus\{a\}}\tau_1)$
6  **else if** type$(t) = join$ **then**
7  $\quad | \quad \tau_t := \dot\Pi_{\chi_t, \{\text{cnt}\leftarrow\text{cnt}\cdot\text{cnt}'\}}(\tau_1 \bowtie_{\bigwedge_{a \in \chi_t} a = a'} \rho_{\bigcup\{a\mapsto a'\}}\tau_2)$
   $\quad\quad\quad\quad\quad\quad\quad\quad\quad\quad\quad\quad\quad\quad\quad\quad {}_{a\in\text{att}(\tau_2)}$

---

with $\text{att}(\tau') \cap \text{att}(\tau) = \emptyset$, we refer to the *cross-join* by $\tau \times \tau' := \{r \cup r' \mid r \in \tau, r' \in \tau'\}$. Further, a *join (using $\varphi$)* corresponds to $\tau \bowtie_\varphi \tau' := \sigma_\varphi(\tau \times \tau')$. We define *renaming* of $\tau$, given a set $A$ of attributes, and a bijective mapping $m : \text{att}(\tau) \to A$ by $\rho_m(\tau) := \{(m(a), v) \mid (a, v) \in \tau\}$. $\tau$ *projected to* $A \subseteq \text{att}(\tau)$ is given by $\Pi_A(\tau) := \{r_A \mid r \in \tau\}$, where $r_A := \{(a, v) \mid (a, v) \in r, a \in A\}$. This is lifted to *extended projection* $\dot\Pi_{A,(a\leftarrow f)}$, assuming attribute $a \in \text{att}(\tau) \setminus A$ and arithmetic function $f : \tau \to \mathbb{N}$. Formally, we define $\dot\Pi_{A,(a\leftarrow f)}(\tau) := \{r_A \cup \{(a, f(r))\} \mid r \in \tau\}$. We use *aggregation by grouping* ${}_AG_{(a\leftarrow g)}$, where we assume $A \subseteq \text{att}(\tau), a \in \text{att}(\tau) \setminus A$ and an *aggregate function* $g : 2^\tau \to \text{dom}(a)$. We define ${}_AG_{(a\leftarrow g)}(\tau) := \{r \cup \{(a, g(\tau[r]))\} \mid r \in \Pi_A(\tau)\}$, where $\tau[r] := \{r' \mid r' \in \tau, r' \supseteq r\}$.

## 3  Towards Nested Dynamic Programming

A solver based on *dynamic programming (DP)* evaluates a given input instance $\mathcal{I}$ in parts along a given TD of a graph representation $G$ of the instance. Thereby, for each node $t$ of the TD, intermediate results are stored in a *table* $\tau_t$. This is achieved by running a so-called *table algorithm*, which is designed for a certain graph representation, and stores in $\tau_t$ results of problem parts of $\mathcal{I}$, thereby considering tables $\tau_{t'}$ for child nodes $t'$ of $t$. DP works for *many problems*:

1. Construct a *graph representation* $G$ of $\mathcal{I}$.
2. Compute (some) tree decomposition $\mathcal{T} = (T, \chi)$ of $G$.
3. Traverse the nodes of $T$ in post-order (bottom-up tree traversal of $T$). At every node $t$ of $T$ during post-order traversal, execute a table algorithm that takes as input bag $\chi(t)$, a certain *bag instance* $\mathcal{I}_t$ depending on the problem, as well as previously computed child tables of $t$. Then, the results of this execution are stored in table $\tau_t$.
4. Finally, interpret table $\tau_n$ for the root node $n$ of $T$ in order to *output the solution* to the problem for instance $\mathcal{I}$.

Having relational algebra and this paradigm at hand, we exemplarily show how to solve $\#\text{SAT}$, required for solving $\#\exists\text{SAT}$ later. To this end, we need the following graph representation for a given formula $\varphi$. The *primal graph* $G_\varphi$ [44]

of a formula $\varphi$ has as vertices its variables, where two variables are joined by an edge if they occur together in a clause of $\varphi$. Given a TD $\mathcal{T} = (T, \chi)$ of $G_\varphi$ and a node $t$ of $T$. Then, we let bag instance $\varphi_t$ of $\varphi$, called *bag formula*, be the clauses $\{\, c \mid c \in \varphi, \mathrm{var}(c) \subseteq \chi(t) \,\}$ entirely covered by the bag $\chi(t)$.

Now, the only ingredient that is still missing for solving #SAT via dynamic programming along a given TD, is the table algorithm #SAT$_t$. For brevity, table algorithm #SAT$_t$ as presented in Listing 1 shows the four cases corresponding to the four node types of a nice TD, as any TD node forms just an overlap of these four cases. Each table $\tau_t$ consists of rows using attributes $\chi(t) \cup \{\mathrm{cnt}\}$, representing an assignment of $\varphi_t$ and cnt is a counter. Then, the table $\tau_t$ for a leaf node $t$, where type$(t) = $ *leaf*, consists of the empty assignment and counter 1, cf., Line 1. For nodes $t$ with introduced variable $a \in \chi(t)$, we guess in Line 3 for each assignment of the child table, whether $a$ is set to true or to false, and ensure that $\varphi_t$ is satisfied. When an atom $a$ is removed in a remove node $t$, we project assignments of child tables to $\chi(t)$, cf., Line 5, and sum up counters of the same assignments. For join nodes, counters of equal assignments are multiplied (Line 7).

*Example 2.* Let $\varphi := \{\overbrace{\{\neg x, y, a\}}^{c_1}, \overbrace{\{x, \neg y, \neg a\}}^{c_2}, \overbrace{\{x, b\}}^{c_3}, \overbrace{\{x, \neg b\}}^{c_4}\}$. Observe that $G$ of Fig. 1 is the primal graph $G_\varphi$ and that there are 6 satisfying assignments of $\varphi$. We discuss selected cases of running algorithm #SAT$_t$ on each node $t$ of TD $\mathcal{T}_{\mathrm{nice}}$ of Fig. 2 in post-order, thereby evaluating $\varphi$ in parts. Observe that type$(t_1) = $ *leaf*. Consequently, $\tau_1 = \{\{(\mathrm{cnt}, 1)\}\}$, cf., Line 1. Nodes $t \in \{t_2, t_3, t_4\}$ are of type$(t) = $ *intr*. Thus, we cross-join table $\tau_1$ with $\{\{(x, 0)\}, \{(x, 1)\}\}$ (two possible truth assignments for $x$), cf., Line 3, which is cross-joined with $\{\{(a, 0)\}, \{(a, 1)\}\}$, and then with $\{\{(y, 0)\}, \{(y, 1)\}\}$. Then, for node $t_4$ we additionally filter, cf., Line 3, those rows, where $\varphi_{t_4} = \{c_1, c_2\}$ is satisfied and obtain table $\tau_4$. Node $t_5$ is of type$(t_5) = $ *rem*, where $a$ is removed, i.e., by the properties of TDs, it is guaranteed that all clauses of $\varphi$ using $a$ are checked below $t_5$ and that no clause involving $a$ will occur above $t_4$. Consequently, $\tau_5$ is obtained from $\tau_4$ by projecting to $\{x, y\}$ and summing up the counters cnt of rows of equal assignments correspondingly, cf., Line 5. Similarly, one proceeds with $\tau_6$ and the right part of the tree, obtaining tables $\tau_7 - \tau_{10}$. In node $t_{11}$, we join common assignments of tables $\tau_6$ and $\tau_{10}$, and multiply counters cnt accordingly. Finally, we obtain 6 satisfying assignments, as expected. In all the tables the corresponding parts of assignment $I$, where $x, y, b$ are set to 1 and $a$ is set to 0 are highlighted.

Although these tables obtained via table algorithms might be exponential in size, the size is bounded by the width of the given TD of the primal graph $G_\varphi$. Still, practical results of such algorithms show competitive behaviour [3,25] up to a certain width. As a result, instances with high (tree-)width seem out of reach. Even further, if we lift the table algorithm #SAT$_t$ in order to solve problem #∃SAT, we are double exponential in the treewidth [23] and suffer from a rather complicated algorithm. To mitigate these issues, we present a novel approach to deal with high treewidth, by nesting of DP on abstractions of $G_\varphi$. As we will see, this not only works for #SAT, but also for #∃SAT with adaptions.

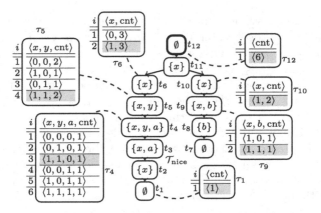

**Fig. 2.** Tables obtained by #SAT$_t$ on $\mathcal{T}_{\text{nice}}$ for $\varphi$ of Example 2.

## 3.1   Essentials for Nested Dynamic Programming

Assume that a set $U$ of variables of $\varphi$, called *nesting variables*, appears *uniquely* in one TD node $t$. Then, one could do DP on the TD as before, but no truth value for any variable in $U$ is stored. Instead, clauses involving $U$ could be evaluated by nested DP within node $t$, since variables $U$ appear uniquely in $t$. Indeed, for DP on the other (non-nesting) variables, only the result of this evaluation is essential. Now, before we can apply nested DP, we need abstractions with room for choosing nesting variables between the empty set and the set of all the variables. Inspired by related work [16,20,26,29], we define the *nested primal graph* $N_\varphi^A$ for a given formula $\varphi$ and a given set $A \subseteq \text{var}(\varphi)$ of *abstraction variables*. To this end, we say a path $P$ in primal graph $G_\varphi$ is a *nesting path* *(between $u$ and $v$)* using $A$, if $P = u, v_1, \ldots, v_\ell, v$ ($\ell \geq 0$), and every vertex $v_i$ is a *nesting variable*, i.e., $v_i \notin A$ for $1 \leq i \leq \ell$. Note that any path in $G_\varphi$ is nesting using $A$ if $A = \emptyset$. Then, the vertices of nested primal graph $N_\varphi^A$ correspond to $A$ and there is an edge between two vertices $u, v \in A$ if there is a nesting path between $u$ and $v$. Observe that the nested primal graph only consists of abstraction variables and, intuitively, "hides" nesting variables in nesting paths of primal graph $G_\varphi$.

*Example 3.* Recall formula $\varphi$ and primal graph $G_\varphi$ of Example 2. Given abstraction variables $A = \{x, y\}$, nesting paths of $G_\varphi$ are, e.g., $P_1 = x$, $P_2 = x, b$, $P_3 = b, x$, $P_4 = x, y$, $P_5 = x, a, y$. However, neither path $P_6 = y, x, b$, nor path $P_7 = b, x, y, a$ is nesting using $A$. Nested primal graph $N_\varphi^A$ contains edge $\{x, y\}$ over vertices $A$ due to paths $P_4, P_5$.

The nested primal graph provides abstractions of needed flexibility for nested DP. Indeed, if we set abstraction variables to $A = \text{var}(\varphi)$, we end up with full DP and zero nesting, whereas setting $A = \emptyset$ results in full nesting, i.e., nesting of all variables. Intuitively, the nested primal graph ensures that clauses subject to nesting (containing nesting variables) can be safely evaluated in exactly one node of a TD of the nested primal graph. To formalize this, we let *nestReach*($U$)

---

**Listing 2:** Algorithm $\text{HybDP}_{\#\exists\text{SAT}}(\text{depth}, \varphi, P', A')$ for hybrid solving of $\#\exists\text{SAT}$ by nested DP with abstraction variables $A'$.

---

**In:** Nesting depth $\geq 0$, formula $\varphi$, projection variables $P' \subseteq \text{var}(\varphi)$, and abstraction variables $A' \subseteq \text{var}(\varphi)$.

**Out:** Number $\#\exists\text{SAT}(\varphi, P')$ of assignments.

```
1  φ, P ← BCP_And_Preprocessing(φ, P')
2  A ← A' ∩ P
3  if φ ∈ dom(cache)   /*Cache Hit occurred*/        then return cache(φ) · 2^|P'\P|

4  if var(φ) ∩ P = ∅ then return SAT(φ) · 2^|P'\P|
5  (T, χ) ← Decompose_via_Heuristics(N_φ^A)     /* Decompose */
6  width ← max_{t in T} |χ(t)| − 1
7  if width ≥ threshold_hybrid or depth ≥ threshold_depth /* Standard Solver */ then
8  │   if var(φ) = P then cache ← cache ∪ {(φ, #SAT(φ))}
9  │   else                  cache ← cache ∪ {(φ, #∃SAT(φ, P))}
10 └   return cache(φ) · 2^|P'\P|

11 if width ≥ threshold_abstr /* Abstract via Heuristics & Decompose */ then
12 │   A      ← Choose_Subset_via_Heuristics(A, φ)
13 └   (T, χ) ← Decompose_via_Heuristics(N_φ^A)

14 n ← root(T)
15 τ ← {}     /* empty mapping */
16 for iterate t in post-order(T, n) /* Nested Dynamic Programming */       do
17 │   {t_1, …, t_ℓ} ← children(T, t)
18 └   τ_t ← #∃SAT_t(depth, χ(t), φ_t, P, φ_t^A, A \ A, ⟨τ_{t_1}, …, τ_{t_ℓ}⟩)
19 cache ← cache ∪ {(φ, c)}  where Π_cnt(τ_n) = {{(cnt, c)}}
20 return cache(φ) · 2^|P'\P|
```

---

for any set $U \subseteq \text{var}(\varphi)$ of variables containing nesting variables ($U \not\subseteq A$), be the set of vertices of all nesting paths of $G_\varphi$ between vertices $a, b$ using $A$ such that (i) both $a, b \in U$, or (ii) $a \in U \setminus A$. Intuitively, this definition ensures that from a given set $U$ of variables, we obtain reachable (i) nesting and (ii) abstraction variables, needed to evaluate clauses over $U$. Then, assuming a TD $T$ of $N_\varphi^A$, we say a set $U \subseteq \text{var}(\varphi)$ of variables (*"compatible set"*) is *compatible* with a node $t$ of $T$, and vice versa, if (I) $U = nestReach(U)$, and (II) $U \cap A \subseteq \chi(t)$.

*Example 4.* Assume again formula $\varphi$, primal graph $G_\varphi$ and abstraction variables $A = \{x, y\}$ of the previous example. Further, consider any TD $(T, \chi)$ of $N_\varphi^A$. Observe that $nestReach(\{b\}) = \{b, x\}$ due to nesting path $b, x$, i.e., $\{b\}$ is not a compatible set. However, $\{b, x\}$ is compatible with any node $t$ of $T$ where $x \in \chi(t)$. Indeed, to evaluate clauses $c_3, c_4 \in \varphi$, we need to evaluate both $b$ and $x$. Similarly, $\{a, x\}$ is not a compatible set due to nesting path $a, y$, but $\{a, x, y\}$ is a compatible set. Also, $\{a, b, x, y\}$ is a compatible set.

By construction any nesting variable is in at least one compatible set. However, (1) a nesting variable could be even in several compatible sets, and (2) a compatible set could be compatible with several nodes of $T$. Hence, to allow nested evaluation, we need to ensure that each nesting variable is evaluated only in one unique node $t$. As a result, we formalize for every compatible set $U$ that is

**Listing 3:** Nested table algorithm $\#\exists\mathrm{SAT}_t(\mathrm{depth}, \chi_t, \varphi_t, P, \varphi_t^A, A', \langle\tau_1, \ldots, \tau_\ell\rangle)$ for solving $\#\exists\mathrm{SAT}$ on node $t$ of a nice TD.

---

**In:** Nesting depth $\geq 0$, bag $\chi_t$, bag formula $\varphi_t$, projection variables $P$, nested bag formula $\varphi_t^A$, abstraction variables $A'$, and child tables $\langle\tau_1, \ldots \tau_\ell\rangle$ of $t$.
**Out:** Table $\tau_t$.

1 **if** type$(t) = leaf$ **then** $\tau_t \leftarrow \{\{(\mathrm{cnt}, 1)\}\}$
2 **else if** type$(t) = intr$, and $a \in \chi_t$ is introduced **then**
3 $\quad\Big|\ \tau_t \leftarrow \tau_1 \bowtie_{\varphi_t} \{\{(a, 0)\}, \{(a, 1)\}\}$
4 $\quad\Big|\ \tau_t \leftarrow \sigma_{\mathrm{cnt}>0}\big(\ddot{\Pi}_{\chi_t, \{\mathrm{cnt} \leftarrow \mathrm{cnt} \cdot \mathrm{HybDP}_{\#\exists\mathrm{SAT}}(\mathrm{depth}+1, \varphi_t^A[\mathrm{ass}], P \cap \mathrm{var}(\varphi_t^A[\mathrm{ass}]), A')\}}\tau_t\big)$
5 **else if** type$(t) = rem$, and $a \notin \chi_t$ is removed **then**
6 $\quad\Big|\ \tau_t \leftarrow {}_{\chi_t}G_{\mathrm{cnt} \leftarrow \mathrm{SUM}(\mathrm{cnt})}\big(\Pi_{\mathrm{att}(\tau_1)\setminus\{a\}}\tau_1\big)$
7 **else if** type$(t) = join$ **then**
8 $\quad\Big|\ \tau_t \leftarrow \ddot{\Pi}_{\chi_t, \{\mathrm{cnt} \leftarrow \mathrm{cnt} \cdot \mathrm{cnt}'\}}\big(\tau_1 \bowtie_{\bigwedge_{a \in \chi_t} a = a'} \rho_{\underset{a \in \mathrm{att}(\tau_2)}{\bigcup\{a \mapsto a'\}}}\tau_2\big)$

---

$\star$) Function ass refers to the respective truth assignment $I: \chi_t \to \{0, 1\}$ of a given row $r \in \tau_t$.

subset-minimal, a *unique* node $t$ compatible with $U$, denoted by comp$(U) := t$. For simplicity of our algorithms, we assume these unique nodes for $U$ are introduce nodes, i.e., type$(t) = intr$. We denote the union of all compatible sets $U$ where comp$(U) = t$, by *nested bag variables* $\chi_t^A$. Then, the *nested bag formula* $\varphi_t^A$ for a node $t$ of $\mathcal{T}$ equals $\varphi_t^A := \{c \mid c \in \varphi, \mathrm{var}(c) \subseteq \chi_t^A\} \setminus \varphi_t$, where formula $\varphi_t$ is defined above.

*Example 5.* Recall formula $\varphi$, TD $\mathcal{T} = (T, \chi)$ of $G_\varphi$, and abstraction variables $A = \{x, y\}$ of Example 3. Consider TD $\mathcal{T}' := (T, \chi')$, where $\chi'(t) := \chi(t) \cap \{x, y\}$ for each node $t$ of $T$. Observe that $\mathcal{T}'$ is $\mathcal{T}$, but restricted to $A$ and that $\mathcal{T}'$ is a TD of $N_\varphi^A$ of width 1. Observe that only for compatible set $U = \{b, x\}$ we have two nodes compatible with $U$, namely $t_2$ and $t_3$. We assume comp$(U) = t_2$. Consequently, nested bag formulas are $\varphi_{t_1}^A = \{c_1, c_2\}$, $\varphi_{t_2}^A = \{c_3, c_4\}$, and $\varphi_{t_3}^A = \emptyset$.

Assume any TD $\mathcal{T}$ of $N_\varphi^A$ using any set $A$ of abstraction variables. Observe that the definitions of nested primal graph and nested bag formula ensure that any set $S$ of vertices connected via edges in $G_\varphi$ will "appear" among nested bag variables of some node of $\mathcal{T}$. Even more stringent, each variable $a \in \mathrm{var}(\varphi) \setminus A$ appears *only* in nested bag formula $\varphi_t^A$ of node $t$ unique for $a$. These unique variable appearances allow to nest evaluating $\varphi_t^A$ under some assignment to $\chi(t)$.

## 3.2 Hybrid Solving Based on Nested DP

Now, we have definitions at hand to discuss nested DP in the context of *hybrid solving*, which combines using both standard solvers and parameterized solvers exploiting treewidth. We first illustrate the ideas for the problem $\#\exists\mathrm{SAT}$ and then discuss possible generalizations in Sect. 3.3; a concrete implementation is presented in Sect. 4.

Listing 2 depicts our algorithm $\mathrm{HybDP}_{\#\exists\mathrm{SAT}}$ for solving $\#\exists\mathrm{SAT}$. Note that the recursion is indirect in Line 18 through Line 4 of Listing 3 (discussed later).

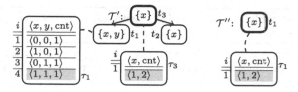

**Fig. 3.** Selected tables obtained by nested DP on TD $\mathcal{T}'$ of $N_\varphi^{\{x,y\}}$ (left) and on TD $\mathcal{T}''$ of $N_\varphi^{\{x\}}$ (right) for $\varphi$ and projection variables $P = \{x, y\}$ of Example 6 via $\text{HybDP}_{\#\exists\text{SAT}_t}$.

Algorithm $\text{HybDP}_{\#\exists\text{SAT}}$ takes formula $\varphi$, projection variables $P'$ and abstraction variables $A'$. The algorithm uses a global, but rather naive and simple cache mapping a formula to an integer, and consists of four subsequent blocks of code, separated by empty lines: (1) Preprocessing & Cache Consolidation, (2) Standard Solving, (3) Abstraction & Decomposition, and (4) Nested DP.

Block (1) spans Lines 1–3 and performs Boolean conflict propagation and preprocessing, thereby obtaining projection variables $P \subseteq P'$ (preserving satisfying assignments w.r.t. $P'$), sets $A$ to $A' \cap P$ in Line 2, and consolidates cache with the updated formula $\varphi$. If $\varphi$ is not cached, we do standard solving if the width is out-of-reach for nested DP in Block (2), spanning Lines 4–10. More concretely, if $\varphi$ does not contain projection variables, we employ a SAT solver returning integer 1 or 0. If $\varphi$ contains projection variables and either the width obtained by heuristically decomposing $G_\varphi$ is above threshold$_{\text{hybrid}}$, or the nesting depth exceeds threshold$_{\text{depth}}$, we use a standard #SAT or #∃SAT solver depending on $\text{var}(\varphi) \cap P$. Block (3) spans Lines 11–13 and is reached if no cache entry was found in Block (1) and standard solving was skipped in Block (2). If the width of the computed decomposition is above threshold$_{\text{abstr}}$, we need to use an abstraction in form of the nested primal graph. This is achieved by choosing suitable subsets $E \subseteq A$ of abstraction variables and decomposing $\varphi_t^E$ heuristically. Finally, Block (4) concerns nested DP, cf., Lines 14–20. This block relies on nested table algorithm $\#\exists\text{SAT}_t$, which takes parameters similar to table algorithm $\#\text{SAT}_t$, but additionally requires the nested bag formula for current node $t$, projection variables $P$ and abstraction variables. Nested table algorithm $\#\exists\text{SAT}_t$ is sketched in Listing 3 and recursively calls for each row $r \in \tau_t$, $\text{HybDP}_{\#\exists\text{SAT}}$ on nested bag formula $\varphi_t^A$ simplified by the assignment $\text{ass}(r)$ of the current row $r$. This is implemented in Line 4 by using extended projection, cf., Listing 1, where the count cnt of the respective row $r$ is updated by multiplying the result of the recursive call $\text{HybDP}_{\#\exists\text{SAT}}$. Notably, as the recursive call $\text{HybDP}_{\#\exists\text{SAT}}$ within extended projection of Line 4 implicitly takes a given current row $r$, the function occurrences ass in Line 4 implicitly take this row $r$ as an argument. As a result, our approach deals with high treewidth by recursively finding and decomposing abstractions of the graph. If the treewidth is too high for some parts, TDs of abstractions are used to guide standard solvers.

*Example 6.* Recall formula $\varphi$, set $A$ of abstraction variables, and TD $\mathcal{T}'$ of nested primal graph $N_\varphi^A$ given in Example 5. Restricted to projection set $P := \{x, y\}$, $\varphi$ has two satisfying assignments, namely $\{x \mapsto 1, y \mapsto 0\}$ and $\{x \mapsto 1, y \mapsto 1\}$.

**Listing 4:** Nested table algorithm $\text{QSAT}_t(\text{depth}, \chi_t, \varphi_t, \varphi_t^A, A', \langle \tau_1, \ldots, \tau_\ell \rangle)$ for solving QSAT on node $t$ of a nice tree decomposition.

---

**In:** Nesting depth $\geq 0$, bag $\chi_t$, bag QBF $\varphi_t = Q\,V.\gamma$, nested bag QBF $\varphi_t^A$, abstraction variables $A'$, and child tables $\langle \tau_1, \ldots \tau_\ell \rangle$ of $t$.
**Out:** Table $\tau_t$.

1  **if** type($t$) = *leaf* **then** $\tau_t \leftarrow \{\emptyset\}$
2  **else if** type($t$) = *intr, and* $a \in \chi_t$ *is introduced* **then**
3  $\quad \big| \tau_t \leftarrow \tau_1 \bowtie_{\varphi_t} \{\{(a,0)\}, \{(a,1)\}\}$
4  $\quad \big| \tau_t \leftarrow \sigma_{(Q=\exists \,\vee\, |\tau_t|=2^{|\chi_t|}) \,\wedge\, \text{HybDP}_{\text{QSAT}}(\text{depth}+1, \varphi_t^A[\text{ass}], A')}(\tau_t)$
5  **else if** type($t$) = *rem, and* $a \notin \chi_t$ *is removed* **then**
6  $\quad \big| \tau_t \leftarrow \Pi_{\text{att}(\tau_1) \setminus \{a\}} \tau_1$
7  **else if** type($t$) = *join* **then**
8  $\quad \big| \tau_t \leftarrow \Pi_{\chi_t} (\tau_1 \bowtie_{\bigwedge_{a \in \chi_t} a = a'} \rho_{\bigcup_{\{a \mapsto a'\}} \tau_2)$
$\qquad\qquad\qquad\qquad\qquad _{a \in \text{att}(\tau_2)}$

---

$\star$) The cardinality of a table $\tau$ can be obtained via relational algebra (sub-expression): $|\tau| := c$, where $\{\{(\text{card}, c)\}\} = {}_\emptyset G_{\text{card} \leftarrow \text{SUM}(1)} \tau$

Consequently, the solution to #∃SAT is 2. Figure 3 (left) shows TD $\mathcal{T}'$ of $N_\varphi^A$ and tables obtained by $\text{HybDP}_{\text{\#∃SAT}_t}(\varphi, P, A)$ for solving projected model counting on $\varphi$ and $P$. Note that the same example easily works for #SAT, where $P = \text{var}(\varphi)$.

Algorithm #∃SAT$_t$ of Listing 3 works similar to algorithm #SAT$_t$, but uses attribute "cnt" for storing (projected) counts accordingly. We briefly discuss executing #∃SAT$_{t_1}$ in the context of Line 18 of algorithm $\text{HybDP}_{\text{\#∃SAT}_t}$ on node $t_1$ of $\mathcal{T}'$, resulting in table $\tau_1$ as shown in Fig. 3 (left). Recall that comp($\{a, x, y\}$) = $t_1$, and, consequently, $\varphi_{t_1}^A = \{\{\neg x, y, a\}, \{x, \neg y, \neg a\}\}$. Then, in Line 4 of algorithm #∃SAT$_t$, for each assignment ass($r$) to $\{x, y\}$ of each row $r$ of $\tau_1$, we compute $\text{HybDP}_{\text{\#∃SAT}_t}(\psi, P \cap \text{var}(\psi), \emptyset)$ using $\psi = \varphi_{t_1}^A[\text{ass}(r)]$. Each of these recursive calls, however, is already solved by BCP and preprocessing, e.g., $\varphi_{t_1}^A[\{x \mapsto 1, y \mapsto 0\}]$ of Row 2 simplifies to $\{\{a\}\}$.

Figure 3 (right) shows TD $\mathcal{T}''$ of $N_\varphi^E$ with $E := \{x\}$, and tables obtained by $\text{HybDP}_{\text{\#∃SAT}_t}(\varphi, P, E)$. Still, $\varphi_{t_1}^E[\text{ass}(r)]$ for a given assignment ass($r$) : $\{x\} \to \{0, 1\}$ of any row $r \in \tau_1$ can be simplified. Concretely, $\varphi_{t_1}^E[\{x \mapsto 0\}]$ evaluates to $\emptyset$ and $\varphi_{t_1}^E[\{x \mapsto 1\}]$ evaluates to two variable-distinct clauses, namely $\{\neg b\}$ and $\{y, a\}$. Thus, there are 2 satisfying assignments $\{y \mapsto 0\}$, $\{y \mapsto 1\}$ of $\varphi_{t_1}^E[\{x \mapsto 1\}]$ restricted to $P$.

**Theorem 1.** *Given formula* $\varphi$, *projection variables* $P \subseteq \text{var}(\varphi)$, *and abstraction variables* $A' \subseteq \text{var}(\varphi)$. *Then,* $\text{HybDP}_{\text{\#∃SAT}}(\varphi, P, A')$ *correctly returns* #∃SAT($\varphi, P$).

*Proof (Sketch).* Observe that (A): $(T, \chi)$ is a TD of nested primal graph $N_\varphi^A$ such that $A \subseteq A' \cap P$. The interesting part of algorithm $\text{HybDP}_{\text{\#∃SAT}}$ lies in Block (3), in particular in Lines 11–13. The proof proceeds by structural induction on $\varphi$. By construction, we have (B): Every variable of $\text{var}(\varphi) \setminus A$ occurs in some nested bag formula $\varphi_t^A$ as used in the call to #∃SAT$_t$ in Line 18 for a unique node $t$ of $T$. Observe that $\#\exists\text{SAT}_t$ corresponds to $\#\text{SAT}_t$, whose correctness is established via invariants, cf., [24,44], only Line 4 differs. In Line 4 of $\#\exists\text{SAT}_t$, $\text{HybDP}_{\#\exists\text{SAT}}$ is called recursively on subformulas $\varphi_t^A[\text{ass}(r)]$ for each $r \in \tau_t$. By induction hypothesis, we have (C): these calls result to $\#\exists\text{SAT}(\varphi_t^A[\text{ass}(r)], P \cap \text{var}(\varphi_t^A[\text{ass}(r)]))$ for each $r \in \tau_t$. By (A), $\#\exists\text{SAT}_t$ as called in Line 18 stores only table attributes in $\chi_t \subseteq A \subseteq P$. Thus, by (C), recursive calls can be subsequently multiplied to cnt for each $r \in \tau_t$.

### 3.3   Generalizing Nested DP to Other Formalisms

Nested DP as proposed above is by far not restricted to (projected) model counting, or counting problems in general. In fact, one can easily generalize nested DP to other relevant formalisms, briefly sketched for the QBF formalism.

**Quantified Boolean Formulas (QBFs).** We assume QBFs of the form $\varphi = \exists V_1.\forall V_2.\dots.\exists V_\ell.\gamma$ using *quantifiers* $\exists, \forall$, where $\gamma$ is a CNF formula and $\text{var}(\varphi) = \text{var}(\gamma) = V_1 \cup V_2 \cdots \cup V_\ell$. Given QBF $\varphi = Q\ V.\psi$ with $Q \in \{\exists, \forall\}$, we let $\text{qvar}(\varphi) := V$. For an assignment $I : V' \rightarrow \{0,1\}$ with $V' \subseteq V$, we let $\varphi[I] := \psi[I]$ if $V' = V$, and $\varphi[I] := Q(V \setminus V').\psi[I]$ if $V' \subsetneq V$. *Validity* of $\varphi$ (QSAT) is recursively defined: $\exists V.\varphi$ is *valid* if there is $I: V \rightarrow \{0,1\}$ where $\varphi[I]$ is valid; $\forall V.\varphi$ is valid if for every $I: V \rightarrow \{0,1\}$, $\varphi[I]$ is valid.

Hybrid solving by nested DP can be extended to problem QSAT. To the end of using this approach for QBFs, we define the primal graph $G_\varphi$ for a QBF $\varphi = \exists V_1.\forall V_2.\dots.\exists V_\ell.\gamma$ analogously to the primal graph of a Boolean formula, i.e., $G_\varphi := G_\gamma$. Consequently, also the nested primal graph is defined for a given set $A \subseteq \text{var}(\varphi)$ by $N_\varphi^A := N_\gamma^A$. Now, let $A \subseteq \text{var}(\varphi)$ be a set of abstraction variables, and $\mathcal{T} = (T, \chi)$ be a TD of $N_\varphi^A$ and $t$ be a node of $T$. Then, the *bag QBF* $\varphi_t$ is given by $\varphi_t := \exists V_1.\forall V_2.\dots.\exists V_\ell.\gamma_t$ and the *nested bag QBF* $\varphi_t^A$ for a set $A \subseteq \text{var}(\varphi)$ amounts to $\varphi_t^A := \exists V_1.\forall V_2.\dots.\exists V_\ell.\gamma_t^A$.

Algorithm $\text{HybDP}_{\text{QSAT}}$ is similar to $\text{HybDP}_{\#\exists\text{SAT}}$ of Listing 2, where the projection variables parameter $P'$ is removed since $P'$ constantly coincides with variables $\text{qvar}(\varphi)$ of the outermost quantifier. Further, Line 4 is removed, Lines 8 and 9 are replaced by calling a QSAT solver and nested table algorithm $\#\exists\text{SAT}_t$ of Line 18 is replaced by nested table algorithm $\text{QSAT}_t$ as presented in Listing 4. Algorithm $\text{QSAT}_t$ is of similar shape as algorithm $\#\exists\text{SAT}_t$, cf., Listing 3, but does not maintain counts cnt. Further, Line 4 of algorithm $\text{QSAT}_t$ intuitively filters $\tau_t$ fulfilling the outer-most quantifier, and keeps those rows $r$ of $\tau_t$, where the recursive call to $\text{HybDP}_{\text{QSAT}}$ on nested bag formula simplified by the assignment $\text{ass}(r)$ of $r$ succeeds. For ensuring that the outer-most quantifier $Q$ is fulfilled, we are either in the situation that $Q = \exists$, which immediately is fulfilled for every row $r$ in $\tau_t$ since $r$ itself serves as a witness. If $Q = \forall$, we need to check that $\tau_t$ contains $2^{|\chi(t)|}$ many (all) rows. The cardinality of table $\tau_t$ can be computed via a sub-expression of relational algebra as hinted in the footnote of Listing 4. Notably, if $Q = \forall$, we do not need to check in Line 8 of Listing 4, whether all rows sustain in table $\tau_t$ since this is already ensured for both child tables $\tau_1, \tau_2$ of $t$. Then, if in the end the table for the root node of $\mathcal{T}$ is not empty, it is guaranteed that either the table contains some (if $Q = \exists$) or all (if $Q = \forall$) rows and that $\varphi$ is valid. Note that algorithm $\text{QSAT}_t$ can be extended to also consider more fine-grained quantifier dependency schemes.

Compared to other algorithms for QSAT using treewidth [9,10], hybrid solving based on nested DP is quite compact without the need of nested tables. Instead of rather involved data structures (nested tables), we use here plain tables that can be handled by modern database systems efficiently.

## 4    Implementation and Preliminary Results

We implemented a hybrid solver nestHDB[1] based on nested DP in Python3 and using table manipulation techniques by means of SQL and the *database management system (DBMS)* Postgres. Our solver builds upon the recently published prototype dpdb [24], which applied a DBMS for plain dynamic programming algorithms. However, we used the most-recent version 12 of Postgres and we let it operate on a tmpfs-ramdisk. In our solver, the DBMS serves the purpose of extremely efficient in-memory table manipulations and query optimization required by nested DP, and therefore nestHDB benefits from database technology.

**Nested DP & Choice of Standard Solvers.** We implemented dedicated nested DP algorithms for solving #SAT and #∃SAT, where we do (nested) DP up to threshold$_{\text{depth}}$ = 2. Further, we set threshold$_{\text{hybrid}}$ = 1000 and therefore we do not "fall back" to standard solvers based on the width (cf., Line 7 of Listing 2), but based on the nesting depth.

Also, the evaluation of the nested bag formula is "shifted" to the database if it uses at most 40 abstraction variables, since Postgres efficiently handles these small-sized Boolean formulas. Thereby, further nesting is saved by executing optimized SQL statements within the TD nodes. A value of 40 seems to be a nice balance between the overhead caused by standard solvers for small formulas and exponential growth counteracting the advantages of the DBMS. For hybrid solving, we use #SAT solver sharpSAT [48] and for #∃SAT we employ the recently published #∃SAT solver projMC [35], solver sharpSAT and SAT solver picosat [4]. Observe that our solver immediately benefits from better standard solvers and further improvements of the solvers above.

**Choosing Non-nesting Variables & Compatible Nodes.** TDs are computed by means of heuristics via decomposition library htd [1]. For finding good abstractions (crucial), i.e., abstraction variables for the nested primal graph, we use encodings for solver clingo [27], which is based on logic programming (ASP) and therefore perfectly suited for solving reachability via nesting paths. There, among a reasonably sized subset of vertices of smallest degree, we aim for a preferably large (maximal) set $A$ of abstraction variables such that at the same time the resulting graph $N_\varphi^A$ is reasonably sparse, which is achieved by minimizing the number of edges of $N_\varphi^A$. To this end, we use built-in (cost) optimization,

---

[1] Source code, instances, and detailed results are available at: tinyurl.com/nesthdb.

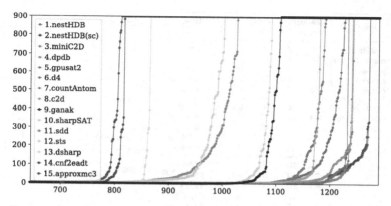

**Fig. 4.** Cactus plot of instances for #SAT, where instances (x-axis) are ordered for each solver individually by runtime[seconds] (y-axis). threshold$_{abstr}$ = 38.

where we take the best results obtained by clingo after running at most 35 s. For the concrete encodings used in nestHDB, we refer to the online repository as stated above. We expect that this initial approach can be improved and that extending by problem-specific as well as domain-specific information might help in choosing promising abstraction variables $A$.

As rows of tables during (nested) DP can be independently computed and parallelized [25], hybrid solver nestHDB potentially calls standard solvers for solving subproblems in parallel using a thread pool. Thereby, the uniquely compatible node for relevant compatible sets $U$, as denoted in this paper by comp($U$), is decided during runtime among compatible nodes on a first-come-first-serve basis.

**Benchmarked Solvers & Instances.** We benchmarked nestHDB and 16 other publicly available #SAT solvers on 1,494 instances recently considered [24]. Among those solvers are single-core solvers miniC2D [41], d4 [34], c2d [13], ganak [46], sharpSAT [48], sdd [14], sts [21], dsharp [39], cnf2eadt [32], cachet [45], sharpCDCL [30], approxmc3 [8], and bdd_minisat [49]. We also included multi-core solvers dpdb [24], gpusat2 [25], as well as countAntom [7]. While nestHDB itself is a multi-core solver, we additionally included in our comparison nestHDB(sc), which is nestHDB, but restricted to a single core only. The instances [24] we took are already preprocessed by pmc [33] using recommended options `-vivification -eliminateLit -litImplied -iterate=10 -equiv -orGate -affine` for preserving model counts. However, nestHDB still uses pmc with these options also in Line 1 of Listing 2.

Further, we considered the problem #∃SAT, where we compare solvers projMC [35], clingo [27], ganak [46], nestHDB (see footnote 1), and nestHDB(sc) on 610 publicly available instances[2] from projMC (consisting of 15 *planning*, 60 *circuit*, and 100 *random* instances) and Fremont, with 170 *symbolic-markov* applications, and 265 *misc* instances. For preprocessing in Line 1 of Listing 2,

---

[2] Sources: tinyurl.com/projmc;tinyurl.com/pmc-fremont-01-2020.

| bench-mark set | solver | tw upper bound | | | Σ | time [h] |
| --- | --- | --- | --- | --- | --- | --- |
| | | max | 0-30 | 31-50 >50 | | |
| planning | nestHDB | 30 | 7 | 0 0 | 7 | **2.88** |
| | nestHDB(sc) | 30 | 7 | 0 0 | 7 | 3.31 |
| | projMC | 26 | 6 | 0 0 | 6 | 3.01 |
| | ganak | 19 | 5 | 0 0 | 5 | 3.36 |
| | clingo | 4 | 1 | 0 0 | 1 | 4.00 |
| circ | nestHDB | 99 | **34** | **10 16** | **60** | 2.10 |
| | nestHDB(sc) | 99 | **34** | 4 14 | 52 | 4.60 |
| | projMC | 91 | 28 | **10** 11 | 49 | 6.23 |
| | ganak | 99 | **34** | **10 16** | **60** | **1.21** |
| | clingo | 99 | 31 | **10 16** | 57 | 4.44 |
| random | nestHDB | 79 | **30** | **20** 17 | **67** | 10.91 |
| | nestHDB(sc) | 79 | **30** | **20** 15 | 65 | 11.29 |
| | projMC | **84** | **30** | **20** 15 | 65 | 11.09 |
| | ganak | 19 | 19 | 0 0 | 19 | 23.18 |
| | clingo | 24 | 25 | 0 0 | 25 | 21.38 |
| markov | nestHDB | 23 | 62 | 0 0 | 62 | 31.98 |
| | nestHDB(sc) | 23 | 61 | 0 0 | 61 | 32.54 |
| | projMC | 8 | 54 | 0 0 | 54 | 33.65 |
| | ganak | **59** | **64** | 0 4 | **68** | **30.32** |
| | clingo | 3 | 38 | 0 0 | 38 | 37.54 |
| misc | nestHDB | 47 | **38** | **17** 0 | **55** | 46.12 |
| | nestHDB(sc) | 47 | **38** | 13 0 | 51 | 48.20 |
| | projMC | 47 | **38** | 13 0 | 51 | 45.90 |
| | ganak | 44 | **38** | 15 0 | 53 | **45.72** |
| | clingo | **63** | **38** | 15 1 | 54 | 44.79 |
| Σ | nestHDB | 99 | 171 | 47 33 | 251 | **93.99** |
| | nestHDB(sc) | 99 | 170 | 37 29 | 236 | 99.95 |
| | projMC | 91 | 156 | 43 26 | 225 | 99.88 |
| | ganak | 99 | 160 | 25 20 | 205 | 103.78 |
| | clingo | 99 | 133 | 25 17 | 175 | 112.15 |

**Fig. 5.** Number of solved #∃SAT insts., grouped by upper bound intervals of treewidth (left), cactus plot (right). time[h] is cumulated wall clock time, timeouts count as 900 s. threshold$_{abstr} = 8$.

nestHDB uses pmc as before, but without options `-equiv -orGate -affine` to ensure preservation of models (equivalence).

**Benchmark Setup.** Solvers ran on a cluster of 12 nodes. Each node of the cluster is equipped with two Intel Xeon E5-2650 CPUs consisting of 12 physical cores each at 2.2 GHz clock speed, 256 GB RAM. For dpdb and nestHDB, we used Postgres 12 on a tmpfs-ramdisk (/tmp) that could grow up to at most 1 GB per run. Results were gathered on Ubuntu 16.04.1 LTS machines with disabled hyperthreading on kernel 4.4.0-139. We mainly compare total wall clock time and number of timeouts. For parallel solvers (dpdb, countAntom, nestHDB) we allow 12 physical cores. Timeout is 900 s and RAM is limited to 16 GB per instance and solver. Results for gpusat2 are taken from [24].

**Benchmark Results.** The results for #SAT showing the best 14 solvers are summarized in the cactus plot of Fig. 4. Overall it shows nestHDB among the best solvers, solving 1,273 instances. The reason for this is, compared to dpdb, that nestHDB can solve instances using TDs of primal graphs of widths larger than 44, up to width 266. This limit is even slightly larger than the width of 264 that sharpSAT on its own can handle. We also tried using minic2d instead

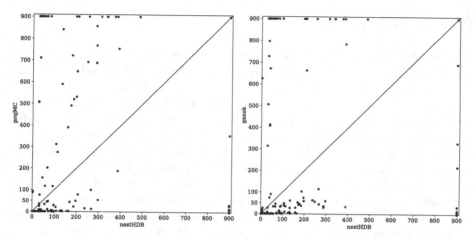

**Fig. 6.** Scatter plot of instances for $\#\exists\mathrm{S}\mathrm{AT}$, where the x-axis shows runtime in seconds of nestHDB compared to the y-axis showing runtime of projMC (left) and of ganak (right). $\mathrm{threshold}_{\mathrm{abstr}} = 8$.

of sharpSAT as standard solver for solvers nestHDB and nestHDB(sc), but we could only solve one instance more. Notably, nestHDB(sc) has about the same performance as nestHDB, indicating that parallelism does not help much on the instances. Further, we observed that the employed simple cache as used in Listing 2, provides only a marginal improvement.

Figure 5 (left) depicts a table of results on $\#\exists\mathrm{S}\mathrm{AT}$, where we observe that nestHDB does a good job on instances with low widths below $\mathrm{threshold}_{\mathrm{abstr}} = 8$ (containing ideas of dpdb), but also on widths well above 8 (using nested DP). Notably, nestHDB is also competitive on widths well above 50. Indeed, nestHDB and nestHDB(sc) perform well on all benchmark sets, whereas on some sets the solvers projMC, clingo and ganak are faster. Overall, parallelism provides a significant improvement here, but still nestHDB(sc) shows competitive performance, which is also visualized in the cactus plot of Fig. 5 (right). Figure 6 shows scatter plots comparing nestHDB to projMC (left) and to ganak (right). Overall, both plots show that nestHDB solves more instances, since in both cases the y-axis shows more black dots at 900 s than the x-axis. Further, the bottom left of both plots shows that there are plenty easy instances that can be solved by projMC and ganak in well below 50 s, where nestHDB needs up to 200 s. Similarly, the cactus plot given in Fig. 5 (right) shows that nestHDB can have some overhead compared to the three standard solvers, which is not surprising. This indicates that there is still room for improvement if, e.g., easy instances are easily detected, and if standard solvers are used for those instances. Alternatively, one could also just run a standard solver for at most 50 s and if not solved within 50 s, the heavier machinery of nested dynamic programming is invoked. Apart from these instances, Fig. 6 shows that nestHDB solves harder instances faster, where standard solvers struggle.

## 5   Conclusion

We presented nested dynamic programming (nested DP) using different levels of abstractions, which are subsequently refined and solved recursively. This approach is complemented with hybrid solving, where (search-intense) subproblems are solved by standard solvers. We provided nested DP algorithms for problems related to Boolean satisfiability, but the idea can be easily applied for other formalisms. We implemented some of these algorithms and our benchmark results are promising. For future work, we plan deeper studies of problem-specific abstractions, in particular for QSAT. We want to further tune our solver parameters (e.g., thresholds, timeouts, sizes), deepen interleaving with solvers like projMC, and to use incremental solving for obtaining abstractions and evaluating nested bag formulas, where intermediate solver references are kept during dynamic programming and formulas are iteratively added and (re-)solved.

**Acknowledgements.** The work has been supported by the Austrian Science Fund (FWF), Grants Y698, and P32830, as well as the Vienna Science and Technology Fund, Grant WWTF ICT19-065.

## References

1. Abseher, M., Musliu, N., Woltran, S.: htd – a free, open-source framework for (customized) tree decompositions and beyond. In: Salvagnin, D., Lombardi, M. (eds.) CPAIOR 2017. LNCS, vol. 10335, pp. 376–386. Springer, Cham (2017). https://doi.org/10.1007/978-3-319-59776-8_30

2. Atserias, A., Fichte, J.K., Thurley, M.: Clause-learning algorithms with many restarts and bounded-width resolution. J. Artif. Intell. Res. **40**, 353–373 (2011)

3. Bannach, M., Berndt, S.: Practical access to dynamic programming on tree decompositions. Algorithms **12**(8), 172 (2019)

4. Biere, A.: PicoSAT essentials. JSAT **4**(2–4), 75–97 (2008)

5. Bliem, B., Charwat, G., Hecher, M., Woltran, S.: D-FLAT$^2$: subset minimization in dynamic programming on tree decompositions made easy. Fundam. Inform. **147**(1), 27–61 (2016)

6. Bodlaender, H., Koster, A.: Combinatorial optimization on graphs of bounded treewidth. Comput. J. **51**(3), 255–269 (2008)

7. Burchard, J., Schubert, T., Becker, B.: Laissez-faire caching for parallel #SAT solving. In: Heule, M., Weaver, S. (eds.) SAT 2015. LNCS, vol. 9340, pp. 46–61. Springer, Cham (2015). https://doi.org/10.1007/978-3-319-24318-4_5

8. Chakraborty, S., Fremont, D.J., Meel, K.S., Seshia, S.A., Vardi, M.Y.: Distribution-aware sampling and weighted model counting for SAT. In: AAAI 2014, pp. 1722–1730. The AAAI Press (2014)

9. Charwat, G., Woltran, S.: Expansion-based QBF solving on tree decompositions. Fundam. Inform. **167**(1–2), 59–92 (2019)

10. Chen, H.: Quantified constraint satisfaction and bounded treewidth. In: ECAI 2004, pp. 161–170. IOS Press (2004)

11. Codd, E.F.: A relational model of data for large shared data banks. Commun. ACM **13**(6), 377–387 (1970)

12. Cygan, M., et al.: Parameterized Algorithms. Springer, Cham (2015). https://doi.org/10.1007/978-3-319-21275-3
13. Darwiche, A.: New advances in compiling CNF to decomposable negation normal form. In: ECAI 2004, pp. 318–322. IOS Press (2004)
14. Darwiche, A.: SDD: a new canonical representation of propositional knowledge bases. In: IJCAI 2011, pp. 819–826. AAAI Press/IJCAI (2011)
15. Dell, H., Komusiewicz, C., Talmon, N., Weller, M.: The PACE 2017 parameterized algorithms and computational experiments challenge: the second iteration. In: IPEC 2017, pp. 30:1–30:13. LIPIcs, Dagstuhl Publishing (2017)
16. Dell, H., Roth, M., Wellnitz, P.: Counting answers to existential questions. In: ICALP 2019. LIPIcs, vol. 132, pp. 113:1–113:15. Schloss Dagstuhl - Leibniz-Zentrum für Informatik (2019)
17. Diestel, R.: Graph Theory, Graduate Texts in Mathematics, vol. 173, 4th edn. Springer, Heidelberg (2012)
18. Downey, R.G., Fellows, M.R.: Fundamentals of Parameterized Complexity. TCS. Springer, London (2013). https://doi.org/10.1007/978-1-4471-5559-1
19. Durand, A., Hermann, M., Kolaitis, P.G.: Subtractive reductions and complete problems for counting complexity classes. Theoret. Comput. Sci. **340**(3), 496–513 (2005). https://doi.org/10.1016/j.tcs.2005.03.012
20. Eiben, E., Ganian, R., Hamm, T., Kwon, O.: Measuring what matters: a hybrid approach to dynamic programming with treewidth. In: MFCS 2019. LIPIcs, vol. 138, pp. 42:1–42:15. Dagstuhl Publishing (2019)
21. Ermon, S., Gomes, C.P., Selman, B.: Uniform solution sampling using a constraint solver as an oracle. In: UAI 2012, pp. 255–264. AUAI Press (2012)
22. Fichte, J.K., Hecher, M., Morak, M., Woltran, S.: Answer set solving with bounded treewidth revisited. In: Balduccini, M., Janhunen, T. (eds.) LPNMR 2017. LNCS (LNAI), vol. 10377, pp. 132–145. Springer, Cham (2017). https://doi.org/10.1007/978-3-319-61660-5_13
23. Fichte, J.K., Hecher, M., Morak, M., Woltran, S.: Exploiting Treewidth for Projected Model Counting and Its Limits. In: Beyersdorff, O., Wintersteiger, C.M. (eds.) SAT 2018. LNCS, vol. 10929, pp. 165–184. Springer, Cham (2018). https://doi.org/10.1007/978-3-319-94144-8_11
24. Fichte, J.K., Hecher, M., Thier, P., Woltran, S.: Exploiting database management systems and treewidth for counting. In: Komendantskaya, E., Liu, Y.A. (eds.) PADL 2020. LNCS, vol. 12007, pp. 151–167. Springer, Cham (2020). https://doi.org/10.1007/978-3-030-39197-3_10
25. Fichte, J.K., Hecher, M., Zisser, M.: An improved GPU-based SAT model counter. In: Schiex, T., de Givry, S. (eds.) CP 2019. LNCS, vol. 11802, pp. 491–509. Springer, Cham (2019). https://doi.org/10.1007/978-3-030-30048-7_29
26. Ganian, R., Ramanujan, M.S., Szeider, S.: Combining treewidth and backdoors for CSP. In: STACS 2017, pp. 36:1–36:17 (2017). https://doi.org/10.4230/LIPIcs.STACS.2017.36
27. Gebser, M., Kaminski, R., Kaufmann, B., Schaub, T.: Multi-shot ASP solving with clingo. TPLP **19**(1), 27–82 (2019). https://doi.org/10.1017/S1471068418000054
28. Giunchiglia, E., Marin, P., Narizzano, M.: Reasoning with quantified Boolean formulas. In: Handbook of Satisfiability, FAIA, vol. 185, pp. 761–780. IOS Press (2009). https://doi.org/10.3233/978-1-58603-929-5-761
29. Hecher, M., Morak, M., Woltran, S.: Structural decompositions of epistemic logic programs. CoRR abs/2001.04219 (2020). http://arxiv.org/abs/2001.04219

30. Klebanov, V., Manthey, N., Muise, C.: SAT-based analysis and quantification of information flow in programs. In: Joshi, K., Siegle, M., Stoelinga, M., D'Argenio, P.R. (eds.) QEST 2013. LNCS, vol. 8054, pp. 177–192. Springer, Heidelberg (2013). https://doi.org/10.1007/978-3-642-40196-1_16

31. Kloks, T. (ed.): Treewidth: Computations and Approximations. LNCS, vol. 842. Springer, Heidelberg (1994). https://doi.org/10.1007/BFb0045375

32. Koriche, F., Lagniez, J.M., Marquis, P., Thomas, S.: Knowledge compilation for model counting: affine decision trees. In: IJCAI 2013. The AAAI Press (2013)

33. Lagniez, J., Marquis, P.: Preprocessing for propositional model counting. In: AAAI 2014, pp. 2688–2694. AAAI Press (2014)

34. Lagniez, J.M., Marquis, P.: An improved decision-DDNF compiler. In: IJCAI 2017, pp. 667–673. The AAAI Press (2017)

35. Lagniez, J., Marquis, P.: A recursive algorithm for projected model counting. In: AAAI 2019, pp. 1536–1543. AAAI Press (2019)

36. Langer, A., Reidl, F., Rossmanith, P., Sikdar, S.: Evaluation of an MSO-solver. In: ALENEX 2012, pp. 55–63. SIAM/Omnipress (2012)

37. Lonsing, F., Egly, U.: Evaluating QBF solvers: quantifier alternations matter. In: Hooker, J. (ed.) CP 2018. LNCS, vol. 11008, pp. 276–294. Springer, Cham (2018). https://doi.org/10.1007/978-3-319-98334-9_19

38. Maniu, S., Senellart, P., Jog, S.: An experimental study of the treewidth of real-world graph data (extended version). CoRR abs/1901.06862 (2019). http://arxiv.org/abs/1901.06862

39. Muise, C., McIlraith, S.A., Beck, J.C., Hsu, E.I.: DSHARP: fast d-DNNF compilation with sharpSAT. In: Kosseim, L., Inkpen, D. (eds.) AI 2012. LNCS (LNAI), vol. 7310, pp. 356–361. Springer, Heidelberg (2012). https://doi.org/10.1007/978-3-642-30353-1_36

40. Niedermeier, R.: Invitation to Fixed-Parameter Algorithms. Oxford Lecture Series in Mathematics and Its Applications, vol. 31. OUP, Oxford (2006)

41. Oztok, U., Darwiche, A.: A top-down compiler for sentential decision diagrams. In: IJCAI 2015, pp. 3141–3148. The AAAI Press (2015)

42. Pulina, L., Seidl, M.: The 2016 and 2017 QBF solvers evaluations (QBFEVAL'16 and QBFEVAL'17). Artif. Intell. **274**, 224–248 (2019). https://doi.org/10.1016/j.artint.2019.04.002

43. Robertson, N., Seymour, P.D.: Graph minors II: algorithmic aspects of tree-width. J. Algorithms **7**, 309–322 (1986)

44. Samer, M., Szeider, S.: Algorithms for propositional model counting. J. Discrete Algorithms **8**(1), 50–64 (2010)

45. Sang, T., Bacchus, F., Beame, P., Kautz, H., Pitassi, T.: Combining component caching and clause learning for effective model counting. In: SAT 2004 (2004)

46. Sharma, S., Roy, S., Soos, M., Meel, K.S.: GANAK: a scalable probabilistic exact model counter. In: IJCAI 2019, pp. 1169–1176. ijcai.org (2019)

47. Tamaki, H.: Positive-instance driven dynamic programming for treewidth. J. Comb. Optim. **37**(4), 1283–1311 (2018). https://doi.org/10.1007/s10878-018-0353-z

48. Thurley, M.: sharpSAT – counting models with advanced component caching and implicit BCP. In: Biere, A., Gomes, C.P. (eds.) SAT 2006. LNCS, vol. 4121, pp. 424–429. Springer, Heidelberg (2006). https://doi.org/10.1007/11814948_38

49. Toda, T., Soh, T.: Implementing efficient all solutions SAT solvers. ACM J. Exp. Algorithmics 21(1.12) (2015). Special Issue SEA 2014

# Reducing Bit-Vector Polynomials to SAT Using Gröbner Bases

Thomas Seed[1]([✉]), Andy King[1], and Neil Evans[2]

[1] University of Kent, Canterbury CT2 7NF, UK
ts495@kent.ac.uk
[2] AWE Aldermaston, Reading RG7 4PR, UK

**Abstract.** We address the satisfiability of systems of polynomial equations over bit-vectors. Instead of conventional bit-blasting, we exploit word-level inference to translate these systems into non-linear pseudo-boolean constraints. We derive the pseudo-booleans by simulating bit assignments through the addition of (linear) polynomials and applying a strong form of propagation by computing Gröbner bases. By handling bit assignments symbolically, the number of Gröbner basis calculations, along with the number of assignments, is reduced. The final Gröbner basis yields expressions for the bit-vectors in terms of the symbolic bits, together with non-linear pseudo-boolean constraints on the symbolic variables, modulo a power of two. The pseudo-booleans can be solved by translation into classical linear pseudo-boolean constraints (without a modulo) or by encoding them as propositional formulae, for which a novel translation process is described.

**Keywords:** Gröbner bases · Bit-vectors · Modulo arithmetic · SMT

## 1 Introduction

Some of the most influential algorithms in algebraic computation, such as Buchberger's algorithm [7] and Collin's Cylindrical Algebraic Decomposition algorithm [8], were invented long before the advent of SMT. SMT itself has evolved from its origins in SAT into a largely independently branch of symbolic computation. Yet the potential of cross-fertilising one branch with the other has been repeatedly observed [1,6,10], and a new class of SMT solvers is beginning to emerge that apply both algebraic and satisfiability techniques in tandem [15,16,23]. The problem, however, is that algebraic algorithms do not readily fit into the standard SMT architecture [22] because they are not normally incremental or backtrackable, and rarely support learning [1].

For application to software verification, the SMT background theory of bit-vectors is of central interest. Solvers for bit-vectors conventionally translate bit-vector constraints into propositional formulae by replacing constraints

Supported by an AWE grant on the verification of AVR microcontroller code.

L. Pulina and M. Seidl (Eds.): SAT 2020, LNCS 12178, pp. 361–377, 2020.
https://doi.org/10.1007/978-3-030-51825-7_26

with propositional circuits that realise them, a technique evocatively called bit-blasting. However, particularly for constraints involving multiplication, the resulting formulae can be prohibitively large. Moreover, bit-blasting foregoes the advantages afforded by reasoning at the level of bit-vectors [3,14].

In this paper we present a new architecture for solving systems of polynomial equalities over bit-vectors. Rather than converting to SAT and bit-blasting, the method sets bits in order of least significance through the addition of certain polynomials to the system. Computing a Gröbner basis [5] for the resulting system effects a kind of high-level propagation, which we have called bit-sequence propagation, in which the values of other bits can be automatically inferred. Furthermore, we show how the procedure can be carried out with symbolic truth values without giving up bit-sequence propagation, thus unifying Gröbner basis calculations that would otherwise be separate.

Once all bits are assigned truth values (symbolic or otherwise), the resulting Gröbner basis prescribes an assignment to the bit-vectors which is a function of the symbolic truth values. The remaining polynomials in the basis relate the symbolic truth values and correspond to non-linear pseudo-boolean constraints modulo a power of two. These constraints can be solved either by translation into classical linear pseudo-boolean constraints (without a modulo) or else by encoding them as propositional formulae, for which a novel translation process is described. Either way, the algebraic Gröbner basis computation is encapsulated in the phase that emits the pseudo-boolean constraints, hence the Gröbner basis engine [5] does not need to be backtrackable, incremental or support learning. The approach can be extended naturally to handle polynomial disequalities since the bit-vectors in the disequalities can be reformulated in terms of the symbolic variables of the equalities, and the disequalities forced to hold as well by virtue of a Tseytin transform [24]. Overall, the architecture provides a principled method for compiling high-level polynomials to low-level pseudo-boolean constraints.

In summary, this paper makes the following contributions:

- We specialise a Gröbner basis algorithm for integers modulo $2^\omega$ [5], using the concept of rank [21], introducing an algorithmic modification to ease the computational burden of computing Gröbner bases;
- We introduce bit-sequence propagation, in which an individual bit is set to a truth value 0 or 1 by adding a suitable (linear) polynomial to the system and then the effect on other bits is inferred by computing a Gröbner basis;
- We show how bit assignments can be handled symbolically in order to unify distinct Gröbner basis computations, eventually yielding a residue system of non-linear pseudo-boolean constraints;
- We show how the resulting pseudo-boolean systems can be solved by employing a novel rewrite procedure for converting non-linear modulo pseudo-booleans to propositional formulae.

The paper is structured as follows: Sect. 2 illustrates bit-sequence propagation through a concrete example. The supporting concepts of Gröbner bases for modulo integers and pseudo-boolean encoding are detailed in Sect. 3. Experimental results are given in Sect. 4 and Sect. 5 surveys related work.

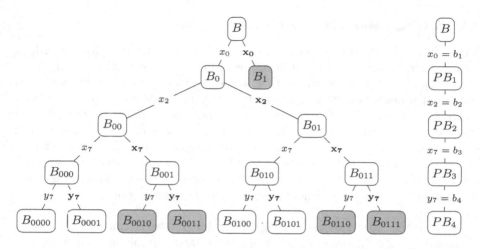

**Fig. 1.** Bit-assignments and word-level propagation: 0/1 bits and symbolic bits

## 2 Bit-Sequence Propagation in a Nutshell

Classically, that is for polynomials over algebraically closed fields, unsatisfiability can be decided by Hilbert's Nullstellensatz [9]. This equates unsatisfiability with the existence of a non-zero constant polynomial in a Gröbner basis for the polynomials. The concept of Gröbner basis is inextricably linked with that of an ideal [9]. The ideal for a given system (set) of polynomials is the least set closed under the addition of polynomials drawn from the set and multiplication of an arbitrary polynomial with a polynomial from the set; an ideal shares the same zeros as the system from which it is derived, but is not finite. A Gröbner basis is merely a finite representation of an ideal, convenient because, among other things, it enables satisfiability to be detected, at least over a field.

Unary bit-vectors constitute a field, but Nullstellensatz does not hold for bit-vectors with multiple bits. To see this, consider the polynomial equation $x^2 + 2 = 0$ where the arithmetic is 3-bit (modulo 8). Any solution $x$ to this equation must be even. But, $0^2 + 2 = 2$, $2^2 + 2 = 6$, $4^2 + 2 = 2$ and $6^2 + 2 = 6$. Hence $x^2 + 2 = 0$ has no solutions, yet the Gröbner basis $\{x^2 + 2\}$ does not contain a non-zero constant polynomial. Moreover, even for a Gröbner basis of a satisfiable system, such as $\{x^2 + 4\}$, the solutions to the system cannot be immediately read off from the basis. The force of these observations is that Gröbner bases need to be augmented with search to test satisfiability and discover models. To illustrate this we consider a more complicated system:

$$B = \begin{cases} y^2 + 120x^2 + 123x + 48 = 0, & yx + 65x^2 + 50x + 32 = 0, \\ 2y + 63x^2 + 59x + 128 = 0, & x^3 + 135x^2 + 100x + 64 = 0, \\ 64x^2 + 192x = 0 \end{cases}$$

where $x, y \in \mathbb{Z}_{256}$. Henceforth we follow convention and omit $= 0$ from systems.

## 2.1 Solving Using 0/1 Truth Values

Since $\mathbb{Z}_{256}$ is finite, this system can be solved by viewing the problem [20] as a finite domain constraint satisfaction problem. In this setting, each bit-vector is associated with a set of values that is progressively pruned using word-level constraint propagation rules. The search tree in the left-hand side of Fig. 1 illustrates how pruning is achieved by setting and inferring bits in the order of least-significance, starting with the bits of $x$ then those of $y$. On a left branch of the tree one bit, $x_i$ or $y_j$, is set to 0; on a right branch the bit is set to 1 (indicated in bold). Each node is labelled with a Gröbner basis that encodes the impact of setting a bit on all other bits. Gröbner bases are indexed by their position in the tree. Grey nodes correspond to the solutions of $B$.

*Computing $B_0$.* Setting the least significant bit of $x$ to 0 can be achieved by imposing $x = 2w$ for some otherwise unconstrained variable $w$. Hence, we add $2w - x$ to $B$ and compute a Gröbner basis with respect to the lexicographical ordering on variables $w \succ y \succ x$, yielding:

$$\begin{cases} wx + 86x + 96, & 2w + 255x, & y^2 + 219x + 48, \\ yx + 134x + 96, & 2y + 231x + 64, & x^2 + 172x + 192, & 64x \end{cases}$$

To eliminate dependence on $w$, polynomials involving $w$ are removed, giving:

$$B_0 = \begin{cases} y^2 + 219x + 48, & yx + 134x + 96, & 2y + 231x + 64, \\ x^2 + 172x + 192, & 64x \end{cases}$$

Note that $B_0$ contains $64x$ (representing $64x = 0$) which indicates that $x$ is a multiple of 4. Thus bit 1 is also 0, although we did not actively impose it.

Now, observe the constraint $64x = 0$ implies $0 = 2^6(x-0)$ hence $x-0 = 2^2 w'$ for some $w'$. To set the next bit to 0, put $w' = 2w$ which gives $x-0 = 8w$ yielding the polynomial $8w - x$. Otherwise, to set the next bit put $w' = 2w + 1$ giving the polynomial $8w - x + 4$.

*Computing $B_{00}$.* Augmenting $B_0$ with $8w - x$, calculating a Gröbner basis, and then eliminating $w$ gives:

$$B_{00} = \begin{cases} y^2 + 219x + 48, & yx + 128, & 2y + 231x + 64, \\ x^2, & 2x + 160 \end{cases}$$

Since $B_{00}$ includes $2x + 160$ (representing $2x + 160 = 0$) it follows that only bit 7 is undetermined. To constrain it, observe $0 = 2(x - 48)$ thus $x - 48 = 2^7 w'$ for some $w'$. Putting $w' = 2w$ gives $x - 48 = 256w = 0$ hence the polynomial $x - 48$. Conversely, putting $w' = 2w + 1$ gives $x - 48 = 256w + 128 = 128$ thus $x - 176$.

*Computing $B_{000}$ and $B_{001}$.* Adding $x - 48$ and $x - 176$ to $B_{00}$, computing a Gröbner basis, and eliminating $w$ (a vacuous step), respectively yields:

$$B_{000} = \{y^2 + 64, \quad 2y + 144, \quad x + 208\} \quad B_{001} = \{y^2 + 192, \quad 2y + 16, \quad x + 80\}$$

Both systems contain a single constraint on $x$ which uniquely determines its value, hence we move attention to $y$. Both $B_{000}$ and $B_{001}$ contain equations with leading terms $2y$ and thus only bit 7 of $y$ must be constrained. Following the same procedure as before, we obtain:

$$B_{0000} = \left\{ \begin{matrix} y + 200, \\ x + 208, \\ 128 \end{matrix} \right\} \quad B_{0001} = \left\{ \begin{matrix} y + 72, \\ x + 208, \\ 128 \end{matrix} \right\} \quad B_{0010} = \left\{ \begin{matrix} y + 136, \\ x + 80 \end{matrix} \right\} \quad B_{0011} = \left\{ \begin{matrix} y + 8, \\ x + 80 \end{matrix} \right\}$$

These Gröbner bases all completely constrain $x$ and $y$, hence are leaf nodes. Note that $B_{0000}$ and $B_{0001}$ contain the non-zero, constant polynomial 128, indicating unsatisfiability. Hence, only $B_{0010}$ and $B_{0011}$ actually yield solutions (highlighted in grey), namely $x \mapsto 176, y \mapsto 120$ and $x \mapsto 176, y \mapsto 248$ respectively.

*Computing $B_*$.* The general principle is that if $2^k(x - \ell)$ is in the basis and $\omega$ is the bit width, then the linear polynomial $2^{\omega-k+1}w - x + \ell$ is added for some fresh $w$ to set the next undermined bit to 0. Conversely, to set the next bit to 1, the polynomial $2^{\omega-k+1}w - x + 2^{\omega-k} + \ell$ is added. We name this tactic bit-sequence propagation. Using this tactic to flesh out the rest of the tree gives the following satisfiable bases (also marked in grey in the figure):

$$B_1 = \{y + 183, x + 91\} \quad B_{0110} = \{y + 158, x + 92\} \quad B_{0111} = \{y + 30, x + 92\}$$

yielding $x \mapsto 165, y \mapsto 73, x \mapsto 164, y \mapsto 98$ and $x \mapsto 164, y \mapsto 226$ respectively.

## 2.2 Solving Using Symbolic Truth Values

To reduce the total number of Gröbner basis calculations, we observe that it is sufficient to work with symbolic bits. The right-hand side of Fig. 1 illustrates how this reduces the number of bases calculated to 4, albeit at the cost of carrying symbolic bits in the basis. Bit-sequence propagation generalises via the single rule: if $2^k(x - \ell)$ is in the basis and $\omega$ is the bit width, then the polynomials $2^{\omega-k+1}w - x + 2^{\omega-k}b + \ell$ and $b^2 - b$ are added to the basis. This sets the next undermined bit to the symbolic value $b$; the polynomial $b^2 - b$ merely asserts that each symbolic $b$ can only be 0 or 1. This construction gives:

$$PB_1 = \left\{ \begin{matrix} y^2 + 219x + 216b_1 + 48, & yx + 6x + 181b_1 + 96, & yb_1 + 183b_1, \\ 2y + 103x + 203b_1 + 64, & x^2 + 44x + 139b_1 + 192, & xb_1 + 91b_1, \\ 64x + 192b_1, & b_1^2 + 255b_1 \end{matrix} \right\}$$

$$\vdots$$

$$PB_4 = \left\{ \begin{matrix} y + 128b_4 + 192b_3 + 214b_2 + 153b_1 + 200, & x + 12b_2 + 255b_1 + 80, \\ b_4^2 + 255b_4, & 128b_4b_1 + 128b_1, & b_3^2 + 255b_3, & 64b_3b_1, \\ 128b_3 + 128b_2 + 128, & b_2^2 + 255b_2, & 2b_2b_1 + 254b_1, & b_1^2 + 255b_1 \end{matrix} \right\}$$

The final $PB_4$ expresses $x$ and $y$ as combinations of $b_4, b_3, b_2$ and $b_1$:

$$y \equiv_{256} -128b_4 - 192b_3 - 214b_2 - 153b_1 - 200 \qquad x \equiv_{256} -12b_2 - 255b_1 - 80$$

Observe that the remaining polynomials are non-linear pseudo-boolean constraints over $b_4, b_3, b_2$ and $b_1$ modulo 256. The polynomials $b_i^2 + 255b_i$, which assert that each $b_i$ is binary, are subsequently ignored.

## 2.3  Solving Using SAT

These pseudo-booleans can be simplified by observing that when all coefficients in the constraint are divisible by a power of 2 then the modulo can be lowered:

$$128b_4b_1 + 128b_1 \equiv_{256} 0 \iff b_4b_1 + b_1 \equiv_2 0$$
$$64b_3b_1 \equiv_{256} 0 \iff b_3b_1 \equiv_4 0$$
$$128b_3 + 128b_2 + 128 \equiv_{256} 0 \iff b_3 + b_2 + 1 \equiv_2 0$$
$$2b_2b_1 + 254b_1 \equiv_{256} 0 \iff b_2b_1 + 127b_1 \equiv_{128} 0$$

Since the reduced versions of the first and third constraints are modulo 2 they can be mapped immediately to the propositional formulae:

$$b_4b_1 + b_1 \equiv_2 0 \iff (b_4 \land b_1) \Leftrightarrow b_1$$
$$b_3 + b_2 + 1 \equiv_2 0 \iff \neg(b_3 \Leftrightarrow b_2)$$

where the negation is introduced because of the constant 1. The second and fourth constraints cannot be handled so directly because the modulus is not 2. However, for the second, we can use the fact that the left-hand side is a single term to infer either $b_3$ or $b_1$ must be 0, yielding the formula $\neg b_3 \lor \neg b_1$. Finally, for the fourth constraint, we do a case split on $b_2$. Setting $b_2 = 0$ simplifies the constraint to $127b_1 \equiv_{128} 0$, from which $b_1 = 0$ is inferred. Conversely, setting $b_2 = 1$ simplifies the constraint to $128b_1 \equiv_{128} 0$ which is vacuous. Overall, we derive the formula $(\neg b_2 \land \neg b_1) \lor b_2$ for the fourth constraint. There are 5 truth assignments for the formula assembled from the above 4 sub-formulae, yielding 5 assignments to $x$ and $y$ that concur with those given previously.

The reasoning exemplified here has been distilled into a series of rules, presented in Sect. 3.8, for encoding non–linear modulo pseudo-booleans into SAT. An alternative approach finds the values for $b_4$, $b_3$, $b_2$ and $b_1$ using a cutting-plane pseudo-boolean solver [19] alongside a modulo elimination transformation [13, Section 3]. Regardless of the particular method employed to solve this system, observe that search has been isolated in the SAT/pseudo-boolean solver; the Gröbner bases are calculated in an entirely deterministic fashion.

## 3  Theoretical Underpinnings

In its classical form, Buchberger's algorithm for computing Gröbner bases is only applicable to polynomials over fields [18], such as rationals and complex numbers, where every non-zero element has a multiplicative inverse. However, integers modulo $2^\omega$ only have this property for $\omega = 1$, as even numbers do not have multiplicative inverses for $\omega > 1$. Nevertheless, a variant of Buchberger's algorithm has been reported that is applicable to modulo integers with respect to arbitrary moduli [5]. This section specialises this variant to integers modulo $2^\omega$, exploiting the concept of rank [21] to efficiently determine divisibility over modulo integers. An algorithmic refinement is also reported that reduces the number of calculated S-polynomials, the key to improving efficiency. The section concludes with a formalisation of the rules for encoding non-linear modulo pseudo-booleans into propositional formulae, first introduced in Sect. 2.3.

## 3.1 Modulo Integers

Let $\mathbb{N}$ (resp., $\mathbb{N}_+$), denote the non-negative (resp., positive) integers, $\omega \in \mathbb{N}_+$ be the bit-width, $m = 2^\omega$ and $\mathbb{Z}_m = \{0, 1, \ldots, m-1\}$ denote the integers modulo $m$. The rank of $x \in \mathbb{Z}_m$ [21] is defined: $\mathrm{rank}_\omega(x) = \max\{j \in \mathbb{N} \mid 2^j \text{ divides } x\}$ if $x > 0$ and $\omega$ otherwise. Rank can be computed by counting the number of trailing zeros in an integer's binary representation [26].

*Example 1.* In $\mathbb{Z}_{256}$ when $\omega = 8$, $\mathrm{rank}_8(0) = 8$, $\mathrm{rank}_8(15) = 0$ and $\mathrm{rank}_8(56) = 3$.

If $x \in \mathbb{Z}_m$, $x \neq 0$ then $x = 2^{\mathrm{rank}_\omega(x)}d$ where $d = x/2^{\mathrm{rank}_\omega(x)}$ is odd. This is referred to as the rank decomposition of $x$. If $x \in \mathbb{Z}_m$ then $x$ is said to be invertible if there exists $x^{-1} \in \mathbb{Z}_m$, necessarily unique, such that $xx^{-1} = 1$. This occurs iff $x$ is odd, in which case $x^{-1}$ can be found as a stationary point of the sequence $y_1 = 1$, $y_{n+1} = y_n(2 - xy_n)$ [21]. For $x_1, x_2 \in \mathbb{Z}_m$, $x_1$ is said to be divisible by $x_2$ if $x_1 = yx_2$ for some $y \in \mathbb{Z}_m$. This occurs iff $\mathrm{rank}_\omega(x_1) \geq \mathrm{rank}_\omega(x_2)$, in which case, letting $x_i = 2^{k_i}d_i$ be the rank decomposition of $x_i$, it follows $y = 2^{k_1-k_2}d_1 d_2^{-1}$.

## 3.2 Polynomials and Ideals

A monomial is an expression $\boldsymbol{x}^{\boldsymbol{\alpha}} = x_1^{\alpha_1} \cdots x_n^{\alpha_n}$ where $\boldsymbol{x} = \langle x_1, \ldots, x_n \rangle$ is a vector of variables and $\boldsymbol{\alpha} = \langle \alpha_1, \ldots, \alpha_n \rangle \in \mathbb{N}^n$. A term is an expression $c\boldsymbol{x}^{\boldsymbol{\alpha}}$ where $c \in \mathbb{Z}_m$. A polynomial is either 0 or an expression $t_1 + \cdots + t_s$ where each $t_i$ is a term. In this expression, we assume all $t_i$ have non-zero coefficients and distinct monomials, since terms with 0 coefficients can be removed and terms with the same monomial can be collected by summing their coefficients. The set of all polynomials over $\mathbb{Z}_m$ is denoted $\mathbb{Z}_m[\boldsymbol{x}]$. For $p \in \mathbb{Z}_m[\boldsymbol{x}]$ and $\boldsymbol{a} \in \mathbb{Z}_m^n$, $[\![p]\!](\boldsymbol{a})$ denotes the result of substituting $a_i$ for each $x_i$ in $p$ and evaluating the result.

An ideal is a set $I \subseteq \mathbb{Z}_m[\boldsymbol{x}]$ such that $\Sigma_{i=1}^s u_i p_i \in I$ for all $s \in \mathbb{N}$, $p_i \in I$ and $u_i \in \mathbb{Z}_m[\boldsymbol{x}]$. If $P \subseteq \mathbb{Z}_m[\boldsymbol{x}]$ then $\langle P \rangle = \{\sum_{i=1}^s u_i p_i \mid s \geq 0, p_i \in P, u_i \in \mathbb{Z}_m[\boldsymbol{x}]\}$ is the ideal generated by $P$; if $I = \langle P \rangle$ then $P$ is said to be a basis for $I$. The solution (zero) set of $P \subseteq \mathbb{Z}_m[\boldsymbol{x}]$ is defined: $\gamma(P) = \{\boldsymbol{a} \in \mathbb{Z}_m^n \mid \forall p \in P.\ [\![p]\!](\boldsymbol{a}) = 0\}$. Note that $\gamma(\langle P \rangle) = \gamma(P)$, hence if $P_1$, $P_2$ are both bases for the same ideal $I$, then $\gamma(P_1) = \gamma(I) = \gamma(P_2)$. Given a set $P_1 \subseteq \mathbb{Z}_m[\boldsymbol{x}]$, it is thus desirable to find a basis $P_2$ for $\langle P_1 \rangle$ which exposes information about the zeros of $P_1$. The concept of Gröbner basis makes this idea concrete.

## 3.3 Leading Terms

Let $\prec$ denote the lexicographical ordering over monomials defined by $\boldsymbol{x}^{\boldsymbol{\alpha}} \prec \boldsymbol{x}^{\boldsymbol{\beta}}$ if $\boldsymbol{\alpha} \neq \boldsymbol{\beta}$ and $\alpha_i < \beta_i$ at the first index $i$ where $\alpha_i \neq \beta_i$. Note that this ordering depends on the order of the variables in $\boldsymbol{x}$. If $p \in \mathbb{Z}_m[\boldsymbol{x}]$ then either $p = 0$ or else $p = c\boldsymbol{x}^{\boldsymbol{\alpha}} + q$ for some polynomial $q$, where all terms $d\boldsymbol{x}^{\boldsymbol{\beta}}$ appearing in $q$ satisfy $\boldsymbol{x}^{\boldsymbol{\beta}} \prec \boldsymbol{x}^{\boldsymbol{\alpha}}$. In the latter case, the leading term, coefficient and monomial of $p$ are defined $\mathrm{lt}(p) = c\boldsymbol{x}^{\boldsymbol{\alpha}}$, $\mathrm{lc}(p) = c$ and $\mathrm{lm}(p) = \boldsymbol{x}^{\boldsymbol{\alpha}}$ respectively.

*Example 2.* Let $p_1 = y^2 + 3yx \in \mathbb{Z}_{256}[y, x]$ and $p_2 = 3xy + y^2 \in \mathbb{Z}_{256}[x, y]$. Note that $p_1$ and $p_2$ consist of the same terms, yet $\mathrm{lt}(p_1) = y^2$, $\mathrm{lc}(p_1) = 1$ and $\mathrm{lm}(p_1) = y^2$ and $\mathrm{lt}(p_2) = 3xy$, $\mathrm{lc}(p_2) = 3$ and $\mathrm{lm}(p_2) = xy$.

## 3.4  Reduction

Leading terms give a rewrite procedure over polynomials. First, note that if $t_1 = c_1 x^{\alpha_1}$, $t_2 = c_2 x^{\alpha_2}$ are terms then there exists a term $t$ such that $t_1 = t t_2$ iff $\alpha_1 \geq \alpha_2$ component-wise and $c_1$ is divisible by $c_2$; in this case, $t = d x^\beta$ where $c_1 = d c_2$ and $\beta = \alpha_1 - \alpha_2$. With divisibility in place, reducibility can be defined:

**Definition 1.** *Let $p, q, r \in \mathbb{Z}_m[x]$, $p, q \neq 0$. Then, $p$ is said to be reducible by $q$ to $r$, denoted $p \rightarrow_q r$, if $\mathsf{lt}(p) = c x^\alpha \, \mathsf{lt}(q)$ and $p = c x^\alpha q + r$ for some term $c x^\alpha$.*

Reducibility lifts to sets $P \subseteq \mathbb{Z}_m[x]$ by $\rightarrow_P = \bigcup_{p \in P} \rightarrow_p$ and $\rightarrow_P^+$ (resp. $\rightarrow_P^*$) is the transitive (resp. transitive, reflexive) closure of $\rightarrow_P$. If $p \rightarrow_P^+ r$ for some $r$ then $p$ is said to be reducible by $P$, otherwise irreducible by $P$, denoted $p \not\rightarrow_P$.

*Example 3.* Let $p = y x^2 + 2 y x + 5 y + x$ and $P = \{p_1, p_2\} \subseteq \mathbb{Z}_{256}[y, x]$ where $p_1 = y x + 3 y$ and $p_2 = 4 y + x$. Then, $\mathsf{lt}(p) = y x^2 = x \, \mathsf{lt}(p_1)$ and $p = x p_1 + r_1$ where $r_1 = 255 y x + 5 y + x$. Similarly, $\mathsf{lt}(r_1) = 255 y x = 255 \, \mathsf{lt}(p_1)$ and $r_1 = 255 p_1 + r_2$ where $r_2 = 8 y + x$. Finally, $\mathsf{lt}(r_2) = 8 y = 2 \, \mathsf{lt}(p_2)$ and $r_2 = 2(4 y + x) + r_3$ where $r_3 = 255 x$. Thus, $p \rightarrow_{p_1} r_1 \rightarrow_{p_1} r_2 \rightarrow_{p_2} r_3 = 255 x$ hence $p \rightarrow_P^+ 255 x$.

Note if $p \rightarrow_P^+ r$ then $r$ has a strictly smaller leading term than $p$. Moreover, if $p \in \mathbb{Z}_m[x]$, $P \subseteq \mathbb{Z}_m[x]$ and $p \rightarrow_P^* 0$ then $p \in \langle P \rangle$. The converse of this does not hold in general, as the following example shows:

*Example 4.* If $p = x$ and $P = \{p_1, p_2\} \subseteq \mathbb{Z}_{256}[y, x]$ where $p_1 = 2 y x^2 + 2 x^2 + 6 y x + x$ and $p_2 = 4 y + 4$ then $p$ is irreducible by $P$, yet $p = (154 y x + 206 y + 154 x + 1) p_1 + (179 y x^3 + 50 y x^2 + 75 y x + 179 x^3 + 25 x^2) p_2 \in \langle P \rangle$.

## 3.5  Gröbner Bases

**Definition 2.** *Let $I \subseteq \mathbb{Z}_m[x]$ be an ideal. A set $P \subseteq I$ is a Gröbner basis for $I$ iff, for all $p \in I$, if $p \neq 0$ then $p$ is reducible by some $q \in P$.*

If $P \subseteq \mathbb{Z}_m[x]$ is a Gröbner basis for the ideal $I \subseteq \mathbb{Z}_m[x]$ and $p \in I$ then $p \rightarrow_P^* 0$. Hence, Gröbner bases allow ideal membership to be tested by reduction. In order to compute Gröbner bases, an auxiliary definition is required:

**Definition 3.** *The S-polynomial of $p_1, p_2 \in \mathbb{Z}_m[x]$ is defined:*

$$\mathsf{spol}(p_1, p_2) = d_2 2^{k - k_1} x^{\alpha - \alpha_1} p_1 - d_1 2^{k - k_2} x^{\alpha - \alpha_2} p_2$$

*where, if $p_i = 0$ then $k_i = \omega$, $d_i = 1$ and $\alpha_i = 0$, else $2^{k_i} d_i$ is the rank decomposition of $\mathsf{lc}(p_i)$ and $x^{\alpha_i} = \mathsf{lm}(p_i)$, $k = \max(k_1, k_2)$ and $\alpha = \max(\alpha_1, \alpha_2)$.*

*Example 5.* Let $p_1 = 2 y x^2 + 2 x^2 + 6 y x + x$ and $p_2 = 4 y + 4$ in $\mathbb{Z}_{256}[y, x]$. Then, $\mathsf{spol}(p_1, p_2) = 2(2 y x^2 + 2 x^2 + 6 y x + x) - x^2(4 y + 4) = 12 y x + 2 x$ and $\mathsf{spol}(p_1, 0) = 128(2 y x^2 + 2 x^2 + 6 y x + x) - y x^2(0) = 128 x$.

The following theorem, adapted from [5], provides an effective criterion for detecting that a finite basis is a Gröbner basis:

**Theorem 1.** *Let $P \subseteq \mathbb{Z}_m[\boldsymbol{x}]$ and suppose for all $p_1, p_2 \in P$, $\mathsf{spol}(p_1, p_2) \to_P^* 0$ and $\mathsf{spol}(p_1, 0) \to_P^* 0$. Then, $P$ is a Gröbner basis for $\langle P \rangle$.*

Note that if $|P| = \ell$ then this criterion takes $\binom{\ell}{2} + \ell$ reductions to verify.

```
function buchberger(F)
begin
    G := F;  S := {{g₁,g₂} | g₁ ∈ G, g₂ ∈ G ∪ {0}, g₁ ≠ g₂}
    while (S ≠ ∅)
        {g₁,g₂} := element(S)
        S := S - {g₁,g₂}
        r := reduce(spol(g₁,g₂), G)
        if (r ≠ 0)
            G := G ∪ {r}
            S := S ∪ {{r,g} | g ∈ G ∪ {0}}
        end if
    end while
    return G
end
```

**Fig. 2.** Buchberger's algorithm over integers modulo $2^\omega$

## 3.6  Buchberger's Algorithm

Theorem 1 also yields a strategy for converting a finite basis $P \subseteq \mathbb{Z}_m[\boldsymbol{x}]$ for an ideal $I \subseteq \mathbb{Z}_m[\boldsymbol{x}]$ to a Gröbner basis for $I$. The strategy works by reducing each S-polynomial of the basis in turn; if some S-polynomial does not reduce to 0 then its reduced form is added to the basis and the procedure continues. Eventually, all S-polynomials of basis elements reduce to 0, at which point the algorithm returns. The algorithm determined by this strategy is called Buchberger's algorithm. Figure 2 presents a version of Buchberger's algorithm, adapted from [5], which utilises a set $S$ to record the set of S-polynomials that have yet to be reduced. It requires an auxiliary function reduce which realises reduction. More specifically, if $p \in \mathbb{Z}_m[\boldsymbol{x}]$ and $P \subseteq \mathbb{Z}_m[\boldsymbol{x}]$ is finite then $p \to_P^* \mathsf{reduce}(p, P)$ and $\mathsf{reduce}(p, P) \not\to_P$.

*Example 6.* The table gives a trace of Buchberger's algorithm on $P = \{p_1, p_2\} \subseteq \mathbb{Z}_{256}[x, y]$ where $p_1 = 2yx^2 + 2x^2 + 6yx + x$ and $p_2 = 4y + 4$. Step $k$ displays the values of $G$ and $S$, and the next reduction, after $k$ iterations of the main loop.

| step | $G$ | $S$ | reduction |
|------|-----|-----|-----------|
| 0 | $\{p_1, p_2\}$ | $\{\{p_2, 0\}, \{p_1, p_2\}, \{p_1, 0\}\}$ | $0 \to_G 0$ |
| 1 | $\{p_1, p_2\}$ | $\{\{p_1, p_2\}, \{p_1, 0\}\}$ | $12yx + 2x \to_G 246x = p_3$ |
| 2 | $\{p_1, p_2, p_3\}$ | $\{\{p_3, 0\}, \{p_2, p_3\}, \{p_1, p_3\},$ $\{p_1, 0\}\}$ | $0 \to_G 0$ |
| 3 | $\{p_1, p_2, p_3\}$ | $\{\{p_2, p_3\}, \{p_1, p_3\}, \{p_1, 0\}\}$ | $236x \to_G 0$ |
| 4 | $\{p_1, p_2, p_3\}$ | $\{\{p_1, p_3\}, \{p_1, p_0\}\}$ | $226yx + 246x^2 + 123x$ $\to_G 123x = p_4$ |
| 5 | $\{p_1, p_2, p_3, p_4\}$ | $\{\{p_4, 0\}, \{p_3, p_4\}, \{p_2, p_4\},$ $\{p_1, p_4\}, \{p_1, 0\}\}$ | $0 \to_G 0$ |
| $\vdots$ | $\vdots$ | $\vdots$ | $\vdots$ |
| 9 | $\{p_1, p_2, p_3, p_4\}$ | $\{\{p_1, 0\}\}$ | $128x \to_G 0$ |
| 10 | $\{p_1, p_2, p_3, p_4\}$ | $\emptyset$ | $-$ |

```
function modifiedBuchberger(F)
begin
    G := ∅;  A := F;  S := ∅
    while (A ≠ ∅ ∨ S ≠ ∅)
        if (A ≠ ∅)
            p := element(A);  A := A − {p}
        else
            {f1, f2} := element(S);  S := S − {f1, f2};  p := spol(f1, f2)
        end if
        r := reduce(p, G)
        if (r ≠ 0)
            H := {g ∈ G | g ↛r};  G := H ∪ {r};  A := A ∪ (G − H)
            S := {{g1, g2} ∈ S | g1, g2 ∈ H ∪ {0}}) ∪ {(p, h) | h ∈ H ∪ {0}}
        end if
    end while
    return G
end
```

**Fig. 3.** Modified Buchberger's algorithm over integers modulo $2^\omega$

### 3.7   Modified Buchberger's Algorithm

Note that, for the Gröbner basis computed in Example 6, only $p_2$ and $p_4$ are necessary to ensure the Gröbner property, since any polynomial reducible by $p_1$ or $p_3$ must be reducible by $p_4$. This observation is formalised in the result:

**Theorem 2.** *Let $P \subseteq \mathbb{Z}_m[\boldsymbol{x}]$ and $P' = \{p \in P \mid p \not\rightarrow_{P-\{p\}}\}$. Suppose for all $p \in P - P'$ that $p \rightarrow^*_{P'} 0$, and for all $p_1, p_2 \in P'$, $spol(p_1, p_2) \rightarrow^*_{P'} 0$ and $spol(p_1, 0) \rightarrow^*_{P'} 0$. Then, $P'$ is a Gröbner basis for $\langle P \rangle$.*

*Proof.* By Theorem 1, $P'$ is a Gröbner basis for $\langle P' \rangle$. However, since $P' \subseteq P$ and $p \rightarrow^*_{P'} 0$ for all $p \in P - P'$ it follows that $\langle P \rangle = \langle P' \rangle$. The result follows.

If $|P| = \ell_1$ and $|P'| = \ell_2$ then this criterion takes $\binom{\ell_2}{2} + \ell_2 + (\ell_1 - \ell_2) = \binom{\ell_2}{2} + \ell_1$ reductions to verify, as opposed to $\binom{\ell_1}{2} + \ell_1$ reductions via the original criterion. Moreover, as for the original criterion, it yields an algorithm for converting a finite basis into a Gröbner basis. The algorithm operates as the original, except whenever a basis element becomes reducible by a newly added element, it is removed from, and then reduced by, the basis. If it reduces to 0 then it is discarded; otherwise, its reduced form is added to the basis. All S-polynomials derived from it are then discarded. Figure 3 presents a modification to Buchberger's algorithm that implements this idea using a third set $A$ to store elements that are removed from the basis. Elements of $A$ are reduced in preference to elements of $S$, and the algorithm terminates when both $A$ and $S$ are empty. To handle the fact that some elements of the input basis could be reducible by other elements of the input basis, the set $G$ is initialised to $\emptyset$ and the set $A$ is initialised to $P$. Hence, the first iterations of the loop effectively add the input polynomials to the basis.

*Example 7.* The following table summarises a trace of the modified Buchberger algorithm on the same input as Example 6.

| step | $G$ | $A$ | $S$ | reduction |
|------|-----|-----|-----|-----------|
| 0 | $\emptyset$ | $\{p_1, p_2\}$ | $\emptyset$ | $p_1 \rightarrow_G p_1$ |
| 1 | $\{p_1\}$ | $\{p_2\}$ | $\{\{p_1, 0\}\}$ | $p_2 \rightarrow_G p_2$ |
| 2 | $\{p_1, p_2\}$ | $\emptyset$ | $\{\{p_2, 0\}, \{p_1, p_2\}, \{p_1, 0\}\}$ | $0 \rightarrow_G 0$ |
| 3 | $\{p_1, p_2\}$ | $\emptyset$ | $\{\{p_1, p_2\}, \{p_1, 0\}\}$ | $12yx + 2x \rightarrow_G 246x = p_3$ |
| 4 | $\{p_2, p_3\}$ | $\{p_1\}$ | $\{\{p_3, 0\}, \{p_2, p_3\}\}$ | $p_1 \rightarrow_G x = p_4$ |
| 5 | $\{p_2, p_4\}$ | $\{p_3\}$ | $\{\{p_4, 0\}, \{p_2, p_4\}\}$ | $p_3 \rightarrow_G 0$ |
| 6 | $\{p_2, p_4\}$ | $\emptyset$ | $\{\{p_4, 0\}, \{p_2, p_4\}\}$ | $0 \rightarrow_G 0$ |
| 7 | $\{p_2, p_4\}$ | $\emptyset$ | $\{\{p_2, p_4\}\}$ | $4x \rightarrow_G 0$ |
| 8 | $\{p_2, p_4\}$ | $\emptyset$ | $\emptyset$ | $-$ |

As noted above, the first two steps of the trace simply add the two input polynomials to the basis, which had already been performed in the original trace. Removing these steps yields a trace length of 6 compared to 10 in the original example. Moreover, by construction, neither $p_2$ nor $p_4$ is reducible by the other.

## 3.8   Encoding Pseudo-Boolean Constraints

Figure 4 presents rules for translating a polynomial in the form $\boldsymbol{c} \cdot \boldsymbol{X} \equiv_{2^r} d$ to a propositional formula such that $\boldsymbol{c} \in \mathbb{Z}_m^\ell$, $d \in \mathbb{Z}_m$ and $\boldsymbol{X} \in \wp(\cup \boldsymbol{x})^\ell$, where

$$\text{true} \frac{}{\epsilon \cdot \epsilon \equiv_{2^r} 0 \;\; \to \;\; \text{true}}$$

$$\text{false} \frac{\forall c_i \in \boldsymbol{c}. \; \text{rank}(c_i) > 0 \quad \text{rank}(d) = 0}{\boldsymbol{c} \cdot \boldsymbol{X} \equiv_{2^r} d \;\; \to \;\; \text{false}}$$

$$\text{xor} \frac{}{\boldsymbol{1} \cdot \boldsymbol{X} \equiv_2 1 \;\; \to \;\; \bigoplus_{X \in \boldsymbol{x}} (\bigwedge X)}$$

$$\text{iff} \frac{\boldsymbol{c} \cdot \boldsymbol{X} \equiv_2 1 \;\; \to \;\; f}{\boldsymbol{c} \cdot \boldsymbol{X} \equiv_2 0 \;\; \to \;\; \neg f}$$

$$\text{scale} \frac{\boldsymbol{c} \cdot \boldsymbol{X} \equiv_{2^r} d \;\; \to \;\; f}{(2^s \boldsymbol{c}) \cdot \boldsymbol{X} \equiv_{2^{r+s}} (2^s d) \;\; \to \;\; f}$$

$$\text{set} \frac{\text{rank}(d) = 0 \quad \exists! c_i \in \boldsymbol{c}. \; \text{rank}(c_i) = 0 \quad (\boldsymbol{c} \cdot \boldsymbol{X} \equiv_{2^r} d)[x \mapsto 1 \mid x \in X_i] \;\; \to \;\; f}{\boldsymbol{c} \cdot \boldsymbol{X} \equiv_{2^r} d \;\; \to \;\; (\bigwedge X_i) \wedge f}$$

$$\text{clear} \frac{\text{rank}(d) > 0 \quad \exists! c_i \in \boldsymbol{c}. \; \text{rank}(c_i) = 0 \quad \forall x \in X_i. \; (\boldsymbol{c} \cdot \boldsymbol{X} \equiv_{2^r} d)[x \mapsto 0] \;\; \to \;\; f_x}{\boldsymbol{c} \cdot \boldsymbol{X} \equiv_{2^r} d \;\; \to \;\; \bigvee_{x \in X_i} (\neg x \wedge f_x)}$$

$$\text{split} \frac{x \in \bigcup \boldsymbol{X} \quad (\boldsymbol{c} \cdot \boldsymbol{X} \equiv_{2^r} d)[x \mapsto 0] \;\; \to \;\; f_0 \quad (\boldsymbol{c} \cdot \boldsymbol{X} \equiv_{2^r} d)[x \to 1] \;\; \to \;\; f_1}{\boldsymbol{c} \cdot \boldsymbol{X} \equiv_{2^r} d \;\; \to \;\; (\neg x \wedge f_0) \vee (x \wedge f_1)}$$

**Fig. 4.** Reduction rules for pseudo-boolean polynomials modulo $2^r$

$\boldsymbol{x}$, recall, is the vector of variables and $\ell$ is the arity of the vectors $\boldsymbol{c}$ and $\boldsymbol{X}$. This form of constraint, although restrictive, is sufficient to express the pseudo-booleans which arise in the final Gröbner basis, as illustrated below:

*Example 8.* Returning to $PB_4$ of Sect. 2 the polynomials $128 b_4 b_1 + 128 b_1$ and $128 b_3 + 128 b_2 + 128$ can be written as $\langle 128, 128 \rangle \cdot \langle \{b_1, b_4\}, \{b_1\} \rangle \equiv_{256} 0$ and $\langle 128, 128 \rangle \cdot \langle \{b_3\}, \{b_2\} \rangle \equiv_{256} 128$ since $128 = -128 \pmod{256}$.

The rules of Fig. 4 collectively reduce the problem of encoding a constraint to that of encoding one or more strictly simpler constraints. For brevity, we limit the commentary to the more subtle rules. The false rule handles constraints which are unsatisfiable because the coefficients $\boldsymbol{c}$ are all even and $d$ is odd. The scale rule reduces the encoding problem to that for an equi-satisfiable constraint obtained by dividing the modulo, coefficients and constant by a power of 2. The set rule handles constraints where $d$ is odd and there is a unique term $c_i X_i$ for which $c_i$ is odd. In this circumstance all the variables of $X_i$ must be 1 for the constraint to be satisfiable. Conversely, clear deals with constraints for which $d$ is even and there exists a unique $c_i X_i$ for which $c_i$ is odd since then one variable of $X_i$ must be 0 for satisfiability. When none of above are applicable, split is applied to reduce to encoding problem to that of two strictly smaller constraints.

## 4   Experimental Results

Our aim is to apply high-level algebraic reasoning to systematically reduce polynomials to compact systems of low-level constraints. Our experimental work thus assesses how the complexity of the low-level constraints relate to that of the input polynomials. Although we provide timings for our Buchberger algorithm, which as far as we know is state-of-the-art, this is not our main concern. Indeed, fast

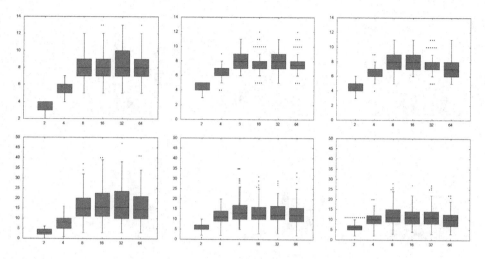

**Fig. 5.** Top row: number of symbolic variables (y) against $\omega$ (x) for $n = 2$, 3 and 4; Bottom row: number of pseudo-booleans (y) against $\omega$ (x) for $n = 2$, 3 and 4

algorithms for calculating Gröbner bases over fields have emerged in the last two decades [11,12], and similar performance gains seem achievable for modulo arithmetic. In light of this, our Buchberger algorithm is implemented in Scala 2.13.0 (compiled to JVM) using BigInt for complete generality. Experiments were performed on a 2.7 GHz Intel i5 Macbook with 16 Gbyte of SDRAM.

*Datasets.* To exercise the symbolic method, polynomial systems were generated for different numbers of bit-vectors $n$ and different bit-widths $\omega$. For each $\omega \in \{2, 4, 8, 16, 32, 64\}$ and $n \in \{2, 3, 4\}$, 100 polynomial systems were constructed by randomly generating points in $\mathbb{Z}_{2^\omega}^n$ and deriving a system with these points as their zeros. First, each point was described as a system of $n$ (linear) polynomials. Second, a single system was then formed with $n$ points as its zeros through the introduction of $n - 1$ fresh variables [2]. Third, the fresh variables were eliminated by calculating a Gröbner base to derive a basis constituting a single datapoint.

*Symbolic Variable and Pseudo-Boolean Count.* Figure 5 presents box and whisker diagrams that summarise the numbers of symbolic variables and pseudo-booleans appearing in the derived pseudo-boolean systems. For each box and whisker, the lower and upper limits of the box indicate the first and third quartiles, the central line the median. The inter-quartile range (IQR) is the distance from the top to the bottom of the box. By convention the whiskers extend to 1.5 times the IQR above and below the median value; any point falling outside of this range is considered to be an outlier and is plotted as an individual point. Figure 5 was derived from datapoints generated from 6 random points. Similar trends are observed with fewer points and appear to be displayed for more points, but variable elimination impedes dataset generation and large-scale evaluation.

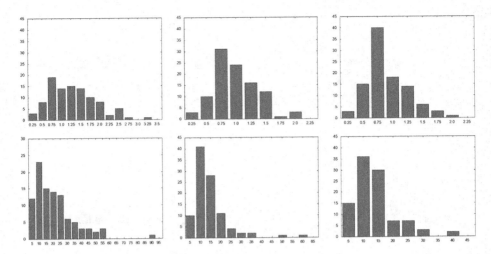

**Fig. 6.** Histograms for the ratio of the number of pseudo-booleans (top row)/logical connectives (bottom row) to the multiplication count for $n = 2, 3, 4$ and $\omega = 32$

For both the number of symbolic variables and the number of pseudo-booleans, the medians level off after an initial increase and then appear to be relatively independent of $\omega$. This surprising result suggests that algebraic methods have a role in reducing the complexity of polynomials for bit-vectors, which is sensitive to $\omega$ for bit-blasting. This implication is that the number of Gröbner base calculations also stabilises with $\omega$ since this tallies with the number of symbolic variables. We also observe that the number of symbolic bits employed is typically only a fraction of the total number of bits occurring in the bit-vectors, hence setting a single symbolic bit is often sufficient to infer values for many other bits.

*Pseudo-Boolean versus Multiplication Count.* The upper row of Fig. 6 presents a fine-grained analysis of the number of pseudo-booleans, comparing this count to the number of bit-vector multiplications in the datasets. Multiplications are counted as follows: a monomial $x^3yz$, say, in a polynomial system contributes $2 + 1 + 1 = 4$ to its multiplication count, irrespective of whether it occurs singly or multiply in the system. The term $42x^3yz$ also contributes 4 to the count, so simple multiplications with constants are ignored. Addition is also not counted, the rationale being to compare the number of pseudo-booleans against the number of bit-vector multiplications, assuming that different occurrences of a monomial are detected and factored together. The $x$-axis of the histograms of Fig. 6 divides the different ratios into bins, the first column giving the number of datasets for which the ratio falls within $[0, 0.25)$. As $n$ increases the ratios bunch more tightly around the bin $[0.5, 0.75)$ and, more significantly, the number of pseudo-booleans rarely exceed twice the multiplication count, at least for $\omega = 32$.

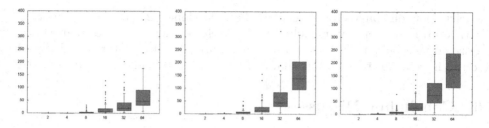

**Fig. 7.** Timings for Buchberger in seconds (y) against $\omega$ (x) for $n = 2, 3, 4$

*Logical Connectives versus Multiplication Count.* The lower row of Fig. 6 examines the complexity of the resulting pseudo-boolean systems from another perspective: the number of logical connectives required to encode them. The pseudo-boolean systems were translated to propositional formulae using the reduction rules of Fig. 4 and their complexity measured by counting the number of logical connectives used within them. The histograms present a frequency analysis of ratios of the number logical connectives to the multiplication count. Remarkably, histograms show that typically no more than 25 logical connectives are required per multiplication for $\omega = 32$.

*Timing.* Although the number of symbolic variables is a proxy for complexity, it ignores that Gröbner basis computations increase in cost with the number of symbolic variables. Figure 7 is intended to add clarity, plotting the time in seconds to calculate the pseudo-booleans against $\omega$. As expected, the median runtimes increase with $\omega$ for any given $n$, though not alarmingly so for an implementation based on Buchberger rather than a modern, fast engine such as F5 [12]. It should be emphasised that the Gröbner basis computations are the dominating overhead: the resulting SAT instances are almost trivial for our datapoints. By way of an initial comparison, our approach is 23-fold slower on average than CVC4 [4] on our 64 bit problems though, no doubt, building on F5 rather than Buchberger would significantly reduce this gap, as would recoding in C++.

## 5    Related Work

Momentum may be growing [1,6,10] for combining algebraic and SMT techniques but work at this intersection has mainly focused on CAD [16,23]. Gröbner bases have been used [3], however, for interpolating non-linear constraints over bit-vectors by use of symbolic conversion predicates. These predicates are used to lazily convert between bit-vectors and rationals, over which Gröbner bases are computed. A closer integration of Gröbner bases and bit-vectors is offered by modifying Buchberger's algorithm [7] to handle modulo arithmetic [5], work which is developed in this paper. This approach has found application in verifying the equivalence of multiplication circuits [17], using signed and unsigned machine arithmetic. Further afield, but also motivated by the desire to bypass bit-blasting, efficient portfolio bit-vector solvers have been developed [25], combining learning with word-level propagators [20] that iteratively restrict the values that can be assigned to a bit-vector. In contrast to our work, the propagators are designed to run in constant time and make use of low-level bit-twiddling operations [26].

## 6    Concluding Discussion

This paper argues for translating polynomial equalities over bit-vectors into pseudo-boolean constraints, the central idea being to use Gröbner bases to expose the consequences of setting an individual bit on the bit-vectors over which a polynomial system is defined. The resulting technique, named bit-sequence propagation, typically infers the values of many bits from setting a single bit, even in the context of symbolic bit assignments. The symbolic bits enable the Gröbner bases to be calculated in a deterministic fashion, with search encapsulated within the pseudo-boolean solver, whether one is employed directly or a reduction to SAT is used. Furthermore, the technique extends to systems of mixed polynomial equalities and disequalities. Disequalities can be handled by expressing each disequality in terms of symbolic variables by rewriting the disequalities using the final basis derived for the equalities. Each disequality can then be converted to propositional formulae, as with an equality, and then a standard transform [24] used to ensure that the negation of each of these formulae holds.

## References

1. Ábrahám, E.: Building bridges between symbolic computation and satisfiability checking. In: International Symposium on Symbolic and Algebraic Computation, pp. 1–6. ACM Press (2015)
2. Adams, W., Loustaunau, P.: An Introduction to Gröbner Bases. American Mathematical Society, Providence (1994)
3. Backeman, P., Rümmer, P., Zeljic, A.: Bit-vector interpolation and quantifier elimination by lazy reduction. In: Formal Methods in Computer Aided Design, pp. 1–10. IEEE (2018)
4. Barrett, C., et al.: CVC4. In: Gopalakrishnan, G., Qadeer, S. (eds.) CAV 2011. LNCS, vol. 6806, pp. 171–177. Springer, Heidelberg (2011). https://doi.org/10.1007/978-3-642-22110-1_14
5. Brickenstein, M., Dreyer, A., Greuel, G., Wedler, M., Wienand, O.: New developments in the theory of Gröbner bases and applications to formal verification. J. Pure Appl. Algebra **213**, 1612–1635 (2009)
6. Brown, C.: Bridging two communities to solve real problems. In: International Symposium on Symbolic and Numeric Algorithms for Scientific Computing, pp. 11–14. IEEE Press (2016)
7. Buchberger, B.: Bruno Buchberger's PhD thesis 1965: an algorithm for finding the basis elements of the residue class ring of a zero dimensional polynomial ideal. J. Symb. Comput. **41**, 475–511 (2006)
8. Collins, G.E.: Quantifier elimination for real closed fields by cylindrical algebraic decompostion. In: Brakhage, H. (ed.) GI-Fachtagung 1975. LNCS, vol. 33, pp. 134–183. Springer, Heidelberg (1975). https://doi.org/10.1007/3-540-07407-4_17

9. Cox, D., Little, J., O'Shea, D.: Ideals, Varieties and Algorithms. Undergraduate Texts in Mathematics. Springer, Heidelberg (1992). https://doi.org/10.1007/978-1-4757-2181-2
10. Davenport, J.H., England, M., Griggio, A., Sturm, T., Tinelli, C.: Symbolic computation and satisfiability checking. J. Symb. Comput. (2019). https://doi.org/10.1016/j.jsc.2019.07.017
11. Faugére, J.: A new efficient algorithm for computing Gröbner bases (F4). J. Pure Appl. Algebra **139**(1–3), 61–88 (1999)
12. Faugére, J.: A new efficient algorithm for computing Gröbner bases without reduction to zero (F5). In: International Symposium on Symbolic and Algebraic Computation, pp. 75–83 (2002)
13. Fekete, Y., Codish, M.: Simplifying pseudo-boolean constraints in residual number systems. In: Sinz, C., Egly, U. (eds.) SAT 2014. LNCS, vol. 8561, pp. 351–366. Springer, Cham (2014). https://doi.org/10.1007/978-3-319-09284-3_26
14. Griggio, A.: Effective word-level interpolation for software verification. In: Formal Methods in Computer-Aided Design, pp. 28–36. IEEE (2011)
15. Horáček, J., Burchard, J., Becker, B., Kreuzer, M.: Integrating algebraic and SAT solvers. In: Blömer, J., Kotsireas, I.S., Kutsia, T., Simos, D.E. (eds.) MACIS 2017. LNCS, vol. 10693, pp. 147–162. Springer, Cham (2017). https://doi.org/10.1007/978-3-319-72453-9_11
16. Jovanović, D., de Moura, L.: Solving non-linear arithmetic. In: Gramlich, B., Miller, D., Sattler, U. (eds.) IJCAR 2012. LNCS (LNAI), vol. 7364, pp. 339–354. Springer, Heidelberg (2012). https://doi.org/10.1007/978-3-642-31365-3_27
17. Kaufmann, D., Biere, A., Kauers, M.: Verifying large multipliers by combining SAT and computer algebra. In: Formal Methods in Computer-Aided Design, pp. 28–36 (2019)
18. Lang, S.: Algebra. Graduate Texts in Mathematics. Springer, Heidelberg (2002). https://doi.org/10.1007/978-1-4613-0041-0
19. Manquinhoand, V.M., Marques-Silva, J.: Using cutting planes in pseudo-boolean optimization. J. Satisf. Boolean Model. Comput. **2**, 199–208 (2006)
20. Michel, L.D., Van Hentenryck, P.: Constraint satisfaction over bit-vectors. In: Milano, M. (ed.) CP 2012. LNCS, pp. 527–543. Springer, Heidelberg (2012). https://doi.org/10.1007/978-3-642-33558-7_39
21. Müller-Olm, M., Seidl, H.: Analysis of modular arithmetic. In: Sagiv, M. (ed.) ESOP 2005. LNCS, vol. 3444, pp. 46–60. Springer, Heidelberg (2005). https://doi.org/10.1007/978-3-540-31987-0_5
22. Nieuwenhuis, R., Oliveras, A., Tinelli, C.: Solving SAT and SAT modulo theories: from an abstract Davis-Putnam-Logemann-Loveland procedure to DPLL(T). J. ACM **53**(6), 937–977 (2006)
23. Viehmann, T., Kremer, G., Ábráham, E.: Comparing different projection operators in the cylindrical algebraic decomposition for SMT solving. In: International Workshop on Satisfiability Checking and Symbolic Computation (2017). http://www.sc-square.org/CSA/workshop2-papers/RP2-FinalVersion.pdf
24. Tseytin, G.S.: On the complexity of derivation in propositional calculus. In: Slisenko, A.O. (ed.) Studies in Constructive Mathematics and Mathematical Logic, pp. 115–125. Steklov Mathematical Institute (1970). Translated from Russian: Zapiski Nauchnykh Seminarov LOMI 8 (1968)
25. Wang, W., Søndergaard, H., Stuckey, P.J.: Wombit: a portfolio bit-vector solver using word-level propagation. J. Autom. Reasoning **63**(3), 723–762 (2018). https://doi.org/10.1007/s10817-018-9493-1
26. Warren, H.: Hacker's Delight. Addison-Wesley, Boston (2012)

# Speeding up Quantified Bit-Vector SMT Solvers by Bit-Width Reductions and Extensions

Martin Jonáš[1] and Jan Strejček[2]([⊠]) [iD]

[1] Fondazione Bruno Kessler, Trento, Italy
mjonas@fbk.eu
[2] Masaryk University, Brno, Czech Republic
strejcek@fi.muni.cz

**Abstract.** Recent experiments have shown that satisfiability of a quantified bit-vector formula coming from practical applications almost never changes after reducing all bit-widths in the formula to a small number of bits. This paper proposes a novel technique based on this observation. Roughly speaking, a given quantified bit-vector formula is reduced and sent to a solver, an obtained model is then extended to the original bit-widths and verified against the original formula. We also present an experimental evaluation demonstrating that this technique can significantly improve the performance of state-of-the-art SMT solvers Boolector, CVC4, and Q3B on quantified bit-vector formulas from the SMT-LIB repository.

## 1 Introduction

We have recently studied the influence of bit-width changes on satisfiability of quantified bit-vector formulas from the SMT-LIB repository. The experiments showed that satisfiability is surprisingly stable under these changes [7]. For example, more than 95% of the considered formulas keep the same satisfiability status after changing the bit-widths of all variables and constants to an arbitrary value between 1 and the original bit-width. Indeed, all these stable formulas have the same satisfiability status even if we replace every bit-vector constant and variable by its least-significant bit and thus transform the formula into a quantified Boolean formula. Moreover, the percentage of stable formulas increased well over 99% if all bit-widths are reduced to any value between 4 and the original bit-width.

The experiments also confirm natural expectation that a formula with smaller bit-widths can be often solved considerably faster than the formula with the same structure but with larger bit-widths. For example, solving the formula

$$\varphi \quad \equiv \quad \forall x \forall y \exists z \, (x \cdot (y + z) = 0)$$

J. Strejček has been supported by the Czech Science Foundation grant GA18-02177S.

L. Pulina and M. Seidl (Eds.): SAT 2020, LNCS 12178, pp. 378–393, 2020.
https://doi.org/10.1007/978-3-030-51825-7_27

takes Boolector [11] several minutes on a standard desktop machine when the variables $x, y, z$ and the constant 0 have the bit-width 32, but it is solved instantly when the bit-width is 2.

This paper presents a new technique for deciding satisfiability of quantified bit-vector formulas that builds on the two mentioned observations: satisfiability status of these formulas is usually stable under bit-width reduction and formulas with reduced bit-widths can be often solved faster. Intuitively, the technique consists of the following steps:

1. Reduce bit-widths of variables and constants in the input formula to a smaller bit-width.
2. Decide satisfiability of the reduced formula using a standard decision procedure. If the reduced formula is satisfiable, obtain its model. If it is unsatisfiable, obtain its countermodel (i.e., a reason for unsatisfiability).
3. Extend the model or the countermodel to the original bit-widths. For example, a model of the formula $\varphi$ defined above reduced to bit-widths 2 has the form $z^{[2]} = -y^{[2]}$, where the superscripts denote bit-widths of the corresponding variables. After extension to the original bit-width, we get $z^{[32]} = -y^{[32]}$.
4. Check whether the extended (counter)model is also a (counter)model of the original formula. If the extended model is a model of the original formula, then the formula is satisfiable. If the extended countermodel is a countermodel of the original formula, then the formula unsatisfiable. In the remaining cases, we increase the bit-widths in the reduced formula and repeat the process.

The technique has some similarities with the approximation framework of Zeljić et al. [12], which also reduces the precision of a given formula, computes a model of the reduced formula, and checks if it is a model of the original formula. However, the framework considers only quantifier-free formulas (and hence models are just elements of the considered domains, while they are interpretations of Skolem functions in our setting) and it does not work with countermodels (it processes unsat cores of reduced formulas instead).

The detailed description of formula reduction and (counter)model extension is given in Sect. 3, preceded by Sect. 2 that recalls all necessary background and notation. The algorithm is precisely formulated in Sect. 4. Section 5 presents our proof-of-concept implementation and it discusses many practical aspects: how to get a counterexample, what to do with an incomplete model etc. Experimental evaluation of the technique can be found in Sect. 6. It clearly shows that the presented technique can improve performance of considered state-of-the-art solvers for quantified bit-vector formulas, namely Boolector [11], CVC4 [1], and Q3B [9], on both satisfiable and unsatisfiable formulas from various subcategories of the relevant SMT-LIB category BV.

## 2   Preliminaries

This section briefly recalls the used logical notions and the *theory of fixed sized bit-vectors* (*BV* or *bit-vector theory* for short). In the description, we assume familiarity with standard definitions of a many-sorted logic, well-sorted terms, atomic formulas, and formulas. To simplify the presentation, we suppose that all formulas are in the *negation normal form*. That is, all formulas use only logical connectives disjunction, conjunction, and negation and the arguments of all negations are atomic formulas. We suppose that sets of all free and all bound variables in the formula are disjoint and that each variable is quantified at most once.

The bit-vector theory is a many-sorted first-order theory with infinitely many sorts denoted $[n]$, where $n$ is a positive integer. Each sort $[n]$ is interpreted as the set of all bit-vectors of length $n$, which is also called their *bit-width*. We denote the set of all bit-vectors of bit-width $n$ as $\mathcal{BV}_n$ and variables of sort $[n]$ as $x^{[n]}$, $y^{[n]}$, etc. The BV theory uses only three predicate symbols, namely *equality* $(=)$, *unsigned inequality* of binary-encoded natural numbers $(\leq_u)$, and *signed inequality* of integers in two's complement representation $(\leq_s)$. The theory also contains various interpreted function symbols. Many of them represent binary operations that produce a bit-vector of the same bit-width as its two arguments. This is the case of *addition* $(+)$, *multiplication* $(\cdot)$, bit-wise *and* $(\&)$, bit-wise *or* $(|)$, bit-wise *exclusive or* $(\oplus)$, *left-shift* $(\ll)$, and *right-shift* $(\gg)$. The theory further contains function symbols for *two's complement negation* $(-)$, *concatenation* (concat), *zero extension* extending the argument with $n$ most-significant zero bits (zeroExt$_n$), *sign extension* extending the argument with $n$ copies of the sign bit (signExt$_n$), and *extraction* of bits delimited by positions $u$ and $l$ (including bits at these positions) from the argument, where position 0 refers to the least-significant bit and $u \geq l$ (extract$_l^u$). The signature of BV theory also contains numerals for constants $m^{[n]}$ for each bit-width $n > 0$ and each number $0 \leq m \leq 2^n - 1$. Each term $t$ has an associated bit-width, which is denoted as $\mathrm{bw}(t)$. The precise definition of the many-sorted logic can be found for example in Barrett et al. [3]. The precise description of bit-vector theory can be found for example in the paper describing complexity of quantified bit-vector theory by Kovásznai et al. [10].

A signature $\Sigma$ is a set of *uninterpreted function symbols*, which is disjoint with the set of all interpreted bit-vector function and predicate symbols. Each function symbol $f \in \Sigma$ has an associated arity $k \in \mathbb{N}_0$ and a sort $(n_1, n_2, \ldots, n_k, n) \in \mathbb{N}^{k+1}$, where the numbers $n_i$ represent bit-widths of the arguments of $f$ and $n$ represents the bit-width of its result. A $\Sigma$-structure $\mathcal{M}$ maps each uninterpreted function $f$ of the sort $(n_1, n_2, \ldots, n_k, n)$ to a function of the type $\mathcal{BV}_{n_1} \times \mathcal{BV}_{n_2} \times \ldots \times \mathcal{BV}_{n_k} \to \mathcal{BV}_n$, and each variable $x^{[n]}$ to a bit-vector value in $\mathcal{BV}_n$.

For a $\Sigma$-structure $\mathcal{M}$, we define the evaluation function $[\![\_]\!]_\mathcal{M}$, which assigns to each term $t$ the bit-vector $[\![t]\!]_\mathcal{M}$ obtained by (i) substituting each variable $x$ in $t$ by its value $\mathcal{M}(x)$ given by $\mathcal{M}$ and (ii) evaluating all interpreted functions and predicates using their given semantics and all uninterpreted functions using their interpretations given by $\mathcal{M}(f)$. Similarly, the function $[\![\_]\!]_\mathcal{M}$ assigns to each formula $\varphi$ the Boolean value $[\![\varphi]\!]_\mathcal{M}$ obtained by substituting free variables in $\varphi$ by values given by $\mathcal{M}$ and evaluating all functions, predicates, logical operators etc. according to $\mathcal{M}$ and the standard semantics. A formula $\varphi$ is *satisfiable* if

$\llbracket \varphi \rrbracket_{\mathcal{M}} = \top$ for some $\Sigma$-structure $\mathcal{M}$; it is *unsatisfiable* otherwise. A $\Sigma$-structure $\mathcal{M}$ is called a *model* of $\varphi$ whenever $\llbracket \varphi \rrbracket_{\mathcal{M}} = \top$.

The *Skolemization* of a formula $\varphi$, denoted $skolemize(\varphi)$, is a formula that is obtained from $\varphi$ by replacing each existentially quantified variable $x^{[n]}$ in $\varphi$ by a fresh uninterpreted function symbol $f_{x^{[n]}}$ that has as arguments all variables that are universally quantified above $x^{[n]}$ in the syntactic tree of the formula $\varphi$. Skolemization preserves the satisfiability of the input formula $\varphi$ [6].

For a satisfiable formula $\varphi$ without uninterpreted functions, a model $\mathcal{M}$ of $skolemize(\varphi)$ assigns a bit-vector $\mathcal{M}(y^{[m]}) \in \mathcal{BV}_m$ to each free variable $y^{[m]}$ in $\varphi$, and a function $\mathcal{M}(f_{x^{[n]}})$ to each Skolem function $f_{x^{[n]}}$, which corresponds to an existentially quantified variable $x^{[n]}$ in the formula $\varphi$. The functions $\mathcal{M}(f_{x^{[n]}})$ may be arbitrary functions (of the corresponding type) in the mathematical sense. To be able to work with the model, we use the notion of a *symbolic model*, in which the functions $\mathcal{M}(f_{x^{[n]}})$ are represented symbolically by terms. Namely, $\mathcal{M}(f_{x^{[n]}})$ is a bit-vector term of bit-width $n$ whose free variables may be only the variables that are universally quantified above $x^{[n]}$ in the original formula $\varphi$. In the further text, we treat the symbolic models as if they assign a term not to the corresponding Skolem function $f_{x^{[n]}}$, but directly to the existentially quantified variable $x^{[n]}$. For example, the formula

$$\forall x^{[32]} \forall y^{[32]} \ \exists z^{[32]} \ \left( x^{[32]} \cdot (y^{[32]} + z^{[32]}) = 0^{[32]} \right)$$

from the introduction has a symbolic model $\{z^{[32]} \mapsto -y^{[32]}\}$.

For a sentence $\varphi$, the dual notion to the symbolic model is a *symbolic countermodel*. The symbolic countermodel of a sentence $\varphi$ is a symbolic model of the negation normal form of $\neg\varphi$, i.e., a $\Sigma$-structure $\mathcal{M}$ that assigns to each *universally* quantified variable $x^{[n]}$ in $\varphi$ a term of bit-width $n$ whose free variables may be only the *existentially* quantified variables that are quantified above $x^{[n]}$ in the original formula $\varphi$.

We can define substitution of a symbolic (counter)model into a given formula. We define this notion more generally to allow substitution of an arbitrary assignment that assigns terms to variables of the formula. For each such assignment $\mathcal{A}$ and a formula $\varphi$, we denote as $\mathcal{A}(\varphi)$ the result of simultaneous substitution of the term $\mathcal{A}(x^{[n]})$ for each variable $x^{[n]}$ in the domain of $\mathcal{A}$ and removing all quantifications of the substituted variables. For example, the value of

$$\mathcal{A}\left( \forall x^{[32]} \forall y^{[32]} \ \exists z^{[32]} \ (x^{[32]} \cdot (y^{[32]} + z^{[32]}) = 0^{[32]}) \right)$$

for $\mathcal{A} = \{z^{[32]} \mapsto -y^{[32]}\}$ is $\forall x^{[32]} \forall y^{[32]} \ (x^{[32]} \cdot (y^{[32]} + (-y^{[32]})) = 0^{[32]})$.

## 3  Formula Reduction and Model Extension

This section describes the basic building blocks of our new technique, namely reduction of bit-widths in a given formula and extension of bit-widths in a given model or countermodel.

## 3.1    Reduction of Bit-Widths in Formulas

The goal of the reduction procedure is to reduce the bit-widths of all variables and constants in a given formula so that they do not exceed a given bit-width. In fact, we reduce bit-widths of all terms in the formula in order to keep the formula type consistent. A similar reduction is defined in our previous paper [7], but only for a simpler fragment of the considered logic.

As the first step, we inductively define a function $rt$ that takes a term and a bit-width $bw \in \mathbb{N}$ and reduces all subterms of the term. The function always cuts off all most-significant bits above the given bit-width $bw$. As the base case, we define the reduction on constants and variables.

$$rt(m^{[n]}, bw) = (m \bmod 2^{\min(n,bw)})^{[\min(n,bw)]}$$

$$rt(x^{[n]}, bw) = x^{[\min(n,bw)]}$$

Further, let $\circ$ range over the set $\{+, \cdot, \&, |, \oplus, \ll, \gg\}$ of binary functions that produce results of the same bit-width as the bit-width of their arguments. To reduce a term $t_1 \circ t_2$, we just need to reduce the arguments.

$$rt(t_1 \circ t_2, bw) = rt(t_1, bw) \circ rt(t_2, bw)$$

$$rt(-t_1, bw) = -rt(t_1, bw)$$

The most interesting cases are the functions that change bit-widths. As the first case, let $ext_n$ be a function that extends its argument with $n$ most-significant zero bits ($\mathsf{zeroExt}_n$) or with $n$ copies of the sign bit ($\mathsf{signExt}_n$). A term $ext_n(t)$ where $t$ has $bw$ or more bits is reduced just to $rt(t, bw)$. Indeed, the function $ext_n$ is completely removed as the bits it would add exceed the maximal bit-width. When $t$ has less than $bw$ bits, we apply the extension function but we decrement its parameter if the bit-width of the resulting term should exceed $bw$. Moreover, we also apply the reduction function to $t$ to guarantee that bit-widths of its subterms do not exceed $bw$.

$$rt(ext_n(t), bw) = \begin{cases} rt(t, bw) & \text{if } \mathrm{bw}(t) \geq bw \\ ext_{\min(n, bw - \mathrm{bw}(t))}(rt(t, bw)) & \text{if } \mathrm{bw}(t) < bw \end{cases}$$

As the second case, consider a term $\mathsf{extract}_l^u(t)$ that represents bits of $t$ between positions $u$ and $l$ (including these positions). The reduction is defined by one of the following three subcases according to the relation of $bw$ and positions $u$ and $l$. Recall that $u \geq l$, the bit-width of the original term is $u - l + 1$, and it has to be reduced to $m = \min(u - l + 1, bw)$.

- If both $u$ and $l$ point to some of the $bw$ least-significant bits of $t$ (i.e., $bw > u$), the positions $u$ and $l$ of $rt(t, bw)$ are defined, and so we just reduce the argument $t$ and do not change the parameters of $\mathsf{extract}$.
- If $l$ points to some of the $bw$ least-significant bits of $t$ but $u$ does not (i.e., $u \geq bw > l$), we reduce the argument $t$, extract its most-significant bits up to the position $l$, and extend the result with most-significant zero bits such that the bit-width of the result is $m$. These additional zero bits correspond to the positions that should be extracted, but are not present in the term $rt(t, bw)$.

– If both positions $u$ and $l$ point outside the $bw$ least-significant bits of $t$ (i.e., $l \geq bw$), we replace the term with the bit-vector of zeroes of the length $m$.

In the following formal definition, we denote by $o$ the bit-width of term $t$ after reduction, i.e., $o = \mathrm{bw}(rt(t, bw)) = \min(\mathrm{bw}(t), bw)$.

$$rt(\mathsf{extract}_l^u(t), bw) = \begin{cases} \mathsf{extract}_l^u(rt(t, bw)) & \text{if } bw > u \\ ext_{m-(o-l)}(\mathsf{extract}_l^{o-1}(rt(t, bw))) & \text{if } u \geq bw > l \\ 0^{[m]} & \text{if } l \geq bw \end{cases}$$

where $ext \in \{\mathsf{signExt}, \mathsf{zeroExt}\}$ can be chosen during the implementation.

Finally, reduction of a term $\mathsf{concat}(t_1, t_2)$ representing concatenation of $t_1$ and $t_2$ is given by one of the following two cases. Note that the reduced term should have the size $m = \min(\mathrm{bw}(t_1) + \mathrm{bw}(t_2), bw)$. If $\mathrm{bw}(t_2) \geq bw$, the term is reduced to $rt(t_2, bw)$ as the bits of $t_1$ exceed the desired maximal bit-width. In the opposite case, we reduce both $t_1$ and $t_2$ and create the term containing all the bits of the reduced term $t_2$ preceded by $m - \mathrm{bw}(t_2)$ least-significant bits of the reduced term $t_1$.

$rt(\mathsf{concat}(t_1, t_2), bw) =$

$$= \begin{cases} rt(t_2, bw) & \text{if } \mathrm{bw}(t_2) \geq bw \\ \mathsf{concat}\left(\mathsf{extract}_0^{m-\mathrm{bw}(t_2)-1}(rt(t_1, bw)), rt(t_2, bw)\right) & \text{if } \mathrm{bw}(t_2) < bw \end{cases}$$

Now we define a function $rf$ that reduces the maximal bit-widths of all terms in a given formula to a given value $bw$. The function is again defined inductively using the function $rt$ in the base case to reduce arguments of predicate symbols. The rest of the definition is straightforward.

$$rf(t_1 \bowtie t_2, bw) = rt(t_1, bw) \bowtie rt(t_2, bw) \qquad \text{for } \bowtie \in \{=, \leq_u, \leq_s\}$$
$$rf(\neg\varphi, bw) = \neg rf(\varphi, bw)$$
$$rf(\varphi_1 \diamond \varphi_2, bw) = rf(\varphi_1, bw) \circ rf(\varphi_2, bw) \qquad \text{for } \diamond \in \{\wedge, \vee\}$$
$$rf(Qx^{[n]}.\varphi, bw) = Qx^{[\min(n,bw)]}.rf(\varphi, bw) \qquad \text{for } Q \in \{\forall, \exists\}$$

## 3.2 Extending Bit-Widths of Symbolic Models

If a reduced formula is satisfiable and its symbolic model $\mathcal{M}$ is obtained, it cannot be directly substituted into the original formula. It first needs to be *extended* to the original bit-widths. Intuitively, for each result $\mathcal{M}(x) = t$, where the original bit-width of the variable $x$ is $n$, we

1. increase bit-widths of all variables in $t$ to match the bit-widths in the original formula $\varphi$,

2. for each operation whose arguments need to have the same bit-width, we increase bit-width of the argument with the smaller bit-width to match the bit-width of the other argument,
3. change the bit-width of the resulting term to match the bit-width of the original variable $x^{[n]}$.

In the formalization, we need to know bit-widths of the variables in the original formula. Therefore, for a formula $\varphi$, we introduce the function $\mathrm{bws}_\varphi$ that maps each variable name $x$ in $\varphi$ to its original bit-width in $\varphi$. For example, $\mathrm{bws}_{x^{[32]}+y^{[32]}=0^{[32]}}(x) = 32$. Further, we use the function $adjust$, which adjusts the bit-width of the given term $t$ to the given bit-width $bw$.

$$adjust(t, bw) = \begin{cases} t & \text{if } \mathrm{bw}(t) = bw \\ ext_{bw-\mathrm{bw}(t)}(t) & \text{if } \mathrm{bw}(t) < bw \\ \mathsf{extract}_0^{bw-1}(t) & \text{if } \mathrm{bw}(t) > bw \end{cases}$$

where $ext \in \{\mathsf{signExt}, \mathsf{zeroExt}\}$ can be chosen during the implementation.

For each term $t$ of the reduced model, we now recursively construct a term $\bar{t}$, which uses only the variables of the original formula and is well-sorted. In other words, this construction implements the first two steps of the symbolic model extension described above.

As the base cases, we keep the bit-width of all constants and extend the bit-width of all variables to their original bit-widths in $\varphi$.

$$\overline{m^{[n]}} = m^{[n]}$$
$$\overline{x^{[n]}} = x^{[\mathrm{bws}_\varphi(x)]}$$

For any operation $\circ \in \{+, \cdot, \&, |, \oplus, \ll, \gg\}$ that requires arguments of the same bit-widths, we may need to extend the shorter of these arguments.

$$\overline{t_1 \circ t_2} = adjust(\overline{t_1}, \max(\mathrm{bw}(\overline{t_1}), \mathrm{bw}(\overline{t_2}))) \circ adjust(\overline{t_2}, \max(\mathrm{bw}(\overline{t_1}), \mathrm{bw}(\overline{t_2})))$$

For the remaining operations, the construction is straightforward.

$$\overline{-t_1} = -\overline{t_1}$$
$$\overline{ext_n(t_1)} = ext_n(\overline{t_1}) \quad \text{for } ext \in \{\mathsf{zeroExt}, \mathsf{signExt}\}$$
$$\overline{\mathsf{extract}_j^i(t_1)} = \mathsf{extract}_j^i(\overline{t_1})$$
$$\overline{\mathsf{concat}(t_1, t_2)} = \mathsf{concat}(\overline{t_1}, \overline{t_2})$$

Now we complete the symbolic model extension with its third step. Formally, for a symbolic model $\mathcal{M}$ we define a model extension $extendM(\mathcal{M})$ that assigns to each variable $x$ in the domain of $\mathcal{M}$ the term $\overline{\mathcal{M}(x)}$ adjusted to the original bit-width of $x$.

$$extendM(\mathcal{M})(x) = adjust(\overline{\mathcal{M}(x)}, \mathrm{bws}_\varphi(x)).$$

*Example 1.* Consider a formula $\varphi$ that contains variables $x^{[8]}$, $y^{[8]}$, $z^{[4]}$, $v^{[8]}$, $w^{[4]}$. Suppose that we have the $\mathcal{M}$ of $rf(\varphi, 4)$ given below. With the parameter *ext* of *adjust* set to signExt, the assignment $extendM(\mathcal{M})$ is defined as follows.

$$\mathcal{M} = \{x^{[4]} \mapsto v^{[4]} + 3^{[4]}, \qquad extendM(\mathcal{M}) = \{x^{[8]} \mapsto v^{[8]} + 3^{[8]},$$
$$y^{[4]} \mapsto w^{[4]}, \qquad\qquad\qquad y^{[8]} \mapsto \text{signExt}_4(w^{[4]}),$$
$$z^{[4]} \mapsto v^{[4]}, \qquad\qquad\qquad z^{[4]} \mapsto \text{extract}_0^3(v^{[8]})\}$$

Note that the numeral $3^{[8]}$ in $extendM(\mathcal{M})$ arises by evaluation of the ground term $\text{signExt}_4(3^{[4]})$.

*Note on additional* SMT-LIB *operations.* The syntax of the BV theory given in SMT-LIB actually contains more predicates and functions than we have defined. The constructions presented in Subsections 3.1 and 3.2 can be extended to cover these additional predicates and functions mostly very easily. One interesting case is the *if-then-else* operator $\text{ite}(\varphi, t_1, t_2)$ where the first argument is a formula instead of a term. To accommodate this operator, the reduction functions $rt$ and $rf$ are defined as mutually recursive, and the symbolic model extension has to be enriched to handle not only terms, but also formulas. All these extensions can be found in the dissertation of M. Jonáš [8]. Note that ite indeed appears in symbolic models in practice.

# 4   Algorithm

In this section, we propose an algorithm that employs bit-width reductions and extensions to decide satisfiability of an input formula. In the first subsection, we describe a simpler approach that can only decide that a formula is satisfiable. The following subsection dualizes this approach to unsatisfiable formulas. We then show how to combine these two approaches in a single algorithm, which is able to decide both satisfiability and unsatisfiability of a formula.

## 4.1   Checking Satisfiability Using Reductions and Extensions

Having defined the functions $rf$ (see Subsect. 3.1), which reduces bit-widths in a formula, and $extendM$ (see Subsect. 3.2), which extends bit-widths in a symbolic model of the reduced formula, it is fairly straightforward to formulate an algorithm that can decide satisfiability of a formula using reduced bit-widths.

This algorithm first reduces the bit-widths in the input formula $\varphi$, thus obtains a reduced formula $\varphi_{red}$, and checks its satisfiability. If the formula is not satisfiable, the algorithm computes a new reduced formula $\varphi_{red}$ with an increased bit-width and repeats the process. If, on the other hand, the reduced formula $\varphi_{red}$ is satisfiable, the algorithm obtains its symbolic model $\mathcal{M}$, which assigns a term to each existentially quantified and free variable of the formula $\varphi_{red}$. The model is then extended to the original bit-widths of the variables in the formula $\varphi$ and the extended model is substituted into the original formula $\varphi$, yielding a formula $\varphi_{sub}$. The formula $\varphi_{sub}$ may not be quantifier-free, but it contains only universally quantified variables and no free variables. The formula $\varphi_{sub}$ may therefore be checked for satisfiability by a solver for quantifier-free bit-vector formulas: the solver can be called on the formula $\varphi_{sub}^{\neg}$ that results from removing all quantifiers from the formula $\neg\varphi_{sub}$ transformed to the negation normal form. Since the formula $\varphi_{sub}$ is closed, the satisfiability of $\varphi_{sub}^{\neg}$ implies unsatisfiability of $\varphi_{sub}$ and vice versa. Finally, if the formula $\varphi_{sub}$ is satisfiable, so is the original formula. If the formula $\varphi_{sub}$ is not satisfiable, the process is repeated with an increased bit-width.

*Example 2.* Consider the formula $\varphi \equiv \forall x^{[32]} \exists y^{[32]} (x^{[32]} + y^{[32]} = 0^{[32]})$. Reduction to 2 bits yields the formula $rf(\varphi, 2) \equiv \forall x^{[2]} \exists y^{[2]} (x^{[2]} + y^{[2]} = 0^{[2]})$. An SMT solver can decide that this formula is satisfiable and its symbolic model is $\{y^{[2]} \mapsto -x^{[2]}\}$. An extended candidate model is then $\{y^{[32]} \mapsto -x^{[32]}\}$. After substituting this candidate model into the formula, one gets the formula $\varphi_{sub} \equiv \forall x^{[32]} (x^{[32]} + (-x^{[32]}) = 0^{[32]})$. Negating the formula $\varphi_{sub}$ and removing all the quantifiers yields the quantifier-free formula $(x^{[32]} + (-x^{[32]}) \neq 0^{[32]})$, which is unsatisfiable. Therefore, the formula $\varphi_{sub}$ is satisfiable and, in turn, the original formula $\varphi$ is satisfiable as well.

The correctness of the approach is guaranteed by the following theorem.

**Theorem 1** ([8, **Theorem 11.1**]). *Let $\varphi$ be a formula in the negation normal form and $\mathcal{A}$ a mapping that assigns terms only to free and existentially quantified variables of $\varphi$. If each term $\mathcal{A}(x)$ contains only universal variables that are quantified in $\varphi$ before the variable $x$, satisfiability of $\mathcal{A}(\varphi)$ implies satisfiability of $\varphi$.*

### 4.2    Dual Algorithm

The algorithm of the previous subsection can improve performance of an SMT solver only for satisfiable formulas. However, its dual version can be used to improve performance on unsatisfiable formulas. In the dual algorithm, one can decide unsatisfiability of a formula by computing a countermodel of a reduced formula and verifying it against the original formula. More precisely, if the solver decides that the reduced formula $\varphi_{red}$ is unsatisfiable, one can extend its countermodel $\mathcal{M}$, substitute the extended countermodel into the original formula, obtaining a formula $\varphi_{sub}$ which contains only existentially quantified variables. Satisfiability of $\varphi_{sub}$ can be again checked by a solver for quantifier-free formulas applied to $\varphi_{sub}$ after removing all its existential quantifiers. If the formula $\varphi_{sub}$ is unsatisfiable, the original formula $\varphi$ must have been unsatisfiable. If the formula $\varphi_{sub}$ is satisfiable, the process is repeated with an increased bit-width.

*Example 3.* Consider the formula $\varphi = \forall y^{[32]} (x^{[32]} + y^{[32]} = 0^{[32]})$. Reduction to one bit yields the formula $rf(\varphi, 1) = \forall y^{[1]} (x^{[1]} + y^{[1]} = 0^{[1]})$. This formula can be decided as unsatisfiable by an SMT solver and its countermodel is $\{y^{[1]} \mapsto$

$-x^{[1]} + 1^{[1]}\}$. The extension of this countermodel to the original bit-widths is then $\{y^{[32]} \mapsto -x^{[32]} + 1^{[32]}\}$. After substituting this candidate countermodel to the original formula, one obtains the quantifier-free formula $\varphi_{sub} = (x^{[32]} + (-x^{[32]} + 1^{[32]}) = 0^{[32]})$, which is unsatisfiable. The original formula $\varphi$ is thus unsatisfiable.

The correctness of the dual algorithm is guaranteed by the following theorem.

**Theorem 2 ([8, Theorem 11.2]).** *Let $\varphi$ be a formula in the negation normal form and $\mathcal{A}$ a mapping that assigns terms only to universally quantified variables of $\varphi$. If each term $\mathcal{A}(x)$ contains only free and existential variables that are quantified in $\varphi$ before the variable $x$, unsatisfiability of $\mathcal{A}(\varphi)$ implies unsatisfiability of $\varphi$.*

### 4.3 Combined Algorithm

Now we combine the two algorithms into one. In the rest of this section, we suppose that there exists a *model-generating solver* that produces symbolic models for satisfiable quantified bit-vector formulas and countermodels for unsatisfiable ones. Formally, let $\mathsf{solve}(\varphi)$ be the function that returns $(\mathsf{sat}, model)$ if $\varphi$ is satisfiable and $(\mathsf{unsat}, countermodel)$ in the opposite case.

Further, we use SMT queries to check the satisfiability of $\varphi_{sub}$. Generally, these queries can be answered by a different SMT solver than the model-generating one. We call it *model-validating solver* and suppose that it has the function $\mathsf{verify}(\psi)$ which returns either $\mathsf{sat}$ or $\mathsf{unsat}$ reflecting the satisfiability of $\psi$.

Using these two solvers, the algorithm presented in Listing 1.1 combines the techniques of the two preceding subsections. This algorithm first reduces the bit-widths in the input formula to 1 and checks satisfiability of the reduced formula $\varphi_{red}$ by the model-generating solver. According to the result, we try to validate either the extended symbolic model or the extended symbolic countermodel with the model-validating solver. If the validation succeeds, the satisfiability of the original formula is decided. Otherwise, we repeat the process but this time we reduce the bit-widths in the input formula to twice the value used in the previous iteration. The algorithm terminates at the latest in the iteration when the value of bw is so high that the formula $\varphi_{red}$ is identical to the input formula $\varphi$. In this case, the model-generating solver provides a model or a countermodel $\mathcal{M}$ of $\varphi$. As $\mathcal{M}$ contains the unchanged variables of $\varphi$, its extension $extendM(\mathcal{M})$ is identical to $\mathcal{M}$ and the model-validating solver has to confirm the result.

## 5 Implementation

We have implemented the proposed algorithm in a proof-of-concept tool. However, our implementation differs in several aspects from the described algorithm. This section explains all these differences and provides more details about the implementation.

**Listing 1.1.** The combined algorithm for checking satisfiability of $\varphi$ using bit-width reductions and extensions.

```
1  bw ← 1
2  while (true) {
3      φ_red ← rf(φ, bw)
4      (result, M) ← solve(φ_red)
5      A ← extendM(M)
6      φ_sub ← A(φ)
7      if (result == sat) {
8          φ⁻_sub ← removeQuantifiers(¬φ_sub)
9          verificationResult ← verify(φ⁻_sub)
10         if (verificationResult == unsat) return SAT
11     }
12     if (result == unsat) {
13         φ_sub ← removeQuantifiers(φ_sub)
14         verificationResult ← verify(φ_sub)
15         if (verificationResult == unsat) return UNSAT
16     }
17     bw ← increaseBW(bw)
18 }
```

## 5.1 Model-Generating Solver

As the model-generating solver, we use Boolector 3.2.0 as it can return symbolically expressed Skolem functions as models of satisfiable quantified formulas, which is crucial for our approach. Unfortunately, Boolector does not satisfy some requirements that we imposed on the model-generating solver.

First, the symbolic model $\mathcal{M}$ returned by Boolector may not contain terms for all existentially quantified variables of the input formula $\varphi$. Therefore, the formula $\varphi_{sub}$ may still contain both existentially and universally quantified variables and we cannot employ an SMT solver for quantifier-free formulas as the model-validation solver. Our implementation thus uses a model-validating solver that supports quantified formulas. An alternative solution is to extend $\mathcal{M}$ to all existentially quantified variables, for example by assigning $0^{[n]}$ to each existentially quantified variable $x^{[n]}$ that is not assigned by $\mathcal{M}$. This allows using a solver for quantifier-free formulas as the model-validating solver. However, our preliminary experiments indicate that this alternative solution does not bring any significant benefit. Moreover, the best performing SMT solvers for the quantifier-free bit-vector formulas can also handle quantified formulas.

Second, Boolector returns symbolic models only for satisfiable formulas and cannot return symbolic countermodels. We alleviate this problem by running two parallel instances of Boolector: one on the original formula $\varphi$ and one on the formula $\neg\varphi'$, where $\varphi'$ arises from $\varphi$ by existential quantification of all free variables. We then use only the result of the solver that decides that the formula is satisfiable; if $\varphi$ is satisfiable, we get its symbolic model, if $\neg\varphi'$ is satisfiable, we get its symbolic model, which is a symbolic countermodel of $\varphi$. Effectively, this is equivalent to running the algorithm of Listing 1.1 without the lines 12–16 in two parallel instances: one on $\varphi$ and the other on $\neg\varphi'$. This is what our implementation actually does.

## 5.2 Portfolio Solver

The aim of our research is to improve the performance of an SMT solver for the BV theory using the bit-width reductions and extensions. The solver is used as the model-validating solver. We investigate two implementations:

- To see real-world benefits, we run the original solver in parallel with the two processes that use bit-width reductions. The result of the first process that decides the satisfiability of the input formula is returned. The schematic overview of our portfolio solver is presented in Fig. 1. In this variant, if the reducing solvers reach the original bit-width of the formula, they return unknown.
- To see the negative overhead of reductions, we also consider a variant of the above-mentioned approach, but without the middle thread with original solver. In this variant, the reducing solvers are additionally executed for the original bit-width in their last iteration.

Our experimental implementation is written in C++ and Python. It utilizes the C++ API of Z3 [5] to parse the input formula in the SMT-LIB format. The Z3 API is also used in the implementation of formula reductions and some simplifications (conversion to the negation normal form and renaming bound variables to have unique names). The only part written in Python is a simple wrapper that executes the three parallel threads and collects their results. As the parameters, we use $ext = $ zeroExt in $rt$, $ext = $ signExt in $adjust$, and increaseBW$(x) = 2 * x$. These parameters had the best performance during our preliminary evaluation, but can be changed. The implementation is available at: https://github.com/martinjonas/bw-reducing-solver.

# 6 Experimental Evaluation

We have evaluated the impact of our technique on the performance of three leading SMT solvers for the BV theory: Boolector 3.2.0 [11], CVC4 1.6 [1], and Q3B 1.0 [9]. Each of these solvers has been employed as the model-validating solver, while the model-generating solver remains the same, namely Boolector. For the evaluation, we have used all 5741 quantified bit-vector formulas from the SMT-LIB benchmark repository [2]. The formulas are divided into 8 benchmark families coming from different sources.

All experiments were performed on a Debian machine with two six-core Intel Xeon E5-2620 2.00 GHz processors and 128 GB of RAM. Each benchmark run was limited to use 16 GB of RAM and 5 min of wall time. All measured times are wall times. For reliable benchmarking we employed BENCHEXEC [4].

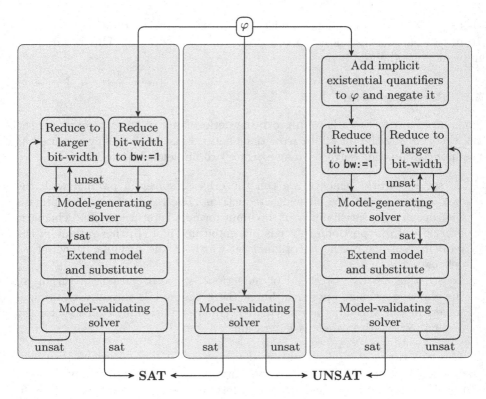

**Fig. 1.** High-level overview of the portfolio solver. The three shaded areas are executed in parallel and the first result is returned.

### 6.1   Boolector

First, we have evaluated the impact of our technique on the performance of Boolector 3.2.0. We have compared the vanilla Boolector (referred to as `btor`), our portfolio solver running Boolector as both model-generating and model-validating solver (`btor-r`), and the portfolio variant without the original solver (`btor-r-no`). The numbers of formulas of individual benchmark families solved by the three solvers can be found in the corresponding columns of Table 1. While `btor-r-no` is not very competitive, the full portfolio solver was able to solve 22 formulas more than Boolector itself. Note that this amounts to 8.6% of the formulas unsolved by Boolector. The scatter plots in Fig. 2 shows the performance of the solvers. With the full portfolio approach, our technique can also significantly reduce the running time of Boolector on a non-trivial number of both satisfiable and unsatisfiable formulas from various benchmark families.

We have also investigated the reduction bit-width that was necessary to improve the performance of Boolector. Among all executions of the full portfolio solver, 475 benchmarks were actually decided by one of the two parallel threads that perform bit-width reductions. From these 475 benchmarks, 193

**Table 1.** The table shows for each benchmark family and each solver the number of benchmarks that were solver by the solver within a given timeout.

| Family | Total | btor | btor-r | btor-r-no | btor\|cvc4 | btor\|cvc4-r | btor\|q3b | btor\|q3b-r |
|---|---|---|---|---|---|---|---|---|
| 2017-Preiner-keymaera | 4035 | 4019 | 4022 | 4020 | 4025 | 4027 | 4025 | 4028 |
| 2017-Preiner-psyco | 194 | 193 | 193 | 129 | 193 | 193 | 193 | 193 |
| 2017-Preiner-scholl-smt08 | 374 | 299 | 304 | 224 | 306 | 306 | 327 | 328 |
| 2017-Preiner-tptp | 73 | 70 | 73 | 69 | 73 | 73 | 73 | 73 |
| 2017-Preiner-ua | 153 | 153 | 153 | 23 | 153 | 153 | 153 | 153 |
| 20170501-Heizmann-ua | 131 | 28 | 30 | 25 | 130 | 130 | 128 | 129 |
| 2018-Preiner-cav18 | 600 | 549 | 554 | 477 | 577 | 577 | 590 | 590 |
| wintersteiger | 181 | 152 | 156 | 125 | 167 | 169 | 172 | 174 |
| Total | 5741 | 5463 | 5485 | 5092 | 5624 | 5628 | 5661 | 5668 |

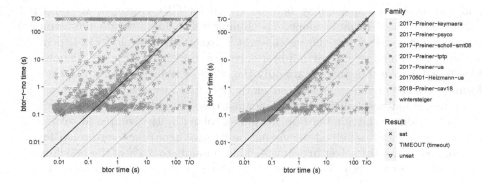

**Fig. 2.** Scatter plots of wall times of the solver `btor` vs the solvers `btor-r` and `btor-r-no`. Each point represents one benchmark, its color shows the benchmark family, and its shape shows its satisfiability.

were decided using the bit-width of 1 bit, 141 using 2 bits, 111 using 4 bits, 23 using 8 bits, and 7 using 16 bits.

## 6.2   CVC4 and Q3B

We have also performed evaluations with CVC4 and Q3B as model-validating solvers. This yields the following four solvers: `cvc4`, `q3b` are the vanilla CVC4 and Q3B, respectively; `cvc4-r`, `q3b-r` are the portfolio solvers using CVC4 and Q3B, respectively, as the model-validating solver.

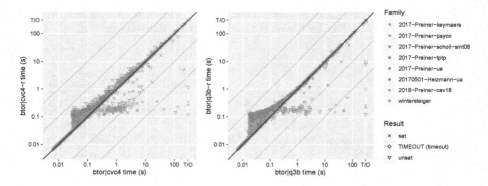

**Fig. 3.** Scatter plots of wall times of the virtual best solvers btor|cvc4 vs. btor|cvc4-r (left) and btor|q3b vs. btor|q3b-r (right).

Whenever the model-generating solver differs from the model-validating solver, the comparison is more involved. For example, the direct comparison of cvc4 and cvc4-r would be unfair and could be biased towards cvc4-r. This happens because models are provided by Boolector as the model-generating solver and the model validation may become trivial for CVC4, even if it could not solve the reduced formula alone. To eliminate this bias, we do not compare cvc4 against cvc4-r, but the virtual-best solver from btor and cvc4, denoted as btor|cvc4, against the virtual-best solver from btor and cvc4-r, denoted as btor|cvc4-r. We thus investigate only the effect of reductions and not the performance of the model-generating solver on the input formula. Similarly, we compare the virtual-best solver btor|q3b against the virtual-best solver btor|q3b-r.

Table 1 shows the number of benchmarks solved by the compared solvers. In particular, reductions helped the virtual-best solver btor|cvc4-r to solve 4 more benchmarks than the solver btor|cvc4. This amounts to 3.4% of the benchmarks unsolved by btor|cvc4. For btor|q3b-r, the reductions help to solve 7 new benchmarks, i.e., 8.8% of unsolved benchmarks.

Similarly to the case of Boolector, reductions also help btor|cvc4-r to decide several benchmarks faster than the solver btor|cvc4 without reductions. This can be seen on the first scatter plot in Fig. 3. As the second scatter plot in this figure shows, reductions also help Q3B to solve some benchmarks faster.

All experimental data, together with additional results and all scripts that were used during the evaluation are available at: https://fi.muni.cz/~xstrejc/sat2020/.

## 7   Conclusions

We have described an algorithm that improves performance of SMT solvers for quantified bit-vector formulas by reducing bit-widths in the input formula. We have shown that if used in a portfolio approach, our proof-of-concept implemen-

tation of this algorithm improves performance of state-of-the art SMT solvers
Boolector, CVC4, and Q3B.

# References

1. Barrett, C., et al.: CVC4. In: Gopalakrishnan, G., Qadeer, S. (eds.) CAV 2011. LNCS, vol. 6806, pp. 171–177. Springer, Heidelberg (2011). https://doi.org/10.1007/978-3-642-22110-1_14
2. Barrett, C., Fontaine, P., Tinelli, C.: The Satisfiability Modulo Theories Library (SMT-LIB) (2016). www.SMT-LIB.org
3. Barrett, C.W., Sebastiani, R., Seshia, S.A., Tinelli, C.: Satisfiability modulo theories. In: Handbook of Satisfiability, pp. 825–885. IOS Press (2009)
4. Beyer, D., Löwe, S., Wendler, P.: Benchmarking and resource measurement. In: Fischer, B., Geldenhuys, J. (eds.) SPIN 2015. LNCS, vol. 9232, pp. 160–178. Springer, Cham (2015). https://doi.org/10.1007/978-3-319-23404-5_12
5. de Moura, L., Bjørner, N.: Z3: an efficient SMT solver. In: Ramakrishnan, C.R., Rehof, J. (eds.) TACAS 2008. LNCS, vol. 4963, pp. 337–340. Springer, Heidelberg (2008). https://doi.org/10.1007/978-3-540-78800-3_24
6. Harrison, J.: Handbook of Practical Logic and Automated Reasoning. Cambridge University Press, Cambridge (2009)
7. Jonáš, M., Strejček, J.: Is satisfiability of quantified bit-vector formulas stable under bit-width changes? In: Barthe, G., Sutcliffe, G., Veanes, M. (eds.) LPAR-22. 22nd International Conference on Logic for Programming, Artificial Intelligence and Reasoning. EPiC Series in Computing, Awassa, Ethiopia, 16–21 November 2018, vol. 57, pp. 488–497. EasyChair (2018)
8. Jonáš, M.: Satisfiability of quantified bit-vector formulas: theory & practice. Ph.D. thesis, Masaryk University (2019)
9. Jonáš, M., Strejček, J.: Q3B: an efficient BDD-based SMT solver for quantified bit-vectors. In: Dillig, I., Tasiran, S. (eds.) CAV 2019. LNCS, vol. 11562, pp. 64–73. Springer, Cham (2019). https://doi.org/10.1007/978-3-030-25543-5_4
10. Kovásznai, G., Fröhlich, A., Biere, A.: Complexity of fixed-size bit-vector logics. Theory Comput. Syst. **59**(2), 323–376 (2015). https://doi.org/10.1007/s00224-015-9653-1
11. Niemetz, A., Preiner, M., Wolf, C., Biere, A.: BTOR2, BtorMC and Boolector 3.0. In: Chockler, H., Weissenbacher, G. (eds.) CAV 2018. LNCS, vol. 10981, pp. 587–595. Springer, Cham (2018). https://doi.org/10.1007/978-3-319-96145-3_32
12. Zeljić, A., Wintersteiger, C.M., Rümmer, P.: An approximation framework for solvers and decision procedures. J. Autom. Reasoning **58**(1), 127–147 (2016). https://doi.org/10.1007/s10817-016-9393-1

# Strong (D)QBF Dependency Schemes via Tautology-Free Resolution Paths

Olaf Beyersdorff, Joshua Blinkhorn$^{(\boxtimes)}$, and Tomáš Peitl

Friedrich-Schiller-Universität, Jena, Germany
joshua.blinkhorn@uni-jena.de

**Abstract.** We suggest a general framework to study dependency schemes for dependency quantified Boolean formulas (DQBF). As our main contribution, we exhibit a *new tautology-free DQBF dependency scheme* that generalises the reflexive resolution path dependency scheme. We establish soundness of the tautology-free scheme, implying that it can be used in any DQBF proof system. We further explore the power of DQBF resolution systems parameterised by dependency schemes and show that our new scheme results in exponentially shorter proofs in comparison to the reflexive resolution path dependency scheme when used in the expansion DQBF system ∀Exp+Res.

On QBFs, we demonstrate that our new scheme is exponentially stronger than the reflexive resolution path dependency scheme when used in Q-resolution, thus resulting in the strongest QBF dependency scheme known to date.

**Keywords:** DQBF · Dependency schemes · Proof complexity · QBF

## 1 Introduction

Quantified Boolean formulas (QBF) have been intensively studied in the past decade, both practically and theoretically. On the practical side, there have been huge improvements in QBF solving [30]. These build on the success of SAT solving [36], but also incorporate new ideas genuine to the QBF domain, such as expansion solving [21] and dependency schemes [32]. Due to its PSPACE completeness, QBF solving is relevant to many application domains that cannot be efficiently encoded into SAT [17,23,26]. On the theoretical side, there is a substantial body of QBF proof complexity results (e.g. [3,6,8–10]), which calibrates the strength of solvers while guiding their development.

In QBF solving, a severe technical complication is that *variable dependencies* stemming from the linear order of quantification[1] must be respected when assigning variables. In contrast, a SAT solver can assign variables in any order, granting

---

[1] The standard input for solvers is a prenex QBF.

This research was supported by grants from FWF (J-4361), the John Templeton Foundation (grant no. 60842), and the Carl Zeiss Foundation.

© Springer Nature Switzerland AG 2020
L. Pulina and M. Seidl (Eds.): SAT 2020, LNCS 12178, pp. 394–411, 2020.
https://doi.org/10.1007/978-3-030-51825-7_28

complete freedom to decision heuristics, which are crucial for performance. As a remedy, QBF researchers have developed *dependency schemes*. Dependency schemes try to determine algorithmically which of the variable dependencies are essential, thereby identifying *spurious* dependencies which can be safely disregarded. The result is greater freedom for decision heuristics.

Practical QBF solving utilises dependency schemes, for example the solvers DepQBF [24] and Qute [27,28], and experiments show dependency-aware solving is particularly competitive on QBFs with high quantifier complexity [20,25].

The performance gains are also underlined by theoretical findings. There is a sequence of results [7,29,35] that establish how and when dependency schemes are sound to use with a QBF proof system, such as the central QBF resolution systems Q-resolution [22] and long-distance Q-resolution [2]. In [6] it is demonstrated that using the reflexive resolution path dependency scheme ($\mathcal{D}^{rrs}$ [35]) in Q-resolution can exponentially shorten proofs.

While dependency schemes aim to algorithmically determine spurious dependencies, *dependency quantified Boolean formulas* (DQBF) allow to directly express variable dependencies by specifying, for each existential variable $x$, a *dependency set* of universal variables on which $x$ depends. This is akin to the use of Henkin quantifiers in first-order logic [18]. Compared to QBFs, DQBFs boost reasoning power and enable further applications (cf. [33] for an overview). The price of succinct encodings is an increase of the complexity of the satisfiability problem from PSPACE (for QBF) to NEXP (for DQBF) [1].

It seems natural that there should be a relationship between dependency schemes and DQBF, and indeed the paper [7] suggests that dependency schemes for QBF should be viewed as truth-preserving mappings from QBF to DQBF.

Now, is there even a need for *dependency schemes for DQBF?* The answer is yes: also for DQBFs it is possible that the dependency sets contain spurious dependencies, which can be safely eliminated [37]. Indeed, Wimmer et al. [37] showed that several dependency schemes for QBF, including $\mathcal{D}^{rrs}$, can be lifted to DQBF. They also demonstrate that using dependency schemes for DQBF preprocessing can have a significant positive impact on solving time.

However, in contrast to QBF, there are currently no results on how DQBF dependency schemes can be incorporated into DQBF proof systems, and how this affects their proof-theoretic strength.

This paper contributes to the theory of DQBF dependency schemes on three main fronts.

**A. A proof complexity framework for DQBF dependency schemes.** We extend the interpretation of QBF dependency schemes proposed in [7] to DQBF. The result is a framework in which a sound DQBF dependency scheme $\mathcal{D}$ can be straightforwardly incorporated into an arbitrary DQBF proof system P, yielding the parametrised system P($\mathcal{D}$). More precisely, in our framework a proof of $\Phi$ in P($\mathcal{D}$) is simply a P proof of $\mathcal{D}(\Phi)$, where $\mathcal{D}$ is a mapping between DQBFs.

A major benefit of this approach is that the rules of the proof system remain independent of the dependency scheme, which essentially plays the role of a preprocessor. Moreover, soundness of a dependency scheme is characterised by the natural property of full exhibition [4,34], independently of proofs. This is a welcome feature, since even *defining* sound parameterisations on the QBF fragment has been fairly non-trivial, e.g. for the long-distance Q-resolution calculus [4,29].

We also extend the notion of *genuine* proof size lower bounds [12,14] to DQBF proof systems. Since DQBF encompasses QBF, proof systems are susceptible to lower bounds from QBF proof complexity. We define a precise condition by which hardness from the QBF fragment is factored out. As such, our framework fosters the first dedicated DQBF proof complexity results.

**B. The tautology-free dependency scheme.** We define and analyse a new DQBF dependency scheme called the *tautology-free dependency scheme* ($\mathcal{D}^{\mathrm{tf}}$). Our scheme builds on the reflexive resolution path dependency scheme ($\mathcal{D}^{\mathrm{rrs}}$) [35], originally defined for QBFs, which prior to this paper was the strongest known DQBF scheme. $\mathcal{D}^{\mathrm{tf}}$ improves on $\mathcal{D}^{\mathrm{rrs}}$ by disallowing certain kinds of tautologies in resolution paths, thereby identifying further spurious dependencies.

We show that $\mathcal{D}^{\mathrm{tf}}$ is fully exhibited, and therefore sound, by reducing its full exhibition to that of $\mathcal{D}^{\mathrm{rrs}}$. For this, we point out that the full exhibition of $\mathcal{D}^{\mathrm{rrs}}$ on DQBF is an immediate consequence of results of Wimmer et al. [37].

**C. Exponential separations of (D)QBF proof systems.** To demonstrate the strength of our new scheme $\mathcal{D}^{\mathrm{tf}}$, we show that it can exponentially shorten proofs in DQBF proof systems. As a case study, we consider the expansion calculus ∀Exp+Res. The choice of ∀Exp+Res is motivated by two considerations: (1) it is a natural calculus, whose QBF fragment models expansion solving [21], and (2) other standard QBF resolution systems such as Q-resolution and long-distance Q-resolution do not lift to DQBF [11].

For ∀Exp+Res parameterised by dependency schemes we show that

$$\forall\mathsf{Exp}+\mathsf{Res} < \forall\mathsf{Exp}+\mathsf{Res}(\mathcal{D}^{\mathrm{rrs}}) < \forall\mathsf{Exp}+\mathsf{Res}(\mathcal{D}^{\mathrm{tf}}) \tag{1}$$

forms a hierarchy of DQBF proof systems of strictly increasing strength.

Since there exist no prior DQBF proof complexity results whatsoever, this entails proving exponential proof-size lower bounds in the first two systems. We obtain these by introducing two new DQBF versions of the equality formulas (originally QBFs [8,13]). Together with the corresponding upper bounds, this yields the separations in (1). We highlight that these are *genuine* separations in the precise sense of our DQBF framework, whereby hardness due to the QBF fragment is factored out.

Finally, we show that our new dependency scheme $\mathcal{D}^{\mathrm{tf}}$ *is also relevant for QBFs*: we prove that Q-resolution parameterised by $\mathcal{D}^{\mathrm{tf}}$ is exponentially stronger than Q-resolution with $\mathcal{D}^{\mathrm{rrs}}$. Thus $\mathcal{D}^{\mathrm{tf}}$ currently constitutes the strongest known dependency scheme for Q-resolution.

**Organisation.** Section 2 defines DQBF preliminaries. In Sect. 3 we explain dependency schemes. Section 4 details how to parameterise DQBF proof systems by dependency schemes. In Sect. 5 we define our new scheme $\mathcal{D}^{\mathrm{tf}}$ and show its soundness. In Sect. 6 we prove the proof complexity upper and lower bounds needed for the strict hierarchy in (1). Section 7 applies $\mathcal{D}^{tf}$ to QBF and shows that it is stronger than $\mathcal{D}^{rrs}$ when used with Q-resolution.

## 2   Preliminaries

**DQBF Syntax.** We assume familiarity with the syntax of propositional logic and the notion of *Boolean formula* (simply *formula*). A *variable* is an element $z$ of the countable set $\mathbb{V}$. A *literal* is a variable $z$ or its *negation* $\bar{z}$. The negation of a literal $a$ is denoted $\bar{a}$, where $\bar{\bar{z}} := z$ for any variable $z$. A *clause* is a disjunction of literals. A *conjunctive normal form formula* (CNF) is a conjunction of clauses. The set of variables appearing in a formula $\psi$ is denoted $\mathrm{vars}(\psi)$. For ease, we often write clauses as sets of literals, and CNFs as sets of clauses. For any clause $C$ and any set of variables $Z$, we define $C\!\restriction_Z := \{a \in C : \mathrm{var}(a) \in Z\}$.

A *dependency quantified Boolean formula* (DQBF) is a sentence of the form $\Psi := \Pi \cdot \psi$, where $\Pi := \forall u_1 \cdots \forall u_m \exists x_1(S_{x_1}) \cdots \exists x_n(S_{x_n})$ is the *quantifier prefix* and $\psi$ is a CNF called the *matrix*. In the quantifier prefix, each existential variable $x_i$ is associated with a *dependency set* $S_{x_i}$, which is a subset of the universal variables $\{u_1, \ldots, u_m\}$. With $\mathrm{vars}_\forall(\Psi)$ and $\mathrm{vars}_\exists(\Psi)$ we denote the universal and existential variable sets of $\Psi$, and with $\mathrm{vars}(\Psi)$ their union. We deal only with *closed* DQBFs, in which $\mathrm{vars}(\psi) \subseteq \mathrm{vars}(\Psi)$. We define a relation $\mathrm{deps}(\Psi)$ on $\mathrm{vars}_\forall(\Psi) \times \mathrm{vars}_\exists(\Psi)$, where $(u, x) \in \mathrm{deps}(\Psi)$ if, and only if, $u \in S_x$.

The set of all DQBFs is denoted DQBF. A QBF is a DQBF whose dependency sets are linearly ordered with respect to set inclusion, i.e. $S_{x_1} \subseteq \cdots \subseteq S_{x_n}$. The prefix of a QBF can be written as a linear order in the conventional way. The set of all QBFs is denoted QBF.

**DQBF Semantics.** An *assignment* $\alpha$ to a set $Z$ of Boolean variables is a function from $Z$ into the set of *Boolean constants* $\{0, 1\}$. The *domain restriction* of $\alpha$ to a subset $Z' \subseteq Z$ is written $\alpha\!\restriction_{Z'}$. The set of all assignments to $Z$ is denoted $\langle Z \rangle$. The *restriction* of a formula $\psi$ by $\alpha$, denoted $\psi[\alpha]$, is the result of substituting each variable $z$ in $Z$ by $\alpha(z)$, followed by applying the standard simplifications for Boolean constants, i.e. $\bar{0} \mapsto 1$, $\bar{1} \mapsto 0$, $\phi \vee 0 \mapsto \phi$, $\phi \vee 1 \mapsto 1$, $\phi \wedge 1 \mapsto \phi$, and $\phi \wedge 0 \mapsto 0$. We say that $\alpha$ *satisfies* $\psi$ when $\psi[\alpha] = 1$, and *falsifies* $\psi$ when $\psi[\alpha] = 0$.

A *model* for a DQBF $\Psi := \Pi \cdot \psi$ is a set of functions $f := \{f_x : x \in \mathrm{vars}_\exists(\Psi)\}$, $f_x : \langle S_x \rangle \to \langle \{x\} \rangle$, for which, for each $\alpha \in \langle \mathrm{vars}_\forall(\Psi) \rangle$, the combined assignment $\alpha \cup \{f_x(\alpha\!\restriction_{S_x}) : x \in \mathrm{vars}_\exists(\Psi)\}$ satisfies $\psi$. A DQBF is called *true* when it has a model, otherwise it is called *false*. When two DQBFs share the same truth value, we write $\Psi \overset{tr}{\equiv} \Psi'$.

**DQBF Expansion.** *Universal expansion* is a syntactic transformation that removes a universal variable from a DQBF. Let $\Psi$ be a DQBF, let $u$ be a universal, and let $y_1, \ldots, y_k$ be the existentials for which $u \in S_{y_i}$. The *expansion* of $\Psi$ by $u$ is obtained by creating two 'copies' of $\Psi$. In the first copy, $u$ is assigned 0 and each $y_i$ is renamed $y_i^{\bar{u}}$. In the second copy, $u$ is assigned 1 and each $y_i$ is renamed $y_i^u$. The two copies are then combined, and $u$ is removed completely from the prefix. Formally, $\exp(\Psi, u) := \Pi' \cdot \psi'$, where $\Pi'$ is obtained from $\Pi$ by removing $\forall u$ and replacing each $\exists y_i(S_{y_i})$ with $\exists y_i^{\overline{u}}(S_{y_i} \setminus \{u\}) \exists y_i^u(S_{y_i} \setminus \{u\})$, and

$$\psi' := \psi[u \mapsto 0, y_1 \mapsto y_1^{\overline{u}}, \ldots, y_k \mapsto y_k^{\overline{u}}] \wedge \psi[u \mapsto 1, y_1 \mapsto y_1^u, \ldots, y_k \mapsto y_k^u].$$

Universal expansion is known to preserve the truth value, i.e. $\Psi \overset{\text{tr}}{\equiv} \exp(\Psi, u)$. Expansion by a set of universal variables $U$ is defined as the successive expansion by each $u \in U$ (the order is irrelevant), and is denoted $\exp(\Psi, U)$. Expansion by the whole set $\text{vars}_\forall(\Psi)$ is denoted $\exp(\Psi)$, and referred to as the *total expansion* of $\Psi$. The superscripts in the renamed existential variables are known as *annotations*. Annotations grow during successive expansions. In the total expansion, each variable is annotated with a total assignment to its dependency set.

# 3    DQBF Dependency Schemes and Full Exhibition

In this section, we lift the 'DQBF-centric' interpretation of QBF dependency schemes [7] to the DQBF domain, and recall the definition of full exhibition.

**How Should We Interpret Variable Dependence?** Dependency schemes [32] were originally introduced to identify so-called *spurious dependencies*: sometimes the order of quantification implies that $z$ depends on $z'$, but forcing $z$ to be independent preserves the truth value. Technically, a dependency scheme $\mathcal{D}$ was defined to map a QBF $\Phi$ to a set of pairs $(z', z) \in \text{vars}(\Phi) \times \text{vars}(\Phi)$, describing an overapproximation of the dependency structure: $(z', z) \in \mathcal{D}(\Phi)$ means that the dependence of $z$ on $z'$ should not be ignored, whereas $(z', z) \notin \mathcal{D}(\Phi)$ means that it can be. The definition was tailored to QBF solving, in which variable dependencies for both true and false formulas come into play.

The DQBF-centric interpretation [7] followed somewhat later. There, the goal was a dependency scheme framework tailored to refutational QBF proof systems. Refutational systems work only with false formulas, and this allows a broad refinement: the dependence of universals on existentials can be ignored. As such, it makes sense to consider merely the effect of deleting some universal variables from the existential dependency sets. Thus, a dependency scheme becomes a mapping from QBF into DQBF.

Likewise, in this work we seek a framework tailored towards refutational proof systems. Hence we advocate the same approach for the *whole domain* DQBF. A DQBF dependency scheme will be viewed as a mapping to and from DQBF, in which the dependency sets may shrink. The notion of shrinking dependency sets is captured by the relation following.

**Definition 1.** *We define the relation $\leq$ on* DQBF $\times$ DQBF *as follows: $\Pi' \cdot \phi \leq \Pi \cdot \psi$ if, and only if, $\phi = \psi$, $\text{vars}_\exists(\Psi') = \text{vars}_\exists(\Psi)$, and the dependency set of each existential in $\Pi'$ is a subset of that of $\Pi$.*

In this paper, we only consider poly-time computable dependency schemes.

**Definition 2 (dependency scheme).** *A dependency scheme is a polynomial-time computable function* $\mathcal{D} : \mathsf{DQBF} \to \mathsf{DQBF}$ *for which* $\mathcal{D}(\Psi) \leq \Psi$ *for all* $\Psi$.

Under this definition, a *spurious dependency according to* $\mathcal{D}$ is a pair $(u, x)$ such that $u$ is in the dependency set for $x$ in $\Psi$, but not in $\mathcal{D}(\Psi)$. A natural property of dependency schemes, identified in [37], is *monotonicity*.[2]

**Definition 3 (monotone (adapted from [37])).** *We call a dependency scheme* $\mathcal{D}$ *monotone when* $\Psi' \leq \Psi$ *implies* $\mathcal{D}(\Psi') \leq \mathcal{D}(\Psi)$, *for all* $\Psi$ *and* $\Psi'$.

A fundamental concept in the DQBF-centric framework, which has strong connections to soundness in related proof systems [6], is *full exhibition*. First used by Slivovsky [34], the name was coined later in [4], describing the fact that there should be a model which 'fully exhibits' all spurious dependencies. 'Full exhibition' is synonymous with 'truth-value preserving'.

**Definition 4 (full exhibition [4,34]).** *A dependency scheme* $\mathcal{D}$ *is called* fully exhibited *when* $\Psi \overset{tr}{\equiv} \mathcal{D}(\Psi)$, *for all* $\Psi$.

# 4    Parametrising DQBF Calculi by Dependency Schemes

In this section we show how to incorporate dependency schemes into DQBF proof systems. In the spirit of so-called 'genuine' lower bounds [12], we also introduce a notion of genuine DQBF hardness.

**Refutational DQBF Proof Systems.** We first define what we mean by a DQBF proof system. With FDQBF we denote the set of false DQBFs. We consider only *refutational* proof systems, which try to show that a given formula is false. Hence, 'proof' and 'refutation' can be considered synonymous.

Following [15], a *DQBF proof system* over an alphabet $\Sigma$ is a polynomial-time computable onto function $\mathsf{P} : \Sigma^* \to \mathsf{FDQBF}$. In practice, we do not always want to define a proof system explicitly as a function on a domain of strings. Instead, we define what constitutes a refutation in the proof system $\mathsf{P}$, and then show: (1) *Soundness:* if $\Psi$ has a refutation, it is false (the codomain of $\mathsf{P}$ is FDQBF); (2) *Completeness:* every false DQBF has a refutation ($\mathsf{P}$ is onto); (3) *Checkability:* refutations can be checked efficiently ($\mathsf{P}$ is polynomial-time computable).

Two concrete examples of DQBF proof systems from the literature are the fundamental expansion-based system $\forall\mathsf{Exp}{+}\mathsf{Res}$ [7], and the more sophisticated instantiation-based system $\mathsf{IR}$-$\mathsf{calc}$ [7].

**Incorporating Dependency Schemes.** A dependency scheme, interpreted as a DQBF mapping as in Definition 2, can be combined with an arbitrary proof system in a straightforward manner.

**Definition 5 ($\mathsf{P}(\mathcal{D})$).** *Let* $\mathsf{P}$ *be a DQBF proof system and let* $\mathcal{D}$ *be a dependency scheme. A* $\mathsf{P}(\mathcal{D})$ *refutation of a DQBF* $\Psi$ *is a* $\mathsf{P}$ *refutation of* $\mathcal{D}(\Psi)$.

---

[2] A different notion of monotonicity for dependency schemes is defined in [29].

The proof system $P(\mathcal{D})$ essentially utilises the dependency scheme as a preprocessing step, mapping its input $\Psi$ to the image $\mathcal{D}(\Psi)$ *before* proceeding with the refutation. In this way, the application of the dependency scheme $\mathcal{D}$ is separated from the rules of the proof system $P$, and consequently the definition of $P$ need not be explicitly modified to incorporate $\mathcal{D}$ (cf. [4, 35]).

Of course, we must ensure that our preprocessing step is correct; we do not want to map a true formula to a false one, which would result in an unsound proof system. Now it becomes clear why full exhibition is central for soundness.

**Proposition 6.** *Given a DQBF proof system $P$ and a dependency scheme $\mathcal{D}$, $P(\mathcal{D})$ is sound if, and only if, $\mathcal{D}$ is fully exhibited.*

*Proof.* Suppose that $\mathcal{D}$ is fully exhibited. Let $\pi$ be a $P(\mathcal{D})$ refutation of a DQBF $\Psi$. Then $\pi$ is a $P$ refutation of $\mathcal{D}(\Psi)$, which is false by the soundness of $P$. Hence $\Psi$ is false by the full exhibition of $\mathcal{D}$, so $P(\mathcal{D})$ is sound.

Suppose now that $\mathcal{D}$ is not fully exhibited. By definition of dependency scheme, for each DQBF $\Psi$ we have $\Psi \geq \mathcal{D}(\Psi)$. It follows that $\mathcal{D}$ preserves falsity, so there must exist a true DQBF $\Psi$ for which $\mathcal{D}(\Psi)$ is false. Then there exists a $P$ refutation of $\mathcal{D}(\Psi)$ by the completeness of $P$, so $P(\mathcal{D})$ is not sound.  $\square$

Note that completeness and checkability of $P$ are preserved trivially by any dependency scheme, so we can even say that $P(\mathcal{D})$ is a DQBF proof system if, and only if, $\mathcal{D}$ is fully exhibited. Thus full exhibition characterises exactly the dependency schemes whose incorporation preserves the proof system.

**Simulations, Separations and Genuine Lower Bounds.** Of course, the rationale for utilising a dependency scheme as a preprocessor lies in the potential for shorter refutations. We first recall the notion of $p$-simulation from [15]. Let $P$ and $Q$ be DQBF proof systems. We say that $P$ *p-simulates* $Q$ (written $Q \leq_p P$) when there exists a polynomial-time computable function from $Q$ refutations to $P$ refutations that preserves the refuted formula.

Since a $p$-simulation is computed in polynomial time, the translation from $Q$ into $P$ incurs at most a polynomial size blow-up. As such, the conventional approach to proving the non-existence of a $p$-simulation is to exhibit a family of formulas $\{\Psi_n\}_{n \in \mathbb{N}}$ that has polynomial-size refutations in $Q$, while requiring super-polynomial size in $P$.

Now, it is of course possible that the hard formulas $\{\Psi_n\}_{n \in \mathbb{N}}$ are QBFs. While this suffices to show that $Q \not\leq_p P$, it is not what we want from a study of DQBF proof complexity; it is rather a statement about the QBF fragments of the systems $P$ and $Q$. In reality the situation is even more complex. The lower bound may stem from QBF proof complexity even when $\{\Psi_n\}_{n \in \mathbb{N}}$ are not QBFs. More precisely, there may exist an 'embedded' QBF family $\{\Phi_n\}_{n \in \mathbb{N}}$ which is already hard for $P$, where 'embedded' means $\Phi_n \leq \Psi_n$. Under the reasonable assumption that decreasing dependency sets cannot increase proof size,[3] *any* DQBF family in which $\{\Phi_n\}_{n \in \mathbb{N}}$ is embedded will be hard for $P$.

---

[3] This holds for all known DQBF proof systems.

For that reason, we introduce a notion of *genuine* DQBF hardness that dismisses all embedded QBF lower bounds.

**Definition 7.** *Let* P *and* Q *be DQBF proof systems. We write* Q $\not\leq_p^*$ P *when there exists a DQBF family* $\{\Psi_n\}_{n\in\mathbb{N}}$ *such that:*

(a) $\{\Psi_n\}_{n\in\mathbb{N}}$ *has polynomial-size* Q *refutations;*
(b) $\{\Psi_n\}_{n\in\mathbb{N}}$ *requires superpolynomial-size* P *refutations;*
(c) *every QBF family* $\{\Phi_n\}_{n\in\mathbb{N}}$ *with* $\Phi_n \leq \Psi_n$ *has polynomial-size* P *refutations.*

*We write* P $<_p^*$ Q *when both* P $\leq_p$ Q *and* Q $\not\leq_p^*$ P *hold.*

Hence, P $<_p^*$ Q means that Q simulates P, but P does not simulate Q, and the hardness result for P is a genuine DQBF lower bound. Prior to this paper, there were no such hardness results in the DQBF literature.

## 5   The Tautology-Free Dependency Scheme

In this section we define the *tautology-free dependency scheme* $\mathcal{D}^{\mathrm{tf}}$ and show that it is fully exhibited.

For any DQBF $\Psi$, we denote by $I_\exists(\Psi)$ the set of independent existential variables, i.e. $I_\exists(\Psi) := \{x \in \mathrm{vars}_\exists(\Psi) : S_x = \emptyset\}$ is the set of existentials whose dependency sets are empty. For any $k \in \mathbb{N}$, we define $[k] := \{n \in \mathbb{N} : n \leq k\}$.

**Definition 8 ($\mathcal{D}^{\mathrm{rrs}}$ [35] and $\mathcal{D}^{\mathrm{tf}}$).**  *The reflexive resolution path dependency scheme ($\mathcal{D}^{\mathrm{rrs}}$) is defined as the mapping* $\Psi \mapsto \Psi'$, *where*

$$\Psi := \forall u_1 \cdots \forall u_m \exists x_1(S_{x_1}) \cdots \exists x_n(S_{x_n}) \cdot \psi,$$
$$\Psi' := \forall u_1 \cdots \forall u_m \exists x_1(S'_{x_1}) \cdots \exists x_n(S'_{x_n}) \cdot \psi,$$

*and* $S'_i$ *is the set of universal variables* $u \in S_i$ *for which there exists a sequence* $C_1, \ldots, C_k$ *of clauses in* $\psi$ *and a sequence* $p_1, \ldots, p_{k-1}$ *of existential literals satisfying the following conditions:*

(a) $u \in C_1$ *and* $\overline{u} \in C_k$;
(b) *for some* $j \in [k-1]$, $x_i = \mathrm{var}(p_j)$;
(c) *for each* $j \in [k-1]$, $p_j \in C_j$, $\overline{p}_j \in C_{j+1}$, *and* $u \in S_{\mathrm{var}(p_j)}$;
(d) *for each* $j \in [k-2]$, $\mathrm{var}(p_j) \neq \mathrm{var}(p_{j+1})$.

*The* tautology-free dependency scheme *($\mathcal{D}^{\mathrm{tf}}$) adds to* $\mathcal{D}^{\mathrm{rrs}}$ *the condition*

(e) *for each* $j \in [k-1]$, $(C_j \cup C_{j+1})\!\restriction_{I_\exists(\Psi)}$ *is non-tautological.*

Let us give an example, illustrating that $\mathcal{D}^{\mathrm{tf}}$ is stronger than $\mathcal{D}^{\mathrm{rrs}}$.

*Example 9.* Consider the DQBF $\Psi = \exists x \forall u \exists z \cdot C_1 \wedge C_2$, where $C_1 = x \vee u \vee z$ and $C_2 = \overline{x} \vee \overline{u} \vee \overline{z}$. The sequence of clauses $C_1, C_2$ and the sequence consisting of the single literal $p_1 = z$ show that $(u, z) \in \mathsf{deps}(\mathcal{D}^{\mathrm{rrs}}(\Psi))$. However, the same sequence of clauses violates condition (e) of Definition 8 because $(C_1 \cup C_2)\!\restriction_{I_\exists(\Psi)}$ is a tautology on $x \in I_\exists(\Psi)$. Since there are no other sequences that satisfy (a), we conclude that $(u, z) \notin \mathsf{deps}(\mathcal{D}^{\mathrm{tf}}(\Psi))$.   □

**Proposition 10.** $\mathcal{D}^{\mathrm{tf}}$ *is a monotone dependency scheme.*

*Proof.* It is easy to see that $\mathcal{D}^{\mathrm{tf}}(\Psi) \leq \Psi$ for each $\Psi$. It remains to verify polynomial-time computability and monotonicity.

*Polynomial-Time Computability.* As there are polynomially many pairs, it suffices to show that whether $(u, x)$ is in $\mathsf{deps}(\Psi)$ can be decided in polynomial time for each pair $(u, x)$. Consider the directed graph $G_\Psi^u = (V_\Psi, E_\Psi^u)$ with the vertex set $V_\Psi = \{(C, a) : C \in \Psi, a \in C\}$ and with an edge from $(C, a)$ to $(D, e)$ if $\overline{e} \in C$, $u \in S_{\mathrm{var}(e)}$, $\mathrm{var}(a) \neq \mathrm{var}(e)$, and $(C \cup D)\!\restriction_{I_\exists(\Psi)}$ is non-tautological.

We claim that $(u, x) \in \mathsf{deps}(\Psi)$ if, and only if, there is a literal $a$, $\mathrm{var}(a) = x$, and clauses $C, C', C''$ such that $\overline{u} \in C''$, $(C', a)$ is reachable from $(C, u)$ and $(C'', e)$ is reachable from $(C', a)$ for some $e$. Indeed, it is easy to verify that the concatenation of a pair of such paths directly translates to the required sequences from Definition 8, and vice versa. Clearly, $G_\Psi^u$ can be constructed in polynomial time, hence we can test all candidates $(C, u)$, compute all middle points $(C', a)$ reachable from them, and check whether some $(C'', e)$ is reachable from any of them, all in polynomial time.

*Monotonicity.* Let $\Psi, \Psi'$ be DQBFs with $\Psi' \leq \Psi$, let $(u, x) \in \mathsf{deps}(\mathcal{D}^{\mathrm{tf}}(\Psi'))$. We show that $(u, x) \in \mathsf{deps}(\mathcal{D}^{\mathrm{tf}}(\Psi))$. It follows that $\mathcal{D}^{\mathrm{tf}}(\Psi') \leq \mathcal{D}^{\mathrm{tf}}(\Psi)$.

There exists a sequence of clauses $C_1, \ldots, C_k$ and a sequence of literals $p_1, \ldots, p_{k-1}$ satisfying conditions (a) to (e) in Definition 8 with respect to $(u, x) \in \mathsf{deps}(\Psi')$. We show that the same sequences satisfy conditions (a) to (e) with respect to $(u, x) \in \mathsf{deps}(\Psi)$, which implies $(u, x) \in \mathsf{deps}(\mathcal{D}^{\mathrm{tf}}(\Psi))$.

Conditions (a), (b) and (d) are satisfied trivially. Since $\Psi' \leq \Psi$, each dependency set $S_{\mathrm{var}(p_i)}$ in $\Psi$ is a superset of the corresponding dependency set $S'_{\mathrm{var}(p_i)}$ in $\Psi'$, so condition (c) is satisfied. Condition (e) is satisfied since the set of independent variables $I_\exists(\Psi)$ is a subset of $I_\exists(\Psi')$. □

Wimmer et al. [37] essentially showed that $\mathcal{D}^{\mathrm{rrs}}$ is fully exhibited, even though they did not use that term. Theorems 3 and 4 in [37] together imply that all spurious dependencies can be removed one by one in any order without changing the truth value (as is remarked at the start of Sect. 3.1 in that paper).

**Theorem 11 (Wimmer et al. [37]).** $\mathcal{D}^{\mathrm{rrs}}$ *is fully exhibited.*

We show full exhibition of $\mathcal{D}^{\mathrm{tf}}$ by reduction to full exhibition of $\mathcal{D}^{\mathrm{rrs}}$.

**Theorem 12.** $\mathcal{D}^{\mathrm{tf}}$ *is fully exhibited.*

*Proof.* Since $\mathcal{D}^{\mathrm{tf}}(\Psi) \leq \Psi$, we only need to show that if $\Psi$ is true, then $\mathcal{D}^{\mathrm{tf}}(\Psi)$ is true. Assume $\Psi$ is true; then there is an assignment $\sigma \in \langle I_\exists(\Psi) \rangle$ such that $\Psi[\sigma]$ is true. We claim that $(u, x) \in \mathsf{deps}(\mathcal{D}^{\mathrm{rrs}}(\Psi[\sigma]))$ implies $(u, x) \in \mathsf{deps}(\mathcal{D}^{\mathrm{tf}}(\Psi))$. Consider sequences $C_1, \ldots, C_k$ and $p_1, \ldots, p_{k-1}$ witnessing $(u, x) \in \mathsf{deps}(\mathcal{D}^{\mathrm{rrs}}(\Psi[\sigma]))$. For each $C_i$ there is $C_i' \in \Psi$, such that $C_i = C_i'[\sigma]$, i.e. $C_i' \subseteq C_i \cup \overline{\sigma}$, where $\overline{\sigma}$ is the largest clause falsified by $\sigma$. It is readily verified that the sequences $C_1', \ldots, C_k'$ and $p_1, \ldots, p_{k-1}$ witness $(u, x) \in \mathsf{deps}(\mathcal{D}^{\mathrm{tf}}(\Psi))$. In particular, no tautologies can appear among $(C_i' \cup C_{i+1}')\!\restriction_{I_\exists(\Psi)}$, because all $C_i'$ agree with $\overline{\sigma}$ on the variables of $I_\exists(\Psi)$. Hence, we get $\mathcal{D}^{\mathrm{rrs}}(\Psi[\sigma]) \leq \mathcal{D}^{\mathrm{tf}}(\Psi)[\sigma]$. By full exhibition of $\mathcal{D}^{\mathrm{rrs}}$, we have that $\mathcal{D}^{\mathrm{rrs}}(\Psi[\sigma])$ is true, which means $\mathcal{D}^{\mathrm{tf}}(\Psi)[\sigma]$ is true, and hence $\mathcal{D}^{\mathrm{tf}}(\Psi)$ is true. □

*Example 13.* Consider $\Psi$ from Example 9. It is easy to see that $\Psi$ is true. As shown in Example 9, $\mathcal{D}^{\mathrm{tf}}(\Psi) = \exists x \exists z \forall u \cdot (x \vee z \vee u) \wedge (\overline{x} \vee \overline{z} \vee \overline{u})$. We can see that the assignment $x \mapsto 1, z \mapsto 0$ is a model of $\mathcal{D}^{\mathrm{tf}}(\Psi)$, which is therefore true, in line with full exhibition of $\mathcal{D}^{\mathrm{tf}}$. □

# 6  Proof Complexity of ∀Exp+Res($\mathcal{D}$)

Among the first DQBF proof systems to be introduced, the expansion based system ∀Exp+Res [7,21] is arguably the most natural. In this section we investigate its proof complexity under parametrisation by dependency schemes; that is, we investigate the proof complexity of P($\mathcal{D}$) where P is ∀Exp+Res. Our main result is the following theorem.

**Theorem 14.** ∀Exp+Res $<^*_p$ ∀Exp+Res($\mathcal{D}^{\mathrm{rrs}}$) $<^*_p$ ∀Exp+Res($\mathcal{D}^{\mathrm{tf}}$).

The simulations present in Theorem 14 follow from two observations, namely (1) $\mathcal{D}^{\mathrm{tf}}(\Psi) \leq \mathcal{D}^{\mathrm{rrs}}(\Psi)$ (by definition), and (2) $\Psi' \leq \Psi$ guarantees that ∀Exp+Res refutations of $\Psi'$ are no larger than those of $\Psi$. Indeed, given a refutation of $\Psi$, restricting the annotations to the dependency sets of $\Psi'$ produces a refutation of $\Psi'$ of the same size. We refer to this property as the *monotonicity* of ∀Exp+Res.

The challenge is to show the genuine separations (Theorems 20 and 26). We note that the QBF analogue of the first separation is known [6]. The question (and indeed the notion) of a genuine separation was not previously considered.

**The DQBF Proof System ∀Exp+Res.** We recall the propositional resolution proof system [31]. A *resolution refutation* of a CNF $\psi$ is a sequence $C_1, \ldots, C_k$ of clauses where $C_k$ is empty and each $C_i$ is derived by one of the following rules:

**A** *Axiom:* $C_i$ is a clause in $\psi$;
**R** *Resolution:* $C_i = A \vee B$, where $C_r = A \vee x$ and $C_s = B \vee \overline{x}$, for some $r, s < i$.

The DQBF proof system ∀Exp+Res, with which we shall concern ourselves for the remainder of the section, is built upon resolution. Perhaps the most obvious way to decide DQBF is to reduce it to propositional logic by expanding out all the universal variables, based on the fact that $\Psi$ is true if, and only if, the matrix of exp($\Psi$) is satisfiable. This is exactly how ∀Exp+Res works. The input DQBF is first expanded, and then refuted in resolution.

**Definition 15** (∀Exp+Res [7,21]). *A* ∀Exp+Res *refutation of a DQBF $\Psi$ is a resolution refutation of the matrix of* exp($\Psi$).

It is known that ∀Exp+Res is sound, complete and checkable on DQBFs [7]. Note that a ∀Exp+Res refutation of $\Psi$ may be small even if its expansion exp($\Psi$) is large, since the underlying resolution refutation of exp($\Psi$) need not necessarily introduce every clause as an axiom.

Given that fully exhibited dependency schemes like $\mathcal{D}^{\mathrm{tf}}$ and $\mathcal{D}^{\mathrm{rrs}}$ (Theorem 12) can be incorporated into an arbitrary DQBF proof system P (Proposition 6), we obtain the DQBF proof systems $\forall\mathsf{Exp}+\mathsf{Res}(\mathcal{D}^{\mathrm{rrs}})$ and $\forall\mathsf{Exp}+\mathsf{Res}(\mathcal{D}^{\mathrm{tf}})$.

Next we show the two genuine separations that together constitute a proof of Theorem 14.

**Separation of $\forall\mathsf{Exp}+\mathsf{Res}$ and $\forall\mathsf{Exp}+\mathsf{Res}(\mathcal{D}^{\mathrm{rrs}})$.** Our separating formulas are DQBFs based on the equality QBFs [8]. Our modification exploits a refined dependency structure and utilises the following notation: the *matrix-clause product* of a CNF $\psi$ and a clause $C$ is the CNF $\psi \otimes C := \{D \cup C : D \in \psi\}$.

**Definition 16 ($\mathrm{EQ}_n^0$ (adapted from [8])).** $\mathrm{EQ}_n^0 := \Pi_n^{\mathrm{EQ}} \cdot \psi_n^{\mathrm{EQ}}$, *where*

$$\Pi_n^{\mathrm{EQ}} := \forall u_1 \cdots \forall u_n \exists x_1(\emptyset) \cdots \exists x_n(\emptyset) \, \exists z_1(u_1) \cdots \exists z_n(u_n) \,,$$

$$\psi_n^{\mathrm{EQ}} := (\overline{z_1} \vee \cdots \vee \overline{z_n}) \wedge \bigwedge_{i=1}^{n} \left( (\overline{x_i} \vee \overline{u_i} \vee z_i) \wedge (x_i \vee u_i \vee z_i) \right).$$

Since the dependency sets of $\mathrm{EQ}_n^0$ are strict subsets of those of the original equality formulas (in which each $z_i$ depends on *each* $u_j$), the QBF lower bound for $\forall\mathsf{Exp}+\mathsf{Res}$ [5] does not suffice for $\mathrm{EQ}_n^0$. Nonetheless, a similar argument works, based on the fact that no small subset of clauses in the expansion is unsatisfiable.

**Theorem 17.** $\{\mathrm{EQ}_n^0\}_{n\in\mathbb{N}}$ *requires exponential-size $\forall\mathsf{Exp}+\mathsf{Res}$ refutations.*

*Proof.* The total expansion of $\mathrm{EQ}_n^0$ is the CNF $\psi \wedge \bigwedge_{i=1}^n \left( (\overline{x_i} \vee z_i^{\overline{u_i}}) \wedge (x_i \vee z_i^{u_i}) \right)$, where $\psi$ is the conjunction of all clauses of the form $(z_1^{a_1} \vee \cdots \vee z_n^{a_n})$ with $\mathrm{var}(a_i) = u_i$. We show that removing any of the $2^n$ clauses from $\psi$ makes the total expansion satisfiable. It follows that any resolution refutation of $\exp(\mathrm{EQ}_n^0)$ must have $2^n$ axiom clauses.

Suppose that some clause $A$ is absent from $\psi$, and let us assume without loss of generality that $A := (z_1^{u_1} \vee \cdots \vee z_n^{u_n})$, i.e. the clause corresponding to $u_i \mapsto 1$ for each $i$ (the general case is symmetrical). Now, assigning each $z_i^{u_i} \mapsto 1$ satisfies every clause in $\psi$ except $A$. Assigning each $z_i^{\overline{u_i}} \mapsto 0$ satisfies each clause $(\overline{x_i} \vee z_i^{u_i})$. Finally, assigning each $x_i \mapsto 1$ satisfies each clause $(x_i \vee z_i^{u_i})$. $\qquad\square$

The corresponding upper bound for $\mathrm{EQ}_n^0$ in $\forall\mathsf{Exp}+\mathsf{Res}(\mathcal{D}^{\mathrm{rrs}})$ *does* follow from that of the original equality QBFs (by the monotonicity of $\mathcal{D}^{\mathrm{rrs}}$ and $\forall\mathsf{Exp}+\mathsf{Res}$). We give a full proof nonetheless, since we will use the details later. The main point is that $\mathcal{D}^{\mathrm{rrs}}$ identifies all pairs as spurious dependencies.

**Proposition 18 ([6]).** *For all $n$, the dependency sets of $\mathcal{D}^{\mathrm{rrs}}(\mathrm{EQ}_n^0)$ are empty.*

*Proof.* Aiming for contradiction, suppose that there exists a sequence of clauses $C_1, \ldots, C_k$ and a sequence of literals $p_1, \ldots, p_{k-1}$ satisfying conditions (a) to (d) of Definition 8 with respect to $(u_i, z_i) \in \mathsf{deps}(\mathrm{EQ}_n^0)$. Since $z_i$ is the unique variable whose dependency set contains $u_i$, we must have $k = 2$, by conditions (c) and (d). By condition (a), we have $u_i \in C_1$, so $C_1 = (x_i \vee u_i \vee z_i)$, and by condition (c) we have $p_1 = z_i$. Also by condition (c) we have $\overline{z_i} \in C_2$, so $C_2 = (\overline{z_1} \vee \cdots \vee \overline{z_n})$. We therefore reach a contradiction, since $\overline{u_i} \notin C_2$ violates condition (a). $\qquad\square$

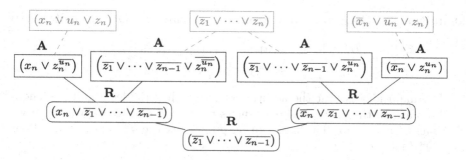

**Fig. 1.** The prelude to a linear-size $\forall$Exp+Res refutation of $\Phi_n^0$. In order to reduce $\exp(\Phi_n^0)$ to $\exp(\mathcal{D}^{\mathrm{rrs}}(\mathrm{EQ}_{n-1}^0))$, we need only derive the clause $(\overline{z_1} \vee \cdots \vee \overline{z_{n-1}})$.

**Theorem 19** ([6]). $\{\mathrm{EQ}_n^0\}_{n \in \mathbb{N}}$ *has linear-size* $\forall$Exp+Res$(\mathcal{D}^{\mathrm{rrs}})$ *refutations.*

*Proof.* By Proposition 18, the total expansion of $\mathcal{D}^{\mathrm{rrs}}(\mathrm{EQ}_n^0)$ is obtained simply by removing the universal literals; that is, the matrix of $\exp(\mathcal{D}^{\mathrm{rrs}}(\mathrm{EQ}_n^0))$ is

$$(\overline{z_1} \vee \cdots \vee \overline{z_n}) \wedge \bigwedge_{i=1}^{n} \left( (\overline{x_i} \vee z_i) \wedge (x_i \vee z_i) \right). \tag{2}$$

It is easy to see that this CNF has linear-size resolution refutations. First, resolve each pair $(x_i \vee z_i), (\overline{x_i} \vee z_i)$ over $x_i$, and resolve the resulting unit clauses $(z_i)$ with the remaining clause to obtain the empty clause. $\qquad\square$

Theorems 17 and 19 together imply that $\forall$Exp+Res does not $p$-simulate $\forall$Exp+Res$(\mathcal{D}^{\mathrm{rrs}})$. It remains to show that the lower bound is genuine.

**Theorem 20.** $\forall$Exp+Res $\not\leq_p^* \forall$Exp+Res$(\mathcal{D}^{\mathrm{rrs}})$.

*Proof.* It is easy to see that the largest QBF $\Phi_n^0$ that is smaller than $\mathrm{EQ}_n^0$ has exactly one non-empty dependency set. Let us assume without loss of generality that this is $S_{z_n} = \{u_n\}$. We will show that $\Phi_n^0$ has a linear-size $\forall$Exp+Res refutation. Hence, by the monotonicity of $\forall$Exp+Res, any family of QBFs smaller than $\{\mathrm{EQ}_n^0\}_{n \in \mathbb{N}}$ has linear-size $\forall$Exp+Res refutations. Thus, by Theorems 17 and 19, $\{\mathrm{EQ}_n^0\}_{n \in \mathbb{N}}$ satisfies all the conditions of Definition 7.

It remains to show that $\Phi_n^0$ has a linear-size $\forall$Exp+Res refutation, or equivalently, that $\exp(\Phi_n^0)$ has a linear-size resolution refutation. It is readily verified that $\exp(\Phi_n^0)$ contains every clause in $\exp(\mathcal{D}^{\mathrm{rrs}}(\mathrm{EQ}_{n-1}^0))$ except $(\overline{z_1} \vee \cdots \vee \overline{z_{n-1}})$. Figure 1 illustrates that this clause can be derived from $\exp(\Phi_n^0)$ in a constant number of resolution steps. Since $\exp(\mathcal{D}^{\mathrm{rrs}}(\mathrm{EQ}_{n-1}^0))$ has a linear-size resolution refutation by Theorem 19, so does $\exp(\Phi_n^0)$. $\qquad\square$

**Separation of** $\forall$Exp+Res$(\mathcal{D}^{\mathrm{rrs}})$ **and** $\forall$Exp+Res$(\mathcal{D}^{\mathrm{tf}})$. For our second separation, we introduce another DQBF family. This time, we refine the prefix of an existing modification of the equality formulas.

**Definition 21 ($EQ_n^1$ (adapted from [13])).** *For each natural number $n$,*

$$EQ_n^1 := \Pi_n^{EQ} \exists r(\emptyset) \exists s(\{u_1, \ldots, u_n\}) \cdot$$
$$\left(\psi_n^{EQ} \otimes (r \vee s)\right) \wedge \left(\psi_n^{EQ} \otimes (\overline{r} \vee \overline{s})\right) \wedge (r \vee \overline{s}) \wedge (\overline{r} \vee s).$$

The main idea is that the addition of the fresh variables $r$ and $s$ is enough to obfuscate all the spurious dependencies for $\mathcal{D}^{rrs}$. As such, preprocessing with $\mathcal{D}^{rrs}$ has no effect, and hardness can be proved via the ∀Exp+Res lower bound for $EQ_n^0$ (Theorem 17).

**Proposition 22.** *For each $n$, $\mathcal{D}^{rrs}(EQ_n^1) = EQ_n^1$.*

*Proof.* To prove the proposition, we must find sequences satisfying conditions (a) to (d) of Definition 8 with respect to both $(u_i, z_i), (u_i, s) \in \mathsf{deps}(EQ_n^1)$ for each $i$. In fact, for both pairs $(u_i, z_i)$ and $(u_i, s)$, the sequence of clauses

$$(r \vee x_i \vee u_i \vee z_i \vee s), (r \vee \overline{z_1} \vee \cdots \vee \overline{z_n} \vee s), (\overline{r} \vee \overline{x_i} \vee \overline{u_i} \vee z_i \vee \overline{s})$$

and the sequence of literals $z_i, s$ suffice.  □

**Theorem 23.** $\{EQ_n^1\}_{n \in \mathbb{N}}$ *requires exponential-size* ∀Exp+Res($\mathcal{D}^{rrs}$) *refutations.*

*Proof.* Consider the assignment $\alpha$ defined by $r \mapsto 0, s \mapsto 0$. It is easy to see that $EQ_n^1[\alpha] = EQ_n^0$. Now consider the 'expanded' assignment $\alpha^U$ defined by $r \mapsto 0, s^\sigma \mapsto 0$ for each $\sigma \in \langle\{u_1, \ldots, u_n\}\rangle$. It is less easy to see, but readily verified, that $\exp(EQ_n^1)[\alpha^U] = \exp(EQ_n^1[\alpha]) = \exp(EQ_n^0)$. Let $\pi$ be a ∀Exp+Res($\mathcal{D}^{rrs}$) refutation of $EQ_n^1$; that is, a resolution refutation of $\exp(\mathcal{D}^{rrs}(EQ_n^1))$. By Proposition 22, $\pi$ is a resolution refutation of $\exp(EQ_n^1)$. Since resolution is closed under restrictions, $\pi[\alpha^U]$ is a resolution refutation of $\exp(EQ_n^1)[\alpha^U] = \exp(EQ_n^0)$ with $|\pi[\alpha^U]| \leq |\pi|$. By Theorem 17, $2^n \leq |\pi[\alpha^U]| \leq |\pi|$.  □

The situation is quite different for the tautology-free dependency scheme $\mathcal{D}^{tf}$. Here, the simple detection of consecutive-clause tautologies in the variable $r$ is enough to identify all spurious dependencies, resulting in linear-size refutations.

**Proposition 24.** *For each $n$, the dependency sets of $\mathcal{D}^{tf}(EQ_n^1)$ are all empty.*

*Proof.* Aiming for contradiction, suppose that there exists a sequence of clauses $C_1, \ldots, C_k$ and a sequence of literals $p_1, \ldots, p_{k-1}$ satisfying conditions (a) to (e) of Definition 8 with respect to $(u_j, y) \in \mathsf{deps}(EQ_n^1)$, for some $y \in \{z_j, s\}$.

By condition (c), none of the $\mathsf{var}(p_i)$ is $r$. Hence, if some $C_i$ is either $(r \vee \overline{s})$ or $\overline{s} \vee r$, we must have $i = 1$ or $i = k$, violating condition (a). Therefore those clauses do not appear in the sequence. It follows that none of the $\mathsf{var}(p_i)$ is $s$, for otherwise we would have consecutive clauses $C_i$ and $C_{i+1}$ whose resolvent over $s$ contains complementary literals in $r$, violating condition (e).

Hence each $\mathsf{var}(p_i) = z_j$, and we must have $k = 2$, by conditions (c) and (d). Now we reach a contradiction as in the proof of Proposition 18, despite the addition of literals in $r$ and $s$. By condition (a), we have $u_i \in C_1$, and we deduce that $\overline{u_i} \notin C_2$, contradicting condition (a).  □

**Theorem 25.** $\{EQ_n^1\}_{n\in\mathbb{N}}$ *has linear-size* $\forall\mathsf{Exp+Res}(\mathcal{D}^{tf})$ *refutations.*

*Proof.* By Proposition 24, the total expansion of $\mathcal{D}^{tf}(EQ_n^1)$ is obtained by removing universal literals, hence $\exp(\mathcal{D}^{tf}(EQ_n^1))$ is the CNF

$$(r \vee \overline{z_1} \vee \cdots \vee \overline{z_n} \vee s) \wedge \bigwedge_{i=1}^{n} \left( (r \vee \overline{x_i} \vee z_i \vee s) \wedge (r \vee x_i \vee z_i \vee s) \right) \wedge (r \vee \overline{s}) \wedge$$

$$(\overline{r} \vee \overline{z_1} \vee \cdots \vee \overline{z_n} \vee \overline{s}) \wedge \bigwedge_{i=1}^{n} \left( (\overline{r} \vee \overline{x_i} \vee z_i \vee \overline{s}) \wedge (\overline{r} \vee x_i \vee z_i \vee \overline{s}) \right) \wedge (\overline{r} \vee s).$$

By resolution of $(r \vee \overline{s})$ over $r$ with each clause containing $\overline{r}$, and likewise of $(\overline{r} \vee s)$ with each clause containing $r$, we obtain all clauses in the CNF

$$(\exp(\mathcal{D}^{rrs}(EQ_n^0)) \otimes (s)) \wedge (\exp(\mathcal{D}^{rrs}(EQ_n^0)) \otimes (\overline{s})),$$

where $\exp(\mathcal{D}^{rrs}(EQ_n^0))$ is the CNF (2) from the proof of Theorem 19. By resolution over $s$ we obtain $\exp(\mathcal{D}^{rrs}(EQ_n^0))$ itself, which has a linear-size resolution refutation by Theorem 19. It is easy to see that the whole refutation of $\exp(\mathcal{D}^{tf}(EQ_n^1))$ is of linear size. $\qquad\square$

**Theorem 26.** $\forall\mathsf{Exp+Res}(\mathcal{D}^{rrs}) \not\leq_p^* \forall\mathsf{Exp+Res}(\mathcal{D}^{tf}).$

*Proof.* It is easy to see that the largest QBF $\varPhi_n^1$ that is smaller than $EQ_n^1$ has $S_s = \{u_1, \ldots, u_n\}$ and exactly one other non-empty dependency set $S_{z_i} = \{u_i\}$, where $i = n$ without loss of generality. We will prove that $\varPhi_n^1$ has linear-size $\forall\mathsf{Exp+Res}(\mathcal{D}^{rrs})$ refutations. We therefore prove the theorem, since by Theorems 23 and 25, and the monotonicity of $\forall\mathsf{Exp+Res}(\mathcal{D}^{rrs})$, $\{EQ_n^1\}_{n\in\mathbb{N}}$ satisfies all the conditions of Definition 7.

A $\forall\mathsf{Exp+Res}(\mathcal{D}^{rrs})$ refutation of $\varPhi_n^1$ is a $\forall\mathsf{Exp+Res}$ refutation of $\mathcal{D}^{rrs}(\varPhi_n^1)$. By definition, $\mathcal{D}^{rrs}(\varPhi_n^1) \leq \varPhi_n^1$. Now, $\varPhi_n^1$ has linear-size $\forall\mathsf{Exp+Res}$ refutations: it is readily verified that $\exp(EQ_n^1)$, which has linear-size resolution refutations, can be derived from $\exp(\varPhi_n^1)$ in a linear number of resolution steps. Hence $\mathcal{D}^{rrs}(\varPhi_n^1)$ has linear-size $\forall\mathsf{Exp+Res}$ refutations, by the monotonicity of $\forall\mathsf{Exp+Res}$; i.e. $\varPhi_n^1$ has linear-size $\forall\mathsf{Exp+Res}(\mathcal{D}^{rrs})$ refutations. $\qquad\square$

# 7   Tautology-Free Dependencies for QBF

We now turn our attention to dedicated QBF proof complexity, in particular to the QBF proof systems Q-Res($\mathcal{D}$) [35] that were introduced to model dependency-aware QBF solving. We show the following.

**Theorem 27.** Q-Res($\mathcal{D}^{tf}$) *is exponentially stronger than* Q-Res($\mathcal{D}^{rrs}$).

Since $\mathcal{D}^{rrs}$ was state-of-the-art for Q-Res($\mathcal{D}$), Theorem 27 shows that $\mathcal{D}^{tf}$ is currently the strongest known dependency scheme applicable to dependency-aware QBF solving. We recall the definition of the QBF proof system Q-Res($\mathcal{D}$).

**Definition 28** (Q-Res($\mathcal{D}$) [22,35]). *A* Q-Res *refutation of a DQBF $\Psi$ is a sequence $C_1, \ldots, C_k$ of clauses in which $C_k$ is empty and each $C_i$ is derived by one of the following rules:*

**A** *Axiom: $C_i$ is a non-tautological clause in the matrix of $\Psi$;*
**R** *Resolution: $C_i = A \vee B$, where $C_r = A \vee x$ and $C_s = B \vee \overline{x}$, for some $r, s < i$ and some $x \in \text{vars}_\exists(\Phi)$, and $C_i$ is not a tautology.*
**U** *Universal reduction: $C_i \vee a = C_r$ for some $r < i$ and some literal $a$ with $\text{var}(a) = u \in \text{vars}_\forall(\Psi)$ and $(u, x) \notin \text{deps}(\Psi)$ for each $x \in \text{vars}(C_i)$.*

*Given a QBF dependency scheme $\mathcal{D}$, a* Q-Res($\mathcal{D}$) *refutation of a QBF $\Phi$ is a* Q-Res *refutation of $\mathcal{D}(\Phi)$.*

Q-Res($\mathcal{D}^{\text{tf}}$) is complete for QBF by [22], and soundness follows by full exhibition.

**Theorem 29.** Q-Res($\mathcal{D}^{\text{tf}}$) *is a QBF proof system.*

**QBF Separation of Q-Res($\mathcal{D}^{\text{rrs}}$) and Q-Res($\mathcal{D}^{\text{tf}}$).** Our separating formulas are the QBFs on which our DQBF modification $\text{EQ}_n^1$ (Definition 21) was based. An exponential lower bound for these formulas in Q-Res($\mathcal{D}^{\text{rrs}}$) was shown in [13].

**Definition 30 ($\text{EQ}_n^2$ [13]).** *For each natural number $n$,*

$$\text{EQ}_n^2 := \exists r \exists x_1 \cdots \exists x_n \forall u_1 \cdots \forall u_n \exists z_1 \cdots z_n \exists s \cdot$$
$$\left( \psi_n^{\text{EQ}} \otimes (r \vee s) \right) \wedge \left( \psi_n^{\text{EQ}} \otimes (\overline{r} \vee \overline{s}) \right) \wedge (r \vee \overline{s}) \wedge (\overline{r} \vee s)$$

**Theorem 31 ([6]).** $\{\text{EQ}_n^2\}_{n \in \mathbb{N}}$ *requires exponential-size* Q-Res($\mathcal{D}^{\text{rrs}}$) *refutations.*

We show that $\text{EQ}_n^2$ have linear-size refutations in Q-Res($\mathcal{D}^{\text{tf}}$). The proof is along similar lines as our upper bound for $\text{EQ}_n^1$ in $\forall\text{Exp}+\text{Res}(\mathcal{D}^{\text{tf}})$. We first show that $\mathcal{D}^{\text{tf}}$ identifies the full set of spurious dependencies, which gives rise naturally to short refutations.

**Proposition 32.** *For each $n$, the dependency sets of $\mathcal{D}^{\text{tf}}(\text{EQ}_n^2)$ are all empty.*

*Proof.* Aiming for contradiction once again, suppose that there exists a sequence of clauses $C_1, \ldots, C_k$ and a sequence of literals $p_1, \ldots, p_{k-1}$ satisfying conditions (a) to (e) of Definition 8 with respect to $(u_i, y) \in \text{deps}(\text{EQ}_n^1)$, for some variable $y \in \{z_1, \ldots, z_n, s\}$.

As in the proof of Proposition 24, we can deduce that variables $r$ and $s$ do not appear in the sequence of literals. By condition (a) we have $u \in C_1$. By condition (c) we have $p_1 = z_i$ and $C_2 = (\overline{z_1} \vee \cdots \vee \overline{z_n} \vee a)$, with $\text{var}(a) = s$. By conditions (c) and (d), we have $p_2 = \overline{z_j}$ for some $j \neq i$. By condition (c) $z_j \in C_2$, and by conditions (c) and (d) we must have $k = 2$. This violates condition (a), since $\overline{u_i} \notin C_2$. $\qquad\square$

**Theorem 33.** $\{\text{EQ}_n^2\}_{n \in \mathbb{N}}$ *has linear-size* Q-Res($\mathcal{D}^{\text{tf}}$) *refutations.*

*Proof.* By Proposition 32, $\mathsf{deps}(\mathcal{D}^{\mathrm{tf}}(\mathrm{EQ}_n^2))$ is the empty relation. It follows that all universal literals in the matrix may be removed by universal reduction. Hence, with a single axiom and universal reduction step per clause, we derive the clauses of $\exp(\mathcal{D}^{\mathrm{tf}}(\mathrm{EQ}_n^1))$ from the proof of Theorem 25. Each step of the linear-size resolution refutation described there is also available in $\mathsf{Q\text{-}Res}(\mathcal{D}^{\mathrm{tf}})$. □

## 8 Conclusions

We conclude with an interesting observation and a question for future research. The family $\{\mathrm{EQ}_n^0\}_{n\in\mathbb{N}}$ from Definition 16 is an adaptation of the equality QBFs $\{\mathrm{EQ}_n\}_{n\in\mathbb{N}}$ from [8], obtained by shrinking the dependency set of each $z_i$ to just $\{u_i\}$. While in QBF $\{\mathrm{EQ}_n\}_{n\in\mathbb{N}}$ requires exponentially long proofs in both $\forall\mathsf{Exp}{+}\mathsf{Res}$ and $\mathsf{Q\text{-}Res}$ [5,8], in DQBF $\{\mathrm{EQ}_n^0\}_{n\in\mathbb{N}}$ remains hard only for $\forall\mathsf{Exp}{+}\mathsf{Res}$. Indeed, even though $\mathsf{Q\text{-}Res}$ is incomplete for DQBF, it is sound, and $\{\mathrm{EQ}_n^0\}_{n\in\mathbb{N}}$ has linear-size $\mathsf{Q\text{-}Res}$ refutations. This suggests that there may be some hidden proof-complexity relationship between $\forall\mathsf{Exp}{+}\mathsf{Res}$ and $\mathsf{Q\text{-}Res}$ in DQBF, even though $\mathsf{Q\text{-}Res}$ is incomplete there.

We have presented the strongest known dependency scheme $\mathcal{D}^{\mathrm{tf}}$. A natural question is whether some even stronger dependency schemes for (D)QBF exist.

## References

1. Azhar, S., Peterson, G., Reif, J.: Lower bounds for multiplayer non-cooperative games of incomplete information. J. Comput. Math. Appl. **41**, 957–992 (2001)
2. Balabanov, V., Jiang, J.R.: Unified QBF certification and its applications. Formal Methods Syst. Des. **41**(1), 45–65 (2012)
3. Balabanov, V., Widl, M., Jiang, J.-H.R.: QBF resolution systems and their proof complexities. In: Sinz, C., Egly, U. (eds.) SAT 2014. LNCS, vol. 8561, pp. 154–169. Springer, Cham (2014). https://doi.org/10.1007/978-3-319-09284-3_12
4. Beyersdorff, O., Blinkhorn, J.: Dependency schemes in qbf calculi: semantics and soundness. In: Rueher, M. (ed.) CP 2016. LNCS, vol. 9892, pp. 96–112. Springer, Cham (2016). https://doi.org/10.1007/978-3-319-44953-1_7
5. Beyersdorff, O., Blinkhorn, J.: Genuine lower bounds for QBF expansion. In: Niedermeier, R., Vallée, B. (eds.) International Symposium on Theoretical Aspects of Computer Science (STACS). Leibniz International Proceedings in Informatics (LIPIcs), vol. 96, pp. 12:1–12:15. Schloss Dagstuhl - Leibniz-Zentrum für Informatik (2018)
6. Beyersdorff, O., Blinkhorn, J.: Dynamic QBF dependencies in reduction and expansion. ACM Trans. Comput. Logic **21**(2), 1–27 (2019)
7. Beyersdorff, O., Blinkhorn, J., Chew, L., Schmidt, R.A., Suda, M.: Reinterpreting dependency schemes: soundness meets incompleteness in DQBF. J. Autom. Reason. **63**(3), 597–623 (2019)
8. Beyersdorff, O., Blinkhorn, J., Hinde, L.: Size, cost, and capacity: a semantic technique for hard random QBFs. Logical Methods Comput. Sci. **15**(1), (2019)
9. Beyersdorff, O., Bonacina, I., Chew, L.: Lower bounds: from circuits to QBF proof systems. In: Sudan, M. (ed.) ACM Conference on Innovations in Theoretical Computer Science (ITCS), pp. 249–260. ACM (2016)

10. Beyersdorff, O., Chew, L., Janota, M.: New resolution-based QBF calculi and their proof complexity. ACM Trans. Comput. Theory **11**(4), 26:1–26:42 (2019)
11. Beyersdorff, O., Chew, L., Schmidt, R.A., Suda, M.: Lifting QBF resolution calculito DQBF. In: Creignou and Berre [16], pp. 490-499
12. Beyersdorff, O., Hinde, L., Pich, J.: Reasons for hardness in QBF proof systems. ACM Trans. Comput. Theory **12**(2), 10:1–10:27 (2020)
13. Blinkhorn, J., Beyersdorff, O.: Shortening QBF proofs with dependency schemes. In: Gaspers, S., Walsh, T. (eds.) SAT 2017. LNCS, vol. 10491, pp. 263–280. Springer, Cham (2017). https://doi.org/10.1007/978-3-319-66263-3_17
14. Chen, H.: Proof complexity modulo the polynomial hierarchy: Understanding alternation as a source of hardness. ACM Trans. Comput. Theory **9**(3), 15:1–15:20 (2017)
15. Cook, S.A., Reckhow, R.A.: The relative efficiency of propositional proof systems. J. Symbol. Logic **44**(1), 36–50 (1979)
16. Creignou, N., Le Berre, D. (eds.) International Conference on Theory and Practice of Satisfiability Testing (SAT). LNCS, vol. 9710. Springer, Cham (2016). https://doi.org/10.1007/978-3-319-40970-2
17. Egly, U., Kronegger, M., Lonsing, F., Pfandler, A.: Conformant planning as a case study of incremental QBF solving. Ann. Math. Artif. Intell. **80**(1), 21–45 (2017)
18. Henkin, L.: Some remarks on infinitely long formulas. Infinitistic Methods, pp. 167–183 (1961)
19. Hooker, J. (ed.): International Conference on Principles and Practice of Constraint Programming (CP). LNCS, vol. 11008. Springer, Cham (2018). https://doi.org/10.1007/978-3-319-98334-9
20. Hoos, H.H., Peitl, T., Slivovsky, F., Szeider, S.: portfolio-based algorithm selection for circuit QBFs. In: Hooker [19], pp. 195–209
21. Janota, M., Marques-Silva, J.: Expansion-based QBF solving versus Q-resolution. Theor. Comput. Sci. **577**, 25–42 (2015)
22. Kleine Büning, H., Karpinski, M., Flögel, A.: Resolution for quantified Boolean formulas. Inf. Comput. **117**(1), 12–18 (1995)
23. Kontchakov, R., et al.: Minimal module extraction from DL-lite ontologies using QBF solvers. In: Boutilier, C. (ed.) International Joint Conference on Artificial Intelligence (IJCAI), pp. 836–841. AAAI Press (2009)
24. Lonsing, F., Egly, U.: DepQBF 6.0: a search-based QBF solver beyond traditional QCDCL. In: de Moura, L. (ed.) CADE 2017. LNCS (LNAI), vol. 10395, pp. 371–384. Springer, Cham (2017). https://doi.org/10.1007/978-3-319-63046-5_23
25. Lonsing, F., Egly, U.: Evaluating QBF solvers: Quantifier alternations matter. In: Hooker [19], pp. 276–294
26. Mangassarian, H., Veneris, A.G., Benedetti, M.: Robust QBF encodings for sequential circuits with applications to verification, debug, and test. IEEE Trans. Comput. **59**(7), 981–994 (2010)
27. Peitl, T., Slivovsky, F., Szeider, S.: Combining resolution-path dependencies with dependency learning. In: International Conference on Theory and Practice of Satisfiability Testing (SAT), pp. 306–318 (2019)
28. Peitl, T., Slivovsky, F., Szeider, S.: Dependency learning for QBF. J. Artif. Intell. Res. **65**, 181–208 (2019)
29. Peitl, T., Slivovsky, F., Szeider, S.: Long-distance Q-resolution with dependency schemes. J. Autom. Reason. **63**(1), 127–155 (2019)
30. Pulina, L., Seidl, M.: The 2016 and 2017 QBF solvers evaluations (QBFEVAL'16 and QBFEVAL'17). Artif. Intell. **274**, 224–248 (2019)

31. Robinson, J.A.: A machine-oriented logic based on the resolution principle. J. ACM **12**(1), 23–41 (1965)

32. Samer, M., Szeider, S.: Backdoor sets of quantified Boolean formulas. J. Autom. Reason. **42**(1), 77–97 (2009)

33. Scholl, C., Wimmer, R.: Dependency quantified boolean formulas: an overview of solution methods and applications. In: Beyersdorff, O., Wintersteiger, C.M. (eds.) SAT 2018. LNCS, vol. 10929, pp. 3–16. Springer, Cham (2018). https://doi.org/10.1007/978-3-319-94144-8_1

34. Slivovsky, F.: Structure in #SAT and QBF. Ph.D. thesis, Vienna University of Technology (2015)

35. Slivovsky, F., Szeider, S.: Soundness of Q-resolution with dependency schemes. Theor. Comput. Sci. **612**, 83–101 (2016)

36. Vardi, M.Y.: Boolean satisfiability: theory and engineering. Commun. ACM **57**(3), 5 (2014)

37. Wimmer, R., Scholl, C., Wimmer, K., Becker, B.: Dependency schemes for DQBF. In: Creignou and Berre [16], pp. 473–489

# Short Q-Resolution Proofs
# with Homomorphisms

Ankit Shukla[1], Friedrich Slivovsky[2], and Stefan Szeider[2(✉)]

[1] JKU, Linz, Austria
ankit.shukla@jku.at
[2] Algorithms and Complexity Group, TU Wien, Vienna, Austria
{fs,sz}@ac.tuwien.ac.at

**Abstract.** We introduce new proof systems for quantified Boolean formulas (QBFs) by enhancing Q-resolution systems with rules which exploit local and global symmetries. The rules are based on homomorphisms that admit non-injective mappings between literals. This results in systems that are stronger than Q-resolution with (injective) symmetry rules. We further strengthen the systems by utilizing a dependency system $D$ in a way that surpasses $Q(D)$-resolution in relative strength.

**Keywords:** Symmetries · Homomorphisms · QBF · $Q(D)$-Resolution · Dependency schemes · Proof complexity

## 1 Introduction

In a 1985 paper, Krishnamurthy [12] introduced *symmetry rules* which strengthen the propositional resolution system to admit exponentially shorter proofs, for instance, linearly sized proofs for the Pigeon Hole Principle. The *global symmetry rule* exploits the automorphisms of the entire input formula. The even stronger *local symmetry rule* exploits the existence of isomorphic images of subsets of clauses within the formula. Szeider [19] further strengthened Krishnamurthy's proof systems, generalizing the symmetry rules to *homomorphism rules* by considering clause-preserving mappings that are not necessarily injective.

Recently, Kauers and Seidl [11] lifted Krishnamurthy's most basic symmetry rule, the global symmetry rule, to *Q-resolution* (Q-Res), the standard resolution-based proof system for quantified Boolean formulas (QBFs) in prenex conjunctive normal form (PCNF). They showed that several families of formulas that require exponentially-sized Q-resolution proofs admit polynomially-sized proofs if the generalized symmetry rule is added.

Our main contribution is the introduction and study of proof systems based on Q-resolution that are even stronger than the one studied by Kauers and

The authors acknowledge the support by the Austrian Science Funds (FWF), projects W1255 and P32441, and by the Vienna Science and Technology Fund (WWTF), projects ICT19-060 and ICT19-065.

L. Pulina and M. Seidl (Eds.): SAT 2020, LNCS 12178, pp. 412–428, 2020.
https://doi.org/10.1007/978-3-030-51825-7_29

Seidl [11]: we lift the local symmetry rule to the quantified setting, as well as the local and global homomorphism rules. A straightforward lifting of the rules from the propositional case to the quantified case insists that the mapping between literals on which the symmetries (or more generally, homomorphisms) operate does not jump between quantifier blocks. A more general version allows a jump between quantifier blocks, as long as the relative position of the variables in the quantifier prefix is preserved. We go even a step further, and parameterize our systems by a *dependency scheme* $D$ [15], and only require the mapping to preserve dependencies according to the chosen dependency scheme. Thus, our systems strengthen Kauers and Seidl's system along three dimensions:

1. from global symmetries to local symmetries,
2. from symmetries to homomorphisms, and
3. from quantifier-block preserving mappings to dependencies preserving mappings with respect to a dependency system.

Each of the three dimensions alone provides an exponential speedup.

Figure 1 gives an overview of the proof systems considered in this paper. In the figure, $D$ stands for the reflexive resolution-path dependency scheme [18], a variant of the resolution-path dependency scheme [17,20], or any stronger dependency scheme. The separations between LH and LS, LH and GH, LS and GS, GH and GS, and GS and Q-Res, as well as the corresponding separations between the systems using a dependency scheme $D$, follows from the propositional case [19].

We show an exponential separation between LH($D$) and LH for the reflexive resolution-path dependency scheme (Theorem 3). This result also provides separations between LS($D$) and LS, GH($D$) and GH, and GS($D$) and GS (see the legend in Fig. 1 for definitions).

## 2   Preliminaries

*Formulas and Assignments.* A *literal* is a negated or unnegated variable. If $x$ is a variable, we write $\overline{x} = \neg x$ and $\overline{\neg x} = x$, and let $var(x) = var(\neg x) = x$. We sometimes call literals $x$ and $\neg x$ the positive and negative *polarity* of variable $x$. If $X$ is a set of literals, we write $\overline{X}$ for the set $\{\,\overline{x} : x \in X\,\}$. A *clause* is a finite disjunction of literals, and a *term* is a finite conjunction of literals. We call a clause *tautological* if it contains the same variable negated as well as unnegated. A *CNF formula* is a finite conjunction of non-tautological clauses. Whenever convenient, we treat clauses and terms as sets of literals, and CNF formulas as sets of sets of literals. We write $var(S)$ for the set of variables occurring (negated or unnegated) in a clause or term $S$, that is, $var(S) = \{\,var(\ell) : \ell \in S\,\}$. Moreover, we let $var(\phi) = \bigcup_{C \in \phi} var(C)$ denote the set of variables occurring in a CNF formula $\phi$.

A *truth assignment* (or simply *assignment*) to a set $X$ of variables is a mapping $\tau : X \to \{0,1\}$. We write $[X]$ for the set of truth assignments to $X$, and extend $\tau : X \to \{0,1\}$ to literals by letting $\tau(\neg x) = 1 - \tau(x)$ for $x \in X$. Let $\tau : X \to \{0,1\}$ be a truth assignment. The restriction $C[\tau]$ of a clause (term)

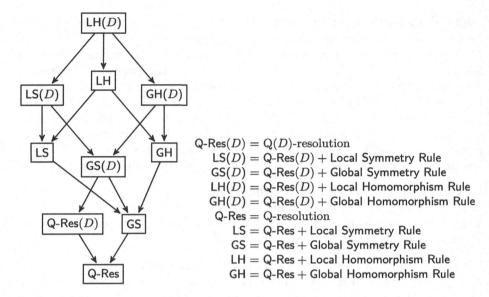

$$Q\text{-Res}(D) = Q(D)\text{-resolution}$$
$$LS(D) = Q\text{-Res}(D) + \text{Local Symmetry Rule}$$
$$GS(D) = Q\text{-Res}(D) + \text{Global Symmetry Rule}$$
$$LH(D) = Q\text{-Res}(D) + \text{Local Homomorphism Rule}$$
$$GH(D) = Q\text{-Res}(D) + \text{Global Homomorphism Rule}$$
$$Q\text{-Res} = Q\text{-resolution}$$
$$LS = Q\text{-Res} + \text{Local Symmetry Rule}$$
$$GS = Q\text{-Res} + \text{Global Symmetry Rule}$$
$$LH = Q\text{-Res} + \text{Local Homomorphism Rule}$$
$$GH = Q\text{-Res} + \text{Global Homomorphism Rule}$$

**Fig. 1.** Proof system map. $A \to B$ indicates that system $A$ p-simulates system $B$, but $B$ cannot p-simulate $A$.

$S$ by $\tau$ is defined as follows: if there is a literal $\ell \in S \cap (X \cup \overline{X})$ such that $\tau(\ell) = 1$ $(\tau(\ell) = 0)$ then $S[\tau] = 1$ $(S[\tau] = 0)$. Otherwise, $S[\tau] = S \setminus (X \cup \overline{X})$. The restriction $\phi[\tau]$ of a CNF formula $\phi$ by the assignment $\tau$ is defined $\phi[\tau] = \{\, C[\tau] : C \in \phi, C[\tau] \neq 1 \,\}$.

*PCNF Formulas.* A *PCNF formula* is denoted by $\Phi = \mathcal{Q}.\phi$, where $\phi$ is a CNF formula and $\mathcal{Q} = Q_1 X_1 \ldots Q_n X_n$ is a sequence such that $Q_i \in \{\forall, \exists\}$, $Q_i \neq Q_{i+1}$ for $1 \leq i < n$, and the $X_i$ are pairwise disjoint sets of variables. We call $\phi$ the *matrix* of $\Phi$ and $\mathcal{Q}$ the *(quantifier) prefix* of $\Phi$, and refer to the $X_i$ as *quantifier blocks*. We require that $var(\phi) = X_1 \cup \cdots \cup X_n$ and write $var(\Phi) = var(\phi)$. We define a partial order $<_\Phi$ on $var(\phi)$ as $x <_\Phi y \Leftrightarrow x \in X_i, y \in X_j, i < j$. We extend $<_\Phi$ to a relation on literals in the obvious way and drop the subscript whenever $\Phi$ is understood. For $x \in var(\Phi)$ we let $R_\Phi(x) = \{\, y \in var(\Phi) : x <_\Phi y \,\}$ and $L_\Phi(x) = \{\, y \in var(\Phi) : y <_\Phi x \,\}$ denote the sets of variables *to the right* and *to the left* of $x$ in $\Phi$, respectively. Relative to the PCNF formula $\Phi$, variable $x$ is called *existential* (*universal*) if $x \in X_i$ and $Q_i = \exists$ $(Q_i = \forall)$. The set of existential (universal) variables occurring in $\Phi$ is denoted $var_\exists(\Phi)$ $(var_\forall(\Phi))$. We define the set $\mathsf{lit}_\exists(\Phi)$ $(\mathsf{lit}_\forall(\Phi))$ as a set of all existential (universal) literals corresponding to $var_\exists(\Phi)$ $(var_\forall(\Phi))$, i.e., if $x \in var_\exists(\Phi)$ then both $x, \neg x \in \mathsf{lit}_\exists(\Phi)$ (resp. for the universal variable). The *length* of a PCNF formula $\Phi = \mathcal{Q}.\phi$ is given by its cardinality $|\Phi|$; the number of clauses in the matrix. The *size* of a PCNF formula $\Phi = \mathcal{Q}.\phi$ is defined as $\|\Phi\| = \sum_{C \in \phi} |C|$. If $\tau$ is an assignment, then $\Phi[\tau]$

denotes the PCNF formula $\mathcal{Q}'.\phi[\tau]$, where $\mathcal{Q}'$ is the quantifier prefix obtained from $\mathcal{Q}$ by deleting variables that do not occur in $\phi[\tau]$. *True* and *false* PCNF formulas are defined in the usual way.

*Proof Systems.* A *proof* of a formula $F$ is a finite object $x$ which certifies falsity of $F$ in the sense that, if $x$ is given, then falsity of $F$ can be verified in polynomial time (proofs of falsity are also called *refutations*). A *proof system* $\Pi$ is a set of proofs such that (i) elements of $\Pi$ can be recognized in polynomial time, and (ii) a formula $F$ is false if and only if $\Pi$ contains a proof of $F$.

Let $\Pi$, $\Pi'$ be proof systems. We say that $\Pi'$ *p-simulates* $\Pi$ if every proof $x \in \Pi$ can be transformed into a proof $x' \in \Pi'$ in polynomial time such that $x$ and $x'$ prove the same formula. If $\Pi$ and $\Pi'$ p-simulate each other, then we say that they are *p-equivalent*.

*Q-Resolution.* Q-resolution is a generalization of propositional resolution to PCNF formulas [6]. Q-resolution is of practical interest due to its relation to search-based QBF solvers that implement Quantified Conflict Driven Constraint Learning (QCDCL) [7,21]: the traces of QCDCL solvers correspond to Q-resolution proofs [9,10].

Q-resolution proof system consists of propositional resolution and the *universal reduction* rule for dealing with universally quantified variables. This system (Fig. 2) was shown to be sound and complete for false PCNF formulas [6].

$$\frac{}{C}\ \text{(input clause)} \qquad \frac{C_1 \vee e \qquad \neg e \vee C_2}{C_1 \vee C_2}\ \text{(resolution)}$$

An *input clause* $C \in \varphi$ can be used as an axiom. From two clauses $C_1 \vee e$ and $\neg e \vee C_2$, where $e$ is an existential variable, the *resolution* rule can derive the clause $C_1 \vee C_2$, provided that it is non-tautological. Here, $e$ is an existential variable called the *pivot*.

$$\frac{C}{C \setminus \{u, \neg u\}}\ \text{($\forall$-reduction)}$$

The $\forall$-*reduction* rule derives the clause $C$ from $C \vee l$ if $var(l)$ is universal and there is no existential variable $e \in var(C)$ with $l <_\Phi e$.

**Fig. 2.** Derivation rules of Q-resolution for a PCNF formula $\Phi = \mathcal{Q}.\phi$.

# 3   Dependency Schemes and Q($D$)-Resolution

QCDCL generalizes the well-known DPLL procedure [8] from SAT to QSAT. In essence, DPLL is a recursive algorithm that picks a variable of its input formula and calls itself for both possible instantiations of that variable. Modern SAT solvers derived from the DPLL algorithm, delegate the choice of which variable to branch on to clever heuristics [16].

In QCDCL, the quantifier prefix imposes constraints on the order of variable assignments: a variable may be assigned only if it occurs in the leftmost quantifier block with unassigned variables. Often, this is more restrictive than necessary. For instance, variables from disjoint subformulas may be assigned in any order. Intuitively, a variable can be assigned as long as it *does not depend* on any unassigned variable. This is the intuition underlying a generalization of QCDCL implemented in the solver DepQBF [13,14]. Dependency schemes are mappings that associate every PCNF formula with a binary relation on its variables that refines the order of variables in the quantifier prefix.[1]

**Definition 1 (Dependency Scheme).** A *dependency scheme* is a mapping $D$ that associates each PCNF formula $\Phi$ with a relation $D_\Phi \subseteq \{ (x,y) : x <_\Phi y \}$ called the *dependency relation* of $\Phi$ with respect to $D$.

The mapping which simply returns the prefix ordering of an input formula can be thought of as a baseline dependency scheme:

**Definition 2 (Trivial Dependency Scheme).** The *trivial dependency scheme* $D^{trv}$ associates each PCNF formula $\Phi$ with the relation $D_\Phi^{trv} = \{ (x,y) : x <_\Phi y \}$.

DepQBF uses a dependency relation to determine the order in which variables can be assigned: if $y$ is a variable and there is no unassigned variable $x$ such that $(x,y)$ is in the dependency relation, then $y$ is considered ready for assignment. DepQBF also uses the dependency relation to generalize the $\forall$-reduction rule used in clause learning [14]. As a result of its use of dependency schemes, DepQBF generates proofs in a generalization of Q-resolution called $Q(D)$-resolution [18], a proof system that takes a dependency scheme $D$ as a parameter.

Dependency schemes can be partially ordered based on their dependency relations: if the dependency relation computed by a dependency scheme $D_1$ is a subset of the dependency relation computed by a dependency scheme $D_2$ for each PCNF formula, then $D_1$ is *more general* than $D_2$. The more general a dependency scheme, the more freedom a solver has in choosing decision variables. Currently, (aside from the trivial dependency scheme) DepQBF supports (a refined version [13, p.49] of) the *standard dependency scheme* [15]. We will work with the more general *reflexive resolution-path dependency scheme* [18], a variant of the resolution-path dependency scheme [17,20]. This dependency scheme computes an overapproximation of variable dependencies based on whether two variables are connected by a (pair of) resolution path(s).

**Definition 3 (Resolution Path).** Let $\Phi = Q.\phi$ be a PCNF formula and let $X$ be a set of variables. A *resolution path* (from $\ell_1$ to $\ell_{2k}$) via $X$ (in $\Phi$) is a sequence $\ell_1, \ldots, \ell_{2k}$ of literals satisfying the following properties:

---

[1] The original definition of dependency schemes [15] is more restrictive than the one given here, but the additional requirements are irrelevant for the purposes of this paper.

1. For all $i \in [k]$, there is a $C_i \in \phi$ such that $\ell_{2i-1}, \ell_{2i} \in C_i$.
2. For all $i \in [k]$, $var(\ell_{2i-1}) \neq var(\ell_{2i})$.
3. For all $i \in [k-1]$, $\{\ell_{2i}, \ell_{2i+1}\} \subseteq X \cup \overline{X}$.
4. For all $i \in [k-1]$, $\overline{\ell_{2i}} = \ell_{2i+1}$.

If $\pi = \ell_1, \ldots, \ell_{2k}$ is a resolution path in $\Phi$ via $X$, then we say that $\ell_1$ and $\ell_{2k}$ are *connected in* $\Phi$ (with respect to $X$). For every $i \in \{1, \ldots, k-1\}$, we say that $\pi$ *goes through* $var(\ell_{2i})$.

One can think of a resolution path as a potential chain of implications: if each clause $C_i$ contains exactly two literals, then assigning $\ell_1$ to 0 requires setting $\ell_{2k}$ to 1. If, in addition, there is such a path from $\overline{\ell_1}$ to $\overline{\ell_{2k}}$, then $\ell_1$ and $\ell_{2k}$ have to be assigned opposite values. Accordingly, the resolution path dependency scheme identifies variables connected by a pair of resolution paths as potentially dependent on each other.

**Definition 4 (Dependency Pair).** Let $\Phi$ be a PCNF formula and $x, y \in var(\Phi)$. We say $\{x, y\}$ is a *resolution-path dependency pair* of $\Phi$ with respect to $X \subseteq var_\exists(\Phi)$ if at least one of the following conditions holds:

- $x$ and $y$, as well as $\neg x$ and $\neg y$, are connected in $\Phi$ with respect to $X$.
- $x$ and $\neg y$, as well as $\neg x$ and $y$, are connected in $\Phi$ with respect to $X$.

**Definition 5.** The *reflexive resolution-path dependency scheme* is the mapping $D^{rrs}$ that assigns to each PCNF formula $\Phi = \mathcal{Q}.\phi$ the relation $D^{rrs}_\Phi = \{ x <_\Phi y : \{x, y\}$ is a resolution-path dependency pair in $\Phi$ with respect to $R_\Phi(x) \setminus var_\forall(\Phi) \}$.

The derivation rules of Q($D$)-resolution are shown in Fig. 3. Here, as in the rest of the paper, $D$ denotes an arbitrary dependency scheme.

---

$$\frac{}{C} \text{ (input clause)} \qquad \frac{C_1 \vee e \qquad \neg e \vee C_2}{C_1 \vee C_2} \text{ (resolution)}$$

An *input clause* $C \in \varphi$ can be used as an axiom. From two clauses $C_1 \vee e$ and $\neg e \vee C_2$, where $e$ is an existential variable, the *resolution* rule can derive the clause $C_1 \vee C_2$, provided that $C_1 \vee C_2$ does not contain a universal variable in both polarities.

$$\frac{C}{C \setminus \{u, \neg u\}} \text{ (}\forall\text{-reduction)}$$

The $\forall$*-reduction* rule derives the clause $C \setminus \{u, \neg u\}$ from $C$, where $u \in var(C)$ is a universal variable such that $(u, e) \notin D_\Phi$ for every existential variable $e \in var(C)$.

---

**Fig. 3.** Derivation rules of Q($D$)-resolution for a PCNF formula $\Phi = \mathcal{Q}.\phi$.

A derivation in a proof system consists of repeated applications of the derivation rules to derive a clause from the clauses of an input formula. A sequence

$S = C_1, \ldots, C_k$ of clauses is a $Q(D)$-resolution derivation of $C_k$ from a PCNF formula $\Phi = \mathcal{Q}.\phi$ if for each $i \in \{1, \ldots, k\}$ at least one of the following holds

1. $C_i \in \phi$ ($C_i$ is an axiom).
2. $C$ can be derived from $C_1$ and $C_2$ by the resolution rule.
3. $C$ can be derived from $C'$ by $\forall$-reduction with respect to the dependency scheme D.

The size $|S|$ of a derivation $S$ is the number $k$ of clauses in the sequence. A *refutation* is a derivation of the empty clause.

**Proposition 1 (Slivovsky and Szeider** [18]**).** $Q(D^{rrs})$*-resolution is a complete proof system for false formulas; i.e., a PCNF formula is false if, and only if, there exists a* $Q(D^{rrs})$*-resolution refutation of it.*

**Definition 6 (Equality formulas** [3]**).** For every $n \in \mathbb{N}$, the $n^{\text{th}}$ equality formula is

$$EQ(n) := \exists x_1 \ldots x_n \forall u_1 \ldots u_n \exists t_1 \ldots t_n. \bigwedge_{i=1}^{n} \left( (x_i \vee u_i \vee \overline{t_i}) \wedge (\overline{x_i} \vee \overline{u_i} \vee \overline{t_i}) \right) \wedge \bigvee_{i=1}^{n} t_i.$$

For every $n \in \mathbb{N}$, the formula $EQ(n)$ is *false*, and any Q-resolution refutation of $EQ(n)$ has size exponential in $n$ [3].

## 4    Homomorphisms

For a finite set $L \subseteq$ lit of literals, a mapping $\rho : L \to$ lit is a *renaming* if $\overline{\rho(\ell)} = \rho(\overline{\ell})$ for every pair $\ell, \overline{\ell} \in L$ of clashing literals in the domain of $\rho$. We generalize renamings to clauses and formulas in the obvious way. For a clause $C$, the image $\rho(C)$ under renaming may be tautological. We define $\rho_{cls}(\phi)$ as the set of all non-tautological $\rho(C)$ with $C \in \phi$. In the propositional case, the image of a resolution derivation under a renaming contains a resolution derivation. For QBFs, we have to take variable dependencies induced by the quantifier prefix into account to make sure universal reduction steps are applicable in the image. We define a notion of renaming that imposes additional restrictions to ensure that the image of a $Q(D)$-resolution derivation is again a $Q(D)$-resolution derivation.

**Definition 7.** Let $\Phi_1 = \mathcal{Q}_1.\phi_1$ and $\Phi_2 = \mathcal{Q}_2.\phi_2$ be PCNF formulas and let $D$ be a dependency scheme for which $Q(D)$-resolution is sound. For any $L \subseteq \text{lit}(\Phi_1)$, a renaming $\rho : L \to \text{lit}(\Phi_2)$ is a *D-renaming* from $\Phi_1$ to $\Phi_2$ if it satisfies the following conditions:

1. For every $\ell \in L$, $qtype_{\Phi_1}(\ell) = qtype_{\Phi_2}(\rho(\ell))$.
2. If $\rho(\ell) = \rho(\ell')$ and $qtype_{\Phi_1}(\ell) = \forall$, then $\ell = \ell'$.
3. If $\ell, \ell' \in L$ and $(var(\rho(\ell)), var(\rho(\ell'))) \in D_{\Phi_2}$ then $(var(\ell), var(\ell')) \in D_{\Phi_1}$.

This definition satisfies several desiderata. First, in the absence of universal variables, it boils down to a previously defined notion of homomorphisms of propositional formulas in CNF. Second, it enables us to transfer $Q(D)$-resolution derivations, as stated in the following lemma.

**Lemma 1.** *Let $\Phi_1 = Q_1.\phi_1$ and $\Phi_2 = Q_2.\phi_2$ be PCNF formulas and let $D$ be a dependency scheme for which $Q(D)$-resolution is sound. If $C_1, \ldots, C_k$ is a $Q(D)$-resolution derivation from $\psi \subseteq \phi_1$ in $\Phi_1$ and $\rho : \mathsf{lit}(\psi) \to \mathsf{lit}(\Phi_2)$ is a $D$-renaming such that $\rho(C_k)$ is non-tautological and $\rho_{cls}(\psi) \subseteq \phi_2$, then $\rho(C_1), \ldots, \rho(C_k)$ contains a $Q(D)$-resolution derivation of $C_k' \subseteq \rho(C_k)$ from clauses $\rho_{cls}(\psi)$ in $\Phi_2$.*

*Proof.* We proceed by induction on the length $k$ of the derivation and distinguish three cases. First, if $C_k \in \phi_1$ is an initial clause and $\rho(C_k)$ is non-tautological then $\rho(C_k) \in \phi_2$ by assumption. Second, if $C_k$ is derived from clause $C_i$ with $1 \le i < k$ by universal reduction, then $C_k = C_i \setminus \{\ell_u\}$ for a universal literal $\ell_u$ with $var(\ell_u) = u$, and $(u, e) \notin D_{\Phi_1}$ for every existential variable $e$ occurring in $C_i$. We argue that $\rho(C_i)$ is non-tautological. Towards a contradiction assume that $\rho(C_i)$ is tautological. Since $\rho(C_k)$ is assumed to be non-tautological the only way for $\rho(C_i) = \rho(C_k) \cup \{\rho(\ell_u)\}$ to be tautological is that $\overline{\rho(\ell_u)} \in \rho(C_i)$. Because $\rho$ preserves quantifier types and $C_i$ is non-tautological, there must be a universal literal $\ell' \in C_i$ with $var(\ell') \ne u$ such that $\rho(\ell') = \overline{\rho(\ell_u)}$. That means $\rho(\overline{\ell'}) = \rho(\ell_u)$ and thus $\ell' = \ell_u$ by Property 2 and in particular $var(\ell') = var(\ell_u) = u$, a contradiction. Thus $\rho(C_i)$ is non-tautological and we can apply the induction hypothesis to conclude that $\rho(C_1), \ldots, \rho(C_i)$ contains a $Q(D)$-resolution derivation of $C_i' \subseteq \rho(C_i)$ from $\rho_{cls}(\psi)$ in $\Phi_2$. By Property 1, every existential literal in $C_i' \subseteq \rho(C_i)$ is the image of an existential literal in $C_i$. Property 3 ensures that $(\rho(u), \rho(e)) \notin D_{\Phi_2}$ for every existential variable $\rho(e)$ occurring in $\rho(C_i)$, so a clause $C_k' \subseteq \rho(C_k)$ can be obtained from $C_i' \subseteq \rho(C_i)$ by universal reduction.

Finally, let $C_k$ be derived by resolution on pivot variable $e$ from $C_i$ and $C_j$ with $1 \le i < j < k$. Assume without loss of generality that $e \in C_i$ and $\neg e \in C_j$, so that $C_i \subseteq C_k \cup \{e\}$ and $C_j \subseteq C_k \cup \{\neg e\}$. If $\rho(C_i)$ and $\rho(C_j)$ are both non-tautological we can apply the induction hypothesis to obtain $Q(D)$-resolution derivations of clauses $C_i' \subseteq \rho(C_i)$ and $C_j' \subseteq \rho(C_j)$ from $\Phi_2$. If the pivot variable is contained in both clauses we obtain $C_k' \subseteq \rho(C_k)$ by resolution, otherwise we choose as $C_k'$ one among the clauses $C_i'$ and $C_j'$ that does not contain the pivot. Otherwise, since $\rho(C_k)$ is non-tautological, the clause $\rho(C_i) \subseteq \rho(C_k) \cup \{\rho(e)\}$ can be tautological only if there is a literal $\ell \in C_i$ such that $\rho(\ell) = \rho(\neg e)$. Symmetrically, the clause $\rho(C_j) \subseteq \rho(C_k) \cup \{\rho(\neg e)\}$ can be tautological only if there is a literal $\ell' \in C_j$ such that $\rho(\ell') = \rho(e)$. It follows that at most one of $\rho(C_i)$ and $\rho(C_j)$ can be tautological. Assume without loss of generality that $\rho(C_i)$ is tautological and let $\ell \in C_i$ such that $\rho(\ell) = \rho(\neg e)$. Then $\rho(C_j) \subseteq \rho(C_k) \cup \rho(\neg e) = \rho(C_k)$ and there is a $Q(D)$-resolution derivation of $C_j' \subseteq \rho(C_k)$ from $\Phi_2$ by induction hypothesis. $\qquad\square$

This result states that if we apply a $D$-renaming to each clause in a $Q(D)$-resolution derivation, a subsequence of the resulting sequence of clauses is a

$Q(D)$-resolution  derivation of a clause subsuming the image of the final clause in the original derivation. In particular, the length of the derivation can only decrease.

If any of the conditions of Definition 7 is dropped, then Lemma 1 no longer holds, in the sense that we might obtain derivations that are not syntactically correct. Mapping existential to universal literals may introduce tautologies that are removed by universal reduction, which is unsound in general and forbidden in $Q(D)$-resolution. The same problem can occur if universal literals are not mapped in an injective way. If universal literals can be mapped to existential literals, or independence according to the dependency scheme $D$ is not preserved, universal reduction may not be applicable in the image.

A $D$-renaming $\rho$ from $\Phi$ to itself is a $D$-*homomorphism* from clause set $\phi$ to clause set $\psi$ with respect to $\Phi$ if $\rho(\phi) \subseteq \psi$. The set of all $D$-homomorphisms from $\phi$ to $\psi$ with respect to $\Phi$ is denoted $\mathsf{Hom}_\Phi^D(\phi, \psi)$. $D$-homomorphisms generalize *symmetries* of PCNF formulas, which are renamings that may only change the order of variables within quantifier blocks [11]: any such mapping is bijective and preserves the type of a variable, as well as dependencies indicated by the trivial dependency scheme.

## 5   The Homomorphism Rule

Let $\Phi = \mathcal{Q}.\phi$ be PCNF formula and let $D$ be a tractable dependency scheme for which $Q(D)$-resolution  is sound. Consider a $Q(D)$-resolution  derivation of a clause $C$ from clauses $\psi \subseteq \phi$ in $\Phi$. If there is a homomorphism $\varphi \in \mathsf{Hom}_\Phi^D(\psi, \phi)$ then the *local homomorphism rule* can derive the clause $\varphi(C)$. We call the restricted form of this rule, which can only be applied if $\psi = \phi$ the *global homomorphism rule*. The proof systems $\mathsf{GH}(D)$ and $\mathsf{LH}(D)$ arise from $Q(D)$-resolution by addition of the global and local homomorphism rule, respectively.

We present an example to illustrate the local homomorphism rule. Consider the PCNF formula $\Phi = \mathcal{Q}.\phi$ where $\mathcal{Q} = \forall a\, b\, \exists x\, \forall c\, \exists y\, z\, w$ and $\phi = \{C_1, C_2, C_3, C_4, C_5\}$ with $C_1 = \{\neg a, \neg y, z\}, C_2 = \{c, y, w\}, C_3 = \{c, \neg z\}, C_4 = \{b, \neg x\}, C_5 = \{\neg a, x\}$. We use trivial dependency scheme $\mathsf{D}_\Phi^{\mathrm{trv}}$ for this illustration. Consider the following $Q(D)$-resolution  derivation $S$ from the formula $\Phi$:

|                      |                                            |
|----------------------|--------------------------------------------|
| $C_1$                | axiom;                                     |
| $C_2$                | axiom;                                     |
| $\{\neg a, c, w, z\}$ | resolution from $C_1$ and $C_2$;           |
| $C_3$                | axiom;                                     |
| $\{\neg a, c, w\}$   | resolution from $\{\neg a, c, w, z\}$ and $C_3$. |

Using the above resolution derivation $S$, we derive the clause $\{\neg a, c, w\}$ from $\Phi$. We define a non-injective mapping $\rho$ over the subset of variables of $\Phi$ as follows; $\rho(a) = a$, $\rho(c) = b$, $\rho(y) = \rho(w) = \neg x$ and $\rho(z) = x$. By the definition of renaming the complement of the literal takes the negation of the value defined by $\rho$, for example, $\rho(\neg z) = \neg x$. Note that the renaming jumps between the quantifier blocks by allowing the mapping of literals from one quantifier block to another. Let $\psi = \{C_1, C_2, C_3\} \subseteq \phi$, the image of $\psi$ under the renaming $\rho$ is $\rho(\psi) = \{C_4, C_5\} \subseteq \phi$. All the three restrictions of Definition 7 are satisfied, hence the renaming $\rho \in \mathsf{Hom}_\Phi^D(\psi, \phi)$. Thus, by using the local homomorphism rule, we can obtain the clause $\rho(\{\neg a, c, w\}) = \{\neg a, b, \neg x\}$ and add it to the matrix $\phi$.

**Proposition 2.** *The systems* $\mathsf{GH}(D)$ *and* $\mathsf{LH}(D)$ *are sound for any dependency scheme* $D$ *such that* $\mathsf{Q}(D)$*-resolution is sound. That is, a PCNF formula that has a refutation in* $\mathsf{GH}(D)$ *or* $\mathsf{LH}(D)$ *is false.*

*Proof.* Let $\Phi = \mathcal{Q}.\phi$ be a PCNF formula and let $S = C_1, \ldots, C_k$ be an $\mathsf{LH}(D)$-refutation of $\Phi$. If $S$ does not use the local homomorphism rule, then $\Phi$ is false by the soundness of $\mathsf{Q}(D)$-resolution. Otherwise, let $C_j$ be derived from $C_i$ by application of the local homomorphism rule, where $1 \leq i < j \leq k$. That is, there is a subset of clauses $\psi \subseteq \phi$ such that $C_i$ can be derived from $\psi$ in $\Phi$ and $\varphi \in \mathsf{Hom}_\Phi^D(\psi, \phi)$ is a homomorphism with $\varphi(C_i) = C_j$. By Lemma 1, the sequence $\varphi(C_1), \ldots, \varphi(C_i)$ contains a $\mathsf{Q}(D)$-resolution derivation of $C_i' \subseteq \varphi(C_i)$ from clauses $\varphi(\psi) \subseteq \phi$ in $\Phi$. We can replace $C_j$ with the corresponding derivation and (possibly) simplify the proof to obtain an $\mathsf{LH}(D)$-refutation of $\Phi$ with one less application of the local homomorphism rule. In this way, we can get rid of all uses of the local homomorphism rule one by one and obtain a $\mathsf{Q}(D)$-resolution refutation of $\Phi$. □

# 6 Lifting Lower Bounds from $\mathsf{Q(D^{rrs})}$-Resolution to $\mathsf{LH(D^{rrs})}$

Let $D$ be an arbitrary but fixed tractable and sound dependency scheme. Let $\Phi = \mathcal{Q}.\phi$ be a PCNF formula with a 3CNF matrix such that each clause contains at least two literals and cannot be simplified by universal reduction. Moreover, we assume that each clause contains at most one universal literal. Observe that any formula not solved by unit propagation can be transformed into this format by applying unit propagation and splitting clauses.

From $\Phi$ we construct a formula $\Phi^\circ = \mathcal{Q}^\circ.\phi^\circ$ as follows. Let $\ell_1, \ldots, \ell_s$ be the sequence of literals appearing in $\phi$. For each existential literal $\ell_j$ we introduce new existential variables $y_{j,1}, \ldots, y_{j,j+9}$ and $z_j$ at the same quantifier depth and create a chain of binary clauses

$$L_j' = \{\{\neg y_{j,1}, y_{j,2}\}, \{\neg y_{j,2}, y_{j,3}\}, \ldots, \{\neg y_{j,j+8}, y_{j,j+9}\}, \{\neg y_{j,j+9}, \ell_j\}\}.$$

We add the variable $z_j$ to all clauses of $L_j'$ except the fourth and (j+7)th one to obtain a formula $L_j$, called the *link* of $\ell_j$. The clause widths of a link yield a sequence

$$3\ 3\ 3\ 2\ \underbrace{3\ \ldots\ 3}_{j+1\ \text{times}}\ 2\ 3\ 3$$

that uniquely identifies an existential literal $\ell_j$.

Next, we replace each existential literal of $\Phi$ by the first literal in its link. More specifically, if $E_i = \{\ell_j, \ell_{j+1}, \ell_{j+2}\}$ is a clause of $\phi$, we let

$$E_i^\circ := \{\, y_{k,1} : \ell_k \text{ is existential, } j \leq k \leq j+2 \,\} \cup \{\, \ell \in E_i : \ell \text{ is universal} \,\}.$$

We combine the above definitions to obtain the formula

$$\phi^\circ := \{E_1^\circ, \ldots, E_m^\circ\} \cup \bigcup_{j=1}^s (L_j \cup \{\{\neg z_j\}\}).$$

We refer to clauses $E_i^\circ$ as *main clauses*, clauses in $L_j$ as *link clauses*, and to unit clauses $\{z_j\}$ as *auxiliary clauses*.

Since link clauses only contain existential variables, homomorphisms from $L_j$ into $\phi^\circ$ coincide with homomorphisms of propositional formulas defined by Szeider [19], and so the following result carries over to our setting.

**Lemma 2 (Szeider [19]).** $\mathrm{Hom}^D_{\Phi^\circ}(L_j, \phi^\circ) = \{id_{L_j}\}$ *for any* $1 \leq j \leq s$.

The formulas $\Phi$ and $\Phi^\circ$ have the same dependencies according to the resolution-path dependency scheme.

**Lemma 3.** *For every existential literal* $\ell_j \in \mathrm{lit}(\Phi)$ *and* $y_{j,i}$ *with* $1 \leq i \leq j+9$, *we have* $\mathrm{D^{rrs}}_{\Phi^\circ}(\ell_j) = \mathrm{D^{rrs}}_{\Phi^\circ}(y_{j,i}) = \mathrm{D^{rrs}}_\Phi(\ell_j)$, *as well as* $\mathrm{D^{rrs}}_{\Phi^\circ}(z_j) = \emptyset$.

*Proof.* There is a natural correspondence between resolution paths of $\Phi$ and $\Phi^\circ$. Each resolution path in $\Phi$ can be extended to a resolution path in $\Phi^\circ$ by using links. Formally, if $\ell_{j_1}, \ell_{j_2}, \ldots, \ell_{j_{2k}}$ is a resolution path of $\Phi$ we obtain a resolution path of $\Phi'$ by replacing each literal $\ell_{j_{2i-1}}$ for $1 \leq i \leq k$ by the sequence

$$\ell_{j_{2i-1}}, \neg y_{j_{2i-1}, j_{2i-1}+9}, y_{j_{2i-1}, j_{2i-1}+9}, \ldots, y_{j_{2i-1}, 2}, \neg y_{j_{2i-1}, 1}, y_{j_{2i-1}}$$

of literals from $L'_{j_{2i-1}}$ in reverse order, and each adjacent literal $\ell_{j_{2i}}$ for $1 \leq i \leq k$ by the sequence

$$y_{j_{2i}, 1}, \neg y_{j_{2i}, 1}, y_{j_{2i}, 2}, \ldots, y_{j_{2i}, j_{2i}+9}, \neg y_{j_{2i}, j_{2i}+9}, \ell_{2i}$$

of literals from $L'_{j_{2i}}$ in order. Conversely, any resolution path of $\Phi^\circ$ with original literals of $\Phi$ as endpoints can be transformed into a resolution path of $\Phi$ by removing sequences of link literals. Since link variables $y_{j,i}$ are introduced at the same quantifier depth as $var(\ell_j)$ and $var_\forall(\Phi) = var_\forall(\Phi^\circ)$, it follows that $\mathrm{D^{rrs}}_{\Phi^\circ}(\ell_j) = \mathrm{D^{rrs}}_\Phi(\ell_j)$. Further, a dependency-inducing resolution path of $\Phi^\circ$ from a universal variable $u$ to its negation $\neg u$ goes through a literal $\ell_j$ if, and only if, it goes through all the link variables $y_{j,i}$, so $\mathrm{D^{rrs}}_{\Phi^\circ}(\ell_j) = \mathrm{D^{rrs}}_{\Phi^\circ}(y_{j,i})$ for $1 \leq i \leq j+9$. Finally, the variables $z_j$ occur negatively exclusively in the unit clauses $(\neg z_j)$, so $\mathrm{D^{rrs}}_{\Phi^\circ}(z_j) = \emptyset$. $\qquad\square$

For a QBF proof system $\Pi$ and a false formula $\Phi$, let $\mathsf{PSize}_\Pi(\Phi)$ denote the size of a shortest $\Pi$-refutation of $\Phi$.

**Corollary 1.** $\mathsf{PSize}_{Q(D^{rrs})\text{-}Res}(\Phi^\circ) \leq \mathsf{PSize}_{Q(D^{rrs})\text{-}Res}(\Phi) + O(\|\Phi\|^2)$.

*Proof.* The original matrix $\phi$ can be obtained from $\phi^\circ$ by resolving each existential literal $\ell_j$ in a main clause $E_i$ with the link clauses in $L_j$ and the auxiliary clause $\{\neg z\}$. This requires $O(j)$ steps for each literal $\ell_j$ with $1 \leq j \leq s$, and $s \in O(\|\Phi\|)$. Let $S$ denote the corresponding $Q(D^{rrs})$-resolution derivation. As resolution-path dependencies are preserved by Lemma 3, a $Q(D^{rrs})$-resolution refutation $S'$ of $\Phi$ can be appended to $S$ so as to obtain a $Q(D^{rrs})$-resolution refutation of $\Phi^\circ$. □

We want to show that any $\mathsf{LH}(D^{rrs})$-refutation of the "rigid" version $\Phi^\circ$ of a PCNF formula $\Phi$ can be mapped back to a $Q(D^{rrs})$-resolution refutation of the original formula $\Phi$. To do this, we introduce a new existential variable $z$ and define a renaming $\rho : \mathsf{lit}(\Phi^\circ) \to \mathsf{lit}(\Phi) \cup \{z\}$ as follows:

$$
\begin{aligned}
\rho(u) &:= u, & &\text{for each universal variable } u; \\
\rho(y_{j,i}) &:= \ell_j, & &\text{for } 1 \leq j \leq s, 1 \leq i \leq j+9; \\
\rho(\ell_j) &:= \ell_j, & &\text{for } 1 \leq j \leq s; \\
\rho(z_j) &:= z, & &\text{for } 1 \leq j \leq s.
\end{aligned}
$$

With this renaming, every link clause becomes tautological, every auxiliary clause $\{\neg z_j\}$ becomes $\rho(\{\neg z_j\}) = \{\neg z\}$, and main clauses $E_i^\circ$ are mapped back to original clauses $\rho(E_i^\circ) = E_i$. Hence $\rho_{cls}(\phi^\circ)$ as defined in Sect. 4 is nothing but $\phi \cup \{\{\neg z\}\}$, and $\neg z$ is a pure literal of $\rho_{cls}(\phi^\circ)$.

**Lemma 4.** *The mapping* $\rho : \mathsf{lit}(\Phi^\circ) \to \mathsf{lit}(\Phi) \cup \{z\}$ *is a* $D^{rrs}$-*renaming from* $\Phi^\circ$ *to* $\Phi_z = \exists z Q.\phi \cup \{\{\neg z\}\}$.

*Proof.* By construction, the mapping preserves quantifier types and is injective with respect to universal variables. Moreover, the new variable $z$ and clause $\{\neg z\}$ do not affect resolution-path dependencies in $\Phi_z$ and $z$ has no dependencies itself, so $D^{rrs}{}_{\Phi^\circ}(v) = D^{rrs}{}_{\Phi_z}(\rho(v))$ holds for every variable $v \in var(\Phi^\circ)$ by Lemma 3. □

The following result establishes that "interesting" $\mathsf{LH}(D)$-derivations using at least two main clauses from $\Phi^\circ$ cannot use the homomorphism rule in a nontrivial way.

**Lemma 5.** *Let* $S = C_1, \ldots, C_k$ *be a* $Q(D)$-resolution *derivation from* $\phi' \subseteq \phi^\circ$ *in* $\Phi^\circ$ *such that no subsequence of* $S$ *is a* $Q(D)$-resolution *derivation of* $C_k$ *in* $\Phi$. *If* $S$ *contains at least two main clauses as input clauses then* $\rho(C_k) = \rho(\varphi(C_k))$ *for any homomorphism* $\varphi \in \mathsf{Hom}_{\Phi^\circ}^D(\phi', \phi^\circ)$.

A formal proof of Lemma 5 is rather tedious and has to be omitted due to space constraints, but the underlying intuition is fairly simple. Since main clauses are only connected through links, two main clauses can take part in a resolution proof only if the two links corresponding to the pivot literal are present, which by Lemma 2 leaves the identity as the only homomorphism that can be applied to the main clauses or the clauses in the links. Having identified such a "rigid" part of a proof, one can then show that any other clause $C$ that participates in the proof has the same image under $\rho$ as its homomorphic image $\varphi(C)$, in symbols $\rho(C) = \rho(\varphi(C))$.

The next result states that $\mathsf{LH}(D)$-derivations that use at most one single main clause cannot be too long.

**Lemma 6.** *For any* $\mathsf{Q}(D)$-*resolution derivation* $S$ *of a clause* $C$ *in* $\Phi^\circ$ *with at most one main clause among its input clauses, there is a* $\mathsf{Q}(D)$-*resolution derivation* $S'$ *of* $C' \subseteq C$ *in* $\Phi^\circ$ *of length* $O(|\Phi^\circ|^3)$.

*Proof.* Let $S = C_1, \ldots, C_k$ be a $\mathsf{Q}(D)$-resolution derivation of $C = C_k$ in $\Phi^\circ$ that contains at most one main clause among its input clauses. Any remaining input clauses are auxiliary or link clauses. We construct the derivation $S'$ by first resolving each link clause containing a literal $z_j$ with the auxiliary clause $\{\neg z_j\}$. This requires at most $|\Phi^\circ|$ resolution steps. We then proceed as in $S$ while (possibly) omitting resolution steps on variables $z_j$. The length of $S'$ can be crudely bounded as follows. After resolving out $z_j$ we are left with binary link clauses $L'_j$ and a single main clause of size at most three. Any $\mathsf{Q}(D)$-resolution derivation starting from these clauses can derive clauses of size at most three, and there are $O(|\Phi^\circ|^3)$ such clauses. □

**Lemma 7.** $\mathsf{PSize}_{\mathsf{Q}(D^{\mathrm{rrs}})\text{-Res}}(\Phi) \leq \mathsf{PSize}_{\mathsf{LH}(D^{\mathrm{rrs}})}(\Phi^\circ) \cdot O(|\Phi^\circ|^3)$.

*Proof.* Let $C_1, \ldots, C_k$ be an $\mathsf{LH}(D^{\mathrm{rrs}})$-refutation of $\Phi^\circ$. By Lemma 4, the mapping $\rho$ is a $D^{\mathrm{rrs}}$-renaming from $\Phi^\circ$ to $\Phi_z$, so if no clause of $C_1, \ldots, C_k$ is derived by the local homomorphism rule, we can apply Lemma 1 and conclude that $\rho(C_1), \ldots, \rho(C_k)$ is a $\mathsf{Q}(D^{\mathrm{rrs}})$-resolution refutation of $\Phi$. Otherwise, we are going to turn $\rho(C_1), \ldots, \rho(C_k)$ into a $\mathsf{Q}(D^{\mathrm{rrs}})$-resolution refutation of $\Phi$ that is not too much larger. Suppose clause $C_j$ is derived from $C_i$ using the local homomorphism rule for some $1 \leq i < j \leq k$. That is, $C_1, \ldots, C_i$ contains a $\mathsf{Q}(D^{\mathrm{rrs}})$-resolution derivation $S$ of $C_i$ from clause set $\phi' \subseteq \phi^\circ$ in $\Phi^\circ$, and there is a homomorphism $\varphi \in \mathsf{Hom}_\Phi^D(\phi', \phi^\circ)$ such that $\varphi(C_i) = C_j$. If the derivation of $C_i$ involves at most one main clause then its size is in $O(|\Phi^\circ|^3)$ by Lemma 6. By Lemma 1, the sequence $\varphi(C_1), \ldots, \varphi(C_i)$ contains a $\mathsf{Q}(D^{\mathrm{rrs}})$-resolution derivation of the clause $\varphi(C_i) = C_j$. We simply replace $\rho(C_j)$ by the image $\rho(\varphi(C_1)), \ldots, \rho(\varphi(C_i))$ of this entire derivation, increasing the proof size by $O(|\Phi^\circ|^3)$. Otherwise, the derivation $S$ uses at least two main clauses. In this case, Lemma 5 tells us that $\rho(C_i) = \rho(\varphi(C_i)) = \rho(C_j)$, so we can simply use $\rho(C_i)$ instead of $\rho(C_j)$. In this manner, we obtain a $\mathsf{Q}(D^{\mathrm{rrs}})$-resolution refutation of $\Phi_z$ of size $k \cdot O(|\Phi^\circ|^3)$. Since $\neg z$ is pure in $\Phi_z$, the refutation cannot contain the clause $\{\neg z\}$, and is in fact a $\mathsf{Q}(D^{\mathrm{rrs}})$-resolution refutation of the original formula $\Phi$. □

## 7   Separating $\mathsf{LH}(D^{\mathrm{rrs}})$ from $\mathsf{LH}$

In this section, we will use Corollary 1 and Lemma 7 to lift known separations of Q-resolution systems *without* the homomorphism rule to systems *with* the homomorphism rule. First, we show that our assumption from the previous section that the formula $\Phi$ has a matrix in 3CNF does not affect certain semantic lower bound techniques. More specifically, we show that long clauses occurring in the equality formulas can be split without affecting the *cost* of these formulas [3].

**Definition 8 (Universal Winning Strategy).** For any set $V$ of variables, let $[V]$ denote the set of assignments of $V$. Let $\Phi = \forall U_1 \exists E_1 \ldots \forall U_n \exists E_n.\phi$ be a false QBF. A *universal strategy* for $\Phi$ is a sequence $S = (S_i)_{1 \leq i \leq n}$ of functions $S_i : [E_1 \cup \cdots \cup E_{i-1}] \rightarrow [U_i]$. The *response* of $S$ to an existential assignment $\tau : var_\exists(\Phi) \rightarrow \{0,1\}$ is the assignment $S(\tau) = \bigcup_{i=1}^n S_i(\tau|_{E_1 \cup \cdots \cup E_{i-1}})$. The universal strategy $S$ is a *universal winning strategy* if the assignment $\tau \cup S(\tau)$ satisfies the matrix $\phi$ for every existential assignment $\tau : var_\exists(\Phi) \rightarrow \{0,1\}$.

**Definition 9 (Cost).** Let $\Phi = \forall U_1 \exists E_1 \ldots \forall U_n \exists E_n.\phi$ be a false QBF and let $S = (S_i)_{1 \leq i \leq n}$ be a universal winning strategy for $\Phi$. The *cost* of $S$ is defined as $cost(S) = \max\{ |rng(S_i)| : 1 \leq i \leq n \}$, where $rng(f)$ denotes the range of function $f$. The *cost* of the QBF $\Phi$ is the minimum cost of any universal winning strategy for $\Phi$.

The cost of a false QBF $\Phi$ is a lower bound on the size of any Q-resolution refutation of $\Phi$.

**Theorem 1 (Beyersdorff, Blinkhorn, and Hinde [3]).** *Let* $C_1, \ldots, C_k$ *be a Q-resolution refutation of a QBF* $\Phi$. *Then* $k \geq cost(\Phi)$.

**Lemma 8.** *Let* $\Phi = Q.\phi \cup \{C\}$ *be a PCNF formula with clause* $C = C_1 \cup C_2$ *and let* $y$ *be a fresh variable. Further, let* $\Phi' = Q \exists y.\phi \cup \{C_1 \cup \{y\}, C_2 \cup \{\neg y\}\}$ *be the formula obtained from* $\Phi$ *by splitting* $C$. *Then* $\Phi$ *and* $\Phi'$ *have the same universal winning strategies.*

**Corollary 2.** *If* $\Phi^*$ *is obtained from* $\Phi$ *by splitting clauses, then* $\Phi^*$ *and* $\Phi$ *have the same cost.*

**Proposition 3 (Beyersdorff, Blinkhorn, and Hinde [3]).** *For each* $n \in \mathbb{N}$, *EQ$(n)$ has cost* $2^n$.

This implies an exponential proof size lower bound by Theorem 1. At the same time, it is known that these formulas have short Q(D$^{rrs}$)-resolution refutations.

**Theorem 2 (Blinkhorn and Beyersdorff [2]).** *For each* $n \in \mathbb{N}$, *EQ$(n)$ has a* Q(D$^{rrs}$)-*resolution refutation of size* $O(n)$.

We are now ready to prove an exponential separation of LH(D$^{rrs}$) from LH(D$^{trv}$).

**Theorem 3.** *There is an infinite sequence* $(\Phi_n)_{n \in \mathbb{N}}$ *of false formulas such that the shortest* LH(D$^{rrs}$)-*refutation of* $\Phi_n$ *is polynomial in* $n$ *but any* LH(D$^{trv}$)-*refutation of* $\Phi_n$ *has length* $2^{\Omega(n)}$.

*Proof.* For each $n \in \mathbb{N}$, let $EQ^*(n)$ denote a QBF obtained from $EQ(n)$ by splitting clauses until each clause contains at most three literals in total and at most one universal literal. By Proposition 3 and Corollary 2, $EQ^*(n)$ has cost $2^n$ and thus requires Q-resolution refutations of size at least $2^n$ by Theorem 1. At the same time, since the original formula can be obtained by resolving on existential variables introduced by splitting, and splitting does not introduce new resolution-path dependencies among the original variables, Theorem 2 implies that $EQ^*(n)$ has a linear-size $Q(D^{rrs})$-resolution refutation. Now, consider the "rigid" versions $EQ^\circ(n)$ of $EQ^*(n)$. Clearly, the size of $EQ^\circ(n)$ is polynomially bounded in the size of $EQ(n)$. By Theorem 2 and Corollary 1, the formulas $EQ^\circ(n)$ have polynomial-size $Q(D^{rrs})$-resolution refutations, and thus also polynomial-size $LH(D^{rrs})$-refutations. On the other hand, Lemma 7 tells us that any $LH(D^{trv})$-refutation of $EQ^\circ(n)$ can be shorter than a Q-resolution refutation of $EQ^*(n)$ by at most a polynomial factor. □

Since the short $LH(D^{rrs})$-refutations in the above theorem do not use the local homomorphism rule, analogous separations hold for weaker systems.

**Corollary 3.** *There is an infinite sequence $(\Phi_n)_{n \in \mathbb{N}}$ of false formulas such that the shortest $\Pi(D^{rrs})$-refutation of $\Phi_n$ is polynomial in $n$ but any $\Pi(D^{trv})$-refutation of $\Phi_n$ has length $2^{\Omega(n)}$, for $\Pi \in \{GH, LS, GS\}$.*

# 8    Concluding Remarks

We have lifted the local and the global homomorphism rule from propositional resolution to the quantified case, introducing several generalizations, including the use of dependency schemes. Although we have established an exponential lower bound for the most general system LH without a dependency scheme, we left open to prove an exponential lower bound for $LH(D^{rrs})$.

The systems introduced here are incomparable with the proof systems LQU+ [1] and IR-calc [4]. Since they are stronger than GS, there are classes of formulas that are easy for our systems and hard for LQU+ and IR-calc [11]. For the converse, we can apply our construction to the QPARITY [4] formulas and make them rigid, so that they are hard for LH. Both LQU+ and IR-calc can derive the original formula and then proceed with short refutations of QPARITY.

There are several possibilities for further strengthening $LH(D^{rrs})$. One possibility is to consider a suitably defined *dynamic homomorphism rule* [19] which considers homomorphisms between sets of *derived* clauses. Neither of the lower bounds established in this paper applies to proof systems that use such a dynamic rule: all the modifications made to the input formula to achieve rigidity can be undone by a polynomial number of resolution steps so that after these steps symmetries and homomorphisms can be exploited to get short proofs.

Another possibility, somewhat related to the dynamic systems discussed above, is based on the idea of *symmetry recomputation*, as considered by Blinkhorn and Beyersdorff [5], which exploits symmetries of the input formula after the application of a partial assignment. We think that this idea can be combined with our homomorphism systems.

All these ideas for even stronger proof systems for QBF give rise to challenging theoretical questions that include separation results, as well as lower and upper bounds. Another interesting line of research is concerned with the possibility of utilizing the strength of the various homomorphism rules considered in this paper within a QBF solver.

# References

1. Balabanov, V., Widl, M., Jiang, J.-H.R.: QBF resolution systems and their proof complexities. In: Sinz, C., Egly, U. (eds.) SAT 2014. LNCS, vol. 8561, pp. 154–169. Springer, Cham (2014). https://doi.org/10.1007/978-3-319-09284-3_12
2. Beyersdorff, O., Blinkhorn, J.: Dynamic QBF dependencies in reduction and expansion. ACM Trans. Comput. Log. **21**(2), 8:1–8:27 (2020)
3. Beyersdorff, O., Blinkhorn, J., Hinde, L.: Size, cost, and capacity: a semantic technique for hard random QBFS. Logical Methods Comput. Sci. **15**(1), (2019)
4. Beyersdorff, O., Chew, L., Janota, M.: New resolution-based QBF calculi and their proof complexity. TOCT **11**(4), 26:1–26:42 (2019)
5. Blinkhorn, J., Beyersdorff, O.: Proof complexity of QBF symmetry recomputation. In: Janota, M., Lynce, I. (eds.) SAT 2019. LNCS, vol. 11628, pp. 36–52. Springer, Cham (2019). https://doi.org/10.1007/978-3-030-24258-9_3
6. Büning, H.K., Karpinski, M., Flögel, A.: Resolution for quantified Boolean formulas. Inf. Comput. **117**(1), 12–18 (1995)
7. Cadoli, M., Schaerf, M., Giovanardi, A., Giovanardi, M.: An algorithm to evaluate quantified Boolean formulae and its experimental evaluation. J. Autom. Reason. **28**(2), 101–142 (2002). https://doi.org/10.1023/A:1015019416843
8. Davis, M., Logemann, G., Loveland, D.: A machine program for theorem-proving. Commun. ACM **5**, 394–397 (1962)
9. Egly, U., Lonsing, F., Widl, M.: Long-distance resolution: proof generation and strategy extraction in search-based QBF solving. In: McMillan, K., Middeldorp, A., Voronkov, A. (eds.) LPAR 2013. LNCS, vol. 8312, pp. 291–308. Springer, Heidelberg (2013). https://doi.org/10.1007/978-3-642-45221-5_21
10. Giunchiglia, E., Narizzano, M., Tacchella, A.: Clause/term resolution and learning in the evaluation of quantified Boolean formulas. J. Artif. Intell. Res. **26**, 371–416 (2006)
11. Kauers, M., Seidl, M.: Short proofs for some symmetric quantified Boolean formulas. Inform. Process. Lett. **140**, 4–7 (2018)
12. Krishnamurthy, B.: Short proofs for tricky formulas. Acta Informatica **22**, 253–275 (1985)
13. Lonsing, F.: Dependency Schemes and Search-Based QBF Solving: Theory and Practice. Ph.D. thesis, Johannes Kepler University, Linz, Austria, April 2012
14. Lonsing, F., Biere, A.: Integrating dependency schemes in search-based QBF solvers. In: Strichman, O., Szeider, S. (eds.) SAT 2010. LNCS, vol. 6175, pp. 158–171. Springer, Heidelberg (2010). https://doi.org/10.1007/978-3-642-14186-7_14
15. Samer, M., Szeider, S.: Backdoor sets of quantified Boolean formulas. J. Autom. Reason. **42**(1), 77–97 (2009)

16. Marques-Silva, J.: The impact of branching heuristics in propositional satisfiability algorithms. In: Barahona, P., Alferes, J.J. (eds.) EPIA 1999. LNCS (LNAI), vol. 1695, pp. 62–74. Springer, Heidelberg (1999). https://doi.org/10.1007/3-540-48159-1_5

17. Slivovsky, F., Szeider, S.: Quantifier reordering for QBF. J. Autom. Reason. **56**(4), 459–477 (2016). https://doi.org/10.1007/s10817-015-9353-1, http://dx.doi.org/10.1007/s10817-015-9353-1

18. Slivovsky, F., Szeider, S.: Soundness of Q-resolution with dependency schemes. Theor. Comput. Sci. **612**, 83–101 (2016). https://doi.org/10.1016/j.tcs.2015.10.020, http://dx.doi.org/10.1016/j.tcs.2015.10.020

19. Szeider, S.: The complexity of resolution with generalized symmetry rules. Theory Comput. Syst. **38**(2), 171–188 (2005)

20. Gelder, A.: Variable independence and resolution paths for quantified Boolean formulas. In: Lee, J. (ed.) CP 2011. LNCS, vol. 6876, pp. 789–803. Springer, Heidelberg (2011). https://doi.org/10.1007/978-3-642-23786-7_59

21. Zhang, L., Malik, S.: The quest for efficient Boolean satisfiability solvers. In: Brinksma, E., Larsen, K.G. (eds.) CAV 2002. LNCS, vol. 2404, pp. 17–36. Springer, Heidelberg (2002). https://doi.org/10.1007/3-540-45657-0_2

# Multi-linear Strategy Extraction for QBF Expansion Proofs via Local Soundness

Matthias Schlaipfer[✉], Friedrich Slivovsky, Georg Weissenbacher, and Florian Zuleger

TU Wien, Vienna, Austria
mschlaipfer@forsyte.at

**Abstract.** In applications, QBF solvers are expected to not only decide whether a given formula is true or false but also return a solution in the form of a strategy. Determining whether strategies can be efficiently extracted from proof traces generated by QBF solvers is a fundamental research task. Most resolution-based proof systems are known to implicitly support polynomial-time strategy extraction through a simulation of the evaluation game associated with an input formula, but this approach introduces large constant factors and results in unwieldy circuit representations. In this work, we present an explicit polynomial-time strategy extraction algorithm for the ∀-Exp+Res proof system. This system is used by expansion-based solvers that implement counterexample-guided abstraction refinement (CEGAR), currently one of the most effective QBF solving paradigms. Our argument relies on a Curry-Howard style correspondence between strategies and ∀-Exp+Res derivations, where each strategy realizes an invariant obtained from an annotated clause derived in the proof system.

## 1 Introduction

Continued improvements in the performance of satisfiability (SAT) solvers [14] are enabling a growing number of applications in areas such as electronic design automation [35]. At the same time, many of the underlying problems are hard for complexity classes beyond NP and as such cannot be expected to have succinct propositional encodings. Super-polynomial growth in encoding size imposes a limit on the problem instances that can be feasibly solved even with extremely efficient SAT solvers. Decision procedures for more succinct languages such as Quantified Boolean Formulas (QBFs) represent a potential solution to this scaling issue. QBFs extend propositional formulas with quantification over truth values and support more succinct encodings for a range of

M. Schlaipfer—Supported by the Vienna Science and Technology Fund (WWTF) through the Heisenbugs project VRG11-005 and the Austrian Science Fund (FWF) through the LogiCS doctoral program W1255-N23.
F. Slivovsky—Supported by the Vienna Science and Technology Fund (WWTF) under grant number ICT19-060.

L. Pulina and M. Seidl (Eds.): SAT 2020, LNCS 12178, pp. 429–446, 2020.
https://doi.org/10.1007/978-3-030-51825-7_30

problems [32]. Recent years have seen significant advancements in QBF solver technology [20, 21, 25, 26, 29, 30, 34, 36], up to a point where reduction to QBF can be more efficient than reduction to SAT [13].

In some applications, QBF solvers are required to not only decide whether a given formula is true or false but also compute a solution in the form of a *strategy*. For example, if a synthesis problem is encoded as a QBF, a solver is expected to either return the synthesized program or an explanation why the specification cannot be satisfied [13]. Determining whether the proof trace of a QBF solver can be efficiently transformed into a strategy—whether the proof system supports polynomial-time *strategy extraction*—is a fundamental research task [2, 3, 6, 10, 27].

One of the most successful QBF solving paradigms relies on partial Shannon expansion [1, 8] of universal variables within a counterexample-guided abstraction refinement (CEGAR) loop, as implemented in RAREQS [21], and, more recently, in IJTIHAD [9] and QFUN [20]. The underlying proof system ∀-Exp+Res [22] offers exponentially shorter proofs for certain classes of formulas than Q-resolution [6], and can polynomially simulate Q-resolution on formulas with few quantifier alternations [4], which includes many practically relevant cases.[1] Polynomial-time strategy extraction follows from the fact that ∀-Exp+Res proofs can be used to guide the universal player in an evaluation game [6, 11], but turning this argument into circuits that compute a winning strategy is rather inefficient. An explicit construction based on this idea for Q-resolution requires the introduction of several gates for each literal in the proof and quantifier level of the input formula [27], leading to unwieldy circuits that are substantially larger than the original proof. In this work, we present a strategy extraction algorithm for ∀-Exp+Res that is multi-linear in the number of proof steps and universal variables. This is asymptotically optimal for a construction that follows the structure of the proof and maintains a circuit for each universal variable.

Our algorithm is inspired by [33], which for the first time has given a local soundness argument for ∀-Exp+Res. [33] constructs partial strategies along the ∀-Exp+Res-proof and provides a semantic abstraction that relates the constructed strategies to the clauses in the proof. In contrast, we associate a full strategy to each node in the ∀-Exp+Res-proof and develop a syntactic argument that ensures the soundness of the construction. For each clause in the proof, we define a propositional invariant that corresponds to a syntactic weakening of the input formula's negated matrix. We then show that strategies satisfying the invariants for the premises of a resolution step can be combined into a strategy that satisfies the invariant for the resolvent. The main technical challenge we had to overcome in deriving this syntactic weakening is that ∀-Exp+Res proofs work over an extended propositional alphabet where multiple versions of the same variable with different annotations may exist simultaneously. Our invariant translates the propositions from the extended alphabet back to formulas over the original vocabulary.

---

[1] Conversely, there are classes of formulas with exponentially shorter Q-resolution proofs [22], so that the systems are mutually separated.

We believe that our syntactic soundness argument is more transparent than the semantic construction from [33]. The clarity of the argument is also what allows us to obtain a concise circuit representation of the resulting strategy. Further, our syntactic argument establishes a Curry-Howard correspondence between proof construction and strategy extraction. For each inference rule combining proof terms, the correspondence provides a rule combining program terms. The result is a program isomorphic to the proof. The widest-known correspondence is between natural deduction proofs and lamba-calculus programs [18]. In this paper we establish a precise correspondence between ∀-Exp+Res-proofs and strategies—the strategy constructed for a node in the proof DAG satisfies the invariant for the clause derived at that node. In contrast, the correspondence stays implicit in the semantic argument from [33]. We expect that our ideas of obtaining such an invariant by weakening the matrix and translating the clauses over the extended alphabet back to a formula over the original variables will have applications in studying further Curry-Howard correspondences for other resolution-based QBF proof systems.

## 2  Preliminaries

*Quantified Boolean Formulas (QBFs).* We consider quantified Boolean formulas (QBFs) with standard propositional connectives $\land, \lor, \neg, \Leftrightarrow, \oplus$, and quantifiers $\forall, \exists$. We denote existentially quantified variables by $x$ and $y$, and universally quantified variables by $u$. Variables range over $\mathbb{B} = \{0, 1\}$. A literal $l$ is a variable $x$ or its negation $\neg x$. We write $\boldsymbol{x}$ for a set of variables or literals. A clause is a disjunction of literals, and a propositional formula in conjunctive normal form is a conjunction of clauses. We write $\square$ for the empty clause. Throughout the paper, QBFs are assumed to be in prenex conjunctive normal form (PCNF). A PCNF formula $\Phi = \Pi.\varphi$ consists of a sequence $\Pi = Q_1 v_1 \ldots Q_n v_n$ with $Q_i \in \{\forall, \exists\}$ for $1 \leq i \leq n$, called the quantifier prefix of $\Phi$, and a propositional formula $\varphi$ in conjunctive normal form, called the matrix of $\Phi$. We define a relation $\prec_\Pi$ on variables from the quantifier prefix as $v_i \prec_\Pi v_j$ whenever $i < j$. We extend $\prec_\Pi$ to a relation on literals in the obvious way and drop the quantifier prefix $\Pi$ from the subscript when it is clear from the context.

*QBF Expansion Proofs.* We consider a proof system for false PCNF formulas known as ∀-Exp+Res [22]. This system combines instantiation of universal variables with propositional resolution. Instantiation leads to existential literals $l^\tau$ that are annotated with an assignment $\tau : \boldsymbol{u}_l \to \mathbb{B}$ of the universal variables $\boldsymbol{u}_l = \{u \mid u \prec l\}$ that precede the variable of $l$ in the quantifier prefix. Following Beyersdorff et al. [6], we write $l^{[\tau]} = l^{\{u \mapsto \tau(u) \mid u \prec l\}}$ to filter out assignments that are not permitted in the annotation of $l$. We sometimes treat an assignment $\tau : \boldsymbol{u} \to \mathbb{B}$ in an annotation as a set of literals and write $l \in \tau$ if $\tau(l) = 1$. We write $C^\tau$ for a clause $C$ with all its literals annotated with $[\tau]$. The proof rules of ∀-Exp+Res are shown in Fig. 1. A ∀-Exp+Res *proof* of a PCNF formula $\Phi$ is a sequence of clauses ending with the empty clause such that each

$$\frac{}{\{l^{[\tau]} \mid l \in C, l \text{ is existential}\}} \; (\forall\text{-exp}) \qquad \frac{C_1 \vee x^\sigma \qquad C_2 \vee \neg x^\sigma}{C_1 \vee C_2} \; (\text{res})$$

Here, $C$ is a clause from the matrix and $\tau$ an assignment to all universal variables falsifying the universal literals of $C$. Both $C_1$ and $C_2$ are annotated clauses and $x^\sigma$ is an annotated variable.

**Fig. 1.** The proof rules of $\forall$-Exp+Res.

clause is derived either by universal expansion ($\forall$-exp) or by resolution (res) from clauses appearing earlier in the sequence.

## 3   Strategies

A PCNF formula can be interpreted as the specification of a game between an existential and a universal player [31]. The game proceeds by the players assigning values to their respective variables in turn, following the order of the quantifier prefix. The goal of the universal player is to falsify the matrix, the goal of the existential player is to satisfy the matrix. Strategies for either player can be conveniently represented as binary trees.

**Definition 1 (Strategy).** *Let* $\Phi = \Pi.\varphi$ *be a PCNF formula. A (universal) strategy for* $\Phi$ *is a labeled, rooted binary tree with the following properties:*

1. *Leaf nodes are labeled with* $\bot$, *inner nodes are labeled with variables of* $\Phi$, *and edges are labeled with* 0 *or* 1.
2. *Nodes labeled with existential variables have exactly two child nodes, and nodes labeled with universal variables have a single child node. Moreover, edges leading to distinct child nodes have distinct labels.*
3. *The sequence of variables encountered as labels on any path from the root to a leaf follows the order* $\prec_\Pi$ *of variables in the quantifier prefix.*

*A strategy* $P$ *for* $\Phi$ *is* complete *if each path from the root of* $P$ *to a leaf contains all variables of* $\Phi$. *Each path from the root to a leaf of a strategy induces a truth assignment in the obvious way. A strategy* $P$ *is a (universal)* winning strategy *for* $\Phi$ *if every such assignment falsifies the matrix* $\varphi$.

We write $P = \mathsf{Str}(v, P^-, P^+)$ for a strategy $P$ with root labeled by variable $v$ and principal subtrees $P^-$ and $P^+$ such that the edge to the root of $P^-$ is labeled with 0 and the edge to the root of $P^+$ is labeled with 1. We use $\emptyset$ to denote the "empty" strategy and write $P = \mathsf{Str}(v, \emptyset, P^+)$ and $P = \mathsf{Str}(v, P^-, \emptyset)$ to denote strategies with root nodes that only have a 1-child and a 0-child, respectively.

In the next section, we will associate each clause $C$ in a $\forall$-Exp+Res proof with a strategy $P$. For clauses $C$ derived by the $\forall$-exp rule with assignment $\tau$, the corresponding strategy simply sets the universal variables according to $\tau$.

**Definition 2.** *Let $\Phi = \Pi.\varphi$ be a PCNF formula and $\tau$ an assignment of the universal variables of $\Phi$. We define* ConstStrat$(\Pi, \tau)$ *as the complete strategy for $\Phi$ where each assignment is consistent with $\tau$.*

*Example 1.* The figure to the right shows the strategy computed by ConstStrat$(\exists x_1 \forall u \exists x_2, \neg u)$. The tree encodes the assignments $\{0/x_1, 0/u, 0/x_2\}$, $\{0/x_1, 0/u, 1/x_2\}$, $\{1/x_1, 0/u, 0/x_2\}$, $\{1/x_1, 0/u, 1/x_2\}$ falsifying $u$.

## 4  Local Soundness

We present a local soundness argument for ∀-Exp+Res using strategies. To this end, we will define a Combine operator that joins strategies along a derivation [33]. For each derived clause $C$, we will show that the strategy created for this clause by the Combine operator satisfies a propositional invariant obtained from $C$. Here, by *a strategy $P$ satisfying a formula $\psi$* we mean that every assignment consistent with $P$ satisfies $\psi$, which we will write as $P \models \psi$.[2] In this notation, we will show that

$$P \models \mathsf{enc}(C) \Rightarrow \neg\varphi,$$

where $\varphi$ denotes the matrix and $\mathsf{enc}(C)$ translates the clause $C$ back into a formula over the original variables of the QBF as

$$\mathsf{enc}(C) \overset{\text{def}}{=} \bigwedge_{x_i^{\tau_i} \in C} \Big( \bigwedge_{u \in \tau_i} u \Big) \Rightarrow \neg x_i.$$

The invariant $\mathsf{enc}(C) \Rightarrow \neg\varphi$ can be understood by considering the evaluation game: if the existential player responds to every universal play in an annotation by setting the literal to false, the current strategy is winning for the universal player. Ultimately, at the empty clause, $\mathsf{enc}(\square) = 1$ and the combined strategy turns into a winning strategy.

### 4.1  Combine

We will now introduce the Combine operator that merges two strategies $P$ and $Q$ in a top-down manner and annotates each clause in a ∀-Exp+Res derivation with a strategy. We write $C\,[P]$ for a clause $C$ annotated with strategy $P$. The definition of Combine as shown in Definition 4 is adapted from the definition of an operator defined by Suda and Gleiss [33]. Since we work with complete strategy trees (rather than partial strategies), the top-most variable remains equivalent between two strategies when recursing on them in lock-step, so it is sufficient to

---

[2] If the strategy $P$ is identified with the disjunction of assignments induced by its root-to-leaf paths, the relation $P \models \psi$ coincides with propositional entailment.

perform a case distinction on the top-most variable encountered in a strategy. Moreover, our definition of Combine is tailored to ∀-Exp+Res.

Clauses derived by (∀-exp) are annotated with the strategy ConstStrat($\Pi, \tau$) that plays the assignment $\tau$. For the (res) rule we have the following cases:

- The top-most variable, say $u$, is universal:

  - If the outgoing edge of $u$ (lit($u$), see Definition 3 below) differs from the annotation $\tau(u)$ of the pivot in at least one of $P$ and $Q$, we select the strategy that differs.
  - If lit($u$) equals the annotation $\tau(u)$ of the pivot in both $P$ and $Q$, we recurse.

- The top-most variable, say $x$, is existential:

  - If $x$ is the pivot of the inference rule, we combine the two strategies.
  - If $x$ is not the pivot, we recurse.

The base cases are when a universal edge differs, or we reach the pivot.

**Definition 3 (lit).** *We define* lit *as the partial function mapping universal strategy nodes to the literal they represent, based on their (unique) child node.*

$$\text{lit}(P) = \begin{cases} \neg u & \text{if } P = \text{Str}(u, P^-, \emptyset) \\ u & \text{if } P = \text{Str}(u, \emptyset, P^+) \end{cases}$$

**Definition 4 (Combine).** *We define* Combine *as a function from two strategies, $P$ and $Q$, and an annotated variable $x^\tau$ to a new strategy inductively on a* ∀-Exp+Res *derivation in Fig. 2. We write* Combine *in infix notation as $P \underset{x^\tau}{\sqcup} Q$.*

Note that in the case where both lit($P$) $\neq l$ and lit($Q$) $\neq l$ there is freedom of which strategy out of $P$ and $Q$ to select. We will use the variant selecting $P$.

*Example 1.* We introduce our running example and use it to demonstrate the combination of two strategies via Combine in Fig. 3.

**Theorem 1.** *Let $C$ be a clause derived by* ∀-Exp+Res *and $P$ be the corresponding strategy annotation computed by* Combine. *Then $P \models \text{enc}(C) \Rightarrow \neg \varphi$.*

*Proof.* By induction on the ∀-Exp+Res derivation.

*Base case.* The base case corresponds to the ∀-exp rule.

$$\frac{}{C^\tau \; [P \in \text{ConstStrat}(\Pi, \tau)]} \; (\text{∀-exp})$$

We need to show that $P \models \text{enc}(C^\tau) \Rightarrow \neg \varphi$. From the definition of ConstStrat we know that $P$ satisfies all universal literals in $\text{enc}(C^\tau)$ following the assignments determined by $\tau$. $P$ similarly satisfies the literals in the corresponding negated clause $\neg C$ in $\neg \varphi$, making both remaining formulas over the existential variables equivalent. The negated matrix $\neg \varphi$ is weaker than just $\neg C$, thus the implication holds.

For a ($\forall$-exp) inference

$$\frac{}{C^\tau \; [\mathsf{ConstStrat}(\Pi, \tau)]} \; (\forall\text{-exp})$$

For a (res) rule with pivot $x^\tau$

$$\frac{[P] \; C_1 \vee \neg x^\tau \quad C_2 \vee x^\tau \; [Q]}{C_1 \vee C_2 \; [P \underset{x^\tau}{\sqcup} Q]} \; (\text{res})$$

**Top-most variable is universal:**

Then $P \in \{\mathsf{Str}(u, P^-, \emptyset), \mathsf{Str}(u, \emptyset, P^+)\}$,
and $Q \in \{\mathsf{Str}(u, Q^-, \emptyset), \mathsf{Str}(u, \emptyset, Q^+)\}$, and $l \in \{u, \neg u\}$.

if $l \in \tau$, and $\mathsf{lit}(P) \neq l$          $P \underset{x^\tau}{\sqcup} Q \overset{\text{def}}{=} P$

if $l \in \tau$, and $\mathsf{lit}(Q) \neq l$          $P \underset{x^\tau}{\sqcup} Q \overset{\text{def}}{=} Q$

if $l \in \tau$, and $\mathsf{lit}(P) = \mathsf{lit}(Q) = l$     $\sigma \overset{\text{def}}{=} \tau - \{l\}$

    $- $ if $l = u$                  $P \underset{x^\tau}{\sqcup} Q \overset{\text{def}}{=} \mathsf{Str}(u, \emptyset, P^+ \underset{x^\sigma}{\sqcup} Q^+)$

    $- $ if $l = \neg u$            $P \underset{x^\tau}{\sqcup} Q \overset{\text{def}}{=} \mathsf{Str}(u, P^- \underset{x^\sigma}{\sqcup} Q^-, \emptyset)$

**Top-most variable is existential:**

if $\tau = \{\}$, $P = \mathsf{Str}(x, P^-, P^+)$
and $Q = \mathsf{Str}(x, Q^-, Q^+)$        $P \underset{x^\tau}{\sqcup} Q \overset{\text{def}}{=} \mathsf{Str}(x, Q^-, P^+)$

if $y \neq x$, $P = \mathsf{Str}(y, P^-, P^+)$
and $Q = \mathsf{Str}(y, Q^-, Q^+)$        $P \underset{x^\tau}{\sqcup} Q \overset{\text{def}}{=} \mathsf{Str}(y, P^- \underset{x^\tau}{\sqcup} Q^-, P^+ \underset{x^\tau}{\sqcup} Q^+)$

**Fig. 2.** Combine defined inductively along a $\forall$-Exp+Res derivation.

*Induction Step.* For a resolution rule with strategy annotations $P$, $Q$ and the combination of $P$ and $Q$, i.e. $P \underset{x^\tau}{\sqcup} Q$

$$\frac{[P] \; C_1 \vee \neg x^\tau \quad C_2 \vee x^\tau \; [Q]}{C_1 \vee C_2 \; [P \underset{x^\tau}{\sqcup} Q]} \; (\text{res})$$

we need to show that

$$P \models \mathsf{enc}(C_1 \vee \neg x^\tau) \Rightarrow \neg\varphi$$

and           $Q \models \mathsf{enc}(C_2 \vee x^\tau) \Rightarrow \neg\varphi$

implies        $P \underset{x^\tau}{\sqcup} Q \models \mathsf{enc}(C_1 \vee C_2) \Rightarrow \neg\varphi$

Let $\pi$ be an arbitrary complete assignment determined by strategy $P \underset{x^\tau}{\sqcup} Q$. We need to show that $\pi \models \mathsf{enc}(C_1 \vee C_2) \Rightarrow \neg\varphi$ given the induction hypothesis. By case distinction:

Consider annotated clauses $C_1 \vee \neg x_2^{u_1}$ $[P]$ and $C_2 \vee x_2^{u_1}$ $[Q]$. The strategies $P$ and $Q$ and their combination along the resolution with pivot $x_2^{u_1}$, i.e., $P \sqcup_{x_2^{u_1}} Q$ are depicted to the right. **Combine** proceeds recursively—top-down—along the trees $P$ and $Q$. At level $x_1$, we simply recurse and proceed by combining the sub-strategies along the paths $0/x_1$ and $1/x_1$ from $P$ and $Q$ because $x_1$ is not the pivot. On the path along $0/x_1$ we detect that $0/u_1$ in $P$ differs from the pivot's annotation $1/u_1$ and we select the sub-strategy anchored in $u_1$ from $P$. On the path along $1/x_1$ the annotation for $u_1$ matches with the values in $P$ and $Q$ and we continue to level $x_2$, which is the pivot. We select the sub-strategy starting in $0/x_2$ from $Q$ and the sub-strategy starting in $1/x_2$ from $P$ and are done.

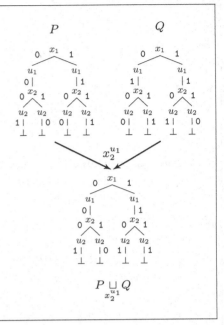

**Fig. 3.** An application of **Combine**.

1. If $\pi \not\models \text{enc}(C_1 \vee C_2)$ the implication is true and we are done.
2. If $\pi \models \text{enc}(C_1 \vee C_2)$ we have two cases:
   (a) $\pi \not\models \bigwedge_{u \in \tau} u$ ($\pi$ differs from the assignment determined by $\tau$):
   Let us assume, w.l.o.g., that $\pi$ is from $P$, then we have the following induction hypothesis:

$$\pi \models \text{enc}(C_1) \wedge \left( \bigwedge_{u \in \tau} u \Rightarrow x \right) \Rightarrow \neg\varphi.$$

   Since we are in case $\pi \models \text{enc}(C_1 \vee C_2)$, by the definition of enc we know that $\pi \models \text{enc}(C_1)$. Furthermore we know that $\pi \not\models \bigwedge_{u \in \tau} u$ satisfying the left-hand side of the outer implication, thus $\pi$ must satisfy $\neg\varphi$ for the IH to be valid. Since, in this case the **Combine** operator evaluates to $P$ and $\pi$ is from $P$, $P \models \text{enc}(C_1 \vee C_2) \Rightarrow \neg\varphi$ is valid.
   (b) $\pi \models \bigwedge_{u \in \tau} u$ ($\pi$ equals the assignment determined by $\tau$):
   Again, since we are in case $\pi \models \text{enc}(C_1 \vee C_2)$, by the definition of enc we know that $\pi \models \text{enc}(C_1)$ and $\pi \models \text{enc}(C_2)$. We also know that $\pi \models \bigwedge_{u \in \tau} u$, so when $\pi \in P$ the IH simplifies to $\pi \models x \Rightarrow \neg\varphi$. Similarly the IH simplifies to $\pi \models \neg x \Rightarrow \neg\varphi$ for $\pi \in Q$. Assume $x = 1$, then $P \models \neg\varphi$. When we assume $x = 0$, then $Q \models \neg\varphi$. In either case, because we assume the IH to be true, we know $\neg\varphi$ needs to be true. **Combine** chooses the respective paths in $P$ and $Q$ and combines them so that $\text{Str}(x, Q^-, P^+) \models \text{enc}(C_1 \vee C_2) \Rightarrow \neg\varphi$ is valid.    $\square$

*Remark on the Curry-Howard correspondence established by Theorem* 1:

- We relate the clauses of a ∀-Exp+Res-proof and the extracted strategies: $P \models$ enc$(C) \Rightarrow \neg\varphi$ signifies that the strategy $P$ is a witness for the validity of the QBF formula $\Pi.$enc$(C) \Rightarrow \neg\varphi$.
- We relate the rules of a ∀-Exp+Res-proof to strategy construction operators: For an expansion-axiom with regard to an assignment $\tau$, the strategy is given by ConstStrat$(\Pi, \tau)$. For a resolution step, the strategy is obtained by applying the Combine operator on the strategies for the parent nodes.

## 5   Implementing Strategies Using Circuits

The strategies we have introduced in the previous section have size exponential in the number of existential variables in the quantifier prefix. Thus, it is impractical to consider strategy extraction using such a data structure. Instead, we will now demonstrate how we can implement the Combine operator on circuits. We will show how we can construct the circuit for $n$ output variables in such a way that the size of the circuit is in the order of $\mathcal{O}(p \cdot n)$, where $p$ is the proof length (number of clauses). This size is asymptotically optimal when constructing circuits locally along the proof derivation for $n$ variables, considering that each inference can potentially manipulate each circuit.

### 5.1   Circuit Construction

We begin by introducing a number of auxiliary circuits. In the following let $L$, $R$, and $B$ (short for "left", "right", and "bottom", according to their respective positions in the inference rule) be tuples of circuits and let $\boldsymbol{y}$ be the input variables. We write $f_{u_i}$ for the circuit with output $u_i$ for $f \in \{L, R, B\}$.

**Definition 5 (Equiv).** *We define the circuits* Equiv$_f^{<i}$ *for* $f \in \{L, R\}$. *These circuits decide if all* $f_{u_i}$ *evaluate to* $\tau(u_i)$ *up to level* $i$, *given input* $\boldsymbol{y}$:

$$\mathsf{Equiv}_f^{<1}(\boldsymbol{y}) \stackrel{\text{def}}{=} 1 \quad and \quad \mathsf{Equiv}_f^{<i}(\boldsymbol{y}) \stackrel{\text{def}}{=} \mathsf{Equiv}_f^{<i-1}(\boldsymbol{y}) \wedge f_{u_{i-1}}(\boldsymbol{y}) \Leftrightarrow \tau(u_{i-1})$$

Next we define the circuits Diff$_L^i$ and Diff$_R^i$ using Equiv. The purpose of the Diff circuits is to choose one of $L$ and $R$ simulating the case of Combine when one of the strategies differs from $\tau$ at a universal edge. We need to consistently select the function values from either $L$ or $R$ starting from some index $i$ for all subsequent outputs $u_j$ with $j \geq i$.

**Definition 6 (Diff).** *We define the circuits* Diff$_L^i$ *and* Diff$_R^i$ *symmetrically to each other. We informally describe the circuit* Diff$_L^i$: Diff$_L^i$ *is true, given input* $\boldsymbol{y}$, *if there has been a difference between an* $L_{u_j}$ *and* $\tau(u_j)$ *for* $j \leq i$ *and when there has been no difference between* $R_{u_k}$ *and* $\tau(u_k)$ *for* $k < j$. *Formally:*

$$\mathsf{Diff}_L^0(\boldsymbol{y}) \stackrel{\text{def}}{=} 0 \quad and \quad \mathsf{Diff}_L^i(\boldsymbol{y}) \stackrel{\text{def}}{=} \mathsf{Diff}_L^{i-1}(\boldsymbol{y}) \vee (\mathsf{Equiv}_R^{<i}(\boldsymbol{y}) \wedge L_{u_i}(\boldsymbol{y}) \oplus \tau(u_i))$$

$$\mathsf{Diff}_R^0(\boldsymbol{y}) \stackrel{\text{def}}{=} 0 \quad and \quad \mathsf{Diff}_R^i(\boldsymbol{y}) \stackrel{\text{def}}{=} \mathsf{Diff}_R^{i-1}(\boldsymbol{y}) \vee (\mathsf{Equiv}_L^{<i}(\boldsymbol{y}) \wedge R_{u_i}(\boldsymbol{y}) \oplus \tau(u_i))$$

**Proposition 1.** *Let $f \in \{L, R\}$ be either the left or right circuit and let $g \in \{L, R\} - \{f\}$ be the opposite one. Once it has been established that $\mathsf{Diff}^i_f$ is true, we know that $\mathsf{Diff}^j_g$ cannot turn true for $j > i$ if it has not been true already at $i$. Formally,*

$$\mathsf{Diff}^i_f(\boldsymbol{y}) \wedge \neg\mathsf{Diff}^i_g(\boldsymbol{y}) \Rightarrow \neg\mathsf{Diff}^j_g(\boldsymbol{y}),$$

*for $j > i$ is a tautology.*

*Proof.* Assume that $f = L$ and $g = R$, with the other case symmetric. It is clear that when $\mathsf{Diff}^i_L(\boldsymbol{y})$ is true, $\mathsf{Equiv}^{<j}_L(\boldsymbol{y})$ must be false for $j > i$. When $\mathsf{Equiv}^{<j}_L(\boldsymbol{y})$ is false, we know that $\mathsf{Diff}^j_R$ will remain false, if $\mathsf{Diff}^i_R$ was false.  □

Note that both $\mathsf{Diff}^i_L$ and $\mathsf{Diff}^i_R$ can be true at the same index $i$. Namely, when there is no difference up to some level $j < i$ ($\mathsf{Equiv}^{<j}_L(\boldsymbol{y}) = \mathsf{Equiv}^{<j}_R(\boldsymbol{y}) = 1$) but both $L_{u_j}(\boldsymbol{y}) \neq \tau(u_j)$ and $R_{u_j}(\boldsymbol{y}) \neq \tau(u_j)$. In this case we have the same freedom as in Combine when both $\mathsf{lit}(P)$ and $\mathsf{lit}(Q)$ differ from $\tau$.

**Definition 7 (Circuit extraction for $\forall$-Exp+Res).** *Let $R$ be a $\forall$-Exp+Res proof. The circuit extraction $\mathsf{Cir}(u_i)$ for output $u_i$ maps vertices in $R$ to circuits as defined in Fig. 4—with the circuits $\mathsf{Comb}_{u_i}$ defined as follows.*
*Let $\diamond \in \{\wedge, \vee\}$. For $u_i \prec x$ we define*

$$\mathsf{Comb}^\diamond_{u_i}(\boldsymbol{y}) \overset{\mathrm{def}}{=} \text{if} \quad \mathsf{Diff}^{i-1}_L(\boldsymbol{y}) \text{ then } L_{u_i}(\boldsymbol{y})$$
$$\text{else if } \mathsf{Diff}^{i-1}_R(\boldsymbol{y}) \text{ then } R_{u_i}(\boldsymbol{y})$$
$$\text{else} \qquad\qquad\qquad L_{u_i}(\boldsymbol{y}) \diamond R_{u_i}(\boldsymbol{y}).$$

*Let $u_m$ be the maximum universal variable with $u_m \prec x$. For $x \prec u_i$, we define*

$$\mathsf{Comb}_{u_i}(\boldsymbol{y}, x) \overset{\mathrm{def}}{=} \text{if} \quad \mathsf{Diff}^m_L(\boldsymbol{y}) \text{ then } L_{u_i}(\boldsymbol{y}, x)$$
$$\text{else if } \mathsf{Diff}^m_R(\boldsymbol{y}) \text{ then } R_{u_i}(\boldsymbol{y}, x)$$
$$\text{else} \qquad (\neg x \vee L_{u_i}(\boldsymbol{y}, x)) \wedge (x \vee R_{u_i}(\boldsymbol{y}, x)).$$

Note that in the case when both $\mathsf{Diff}^i_L$ and $\mathsf{Diff}^i_R$ are true for $i$, we prefer $L$ (like we have preferred the left strategy $P$ in Combine), due to the order of appearance in the if-then-else cascade.

*Example 2.* Consider again the strategies $P$ and $Q$ introduced in Example 1. Strategy $P$ encodes the circuits $L_{u_1}(x_1) = x_1$ and $L_{u_2}(x_1, x_2) = x_1 \Leftrightarrow x_2$. Strategy $Q$ encodes the circuits $R_{u_1}(x_1) = 1$ and $R_{u_2}(x_1, x_2) = x_1 \oplus x_2$. We will show that combining the circuits $L$ and $R$ results in circuits $B$ encoded by $P \underset{x_2^{u_1}}{\sqcup} Q$, i.e. $B_{u_1}(x_1) = x_1$ and $B_{u_2}(x_1, x_2) = x_1 \vee \neg x_2$.

We will demonstrate that our circuit construction yields the same circuits: For $B_{u_1}$ we are in the case $u_1 \prec x_2$ and $\mathsf{Diff}^0_L$ and $\mathsf{Diff}^0_R$ are false by definition, $\diamond = \wedge$ because the annotation $u_1$ of the pivot is 1 so Definition 7 evaluates to

$$B_{u_1}(x_1) = \text{if} \quad \mathsf{Diff}^0_L(x_1) \text{ then } x_1$$
$$\text{else if } \mathsf{Diff}^0_R(x_1) \text{ then } 1$$
$$\text{else} \qquad\qquad x_1 \wedge 1$$
$$= x_1 \wedge 1 = x_1.$$

For a (∀-exp) inference

$$\frac{}{C^{\neg u_i \in \tau} \, [0]} \text{ (∀-exp)} \qquad \frac{}{C^{u_i \in \tau} \, [1]} \text{ (∀-exp)}$$

For a (res) rule with pivot $x^\tau$

$$\frac{[L_{u_i}] \, C_1 \vee \neg x^\tau \quad C_2 \vee x^\tau \, [R_{u_i}]}{C_1 \vee C_2 \, [B_{u_i}]} \text{ (res)}$$

if $u_i \prec x$, and $\neg u_i \in \tau$ $\quad B_{u_i} \stackrel{\text{def}}{=} \text{Comb}^\vee_{u_i}(\boldsymbol{y})$

if $x \prec u_i$ $\qquad\qquad\qquad B_{u_i} \stackrel{\text{def}}{=} \text{Comb}_{u_i}(\boldsymbol{y}, x)$

if $u_i \prec x$, and $u_i \in \tau$ $\quad B_{u_i} \stackrel{\text{def}}{=} \text{Comb}^\wedge_{u_i}(\boldsymbol{y})$

**Fig. 4.** Circuit extraction for ∀-Exp+Res proofs.

For $B_{u_2}$ we are in case $x \prec u_2$, and $u_1$ is the maximum $u_i \prec x$ so we have

$$B_{u_2}(x_1, x_2) = \begin{array}{ll} \text{if} & \text{Diff}^1_L(x_1) \text{ then } x_1 \Leftrightarrow x_2 \\ \text{else if } \text{Diff}^1_R(x_1) \text{ then } x_1 \oplus x_2 \\ \text{else} & (\neg x_2 \vee (x_1 \Leftrightarrow x_2)) \wedge (x_2 \vee (x_1 \oplus x_2)). \end{array}$$

$\text{Diff}^1_L(x_1)$ evaluates to $L_{u_1}(x_1) \oplus \tau(u_1) = x_1 \oplus 1 = \neg x_1$ indicating a difference in $L$ when $x_1 = 0$ leading us to choose the "if-then" branch: $0 \Leftrightarrow x_2$, which is true when $x_2 = 0$. $\text{Diff}^1_R(x_1)$ evaluates to $R_{u_1}(x_1) \oplus \tau(u_1) = 1 \oplus 1 = 0$ indicating no difference in $R$. So when $x_1 = 1$, we reach the "else" branch, which evaluates to 1 for both $x_2 = 0$ and $x_2 = 1$. Overall, we know that only the assignment $x_1 = 0, x_2 = 1$ makes $B_{u_2}$ false, thus we determine that $B_{u_2}(x_1, x_2) = x_1 \vee \neg x_2$.

## 5.2 Correctness and Running Time

**Lemma 1.** *Let $P$ and $Q$ be strategies and let $L$ and $R$ be families of circuits representing $P$ and $Q$, respectively. Then the family $B$ of circuits as specified in Definition 7 represents $P \underset{x^\tau}{\sqcup} Q$.*

*Proof (Sketch).* When a function value for an output differs from the annotation in circuit $L$ we select the circuits from $L$ for all consecutive outputs. While this operation is implicit in Combine by selecting whole sub-trees of a strategy, we need to make this operation explicit for each output in the circuit construction, by looking at all preceding outputs, which we do in the Diff circuits.

If all preceding outputs equal the annotation, then we compute the new function value for the current output as a disjunction or conjunction, depending on the assignment to the output in the annotation. This operation mimics Combine, both in selecting the differing edge, if an edge differs, and keeping the equivalent edge, if both function values equal the annotation.

The case where we reach the pivot variable in Combine and select sub-strategies from both input strategies, again needs to be made explicit in the circuit construction: We need to check that we are in this case, by inspecting whether one of the preceding outputs differs, like described above. However, we need to check only the outputs up to the level of the pivot variable. Beyond that, the selection of the sub-strategies at the pivot needs to be simulated, which we do by adding a multiplexer with the pivot being the selector input.

The case where the top-most existential variable differs from the pivot in Combine and we recurse is implicit in the circuit construction: The function values depend on these variables, but we do not need to handle existential variables beyond the multiplexer construction.

The case where we recurse in Combine when both universal edges adhere to the annotation is implicit in the circuit construction as well: it amounts to iterative computation of the functions according to the quantifier level.    □

**Lemma 2.** *Given a ∀-Exp+Res derivation of length $p$ from a PCNF formula $\Phi$ with $n$ universal variables, the circuits as defined in Definition 7 can be computed in time $\mathcal{O}(p \cdot n)$.*

*Proof.* For each output $u_i$, we need a circuit $\text{Diff}_L^{i-1}$. To compute that circuit we reuse the circuits computing $\text{Equiv}_R^{<i-1}$ and $\text{Diff}_L^{i-2}$, which we have already computed for $u_{i-1}$, so for output $u_i$ we only have to add the checks $R_{u_{i-1}} \Leftrightarrow \tau(u_{i-1})$ and $L_{u_{i-1}} \oplus \tau(u_{i-1})$ of constant size, and gates connecting these circuits, also of constant size. Thus, the number of gates of the $\text{Diff}_L$ circuits for all $n$ outputs is in the order of $\mathcal{O}(n)$. The same analysis applies to the $\text{Diff}_R$ circuits, adding another $\mathcal{O}(n)$. The if-then-else cascade adds another constant, but the overall circuit complexity at a proof node remains $\mathcal{O}(n)$. Thus, overall we have a circuit size and running time of $\mathcal{O}(p \cdot n)$.    □

In combination with Theorem 1, the preceding lemmas imply the following.

**Theorem 2.** *Given a ∀-Exp+Res derivation of length $p$ from a PCNF formula $\Phi$ with $n$ universal variables, a family of circuits implementing a universal winning strategy for $\Phi$ can be computed in time $\mathcal{O}(p \cdot n)$.*

*Similarity to Craig Interpolation.* When the circuit has a single output, note that the Diff circuits are always false and we only use the "else" branches. In this case, our system resembles a *symmetric* Craig interpolation system, cf. [19, 24, 28].

## 6    Circuit Extraction for QParity

We demonstrate our strategy extraction algorithm with the QPARITY formulas. Each formula $\text{QPARITY}_n$ has a single universal variable with the parity function on $n$ variables as the unique universal winning strategy. Since Q-resolution proofs can be efficiently turned into bounded-depth circuits computing a universal winning strategy, QPARITY is known to be hard for Q-resolution [6]. At the same time, it has short (even tree-like) ∀-Exp+Res proofs, and our strategy extraction algorithm obtains a small circuit representing the $n$-bit parity function.

*Example 3 (*QPARITY*).* The formula QPARITY says that there exists an assignment of $x_1, \ldots, x_n$ such that $u \neq x_1 \oplus \cdots \oplus x_n$ for all assignments of $u$. Clearly, this formula is false, and the (unique) winning strategy for the universal player is to assign $u = x_1 \oplus \cdots \oplus x_n$. A PCNF encoding is obtained by introducing auxiliary variables satisfying $y_i \Leftrightarrow \bigoplus_{j=1}^{i} x_j$ as follows:

$$\text{QPARITY}_n := \exists x_1 \ldots x_n \forall u \exists y_0 \ldots y_n . (\neg y_0) \wedge (u \Leftrightarrow y_n) \wedge \bigwedge_{i=1}^{n} (y_i \Leftrightarrow (y_{i-1} \oplus x_i))$$

The biconditional $u \Leftrightarrow y_n$ yields the clauses $(\neg u \vee \neg y_n)$ and $(u \vee y_n)$, and each formula $(y_i \Leftrightarrow (y_{i-1} \oplus x_i))$ translates to clauses $(\neg y_{i-1} \vee x_i \vee y_i)$, $(y_{i-1} \vee \neg x_i \vee y_i)$, $(y_{i-1} \vee x_i \vee \neg y_i)$, and $(\neg y_{i-1} \vee \neg x_i \vee \neg y_i)$. Beyersdorff et al. show how to construct short tree-like proofs for QPARITY in ∀-Exp+Res [4, Theorem 2]. We illustrate their construction for the case $n = 2$. By expanding the universal variable $u$ (applying the ∀-exp rule), we obtain the following initial clauses:

$$\underbrace{(y_0^{\neg u} x_1 \neg y_1^{\neg u})}_{C_1} \wedge \underbrace{(y_0^u \neg x_1 y_1^u)}_{C_2} \wedge \underbrace{(y_1^{\neg u} x_2 \neg y_2^{\neg u})}_{C_3} \wedge \underbrace{(y_1^u \neg x_2 y_2^u)}_{C_4} \wedge \underbrace{(y_0^{\neg u} \neg x_1 y_1^{\neg u})}_{C_5} \, .$$

$$\wedge \underbrace{(y_0^u x_1 \neg y_1^u)}_{C_6} \wedge \underbrace{(\neg y_1^{\neg u} \neg x_2 \neg y_2^{\neg u})}_{C_7} \wedge \underbrace{(\neg y_1^u x_2 y_2^u)}_{C_8} \wedge \underbrace{(\neg y_0^u)}_{C_9} \wedge \underbrace{(\neg y_0^{\neg u})}_{C_{10}} \wedge \underbrace{(y_2^{\neg u})}_{C_{11}} \wedge \underbrace{(\neg y_2^u)}_{C_{12}}$$

A resolution refutation completing the ∀-Exp+Res proof is shown in Fig. 5, where each clause is annotated with the circuit computed for $u$ according to Def. 7. The empty clause is annotated $(x_1 \vee x_2) \wedge (\neg x_1 \vee \neg x_2) = x_1 \oplus x_2$, which is a winning strategy.

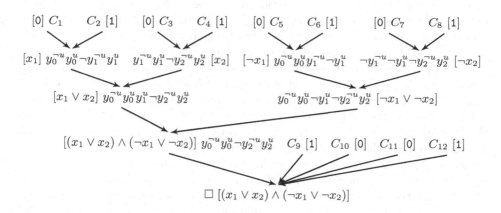

**Fig. 5.** ∀-Exp+Res proof of QPARITY$_2$. We compress the last proof steps since they do not affect the extracted circuit—either we add a conjunction with 1 or a disjunction with 0.

# 7    Related Work

Suda and Gleiss present a local soundness argument for several resolution-based QBF calculi, including a generalization of ∀-Exp+Res [33]. They interpret clauses derived in these systems as abstractions of partial strategies (a partial strategy does not need to be defined for all moves of the existential player), and show that resolution can be understood in terms of combining partial strategies. Soundness of a proof system is obtained by showing that partial strategies with the premises of a resolution step as their abstractions result in a partial strategy that abstracts to the resolvent. The statement that the partial strategy constructed at a particular node of a proof DAG abstracts to the clause derived at that node is proved only indirectly, through observing that there are simple partial strategies abstracting to initial clauses.

By contrast, we define a syntactic weakening of the matrix for each node in the proof DAG and show that the strategy extracted at that node satisfies the weakened matrix. Moreover, we manipulate complete strategies, which are defined for all moves of the existential player. We believe that our use of complete strategies and an explicit syntactic construction offer a considerably simpler and clearer local soundness argument for ∀-Exp+Res.

A correspondence between Q-resolution proofs and strategies was first observed by Goultiaeva et al. [15] and later extended to long-distance Q-resolution by Egly et al. [12]. Balabanov and Jiang [2] present a linear-time strategy extraction algorithm for Q-resolution that was generalized to long-distance Q-resolution by Balabanov et al. [3]. Beyersdorff et al. [5] prove a correspondence between strategies and proofs in IRM-calc, a system that generalizes ∀-Exp+Res. The notion that efficient extraction of winning moves from proofs leads to polynomial-time strategy extraction is folklore. Peitl et al. [27] give an explicit construction for Q-resolution with a dependency scheme. Chew and Clymo [11] provide a general argument for QBF proof systems that combine a propositional proof system with universal expansion. Surprisingly, they also identify *feasible interpolation* of the underlying propositional proof system (i.e, the property that interpolants can be computed from refutation proofs in polynomial time [24]) as a *necessary* condition for such systems to have polynomial-time strategy extraction. They further show that the QRAT proof system does not have polynomial-time strategy extraction unless P=PSPACE. By contrast, Heule et al. [16] proved that the (almost) dual proof system for true formulas does have polynomial-time strategy extraction.

Jiang et al. [23] synthesize Boolean functions with a single output using propositional Craig interpolation [19,24,28]. Given a Boolean relation $\varphi : \mathbb{B}^n \times \mathbb{B} \mapsto \mathbb{B}$, the authors of [23] derive a circuit $f(\boldsymbol{x})$ such that $\forall \boldsymbol{x}.\exists u.\varphi(\boldsymbol{x}, u) \equiv \forall \boldsymbol{x}.\varphi(\boldsymbol{x}, f(\boldsymbol{x}))$ holds. They derive a *resolution refutation* from $\forall \boldsymbol{x}.\exists u.\varphi(\boldsymbol{x}, u)$ by negating it first and then expanding the universal quantifier to obtain an unsatisfiable CNF instance $\neg\varphi(\boldsymbol{x}, 0) \wedge \neg\varphi(\boldsymbol{x}, 1)$, which is then split into two partitions $A^{\neg u} \stackrel{\text{def}}{=} \neg\varphi(\boldsymbol{x}, 0)$ and $B^u \stackrel{\text{def}}{=} \neg\varphi(\boldsymbol{x}, 1)$. An interpolant $I(\boldsymbol{x})$ for these partitions satisfies $(A^{\neg u} \to I)$ and $(B^u \to \neg I)$, hence the circuit $f(\boldsymbol{x}) \stackrel{\text{def}}{=} I(\boldsymbol{x})$ yields 1 if

$\neg\varphi(\boldsymbol{x}, 0)$ and $0$ if $\neg\varphi(\boldsymbol{x}, 1)$, satisfying the requirement above. $I(\boldsymbol{x})$ is obtained by annotating all clauses $C$ in the resolution refutation by partial interpolants $I_C$, where $I_C \stackrel{\text{def}}{=} 0$ if $C \in A^{\neg u}$, $I_C \stackrel{\text{def}}{=} 1$ if $C \in B^u$, and $I_c \stackrel{\text{def}}{=} (x \vee I_1) \wedge (\neg x \vee I_2)$ if $C$ is the result of a resolution of $C_1$ and $C_2$ with partial interpolants $I_1$ and $I_2$, respectively, on the pivot literal $x$.

The construction of the partitions $A^{\neg u}$ and $B^u$ in [23] is analogous to QBF Expansion, and propositional interpolation is a (less general) version of the circuit extraction in Fig. 4. Consequently, [23] can be seen as a special case of our framework that is limited to a single universally quantified variable. In fact, [23] proposes an iterative approach to deal with multiple outputs (universal quantifiers, respectively), requiring the repeated construction of refutations and interpolants and the substitution of outputs one at a time.

Hofferek et al. [17] extend the approach of Jiang et al. [23] to $n$ universally quantified Boolean variables by (syntactically) expanding the quantified formula into $2^n$ partitions and adapting the interpolation system to multiple partitions accordingly. Their approach targets the theory of uninterpreted functions with equality, which is a more expressive logic, but is limited to ∀∃∀-prefixes and imposes an order on the resolution steps in the propositional part of the refutation.

Beyersdorff et al. [7] present a feasible interpolation technique for the calculi LQU+-Res and IRM-calc. Their approach is restricted to instances of the form $\exists \boldsymbol{p}.Q\boldsymbol{q}.Q\boldsymbol{r}.A(\boldsymbol{p}, \boldsymbol{q}) \wedge B(\boldsymbol{p}, \boldsymbol{r})$ (where $\boldsymbol{q}$ and $\boldsymbol{r}$ can be quantified arbitrarily) and yields an interpolant $I(\boldsymbol{p})$. They show that for instances of the form $\exists \boldsymbol{p}.\forall u.Q\boldsymbol{q}.Q\boldsymbol{r}.(A(\boldsymbol{p}, \boldsymbol{q}) \vee u) \wedge (B(\boldsymbol{p}, \boldsymbol{r}) \vee \neg u)$ the resulting interpolant $I(\boldsymbol{p})$ represents a strategy for instantiating $u$. While this approach extends Jiang et al. [23] to arbitrarily quantified partitions, it is still limited to a single output $u$.

## 8    Conclusion

We presented a polynomial-time strategy extraction algorithm for ∀-Exp+Res with a running time that is multi-linear in the number of universal variables and resolution steps in the proof. It is based on a local soundness argument showing that each intermediate strategy constructed for a derived clause satisfies a propositional invariant obtained from that clause. This invariant translates annotated literals back to the vocabulary of the original formula and gives them a clear semantics based on the evaluation game: if the existential player responds to the universal play in the annotation by setting the literal to false, the current strategy is winning for the universal player. We believe that this idea can be extended to more general proof systems such as IRM-calc [5]. Moreover, our interpretation of annotated clauses in terms of the original variables may open up new ways of integrating search-based (QCDCL) solvers with expansion.

# References

1. Ayari, A., Basin, D.: QUBOS: deciding quantified boolean logic using propositional satisfiability solvers. In: Aagaard, M.D., O'Leary, J.W. (eds.) FMCAD 2002. LNCS, vol. 2517, pp. 187–201. Springer, Heidelberg (2002). https://doi.org/10.1007/3-540-36126-X_12
2. Balabanov, V., Jiang, J.R.: Unified QBF certification and its applications. Formal Methods Syst. Des. **41**(1), 45–65 (2012)
3. Balabanov, V., Jiang, J.R., Janota, M., Widl, M.: Efficient extraction of QBF (counter)models from long-distance resolution proofs. In: Bonet, B., Koenig, S. (eds.) Proceedings of the Twenty-Ninth AAAI Conference on Artificial Intelligence (AAAI-15), pp. 3694–3701. AAAI Press (2015)
4. Beyersdorff, O., Chew, L., Clymo, J., Mahajan, M.: Short proofs in QBF expansion. In: Janota, M., Lynce, I. (eds.) SAT 2019. LNCS, vol. 11628, pp. 19–35. Springer, Cham (2019). https://doi.org/10.1007/978-3-030-24258-9_2
5. Beyersdorff, O., Chew, L., Janota, M.: On unification of QBF resolution-based calculi. In: Csuhaj-Varjú, E., Dietzfelbinger, M., Ésik, Z. (eds.) MFCS 2014. LNCS, vol. 8635, pp. 81–93. Springer, Heidelberg (2014). https://doi.org/10.1007/978-3-662-44465-8_8
6. Beyersdorff, O., Chew, L., Janota, M.: New resolution-based QBF calculi and their proof complexity. TOCT **11**(4), 26:1–26:42 (2019)
7. Beyersdorff, O., Chew, L., Mahajan, M., Shukla, A.: Feasible interpolation for QBF resolution calculi. Logical Methods Comput. Sci. **13**(2), 1–20 (2017). https://lmcs.episciences.org/3702/pdf
8. Biere, A.: Resolve and expand. In: Hoos, H.H., Mitchell, D.G. (eds.) SAT 2004. LNCS, vol. 3542, pp. 59–70. Springer, Heidelberg (2005). https://doi.org/10.1007/11527695_5
9. Bloem, R., Braud-Santoni, N., Hadzic, V., Egly, U., Lonsing, F., Seidl, M.: Expansion-based QBF solving without recursion. In: Bjørner, N., Gurfinkel, A. (eds.) 2018 Formal Methods in Computer Aided Design, FMCAD 2018, pp. 1–10. IEEE (2018)
10. Chew, L., Clymo, J.: The equivalences of refutational QRAT. In: Janota, M., Lynce, I. (eds.) SAT 2019. LNCS, vol. 11628, pp. 100–116. Springer, Cham (2019). https://doi.org/10.1007/978-3-030-24258-9_7
11. Chew, L., Clymo, J.: How QBF expansion makes strategy extraction hard. In: International Joint Conference on Automated Reasoning, IJCAR 2020 (2020)
12. Egly, U., Lonsing, F., Widl, M.: Long-distance resolution: proof generation and strategy extraction in search-based QBF solving. In: McMillan, K., Middeldorp, A., Voronkov, A. (eds.) LPAR 2013. LNCS, vol. 8312, pp. 291–308. Springer, Heidelberg (2013). https://doi.org/10.1007/978-3-642-45221-5_21
13. Faymonville, P., Finkbeiner, B., Rabe, M.N., Tentrup, L.: Encodings of bounded synthesis. In: Legay, A., Margaria, T. (eds.) TACAS 2017. LNCS, vol. 10205, pp. 354–370. Springer, Heidelberg (2017). https://doi.org/10.1007/978-3-662-54577-5_20
14. Gomes, C.P., Kautz, H.A., Sabharwal, A., Selman, B.: Satisfiability solvers. In: van Harmelen, F., Lifschitz, V., Porter, B.W. (eds.) Handbook of Knowledge Representation, Foundations of Artificial Intelligence, vol. 3, pp. 89–134. Elsevier (2008)
15. Goultiaeva, A., Gelder, A.V., Bacchus, F.: A uniform approach for generating proofs and strategies for both true and false QBF formulas. In: Walsh, T. (ed.) IJCAI 2011, Proceedings of the 22nd International Joint Conference on Artificial Intelligence, pp. 546–553. IJCAI/AAAI (2011)

16. Heule, M.J.H., Seidl, M., Biere, A.: Solution validation and extraction for QBF preprocessing. J. Autom. Reason. **58**(1), 97–125 (2017)

17. Hofferek, G., Gupta, A., Könighofer, B., Jiang, J.R., Bloem, R.: Synthesizing multiple Boolean functions using interpolation on a single proof. In: Formal Methods in Computer-Aided Design, FMCAD 2013. pp. 77–84. IEEE (2013)

18. Howard, W.A.: The formulas-as-types notion of construction. In: Seldin, J.P., Hindley, J.R. (eds.) To H. B. Curry: Essays on Combinatory Logic, Lambda Calculus, and Formalism, pp. 479–490. Academic Press (1980)

19. Huang, G.: Constructing craig interpolation formulas. In: Du, D.-Z., Li, M. (eds.) COCOON 1995. LNCS, vol. 959, pp. 181–190. Springer, Heidelberg (1995). https://doi.org/10.1007/BFb0030832

20. Janota, M.: Towards generalization in QBF solving via machine learning. In: McIlraith, S.A., Weinberger, K.Q. (eds.) Proceedings of the Thirty-Second AAAI Conference on Artificial Intelligence, (AAAI-2018), pp. 6607–6614. AAAI Press (2018)

21. Janota, M., Klieber, W., Marques-Silva, J., Clarke, E.M.: Solving QBF with counterexample guided refinement. Artif. Intell. **234**, 1–25 (2016)

22. Janota, M., Marques-Silva, J.: Expansion-based QBF solving versus Q-resolution. Theor. Comput. Sci. **577**, 25–42 (2015)

23. Jiang, J.R., Lin, H., Hung, W.: Interpolating functions from large Boolean relations. In: Roychowdhury, J.S. (ed.) 2009 International Conference on Computer-Aided Design, ICCAD 2009, pp. 779–784. ACM (2009)

24. Krajícek, J.: Interpolation theorems, lower bounds for proof systems, and independence results for bounded arithmetic. J. Symb. Logic **62**(2), 457–486 (1997)

25. Lonsing, F., Bacchus, F., Biere, A., Egly, U., Seidl, M.: Enhancing search-based QBF Solving by dynamic blocked clause elimination. In: Davis, M., Fehnker, A., McIver, A., Voronkov, A. (eds.) LPAR 2015. LNCS, vol. 9450, pp. 418–433. Springer, Heidelberg (2015). https://doi.org/10.1007/978-3-662-48899-7_29

26. Peitl, T., Slivovsky, F., Szeider, S.: Dependency learning for QBF. J. Artif. Intell. Res. **65**, 180–208 (2019)

27. Peitl, T., Slivovsky, F., Szeider, S.: Long-distance Q-resolution with dependency schemes. J. Autom. Reason. **63**(1), 127–155 (2019)

28. Pudlák, P.: Lower bounds for resolution and cutting plane proofs and monotone computations. J. Symb. Logic **62**(3), 981–998 (1997)

29. Rabe, M.N., Seshia, S.A.: Incremental determinization. In: Creignou, N., Le Berre, D. (eds.) SAT 2016. LNCS, vol. 9710, pp. 375–392. Springer, Cham (2016). https://doi.org/10.1007/978-3-319-40970-2_23

30. Rabe, M.N., Tentrup, L.: CAQE: A certifying QBF solver. In: Kaivola, R., Wahl, T. (eds.) Formal Methods in Computer-Aided Design, FMCAD 2015, pp. 136–143. IEEE (2015)

31. Schaefer, T.J.: On the complexity of some two-person perfect-information games. J. Comput. Syst. Sci. **16**(2), 185–225 (1978)

32. Shukla, A., Biere, A., Pulina, L., Seidl, M.: A survey on applications of quantified Boolean formulas. In: 31st IEEE International Conference on Tools with Artificial Intelligence, ICTAI 2019, pp. 78–84. IEEE (2019)

33. Suda, M., Gleiss, B.: Local soundness for QBF calculi. In: Beyersdorff, O., Wintersteiger, C.M. (eds.) SAT 2018. LNCS, vol. 10929, pp. 217–234. Springer, Cham (2018). https://doi.org/10.1007/978-3-319-94144-8_14

34. Tentrup, L.: Non-prenex QBF Solving using abstraction. In: Creignou, N., Le Berre, D. (eds.) SAT 2016. LNCS, vol. 9710, pp. 393–401. Springer, Cham (2016). https://doi.org/10.1007/978-3-319-40970-2_24

35. Vizel, Y., Weissenbacher, G., Malik, S.: Boolean satisfiability solvers and their applications in model checking. Proc. IEEE **103**(11), 2021–2035 (2015)
36. Wimmer, R., Reimer, S., Marin, P., Becker, B.: HQSpre – an effective preprocessor for QBF and DQBF. In: Legay, A., Margaria, T. (eds.) TACAS 2017. LNCS, vol. 10205, pp. 373–390. Springer, Heidelberg (2017). https://doi.org/10.1007/978-3-662-54577-5_21

# Positional Games and QBF: The *Corrective* Encoding

Valentin Mayer-Eichberger[1]([⊠]) and Abdallah Saffidine[2]

[1] Technische Universität Berlin, Berlin, Germany
valentin@mayer-eichberger.de
[2] University of New South Wales, Sydney, Australia
abdallah.saffidine@gmail.com

**Abstract.** Positional games are a mathematical class of two-player games comprising Tic-tac-toe and its generalizations. We propose a novel encoding of these games into Quantified Boolean Formulas (QBFs) such that a game instance admits a winning strategy for first player if and only if the corresponding formula is true. Our approach improves over previous QBF encodings of games in multiple ways. First, it is generic and lets us encode other positional games, such as Hex. Second, structural properties of positional games together with a careful treatment of illegal moves let us generate more compact instances that can be solved faster by state-of-the-art QBF solvers. We establish the latter fact through extensive experiments. Finally, the compactness of our new encoding makes it feasible to translate realistic game problems. We identify a few such problems of historical significance and put them forward to the QBF community as milestones of increasing difficulty.

**Keywords:** QBF encodings · Positional games · Quantified Boolean Formula

## 1 Introduction

In a *positional game* [3,12], two players alternately claim unoccupied elements of the board of the game. The goal of a player is to claim a set of elements that form a winning set, and/or to prevent the other player from doing so. TIC-TAC-TOE, and its competitive variant played on a 15 × 15 board, GOMOKU, as well as HEX are the most well-known positional games. When the size of the board is not fixed, the decision problem, whether the first player has a winning strategy from a given position in the game is PSPACE-complete for many such games. The first result was established for GENERALIZED HEX, a variant played on an arbitrary graph [8]. Reisch [24] soon followed up with results for GOMOKU [24] and HEX played on a board [25].

Recent work on the classical and parameterized computational complexity of positional games provides us with elegant first-order logic formulations of such domains [3,4]. We draw inspiration from this approach and introduce a

© Springer Nature Switzerland AG 2020
L. Pulina and M. Seidl (Eds.): SAT 2020, LNCS 12178, pp. 447–463, 2020.
https://doi.org/10.1007/978-3-030-51825-7_31

practical implementation for such games into QBF. We believe that Positional Games exhibit a class of games large enough to include diverse and interesting games and benchmarks, yet allow for a specific encoding exploiting structural properties.

Our contributions are as follows: (1) Introduce the *Corrective Encoding*: a generic translation of positional games into QBF; (2) We identify a few positional games of historical significance and put them forward to the QBF community as milestones of increasing difficulty; (3) Demonstrate on previously published benchmark instances that our encoding leads to more compact instances that can be solved faster by state-of-the-art QBF solvers; (4) Establish that the Corrective Encoding enables QBF solving of realistic small scale puzzles of interest to human players.

After a formal introduction to QBF and to positional games (Sect. 2), we describe the contributed translation of positional games into QBFs (Sect. 3). We then describe the selected benchmark game problems, including the proposed milestones (Sect. 4), before experimentally evaluating the quality of our encoding and comparing it to previous work (Sect. 5). We conclude with a discussion contrasting our encoding with related work (Sect. 6).

## 2    Preliminaries on QBF and Positional Games

We assume a finite set of propositional variables $X$. A literal is a variable $x$ or its negation $\neg x$. A clause is a disjunction of literals. A Conjunctive Normal Form (CNF) formula is a conjunction of clauses. An assignment of the variables is a mapping $\tau : X \to \{\bot, \top\}$. A literal $x$ (resp. $\neg x$) is satisfied by the assignment $\tau$ if $\tau(x) = \top$ (resp. $\tau(x) = \bot$). A clause is satisfied by $\tau$ if at least one of the literals is satisfied. A CNF formula is satisfied if all the clauses are satisfied.

A QBF formula (in *Prenex*-CNF) is a sequence of alternating blocks of existential ($\exists$) and universal ($\forall$) quantifiers over the propositional variables followed by a CNF formula. A QBF formula may be interpreted as a game where an $\exists$ and $\forall$ player take turns building a variable assignment $\tau$ selecting the variables in the order of the quantifier prefix. The objective of $\exists$ (resp. $\forall$) is that $\tau$ satisfies (resp. falsifies) the formula.

*Positional games* are played by two players on a hypergraph $G = (V, E)$. The vertex set $V$ indicates the set of available positions, while the each hyperedge $e \in E$ denotes a winning configuration. For some games, the hyperedges are implicitly defined, instead of being explicitly part of the input. The two players alternatively claim unclaimed vertices of $V$ until either all elements are claimed or one player wins. A *position* in a positional game is an allocation of vertices to the players who have already claimed these vertices. The *empty position* is the position where no vertex is allocated to a player. The notion of winning depends on the game type. In a *Maker-Maker game*, the first player to claim all vertices of some hyperedge $e \in E$ wins. In a *Maker-Breaker game*, the first player (*Maker*) wins if she claims all vertices of some hyperedge $e \in E$. If the game ends and player 1 has not won, then the second player (*Breaker*) wins. The class of $(p, q)$-positional games is defined similarly to that of positional games, except that on the first move, Player 1 claims $q$ vertices and then each move after the first, a player claims $p$ vertices instead of 1. A *winning strategy* for player 1 is a move for player 1 such that for all moves of player 2 there exists a move of player 1... such that player 1 wins.

To illustrate these concepts, Fig. 1 displays a position from a well-known Maker-Maker game, TIC-TAC-TOE, and a position from a Maker-Breaker game, HEX. Although the rules of HEX are typically stated as Player 1 trying to create a path from top left to bottom right and Player 2 trying to connect the top right to bottom left, the objective of Player 2 is equivalent to preventing Player 1 from connecting their edges [20]. Therefore, HEX can indeed be seen as a Maker-Breaker positional game.

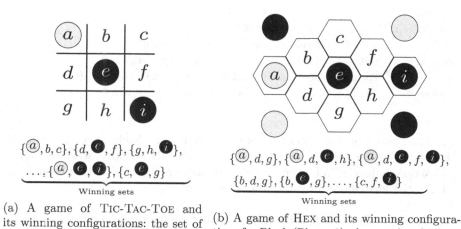

(a) A game of TIC-TAC-TOE and its winning configurations: the set of aligned triples.

(b) A game of HEX and its winning configurations for Black (Player 1): the set of paths from the top left edge to the bottom right edge.

**Fig. 1.** Two positional games played on the same vertices: $a$–$i$ where vertex $a$ has been claimed by Player 2 and vertices $e$ and $i$ have been claimed by Player 1.

# 3   The Corrective Encoding

In this section we present the Corrective encoding ($COR$). First we define positional games formally, describe the set of variables and the clauses in detail and analyse the size of the encoding.

## 3.1   Description

A positional game is a tuple $\prod = \langle T_\mathrm{B}, T_\mathrm{W}, \mathsf{F}, \mathsf{N}, E_\mathrm{B}, E_\mathrm{W} \rangle$ consisting of:

- Disjoint sets $T_\mathrm{B}$ and $T_\mathrm{W}$ of time points in which Black and White make moves. We denote $T = T_\mathrm{B} \cup T_\mathrm{W}$ as the set of all time points that range from $\{1, 2, \ldots \mathsf{F}\}$. For example, in a positional game where black starts and $p = q = 1$, $T_\mathrm{B}$ contains all odd and $T_\mathrm{W}$ all even numbers of $T$.

- $F \in \mathbb{N}$ the depth (or length) of the game.
- A set of vertices $V = \{1 \ldots N\}$ and two sets of hyperedges $E_B$ and $E_W$ of winning configurations for Black and White, respectively.

The remainder of this section defines a translation of a positional game configuration $\prod$ into a prenex QBF in CNF. For this we introduce variables defined in the following table. For readability we use a function style notation instead of variable subscripts. Let $A$ denote the set of the two players $\{B, W\}$.

| Variable | Description |
|---|---|
| time$(t)$ | The game is still running by time point $t \in T$ |
| board$(a, v, t)$ | Player $a \in A$ owns vertex $v \in V$ at time point $t \in T$ |
| occupied$(v, t)$ | Vertex $v \in V$ is occupied at $t \in T$ |
| win$(e)$ | Black has claimed winning configuration $e \in E_B$ |
| move$(v, t)$ | Vertex $v \in V$ is chosen at $t \in T_a$ by player $a \in A$ |
| moveL$(i, t)$ | The moves for White encoded *logarithmically*, $0 \le i < \lceil \log_2(N) \rceil\}$, $t \in T_W$ |
| ladder$(i, t)$ | Auxiliary variables for the ladder encoding, $i \in V$ and $t \in T$ |

The last two sets of variables are of technical nature; ladder is used to encode that a player must claim 1 vertex when it is their time $t$ to move, whereas moveL encodes the choices of White in a way that prevents the universal player from falsifying the formula by breaking the rules of the game.

*Quantification.* Here we specify the quantifier prefix of our encoding. In the order of the time point $t = \{1 \ldots F\}$ we introduce a level of quantifier blocks as follows:

$$\exists \text{time}(t)$$
$$\text{if } t \in T_W \text{ then for all } 0 \le i < \lceil \log_2(N) \rceil \quad \forall \text{moveL}(i, t)$$
$$\text{for } v \in V \quad \exists \text{move}(v, t)$$
$$\text{for } v \in V, a \in A \quad \exists \text{board}(a, v, t) \tag{1}$$
$$\text{for } v \in V \quad \exists \text{occupied}(v, t)$$
$$\text{for } i \in V \quad \exists \text{ladder}(i, t)$$

On the innermost level we have

$$e \in E_B \quad \exists \text{win}(e) \tag{2}$$

In the remainder of this section we list the clauses that make the body of the generated QBF instance. This body is constituted of sets of clauses encoding different aspects of the game. The encoding is almost entirely symmetric for both players apart from clauses which specify the interaction of the universal variables.

*Time Handling.* Variable time($t$) holds if the game is not over at time point $t$.

$$\{\neg\text{time}(t) \vee \text{time}(t-1) \mid t \in T\} \tag{3}$$

*Structure of the Board.* Clauses (5) encode that both players cannot own the same vertex. One fundamental property of positional games is that claimed vertices never change owner. This basic property is captured in clause (6). Note that these two clauses are independent from the time variable and act also when the game is over. Once a vertex is claimed and board($a, v, t$) is true the implication chain in (6) sets all board variables for that vertex in the future, in particular the last one board($a, v, \mathsf{F}$). Once the game is over (i.e. time($t$) is set to false), then all unclaimed vertices stay unclaimed (7) and the situation of the board at the last active move of the game is propagated through to the final time point.

$$\{\neg\text{board}(a, v, 0) \mid a \in A, v \in V\} \tag{4}$$

$$\{\neg\text{board}(\text{B}, v, t) \vee \neg\text{board}(\text{W}, v, t) \mid v \in V, t \in T\} \tag{5}$$

$$\{\neg\text{board}(a, v, t-1) \vee \text{board}(a, v, t) \mid a \in A, v \in V, t \in T\} \tag{6}$$

$$\{\text{time}(t) \vee \text{board}(a, v, t-1) \vee \neg\text{board}(a, v, t) \mid a \in A, v \in V, t \in T\} \tag{7}$$

The following clauses (8)–(10) define the meaning of occupied. Initially all vertices are unoccupied (8) and by definition a vertex is occupied if it is owned by Black or White.

$$\{\neg\text{occupied}(v, 0) \mid v \in V\} \tag{8}$$

$$\{\text{occupied}(v, t) \vee \neg\text{board}(a, v, t) \mid a \in A, v \in V, t \in T\} \tag{9}$$

$$\{\neg\text{occupied}(v, t) \vee \text{board}(\text{B}, v, t) \vee \text{board}(\text{W}, v, t) \mid v \in V, t \in T\} \tag{10}$$

*Player's Actions.* The action clauses specify how the moves of the players affect the board. If player $a$ claims vertex $v$ at time $t \in T_a$, i.e. move($v, t$) is true, then the game still has to be running (11). This clause can also be understood as if the game is over no moves are allowed anymore. Moreover, when move($v, t$) is true, then vertex $v$ was not occupied at the previous time point (12) and as a result of the action the vertex is occupied (13).

$$\{\text{time}(t) \vee \neg\text{move}(v, t) \mid v \in V, t \in T\} \tag{11}$$

$$\{\neg\text{occupied}(v, t-1) \vee \neg\text{move}(v, t) \mid v \in V, t \in T\} \tag{12}$$

$$\{\text{board}(a, v, t) \vee \neg\text{move}(v, t) \mid v \in V, a \in A, t \in T_a\} \tag{13}$$

*White's Choice.* The core of this encoding is how the universal variables interact with the rest of the encoding without having to prevent illegal moves by White. To avoid that White chooses too many vertices we encode the move logarithmically through variables moveL($i, t$). Moreover, these variables only actually imply a move of White in case the game is still running and the vertex was unoccupied before. In case one of the prerequisites is not given, then no move will be forced by this clause. Let $L_1(v)$ denote to the set of indices that are 1 in the binary representation of $v$, likewise $L_0(v)$ where there is a 0. The following equality holds: $v = \sum_{j \in L_1(v)} 2^j$. For example, for $13 = 1101|_2$ the respective sets are $L_1(13) = \{0, 2, 3\}$ and $L_1(13) = \{1\}$.

$$
\left\{
\begin{array}{l}
\bigvee_{i \in L_1(v)} \neg\mathsf{moveL}(i, t) \vee \bigvee_{i \in L_0(v)} \mathsf{moveL}(i, t) \\[2mm]
\vee\neg\mathsf{time}(t) \vee \mathsf{occupied}(v, t-1) \vee \mathsf{move}(v, t) \mid v \in V, t \in T_{\mathrm{W}}
\end{array}
\right\}
\tag{14}
$$

Notice how these clauses interact with (11) and (12) such that any choice for the universal variables is not able to cause a contradiction. In case White chooses a combination of moveL that does not imply an existing vertex, then still a move is selected for White to satisfy the ladder encoding (see (21) to (25)).

*Frame Axioms.* The following two clauses specify what happens when no action is performed on a position and the board variable is unchanged. Clause (15) says that in time points where player $s$ does not claim vertex $v$ and the vertex has not been previously owned by $s$ then it will also not be owned in the following step. In time points $t$ of the opponent to $s$, all unclaimed vertices by $s$ will be unclaimed in the next time point. Clause (16) forces this.

$$
\{\mathsf{move}(v, t) \vee \mathsf{board}(a, v, t-1) \vee \neg\mathsf{board}(a, v, t) \mid a \in A, v \in V, t \in T_a\}
\tag{15}
$$

$$
\{\mathsf{board}(a, v, t-1) \vee \neg\mathsf{board}(a, v, t) \mid a \in A, v \in V, t \in T \setminus T_a\}
\tag{16}
$$

*Winning Configuration.* For each winning configuration $e \in E_{\mathrm{B}}$ we have introduced a variable $\mathsf{win}(e)$ and clauses (17) specifies that at least one of the winning configuration have to be reached and (17) defines which vertices belong to it.

White should never reach a winning position, for this we introduce a clause for each winning positions specified in clauses (19). We only need to encode this for the last time point $F$ due to the implication chain (6). This looks straightforward from the definition, but we need to make sure with other clauses that White is unable to play illegal moves to reach a winning position.

$$
\left\{ \bigvee_{e \in E_{\mathrm{B}}} \mathsf{win}(e) \right\}
\tag{17}
$$

$$
\{\neg\mathsf{win}(e) \vee \mathsf{board}(\mathrm{B}, v, \mathrm{F}) \mid v \in e, e \in E_{\mathrm{B}}\}
\tag{18}
$$

$$
\left\{ \bigvee_{v \in e} \neg\mathsf{board}(\mathrm{W}, v, \mathrm{F}) \;\middle|\; e \in E_{\mathrm{W}} \right\}
\tag{19}
$$

$$
\left\{ \mathsf{win}(e) \vee \bigvee_{v \in e} \neg\mathsf{board}(\mathrm{B}, v, \mathrm{F} \;\middle|\; e \in E_{\mathrm{B}} \right\}
\tag{20}
$$

*Number of Moves.* To restrict the number of moves we apply the *ladder encoding* [9] to translate the cardinality constraint specifying the number of moves that a player can make in a round.

The ladder essentially encodes the equivalence move$(i+1, t) \Leftrightarrow \neg$ladder$(i, t) \wedge$ ladder$(i + 1, t)$. As soon as a move variable is set to true, all following ladder variables are forced to true (21, 22) and all previous are forced to false (21, 23). This ensures that no two move variables can be set to true. Clauses (25, 26) ensure that at least one move variable is true.

$$\{\neg \text{ladder}(i, t) \vee \text{ladder}(i + 1, t) \mid i \in V, i < \mathsf{N}, t \in T\} \qquad (21)$$

$$\{\neg \text{move}(i, t) \vee \text{ladder}(i, t) \mid i \in V, t \in T\} \qquad (22)$$

$$\{\neg \text{move}(i + 1, t) \vee \neg \text{ladder}(i, t) \mid i \in V, i < \mathsf{N}, t \in T\} \qquad (23)$$

$$\{\text{move}(1, t) \vee \neg \text{ladder}(1, t) \mid t \in T\} \qquad (24)$$

$$\{\text{move}(i + 1, t) \vee \text{ladder}(i, t) \vee \neg \text{ladder}(i + 1, t) \mid i \in V, i < \mathsf{N}, t \in T\} \qquad (25)$$

$$\{\neg \text{time}(t) \vee \text{ladder}(\mathsf{N}, t) \mid t \in T\} \qquad (26)$$

These clauses also enforce a move by White even if White had chosen an already claimed vertex or no vertex with the moveL variables and clause (14) does not fire. This is a crucial property of the encoding. An arbitrary vertex for White is chosen by, the game continues and there is no backtracking even though the universal player acted illegal.

*Initial Positions.* The QBF generator can also translate positional games that contain initial positions of White and Black, i.e. vertices that players own before the actual game starts. It is straight forward to turn this into an equivalent description without initial positions: For each initial position $v$ of one player remove this vertex from all its winning configurations and remove all winning configurations of the opposing player that contain $v$. After this operation we can remove $v$ from $V$ and have an equivalent game.

*Symmetry Breaking.* We employ a simple form of manual symmetry breaking by restricting the set of vertices from which the first move can be chosen. For instance in Generalized Tic-tac-toe (GTTT) and a $n \times n$ board, if this set contains the upper left triangle of the board (the set of coordinates $(i, j)$ such that $1 \leq i \leq j \leq n/2$), the symmetries of the squared board are broken. Typically, for other games with some initial positions for White and Black there is not much need for symmetry breaking since row or column symmetries are usually already broken by such a position.

*Consecutive Moves.* The positional game description need not have Black and White alternate moves: a description may allow a player to select several vertices consecutively. For instance, when $q = 2$ players claim two vertices in each round. For the sake of simplicity, our presentation of *COR* above does not break this symmetry. Our implementation avoids such symmetries by merging consecutive moves into a single turn where a subset of vertices of the right cardinality must be chosen. We implemented this cardinality constraint as a *sequential counter* [26]. In coincides with the ladder encodings in the single move case.

## 3.2   Size of the *COR* Encoding

It is straightforward to estimate the number of variables and clauses of the encoding. For instance, clauses with a description containing $v \in V, t \in T$ are generated at most $N \cdot F$ times. Since the depth of a game is limited by the number of vertices, the number of clauses is roughly $20N^2 + N \cdot |E_B| + |E_W|$.

| ∃ variables | ∀ variables | binary clauses | ternary clauses | long clauses |
|---|---|---|---|---|
| $4FN$ | $F \log_2(N)$ | $3N + 12NF + N|E_B|$ | $7NF$ | $NF + |E_B| + |E_W|$ |

# 4   Instances

We used the encoding above to generate three sets of QBF instances based on some well-known positional games.[1] The first two sets consist of positions of HEX and of a generalization of TIC-TAC-TOE on boards that are relatively small by human playing standards. Positions from that benchmark are fairly easy to solve even for relatively inexperienced human players of these games, and they can be solved almost instantaneously by specialized solvers. Our encoding of these positions should provide a reasonable challenge for QBF solvers as of 2019.

The third set contains the starting position of 4 positional games that are of interest to experienced human players and to mathematicians. At least 3 of these positions can be solved by specialized game algorithms developed in the 1990s and 2000s albeit with a non-trivial programming effort. The instances in this third set are out of reach of current QBF solvers and we believe that solving these positions with a QBF solver—via our encoding or a better one— can constitute a good milestone for the field.

## 4.1   Harary's Tic-Tac-Toe and GTTT($p, q$)

HARARY'S TIC-TAC-TOE is a Maker-Maker generalization of TIC-TAC-TOE where instead of marking 3 aligned stones, the players are trying to mark a set of cells congruent to a given polyomino. This type of game has received accrued interest from the mathematical community which was able to show the existence of a winning strategy or lack thereof for most polyomino shapes. GTTT($p, q$) is a further generalization of HARARY'S TIC-TAC-TOE along the principle of ($p, q$)-positional games [7].

Previous work has already proposed an encoding of GTTT($p, q$) played on small boards to QBF [7]. We refer to this existing in encoding as *DYS*. In our first set of benchmark, we encode the exact same GTTT($p, q$) configurations as previous work. However, since our encoding is different, with obtain a different set of QBF instances. This provides us with an opportunity to directly compare our approach with existing work. We report results on the 96 instances

---

[1] All our generated instances are available at github.com/vale1410/positional-games-qbf-encoding.

of GTTT(1, 1) played on a 4 × 4 board and compare formula size and solving performance with the DYS encoding.[2]

## 4.2 Hex

We use 20 hand-crafted HEX puzzles of board size 4×4 up to 7×7 that all have a winning strategy for Player 1. The first 19 of these HEX instances are of historical significance. Indeed, they were created by Piet Hein, one of the inventors of HEX and first appeared in the Danish newspaper *Politken* [14,20] during World War II [13].[3] The remaining puzzle is a 5 × 5 position proposed by Cameron Browne, it arose during standard play and offers a significant challenge for the Monte Carlo Tree Search (MCTS) algorithm and the associated RAVE enhancement [5]. This is noteworthy because MCTS is the foundational algorithm behind the top artificial players for numerous games including HEX and GO [6].

## 4.3 Challenges

We now put foward a few positional games that have attracted the attention of board game players as well as AI or mathematics researchers. Table 1 summarizes the proposed challenges together with the size of their *COR* encoding.

QUBIC, also known as 3-DIMENSIONAL TIC-TAC-TOE, is played on a 4×4×4 cube and the goal is to mark 4 aligned cells, horizontally, vertically, or diagonally. Our first domain was solved for the first time in 1980 by combining depth-first search with expert domain knowledge [21]. A second time in the 1990s using Proof Number Search (PNS), a tree search algorithm for two-player games [27].

The second domain, *freestyle* GOMOKU, is played on a 15 × 15 board and the goal is to mark 5 aligned cells, horizontally, vertically, or diagonally. Already in the 1930s, GOMOKU was perceived to be giving an overwhelming advantage to Black [27], the starting player, and by the 1980s professional GOMOKU players from Japan had claimed that the initial position admitted a Black winning strategy [1]. This was confirmed in 1993 using the PNS algorithm, a domain heuristic used to dramatically reduce the branching factor, and a database decomposing the work in independent subtasks [1].

CONNECT6 is akin to GOMOKU but the board is 19 × 19, the goal is 6 aligned cells, and players place 2 stones per move [16]. The *Mickey Mouse* setup once was among the most popular openings of CONNECT6 until it was solved in 2010 [30]. The resolution of CONNECT6 was based on PNS distributed over a cluster.

Our last challenge is an open-problem in HARARY'S TIC-TAC-TOE which corresponds to achieving shape SNAKY on a 9 × 9 board. This problem was recently put forward as an intriguing challenge for QBF solvers [7] and we offer here an alternative, more compact, encoding.

---

[2] We also generated the instances for larger values of $p$ and $q$ but the formulas are much easier to solve and provide less insight.

[3] We are grateful to Ryan Hayward and Bjarne Toft for providing us with this collection of puzzles.

**Table 1.** Selected problems put forward to the QBF community and size of their Corrective encoding. No preprocessing has been applied to these instances.

| Challenge problem | | First systematic solution | Size in QBF | | | | |
|---|---|---|---|---|---|---|---|
| Domain | Variant | | #qb | # ∀ | # ∃ | #cl | #lits |
| QUBIC | $4 \times 4 \times 4$ | 1980 | 65 | 192 | 29275 | 80343 | 245723 |
| SNAKY | $9 \times 9$ | Open | 81 | 280 | 47137 | 130702 | 403595 |
| GOMOKU | $15 \times 15$ freestyle | 1993 | 225 | 896 | 357097 | 991430 | 3078404 |
| CONNECT6 | $19 \times 19$ Mickey Mouse | 2010 | 179 | 1602 | 510651 | 1527064 | 5031059 |

## 5   Analysis

### 5.1   Setup of Experiments

When solving problems encoded in QBF, the ideas underlying the encoding of a problem are only a factor in whether the instances can be solved withing reasonable resources. Two other important factors are the specific solver invoked and the kind of preprocessing performed on the instance before solving, if any. In our experiments, we chose four state-of-the-art QBF solvers, including the top three solvers of the latest QBF Competition[4] and three preprocessors, as indicated in Table 2a.[5] All software was called with default command line parameters.

**Table 2.** Software used and resulting performance on the first benchmark.

(a) Solvers and preprocessors used in the experiments.

| | Software | Shorthand |
|---|---|---|
| Solver | DepQbf 6.03 [18] | depqbf |
| | Caqe 4.0.1 [23] | caqe |
| | Qesto 1.0 [17] | qesto |
| | Qute 1.1 [22] | qute |
| Preprocessor | QratPre+ 2.0 [19] | Q |
| | HQSPRE 1.4 [29] | H |
| | Bloqqer v37 [15] | B |
| | None | N |

(b) Solver performance depending on the encoding, always using the best preprocessor.

| | Solver | Preproc. | S | ⊤ | ⊥ | U | time(s) |
|---|---|---|---|---|---|---|---|
| DYS | caqe | B | 92 | 31 | 61 | 4 | **11468** |
| | depqbf | Q | 82 | 28 | 54 | 14 | 27211 |
| | qesto | BQ | 74 | 27 | 47 | 22 | 34887 |
| | qute | BQ | 69 | 27 | 42 | 27 | 35837 |
| COR | caqe | Q | 96 | 34 | 62 | 0 | 4099 |
| | depqbf | N | 96 | 34 | 62 | 0 | **602** |
| | qesto | BQ | 96 | 34 | 62 | 0 | 3404 |
| | qute | B | 82 | 30 | 52 | 14 | 23266 |

The experiments have been running on a i7-7820X CPU @ 3.60 GHz with 8 cores, 24 GB RAM. All solvers have been running with a dedicated single core.

---

[4] http://www.qbflib.org/eval19.html.

[5] We also attempted to use the `rareqs` QBF solver, but it timed out on almost all instances. Preprocessor H was omitted due to large timeouts.

## 5.2   Experimental Comparison of Our New Encoding to the *DYS*

Before attempting to solve positional games, let us first examine how large and amenable to preprocessing the generated encodings are. We use an approach inspired by recent work on QBF preprocessors [19] and report in Table 3 the number of quantifier blocks, universal and existential variables, clauses, and literals, as well the time time needed for the preprocessing of a representative instance of GTTT(1, 1). On both the existing *DYS* encoding and our proposed *COR*, we test each preprocessor individually as well as the outcome of running one preprocessor then another. No preprocessor timed out.

**Table 3.** Preprocessing on $5 \times 5$ instance gttt_1_1_00101121_5x5_b

|       |         | N      | Q      | H      | B      | QB     | BQ     | HQ     | QH     |
|-------|---------|--------|--------|--------|--------|--------|--------|--------|--------|
| *DYS* | #qb     | 25     | 25     | 25     | 25     | 25     | 25     | 25     | 25     |
|       | #∀      | 300    | 300    | 300    | 299    | 299    | 299    | 300    | 300    |
|       | #∃      | 21056  | 12058  | 7553   | 2750   | 2605   | 2750   | 7553   | 7545   |
|       | #cl     | 53589  | 35875  | 32978  | 21434  | 19625  | 19257  | 30191  | 30444  |
|       | #lits   | 191485 | 127366 | 145480 | 120237 | 106590 | 105697 | 103312 | 135061 |
|       | time(s) | 0      | 46     | 1210   | 9      | 55     | 22     | 1233   | 2030   |
| *COR* | #qb     | 25     | 25     | 25     | 25     | 25     | 25     | 25     | 25     |
|       | #∀      | 60     | 60     | 58     | 58     | 58     | 58     | 58     | 58     |
|       | #∃      | 4649   | 3127   | 3433   | 1396   | 1360   | 1396   | 3432   | 2981   |
|       | #cl     | 12490  | 8183   | 28918  | 8245   | 7943   | 7394   | 14899  | 19757  |
|       | #lits   | 28544  | 19672  | 117953 | 35457  | 34412  | 30732  | 52381  | 88712  |
|       | time(s) | 0      | 0      | 275    | 2      | 2      | 2      | 277    | 20     |

Two observations stand out when looking at Table 3. First, *DYS* is much larger than *COR* across most of the size dimensions. Second, preprocessors seem to be much more capable or reducing the size of the *DYS* instance than the size of the *COR* instance. Our interpretation is that it is a direct consequence of the effort we have put in crafting the proposed new encoding: there is relatively little improvement room left for the preprocessors to improve the formulas. Since the size of a formula directly impacts how hard it is to solve, we expect QBF solvers to struggle much more with *DYS*-encoded game instances than with *COR*-encoded ones.

In our next experiment, we compare how well QBF solvers manage to solve GTTT game instances when encoded with *DYS* and with *COR*. Since different preprocessors tend to play to the strength of different solvers, we report the preprocessor that lead to the best performance for each solver separately. We compare the solvers and the encodings using 96 GTTT(1, 1) $4 \times 4$ game instances and assuming a timeout of 1000 s, Table 2b displays for each configuration the number of formulas solved (S), proven satisfiable ($\top$), proven unsatisfiable ($\bot$), and unsolved (U), as well as the cumulative time spent by the solver.

The data in Table 2b confirms our intuition. GTTT games can be more effectively solved through our encoding: 3 out of 4 solvers solve all *COR* instances whereas none solve all *DYS* instances, and *COR* instances are solved between up to two orders of magnitude faster. Furthermore, our results demonstrate that the choice of encoding has a bigger impact than the choice of solver and preprocessor.

## 5.3   Solving Increasingly Realistic Games

Iterative deepening is an algorithmic principle in game search recommending to search for a *d*-move strategy before attempting to find a deeper one. This principle lets one benefit from the memory-efficiency of depth-first search and from the completeness of breadth-first search. It is easily adapted to solving games via QBF: encode one formula per depth and attempt to solve them one by one in order. We demonstrate the benefits of this adaptation in Fig. 2: the position admits a depth 5 winning strategy. Proving the existence of a strategy of depth $\leq 5$ needs 0.1 s, but the formula stating the existence of a strategy of depth $\leq 13$ needs 2 hours to be proven. Although the outcome of searches at short depths is subsumed by that of deeper searches, the exponential growth of the required solving time makes iterative deepening a worthy trade-off.

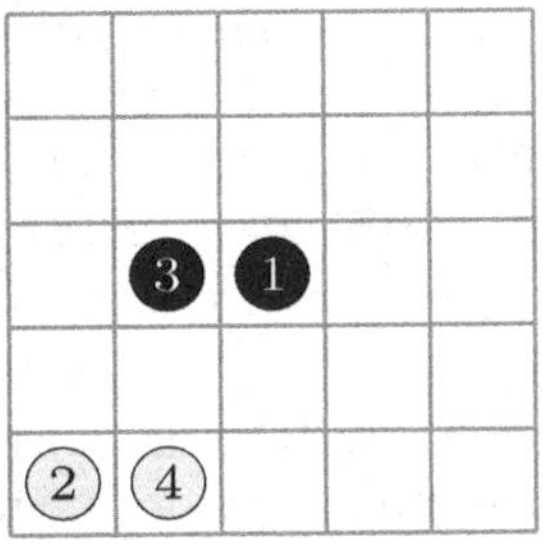

| $d$ | $\not\models \phi_d$ | $\models \phi_d$ |
|----|------|------|
| 1  | 0.01 |      |
| 3  | 0.01 |      |
| 5  |      | 0.10 |
| 7  |      | 2.33 |
| 9  |      | 22.0 |
| 11 |      | 334. |
| 13 |      | 7753.|

(a) Position after White's mistaken second move.

(b) Time (s) needed by depqbf-N-*COR* to establish whether Black can win within depth $\leq d$.

**Fig. 2.** GTTT $5 \times 5$ L game where Black can force a 5-move win.

Our final benchmark is the set of 20 historical HEX puzzles described in Sect. 4.2. Except for one puzzle on which `qesto` needed more than 10 GB of memory, all positions on board sizes $5 \times 5$ or less can be solved by state of the art QBF solvers (Table 4). The 5 remaining puzzles remain out of reach at this stage. This is a remarkable feat: for the first time, the QBF technology can address game situations considered of interest to human players.

**Table 4.** Solving classic HEX puzzles by encoding them through *COR*.

| Puzzle | size | depth | caqe-Q | | depqbf-N | | questo-BQ | |
|--------|------|-------|--------|--------|----------|--------|-----------|--------|
| | | $d$ | $\not\models \phi_{d-2}$ | $\models \phi_d$ | $\not\models \phi_{d-2}$ | $\models \phi_d$ | $\not\models \phi_{d-2}$ | $\models \phi_d$ |
| Hein 04 | $3 \times 3$ | 05 | 0.01 | 0.01 | 0.02 | 0.02 | 0.01 | 0.00 |
| Hein 09 | $4 \times 4$ | 07 | 0.01 | 0.11 | 0.03 | 0.15 | 0.01 | 0.06 |
| Hein 12 | $4 \times 4$ | 07 | 0.02 | 0.10 | 0.05 | 0.22 | 0.00 | 0.02 |
| Hein 07 | $4 \times 4$ | 09 | 0.30 | 4.31 | 0.33 | 5.69 | 0.09 | 1.66 |
| Hein 06 | $4 \times 4$ | 13 | 10.2 | 15.5 | 2.95 | 17.7 | 3.92 | 9.79 |
| Hein 13 | $5 \times 5$ | 09 | 0.24 | 15.6 | 0.72 | 17.1 | 0.06 | 4.61 |
| Hein 14 | $5 \times 5$ | 09 | 0.38 | 19.0 | 1.24 | 42.4 | 0.18 | 4.40 |
| Hein 11 | $5 \times 5$ | 11 | 5.17 | 240. | 21. | 0.457 | 1.84 | 23.6 |
| Hein 19 | $5 \times 5$ | 11 | 2.29 | 44.4 | 3.60 | 80.8 | 0.91 | 13.1 |
| Hein 08 | $5 \times 5$ | 11 | 4.13 | 104. | 6.84 | 247.0 | 1.98 | 34.4 |
| Hein 10 | $5 \times 5$ | 13 | 367. | 4906. | 443. | 10259. | 74.3 | 1543. |
| Hein 16 | $5 \times 5$ | 13 | 651. | 8964. | 1794. | 8506. | 278. | 4406. |
| Hein 02 | $5 \times 5$ | 13 | 719. | 22526. | 1258. | 10876. | 317. | 2957. |
| Hein 15 | $5 \times 5$ | 15 | 3247. | 26938. | 2928. | 19469. | 767. | MO |
| Browne | $5 \times 5$ | 09 | 0.87 | 57.45 | 0.91 | 21.2 | 0.25 | 2.89 |

## 6 Comparison to Related Encodings

Despite the similarity of QBF solving with systematic search for winning strategies in games with perfect information, to the best of our knowledge there are not many encodings published that have attempted translation from games to QBF. After Walsh [28] challenged the QBF community to solve CONNECT4 on a $7 \times 6$ board using QBF techniques in 2003, there was some activity in this direction but with rather little success in experiments.

The first concrete and implemented encoding of a game is the work by Gent and Rowley that presents a translation from CONNECT4 to QBF [11]. Building upon Gent's encoding [2] presents a QBF encoding of an *Evader/Pursuer* game that resembles simpler chess-like endgames on boards of size $4 \times 4$ and $8 \times 8$. Both papers analyse problems that are not positional games, but the authors do report similar challenges in the construction of QBF formulas.

The closest to our encoding is the *DYS* encoding of GTTT [7] that has been proposed recently which is an adaptation of the encoding for [11]. The structure and clauses for these two encodings are similar so our more detailed comparison to *DYS* also applies to the encoding in [11].

Apart from these games and encodings to the best of our knowledge we are not aware of any other QBF formulations of games that improve upon them. In the remainder of this section we will go into various properties regarding *COR* and the existing encodings.

*Generalisation.* Although we presume that the ideas behind $DYS$ could be extended to arbitrary positional games, the description of the encoding was tailored to the GTTT domain. We chose positional game as an input formalism to reach a reasonable level of generalisation, i.e. many two-player can be formulated as positional games, but enough structural properties from the description to create neat encodings.

*Clausal Description.* The description of the clauses in $DYS$ is not purely clausal and contains many equivalences. The general translation of equivalences into CNF introduces auxiliary (Tseytin) variables of which some can be avoided through better techniques. Our description consists only of clauses and much work has gone to reduce the number of variables.

*Size.* The size of our encoding is quadratic in the number of vertices and linear in the number of winning configurations. Even for smaller boards these scale effects materialize and we demonstrate and discuss our observations in the following Section.

*Binary Clauses.* Binary clauses enjoy many theoretical and practical advantages. A purely 2CNF problem can be solved in polynomial time, SAT solvers invest in the special treatment of binary clauses to speed up propagation and learning. This should also apply to $QBF$ solving. Discovering a binary clause structure of a certain aspect of a problem description might be the key to crafting encodings that also solve fast. Our encoding demonstrates that many aspects of positional games can be captured through sets of binary clauses forming chains.

*Timing.* The variable $gameover_z$ in $DYS$ has the same meaning as time in our encoding, it marks the end of the game and is crucial to prevent white from reaching a winning position after black has already won. In $DYS$ this variable is added to almost all clauses such that when the game is finished the universal variables cannot falsify these clauses anymore. This technique produces correct encodings, but affects propagation negatively and weakens clause learning. We avoid such weakening of the other clauses by de-coupling time and the board structure of a game. When the game is finished (i.e. time is set to false) independently of the choices of the players no moves are allowed anymore and all empty board positions are propagated through to the end. We expect that $COR$ makes it easier for the solver to exploit transposition of sequence of moves leading to the same board configuration than previous encodings.

*Monotonicity.* Using the property of monotonicity of positional games—the set of claimed vertices only grows throughout the game—our clauses manage to captures this property more directly than the previous encodings. For instance, $DYS$ introduces variables that are true if a player wins in time point $t$ by reaching a winning configuration. We avoid the need to know explicitly by which time point a player won via propagating the claimed vertices through to the last time point and only need to test for the winning configuration of both players there. Through de-coupling the time aspect of games from checking winning positions our encoding has fewer variables and shorter clauses, that again benefit propagation.

*Adapted Log Encoding.* This concept was first introduced for the encoding of quantified constraint satisfaction problems (CSPs) into QBF [10]. There, a logarithmic number of universal variables encode the binary representation of a CSP variable. This technique was also applied with success in a game encoding to QBF [2].

*Indicator Variables.* To the best of our knowledge all translations of games (including non-positional ones) to QBF introduce variables that indicate types of illegal moves by white. Such variables again weaken the encoding due to longer clauses. The encoding $DYS$ has the following types of illegal moves; white claims too many vertices, too few vertices, already occupied vertices. All of these are explicitly encoded in $DYS$ whereas $COR$ corrects white's move. The key insight in the work of [2] raises the question how to address White's illegal behavior without weakening the encoding or introducing auxiliary variables.

## 7    Conclusion and Future Work

We consider the craft of finding efficient translations of a problem description to the clausal representation an important step towards better performances of $QBF$ solvers. Our investigation regarding the class of positional games demonstrates that a carefully crafted translation using structural properties to decrease and shorten clauses and decrease the number of variables improves the applicability beyond trivial problems. We list some key insights:

- Binary implication chains capturing monotone structural properties of problem are crucial.
- Variables representing illegal moves by the universal player can be avoided.
- Encoding the choices of the choices by the universally player logarithmically helps in this problem description.
- Preprocessing is crucial for previous encodings to perform, whereas on our encoding has minor impact.

Our investigation focused on clause representation and we have yet to extend to non-clausal description languages to QBF such as QCIR.

The insights from our investigation can be applied to translation of other almost-positional games and to planning problems with similar structures.

Although HEX is a positional game, its hypergraph representation is exponential in the board size because it needs to account for all paths between a pair of sides. For larger boards one will need an implicit representation of the paths between the two sides, possibly drawing inspiration from existing first-order logic modeling [4].

# References

1. Allis, L.V., van den Herik, H.J., Huntjens, M.P.: Go-Moku solved by new search techniques. Comput. Intell. **12**(1), 7–23 (1996)
2. Ansótegui, C., Gomes, C.P., Selman, B.: The achilles' heel of QBF. In: Veloso, M.M., Kambhampati, S. (eds.) Proceedings, The Twentieth National Conference on Artificial Intelligence and the Seventeenth Innovative Applications of Artificial Intelligence Conference, Pittsburgh, Pennsylvania, USA, 9–13 July 2005, pp. 275–281. AAAI Press/The MIT Press (2005)
3. Bonnet, É., Gaspers, S., Lambilliotte, A., Rümmele, S., Saffidine, A.: The parameterized complexity of positional games. In: International Colloquium on Automata, Languages and Programming (ICALP), July 2017
4. Bonnet, É., Jamain, F., Saffidine, A.: On the complexity of connection games. Theor. Comput. Sci. (TCS) **644**, 2–28 (2016)
5. Browne, C.: A problem case for UCT. IEEE Trans. Comput. Intell. AI Games **5**(1), 69–74 (2013)
6. Browne, C., et al.: A survey of monte carlo tree search methods. IEEE Trans. Comput. Intell. AI Games **4**(1), 1–43 (2012)
7. Diptarama, R.Y., Shinohara, A.: QBF encoding of generalized tic-tac-toe. In: 4th International Workshop on Quantified Boolean Formulas(QBF) Co-located with 19th International Conference on Theory and Applications of Satisfiability Testing (SAT), Bordeaux, France, pp. 14–26, July 2016
8. Even, S., Tarjan, R.E.: A combinatorial problem which is complete in polynomial space. J. ACM **23**(4), 710–719 (1976)
9. Gent, I.P., Nightingale, P.: A new encoding of all different into SAT. In: Third International Workshop on CP 2004 Workshop on Modelling and Reformulating CSPs, pp. 95–110 (2004)
10. Gent, I.P., Nightingale, P., Rowley, A.: Encoding quantified CSPs as quantified Boolean formulae. In: Proceedings of ECAI-2004, pp. 176–180 (2004)
11. Gent, I.P., Rowley, A.G.: Encoding connect-4 using quantified Boolean formulae. In: Modelling and Reformulating Constraint Satisfaction Problems, pp. 78–93 (2003)
12. Hales, A.W., Jewett, R.I.: Regularity and positional games. Trans. Am. Math. Soc. **106**, 222–229 (1963)
13. Hayward, R.B., Toft, B.: Hex, the Full Story. AK Peters/CRC Press/Taylor Francis (2019)
14. Hein, P.: Polygon. In: Politiken, 26 December 1942–11 August 1943
15. Heule, M., Järvisalo, M., Lonsing, F., Seidl, M., Biere, A.: Clause elimination for SAT and QSAT. J. Artif. Intell. Res. (JAIR),**53**, 127-168 (2015). fmv.jku.at/bloqqer
16. Hsieh, M.Y., Tsai, S.C.: On the fairness and complexity of generalized k-in-a-row games. Theor. Comput. Sci. **385**(1–3), 88–100 (2007)
17. Janota, M., Marques-Silva, J.: Solving QBF by clause selection. In: Twenty-Fourth International Joint Conference on Artificial Intelligence (IJCAI), Buenos Aires, Argentina, pp. 325–331, July 2015. sat.inesc-id.pt/mikolas/sw/qesto/
18. Lonsing, F., Egly, U.: DepQBF 6.0: a search-based QBF solver beyond traditional QCDCL. In: de Moura, L. (ed.) CADE 2017. LNCS (LNAI), vol. 10395, pp. 371–384. Springer, Cham (2017). https://doi.org/10.1007/978-3-319-63046-5_23. lonsing.github.io/depqbf/

19. Lonsing, F., Egly, U.: QRATPre +: effective QBF preprocessing via strong redundancy properties. In: Janota, M., Lynce, I. (eds.) SAT 2019. LNCS, vol. 11628, pp. 203–210. Springer, Cham (2019). https://doi.org/10.1007/978-3-030-24258-9_14. lonsing.github.io/qratpreplus/

20. Maarup, T.: Everything you always wanted to know about Hex but were afraid to ask. Master's thesis, University of Southern Denmark (2005)

21. Patashnik, O.: Qubic: 4 × 4 × 4 tic-tac-toe. Math. Mag. **53**(4), 202–216 (1980)

22. Peitl, T., Slivovsky, F., Szeider, S.: Dependency learning for QBF. In: Gaspers, S., Walsh, T. (eds.) SAT 2017. LNCS, vol. 10491, pp. 298–313. Springer, Cham (2017). https://doi.org/10.1007/978-3-319-66263-3_19. www.ac.tuwien.ac.at/research/qute/

23. Rabe, M.N., Tentrup, L.: CAQE: a certifying QBF solver. In: Formal Methods in Computer-Aided Design (FMCAD), Austin, Texas, USA, pp. 136–143, September 2015.github.com/ltentrup/caqe

24. Reisch, S.: Gobang ist PSPACE-vollständig. Acta Informatica **13**(1), 59–66 (1980)

25. Reisch, S.: Hex ist PSPACE-vollständig. Acta Informatica **15**, 167–191 (1981)

26. Sinz, C.: Towards an optimal CNF encoding of Boolean cardinality constraints. In: CP, pp. 827–831 (2005)

27. Van Den Herik, H.J., Uiterwijk, J.W., Van Rijswijck, J.: Games solved: now and in the future. Artif. Intell. **134**(1–2), 277–311 (2002)

28. Walsh, T.: Challenges for SAT and QBF. In: Keynote, SAT-03 (2003)

29. Wimmer, R., Reimer, S., Marin, P., Becker, B.: HQSpre – an effective preprocessor for QBF and DQBF. In: Legay, A., Margaria, T. (eds.) TACAS 2017. LNCS, vol. 10205, pp. 373–390. Springer, Heidelberg (2017). https://doi.org/10.1007/978-3-662-54577-5_21. projects.informatik.uni-freiburg.de/

30. Wu, I.-C., et al.: Job-level proof number search. IEEE Trans. Comput. Intell. AI Games **5**(1), 44–56 (2013)

# Matrix Multiplication: Verifying Strong Uniquely Solvable Puzzles

Matthew Anderson[(⊠)], Zongliang Ji, and Anthony Yang Xu

Department of Computer Science, Union College,
Schenectady, NY, USA
{andersm2,jiz,xua}@union.edu

**Abstract.** Cohn and Umans proposed a framework for developing fast matrix multiplication algorithms based on the embedding computation in certain groups algebras [9]. In subsequent work with Kleinberg and Szegedy, they connected this to the search for combinatorial objects called strong uniquely solvable puzzles (strong USPs) [8]. We begin a systematic computer-aided search for these objects. We develop and implement algorithms based on reductions to SAT and IP to verify that puzzles are strong USPs and to search for large strong USPs. We produce tight bounds on the maximum size of a strong USP for width $k < 6$, and construct puzzles of small width that are larger than previous work. Although our work only deals with puzzles of small-constant width and does not produce a new, faster matrix multiplication algorithm, we provide evidence that there exist families of strong USPs that imply matrix multiplication algorithms that are more efficient than those currently known.

**Keywords:** Matrix multiplication · Strong uniquely solvable puzzle · Arithmetic complexity · Integer programming · Satisfiability · Reduction · Application

## 1 Introduction

An optimal algorithm for matrix multiplication remains elusive despite substantial effort. We focus on the square variant of the matrix multiplication problem, i.e., given two $n$-by-$n$ matrices $A$ and $B$ over a field $\mathcal{F}$, the goal is to compute the matrix product $C = A \times B$. The outstanding open question is: How many field operations are required to compute $C$? The long thought-optimal naïve algorithm based on the definition of matrix product is $O(n^3)$ time. The groundbreaking work of Strassen showed that it can be done in time $O(n^{2.808})$ [24] using a divide-and-conquer approach. A long sequence of work concluding with Coppersmith and Winograd's algorithm (CW) reduced the running time to $O(n^{2.376})$ [10,21,22,25]. Recent computer-aided refinements of CW by others reduced the exponent to $\omega \leq 2.3728639$ [13,18,26].

© Springer Nature Switzerland AG 2020
L. Pulina and M. Seidl (Eds.): SAT 2020, LNCS 12178, pp. 464–480, 2020.
https://doi.org/10.1007/978-3-030-51825-7_32

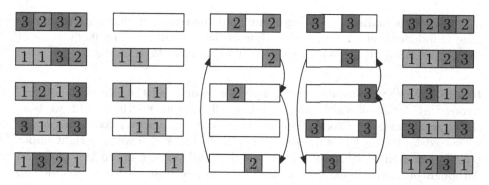

**Fig. 1.** The leftmost diagram is a width-4 size-5 puzzle $P$. The middle three diagrams are the three sets of subrows of $P$. The rightmost diagram is the puzzle $P'$ resulting from reordering the subrows of $P$ as indicated by the arrows and then recombining them. Since $P$ can be rearranged as $P' \neq P$ without overlap, $P$ is not uniquely solvable.

**Approach.** Cohn and Umans [9] introduced a framework for developing faster algorithms for matrix multiplication by reducing this to a search for groups with subsets that satisfy an algebraic property called the *triple-product property* that allows matrix multiplication to be embedded in the group algebra. Their approach takes inspiration from the $O(n \log n)$ algorithm for multiplying degree-$n$ univariate polynomials by embedding in the group algebra of the fast Fourier transform, c.f., e.g., [11, Chapter 30]. Subsequent work [8] elaborated on this idea and developed the notion of combinatorial objects called *strong uniquely solvable puzzles* (strong USPs). These objects imply a group algebra embedding for matrix multiplication, and hence give a matrix multiplication algorithm as well.

A *width-k* puzzle $P$ is a subset of $\{0, 1, 2\}^k$, and the cardinality of $P$ is the puzzle's *size*. Each element of $P$ is called a *row* of $P$, and each row consists of three *subrows* that are elements of $\{0, *\}^k$, $\{1, *\}^k$, $\{2, *\}^k$ respectively. Informally, a puzzle $P$ is a *uniquely solvable puzzle* (USP) if there is no way to permute the subrows of $P$ to form a distinct puzzle $P'$ without cells with numbers overlapping. Figure 1 demonstrates a puzzle that is not a USP. A uniquely solvable puzzle is *strong* if a tighter condition for non-overlapping holds (see Definition 2). For a fixed width $k$, the larger the size of a strong USP, the faster matrix multiplication algorithm it gives [8]. In fact Cohn et al. show that there exist an infinite family of strong USPs that achieves $\omega < 2.48$.

We follow Cohn et al.'s program by: (i) developing **verification** algorithms to determine whether a puzzle is a strong USP, (ii) developing **search** algorithms to find large strong USPs, and (iii) implementing and running practical **implementations** of these algorithms. The most successful of the verification algorithms function by reducing the problem through 3D matching to SAT and IP which are then solved with existing tools. The algorithms we develop are not efficient—they run in worst-case exponential time in the natural parameters. However, the goal is to find a sufficiently large strong USP that would provide a faster matrix multiplication algorithm, and the resulting algorithm's running time is independent of the running time of our algorithms. The inefficiency of our algorithms limit the search space that we can feasibly examine.

**Results.** Our experimental results give new bounds on the size of the largest strong USP for small-width puzzles. For small-constant width, $k \leq 12$, we beat the largest sizes of [8, Proposition 3.8]. Our lower bounds on maximum size are witnessed by strong USPs we found via search. For $k \leq 5$ we give tight upper bounds determined by exhaustively searching all puzzles up to isomorphism. Although our current experimental results do not beat [8] for unbounded $k$, they give evidence that there may exist families of strong USPs that give matrix multiplication algorithms that are more efficient than those currently known.

**Related Work.** There are a number of negative results known. Naïvely, the dimensions of the output matrix $C$ implies that the problem requires at least $\Omega(n^2)$ time. Slightly better lower bounds are known in general and also for specialized models of computation, c.f., e.g., [16,23]. There are also lower bounds known for a variety of algorithmic approaches to matrix multiplication. Ambainis et al. showed that the laser method cannot alone achieve an algorithm with $\omega \leq 2.3078$ [4]. A recent breakthrough on arithmetic progressions in cap sets [12] combined with a conditional result on the Erdös-Szemeredi sunflower conjecture [3] imply that Cohn et al.'s strong USP approach cannot achieve $\omega = 2 + \epsilon$ for some $\epsilon > 0$ [7]. Subsequent work has generalized this barrier [1,2] to a larger class of algorithmic techniques. Despite this, we are unaware of a concrete lower bound on $\epsilon$ implied by these negative results. There remains a substantial gap in our understanding between what has been achieved by the positive refinements of LeGall, Williams, and Stothers, and the impossibility of showing $\omega = 2$ using the strong USP approach.

**Organization.** Section 2 begins with the formal definition of a strong USP. Sections 3 and 4, respectively, discuss our algorithms and heuristics for verifying that and searching for a puzzle that is a strong USP. Section 5 discusses our experimental results.

## 2   Preliminaries

For an integer $k$, we use $[k]$ to denote the set $\{0, 1, 2, \ldots, k-1\}$. For a set $Q$, $\mathrm{Sym}_Q$ denotes the symmetric group on the elements of $Q$, i.e., the group of permutations acting on $Q$. Cohn et al. introduced the idea of a *puzzle* [8].

**Definition 1 (Puzzle).** *For $s, k \in \mathcal{N}$, an $(s, k)$-puzzle is a subset $P \subseteq [3]^k$ with $|P| = s$. We call $s$ the size of $P$, and $k$ the width of $P$.*

We say that an $(s, k)$-puzzle has $s$ rows and $k$ columns. The columns of a puzzle are inherently ordered and indexed by $[k]$. The rows of a puzzle have no inherent ordering, however, it is often convenient to assume that they are ordered and indexed by the set of natural numbers $[s]$.

Cohn et al. establish a particular combinatorial property of puzzles that allows one to derive group algebras that matrix multiplication can be efficiently embedded into. Such puzzles are called *strong uniquely solvable puzzles*.

**Definition 2 (Strong USP).** *An $(s, k)$-puzzle $P$ is* strong uniquely solvable *if for all $\pi_0, \pi_1, \pi_2 \in \mathrm{Sym}_P$: Either (i) $\pi_0 = \pi_1 = \pi_2$, or (ii) there exists $r \in P$ and $i \in [k]$ such that exactly two of the following hold: $(\pi_0(r))_i = 0$, $(\pi_1(r))_i = 1$, $(\pi_2(r))_i = 2$.*

Note that strong uniquely solvability is invariant to the (re)ordering of the rows or columns of a puzzle. We use this fact implicitly.

Cohn et al. show the following connection between the existence of strong USPs and upper bounds on the exponent of matrix multiplication $\omega$.

**Lemma 1 ([8, Corollary 3.6]).** *Let $\epsilon > 0$, if there is a strong uniquely solvable $(s, k)$-puzzle, there is an algorithm for multiplying $n$-by-$n$ matrices in time $O(n^{\omega+\epsilon})$ where*

$$\omega \leq \min_{m \geq 3, m \in \mathcal{N}} \frac{3 \log m}{\log(m-1)} - \frac{3 \log s!}{sk \log(m-1)}.$$

This result motivates the search for large strong USPs that would result in faster algorithms for matrix multiplication. In the same article, the authors also demonstrate the existence of an infinite family of strong uniquely solvable puzzles, for width $k$ divisible by three, that achieves a non-trivial bound on $\omega$.

**Lemma 2 ([8, Proposition 3.8]).** *There is an infinite family of strong uniquely solvable puzzles that achieves $\omega < 2.48$.*

## 3    Verifying Strong USPs

The core focus of this article is the problem of verifying strong USPs, i.e., given an $(s, k)$-puzzle $P$, output YES if $P$ is a strong USP, and NO otherwise. In this section we discuss the design of algorithms to solve this computational problem as a function of the natural parameters $s$ and $k$. Along the way we also discuss some aspects of our practical implementation that informed or constrained our designs. All the exact algorithms we develop in this section have exponential running time. However, asymptotic worst-case running time is not the metric we are truly interested in. Rather we are interested in the practical performance of our algorithms and their capability for locating new large strong USPs. The algorithm that we ultimately develop is a hybrid of a number of simpler algorithms and heuristics.

---

**Algorithm 1:** Brute Force

---

**Input:** An $(s, k)$-puzzle $P$.
**Output:** YES, if $P$ is a strong USP and NO otherwise.
  1: **function** VERIFYBRUTEFORCE($P$)
  2:   **for** $\pi_1 \in \mathrm{Sym}_P$ **do**
  3:     **for** $\pi_2 \in \mathrm{Sym}_P$ **do**
  4:       **if** $\pi_1 \neq 1 \vee \pi_2 \neq 1$ **then**
  5:         $found = false$.
  6:         **for** $r \in P$ **do**
  7:           **for** $i \in [k]$ **do**
  8:             **if** $\delta_{r_i,0} + \delta_{(\pi_1(r))_i,1} + \delta_{(\pi_2(r))_i,2} = 2$ **then** $found = true$.
  9:         **if not** $found$ **then return** NO.
 10:   **return** YES.

---

## 3.1 Brute Force

The obvious algorithm for verification comes directly from the definition of a strong USP. Informally, we consider all ways of permuting the ones and twos pieces relative to the zeroes pieces and check whether the non-overlapping condition of Definition 2 is met. A formal description of the algorithm is found in Algorithm 1.

The ones in Line 4 of Algorithm 1 denote the identity in $\mathrm{Sym}_P$, and $\delta_{a,b}$ is the Kronecker delta function which is one if $a = b$ and zero otherwise. Observe that Algorithm 1 does not refer to the $\pi_0$ of Definition 2. This is because the strong USP property is invariant to permutations of the rows and so $\pi_0$ can be thought of as an arbitrary phase. Hence, we fix $\pi_0 = 1$ to simplify the algorithm. Seeing that $|\mathrm{Sym}_P| = s!$, we conclude that the algorithm runs in time $O((s!)^2 \cdot s \cdot k \cdot \mathrm{poly}(s))$ where the last factor accounts for the operations on permutations of $s$ elements. The dominant term in the running time is the contribution from iterating over pairs of permutations. Finally, notice that if $P$ is a strong USP, then the algorithm runs in time $\Theta((s!)^2 \cdot s \cdot k \cdot \mathrm{poly}(s))$, and that if $P$ is not a strong USP the algorithm terminates early. The algorithm's poor performance made it unusable in our implementation, however, its simplicity and direct connection to the definition made its implementation a valuable sanity check against later more elaborate algorithms (and it served as effective onboarding to the undergraduate students collaborating on this project).

Although Algorithm 1 performs poorly, examining the structure of a seemingly trivial optimization leads to substantially more effective algorithms. Consider the following function on triples of rows $a, b, c \in P$: $f(a, b, c) = \vee_{i \in [k]}(\delta_{a_i,0} + \delta_{b_i,1} + \delta_{c_i,2} = 2)$. We can replace the innermost loop in Lines 7 & 8 of Algorithm 1 with the statement $found = found \vee f(r, \pi_1(r), \pi_2(r))$. Observe that $f$ neither depends on $P$, $r$, nor the permutations, and that Algorithm 1 no longer depends directly on $k$. To slightly speed up Algorithm 1 we can precompute and cache $f$ before the algorithm starts and then look up values as the algorithm runs. We precompute $f$ specialized to the rows in the puzzle $P$, and call it $f_P$.

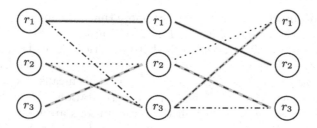

**Fig. 2.** An example hypergraph $G$ with edges $E = \{(r_1, r_1, r_2), (r_1, r_3, r_3), (r_2, r_2, r_1),$ $(r_2, r_3, r_1), (r_3, r_2, r_3)\}$. The highlighted edges are a non-trivial 3D matching $M = \{(r_1, r_1, r_2), (r_2, r_3, r_1), (r_3, r_2, r_3)\}$ of $G$.

### 3.2  Strong USP Verification to 3D Matching

It turns out to be more useful to work with $f_P$ than with $P$. It is convenient to think of $f_P$ as a function $f_P : P \times P \times P \to \{0, 1\}$ that is the complement of the characteristic function of the relations of a tripartite hypergraph $H_P = \langle P \sqcup P \sqcup P, \bar{f}_P \rangle$ where the vertex set is the disjoint union of three copies of $P$ and $f_P$ indicates the edges that are not present in $H_P$.

Let $H = \langle P \sqcup P \sqcup P, E \subseteq P^3 \rangle$ be a tripartite 3-hypergraph. We say $H$ has a *3D matching* (3DM) iff there exists a subset $M \subseteq E$ with $|M| = |P|$ and for all distinct edges $e_1, e_2 \in M$, $e_1$ and $e_2$ are *vertex disjoint*, i.e., $e_1 \cap e_2 = \emptyset$. Determining whether a hypergraph has a 3D matching is a well-known NP-complete problem (c.f., e.g., [14]). We say that a 3D matching is *non-trivial* if it is not the set $\{(r, r, r) \mid r \in P\}$. Figure 2 demonstrates a 3-hypergraph with a non-trivial 3D matching.

The existence of non-trivial 3D matchings in $H_P$ is directly tied to whether $P$ is a strong USP.

**Lemma 3.** *A puzzle $P$ is a strong USP iff $H_P$ has no non-trivial 3D matching.*

*Proof.* We first argue the reverse. Suppose that $H_p$ has a non-trivial 3D matching $M$. We show that $P$ is not a strong USP by using $M$ to construct $\pi_0, \pi_1, \pi_2 \in \mathrm{Sym}_P$ that witness this. Let $\pi_0$ be the identity permutation. For each $r \in P$, define $\pi_1(r) = q$ where $(r, q, *) \in M$. Note that $q$ is well defined and unique because $M$ is 3D matching and so has vertex disjoint edges. Similarly define $\pi_2(r) = q$ where $(r, *, q) \in M$. Observe that by construction

$$M = \{(\pi_0(r), \pi_1(r), \pi_2(r)) \mid r \in P\}.$$

Since $M$ is a matching of $H_P$, $M \subseteq \bar{f}_P$. Because $M$ is a non-trivial matching at least one edge in $(a, b, c) \in M$ has either $a \neq b$, $a \neq c$, or $b \neq c$. This implies, respectively, that as constructed $\pi_0 \neq \pi_1$, $\pi_0 \neq \pi_2$, or $\pi_1 \neq \pi_2$. In each case we have determined that $\pi_0$, $\pi_1$, and $\pi_2$ are not all identical. Thus we determined permutations such that for all $r \in P$, $f(\pi_0(r), \pi_1(r), \pi_2(r)) = 0$. This violates Condition (ii) of Definition 2, hence $P$ is not a strong USP.

The forward direction is symmetric. Suppose that $P$ is not a strong USP. We show that $H_P$ has a 3D matching. For $P$ not to be a strong USP there must exist $\pi_0, \pi_1, \pi_2 \in \mathrm{Sym}_P$ not all identical such that Condition (ii) of Definition 2 fails. Define $e(r) = (\pi_0(r), \pi_1(r), \pi_2(r))$ and $M = \{e(r) \mid r \in P\}$. Since Condition (ii) fails, we have that $f_P(e(r)) = false$ for all $r \in P$. This means that for all $r \in P$, $e(r) \in \bar{f}_P$ and hence $M \subseteq \bar{f}_P$. Since $\pi_0$ is a permutation, $|M| = |P|$. Observe that $M$ is non-trivial because not all the permutations are identical and there must be some $r \in P$ with $e(r)$ having non-identical coordinates. Thus $M$ is a non-trivial 3D matching.                                                              □

Note that although 3D matching is an NP-complete problem, Lemma 3 does not immediately imply that verification of strong USPs is coNP-complete because $H_P$ is not an arbitrary hypergraph. As a consequence of Definition 2, verification is in coNP. It remains open whether verification is coNP-complete. Lemma 3 implies that to verify $P$ is a strong USP it suffices to determine whether $H_P$ has a 3D matching. In the subsequent sections we examine algorithms for the later problem. We can, in retrospect, view Algorithm 1 as an algorithm for solving 3D matching.

The realization that verification of strong USPs is a specialization of 3D matching leads to a dynamic programming algorithm for verification that runs in linear-exponential time $O(2^{2s}\mathrm{poly}(s) + \mathrm{poly}(s, k))$. Applying more advanced techniques like those of Björklund et al. can achieve a better asymptotic time of $O(2^s\mathrm{poly}(s) + \mathrm{poly}(s, k))$ [6]. For brevity, we defer the details of our algorithm to the long version of this article.

### 3.3   3D Matching to Satisfiability

By Lemma 3, one can determine whether a puzzle $P$ is a strong USP by constructing the graph $H_P$ and deciding whether it has a non-trivial 3D matching. Here we reduce our 3D matching problem to the satisfiability (SAT) problem on conjunctive normal form (CNF) formulas and then use a state-of-the-art SAT solver to resolve the reduced problem. To perform the reduction, we convert the graph $H_P$ into a CNF formula $\Psi_P$, a depth-2 formula that is the AND of ORs of Boolean literals. We construct $\Psi_P$ so that $\Psi_P$ is satisfiable iff $H_P$ has a non-trivial 3D matching.

Let $H_P = \langle V = P \sqcup P \sqcup P, E \subseteq P^3 \rangle$ be the 3D matching instance associated with the puzzle $P$. Our goal is to determine whether there is a non-trivial 3D matching $M \subseteq E$. A naïve reduction would be to have variables $M_{u,v,w}$ indicating inclusion of each edge $(u, v, w) \in P^3$ in the matching. This results in a formula $\Psi_P$ with $s^3$ variables and size $\Theta(s^5)$ because including an edge $e \in P^3$ excludes the $\Theta(s^2)$ edges $e'$ with $e \cap e' \neq \emptyset$. To decrease the size of $\Psi_P$ we instead use sets of variables to indicate which vertices in the second and third part of $V$ are matched with each vertex in the first part. In particular we have Boolean variables $M_{u,v}^1$ and $M_{u,w}^2$ for all $u, v, w \in P$, and these variable map to assignments in the naïve scheme in the following way: $M_{u,v}^1 \wedge M_{u,w}^2 \Leftrightarrow M_{u,v,w}$.

We now write our CNF formula for 3D matching. First, we have clauses that prevents non-edges from being in the matching:

$$\Psi_P^{\text{non-edge}} = \bigwedge_{(u,v,w)\in\overline{E}} (\neg M_{u,v}^1 \vee \neg M_{u,w}^2). \tag{1}$$

Second, we add clauses require that every vertex in $H_P$ is matched with some edge:

$$\Psi_P^{\geq 1} = \left( \bigwedge_{u\in P} (\vee_{v\in P} M_{u,v}^1) \wedge (\vee_{w\in P} M_{u,w}^2) \right)$$
$$\wedge \left( \bigwedge_{v\in P} (\vee_{u\in P} M_{u,v}^1) \right) \wedge \left( \bigwedge_{w\in P} (\vee_{u\in P} M_{u,w}^2) \right). \tag{2}$$

Third, we require that each vertex be matched with at most one edge and so have clauses that exclude matching edges that overlap on one or two coordinates.

$$\Psi_P^{\leq 1} = \bigwedge_{i\in\{1,2\}} \bigwedge_{(u,v),(u',v')\in P^2} (u = u' \vee v = v') \wedge (u,v \neq u',v') \Rightarrow \neg M_{u,v}^i \vee \neg M_{u',v'}^i. \tag{3}$$

Fourth, we exclude the trivial 3D matching by requiring that at least one of the diagonal edges not be used: $\Psi_P^{\text{non-trivial}} = \bigvee_{u\in P} \neg M_{u,u}^1 \vee \neg M_{u,u}^2$. Finally, we AND these into the overall CNF formula: $\Psi_P = \Psi_P^{\text{non-edge}} \wedge \Psi_P^{\leq 1} \wedge \Psi_P^{\geq 1} \wedge \Psi_P^{\text{non-trivial}}$. The size of the CNF formula $\Psi_P$ is $\Theta(s^3)$, has $2s^2$ variables, and is a factor of $s^2$ smaller than the naïve approach. Thus we reduce 3D matching to satisfiability by converting the instance $H_P$ into the CNF formula $\Psi_P$.

To solve the reduced satisfiability instance we used the open-source solver MapleCOMPSPS from the 2016 International SAT Competition [5]. This solver is conflict driven and uses a learning rate branching heuristic to decide which variables are likely to lead to conflict and has had demonstrable success in practice [19]. We chose MapleCOMPSPS because it was state of the art at the time our project started. It is likely that more recently-developed solvers would achieve similar or better performance on our task.

### 3.4   3D Matching to Integer Programming

In parallel to the previous subsection, we use the connection between verification of strong USPs and 3D matching to reduce the former to integer programming, another well-known NP-complete problem (c.f., e.g., [17]). Again, let $H_P = \langle V, E \rangle$ be the 3D matching instance associated with $P$. We construct an integer program $Q_P$ over $\{0, 1\}$ that is infeasible iff $P$ is a strong USP. Here the reduction is simpler than the previous one because linear constraints naturally capture matching.

We use $M_{u,v,w}$ to denote a variable with values in $\{0, 1\}$ to indicate whether the edge $(u, v, w) \in P^3$ is present in the matching. To ensure that $M$ is a subset of $E$ we add the following edge constraints to $Q_P$: $\forall u, v, w \in P, \forall (u, v, w) \notin$

$E, M_{u,v,w} = 0$. We also require that each vertex in each of the three parts of the graph is incident to exactly one edge in $M$. This is captured by the following vertex constraints in $Q_P$: $\forall w \in P, \sum_{u,v \in P} M_{u,v,w} = \sum_{u,v \in P} M_{u,w,v} = \sum_{u,v \in P} M_{w,u,v} = 1$. Lastly, since we need that the 3D matching be non-trivial we add the constraint: $\sum_{u \in P} M_{u,u,u} < |P|$.

To check whether $P$ is a strong USP we determine whether $Q_P$ is not feasible, i.e., that no assignment to the variables $M$ satisfy all constraints. In practice this computation is done using the commercial, closed-source, mixed-integer programming solver Gurobi [15]. We note that reduction from 3D matching to IP is polynomial time and that there are $s^3$ variables in $Q_P$, and that the total size of the constraints is $s^3 \cdot \Theta(1) + 3s \cdot \Theta(s^2) + 1 \cdot \Theta(s^3) = \Theta(s^3)$, similar to size of $\Psi_P$ in the SAT reduction.

### 3.5   Heuristics

Although the exact algorithms presented in the previous sections make substantial improvements over the brute force approach, the resulting performance remains impractical. To resolve this, we also develop several fast verification heuristics that may produce the non-definitive answer MAYBE in place of YES or NO. Then, to verify a puzzle $P$ we run this battery of fast heuristics and return early if any of the heuristics produce a definitive YES or NO. When all the heuristics result in MAYBE, we then run one of the slower exact algorithms that were previously discussed. The heuristics have different forms, but all rely on the structural properties of a strong USP. Here we discuss the two most effective heuristics, downward closure and greedy, and defer a deeper discussion of these and several less effective heuristics, including projection to 2D matching, to the full version of this article.

**Downward Closed.** The simplest heuristics we consider is based on the fact that strong USPs are downward closed.

**Lemma 4.** *If $P$ is a strong USP, then so is every subpuzzle $P' \subseteq P$.*

*Proof.* Let $P$ be a strong USP and $P' \subseteq P$. By Definition 2, for every $(\pi_1, \pi_2, \pi_3) \in \mathrm{Sym}_P^3$ not all identity, there exist $r \in P$ and $i \in [k]$ such that exactly two of the following hold: $(\pi_0(r))_i = 0, (\pi_1(r))_i = 1, (\pi_2(r))_i = 2$. Consider restricting the permutations to those that fix the elements of $P \backslash P'$. For these permutations it must be the case that $r \in P'$ because otherwise $r \in P \backslash P'$ and there is exactly one $j \in [3]$ for which $(\pi_j(r))_i = j$ holds. Thus we can drop the elements of $P \backslash P'$ and conclude that for every tuple of permutations in $\mathrm{Sym}_{P'}$ the conditions of Definition 2 hold for $P'$, and hence that $P'$ is a strong USP. □

This leads to a polynomial-time heuristic that can determine that a puzzle is not a strong USP. Informally, the algorithm takes an $(s, k)$-puzzle $P$ and $s' \leq s$, and verifies that all subsets $P' \subseteq P$ with size $|P'| = s'$ are strong USPs. If any subset $P'$ is not a strong USP, the heuristic returns NO, otherwise it returns MAYBE. This algorithm runs in time $O(\binom{s}{s'} \cdot T(s', k))$ where $T(s', k)$ is the runtime for verifying an $(s', k)$-puzzle. In practice we did not apply this heuristic for $s'$ larger than 3, so the effective running time was $O(s^3 \cdot T(3, k))$, which is polynomial in $s$ and $k$ using the verification algorithms from the previous subsections that eliminate dependence on $k$ for polynomial cost. This heuristic can be made even more practical by caching the results for puzzles of size $s'$, reducing the verification time per iteration to constant in exchange for $\Theta(\binom{3^k}{s'}) \cdot T(s', k))$ time and $\Theta(\binom{3^k}{s'})$ space to precompute the values for all puzzles of size $s'$. From a practical point of view, running this heuristic is free for small constant $s' \leq 3$, as even the reductions in the exact verification algorithms have a similar or higher running time.

**Greedy.** This heuristic attempts to greedily solve the 3D matching instance $H_P$. The heuristic proceeds iteratively, determining the vertex of the first part of the 3D matching instance with the least edges and randomly selecting an edge of that vertex to put into the 3D matching. If the heuristic successfully constructs a 3D matching it returns NO indicating that the input puzzle $P$ is not a strong USP. If the heuristic reaches a point were prior commitments have made the matching infeasible, the heuristic starts again from scratch. This process is repeated some number of times before it gives up and returns MAYBE. In our implementation we use $s^3$ attempts because it is similar to the running time of the reductions and it empirically reduced the number of instances requiring full verification in the domain of puzzles with $k = 6, 7, 8$ while not increasing the running time by too much.

## 3.6  Hybrid Algorithm

Our final verification algorithm (Algorithm 2) is a hybrid of several exact algorithms and heuristics. The size thresholds for which algorithm and heuristic to apply were determined experimentally for small $k$ and were focused on the values were our strong USP search algorithms were tractable $k \leq 6$ (or nearly tractable $k \leq 8$). We decided to run both the reduction to SAT and IP in parallel because it was not clear which algorithm performed better. Since verification halts when either algorithm completes, the wasted effort is within a factor of two of what the better algorithm could have done alone. We also chose to do this because we experimentally observed that there were many instances that one of the algorithms struggled with that the other did not—this resulted in a hybrid algorithm that out performed the individual exact algorithms on average.

## 4  Searching for Strong USPs

In some ways the problem of constructing a large strong USP is similar to the problem of constructing a large set of linearly independent vectors. In both cases, the object to be constructed is a set, the order that elements are added does not

---

**Algorithm 2:** Hybrid Verification Algorithm

---

**Input:** An $(s, k)$-puzzle $P$.
**Output:** YES, if $P$ is found to be strong USP, and NO otherwise.
1: **if** $s \leq 2$ **then return** VERIFYBRUTEFORCE($P$).
2: **if** $s \leq 7$ **then return** VERIFYDYNAMICPROGRAMMING($P$).
3: **if** $s \leq 10$ **then**
4:     Return result if HEURISTICDOWNWARDCLOSED($P, 2$) is not MAYBE.
5:     **return** VERIFYDYNAMICPROGRAMMING($P$).
6: Return result if HEURISTICDOWNWARDCLOSED($P, 3$) is not MAYBE.
7: Return result if HEURISTICGREEDY($P$) is not MAYBE.
8: Run VERIFYSAT($P$) and VERIFYIP($P$) in parallel and return the first result.

---

matter, the underlying elements are sequences of numbers, and there is a notion of (in)dependence among sets of elements. There are well-known polynomial-time algorithms for determining whether a set of vectors are independent, e.g., Gaussian elimination, and we have a practical implementation for deciding whether a puzzle is a strong USP.

There is a straightforward greedy algorithm for constructing maximum-size sets of independent vectors: Start with an empty set $S$, and repeatedly add vectors to $S$ that are linearly independent of $S$. After this process completes $S$ is a largest set of linearly independent vectors. This problem admits such a greedy algorithm because the family of sets of linearly independent vectors form a matroid. The vector to be added each step can be computed efficiently by solving a linear system of equations for vectors in the null space of $S$.

Unfortunately this same approach does not work for generating maximum-size strong USPs. The set of strong USPs does not form a matroid, rather it is only an independence system, c.f., e.g., [20]. In particular, (i) the empty puzzle is a strong USP and (ii) the set of strong USP are downward closed by Lemma 4. The final property required to be a matroid, the augmentation property, requires that for every pair of strong USPs $P_1, P_2$ with $|P_1| \leq |P_2|$ there is a row of $r \in P_2 \backslash P_1$ such that $P_1 \cup \{r\}$ is also a strong USP. A simple counterexample with the strong USPs $P_1 = \{32\}$ and $P_2 = \{12, 23\}$ concludes that neither $P_1 \cup \{12\} = \{12, 32\}$ nor $P_1 \cup \{23\} = \{23, 32\}$ are strong USPs, and hence the augmentation property fails. One consequence is that naïve greedy algorithms will likely be ineffective for finding maximum-size strong USPs. Furthermore, we do not currently have an efficient algorithm that can take a strong USP $P$ and determine a row $r$ such that $P \cup \{r\}$ is a strong USP aside from slight pruning of the $\leq 3^k$ possible next rows $r$.

That said, we have had some success applying general purpose search techniques together with our practical verification algorithm to construct maximum-size strong USPs for small $k$. In particular, we implemented variants of depth-first search (DFS) and breadth-first search (BFS). We defer the details of this to the full version of this article.

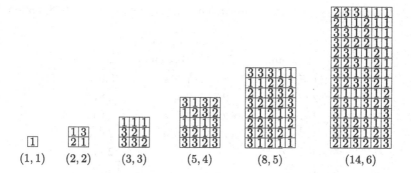

**Fig. 3.** Representative maximum-size strong USPs found for width $k = 1, 2, \ldots, 6$.

The actual running times of both of these algorithms are prohibitive even for $k > 5$, and the greater memory usage of BFS to store the entire search frontier is in the tens of terabytes even for $k = 6$. There are some silver linings, DFS can report intermediate results which are the maximally strong USPs that it has discovered so far. Both algorithms admit the possibility of eliminating puzzles from the search that are equivalent to puzzles that have already been searched, though it is easier to fit into the structure BFS as the puzzles are already being stored in a queue.

## 5    Experimental Results

Our experimental results come in three flavors for small-constant width $k$: (i) constructive lower bounds on the maximum size of width-$k$ strong USPs witnessed by found puzzles, (ii) exhaustive upper bounds on the maximum size of width-$k$ strong USPs, and (iii) experimental run times comparing the algorithms for verifying width-$k$ strong USPs. BFS and DFS when able to run to completion search the entire space, up to puzzle isomorphism, and provide tight upper and lower bounds. When unable to run to completion they provide only results of form (i) and are not guaranteed to be tight.

### 5.1    New Bounds on the Size of Strong USPs

Figure 3 contains representative examples of maximum-size strong USPs we found for $k \leq 6$. Table 1 summarizes our main results in comparison with [8]. The lower bounds of [8] are from the constructions in their Propositions 3.1 and 3.8 that give families of strong USPs for $k$ even or $k$ divisible by three. The upper bounds of [8] follow from their Lemma 3.2 (and the fact that the capacity of strong USPs is bounded above by the capacity of USPs). The values of $\omega$ in this table are computed by plugging $s$ and $k$ into Lemma 1 and optimizing over $m$. For clarity we omit $\omega$'s that would be larger than previous lines.

We derive tight bounds for all $k \leq 5$ and constructively improve the known lower bounds for $4 \leq k \leq 12$. The strong uniquely solvable (14, 6)-puzzles we

**Table 1.** Comparison of bounds on maximum size of strong USPs with [8] for small $k$.

| $k$ | [8] Maximum $s$ | $\omega$ | This work Maximum $s$ | $\omega$ |
|---|---|---|---|---|
| 1 | $1 = s$ | 3.000 | $1 = s$ | 3.000 |
| 2 | $2 \leq s \leq 3$ | 2.875 | $2 = s$ | 2.875 |
| 3 | $3 \leq s \leq 6$ | 2.849 | $3 = s$ | 2.849 |
| 4 | $4 \leq s \leq 12$ | 2.850 | $5 = s$ | 2.806 |
| 5 | $4 \leq s \leq 24$ | | $8 = s$ | 2.777 |
| 6 | $10 \leq s \leq 45$ | 2.792 | $14 \leq s$ | 2.733 |
| 7 | $10 \leq s \leq 86$ | | $21 \leq s$ | 2.722 |
| 8 | $16 \leq s \leq 162$ | | $30 \leq s$ | 2.719 |
| 9 | $36 \leq s \leq 307$ | 2.739 | $42 \leq s$ | 2.718 |
| 10 | $36 \leq s \leq 581$ | | $64 \leq s$ | 2.706 |
| 11 | $36 \leq s \leq 1098$ | | $112 \leq s$ | 2.678 |
| 12 | $136 \leq s \leq 2075$ | 2.696 | $196 \leq s$ | 2.653 |

found represent the greatest improvement in $\omega$ versus the construction of [8]. Further, our puzzle for $k = 12$ is the result of taking the Cartesian product of two copies of a strong uniquely solvable (14, 6)-puzzle. Repeating this process with more copies of the puzzle gives a strong USP implying $\omega < 2.522$. Note that Proposition 3.8 of [8] gives an infinite family of strong USPs that achieves $\omega < 2.48$ in the limit.

Based on the processing time we spent on $k = 6$, we conjecture that $s = 14$ is tight for $k = 6$ and that our lower bounds for $k > 6$ are not. Our results suggests there is considerable room for improvement in the construction of strong USPs, and that it is likely that there exist large puzzles for $k = 7, 8, 9$ that would beat [8]'s construction and perhaps come close to the Coppersmith-Winograd refinements. It seems that new insights into the search problem are required to proceed for $k > 6$.

## 5.2　Algorithm Performance

We implemented our algorithms in C++ (source code to be made available on github) and ran them on a 2010 MacPro running Ubuntu 16.04 with dual Xeon E5620 2.40 Ghz processors and 16 GB of RAM. Figure 4 contains log plots that describe the performance of our algorithms on sets of 10000 random puzzles at each point on a sweep through parameter space for width $k = 5 \ldots 10$ and size $s = 1 \ldots 30$. We chose to test performance via random sampling because we do not have access to a large set of solved instances. This domain coincides with the frontier of our search space, and we tuned the parameters of the heuristics and algorithms in the hybrid algorithm to perform well in this domain. We did not deeply investigate performance characteristics outside of this domain.

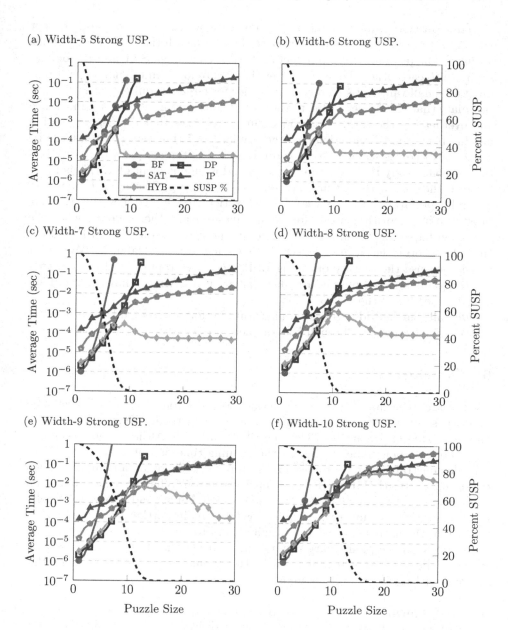

**Fig. 4.** Log plots of the average running times for verifying 10000 random puzzles of widths five to ten. Note that the legend in (a) applies to all six plots, and that the axes are named and labeled only on the edges of the page. Each plot describes the behavior of five algorithms brute force (BF), dynamic programming (DP), reduction to satisfiability (SAT), reduction to integer programming (IP), and the hybrid algorithm (HYB). The final dashed line indicates the percentage of strong USP found among the 10000 random puzzles.

The brute force and dynamic programming algorithms perform poorly except for very small size, $s \leq 8$, and their curves loosely match the $2^{\Omega(n)}$ time bounds we have. The plots for the two reduction-based algorithms (SAT and IP) behave similarly to each other. They are slower than brute force and dynamic programming for small values of $s$, and their behavior for large $s$ is quite a bit faster. We speculate that the former is due to the cost of constructing the reduced instance and overhead of the third party tools. Further observe that the SAT reduction handily beats the IP reduction on large size for $k = 5$, but as $k$ increases, the IP reduction becomes faster. We also note that across the six plots the IP reduction has effectively the same running time and is independent of $k$, this is likely because the size of the IP instance depends only on $s$. The hybrid algorithm generally performs best or close to best. Notice that it matches the dynamic programming algorithm closely for small values of $s$ and then diverges when the reduction-based algorithms and heuristics are activated around $s = 10$. Observe that the hybrid algorithm is effectively constant time for large $s$. We expect this is because the density of strong USPs decreases rapidly with $s$, and that the randomly selected puzzles are likely far from satisfying Definition 2 and are quickly rejected by the downward closure heuristic.

Overall, our hybrid verification algorithm performs reasonably well in practice, despite reductions through NP-complete problems.

## 6    Conclusions

We initiated the first study of the verification of strong USPs and developed practical software for both verifying and searching for them. We give tight results on the maximum size of width-$k$ strong USPs for $k \leq 5$. Although our results do not produce a new upper bound on the running time of matrix multiplication, they demonstrate there is promise in this approach. There are a number of open questions. Is strong USP verification coNP-complete? What are tight bounds on maximum size strong USPs for $k \geq 6$ and do these bound lead to asymptotically faster algorithms for matrix multiplication? The main bottleneck in our work is the size of the search space—new insights seem to be required to substantially reduce it. We have preliminary results that indicate that the size of the search space can be reduced by modding out by the symmetries of puzzles, though this has not yet led to new lower bounds.

**Acknowledgments.** The second and third authors thank Union College for Undergraduate Summer Research Fellowships funding their work. The authors thank the anonymous reviewers for their detailed and thoughtful suggestions for improving this work.

## References

1. Alman, J., Williams, V.V.: Further limitations of the known approaches for matrix multiplication. In: Leibniz International Proceedings Information LIPIcs of the 9th

Innovations in Theoretical Computer Science (ITCS), vol. 94, p. 15, Article no. 25. Schloss Dagstuhl. Leibniz-Zent. Inform., Wadern (2018)

2. Alman, J., Williams, V.V.: Limits on all known (and some unknown) approaches to matrix multiplication. In: 59th Annual IEEE Symposium on Foundations of Computer Science (FOCS), pp. 580–591, October 2018. https://doi.org/10.1109/FOCS.2018.00061

3. Alon, N., Shpilka, A., Umans, C.: On sunflowers and matrix multiplication. Comput. Complex. **22**(2), 219–243 (2013)

4. Ambainis, A., Filmus, Y., Le Gall, F.: Fast matrix multiplication: limitations of the coppersmith-winograd method. In: 47th Annual ACM Symposium on Theory of Computing (STOC), pp. 585–593. ACM (2015)

5. Balyo, T., Heule, M.J., Jarvisalo, M.: SAT competition 2016: recent developments. In: 31st AAAI Conference on Artificial Intelligence (AAAI) (2017)

6. Björklund, A., Husfeldt, T., Kaski, P., Koivisto, M.: Narrow sieves for parameterized paths and packings. J. Comput. Syst. Sci. **87**, 119–139 (2017)

7. Blasiak, J., Church, T., Cohn, H., Grochow, J.A., Umans, C.: Which groups are amenable to proving exponent two for matrix multiplication? arXiv preprint arXiv:1712.02302 (2017)

8. Cohn, H., Kleinberg, R., Szegedy, B., Umans, C.: Group-theoretic algorithms for matrix multiplication. In: 46th Annual IEEE Symposium on Foundations of Computer Science (FOCS), pp. 379–388, October 2005. https://doi.org/10.1109/SFCS.2005.39

9. Cohn, H., Umans, C.: A group-theoretic approach to fast matrix multiplication. In: 44th Annual IEEE Symposium on Foundations of Computer Science (FOCS), pp. 438–449, October 2003. https://doi.org/10.1109/SFCS.2003.1238217

10. Coppersmith, D., Winograd, S.: Matrix multiplication via arithmetic progressions. J. Symbolic Comput. **9**(3), 251–280 (1990)

11. Cormen, T.H., Leiserson, C.E., Rivest, R.L., Stein, C.: Introduction to Algorithms, 3rd edn. The MIT Press, Cambridge (2009)

12. Croot, E., Lev, V.F., Pach, P.P.: Progression-free sets in are exponentially small. Ann. Math. **185**, 331–337 (2017)

13. Davie, A.M., Stothers, A.J.: Improved bound for complexity of matrix multiplication. Proc. R. Soc. Edinb. Sect. A Math. **143**(2), 351–369 (2013)

14. Garey, M.R., Johnson, D.S.: Computers and Intractability: A Guide to the Theory of NP-Completeness (1979)

15. Gurobi Optimization LLC: Gurobi optimizer reference manual (2018). http://www.gurobi.com

16. Kaminski, M.: A lower bound on the complexity of polynomial multiplication over finite fields. SIAM J. Comput. **34**(4), 960–992 (2005)

17. Korte, B., Vygen, J.: Combinatorial Optimization, vol. 2. Springer, Heidelberg (2012). https://doi.org/10.1007/978-3-642-24488-9

18. Le Gall, F.: Powers of tensors and fast matrix multiplication. In: 39th International Symposium on Symbolic and Algebraic Computation (ISSAC), pp. 296–303. ACM (2014)

19. Liang, J.H., Ganesh, V., Poupart, P., Czarnecki, K.: Learning rate based branching heuristic for SAT solvers. In: Creignou, N., Le Berre, D. (eds.) SAT 2016. LNCS, vol. 9710, pp. 123–140. Springer, Cham (2016). https://doi.org/10.1007/978-3-319-40970-2_9

20. Oxley, J.G.: Matroid Theory, vol. 3. Oxford University Press, USA (2006)

21. Pan, V.Y.: Strassen's algorithm is not optimal trilinear technique of aggregating, uniting and canceling for constructing fast algorithms for matrix operations. In: 19th Annual Symposium on Foundations of Computer Science (FOCS), pp. 166–176. IEEE (1978)
22. Schönhage, A.: Partial and total matrix multiplication. SIAM J. Comput. **10**(3), 434–455 (1981)
23. Shpilka, A.: Lower bounds for matrix product. SIAM J. Comput. **32**(5), 1185–1200 (2003)
24. Strassen, V.: Gaussian elimination is not optimal. Numer. Math. **13**(4), 354–356 (1969)
25. Strassen, V.: The asymptotic spectrum of tensors and the exponent of matrix multiplication. In: 27th Annual Symposium on Foundations of Computer Science (FOCS), pp. 49–54. IEEE (1986)
26. Williams, V.V.: Multiplying matrices faster than Coppersmith-Winograd. In: 44th Annual ACM Symposium on Theory of Computing (STOC), pp. 887–898. ACM (2012)

# Satisfiability Solving Meets Evolutionary Optimisation in Designing Approximate Circuits

Milan Češka, Jiří Matyáš[✉], Vojtech Mrazek, and Tomáš Vojnar

FIT, Brno University of Technology, Brno, Czech Republic
imatyas@fit.vutbr.cz

**Abstract.** Approximate circuits that trade the chip area or power consumption for the precision of the computation play a key role in development of energy-aware systems. Designing complex approximate circuits is, however, very difficult, especially, when a given approximation error has to be guaranteed. Evolutionary search algorithms together with SAT-based error evaluation currently represent one of the most successful approaches for automated circuit approximation. In this paper, we apply satisfiability solving not only for circuit evaluation but also for its minimisation. We consider and evaluate several approaches to this task, both inspired by existing works as well as novel ones. Our experiments show that a combined strategy, integrating evolutionary search and SMT-based sub-circuit minimisation (using quantified theory of arrays) that we propose, is able to find complex approximate circuits (e.g. 16-bit multipliers) with considerably better trade-offs between the circuit precision and size than existing approaches.

## 1 Introduction

*Approximate circuits* are digital circuits that trade functional correctness (precision of computation) for other design objectives such as chip area or power consumption. Such circuits play an important role in development of resource-efficient systems, including applications such as image and video processing [10] or neural networks [14,17]. Designing approximate systems, i.e. finding optimal trade-offs between the approximation error and resource savings is, however, a complex and time-demanding process. Automated methods allowing one to develop high-quality approximate circuits are thus in high demand, especially when a bound on the approximation error is to be guaranteed.

There exists a vast body of literature (see, e.g. [13,16,18,19,26]) demonstrating that evolutionary-based algorithms are able to automatically design innovative approximate circuits providing high-quality trade-offs among the different design objectives. There are two main challenges related to the

This work was supported by the Czech Science Foundation grant GJ20-02328Y, the JCMM Brno Ph.D. Talent scholarship program, and the BUT project FIT-S-20-6427.

L. Pulina and M. Seidl (Eds.): SAT 2020, LNCS 12178, pp. 481–491, 2020.
https://doi.org/10.1007/978-3-030-51825-7_33

evolutionary-driven circuit approximation: (1) Finding a fast and reliable evaluation of candidate solutions. (2) Designing a quickly converging search strategy that drives the exploration towards high-quality solutions.

Concerning the first challenge, several circuit evaluation techniques have been proposed including parallel circuit simulation [27] and various formal methods [5,9,25,28]. In our recent work [3], we proposed and implemented a new *miter* construction together with a resource-limited verifier for SAT-based evaluation of the worst-case error. This approach has made feasible approximation of complex circuits, going beyond 16-bit adders and 12-bit multipliers, which were the limits of previously known techniques. In this paper, we aim at the second challenge. Inspired by recent advances in SAT-based exact synthesis [11,24] (the problem of finding the optimum logic representation of a given Boolean function), we investigate whether a search strategy based on *satisfiability solving* (StS)—i.e. SAT or SMT solving—can improve state-of-the-art methods for designing complex approximate circuits.

We emphasize that complex circuits typically have more than a thousand gates and thus a monolithic approach, i.e. representing the circuit approximation problem as a single StS query, is not tractable. Instead, we build on an iterative approach where sub-circuits are optimised (i.e. the sub-circuit logic is minimised while the original functionality is preserved) [23] or approximated (i.e. the functionality of the sub-circuit is not preserved and the error of the whole circuit is increased—to our best knowledge the iterative approximation has not been considered yet).

Despite the enormous progress in satisfiability solving, our experiments clearly show that the purely StS-based approximation significantly lags behind the standard evolutionary approximation. Although the StS-based approximation performs informed (and thus in some sense more useful) changes in the candidate circuits, the overhead caused by calling the solver does not pay off compared to the uninformed but very cheap genetic mutations. Put differently, the evolution can perform over 100-times more approximation attempts which is enough to overcome the benefit of the informed changes.

In order to leverage the benefits of the informed changes, we propose a *combined approach*. We interleave the evolutionary approximation and the StS-based optimisation. The evolutionary approximation typically quickly converges to a sub-optimal solution. After the progress of the evolution decreases below a certain threshold, we run the StS-based optimisation. It further reduces the circuit size, but, more importantly, it introduces new reconnections in the circuit causing that the subsequent evolution is able to escape a local minimum and further explore the design space.

## 2    Designing Approximate Circuits

Technology-independent functional approximation is the most preferred approach to approximation of digital circuits. The goal is to replace the original accurate circuit (further denoted as the *golden circuit*) by a less complex circuit which exhibits some errors but improves non-functional circuit parameters such as power, delay, or chip area. Fully-automated functional approximation methods typically employ various heuristics to simplify the circuit logic and reduce its area approximated by the sum of the sizes of the gates used—this sum is further denoted as the *circuit size*.

The circuit size can be reduced either by replacing a gate by a smaller one or by disconnecting a gate. The gate is disconnected if there is no connection between its output and the primary outputs of the circuit. The essential operation in the approximation process is thus gate reconnection allowing one to disconnect some gates. We stress that the space of possible reconnections grows exponentially with the circuit size, and each reconnection typically causes a non-trivial change in the overall circuit functionality.

To overcome this complexity, existing approximation techniques leverage various forms of greedy algorithms, such as ABACUS [18], or genetic algorithms, such as Cartesian Genetic Programming [13,26], to identify suitable reconnections. The approximation then boils down to iteratively generating candidate solutions and evaluating their quality, i.e. the obtained trade-off between the circuit area and error. Circuit approximation can be naturally formulated as multi-objective optimisation, but most works consider single-objective optimisation of the circuit size for several predefined target errors—this is motivated practically as the required error levels are typically known in advance and single-objective optimisation is computationally less demanding.

There exist several metrics to quantify the error [8] and different techniques allowing one to evaluate these metrics. For small circuits (up to 12-bit inputs), parallel circuit simulation [27] provides the best performance. For larger circuits, various formal verification techniques have been proposed [5,9,25,28]. In this paper, we build on the SAT-based technique we proposed in [3] allowing one to verify whether a given candidate circuit meets the required bound $T$ on the worst-case absolute error (WCAE), i.e. whether the difference between the candidate and the golden circuit is smaller than $T$ for every input. The technique constructs a *miter* [28], an auxiliary circuit interconnecting the golden and candidate circuit and allowing one to check their approximate equivalence given by the bound $T$. The technique allows us to approximate complex circuits (16-bit multipliers and beyond).

Recent advances in exact SAT-based synthesis of Boolean chains [24], providing efficient implementation of a given Boolean function, have opened new avenues for automated circuit design and optimisation. In this paper, we investigate whether these advances can improve circuit approximation too.

## 3    SAT-based Circuit Approximation

We propose three different approaches for SAT-based circuit approximation.

### 3.1    A Monolithic Approach

The monolithic approach builds a single formula encoding the following synthesis problem: *For a given golden circuit GC, the size S of its currently* *best-known approximation, and an error bound $T$, synthesize an approximating circuit $AC$ whose size is smaller than $S$ and that satisfies the constraint that error$(GC, AC) < T$.*

The formula has to encode the following features: (1) possible designs of the circuit (i.e. possible interconnections and functionality of the gates), which must be encoded using free variables whose suitable values are to be found by the solver, thus fixing a certain design of the circuit; (2) the way the error of the circuit is to be checked; and (3) the way the circuit size is to be evaluated.

In our approach, we use a forward-propagating network of two-input gates to represent the designed circuit. We represent each gate by three integers. The first two represent the inputs of the gate, and they can refer to some of the primary inputs or to the output of one of the gates (which we identify with the gate itself). The third integer then encodes the gate's functionality that is chosen from a predefined set of operations. A gate implementing each possible operation has a predefined size given by the target chip architecture. To ensure that the size of the synthesized circuit $AC$ is smaller than the size $S$ of the currently best approximation, we add a constraint on the sum of the sizes of the gates forming $AC$. We investigate and compare (cf. Section 4) the below presented three ways of encoding the structure and functionality of $C$.

The first encoding is purely *SAT-based* although we present it using both Boolean and integer variables—those are, however, bit-blasted away. Assume we have $k$ types of (binary) gates, use $l$ gates, and have $m/n$ primary input/output bits, respectively. For each gate $g \in G = \{1, ..., l\}$, we use the integer variables $in_{g,1}$ and $in_{g,2}$ to denote the first and second input of $g$. These variables range over the domain $W = \{1, ..., m + l\}$ of all wires in the circuit where the first $m$ wires carry the primary inputs and the next $l$ wires carry the outputs of the different gates. For $g \in G$, we also use the integer variable $f_g$ to denote its type with the domain $F = \{1, ..., k\}$. Let $\mathbb{I} = \{0, 1\}^m$ denote the different input combinations. For $u \in W$, we use the Boolean variable $b_u^I$ to hold the value of the wire $u$ for a primary input $I \in \mathbb{I}$. We encode all possible circuits by the conjunction of the formulae $(in_{g,1} = u \wedge in_{g,2} = v \wedge f_g = f) \rightarrow \bigwedge_{I \in \mathbb{I}}(b_{m+g}^I = b_u^I \ op_f \ b_v^I)$ that are generated for all gates $g \in G$, all possible types $f \in F$ of $g$, and all wires $u, v \in W$ that may be used as the inputs of $g$. In particular, we require that $u < g$ and $v < g$ to prevent backward connections in the circuit (e.g. the input of $g_4$ cannot be connected to the output of $g_6$). In the formula, $op_f$ denotes the Boolean operation implemented by gates of the type $f \in F$.

We also need to link the concrete input combinations with the input wires, which is done by the conjunction $\bigwedge_{I \in \mathbb{I}} \bigwedge_{j \in \{1,...,m\}} b_j^I = I[j]$ where $I[j]$ denotes the $j$-th bit of $I$. Finally, for each output $o \in O = \{1, ..., n\}$, we introduce the integer variable $out_o$, which ranges over the domain of wires $W$ and says from where the output $o$ is taken, and the Boolean variable $out_o^I$ carrying the value of the output $o$ for the primary input $I \in \mathbb{I}$ (this variable will be compared with the appropriate output of the golden circuit). These variables are connected with the rest of the circuit using the conjunction of the formulae $out_o = u \rightarrow \bigwedge_{I \in \mathbb{I}} out_o^I = b_u^I$ generated for every output $o \in O$ and every wire $u \in W$. The solver then chooses a concrete circuit by fixing the values of the variables $in_{g,1}$, $in_{g,2}$, and $f_g$ for every $g \in G$ as well as the values of the variables $out_o$ for every $o \in O$.

Second, using a *theory of arrays*, we simplify the above encoding by using an array $\bar{b}^I : W \to \{0,1\}$ for each $I \in \mathbb{I}$ to hold the values of the wires in $W$ for the input $I$. Then, the conjuncts describing the structure of the circuit may be simplified to $f_g = f \to \bigwedge_{I \in \mathbb{I}} (\bar{b}^I[m + g] = \bar{b}^I[in_{g,1}] \; op_f \; \bar{b}^I[in_{g,2}])$. The input formula is changed to $\bigwedge_{I \in \mathbb{I}} \bigwedge_{j \in \{1,...,m\}} \bar{b}^I[j] = I[j]$ and similarly for the output. Finally, third, using a *theory of arrays with quantifiers*, one suffices with a single array $\bar{b}$, simplifying the formulae describing the structure of the circuit to $f_g = f \to \bar{b}[m + g] = \bar{b}[in_{g,1}] \; op_f \; \bar{b}[in_{g,2}])$, adding the universal quantification $\forall i_1, ..., i_m$ over the entire formula, using the input formula $\bigwedge_{j \in \{1,...,m\}} \bar{b}[j] = i_j$, and handling the output accordingly.

Using our encodings of $AC$, we can easily add a constraint on the required error that compares the WCAE between the result coming from $AC$ (using the $out_o$ variables) and the expected result for all input combinations.

*A comparison to existing encoding schemes for exact synthesis.* Our encoding of circuits is quite similar to other works such as [24]. The authors of [24] do not consider a predefined set of gates. Instead, they synthesize the internal function-ality of the gates too. The work [24] and other existing approaches use SAT based encodings only. Further, they consider uniform gate sizes only (the circuit size is given by the number of gates). Our more general formulation using non-uniform gate sizes leads to more complex problems. As in [22], we use simplifications and symmetry-pruning to reduce the complexity of the StS queries.

## 3.2   Sub-circuit Approximation

As discussed in [11], the monolithic approach for exact synthesis is feasible only for small circuits up to 8 input bits (depending on the complexity of the syn-thesized function). Our experiments confirm similar scalability limits also for circuit approximation (cf. Sect. 4), and thus we focus on an iterative approach that approximates selected sub-circuits. We focus on approximation wrt. Ham-ming Distance as arithmetic metrics are not suitable for sub-circuits. Note that there is no effective method allowing us to determine how the error introduced in the sub-circuit affects the overall circuit error.

In every iteration, we select a single gate (either randomly or by enumeration, depending on the circuit size) and perform a breadth-first search starting from the selected gate to identify a sub-circuit of a suitable size. Note that, in our approach, we consider multi-input and multi-output sub-circuits. The size of the sub-circuits is indeed essential: Considering only very small sub-circuits prevents the approximation from doing more complicated and non-local changes that are crucial for finding high-quality approximate circuits. On the other hand, approx-imation of larger sub-circuits introduces a significant overhead causing that only a small number of iterations can be done within the given time limit. Regarding the encoding of sub-circuit approximation, we consider the same schemes as in the monolithic approach discussed above. After every sub-circuit approximation, we need to evaluate the error of the whole circuit. If it satisfies the error bound, we accept the circuit as the new candidate solution, otherwise the next iteration continues with the circuit before the approximation.

### 3.3 Evolutionary Approximation with StS-Based Optimisation

Evolutionary algorithms, in particular Cartesian Genetic Programming (CGP), have achieved excellent results in approximation of complex circuits [3]. The key idea is similar to sub-circuit approximation, but here CGP performs random changes in the candidate solution instead of utilising satisfiability solving. Unlike finding an optimal sub-circuit approximation, random changes are very fast, and the success of CGP is typically achieved by a large number of small changes. We emphasize that CGP is also able to accumulate a large change in the candidate circuit via so-called *inactive mutations* [15]—a chain of changes where only the last change directly affects the circuit functionality. Although CGP usually quickly converges to a sub-optimum solution, it can get stuck in this solution for a long time. On the other hand, the StS-based approach is able to systematically search for improvements that are hard to find for CGP.

We hence propose a *combined approach* leveraging the benefits of both techniques. In particular, we interleave the evolutionary search by iterative StS optimisation. In contrast to StS-based approximation, StS-based optimisation minimises the size of the selected sub-circuit by changing the internal structure while preserving its functionality. The rationale behind this is based on the observation that a large portion of approximated sub-circuits are rejected as they cause that the WCAE error of the whole circuit gets above the allowed bound. Compared with CGP, the cost of each approximation operation is too high—in our scenarios, CGP is about 100-times faster. Therefore, the combined approach uses CGP to introduce changes affecting the functionality, and the StS-based optimisation to minimise the logic.Further, we also explore different encoding schemes for the optimisation problem.

The interleaving is controlled in the following way. If CGP gets stuck in a local optimum, we switch to the iterative StS optimisation that has a time budget depending on the given overall time for the approximation. Once the budget is spent, we continue with again CGP. Our experiments show that the optimisation helps CGP to escape the local optimum and to further effectively explore the space of candidate circuits.

## 4   Experimental Results

We ran all our experiments on a server with an Intel(R) Xeon(R) CPU at 2.40 GHz. Although search-based approximation can naturally benefit from a simple task parallelisation, we use a single-core computation to simplify the interpretation of the results.

*A Comparison of different encoding schemes and satisfiability solvers.* We consider a set of formulae relevant for the monolithic as well as for the iterative approach. The set includes both SAT and UNSAT instances including a full adder, 2-bit adder, 2-bit multiplier, 4-1 multiplexor, and some randomly generated 4-input functions. We compare the total time needed to solve all the formulae with a 3 hour time limit. We do not include any additional penalty for timing out. The Z3 solver [6] and the quantified array encoding proved to be the fastest combination of the encoding and the solver. Z3 with separate arrays for different input combinations is about 2 times slower, and the Glucose solver [1] with the purely SAT-based encoding is about 3 times slower. Z3 with the purely SAT-based encoding as well as its SMT variant without bit-blasting were roughly 5 times slower. Other tested solvers—MathSAT [2], Minisat [7], Sadical [12], and Vampire [20]—were all more than 5 times slower. Based on these observations, we use Z3 with the quantified array encoding in all further StS-based queries.

*The monolithic approach.* The monolithic approach was able to find optimal approximations of 2-bit adders and multipliers as well as randomly generated functions with 4 inputs and 2 outputs. Approximation of larger circuits proved to be infeasible, i.e. most instances timed out within the given limit of 3 h.

We compare the performance of our monolithic approach with Cirkit [21], a state-of-the-art tool for exact synthesis. As expected, Cirkit is able to achieve a better performance and scalability: It is significantly faster on 4-bit functions and it can also synthesize optimal solutions for some 6-bit and 8-bit functions. However, there are also some hard 6-bit instances that are infeasible for Cirkit.

The better performance of Cirkit is mainly caused by the following factors: (1) Our formulation of circuit approximation is more complicated due to the non-uniform gate sizes and the error quantification. (2) Cirkit uses different circuit representations (such as AIGs, MIGs, or n-bit look-up tables) that proved to be more efficient for some exact synthesis problems [22]. (3) Cirkit implements various optimisations and symmetry breaking methods [11]. Some of these methods are problem- and representation-specific and thus not directly applicable to our approximation problem. We are, however, aware that our current prototype implementation could be improved by adapting some of the methods. However, the improvements would not change the practical limits of the monolithic approach.

In the following subsections, we will examine three strategies for approximation of complex circuits: (1) CGP: the state-of-the-art evolutionary approximation [4]. (2) SMT: the sub-circuit approximation from Sect. 3.2. (3) COMB: the combined approach from Sect. 3.3 using the following interleaving strategy:

In each iteration, we run the CGP-based approximation until no improvement is found for 100 K generations. Then, for 10% of the overall time limit, we switch to the SMT-based optimisation. Afterwards, a new iteration starts.

Based on our preliminary experiments, we use sub-circuits with 5 gates (recall the discussion in Sect. 3.2) in all SMT-based sub-circuit approximation and optimisation queries. We also introduce a hard time limit on every such query.

**Table 1.** The resulting size of the approximate circuits, obtained using the proposed approximate strategies, expressed as the percentage of the size of the golden circuits (left) and of the size of the best known approximations presented in [3] (right).

| | 8-bit adders | | | 4-bit multipliers | | |
|---|---|---|---|---|---|---|
| Err | CGP | SMT | COMB | CGP | SMT | COMB |
| 1 % | 64.8 | 83.5 | **54.5** | 78.4 | 90.5 | **74.6** |
| 2 % | 52.6 | 78.0 | **44.9** | 69.3 | 82.6 | **67.1** |
| 5 % | 37.1 | 57.4 | **32.3** | 53.4 | 77.0 | **49.7** |

| | 32-bit adders | | | 16-bit multipliers | |
|---|---|---|---|---|---|
| Err[%] | CGP | COMB | Err[%] | CGP | COMB |
| $10^{-5}$ | 100.0 | **81.5** | $10^{-3}$ | 97.9 | **91.4** |
| $10^{-4}$ | 100.0 | **81.3** | 0.01 | 97.6 | **91.1** |
| $10^{-3}$ | 100.0 | **81.1** | 0.1 | 95.0 | **90.1** |

It prevents the SMT solver to spend a prohibitively long time in complex queries and thus to significantly slow down the approximation process.

## 4.1   Performance on Small Circuits

We first consider small circuits (a 4-bit multiplier with 67 gates and an 8-bit adder with 49 gates) to understand performance aspects of the search strategies. We report the area savings (as the percentage of the size of the golden circuit) for selected WCAE error bounds and the approximation time limit of 1 h.

Table 1 (left) shows the results obtained from 15 independent approximation runs for each combination of the approximation method, circuit, and target error. For the 8-bit adder, the combined approach wins in all 45 evolutionary runs. On average, the combined approach saves 7.6% more than the pure CGP and 29% more than the pure SMT-based approach. For the 4-bit multiplier, the combined strategy provides 3.27% better savings than the pure CGP and 19,6% more than the pure SMT-based approach. It also wins 37 out of 45 comparisons.

These experiments show that the pure SMT approximation is not competitive, and it is not considered in the following approximation of complex circuits.

## 4.2   Performance on Complex Circuits

In this subsection, we focus on our key research question: *Can the combined strategy improve the performance of the approximation of complex circuits?*

We consider approximation of (1) a 32-bit adder (the golden model has 235 gates), and (2) 16-bit multiplier (the golden model has 1,534 gates). To evaluate the potential of the combined strategy, we start with state-of-art approximate circuits we obtained in our previous work by a pure evolutionary search strategy [3]. For each target error, we choose the best 32-bit adders and 16-bit multipliers, obtained by 2 and 8-h approximation runs respectively. From each of these circuits (seeds), we continue the approximation using pure CGP and combined strategy for 10 h (adders) and 75 h (multipliers).

*32-bit adders.* Each pure CGP run performs around 10 million iterations within the given 10 h but achieves no improvements at all. The sub-circuit optimisation, however, introduces changes in the circuit structure, which allow CGP to escape the local optimum and perform further improvements. In total, the combined strategy saves roughly 19% of the seeding circuit area—11% was achieved by the CGP approximation and 8% by SMT optimisation.

*16-bit multipliers.* As illustrated in Fig. 1, which shows the progress of the two approximation strategies for different target errors, the pure CGP approximation improves the candidate slowly and achieves only marginal improvements after 45 h. The combined strategy is able to improve the candidate solution during the whole 75-h run—after this time, it saves 4–6% more than the pure CGP. Recall that, compared to

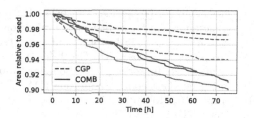

**Fig. 1.** Progress of the area reduction for the 16-bit multiplier and target WCAEs: red = $10^{-1}$%, green = $10^{-2}$%, blue = $10^{-3}$%.

32-bit adders, the approximation of the 16-bit multipliers is significantly more complex. The 8-h CGP run computing the seed performs around 230 K iterations, which is around 13-times less than the 2-h run for the 32-bit adder. Hence, the pure CGP run requires much more time to reach the local optimum.

*Conclusion.* The proposed fusion of satisfiability solving and evolutionary optimisation leads to a new circuit approximation strategy that is able to effectively escape local optima and thus to explore the design space more effectively than pure evolutionary search strategies. The obtained approximate circuits provide the best known trade-offs between the precision and the chip area.

# References

1. Audemard, G., Simon, L.: On the glucose SAT solver. Int. J. Artif. Intell. Tools **27**, 1840001 (2018)
2. Bruttomesso, R., Cimatti, A., Franzén, A., Griggio, A., Sebastiani, R.: The MATH-SAT 4 SMT Solver. In: Gupta, A., Malik, S. (eds.) CAV 2008. LNCS, vol. 5123, pp. 299–303. Springer, Heidelberg (2008). https://doi.org/10.1007/978-3-540-70545-1_28
3. Češka, M., Matyáš, J., et al.: Approximating complex arithmetic circuits with formal error guarantees: 32-bit multipliers accomplished. In: International Conference on Computer Aided Design (ICCAD'2017), pp. 416–423. IEEE (2017)
4. Češka, M., Matyáš, J., Mrazek, V., Sekanina, L., Vasicek, Zdenek, Vojnar, Tomáš: ADAC: Automated design of approximate circuits. In: Chockler, H., Weissenbacher, G. (eds.) CAV 2018. LNCS, vol. 10981, pp. 612–620. Springer, Cham (2018). https://doi.org/10.1007/978-3-319-96145-3_35
5. Chandrasekharan, A., Soeken, M., et al.: Precise error determination of approximated components in sequential circuits with model checking. In: Design Automation Conference (DAC'2016), pp. 129:1–129:6. ACM (2016)
6. de Moura, L., Bjørner, N.: Z3: An efficient SMT solver. In: Ramakrishnan, C.R., Rehof, J. (eds.) TACAS 2008. LNCS, vol. 4963, pp. 337–340. Springer, Heidelberg (2008). https://doi.org/10.1007/978-3-540-78800-3_24
7. Eén, N., Sörensson, N.: An extensible SAT-solver. In: Giunchiglia, E., Tacchella, A. (eds.) SAT 2003. LNCS, vol. 2919, pp. 502–518. Springer, Heidelberg (2004). https://doi.org/10.1007/978-3-540-24605-3_37

8. Froehlich, S., Große, D., Drechsler, R.: One method - all error-metrics: a three-stage approach for error-metric evaluation in approximate computing. In: Design, Automation Test in Europe Conference Exhibition (2019)

9. Froehlich, S., Grosse, D., Drechsler, R.: Approximate hardware generation using symbolic computer algebra employing grobner basis. In: Design, Automation Test in Europe Conference Exhibition (DATE'2018), pp. 889–892. IEEE (2018)

10. Gupta, V., Mohapatra, D., et al.: Low-power digital signal processing using approximate adders. IEEE Trans. Comput. Aided Des. Integr. Circuits Syst. **32**(1), 124–137 (2013)

11. Haaswijk, W., Soeken, M., et al.: SAT based exact synthesis using DAG topology families. In: Design Automation Conference (DAC'2018), pp. 1–6 (2018)

12. Heule, M.J.H., Kiesl, B., Biere, A.: Encoding redundancy for satisfaction-driven clause learning. In: Vojnar, T., Zhang, L. (eds.) TACAS 2019. LNCS, vol. 11427, pp. 41–58. Springer, Cham (2019). https://doi.org/10.1007/978-3-030-17462-0_3

13. Lotfi, A., Rahimi, A., et al.: Grater: an approximation workflow for exploiting data-level parallelism in FPGA acceleration. In: Design, Automation Test in Europe Conference Exhibition (DATE'2016), pp. 1279–1284. EDA Consortium (2016)

14. Mahdiani, H.R., Ahmadi, A., et al.: Bio-inspired imprecise computational blocks for efficient VLSI implementation of soft-computing applications. IEEE Trans. Circuits Syst. I Regul. Pap. **57**(4), 850–862 (2010)

15. Miller, J.F., Thomson, P.: Cartesian genetic programming. In: Poli, R., Banzhaf, W., Langdon, W.B., Miller, J., Nordin, P., Fogarty, T.C. (eds.) EuroGP 2000. LNCS, vol. 1802, pp. 121–132. Springer, Heidelberg (2000). https://doi.org/10.1007/978-3-540-46239-2_9

16. Mrazek, V., Hrbacek, R., et al.: EvoApprox8b: library of approximate adders and multipliers for circuit design and benchmarking of approximation methods. In: Design, Automation Test in Europe Conference Exhibition (DATE'2017) (2017)

17. Mrazek, V., Sarwar, S.S., et al.: Design of power-efficient approximate multipliers for approximate artificial neural networks. In: International Conference on Computer Aided Design (ICCAD'2016), pp. 811–817. ACM (2016)

18. Nepal, K., Hashemi, S., et al.: Automated high-level generation of low-power approximate computing circuits. IEEE Trans. Emerg. Top. Comput. **7**, 18–30 (2018)

19. Reda, S., Shafique, M. (eds.): Approximate Circuits. Springer, Cham (2019). https://doi.org/10.1007/978-3-319-99322-5

20. Riazanov, A., Voronkov, A.: The design and implementation of vampire. AI Commun. **15**, 91–110 (2002)

21. Soeken, M.: Cirkit (version 3). https://github.com/msoeken/cirkit (2019)

22. Soeken, M., Amarù, L.G., et al.: Exact synthesis of majority-inverter graphs and its applications. IEEE Trans. Comput. -Aided Des. Integr. Circuits Syst. **36**(11), 1842–1855 (2017)

23. Soeken, M., De Micheli, G., Mishchenko, A.: Busy man's synthesis: combinational delay optimization with sat. In: Design, Automation Test in Europe Conference Exhibition (DATE'2017), pp. 830–835 (2017)

24. Soeken, M., Haaswijk, W., et al.: Practical exact synthesis. In: Design, Automation Test in Europe Conference Exhibition (DATE'2018), pp. 309–314 (2018)

25. Vasicek, Z., Mrazek, V.: Towards low power approximate DCT architecture for HEVC standard. In: Design, Automation Test in Europe Conference Exhibition (DATE'2017) (2017)

26. Vasicek, Z., Sekanina, L.: Evolutionary approach to approximate digital circuits design. IEEE Trans. Evol. Comput. **19**(3), 432–444 (2015)

27. Vašíček, Z., Slaný, K.: Efficient phenotype evaluation in cartesian genetic programming. In: Moraglio, A., Silva, S., Krawiec, K., Machado, P., Cotta, Carlos (eds.) EuroGP 2012. LNCS, vol. 7244, pp. 266–278. Springer, Heidelberg (2012). https://doi.org/10.1007/978-3-642-29139-5_23
28. Venkatesan, R., Agarwal, A., et al.: MACACO: Modeling and analysis of circuits for approximate computing. In: International Conference on Computer Aided Design (ICCAD'2011), pp. 667–673. ACM(2011)

# SAT Solving with Fragmented Hamiltonian Path Constraints for Wire Arc Additive Manufacturing

Rüdiger Ehlers[1](✉), Kai Treutler[2], and Volker Wesling[2]

[1] Institute of Software and Systems Engineering, Clausthal University of Technology,
Clausthal-Zellerfeld, Germany
`ruediger.ehlers@tu-clausthal.de`
[2] Institute of Welding and Machining, Clausthal University of Technology,
Clausthal-Zellerfeld, Germany

**Abstract.** In Wire Arc Additive Manufactoring (WAAM), an object is welded from scratch. Finding feasible welding paths that make use of the potential of the technology is a computationally complex problem as it requires planning paths in 3D. All parts of the object to be manufactured have to be visited in few welding paths. The search for such welding paths in 3D can be mapped to searching for a fragmented Hamiltonian path in a mathematical graph.

We propose a SAT-based approach to finding such fragmented Hamiltonian paths that is suitable for planning WAAM paths. We show how to encode the search for such paths as a mix of SAT clauses and one non-clausal constraint that can be integrated into the SAT solver itself. The reasoning power of the solver enables us to impose additional constraints coming from the application domain on the planned paths, and we show experimentally that in this way, we can find welding paths for relatively complex object geometries.

## 1 Introduction

Modern additive manufacturing approaches hold a potential to add substantial flexibility to manufacturing processes. In additive manufacturing, an object is built step by step and ground up from raw material. While for plastics, 3D printing is already established, for metal, additive manufacturing is more complex, leading to a plethora of manufacturing approaches with different properties.

A notable approach in this context is wire arc additive manufacturing (WAAM), where a metal object is welded from scratch onto a metal ground body using metal from a wire roll [5]. Utilizing an industrial welding robot, the approach enables the processing of relatively high volumes of metal in short time frames with a large number of possible materials (e.g. Titanium, Aluminum, Steels, Copper, . . .) [9,14]. WAAM is based on arc welding, where an electric current induces the heat necessary to melt both the wire and the surface to which wire material gets attached. Different shielding gases are used to protect the

L. Pulina and M. Seidl (Eds.): SAT 2020, LNCS 12178, pp. 492–500, 2020.
https://doi.org/10.1007/978-3-030-51825-7_34

**Fig. 1.** Welding robot for WAAM (Institute of Welding and Machining - Clausthal University of Technology)

weld from interacting with the air. A typical manufacturing system can be seen in Fig. 1. The robot welds along a path that needs to be planned in advance. Traditionally, additive manufacturing is done layer by layer, just like for most 3D printers for plastics, so that these paths only need to be planned in 2D. The material properties of metals however depend on how quickly it cools down. For example, for low alloyed steel, the final material properties mainly depend on the time it takes for the material to cool down from 800 °C to 500 °C [4]. Shorter such so-called *t8/5 times* lead to a stronger and brittle material, while longer times lead to a tougher but weaker material with less residual stresses [4]. This can be exploited by making use of the possibility to stack up material locally during the welding process and to thus deviate from welding layer by layer, so that the material stays warm for a bit longer. The resulting local higher toughness can for instance be useful in the region of highly loaded notches [15]. On the other hand, far-sweeping welding paths speed up cooling, which is useful when a higher strength of the material is locally needed. Such material properties can at the same time be less important in other parts of the object. This gives the planning process some flexibility to attain local material properties, which is a clear advantage of additive manufacturing over traditional manufacturing processes. Planning in full 3D increases the size of the search space for possible welding paths dramatically, making the planning problem combinatorially complex. This observation asks for algorithmic support from computational engines such as SAT solvers. The planning problem has many side-constraints such as gravity (it is not possible to weld underneath a part of the object already welded), which can be encoded as clauses provided to the solver. The object to be welded can be composed of several welding paths, but the number of paths is typically low, as the disturbances caused by the ignition of the welding arc

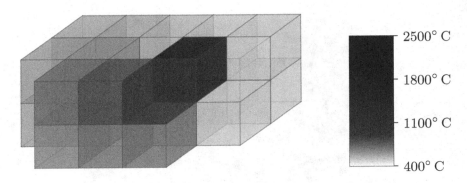

**Fig. 2.** Visualization of the block temperatures during a welding process. All blocks including those not yet welded are write-framed, while blocks already welded are filled with a semi-transparent color representing their temperatures. (Color figure online)

cause the local material properties to be worse and moving the robot head while not welding leads to an additional loss of shielding gas and material.

We show in this paper how to search for welding paths with a satisfiability (SAT) solver. We discretize the object to be manufactured into blocks and represent the connections between these blocks in the form of a graph. Two blocks that can be manufactured in succession are connected by an edge in the graph. For this initial study of planning welding paths under cooling time constraints in 3D, we discretize the object parts into cubes, but the approach presented is not restricted to cubes. By searching for a *fragmented* Hamiltonian path in the graph, i.e., one that consists of multiple independent paths that together visit all vertices, we encode the search for welding paths that together implement the complete object manufacturing process. Additional constraints on the paths encode the process-induced requirements. In our approach, checking if the planned fragmented path satisfies the requested t8/5 times is done after finding a satisfying assignment to all variables. A simplified simulator calculates the temperatures in all blocks during the welding process and implements abstract versions of heat loss due to thermal conduction and radiation. Once in any block with upper and lower t8/5 time limits a t8/5 time span has been observed that lies outside of the specified range, the simulation stops and a clause is generated that requires some part of the welding paths simulated until then to be different. To illustrate a simulation, Fig. 2 shows a visualization the simulator state in the middle of a welding process.

This paper is structured as follows: In the next section, we discuss the general approach to encoding the search for a fragmented Hamiltonian path into the SAT problem, including the special considerations from the application domain. Section 3 contains the architecture of our prototype solver, including details on the simulator for the evolution of the temperatures. Section 4 reports some experimental results, followed by a discussion of related work. We conclude with an outlook on future work and explain what role we expect SAT solving to play in additive manufacturing in the future.

# 2  Encoding Fragmented Hamiltonian Path Constraints for WAAM

We assume familiarity with the basics of satisfiability solving (see, e.g., [3] for an introduction) in the following. To describe the encoding of the welding path planning problem as a fragmented Hamiltonian path problem, we first need to define the latter.

**Definition 1.** *Let $G = (V, E)$ be a directed graph. We say that a set of sequences $S \subseteq V^*$ is a* fragmented Hamiltonian path *with $k \in \mathbb{N}$ fragments if $|S| = k$, every sequence $(s_0, \ldots, s_n) \in S$ is a path in $G$ (i.e., such that for all $0 \le i < n$, we have $(s_i, s_{i+1}) \in E$), and after adding at most $k-1$ edges to $E$, the sequences in $S$ can be connected by the additional edges to a Hamiltonian path in $G$.*

The fragmented Hamiltonian path problem is NP-hard since the Hamiltonian path problem (its special case for $k = 1$) is also NP-hard. We want to use a SAT solver to tackle this problem despite its NP-hardness. Since the fragments of a fragmented Hamiltonian path cannot share vertices, they cannot share edges. We exploit this by defining one Boolean variable $x_e$ for every edge $e \in E$ to encode whether the edge is part of a fragment or not. The values of these variables can then together completely represent a fragmented Hamiltonian path.

Since every node can only be visited once, for every node and every pair of incoming edges of the node, we use a SAT clause requiring one of the respective Boolean edge variables to have a value of **false**. If a fragment originates in a node, there does not actually need to be an active incoming edge. To encode that there are at most $k$ such nodes, we first allocate variables $y_v$ for every node $v \in V$ to represent whether a fragment originates from the node. Then, we add a clause $y_v \vee \bigvee_{(v',v) \in E} x_{(v',v)}$ for every vertex $v$ to ensure that all vertices in which no fragment starts have incoming edges. Finally, we use some type of cardinality constraint [13] to ensure that at most $k$ variables in $\{y_v\}_{v \in V}$ have values of **true**.

So far, such an encoding does not guarantee that the edges selected cannot contain cycles, which is disallowed by Definition 1. Pandey and Rintanen [12] showed that *acyclicity* can be efficiently taken into account by *non-clausal constraint* that the SAT solver evaluates before deciding next variable values (and after unit propagation), as a clausal encoding needs large numbers of clauses and variables. Following their observations, we use the same approach in this paper.

For the WAAM application, the graph represents the discretized blocks of the object to be welded and the order in which successive blocks can be welded. The fragments of a fragmented Hamiltonian path in the graph represent the welding paths that together build up the object. They cannot contain cycles since a block would then be welded twice. As the t8/5 times at the blocks depend on in which order the fragments are welded, we also need to encode this order. Using the assumption that $k$ will normally be small in this application domain, we chose to do so using $|V| \times k$ many variables $\{z_{v,i}\}_{v \in V, 1 \le i \le k}$. A variable $z_{v,i}$ should have a value of **true** iff vertex $v$ belongs to the $i$th path to be welded.

To ensure that these variables have meaningful values, we first encode that every vertex has to be part of at least one fragment number, using $|V|$

many clauses of length $k$. Then, we encode that all (successor) elements of a path are marked as belonging to the same order. We do so with the clauses $\bigwedge_{(v,v')\in E, 1\leq i\leq k} \neg z_{i,v} \vee \neg x_{(v,v')} \vee z_{i,v'}$. Also, to exclude that all paths are marked as having the same index, we add the clauses $\bigwedge_{v,v'\in V, 1\leq i\leq k} \neg y_v \vee \neg y_{v'} \vee \neg z_{v,i} \vee \neg z_{v',i}$ which make sure that all path starting points have different fragment numbers. This constraint will be the only one in our encoding that requires a number of clauses quadratic in the number of vertices.

The relatively variable-intensive encoding of which vertex belongs to which path number has the benefit that it enables us to concisely encode another important requirement for welding: vertices $v'$ that lie above vertices $v$ in the three-dimensional grid cannot be welded in earlier fragments. It suffices to add the clauses $\bigwedge_{v,v' \text{ s.t. } v \text{ is directly above } v', 1\leq i<i'\leq k} \neg x_{i,v} \vee \neg x_{i',v'}$ to ensure this. We do not encode any downwards edges into $E$ (and only the transitions back, front, left, right, and up from each block), so we can ignore the $i = i'$ case.

## 3   Solver and Simulator Engineering

To evaluate the encoding for fragmented Hamiltonian paths presented in the preceding section, we implemented the acyclicity non-clausal constraint on top of the SAT solver `Glucose 3.0` [2], which bases on `Minisat` [6]. The solver is called by a host tool that reads a geometry description of the object to be manufactured, constraints on t8/5 times for some blocks, and two constants (described below) detailing the cooling properties of the welding process. The tool then computes the underlying graph and performs the encoding presented above. For encoding the cardinality constraints, we use cardinality networks [1] as implemented in `PySAT` [10]. The SAT solver is used in incremental mode, so that after each found fragmented Hamiltonian path satisfying the encoded side constraints, we can run a simulator to check if welding the path would lead to the satisfaction of the t8/5 time constraints, and rule out this path afterwards.

The simulator tracks the temperatures of all blocks during the welding process. Initially, no block has been welded, meaning that no block has a temperature. When a block is welded, its temperature is initialized to 2500 °C. Between welding the blocks, the simulator calculates temperature changes due to two physical phenomena:

– Neighbouring blocks that have both been already welded exchange heat at a rate that is proportional to their temperature difference (and the surface area, which is however the same for all block faces in this work).
– Hot objects emit radiation from their outer surfaces (which can be detected by infrared heat cameras). The intensity of radiation is proportional to the difference between the fourth powers of the temperatures of the block and the environment.

For very hot object parts (e.g., the weld), the temperature loss/exchange due to radiation typically dominates the former physical phenomenon. In both phenomena, temperature differences are driving the heat exchange, which means that the dynamics of the system are represented as differential equations. Our simulator

**Table 1.** Overview of the experimental results. computation times are taken on a Linux-based computer with a i5-4200U processor running at 1.6 GHz (8 GB RAM).

| Example | Example 1 | Example 2 | Example 3 |
|---|---|---|---|
| # Blocks to be welded | 47 | 113 | 168 |
| # t8/5 constraints | 2 | 6 | 15 |
| # Allowed path fragments | 1 | 2 | 3 |
| Comp. time to first solution | 0.008 s | 0.603 s | 0.661 s |
| Comp. time enumerating all solutions | 0.442 s | 48.396 s | 680.28 s |
| Comp. time enum. all frag. Ham. paths | 0.228 s | 26.552 s | 422.40 s |
| # Solutions (satisfying t8/5 times) | 64 | 256 | 2272 |
| # Frag. Hamiltonian paths | 728 | 31744 | 162624 |
| Min/max/mean length of clauses added after t8/5 time violation | 34/40/ 37.1084 | 32/116/ 55.1545 | 30/162/ 79.961 |

solves them approximately using the Euler method and 20 sub-steps per welded block. While the simulation is of relatively low precision, its computational cost is already substantial. During the simulation, we keep track of how quickly a block in the object to be welded cools down from 800 °C to 500 °C. Since research on the expected material properties in case of crossing the 500 °C barrier multiple times (without exceeding 800 °C) has not converged to a well-usable model yet, we require that for every crossing of the 500 °C limit, the time spent in temperatures between 800 °C and 500 °C is within the boundaries imposed in the object description. As soon as one boundary is crossed, the simulator computes a clause ruling out the choice of fragments already welded until that point (including their order). This clause is then added to the SAT solver's clause database, so that all fragmented paths with the same simulated prefix are ruled out.

The simulator requires two constants that define the magnitude of the two physical phenomena causing temperature changes listed above. We chose those values so that a visualization of the temperature evolution shows reasonable behaviour and leads to non-trivial path planning problems that help us to evaluate our SAT-based approach. Similarly, the minimal and maximal t8/5 times are chosen to lead to computationally interesting path planning problems to help with our experimental evaluation in the next section. Fine-tuning the parameters to conform precisely to the actual welding process is left for future work after extensive experimentation with welding the planned paths.

## 4   Experiments

We evaluate the approach for encoding fragmented Hamiltonian path constraints on a few WAAM welding path planning problem instances. The aim of the experiments is to determine (a) how big the path-planned objects can be with our approach, i.e., scalability, (b) how much solving time is spent in the CDCL-part of the SAT solver, the acyclicity constraint program code, and the simulator, and (c) how long the conflict clauses computed by the temperature evolution simulator are, i.e, how early in a fragmented path it can be detected that some t8/5 time is not in the allowed range. With the experimental evaluation, we want to shed light on the principal applicability of SAT solvers for the welding path planning problem and what the most pressing further algorithmic improvements necessary to support solving this problem for large objects to be manufactured are. Table 1 contains experimental results on three example objects of different sizes. To get an idea of how much computation time is spent in the core SAT solver, in the non-clausal constraint code, and in the simulator, we evaluated this on the second example using the profiling tool `callgrind/valgrind`. This yielded 36.94% of time for the core parts of the SAT solver, 5.02% of time for the non-clausal constraint, and 58.03% of time for the simulator. It can be seen that the SAT solving part (including the non-clausal constraint) is already quite efficient, and the simulator is the bottleneck. The clauses generated after a simulation are quite long, so that the t8/5 time constraints (currently) prune the search space only little.

## 5   Related Work

We are not aware of any published previous approach to encoding fragmented Hamiltonian paths in SAT. However, Hamiltonian circuit constraints have been considered in the context of constraint programming. One of the latest works in this area is the one by Francis et al. [7], who distill previous approaches to incorporating such constraints to a new strong technique with a very high *propagation strength*, i.e., such that for solving the problem instances, fewer decisions need to be made by the solver. While their approach can be extended to Hamiltonian paths using artificial cycle closing edges, we observed that propagation in their approach is mostly dependent on which of these artificial edges are chosen by the solver. This leads to unnecessary solver decisions.

Francis et al. [7] also argue that strong propagators based on computing flows in the graphs consisting of the edges not yet ruled out by the solver, as previously proposed and used for the *all-different* constraint in constraint programming [8], is too costly in an efficient solving approach. Since this may change when dealing with fragmented Hamiltonian paths, it makes sense to re-evaluate this statement in this context in the future.

## 6   Conclusion and Outlook

In this paper, we described how SAT solving can help with new additive manufacturing approaches such as Wire Arc Additive Manufacturing (WAAM). From an application point of view, the contribution of the paper is a way to handle the complex combinational search problem for feasible welding paths. From a SAT solving perspective, this paper describes a new encoding (using an acyclicity non-clausal constraint) for finding fragmented Hamiltonian paths. It should be noted that there is an alternative to the non-clausal constraints employed in our approach. We also experimented with an adaptation of the linear feedback-shift register acyclicity SAT encoding by Johnson [11] to the case of fragmented Hamiltonian paths. Using it resulted in only slightly longer computation times. Adding further clausal constraints on welding paths may flip the advantage to the the pure SAT encoding in the future.

The experiments performed with a first approximate simulator for the local temperatures during welding show two results. First of all, SAT solvers are a suitable computational engine for this application domain. Future side-constraints on feasible curve radius sizes or similar requirements on the welded paths that are yet to be identified in WAAM research can easily be integrated as clauses. Hence, SAT solvers can serve as a reasoning platform for such planning problems. Secondly, during simulation, the violation of t8/5 time constraints becomes apparent relatively late, leading to long added clauses.

To solve this, we plan to investigate whether for complex system dynamics, some sort of specialized simulation theory can be developed to exploit the fact that while temperatures evolve non-linearly, they still evolve monotonely with the heat applied in the system. At the same time, the fact that every block is visited by the welding torch exactly once can be exploited to detect conflicts earlier, suggesting that a full SMT approach with such a specialized theory solver should be evaluated.

# References

1. Asín, R., Nieuwenhuis, R., Oliveras, A., Rodríguez-Carbonell, E.: Cardinality networks and their applications. In: Kullmann, O. (ed.) SAT 2009. LNCS, vol. 5584, pp. 167–180. Springer, Heidelberg (2009). https://doi.org/10.1007/978-3-642-02777-2_18
2. Audemard, G., Simon, L.: Glucose and syrup in the SAT race 2015. In: Reports on the SAT 2015 Competition (2015)
3. Biere, A., Heule, M., van Maaren, H., Walsh, T. (eds.): Handbook of Satisfiability, Frontiers in Artificial Intelligence and Applications, vol. 185. IOS Press, Amsterdam (2009)
4. Schröpfer, D., Kromm, A., Hannemann, A., Kannengießer, T.: In-situ determination of critical welding stresses during assembly of thick-walled components made of high-strength steel. In: Materials Research Proceedings, vol. 6, pp. 191–196 (2018)
5. Dahat, S., Hurtig, K., Andersson, J., Scotti, A.: A methodology to parameterize wire + arc additive manufacturing: a case study for wall quality analysis. J. Manuf. Mater. Process. 4(1), 14 (2020)
6. Eén, N., Sörensson, N.: An extensible SAT-solver. In: Giunchiglia, E., Tacchella, A. (eds.) SAT 2003. LNCS, vol. 2919, pp. 502–518. Springer, Heidelberg (2004). https://doi.org/10.1007/978-3-540-24605-3_37
7. Francis, K.G., Stuckey, P.J.: Explaining circuit propagation. Constraints 19(1), 1–29 (2013). https://doi.org/10.1007/s10601-013-9148-0
8. van Hoeve, W.: The all different constraint: a survey. In: Annual Workshop of the ERCIM Working Group on Constraints (2001)

9. A Hosseini, V., Högström, M., Hurtig, K., Valiente Bermejo, M.A., Stridh, L.-E., Karlsson, L.: Wire-arc additive manufacturing of a duplex stainless steel: thermal cycle analysis and microstructure characterization. Weld. World **63**(4), 975–987 (2019). https://doi.org/10.1007/s40194-019-00735-y

10. Ignatiev, A., Morgado, A., Marques-Silva, J.: PySAT: a python toolkit for prototyping with SAT oracles. In: Beyersdorff, O., Wintersteiger, C.M. (eds.) SAT 2018. LNCS, vol. 10929, pp. 428–437. Springer, Cham (2018). https://doi.org/10.1007/978-3-319-94144-8_26

11. Johnson, A.: Stedman and Erin triples encoded as a SAT problem. In: Federated Logic Conference 2018: Pragmatics of SAT. Technical report (2018)

12. Pandey, B., Rintanen, J.: Planning for partial observability by SAT and graph constraints. In: ICAPS, pp. 190–198. AAAI Press (2018)

13. Roussel, O., Manquinho, V.: Pseudo-Boolean and Cardinality Constraints, chap. 22, pp. 695–733. Vol. 185 of Biere et al. [3] (2009)

14. Szost, B.A., et al.: A comparative study of additive manufacturing techniques: residual stress and microstructural analysis of CLAD and WAAM printed Ti-6Al-4V components. Mater. Des. **89**, 559–567 (2016)

15. Treutler, K., Kamper, S., Leicher, M., Bick, T., Wesling, V.: Multi-material design in welding arc additive manufacturing. Metals **9**(7), 809 (2019)

# SAT-Based Encodings for Optimal Decision Trees with Explicit Paths

Mikoláš Janota[1,2]([⊠])[iD] and António Morgado[1][iD]

[1] INESC-ID/IST, Universidade de Lisboa, Lisbon, Portugal
Mikolas.Janota@gmail.com
[2] Czech Technical University in Prague, Prague, Czech Republic

**Abstract.** Decision trees play an important role both in Machine Learning and Knowledge Representation. They are attractive due to their immediate interpretability. In the spirit of Occam's razor, and interpretability, it is desirable to calculate the smallest tree. This, however, has proven to be a challenging task and greedy approaches are typically used to learn trees in practice. Nevertheless, recent work showed that by the use of SAT solvers one may calculate the optimal size tree for real-world benchmarks. This paper proposes a novel SAT-based encoding that explicitly models paths in the tree, which enables us to control the tree's depth as well as size. At the level of individual SAT calls, we investigate splitting the search space into tree topologies. Our tool outperforms the existing implementation. But also, the experimental results show that minimizing the depth first and then minimizing the number of nodes enables solving a larger set of instances.

## 1 Introduction

Decision trees play an important role in machine learning either on their own [6] or in the context of ensembles [5]. Learning decision trees is especially attractive in the context of interpretable machine learning due to their simplicity. However, despite this simplicity, minimization of decision trees is well-known to be an NP-hard problem [10,16]. Yet, smaller trees are likely to generalize better.

To learn trees, suboptimal, greedy algorithms are used in practice. With the rise of powerful reasoning engines, recent research has tackled the problem by the use of SAT, CSP, or MILP solvers [1,27,37,38]. Indeed, the state-of-the-art technology shows that many (NP) hard problems are often successfully solved. Conversely, such applications drive the reasoning technology by providing interesting benchmarks.

This paper, follows this line of research and proposes a novel SAT-based encoding. This encoding enables finding a decision tree conforming to the given set of examples with a given depth and number of nodes. A minimal tree is found by iterative calls to a SAT solver while minimizing size and depth.

Focusing not only on size but also on depth of the tree brings about opportunities for further analysis. Intuitively, more shallow trees are less likely to over-fit.

© Springer Nature Switzerland AG 2020
L. Pulina and M. Seidl (Eds.): SAT 2020, LNCS 12178, pp. 501–518, 2020.
https://doi.org/10.1007/978-3-030-51825-7_35

Indeed, modern packages such as Scikit [30] enable imposing a threshold on the depth, which users have to set manually. Also, a shallow tree is more likely to be interpretable by a human because less memory is required to keep track of a single branch.

The problem at hand is of challenging complexity. In practice, we may need to deal with a high number of features and examples, which brings the search-space of possible trees into extreme dimensions. Looking for an optimal tree means not only finding such tree but also proving that no smaller tree exists.

The SAT technology has recently shown a lot of promise in tackling difficult combinatorial questions, e.g. Erdős' discrepancy [22] or the Boolean Pythagorean triples problem [13], among others. Inspired by these results we also investigate the splitting of search-space based on the topology of the decision tree. The paper has the following main contributions.

1. It proposes a novel SAT-based encoding for decision trees, along with a number of optimizations.
2. Compared to existing encoding, rather than representing nodes it represents *paths* of the tree. This enables natively controlling not only the tree's size but also the tree's depth.
3. It shows that minimizing depth first and then size enables tackling harder instances.
4. It shows that search-space splitting by topologies enables tackling harder instances.
5. The implemented tool outperforms existing work [27].

## 2    Preliminaries

Standard notions and notation for propositional logic are assumed [36]. A *literal* is a Boolean variable ($x$) or its negation (denoted $\neg x$); a *clause* is a disjunction of literals a *cube* is a conjunction of literals. A formula is in *conjunctive normal form (CNF)* if it is a conjunction of clauses. General Boolean formulas are also considered constructed by using the standard connectives conjunction ($\wedge$), disjunction ($\vee$), implication ($\rightarrow$), bi-implication ($\leftrightarrow$). State-of-the-art SAT solvers typically accept input in CNF. Non-CNF formulas are converted to CNF by standard equisatisfiable clausification methods [31].

Several constraints in the paper also rely on *cardinality constraints* [34]. These are also turned into CNF through standard means, the implementation avails of the cardinality encodings in the tool PySAT [18,26].

### 2.1    Training Data

Standard setting of supervised learning is assumed [35]. Following notation and concepts of [27] we expect features to be binary (with values 0, 1). Non-binary features can be reduced to binary by unary or binary encoding. Analogously, classes are also binary (positive, negative).

Examples are defined on a fixed set of features 1..F given as two sets, one containing the negative examples $(\mathcal{E}^-)$ and second containing positive examples $(\mathcal{E}^+)$. The examples are assumed consistent, i.e. $\mathcal{E}^- \cap \mathcal{E}^+ = \emptyset$. We write $\mathcal{E}$ for the whole set of examples, i.e. $\mathcal{E} = \mathcal{E}^- \cup \mathcal{E}^+$. Each example consists of feature-value pairs. We write $\sigma(q, f)$ for the value of a feature $f$ in an example $q$. We assume that all the examples are complete, i.e. $\sigma(q)$ is total on 1..F.

# 3    SAT-Based Optimization of Decision Trees

The objective is to develop a propositional formula whose models are decision trees congruent with the given set of samples. Such model then is found by a call to an off-the-shelf SAT solver. As customary, we take the approach of optimizing by solving a series of decision problems. This means finding a decision tree with a certain size and diminishing the size until no such tree exists. Alternatively, other type of search can be used, e.g., binary or progression.

This paper targets *two* optimization criteria: *size* and *depth*. Minimizing any combination of the two may be potentially be of interest. Section 5 discusses the exact type of search used in the implementation.

The structure of binary trees guarantees a number of well-known properties. Any tree with $n$ nodes has $(n + 1)/2$ leaves and $(n - 1)/2$ internal nodes. Further, $n$ is always odd and the number of leaves is equal to the number of paths going from the root to a leaf. Our encoding heavily exploits this property:
*Rather than modeling nodes of a tree, we model the set of unique paths from the root to leaves.*

The optimization algorithm has two levels. At the first level, search is being carried out on the tree's size and depth. At the second level, the decision problem of finding a tree with such depth and size is solved via a SAT solver. The SAT solver is used in a black-box fashion, i.e., the problem is encoded into its propositional form and any off-the-shelf SAT solver may be used to solve it.

In the remainder of this section we focus on the decision problem, which is invoked with a given number of paths P (controlling size) and maximum allowed number of steps in a path S (controlling depth).

The steps in a path are numbered in the following way. In the first step, each path is in the root. In the last step of a path, the path goes from an internal node to a leaf. This means that if we are looking for a tree with a particular depth and particular number of nodes we set S and P accordingly. If we are looking only for a tree with minimal number of nodes but with an arbitrary depth, the value of S is set to P − 1, which corresponds to the number of internal nodes.

## 3.1    Path-Based Encoding

The encoding we propose models each path from the root to a leaf separately while imposing relations between them that guarantee that the paths form a binary tree. Throughout the paper, we use the convention that for a node labeled

**Table 1.** Variables used in the encoding

| Variable | Semantics | Range |
|---|---|---|
| $g_s^p$ | Path $p$ at step $s$ goes right=1/left=0 | $p \in 1..P, s \in 1..S$ |
| $t_s^p$ | Path $p$ at step $s$ is terminated | $p \in 1..P, s \in 1..S+1$ |
| $e_s^p$ | Path $p$ at step $s$ is equal to path $p-1$ | $p \in 2..P, s \in 1..S+1$ |
| $a_{s,f}^p$ | Path $p$ at step $s$ is assigned feature $f$ | $p \in 1..P, s \in 1..S, f \in 1..F$ |
| $m_q^p$ | Path $p$ matches an example $q$ | $p \in 1..P, q \in \mathcal{E}$ |
| $m_{f,v}^p$ | Path $p$ matches on value $v$ for feature $f$ | $p \in 1..P, f \in 1..F, v \in \{0,1\}$ |
| $c^p$ | Path $p$ is classified as positive | $p \in 1..P$ |

by a feature $f$, the left child corresponds to the value 0 of $f$ and the right child corresponds to the value 1 of $f$.

To model the tree, introduce a matrix of variables, where each row represents a path and each column represents a step in the path. The first row (the first path) is a path that only goes to the left—it is the leftmost path in the tree. Analogously, the last row (the last path) is a path that only goes to the right—it is the rightmost path in the tree. In general, the paths are ordered in the way they would be obtained by running DFS that goes to the left first.

Each path corresponds to a sequence of 0's and 1's so that 0 is a step to the left and 1 is a step to the right. Then, we consider these paths in a lexicographic order. Each path is represented by a sequence of variables, one for each step, where the variable represents whether the path goes left or right in that step. Additionally, for each step we need to remember whether the path has already terminated and which prefix is shared with the previous path.

Table 1 summarizes the main variables of the encoding. The direction of each step $s$ in a path $p$ is determined by the variable $g_s^p$. What is somewhat unusual about this encoding is that paths may share prefixes. To that effect, the variable $e_s^p$ represents that the path $p$ in step $s$ is in the same node as the preceding path $p-1$. The semantics of the variables $e_s^p$ is defined inductively. All paths share the root and therefore $e_1^p$ must be always true. In further steps, paths $p$ and $p-1$ remain equal as long as both paths take steps in the same direction.

$$e_1^p, p \in 2..P \tag{1}$$

$$e_{s+1}^p \leftrightarrow \left((g_s^p \leftrightarrow g_s^{p-1}) \wedge e_s^p\right), p \in 2..P, s \in 1..S \tag{2}$$

Since it is unknown beforehand how many steps are in either path, the variables $t_s^p$ determine whether the path has already terminated or not. Observe that the variables $t_s^p$ go up to step $S+1$, whereas the variables $g_s^p$ go only to step $S$. This is because the $g_s^p$ variables correspond to edges in the path while termination is tracked for nodes (as well as equality). A terminated path remains terminated and cannot terminate if it is still equal to the previous one. Any path

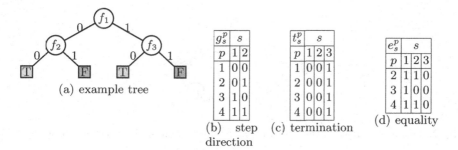

**Fig. 1.** Assignment to the variables determining the tree's topology

must be terminated after the last step.

$$t_s^p \rightarrow t_{s+1}^p, p \in 1..\mathrm{P}, s \in 1..\mathrm{S} \tag{3}$$

$$t_s^p \rightarrow \neg e_s^p, p \in 2..\mathrm{P}, s \in 1..\mathrm{S}+1 \tag{4}$$

$$t_{\mathrm{S}+1}^p, p \in 1..\mathrm{P} \tag{5}$$

*Example 1.* Figure 1 shows a binary tree along with the values of the topology variables ($g_s^p$, $t_s^p$, and $e_s^p$). The tree is comprising 4 leaves, therefore 4 paths. In this simple example each path makes two steps and then it terminates. The second path shares everything with the first one except for the leaf. The third path only shares the root with the second path. The last path shares everything with the third path, except for the leaf. Observe that since this is a full binary tree, the $g_s^p$ variables represent the binary numbers from 0 to 3.

Now it is necessary to ensure that the paths are lexicographically ordered. The first path always goes left and the last one always goes right. If a path $p$ in step $s$ is in the same node as path $p-1$, the path $p$ can go left only if $p-1$ also went left (otherwise they would cross).

$$\neg g_s^1 \wedge g_s^{\mathrm{P}}, s \in 1..\mathrm{S} \tag{6}$$

$$e_s^p \rightarrow (g_s^{p-1} \rightarrow g_s^p), p \in 2..\mathrm{P}, s \in 1..\mathrm{S}+1 \tag{7}$$

The lexicographic order alone does not guarantee a correct topology. Since the tree is binary, any path must adhere to the following pattern. For a certain number of steps it shares the prefix with the preceding path until it breaks off. Once it breaks off, it has to go only to the left (or terminate). At the same time, the preceding path can only go right after the break-off point (or terminate). Otherwise, there would be a gap in the tree.

$$(\neg t_s^p \wedge \neg e_s^p) \rightarrow \neg g_s^p, p \in 2..\mathrm{P}, s \in 1..\mathrm{S} \tag{8}$$

$$(\neg t_s^p \wedge \neg e_s^p) \rightarrow g_s^{p-1}, p \in 2..\mathrm{P}, s \in 1..\mathrm{S} \tag{9}$$

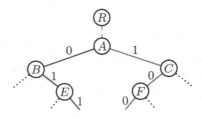

**Fig. 2.** Two consecutive paths $R$–$E$ and $R$–$F$ diverging in node $A$. (Color figure online)

Figure 2 illustrates these constraints. Consider the blue path, $R \to A \to B \to E$ and the red path, $R \to A \to C \to F$, where the blue one is lexicographically smaller. The paths diverge in node $A$—blue goes left, the red goes right. Afterwards, the blue path may only go right or terminate. In contrast, the red path may only go left or terminate. The reason why this has to be the case is that for the red one to follow blue in our ordering, the blue one has to contain the *last* path for the subtree rooted in $B$ while the red one has to contain the *first* path for the subtree rooted in $C$.

**Assigning Features and Their Semantics.** The encoding of semantics of the training data is similar to the one in [27] but with two major differences:

1. Here classification is only per *path*, while in [27] it is per *node* because any node can potentially be a leaf, which means semantics of the examples in our approach need only to be repeated $(n + 2)/2$ times rather than $n$ times.
2. Our encoding introduces explicit variables to track whether a given training example is matched for a given path, this is useful for one of the optimizations (see Sect. 3.2).

We make sure that each step is assigned exactly one feature and that no feature appears more than once on any path.

$$\sum_{f \in 1..F} a^p_{s,f} = 1, p \in 1..P, s \in 1..S \tag{10}$$

$$\sum_{p \in 1..P, s \in 1..S} a^p_{s,f} \leq 1, f \in 1..F \tag{11}$$

Recall that an example is seen as a set of feature-value pairs. We say that a feature-value pair $f, v$ is *matched* on a path if the path makes a step in the direction of $v$ in the node that is assigned the feature $f$. An example is matched if all its feature-value pairs are matched. These two concepts are modeled by the variables $m^p_{f,v}$ and $m^p_q$, respectively. Observe that $f, v$ is also matched on any path that does not contain $f$ at all. Finally, once a path matches any positive example, it must be classified as positive and the other way around.

$$m^p_{f,0} \leftrightarrow \bigwedge_{s \in 1..S} (\neg t^p_s \wedge a^p_{s,f} \rightarrow \neg g^p_s) \qquad\qquad p \in 1..\mathsf{P}, f \in 1..\mathsf{F} \qquad (12)$$

$$m^p_{f,1} \leftrightarrow \bigwedge_{s \in 1..S} (\neg t^p_s \wedge a^p_{s,f} \rightarrow g^p_s) \qquad\qquad p \in 1..\mathsf{P}, f \in 1..\mathsf{F} \qquad (13)$$

$$m^p_q \leftrightarrow \bigwedge_{f \in 1..F} m^p_{f,\sigma(q,f)} \qquad\qquad\qquad p \in 1..\mathsf{P}, q \in \mathcal{E} \qquad\quad (14)$$

$$m^p_q \rightarrow c^p \qquad\qquad\qquad\qquad\qquad\qquad q \in \mathcal{E}^+, p \in 1..\mathsf{P} \qquad (15)$$

$$m^p_q \rightarrow \neg c^p \qquad\qquad\qquad\qquad\qquad\quad q \in \mathcal{E}^-, p \in 1..\mathsf{P} \qquad (16)$$

*Summary of the Encoding.* The constraints (1)–(16) are parameterized by natural numbers $\mathsf{P}$ and $\mathsf{S}$ and their satisfying assignments represent a sequence of $\mathsf{P}$ paths in a binary tree from the root to a leaf, where each path has at most $\mathsf{S}$ edges. The paths are lexicographically ordered, starting from the leftmost path and ending in the rightmost one. Additionally, the encoding ensures that there are no gaps between paths and therefore these represent the whole binary tree. Each node in a path is labeled by a feature in a way that shared prefixes among paths are labeled by the same features. Each path is assigned a classification class that must be congruent with the training examples given on the input.

### 3.2 Path Encoding Optimizations

The encoding described above permits constructing any decision tree conforming to the given set of examples. However, certain optimizations can be made if we assume that we are not interested in superfluous nodes.

**Enforcing Example Matching.** We make sure that any path (equivalently any leaf), *matches* at least one of the given examples. If it does not, its classification does not come from the examples and may therefore be arbitrary, which means it can be removed from any tree without violating the classification of the examples. At the formula level, additional constraints are added.

$$\bigvee_{q \in \mathcal{E}} m^p_q \qquad p \in 1..\mathsf{P} \qquad\qquad\qquad (17)$$

**Pure Features.** Features with the same value in all the examples can be ignored as they never permit distinguishing between two examples of a different class. This is done at the preprocessing level, so the encoder never sees them.

**Quasi-pure Features.** A feature may appear with a fixed value $v$ within all the examples of one of the classes $c$. If such feature is assigned with the direction $v$, then the tree can immediately terminate with a leaf classified with $c$. As such, we

**Table 2.** Number of topologies (t) for tree size $\mathbf{n} \in 3..31$ (Catalan numbers)

| n | 3 | 5 | 7 | 9 | 11 | 13 | 15 | 17 | 19 | 21 | 23 | 25 | 27 | 29 | 31 |
|---|---|---|---|---|----|----|----|----|----|----|----|----|----|----|----|
| t | 1 | 2 | 5 | 14 | 42 | 132 | 429 | 1,430 | 4,862 | 16,796 | 58,786 | 208,012 | 742,900 | 2,674,440 | 9,694,845 |

enforce that the child of a node assigned in the direction of the value $v$ is a leaf classified with the class $c$. At the formula level, we add the following constraint for $s \in 1..S$ and $p \in 1..P$.

$$(a_{s,f}^p \wedge \mathrm{lit}(v, g_s^p)) \to (t_{s+1}^p \wedge \mathrm{lit}(c, c^p)) \text{ where } \mathrm{lit}(0, x) = \neg x, \mathrm{lit}(1, x) = x \quad (18)$$

**Path Lower Bounds.** We propose to use MaxSAT to obtain lower bounds on the length of a path. The question we ask is what is the shortest possible path that separates positive and negatives examples. Since the lower bound considers only one path at a time, the order of features on that path is irrelevant. In preliminary experiments we have observed rather small lower bounds. However, the bound can be improved for the leftmost and rightmost branches. This gives us three types of bounds: for the leftmost and rightmost branches, and for any branch in between. In any path a feature either does not appear, or appears on a step that goes left or on a step that goes right. To model this behavior we introduce two variables for each feature $x_f^0$ and $x_f^1$ (similar to the dual rail encoding [25]). This corresponds to the following hard and soft constraints.

| | | |
|---|---|---|
| hard: | $\neg x_f^0 \vee \neg x_f^1$ | $f \in 1..F$ |
| hard: | $\neg x_f^1/\neg x_f^0$ | $f \in 1..F$, for leftmost/rightmost branch |
| hard: | $\bigvee_{f \in 1..F} x_f^0 \wedge \bigvee_{f \in 1..F} x_f^1$ | for general branch |
| hard: | $m_q^p \leftrightarrow \bigwedge_{f \in 1..F} \neg x^{1-\sigma(q,f)}$ | $p \in 1..P, q \in \mathcal{E}$ |
| hard: | $\bigwedge_{q \in \mathcal{E}^+} \neg m_q^p \vee \bigwedge_{q \in \mathcal{E}^-} \neg m_q^p$ | |
| soft: | $\neg x_f^v$ | $f \in 1..F, v \in \{0, 1\}$ |

## 4    Search-Space Splitting by Topologies

Upon initial experiments, we observed that the SAT solver may struggle even on decision trees of modest size, e.g. 9 nodes. This is somewhat surprising because the number of topologies does not initially grow that much; see Table 2.

This suggests splitting the search space into individual topologies and call the SAT solver for each one of them separately. Like so, the SAT solver only needs to find the labeling of the tree. Intuitively this should be an easier problem because the SAT solver only needs to deal with one type of decisions.

This approach is not generally viable because eventually the number of topologies is too large. To which we propose the following approach. The upper part of the topology is fixed—until a certain depth—and the rest is left for the SAT solver to complete. This gives rise to *topology templates*. Each topology template is a tree, where each leaf is an actual leaf ($\square$) of the topology or an incomplete subtree ($\triangle$).

---

**Algorithm 1.** Topology enumeration, with $\square$ - leaf, $\triangle$ - subtree

---

1   **Function** TE $(n, d)$ **begin**
2      **if** $n = 1$ **then return** $\{\square\}$             // leaf
3      **if** $d = 0$ **then return** $\{\triangle\}$       // incomplete subtree
4      **if** $d = 1$ **then**
5          **if** $n = 3$ **then return** { tree ($\square$, $\square$) }
6          **else if** $n = 5$ **then return** { tree ($\triangle$, $\square$), tree ($\square$, $\triangle$) }
7          **else return** { tree ($\square$, $\triangle$), tree ($\triangle$, $\square$), tree ($\triangle$, $\triangle$) }
8      **return** $\{\text{tree}(l, r) \mid l \in \text{TE}(i, d-1), r \in \text{TE}(n-i-1, d-1), i \in 1..n-1\}$

---

Algorithm 1 recursively enumerates incomplete topologies on $n$ nodes with the cut-off parameter $d$. In order to avoid repetitions in enumeration, certain cases need to be treated separately. If the cut-off parameter reaches 1, the children of the current node will either be leaves ($\square$) or incomplete subtrees ($\triangle$). This, in general gives three scenarios where either the left or the right child is a leaf and the second child is a subtree, or both are subtrees. However, in the case of $n = 3$, $n = 5$ the scenarios are different. Observe that because of the cut-off parameter, the generated topology template may have less than $n$ nodes.

We study topology enumeration both for our encoding as well as the encoding of Narodytska et al. [27]. A given topology template in the encoding of Narodytska et al. is enforced by a cube corresponding to the child relation and the information whether a node is a leaf or not. An important property of our generation procedure is that the cut-off parameter is equal on all branches. This means that numbering the topology template by BFS gives the same numbers as a BFS on any topology corresponding to it. Since the encoding of mindt relies on BFS, this property lets us directly translate the relation into a cube.

Our path-based encoding does not allow easily encoding a topology template because the number of paths in an incomplete subtree ($\triangle$) is unknown. To this

---

**Algorithm 2.** Topology enumeration with cardinalities

---

1   **Function** TE$^{\#}$ $(n, d)$ **begin**
2      **if** $n = 1$ **then return** $\{\square\}$              // leaf
3      **if** $d = 0$ **then return** $\{\#n\}$     // incomplete subtree of size $n$
4      **return** $\{\text{tree}(l, r) \mid l \in \text{TE}^{\#}(i, d-1), r \in \text{TE}^{\#}(n-i-1, d-1), i \in 1..n-1\}$

---

---

**Algorithm 3.** Measuring difference between topologies

---

1  **Function** TDiff $(t_1, t_2, w)$ **begin**
2  |     **if** $|t_1| = 0$ **then return** $w|t_2|$
3  |     **else if** $|t_2| = 0$ **then return** $w|t_1|$
4  |     **else return** TDiff$(t_1.\text{left}, t_2.\text{left}, w\Delta)$ + TDiff$(t_1.\text{right}, t_2.\text{right}, w\Delta)$

---

effect, we introduce a variation on the topology template where the leaves of the topology template are actual leaves ($\square$) or an incomplete subtree with a given cardinality ($\#k$). These topology templates can be easily enumerated as shown by Algorithm 2. Observe that the number of these topology templates may be larger than in the previous version. Such topology template is encoded into our path-based model in a straightforward fashion. Each path in the topology template fixes the direction in prefixes in a certain number of paths. The number of these paths corresponds to the $\#k$ node at the end of the path. Any path terminating in $\square$ corresponds exactly to one path in the path-based model.

### 4.1   Topology Enumeration

A cube describing a topology template can either be encoded into assumptions to enable *incremental SAT solving* [7] or appended as a set of unit clauses. We observed that in our case incremental solving does not pay off for hard instances. However, at the same time, if a large number of topology templates need to be examined, initializing a new SAT solver for each one of them is too costly. Therefore, the implementation employs both modes, incremental and non-incremental, depending on the number of topology templates to be examined.

Another point of interest is the order in which the topology templates are examined. In the case of non-incremental SAT solving and unsatisfiable instances, the order does not matter because all formulas need to be solved independently of one another. Hence, the order plays mainly a role in the case of satisfiable instances. The order heuristics we propose is the following.

We start with the assumption that we already have a suboptimal solution to the problem from a greedy (fast) algorithm. We would like to first focus on topologies that are similar to the topology of this suboptimal solution. In order to do so, we need some notion of *difference* between topologies (and topology templates). For this purpose we define a simple function that recursively compares the two topologies and accumulates a penalty once they are different. Additionally, subtrees with lower depth are accounted with less weight.

Algorithm 3 shows the function. If one of the given trees is empty, the penalty is the size of the other tree weighted by the factor $w$. Otherwise, the penalties are calculated as a sum of the left and the right subtrees, respectively. As the recursion descends, the weight is gradually decayed by the factor $\Delta \in (0, 1]$. In the implementation we chose the ad-hoc value of 0.75.

When partitioning the search space, the topology templates are enumerated in the increasing order of the difference from the suboptimal solution.

## 5    Experimental Evaluation

The tool was implemented on top of the PySAT package [18], which interfaces with a number of modern SAT solvers and provides a number of implementations of cardinality encodings. We used the *CaDiCaL solver* [3] and the *k-Cardinality Modulo Totalizer* [26]. This configuration was chosen after some careful preliminary experiments. We show that this configuration performs significantly better than the configuration used in the evaluation of Narodytska et al.

Our preliminary experiments also informed other ad-hoc choices that had to be made as the search and encodings can be configured in a large number of ways. An alternative would be to employ automated parameter tuning in the spirit of ParamILS [15]; we leave this as future work.

The SAT solver is used in a non-incremental fashion, i.e., every decision problem is solved independently of the other ones. The exception is topology enumeration: if the number of topologies is larger than 500, the incremental mode is employed (see Sect. 4).

A suboptimal greedy solution is obtained by the popular modern machine learning library Scikit-learn [30], which also enables a seamless integration with the Python implementation. The greedy solution is used in two scenarios: 1) to obtain an upper bound on the number of nodes in the solution 2) to inform the ordering of topologies during enumeration (see Sect. 4).

The experiments were performed on servers with Intel(R) Xeon(R) CPU at 2.60 GHz, 24 cores, 64 GB RAM, while always running 4 tasks in parallel. The time limit was set to 1000 s and the memory limit to 3 GB. The experimental results report on the following search modes:

(1) binary search on the number of nodes with no restriction on the depth without topology enumeration (with `sklearn` upper-bound)
(2) linearly increasing the number of nodes with no restriction on the depth with topology enumeration (linear UNSAT-SAT search)
(3) linearly increasing depth and linearly increasing number of nodes for each considered depth.

Searches (1) and (2) find the smallest tree just as in [27]. The search (3) finds the smallest tree in the lexicographic ordering of the pair depth-size.

The evaluation was carried out on the benchmarks used in [27], kindly provided by Narodytska. These benchmarks were originally obtained by sampling a large set of instances [29], with sampling percentages of 20%, and 50% (we have used the same exact sampled benchmarks as Narodytska et al.). The reader is referred to [27] for the details of the sampling procedure.

We compare our tools with the state-of-the-art tool `mindt` [27]. Our tool is run in the following configurations. The configuration `dtfinder` corresponds to the search (1), i.e. size minimization via binary search and path-based encoding. The configuration `dtfinder-DT1` is the same type of search but with encoding of Narodytska et al. The suffix `-T` in a configuration indicates topology-based

**Table 3.** Results on all the benchmarks divided by the percentage of random sampling.

| % | **0.2** | | | | **0.5** | | | |
|---|---|---|---|---|---|---|---|---|
| nf./ns | 447/136 | | | | 473/357 | | | |
| #I | 754 | | | | 709 | | | |
| | #nd | depth | cpu-time | #slv | #nd | depth | cpu-time | #slv |
| mindt | 6 | 3 | 58 | 394 | 5 | 3 | 52 | 249 |
| dtfinder-DT1 | 7 | 3 | 14 | 457 | 6 | 3 | 43 | 337 |
| dtfinder-DT1-T | 7 | 3 | 28 | 473 | 7 | 3 | 41 | 345 |
| dtfinder-DT1-T-O | 7 | 3 | 27 | 473 | 7 | 3 | 39 | 345 |
| dtfinder | 7 | 3 | 14 | 458 | 7 | 3 | 60 | 339 |
| dtfinder-T | 7 | 3 | 29 | 470 | 7 | 3 | 42 | 342 |
| dtfinder-T-O | 7 | 3 | 30 | 471 | 7 | 3 | 39 | 341 |
| d-dtfinder | 8 | 3 | 65 | **519** | 7 | 3 | 73 | **352** |
| d-dtfinder-T-O | 8 | 3 | 49 | 486 | 7 | 3 | 57 | 345 |
| vbs | 8 | – | 69 | 528 | 7 | – | 46 | 355 |

search (search (2)). The suffix -T-O topology-based is search with heuristic ordering. The configuration d-dtfinder corresponds to the search (3), i.e. depth-size minimization.

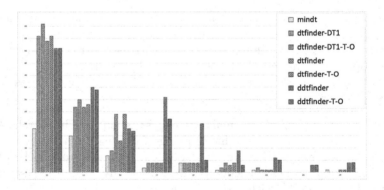

**Fig. 3.** Distribution of the sizes of calculated optimal trees

Table 3 summarizes results for all the considered benchmarks and tools. The first row (%) shows the percentage of random samplings used to construct the instance, the second row the average number of features (nf.) and samples (ns.). The third row (#I) shows the number of benchmarks in that category. The remaining rows are grouped according to the tool they represent.

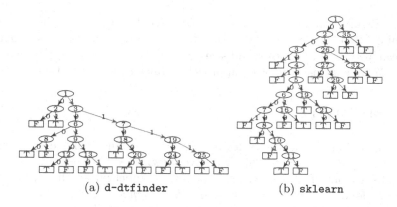

(a) d-dtfinder                    (b) sklearn

**Fig. 4.** postoperative-patient-data-un_1-un with 50% sampling

For each of the tools we present four values: the average number of nodes discovered (#nd); the average depth of the tree reported (depth); the average CPU time taken in solved instances (cpu-time); and the number of instances solved (#slv).

Figure 3 shows a histogram of the sizes of the optimal trees per solver. The vertical axis shows the number of solved instances and the horizontal groups the solvers according to the number of nodes of the reported decision trees. More detailed overview of the data can be found here on the authors' website [20].

Table 3 enables the following conclusions. Our implementation (dtfinder) outperforms the tool by Narodytska et al. (mindt) in all cases. This is also the case for their encoding; We attribute this to the choice of cardinality encoding and the SAT solver. We used $k$-Cardinality Modulo Totalizer and CaDiCaL while mindt uses sequential counter and glucose-0.3.

Comparing dtfinder with d-dtfinder, we can see that d-dtfinder is faster to compute a minimum depth solution than dtfinder is to compute a minimum size solution and even more interestingly, again solves even more instances.

The topology search-space splitting is beneficial in all encodings except for depth minimization. Both our encoding and encoding of Narodytska et al. solves more instances with topology enumeration. Not always this helps the average CPU time; however, it went from 60s to 39s in path-based encoding for the 0.5 instances. The ordering of topologies enables a minor speed-up but overall the effect is small.

The distribution of sizes of solved instances (Fig. 3) shows that the hardness of an instance grows drastically with the size. While depth minimization is able to solve a handful of instances of size 29, the path-based encoding solves just 1, our implementation of Narodytska et al. none and, surprisingly mindt 1. This can be attributed to the number of topologies (see Table 2).

Overall, focusing on minimizing depth first is computationally advantageous, yet yielding decision trees of good quality. We illustrate this on a particular instance. Figure 4 shows decision trees calculated by our approach minimizing depth first (`d-dtfinder`) and calculated by the greedy approach (`sklearn`). The optimal tree gives depth 6 and size 29, the greedy approach gives depth 11 and size 37. In contrast, the other approaches timeout on this instance in 1000 s.

## 6    Related Work

Greedy algorithms for learning decision trees based on recursive splitting are well-known [6,32,33]; see also [8] for an overview.

Various notions of optimality of decision trees appear in the literature. Some approaches focus on finding a tree with a *fixed depth* but with the best *accuracy* [1,37,38]. These approaches assume a full (perfectly balanced) binary tree of the fixed depth whose accuracy is to be optimized. While the problem is still very hard, it is in some sense easier because the topology is fixed and only the labeling needs to be calculated. However, combinations of these approaches in our approach is an interesting line of research.

Another approach is taken by [14], which optimizes a linear combination of accuracy and size. However, this approach is based on brute force search and in our experiments we were only able to synthesize trees with a handful of features while the considered benchmarks contain hundreds of features.

Closest to our work is [27], which uses SAT encoding to construct a size-optimal decision tree for a given set of consistent samples. In contrast to our work, individual nodes and their children relation are modeled explicitly. This means that a path from the root to a leaf is implicit. In principle, one could also restrict the depth of these implicit paths by adding additional counters or some other form of cardinality constraints. This is bound to be less efficient. Further, our encoding is closer to the idea of a tree. If the tree is modeled through nodes, it must be ensured that is in fact a tree via cardinality constraints—ensuring that each node has one and only one parent (except for the root) and that each internal node has two children. These cardinality constraints are not needed in our encoding. Since in our case, classes are per path rather than node we save half of the semantic constraint (see Sect. 3.1). It is interesting to compare how symmetries are broken in [27], where restrictions are imposed on the possible children nodes. In our approach paths are ordered lexicographically rather than in an arbitrary order. This order lets us single out the leftmost in the rightmost branches, which turned out to be useful in lower-bounding the depth (Sect. 3.2). We remark that lexicographic order is a popular means of breaking symmetries in general graphs, cf. [12].

Earlier work for minimization of decision tree using Constraint Programming (CP) exists [2]. It was shown in [27], that the approach by Narodytska et al. strictly outperforms the approach of Bessiere at al. this is most likely to be attributed to the fact that the CP encoding is asymptotically much larger.

Synthesis by calls to a SAT/SMT solver has seen increased interest in the recent years, cf. [19,21,28]. Haaswi et al. used topology enumeration to synthesize Boolean circuits [9]. The general idea is analogous to our approach (see Sect. 4).

However, the set of possible topologies is partitioned differently. The possible topologies are DAGs, whereas they are trees in our case. Topologies in their approach belong to the same partition if they have the same number of nodes at each level (levels are obtained by BFS). This approach is unlikely to give good partitioning for binary trees and is more expensive to encode than our approach. Further, in our approach, the enumeration of topologies simply goes over all possible topologies if the number of nodes is small.

The well-known technique of *cube-and-conquer* (CnC) splits the search-space by a lookahead solver [11,17]. The lookahead solver is run with a bound, which yields cubes to be decided by a traditional CDCL solver. Compared to our approach, CnC is much more general since it is applicable to any SAT instance, and, the lookahead solver is less likely to generate cubes that will be decided trivially. The downside is that CnC may not come up with a splitting as a human would. Further, the lookahead solver can be very costly. In our preliminary experiments, CnC performs much more poorly than a plain SAT solver on our instances. The order in which cubes are decided is also investigated by Heule et al. [11].

# 7    Conclusions and Future Work

This paper proposes a novel SAT-based encoding for decision trees, which enables natively controlling both the tree's size and depth. We also study search-space splitting by topology enumeration. Our implementation outperforms existing work of [27] but also enables a finer control due to the explicit representation of paths of the tree. This finer control lets us optimize practically interesting instances that had been out of reach.

The proposed approaches open a number of avenues for future research. The solving itself could be further improved by better splitting, parallelization, and combining with cube-and-conquer [11]. While some preprocessing of the examples was already used in our optimization techniques (Sect. 3.2), further inspection could be used to draw more information from them, e.g. introduction of extended variables in the spirit of [23]. The proposed techniques could also be integrated into more expressive approaches, e.g. SMT-based synthesis [21].

At the application level, we are investigating the integration of our tool with some greedy approaches, e.g. ensembles, where only limited depth is considered. Or, consider a hybrid between a greedy approach and an exact approach where an exact approach is invoked on smaller sub-problems. It would be interesting to investigate whether trees with a smaller depth are really easier to understand and interpret, and, what is the trade-off between depth and size. Our approach provides the means to exactly quantify these metrics.

The experimental evaluation shows that SAT solvers poorly handle a search-space with many topologies. We believe that this represents an important challenge for the SAT community.

**Acknowledgements.** This work was supported by national funds through FCT, Fundação para a Ciência e a Tecnologia, under project UIDB/50021/2020, the project

INFOCOS with reference PTDC/CCI-COM/32378/2017. The results were supported by the Ministry of Education, Youth and Sports within the dedicated program ERC CZ under the project POSTMAN with reference LL1902.

# References

1. Bertsimas, D., Dunn, J.: Optimal classification trees. Mach. Learn. **106**(7), 1039–1082 (2017). https://doi.org/10.1007/s10994-017-5633-9
2. Bessiere, C., Hebrard, E., O'Sullivan, B.: Minimising decision tree size as combinatorial optimisàtion. In: Gent, I.P. (ed.) CP 2009. LNCS, vol. 5732, pp. 173–187. Springer, Heidelberg (2009). https://doi.org/10.1007/978-3-642-04244-7_16
3. Biere, A.: CaDiCaL, Lingeling, PLingeling, Treengeling and YalSAT entering the SAT competition 2017. In: Balyo,T., Heule, M., Järvisalo, M. (eds.) SAT Competition 2017: Solver and Benchmark Descriptions, pp. 14–15. University of Helsinki (2017)
4. Biere, A., Heule, M., van Maaren, H., Walsh, T. (eds.): Handbook of Satisfiability, Frontiers in Artificial Intelligence and Applications, vol. 185, p. 980. IOS Press, Amsterdam (2009)
5. Breiman, L.: Random forests. Mach. Learn. **45**(1), 5–32 (2001). https://doi.org/10.1023/A:1010933404324
6. Breiman, L., Friedman, J.H., Olshen, R.A., Stone, C.J.: Classification and Regression Trees. Wadsworth, Belmont (1984)
7. Eén, N., Sörensson, N.: Temporal induction by incremental SAT solving. Electron. Notes Theor. Comput. Sci. **89**(4), 543–560 (2003). https://doi.org/10.1016/S1571-0661(05)82542-3
8. Fürnkranz, J.: Decision tree. In: Sammut, C., Webb, G.I. (eds.) Encyclopedia of Machine Learning and Data Mining, pp. 330–335. Springer, Boston (2017). https://doi.org/10.1007/978-1-4899-7687-1_66
9. Haaswijk, W., Mishchenko, A., Soeken, M., Micheli, G.D.: SAT based exact synthesis using DAG topology families. In: Proceedings of the 55th Annual Design Automation Conference, DAC, pp. 53:1–53:6 (2018). https://doi.org/10.1145/3195970.3196111
10. Hancock, T.R., Jiang, T., Li, M., Tromp, J.: Lower bounds on learning decision lists and trees. Inf. Comput. **126**(2), 114–122 (1996). https://doi.org/10.1006/inco.1996.0040
11. Heule, M.J.H., Kullmann, O., Wieringa, S., Biere, A.: Cube and conquer: guiding CDCL SAT solvers by lookaheads. In: Eder, K., Lourenço, J., Shehory, O. (eds.) HVC 2011. LNCS, vol. 7261, pp. 50–65. Springer, Heidelberg (2012). https://doi.org/10.1007/978-3-642-34188-5_8
12. Heule, M.J.H.: Optimal symmetry breaking for graph problems. Math. Comput. Sci. **13**(4), 533–548 (2019). https://doi.org/10.1007/s11786-019-00397-5
13. Heule, M.J.H., Kullmann, O., Marek, V.W.: Solving and verifying the boolean pythagorean triples problem via cube-and-conquer. In: Creignou, N., Le Berre, D. (eds.) SAT 2016. LNCS, vol. 9710, pp. 228–245. Springer, Cham (2016). https://doi.org/10.1007/978-3-319-40970-2_15
14. Hu, X., Rudin, C., Seltzer, M.: Optimal sparse decision trees. In: Neural Information Processing Systems NeurIPS (2019). http://papers.nips.cc/paper/8947-optimal-sparse-decision-trees
15. Hutter, F., Hoos, H.H., Leyton-Brown, K., Stützle, T.: ParamILS: an automatic algorithm configuration framework. J. Artif. Intell. Res. **36**, 267–306 (2009)

16. Hyafil, L., Rivest, R.L.: Constructing optimal binary decision trees is NP-complete. Inf. Process. Lett. **5**(1), 15–17 (1976). https://doi.org/10.1016/0020-0190(76)90095-8
17. Hyvärinen, A.E.J., Junttila, T., Niemelä, I.: Partitioning SAT instances for distributed solving. In: Fermüller, C.G., Voronkov, A. (eds.) LPAR 2010. LNCS, vol. 6397, pp. 372–386. Springer, Heidelberg (2010). https://doi.org/10.1007/978-3-642-16242-8_27
18. Ignatiev, A., Morgado, A., Marques-Silva, J.: PySAT: a python toolkit for prototyping with SAT oracles. In: Beyersdorff, O., Wintersteiger, C.M. (eds.) SAT 2018. LNCS, vol. 10929, pp. 428–437. Springer, Cham (2018). https://doi.org/10.1007/978-3-319-94144-8_26
19. Ignatiev, A., Pereira, F., Narodytska, N., Marques-Silva, J.: A SAT-based approach to learn explainable decision sets. In: Galmiche, D., Schulz, S., Sebastiani, R. (eds.) IJCAR 2018. LNCS (LNAI), vol. 10900, pp. 627–645. Springer, Cham (2018). https://doi.org/10.1007/978-3-319-94205-6_41
20. Janota, M., Morgado, A.: (2020). http://sat.inesc-id.pt/%7Emikolas/dectrees
21. Kolb, S., Teso, S., Passerini, A., Raedt, L.D.: Learning SMT(LRA) constraints using SMT solvers. In: Lang [24]. https://doi.org/10.24963/ijcai.2018/323, http://www.ijcai.org/proceedings/2018/
22. Konev, B., Lisitsa, A.: Computer-aided proof of Erdős discrepancy properties. Artif. Intell. **224**, 103–118 (2015). https://doi.org/10.1016/j.artint.2015.03.004
23. Lagniez, J.-M., Biere, A.: Factoring out assumptions to speed up MUS extraction. In: Järvisalo, M., Van Gelder, A. (eds.) SAT 2013. LNCS, vol. 7962, pp. 276–292. Springer, Heidelberg (2013). https://doi.org/10.1007/978-3-642-39071-5_21
24. Lang, J. (ed.): Proceedings of the Twenty-Seventh International Joint Conference on Artificial Intelligence (IJCAI 2018), Stockholm, Sweden, 13–19 July 2018. ijcai.org (2018). http://www.ijcai.org/proceedings/2018/
25. Manquinho, V.M., Flores, P.F., Silva, J.P.M., Oliveira, A.L.: Prime implicant computation using satisfiability algorithms. In: 9th International Conference on Tools with Artificial Intelligence ICTAI, pp. 232–239 (1997). https://doi.org/10.1109/TAI.1997.632261
26. Morgado, A., Ignatiev, A., Marques-Silva, J.: MSCG: robust core-guided MaxSAT solving. JSAT **9**, 129–134 (2015)
27. Narodytska, N., Ignatiev, A., Pereira, F., Marques-Silva, J.: Learning optimal decision trees with SAT. In: Lang [24], pp. 1362–1368. https://doi.org/10.24963/ijcai.2018/189, http://www.ijcai.org/proceedings/2018/
28. Narodytska, N., Shrotri, A., Meel, K.S., Ignatiev, A., Marques-Silva, J.: Assessing heuristic machine learning explanations with model counting. In: Janota, M., Lynce, I. (eds.) SAT 2019. LNCS, vol. 11628, pp. 267–278. Springer, Cham (2019). https://doi.org/10.1007/978-3-030-24258-9_19
29. Olson, R.S., Cava, W.G.L., Orzechowski, P., Urbanowicz, R.J., Moore, J.H.: PMLB: a large benchmark suite for machine learning evaluation and comparison. BioData Min. **10**(1), 36:1–36:13 (2017). https://doi.org/10.1186/s13040-017-0154-4
30. Pedregosa, F., et al.: Scikit-learn: machine learning in Python. J. Mach. Learn. Res. **12**, 2825–2830 (2011). https://scikit-learn.org/
31. Plaisted, D.A., Greenbaum, S.: A structure-preserving clause form translation. J. Symb. Comput. **2**(3), 293–304 (1986)
32. Quinlan, J.R.: Induction of decision trees. Mach. Learn. **1**(1), 81–106 (1986)
33. Quinlan, J.R.: C4.5: Programs for Machine Learning. Morgan Kaufmann, San Mateo (1993)

34. Roussel, O., Manquinho, V.M.: Pseudo-boolean and cardinality constraints. In: Biere, A., et al. (eds.) Handbook of Satisfiability, Frontiers in Artificial Intelligence and Applications, vol. 185, pp. 695–733. IOS Press, Amsterdam (2009). https://doi.org/10.3233/978-1-58603-929-5-695

35. Russell, S.J., Norvig, P.: Artificial Intelligence: A Modern Approach. Prentice Hall, Upper Saddle River (2010)

36. Silva, J.P.M., Lynce, I., Malik, S.: Conflict-driven clause learning SAT solvers. In: Biere, A., et al. (eds.) Handbook of Satisfiability, Frontiers in Artificial Intelligence and Applications, vol. 185, pp. 131–153. IOS Press, Amsterdam (2009). https://doi.org/10.3233/978-1-58603-929-5-131

37. Verhaeghe, H., Nijssen, S., Pesant, G., Quimper, C., Schaus, P.: Learning optimal decision trees using constraint programming. In: 31st Benelux Conference on Artificial Intelligence (BNAIC) (2019). http://ceur-ws.org/Vol-2491/abstract109.pdf

38. Verwer, S., Zhang, Y.: Learning optimal classification trees using a binary linear program formulation. In: The Thirty-Third AAAI Conference on Artificial Intelligence (AAAI), pp. 1625–1632. AAAI Press (2019). https://doi.org/10.1609/aaai.v33i01.33011624, https://www.aaai.org/Library/AAAI/aaai19contents.php

# Incremental Encoding of Pseudo-Boolean Goal Functions Based on Comparator Networks

Michał Karpiński[(✉)][iD] and Marek Piotrów[iD]

Institute of Computer Science, University of Wrocław,
Joliot-Curie 15, 50-383 Wrocław, Poland
{karp,mpi}@cs.uni.wroc.pl

**Abstract.** Incremental techniques have been widely used in solving problems reducible to SAT and MaxSAT instances. When an algorithm requires making subsequent runs of a SAT-solver on a slightly changing input formula, it is usually beneficial to change the strategy, so that the algorithm only operates on a single instance of a SAT-solver. One way to do this is via a mechanism called *assumptions*, which allows to accumulate and reuse knowledge from one iteration to the next and, in consequence, the provided input formula need not to be rebuilt during computation. In this paper we propose an encoding of a Pseudo-Boolean goal function that is based on sorting networks and can be provided to a SAT-solver only once. Then, during an optimization process, different bounds on the value of the function can be given to the solver by appropriate sets of assumptions. The experimental results show that the proposed technique is sound, that is, it increases the number of solved instances and reduces the average time and memory used by the solver on solved instances.

**Keywords:** Incremental encoding · CNF encoding · Pseudo-Boolean constraints · Comparator networks · SAT-solvers

## 1 Introduction

A Pseudo-Boolean constraint (a PB-constraint, in short) is of the form $a_1 x_1 + a_2 x_2 + \cdots + a_n x_n \# k$, where $n, k \in \mathbb{N}$, $\{x_1, \ldots, x_n\}$ is a set of propositional literals (that is, variables or their negations), $\{a_1, \ldots, a_n\}$ is a set of integer coefficients, and $\# \in \{<, \leq, =, \geq, >\}$. PB-constraints are more expressive and more compact than clauses when representing some Boolean formulas, especially for optimization problems. PB-constraints are used in many real-life applications, for example, in cumulative scheduling [31], logic synthesis [3] or verification [9]. There have been many approaches for handling PB-constraints in the past, for example, extending existing SAT-solvers to support PB-constraints natively [14,21]. One of the most successful ideas was introduced by Eén and Sörensson [13], who show how PB-constraints can be handled through translation to SAT.

L. Pulina and M. Seidl (Eds.): SAT 2020, LNCS 12178, pp. 519–535, 2020.
https://doi.org/10.1007/978-3-030-51825-7_36

The algorithm, implemented in a tool called MINISAT+, incrementally strengthens the constraint on a goal function to find the optimum value, rebuilding partially a formula on each iteration, and making a new call to the underlying SAT-solver.

A typical SAT-solver accepts a problem instance as an input and outputs a satisfying assignment or an **Unsatisfiable** statement as a result. This can be inefficient if we want to minimize a value of a given goal function by solving many similar SAT instances (like in the aforementioned PB-solving algorithm of MINISAT+). Parsing almost the same constraint sets, and then applying the same inferences could be costly, therefore a more preservative approach is recommended.

An incremental approach for solving a series of related SAT instances was introduced, for example, in [12], as the means of checking safety properties on finite state machines. Later, the same authors implemented this technique in MINISAT [11] as a general tool, which they simply called *assumptions*. Assumptions are propositions that hold solely for one specific invocation of the solver. The goal of this paper is to propose an incremental algorithm for solving PB-constraint optimization problems by modifying an iterative SAT-based algorithm of KP-MINISAT+ [17], such that the input instance is encoded only once, and later, a set of assumptions is changed from one iteration to another, such that the encoding of the new constraint (on the goal function) is preserved, without the need to rebuild the CNF formula.

## 1.1   Related Work

One way to solve a PB-constraint is to transform it to a SAT instance (via Binary Decision Diagrams (BDDs), adders or sorting networks [7,13]) and process it using – increasingly improving – state-of-the-art SAT-solvers. Recent research have favored the approach that uses BDDs, which is evidenced by several new constructions and optimizations [2,30]. In our previous paper we showed that encodings based on comparator networks can still be very competitive [17]. Comparator networks have been successfully applied to construct very efficient encodings of cardinality and Pseudo-Boolean constraints. Codish and Zazon-Ivry [10] introduced pairwise selection networks. We have later improved their construction [16]. In [1] the authors proposed a mixed parametric approach to the encodings, where the direct encoding is chosen for small sub-problems and the splitting point is optimized when large problems are divided into two smaller ones. They proposed to minimize the function $\lambda \cdot num\_vars + num\_clauses$ in the encodings, where lambda is a constant chosen empirically. The constructed encodings are small and efficient. Most encodings based on comparator networks use variations of the Batcher's Odd-Even Sorting Network [1,4,5,18].

Incremental usage of SAT-solvers has been studied extensively in the past years, which allowed for the huge increase in the performance of SAT-based algorithms [12,25,32,33]. Recently, incremental algorithms for MaxSAT instances have appeared [24,27,34], and the experimental results show that the performance of MaxSAT-solvers can be greatly improved by maintaining the learned information and the internal state of the SAT-solver between iterations. Some incremental SAT algorithms also exist for solving PB-constraint instances. For example, Manolios and Papavasileiou [22] proposed an algorithm for PB-solving that uses a SAT-solver for the efficient exploration of the search space, but at the same time exploits the high-level structure of the PB-constraints to simplify the problem and direct the search. Some popular solvers also implement incremental methods, for example, QMaxSAT [20] or Glucose [6].

## 1.2 Our Contribution

Even though MaxSAT problems and Pseudo-Boolean constraint satisfaction problems have a very close relation with each other (by a simple reduction), the notion of incrementality for encoding PB-constraints has not yet been fully exploited. In this paper we show how sorter-based algorithm of KP-MINISAT+ can be extended to solve, even more efficiently, optimization problems involving PB-constraints.

MINISAT+ has served as a base for many new solvers and has been extended to test new constructions and optimizations in the field of PB-solving. Similarly, we have developed a system based on it which encodes PB-constraints using a new sorter-based algorithm [17], efficiently finds good mixed-radix bases for the encoding (see Subsect. 2.2 for a definition) and incorporates a few other optimizations. The underlying comparator network is called a 4-Way Merge Selection Network [18], and experiments showed that on many instances of popular benchmarks our technique outperformed other state-of-the-art PB-solvers. Furthermore, our solver has been recently extended to MaxSAT problems and can successfully compete with state-of-the-art MaxSAT-solvers, which is evidenced by achieving high places in MaxSAT Evaluation 2019. The new MaxSAT-solver, called UWrMaxSat [29], took second place in both Weighted Complete Track and Unweighted Complete Track of the competition.

In this paper we show how the encoding algorithm of the PB-solver can be further improved by extending the usage of assumptions in the comparator network encoding scheme. The new technique is a modification of the idea found in NAPS [30] for simplifying inequality assertions in a constraint. It is applied when a mixed-radix base is used to encode a constraint as an interconnected sequence of sorting networks. The idea is to add a certain integer constant to both sides of the constraint, such that the representation of right side constant (in the base) contains only one non-zero digit. Now, in order to enforce the inequality, one only needs to assert a single output variable of the encoding of the last network. This simplification allows for a reduction of the number of clauses in the resulting CNF encoding, as well as allows better propagation. We have successfully implemented the technique in KP-MINISAT+. In the process of minimizing the value of a goal function, the solver has to try a series of bounds on it. The main purpose of our new construction is to avoid adding new variables and clauses to the encoding after each bound change. In this paper we show how to remedy this situation by adding a certain number of fresh variables to the encoded networks and then using them as assumptions to set a value of the changing constant.

We experimentally compare our solver with other state-of-the-art general constraints solvers like PBLIB [28] and NAPS [30] to prove that our techniques are good in practice. We use COMINISATPS [26] by Chanseok Oh as the underlying SAT-solver, as it has been observed to perform better than the original MINISAT [11] for many instances.

Since more than a decade there have been organized a series of Pseudo-Boolean Evaluations [23] which aim to assess the state-of-the-art in the field of PB-solvers. We use the competition problems from the PB 2016 Competition as benchmarks for the solver proposed in this paper.

### 1.3   Structure of the Paper

In Sect. 2 we briefly describe our comparator network algorithm, then we explain the Mixed Radix Base technique used in MINISAT+ and we show how it is applied to encode a PB-constraint by constructing a series of comparator networks. In Sect. 3 we show how to leverage assumptions in order to build an incremental algorithm on top of KP-MINISAT+'s PB-solving algorithm. We present results of our experiments in Sect. 4, and we give concluding remarks in Sect. 5.

## 2   Background

The main tool in our encoding algorithms is a comparator network. Traditionally comparator networks are presented as circuits that receive $n$ inputs and permute them using comparators (2-sorters) connected by "wires". Each comparator has two inputs and two outputs. The "lower" output is the maximum of inputs, and "upper" one is the minimum. Their standard definitions and properties can be found, for example, in [19].

### 2.1   4-Way Merge Selection Network

MINISAT+ uses Batcher's original construction [8] – the 2-Odd-Even Sorting Network. Later, it has been proposed to replace it with a *selection network*. A selection network of order $(n, k)$ is a comparator network such that for any 0–1 input of length $n$ it outputs its $k$ largest elements, where $k$ is the RHS of a constraint. Those $k$ elements must also be sorted in order to easily assert the given constraint, by asserting only the $k$-th output. In this paper we use sorting networks as black-boxes, therefore we describe the algorithm in a brief manner.

The main building block of our encoding is a direct selection network, which is a certain generalization of a comparator. Encoding of the direct selection network of order $(n, k)$ with inputs $\langle x_1, \ldots, x_n \rangle$ and outputs $\langle y_1, \ldots, y_k \rangle$ is the set of clauses $\{x_{i_1} \wedge \cdots \wedge x_{i_p} \Rightarrow y_p : 1 \leq p \leq k, 1 \leq i_1 < \cdots < i_p \leq n\}$. The direct $n$-sorter is a direct selector of order $(n, n)$, therefore we need $n$ auxiliary variables and $2^n - 1$ clauses to encode it. This shows that $n$ should be small in order to avoid an exponential blowup in the number of clauses.

It has already been observed that using selection networks instead of sorting networks is more efficient for the encoding of constraints [10], as the resulting encodings are smaller and can achieve faster SAT-solver run-time. This fact has been successfully used to encode cardinality constraints, and we have applied this technique to PB-constraints using a construction called a 4-Way Merge Selection Network. A detailed description of the algorithm, a proof of its correctness and the corresponding analysis can be found in our previous paper [18]. We extended our construction by mixing our network with the direct encoding for small values of parameters $n$ and $k$ – the technique which was first described by Abío et al. [1].

## 2.2 Mixed Radix Base Technique

The authors of MiniSat+ devised a method to decompose a PB-constraint into a number of interconnected sorting networks, where sorters play the role of adders on unary numbers in a *mixed radix representation*.

In the classic base $r$ radix system, positive integers are represented as finite sequences of digits $\mathbf{d} = \langle d_0, \ldots, d_{m-1} \rangle$ where for each digit $0 \leq d_i < r$, and for the most significant digit, $d_{m-1} > 0$. The integer value associated with $\mathbf{d}$ is $v = d_0 + d_1 r + d_2 r^2 + \cdots + d_{m-1} r^{m-1}$. A mixed radix system is a generalization where a base $\mathbf{B}$ is a sequence of positive integers $\langle r_0, \ldots, r_{m-1} \rangle$. The integer value associated with $\mathbf{d}$ is $v = d_0 w_0 + d_1 w_1 + d_2 w_2 + \cdots + d_m w_m$ where $w_0 = 1$ and for $i \geq 0$, $w_{i+1} = w_i r_i$. For example, the number $\langle 2, 4, 10 \rangle_{\mathbf{B}}$ in base $\mathbf{B} = \langle 3, 5 \rangle$ is interpreted as $2 \times \mathbf{1} + 4 \times \mathbf{3} + 10 \times \mathbf{15} = 164$ (values of $w_i$'s in boldface).

The decomposition of a PB-constraint into sorting networks is roughly as follows: first, find a "suitable" finite base $\mathbf{B}$ for the given set of coefficients, for example, in MiniSat+ the base is chosen so that the sum of all the digits of the coefficients written in that base is as small as possible. Then for each element $r_i$ of $\mathbf{B}$ construct a sorting network where the inputs of the $i$-th sorter will be those digits $\mathbf{d}$ (from the coefficients) where $d_i$ is non-zero, plus the potential carry bits from the $(i-1)$-th sorter.

$$
\begin{pmatrix}
0 & 1 & 0 & 0 \\
0 & 1 & 0 & 0 \\
0 & 1 & 0 & 0 \\
0 & 1 & 0 & 0 \\
1 & 2 & 0 & 0 \\
0 & 0 & 0 & 1
\end{pmatrix}
$$

**Fig. 1.** Coefficients of $\psi$ in base $\mathbf{B}$

We show a construction of a sorting network system using an example. We present a step-by-step process of translating a PB-constraint $\psi = 2x_1 + 2x_2 + 2x_3 + 2x_4 + 5x_5 + 18x_6 \leq 22$. Let $\mathbf{B} = \langle 2, 3, 3 \rangle$ be the considered mixed radix base. The representation of the coefficients of $\psi$ in base $\mathbf{B}$ may be illustrated by a $6 \times 4$ matrix (see Fig. 1). The rows of the matrix correspond to the representation of the coefficients in base $\mathbf{B}$. Weights of the digit positions of base $\mathbf{B}$ are $\bar{w} = \langle 1, 2, 6, 18 \rangle$. Thus, the decomposition of the LHS (left-hand side) of $\psi$ is:

$$\mathbf{1} \cdot (x_5) + \mathbf{2} \cdot (x_1 + x_2 + x_3 + x_4 + 2x_5) + \mathbf{6} \cdot (0) + \mathbf{18} \cdot (x_6)$$

Now we construct a series of four sorting networks in order to encode the sums at each digit position of $\bar{w}$. Given values for the variables, the sorted outputs

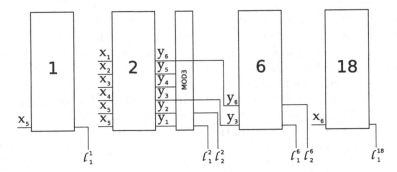

**Fig. 2.** Decomposition of a PB-constraint into a series of interconnected sorting networks. Outputs of sorting networks are ordered such that the bottom bit is the largest.

from these networks represent unary numbers $d_1$, $d_2$, $d_3$, $d_4$ such that the LHS of $\psi$ takes the value $\mathbf{1} \cdot d_1 + \mathbf{2} \cdot d_2 + \mathbf{6} \cdot d_3 + \mathbf{18} \cdot d_4$.

The final step is to encode the carry operation from each digit position to the next. The first three outputs must represent valid digits (in unary) for **B**. In our example the single potential violation to this is $d_2$, which is represented in 6 bits. To this end we add two components to the encoding: (1) each third output of the second network is fed into the third network as carry input; and (2) a *normalizer* **MOD3** is added to encode that the output of the second network is to be considered modulo 3. The full construction is illustrated in Fig. 2.

The outputs from these four sorting networks now specify a number in base **B**, i.e., bits representing LHS of the constraint, each digit represented in unary. To enforce the constraint, we have to add clauses representing the relation $\leq 22$ (in base **B**). It is done by lexicographical comparison of digits representing LHS to digits representing $22 = \langle 0,2,0,1 \rangle_{\mathbf{B}}$. Let $l_1^1$, $l_1^2$, $l_2^2$, $l_1^6$, $l_2^6$, $l_1^{18}$ represent the outputs of the networks, like in Figure 2. Then the following set of clauses enforce the $\leq 22$ constraint: $l_1^{18} \Rightarrow \neg l_1^6$ and $l_1^{18} \Rightarrow \neg(l_2^2 \wedge l_1^1)$.

Could we eliminate the clauses and the **MOD** sub-networks as well? Consider the following scheme. If we add $13 = \langle 1,0,2,0 \rangle_{\mathbf{B}}$ to both sides of $\psi$, then we get $\psi' = 2x_1 + 2x_2 + 2x_3 + 2x_4 + 5x_5 + 18x_6 + 13 < 36$. Observe that $36 = \langle 0,0,0,2 \rangle_{\mathbf{B}}$ and the new decomposition of the LHS is:

$$\mathbf{1} \cdot (1 + x_5) + \mathbf{2} \cdot (x_1 + x_2 + x_3 + x_4 + 2x_5) + \mathbf{6} \cdot (1 + 1) + \mathbf{18} \cdot (x_6)$$

After this change we virtually add 1s as additional outputs to the corresponding networks (one to the first network and two to the third network, as indicated by the new decomposition). This will change the number of inputs to some networks, that is, $l_1^1$ will be an additional input to the second network (as a carry) and $l_1^6$ will be a similar input to the fourth one. Thus, the fourth network will now have 2 inputs and an additional literal $l_2^{18}$ representing its second output needs to be created.

Observe that $\psi$ and $\psi'$ are equivalent, but in the representation of 36 (in base **B**) only the most significant bit has a non-zero value, therefore enforcing the $< 36$ constraint is as easy as adding a singleton clause $\neg l_2^{18}$ (or setting it as an assumption). In consequence, the only relevant outputs of the networks (except the last one) are the ones that represent the carry bits, therefore there is no need to use normalizers. This optimization was first proposed in NaPS [30] and we have already implemented it in our previous solver [17]. Notice that after changing the RHS, we need to rebuild most of the construction in order to account for the increased number of inputs and outputs of each network and the new carry bit positions. What follows is an improvement of this strategy, such that we do not need to do the rebuilding step.

# 3   The Incremental Algorithm

We now show how we can better encode the goal function of a PB-constraint optimization instance by adding assumptions to the previous construction. To demonstrate each step of the algorithm, we will be using our running example, i.e., the goal function is $2x_1 + 2x_2 + 2x_3 + 2x_4 + 5x_5 + 18x_6$ and $\mathbf{B} = \langle 2, 3, 3 \rangle$ is the chosen base.

*Code Notation.* The pseudo-code is presented in Algorithms 1 and 2. The only non-trivial data structure used is a vector, i.e., a dynamic array, which in our case can store either numbers or literals, depending on the context. Vectors are indexed starting from 0, and $x_i$ is the $i$-th element of a vector $\bar{x}$. The vector structure supports three straightforward operations:

- *pushBack* – appends a given element to the end of the vector.
- *size* – returns the number of elements currently stored in the vector.
- *clear* – removes all elements of the vector.

A special SAT-solver object $ss$ is also available. It supports the following set of operations:

- *newVar* – creates a fresh variable and adds it to the solver instance.
- *addClause* – adds a clause to the solver instance (a clause is given as a sequence of literals).
- *encodeBySorter* – given a sequence of input literals of size $n$, it constructs a sorter with $n$ inputs and $n$ outputs and transforms it to a CNF formula (for example, using our 4-Way Merge Selection Network). The formula is added to the SAT-solver and the operation returns a sequence of literals representing the output of the sorter.
- *solve* – takes a set of assumptions as input and returns a model if the solver instance is satisfiable under given assumptions, otherwise returns UNSAT.

We now describe our algorithm and show how it works using our running example. We do this in a bottom-up manner, starting with the *encodeGoal* procedure (Algorithm 1).

---

**Algorithm 1.** encodeGoal

---

**Input:** A PB function $g(\bar{x}) = a_1 x_1 + a_2 x_2 + \cdots + a_n x_n$, where $a_1, a_2, \ldots, a_n > 0$, and a SAT-solver $ss$.

**Output:** A tuple $\langle \bar{r}, \bar{z}, \bar{y} \rangle$, where $\bar{r}$ is a mixed radix base, $\bar{z}$ are new assumption variables and $\bar{y}$ is a sequence of variables representing output of the encoding. The output is used in Algorithm 2 to force any upper bound on $g(\bar{x})$ by setting assumptions to $ss$.

1: $\bar{r} \leftarrow findGoodBase(a_1, a_2, \ldots, a_n)$
2: **let** $\bar{w}$ be the weight vector of $\bar{r}$
3: **let** $\bar{z}, \bar{y}, carry, in$ and $out$ be empty vectors
4: **for** $i = 0$ **to** $\bar{r}.size() - 1$ **do**
5:     $in \leftarrow carry$
6:     **for** $j = 1$ **to** $r_i - 1$ **do**
7:         $z_j^{w_i} \leftarrow ss.newVar()$
8:         $in.pushBack(z_j^{w_i}), \bar{z}.pushBack(z_j^{w_i})$
9:     **for** $j = 2$ **to** $r_i - 1$ **do** $ss.addClause(\neg z_j^{w_i} \vee z_{j-1}^{w_i})$
10:     **for** $j = 1$ **to** $n$ **do**
11:         **repeat** $a_j \bmod r_i$ **times** $in.pushBack(x_j)$
12:         $a_j \leftarrow a_j / r_i$
13:     $out \leftarrow ss.encodeBySorter(in)$
14:     $carry.clear()$
15:     **for** $j = r_i - 1$ **while** $j < out.size()$ **step** $r_i$ **do** $carry.pushBack(out_j)$
16: $in \leftarrow carry$
17: **for** $j = 1$ **to** $n$ **do**
18:     **repeat** $a_j$ **times** $in.pushBack(x_j)$
19: $\bar{y} \leftarrow ss.encodeBySorter(in)$
20: **return** $\langle \bar{r}, \bar{z}, \bar{y} \rangle$

---

*encodeGoal.* Find a mixed radix base $\langle r_0, \ldots, r_{m-1} \rangle$ (for some $m \geq 0$) and its weight vector $\langle w_0, \ldots, w_m \rangle$ (lines 1–2). Next, decompose the goal function as shown in the previous section with the following modifications. The $i$-th iteration creates the $i$-th sorter for which inputs are stored in vector $in$. New assumption variables are created and passed to the sorter as additional input. A more detailed description is as follows, given we are in the $i$-th iteration of the main loop ($0 \leq i \leq m - 1$).

In lines 6–8, create a new variable $z_j^{w_i}$ for each $0 < j < r_i$ (line 7). Add the new variables as input to the current sorter (line 8). Next, for $1 < j < r_i$ add the clause $z_j^{w_i} \Rightarrow z_{j-1}^{w_i}$ to the instance (line 9).

The purpose for this step is as follows. The new variables allow to represent any number between 0 and $r_i - 1$ (for a given $0 \leq i \leq m - 1$) in unary, and the new clauses enforce the order of the bits. Now, if we would like to set the variables $\langle z_1^{w_i}, \ldots, z_{r_i-1}^{w_i} \rangle$ such that they represent a number $0 < j < r_i - 1$, we need to only set $z_j^{w_i} = 1$ and $z_{j+1}^{w_i} = 0$, and the unit propagation will set all other $z_{j'}^{w_i}$'s such that exactly $j$ of them will be set to true. If $j = 0$, then we only

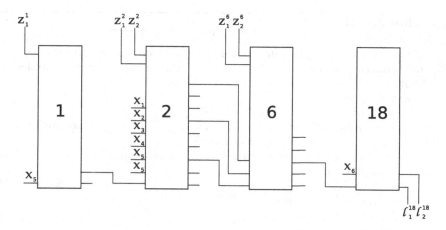

**Fig. 3.** An example of a novel PB-constraint decomposition. The top variables are stored as assumptions and their values are adjusted after each iteration of the algorithm.

need to set $z_1^{w_i} = 0$, and if $j = r_i - 1$ we set $z_{r_i-1}^{w_i} = 1$, and similarly the unit propagation correctly sets the rest of the variables.

Let $a_j^i$, $1 \le j \le n$, denote the value of $a_j$ in line 11 in the $i$-th iteration of the loop 4–15. In lines 10–12, add multiple copies of the variables $x_1, \ldots, x_n$ to $in$, in such a way that they represent (in unary) terms of the sum $\sum_{j=1}^{n}(a_j^i \bmod r_i)x_j$. Since each sorter acts as an adder, the sequence $out$ in line 13 represents the value of the sum (plus $carry$ and $\langle z_1^{w_i}, \ldots, z_{r_i-1}^{w_i} \rangle$) in unary.

In our running example we create five new variables: $z_1^1$, $z_1^2$, $z_2^2$, $z_1^6$, $z_2^6$, and the set of clauses consists of $z_2^2 \Rightarrow z_1^2$ and $z_2^6 \Rightarrow z_1^6$.

Remember that we represent the value of the LHS as an expression $w_0 \cdot d_0 + \cdots + w_m \cdot d_m$, as explained in the previous section (each $d_i$ is a sum of some input variables). For each $0 \le i \le m-1$ and $0 < j < r_i$ we add $z_j^{w_i}$ to $d_i$. In our running example the decomposition will look like this:

$$\mathbf{1} \cdot (z_1^1 + x_5) + \mathbf{2} \cdot (z_1^2 + z_2^2 + x_1 + x_2 + x_3 + x_4 + 2x_5) + \mathbf{6} \cdot (z_1^6 + z_2^6) + \mathbf{18} \cdot (x_6)$$

In lines 10–15 a single sorter is created and the carry bits are set for the next one. Notice that the output of the current sorter (stored in the $out$ vector) is only needed for calculating the carry bits passed to the next sorter (line 15). This is because the only necessary output variable which enforces constraints belongs to the last sorter (created in lines 16–19). Therefore no additional normalizers are required, which is another advantage of using our construction.

We show in Fig. 3 how such a construction looks for our running example. The new assumption variables are shown on the top. Compared to the example from Fig. 2 we added some new inputs, therefore we needed to also create additional outputs for each network. Notice that this changed the carry bit positions but no normalizers were constructed.

---

**Algorithm 2.** optimizeGoal

---

**Input:** A PB function $g(\bar{x}) = a_1 x_1 + a_2 x_2 + \cdots + a_n x_n$, where $a_1, a_2, \ldots, a_n > 0$, and
a SAT-solver $ss$ that is filled already with encodings of other constraints.
**Output:** A model that minimizes the value of $g(\bar{x})$ and satisfies other constraints or
$UNSAT$.

1:  $\langle res, model \rangle \leftarrow ss.solve(\varnothing)$
2:  **if** $res = UNSAT$ **then return** $UNSAT$
3:  $\langle \bar{r}, \bar{z}, \bar{y} \rangle \leftarrow encodeGoal(g, ss)$                                    # see Algorithm 1
4:  $UB \leftarrow g(model),\ LB \leftarrow 0,\ optModel \leftarrow model$
5:  **let** $assump$ be an empty vector and $\bar{w}$ be the weight vector of $\bar{r}$
6:  **while** $UB > LB$ **do**
7:      $bound \leftarrow \lceil 0.65 \cdot UB + 0.35 \cdot LB \rceil$                              # $LB < bound \leq UB$
8:      $assump.clear(),\ b \leftarrow bound$
9:      **for** $i = 0$ **to** $r.size() - 1$ **do**
10:         $j \leftarrow b \bmod r_i,\ b \leftarrow b/r_i$
11:         **if** $j \neq 0$ **then** $j \leftarrow r_i - j,\ b \leftarrow b + 1$
12:         **if** $j = 0$ **then** $assump.pushBack(\neg z_1^{w_i})$
13:         **else if** $j = r_i - 1$ **then** $assump.pushBack(z_{r_i-1}^{w_i})$
14:         **else** $assump.pushBack(z_j^{w_i}),\ assump.pushBack(\neg z_{j+1}^{w_i})$
15:      $assump.pushBack(\neg y_b)$
16:      $\langle res, model \rangle \leftarrow ss.solve(assump)$           # $g(\bar{x}) < bound$ is enforced by $assump$
17:      **if** $res = SAT$ **then**
18:         $UB \leftarrow g(model)$                                    # $g(model) < bound$
19:         $optModel \leftarrow model$
20:         $ss.addClause(\neg y_b)$
21:      **else**
22:         $LB \leftarrow bound$
23:  **return** $optModel$

---

*optimizeGoal.* The optimization procedure is presented in Algorithm 2. Notice
that we assume $a_1, a_2, \ldots, a_n > 0$. The goal function can be easily normalized
to satisfy this condition (see [13]). After encoding every constraint into CNF
formulas we first check if the given set of constraints is satisfiable (lines 1–2). If
it is, then we can optimize the goal function given the constraints. We encode the
goal function using the *encodeGoal* procedure (line 3). The optimization strategy
used is the binary search with the 65/35 split ratio. The detailed description
follows.

For the current bound on the constraint (stored in the *bound* variable) compute how many 1s need to be added to both sides of the inequality, such that the
RHS has only the most significant position set in the base $\langle r_0, \ldots, r_{m-1} \rangle$. Let $c$
be that number, that is, if *bound* is divisible by $w_m$ then $c$ is zero, otherwise $c$
is set to $w_m - bound \bmod w_m$, and let $\langle c_0, \ldots, c_{m-1} \rangle$ be the representation of $c$
in base $\langle r_0, \ldots, r_{m-1} \rangle$. Notice that $c_m$ is omitted since it is equal to 0. For each
$0 \leq i \leq m - 1$, let $j = c_i$ and do:

- if $j = 0$, set $z_1^{w_i} = 0$,
- if $j = r_i - 1$, set $z_{r_i-1}^{w_i} = 1$,
- otherwise set $z_j^{w_i} = 1$ and $z_{j+1}^{w_i} = 0$.

This is done in lines 9–14, where variable $b$ in the $i$-th iteration is set to the value of $\lceil bound/w_{i+1} \rceil$. Thus, $j$ is the $i$-th digit (in base $\bar{r}$) of $bw_{i+1} - bound$. Next, add a singleton clause enforcing the constraint (line 15) to the set of assumptions.

In our running example let us assume that the current bound is 23, therefore $c = 13$, so the assumptions are $z_1^1 = 1$, $z_1^2 = 0$, $z_2^6 = 1$ ($z_2^2$ and $z_1^6$ will be set by unit propagation), which means that in order to enforce a constraint $<36$, we only need to add $\neg y_2$ as another assumption. Note that $\bar{y}$ is the output of the last sorter created by the $encodeGoal$ procedure, so $y_2$ is equivalent to the $l_2^{18}$ in Fig. 3.

Finally, we run the underlying SAT-solver under the current set of assumptions (line 16) and based on the answer we strengthen the bounds on the goal function (lines 17–22). The binary search continues until the optimum is found. For example, if the algorithm determines that the next bound to check for our running example is 19, then we revert the assignment of the assumptions and now we set $z_1^1 = 1$, $z_2^2 = 1$ and $z_2^6 = 1$, since now we need to add 17 to both sides of the inequality so that the encoding is still equisatisfiable with the $< 36$ constraint. Notice that no other operation is necessary. As we will see in the next section, the fact that we are building the sorting networks structure only once for the goal function leads to a performance increase in both running time and memory use, compared to other state-of-the-art methods. Let us now prove the correctness of our algorithm, for the sake of completeness.

**Theorem 1.** *Let $g(\bar{x}) = a_1 x_1 + a_2 x_2 + \cdots + a_n x_n$, where $a_1, a_2, \ldots, a_n > 0$ are integer coefficients and $x_1, x_2, \ldots, x_n$ are propositional literals. Let $\phi$ be a CNF formula. Algorithm 2 returns a model of $\phi$ which minimizes the value of $g(\bar{x})$ or UNSAT, if $\phi$ is unsatisfiable.*

*Proof* (sketch). If $\phi$ is unsatisfiable, then the algorithm terminates on line 2. Assume that $\phi$ is satisfiable. The binary search of $optimizeGoal$ will find the optimal model of $\phi$ with respect to the goal function $g(\bar{x})$, if the distance between upper and lower bounds decreases in each iteration of the algorithm. It is obviously true if $g(\bar{x}) < bound$ is enforced on SAT solver by $assump$ set in lines 8–15. To prove this, let $\langle \bar{r}, \bar{z}, \bar{y} \rangle$ be the result of line 3, $m$ be the size of $\bar{r}$ and let $\bar{w}$ be the weight vector of base $\bar{r}$ (see Subsect. 2.2). Fix the value of $bound$ and let $b_0 = bound$ and define $b_{i+1}$ and $j_i$ to be the values of variables $b$ and $j$ after line 11 in the $i$-th iteration of the loop in lines 9–14. Notice that $b_{i+1} = \left\lceil \frac{b_i}{r_i} \right\rceil$ and $j_i = r_i b_{i+1} - b_i$. By induction one can prove the following invariants of the loop: $b_i = \left\lceil \frac{bound}{w_i} \right\rceil$ and $\sum_{s=0}^{i-1} j_s w_s = b_i w_i - bound$.

Therefore, after the loop, we have $bound = b_m w_m - \sum_{s=0}^{m-1} j_s w_s$. It follows that the inequality $g(\bar{x}) < bound$ is equivalent to $g(\bar{x}) + \sum_{s=0}^{m-1} j_s w_s < b_m w_m$ (1). Each value $j_s$ is set (in unary) on variables $z_1^{w_s}, \ldots, z_{r_i-1}^{w_s}$ in lines 12–14 (see also line 9 in Algorithm 1). In this way the LHS of (1) is set in the encoding generated by $encodeGoal$. The sequence $\bar{y} = (y_1, y_2, \ldots)$ represents (in unary) a value that is multiplied by $w_m$, thus, by adding $\neg y_{b_m}$ to $assump$ in line 15, we enforce the value to be less than $b_m$. That ends the proof that the SAT-solver call in line 16 returns SAT if and only if both $g(\bar{x}) < bound$ and $\phi$ are satisfied.

# 4    Experimental Evaluation

Our extension of MiniSat+, based on the features explained in this paper and in the previous one [17], is available online[1]. We call it KP-MiniSat+ (**KP-MSP**, in short). It should be linked with a slightly modified COMiniSatPS[2], where the patch is also given at the link above (See footnote 1). The latest addition to the patch is an assumptions processing improvement due to Hickey and Bacchus [15]. Detailed results of the experimental evaluation are also available online[3].

The set of instances we use is from the Pseudo-Boolean Competition 2016[4]. We use instances with linear Pseudo-Boolean constraints that encode optimization problems. To this end, two categories from the competition have been selected:

- **OPT-BIGINT-LIN** - 1109 instances of optimization problems with big coefficients in the constraints (at least one constraint with a sum of coefficients greater than $2^{20}$). An objective function is present. The solver must find a solution with the best possible value of the objective function.
- **OPT-SMALLINT-LIN** - 1600 instances of optimization problems. Like OPT-BIGINT-LIN but with small coefficients in the constraints (no constraint with sum of coefficients greater than $2^{20}$).

We compare our solver with two state-of-the-art general purpose constraint solvers. The first one is PBSOLVER from PBLib ver. 1.2.1, by Tobias Philipp and Peter Steinke [28] (abbreviated to **PBLib** in the results). This solver implements a plethora of encodings for three types of constraints: at-most-one, at-most-k (cardinality constraints) and Pseudo-Boolean constraints. PBLib automatically normalizes the input constraints and decides which encoder provides the most effective translation. We have launched the program `./BasicPBSolver/pbsolver` of PBLib on each instance with the default parameters.

The second solver is NaPS ver. 1.02b by Masahiko Sakai and Hidetomo Nabeshima [30] which implements improved ROBDD structure for encoding constraints in band form, as well as other optimizations. This solver is also built on the top of MiniSat+. NaPS won two of the optimization categories in the Pseudo-Boolean Competition 2016: OPT-BIGINT-LIN and OPT-SMALLINT-LIN. We have launched the main program of NaPS on each instance, with parameters `-a -s -nm`.

We also compare our solver with the original MiniSat+ in two different versions, one using the original MiniSat SAT-solver and the other using COMiniSatPS. We label these **MS+** and **MS+COM** in the results. We present results for **MS+COM** in order to show that the advantage of using our solver does not come simply from changing the underlying SAT-solver.

---

[1] See https://github.com/karpiu/kp-minisatp.
[2] See https://baldur.iti.kit.edu/sat-competition-2016/solvers.
[3] See http://www.ii.uni.wroc.pl/~karp/sat/2020.html.
[4] See http://www.cril.univ-artois.fr/PB16/.

**Table 1.** Results summary for the OPT-BIGINT-LIN category

| solver | solved | Opt | UnSat | cpu (s) | scpu (s) | avg(scpu) | smem (MB) | avg(smem) |
|---|---|---|---|---|---|---|---|---|
| KP-MSP++ | **468** | **395** | **73** | 1046518 | 44424 | 94.9 | 208035 | 444.5 |
| KP-MSP+- | 467 | **395** | 72 | **1037085** | 44886 | 96.1 | 213973 | 458.2 |
| KP-MSP-- | 461 | 389 | 72 | 1039499 | **37672** | **81.7** | 283681 | 615.4 |
| NaPS | 383 | 314 | 69 | 1314536 | 51557 | 134.6 | 245533 | 641.1 |
| MS+ | 220 | 149 | 71 | 1647958 | 47759 | 217.1 | **42181** | 191.7 |
| MS+COM | 245 | 174 | 71 | 1609433 | 54234 | 221.4 | 46336 | **189.1** |

**Table 2.** Results summary for the OPT-SMALLINT-LIN category

| solver | solved | Opt | UnSat | cpu (s) | scpu (s) | avg(scpu) | smem (MB) | avg(smem) |
|---|---|---|---|---|---|---|---|---|
| KP-MSP++ | **894** | 808 | 86 | 1282788 | 43556 | 48.7 | 164223 | 183.7 |
| KP-MSP+- | 893 | 806 | **87** | 1278926 | 38474 | 43.1 | 162405 | 181.9 |
| KP-MSP-- | 893 | **809** | 84 | **1278722** | **37747** | **42.3** | 153619 | 172.0 |
| NaPS | 887 | 803 | 84 | 1310006 | 40376 | 45.5 | 186760 | 210.6 |
| PBLib | 747 | 691 | 56 | 1611247 | 74993 | 100.4 | 112993 | 151.3 |
| MS+ | 788 | 715 | 73 | 1515166 | 53566 | 68.0 | 113606 | 144.2 |
| MS+COM | 805 | 734 | 71 | 1491269 | 60270 | 74.9 | **106886** | **132.8** |

We are providing results for three versions of KP-MSP: (1) **KP-MSP++** that contains our algorithms and the latest modification to COMiniSatPS, (2) **KP-MSP+-** that also contains the algorithms but not the modification, and (3) **KP-MSP--** - without the algorithms and the modification, but still with optimizations of KP-MSP described in [17] (in particular, in encodings of constraints on a goal function, it reuses clauses from previous encoding by the "shared-formulas" original technique of MINISAT+). We would like to see what is the impact of new techniques on the number of solved instances and the average times and spaces used.

All the three versions of KP-MSP used default parameters, except for the parameter **-gs**, which forces the algorithm to always encode the goal function using our selection network (and the direct encoding for small sub-networks). This means that other constraints can sometimes be encoded using either BDDs or adder networks, and the original MINISAT+'s heuristics (slightly modified by us to strongly prefer encoding by sorters) decide which method is used. For example, for OPT-BIGINT-LIN instances, in all encoded non-goal constraints: 99.58% were encoded by sorters, 0.34% by BDDs and 0.08% by adders. If we consider only the successfully solved instances then the corresponding numbers are: 99.73%, 0.02% and 0.25%.

All experiments were carried out on machines with Intel(R) Core(TM) i7-2600 CPU @ 3.40 GHz. The timeout limit is set to 1800 s and the memory limit is 15 GB, which are enforced with the following commands: ulimit -Sv 15000000 and timeout  − k 20 1809  < solver > < parameters > < instance >.

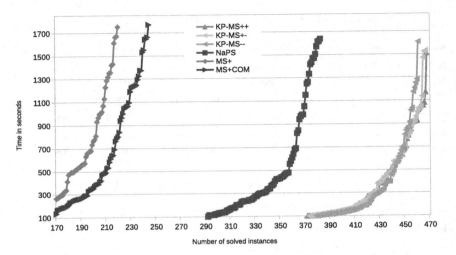

**Fig. 4.** Cactus plot for OPT-BIGINT-LIN division from the PB16 suite

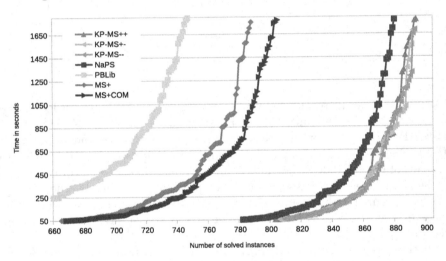

**Fig. 5.** Cactus plot for OPT-SMALLINT-LIN division from the PB16 suite

In Tables 1 and 2 we present results for categories OPT-BIGINT-LIN and OPT-SMALLINT-LIN, respectively. In the **solved** column we show the total number of solved instances, which is the sum of the number of instances where the optimum was found (the **Opt** column) and the number of unsatisfiable instances found (the **UnSat** column). In the **cpu** column we show the total solving time (in seconds) of the solver over all instances of a given category, and **scpu** is the total solving time over solved instances only. Similarly, **smem** is the total memory space used (in megabytes) during the computation of the solved instances. Averages have been computed as follows: **avg(scpu) = scpu/solved** and **avg(smem) = smem/solved**.

Looking at the results, one can observe that new algorithms increase the number of solved instances to 468 in the OPT-BIGINT-LIN category. It is now almost equal to the number of 470 instances solved together by all the competitors of PB Competition 2016. The modification of COMiniSatPS add 1 solved instance and reduces the average time (by 1.2 s) and memory use (by 13.7 MB). The algorithms reduce the average memory use by 157.2 MB (KP-MSP+- versus KP-MSP--). Moreover, one can observe significant improvement in the number of solved instances in comparison to **NaPS** in this category.

In case of OPT-SMALLINT-LIN category, the differences among the results of all three versions of KP-MSP are small. It is understandable, as the coefficients of goal functions are not big in this category, thus, the sizes of mixed-radix bases are small, so the optimization techniques of [17] are equivalently efficient to the new algorithms.

In terms of memory usage **MS+** and **MS+COM** are the most efficient in this evaluation, but their overall performance is poor. Observe also that their average values are computed over much smaller sets of solved instances. Solver **PBLib** had the worst performance in this evaluation. Notice that the results of **PBLib** for OPT-BIGINT-LIN division are not available. This is because **PBLib** is using 64-bit integers in calculations, thus could not be launched with all OPT-BIGINT-LIN instances.

Figures 4 and 5 show cactus plots of the results, which indicate the number of solved instances within the time. We see a clear advantage of our solvers over the competition in the OPT-BIGINT-LIN category.

## 5    Conclusions

In this paper we showed that comparator networks are still competitive when used in encoding Pseudo-Boolean constraints to SAT. The popular idea of incremental encoding applied to the sorting network encoding of a pseudo-Boolean goal function leads to an increase in the number of solved instances in the OPT-BIGINT-LIN category and reduces the memory use compared to other state-of-the-art methods. The proposed modification is short and easy to implement using any modern SAT-solver which supports *assumptions*.

## References

1. Abío, I., Nieuwenhuis, R., Oliveras, A., Rodríguez-Carbonell, E.: A parametric approach for smaller and better encodings of cardinality constraints. In: Schulte, C. (ed.) CP 2013. LNCS, vol. 8124, pp. 80–96. Springer, Heidelberg (2013). https://doi.org/10.1007/978-3-642-40627-0_9
2. Abío, I., Nieuwenhuis, R., Oliveras, A., Rodríguez-Carbonell, E., Mayer-Eichberger, V.: A new look at bdds for pseudo-boolean constraints. J. Artif. Intell. Res. **45**, 443–480 (2012)
3. Aloul, F.A., Ramani, A., Markov, I.L., Sakallah, K.A.: Generic ILP versus specialized 0–1 ILP: an update. In: Proceedings of the 2002 IEEE/ACM International Conference on Computer-aided Design, pp. 450–457. ACM (2002)

4. Asín, R., Nieuwenhuis, R., Oliveras, A., Rodríguez-Carbonell, E.: Cardinality networks and their applications. In: Kullmann, O. (ed.) SAT 2009. LNCS, vol. 5584, pp. 167–180. Springer, Heidelberg (2009). https://doi.org/10.1007/978-3-642-02777-2_18

5. Asín, R., Nieuwenhuis, R., Oliveras, A., Rodríguez-Carbonell, E.: Cardinality networks: a theoretical and empirical study. Constraints **16**(2), 195–221 (2011)

6. Audemard, G., Lagniez, J.-M., Simon, L.: Improving glucose for incremental SAT solving with assumptions: application to MUS extraction. In: Järvisalo, M., Van Gelder, A. (eds.) SAT 2013. LNCS, vol. 7962, pp. 309–317. Springer, Heidelberg (2013). https://doi.org/10.1007/978-3-642-39071-5_23

7. Bailleux, O., Boufkhad, Y., Roussel, O.: A translation of pseudo-boolean constraints to SAT. J. Satisfiability Boolean Model. Comput. **2**, 191–200 (2006)

8. Batcher, K.E.: Sorting networks and their applications. In: Proceedings of the 30 April–2 May 1968, Spring Joint Computer Conference, pp. 307–314. ACM (1968)

9. Bryant, R.E., Lahiri, S.K., Seshia, S.A.: Deciding CLU logic formulas via boolean and pseudo-boolean encodings. In: Proceedings of the International Workshop on Constraints in Formal Verification (CFV 2002). Citeseer (2002)

10. Codish, M., Zazon-Ivry, M.: Pairwise cardinality networks. In: Clarke, E.M., Voronkov, A. (eds.) LPAR 2010. LNCS (LNAI), vol. 6355, pp. 154–172. Springer, Heidelberg (2010). https://doi.org/10.1007/978-3-642-17511-4_10

11. Eén, N., Sörensson, N.: An extensible SAT-solver. In: Giunchiglia, E., Tacchella, A. (eds.) SAT 2003. LNCS, vol. 2919, pp. 502–518. Springer, Heidelberg (2004). https://doi.org/10.1007/978-3-540-24605-3_37

12. Eén, N., Sörensson, N.: Temporal induction by incremental SAT solving. Electron. Notes Theor. Comput. Sci. **89**(4), 543–560 (2003)

13. Eén, N., Sörensson, N.: Translating pseudo-boolean constraints into SAT. J. Satisfiability Boolean Model. Comput. **2**, 1–26 (2006)

14. Elffers, J., Nordström, J.: Divide and conquer: towards faster pseudo-boolean solving. In: IJCAI, pp. 1291–1299 (2018)

15. Hickey, R., Bacchus, F.: Speeding up assumption-based SAT. In: Janota, M., Lynce, I. (eds.) SAT 2019. LNCS, vol. 11628, pp. 164–182. Springer, Cham (2019). https://doi.org/10.1007/978-3-030-24258-9_11

16. Karpiński, M., Piotrów, M.: Smaller selection networks for cardinality constraints encoding. In: Pesant, G. (ed.) CP 2015. LNCS, vol. 9255, pp. 210–225. Springer, Cham (2015). https://doi.org/10.1007/978-3-319-23219-5_16

17. Karpiński, M., Piotrów, M.: Competitive sorter-based encoding of PB-constraints into SAT. In: Berre, D.L., Järvisalo, M. (eds.) Proceedings of Pragmatics of SAT 2015 and 2018. EPiC Series in Computing, vol. 59, pp. 65–78. EasyChair (2019). https://doi.org/10.29007/hh3v. https://easychair.org/publications/paper/tsHw

18. Karpiński, M., Piotrów, M.: Encoding cardinality constraints using multiway merge selection networks. Constraints **24**(3–4), 234–251 (2019). https://doi.org/10.1007/s10601-019-09302-0

19. Knuth, D.E.: The art of computer programming. In: Sorting and Searching, 2nd edn, vol. 3. Addison Wesley Longman Publishing Co., Inc., Redwood City (1998)

20. Koshimura, M., Zhang, T., Fujita, H., Hasegawa, R.: QMaxSAT: a partial Max-SAT solver. J. Satisfiability Boolean Model. Comput. **8**, 95–100 (2012)

21. Le Berre, D., Parrain, A.: The sat4j library, release 2.2. J. Satisfiability Boolean Model. Comput. **7**(2–3), 59–64 (2010)

22. Manolios, P., Papavasileiou, V.: Pseudo-boolean solving by incremental translation to SAT. In: 2011 Formal Methods in Computer-Aided Design (FMCAD), pp. 41–45. IEEE (2011)

23. Manquinho, V.M., Roussel, O.: The first evaluation of pseudo-boolean solvers (PB'05). J. Satisfiability Boolean Model. Comput. **2**, 103–143 (2006)
24. Martins, R., Joshi, S., Manquinho, V., Lynce, I.: On using incremental encodings in unsatisfiability-based MaxSAT solving. J. Satisfiability Boolean Model. Comput. **9**(1), 59–81 (2014)
25. Nadel, A., Ryvchin, V.: Efficient SAT solving under assumptions. In: Cimatti, A., Sebastiani, R. (eds.) SAT 2012. LNCS, vol. 7317, pp. 242–255. Springer, Heidelberg (2012). https://doi.org/10.1007/978-3-642-31612-8_19
26. Oh, C.: Improving SAT Solvers by Exploiting Empirical Characteristics of CDCL. Ph.D. thesis, New York University (2016)
27. Paxian, T., Reimer, S., Becker, B.: Dynamic polynomial watchdog encoding for solving weighted MaxSAT. In: Beyersdorff, O., Wintersteiger, C.M. (eds.) SAT 2018. LNCS, vol. 10929, pp. 37–53. Springer, Cham (2018). https://doi.org/10.1007/978-3-319-94144-8_3
28. Philipp, T., Steinke, P.: PBLib – a library for encoding pseudo-boolean constraints into CNF. In: Heule, M., Weaver, S. (eds.) SAT 2015. LNCS, vol. 9340, pp. 9–16. Springer, Cham (2015). https://doi.org/10.1007/978-3-319-24318-4_2
29. Piotrów, M.: UWrMaxSAT-a new minisat+-based solver in maxsat evaluation 2019. MaxSAT Eval. **2019**, 11 (2019)
30. Sakai, M., Nabeshima, H.: Construction of an ROBDD for a PB-constraint in band form and related techniques for PB-solvers. IEICE Trans. Inf. Syst. **98**(6), 1121–1127 (2015)
31. Schutt, A., Feydy, T., Stuckey, P.J., Wallace, M.G.: Why decomposition is not as bad as it sounds. In: Gent, I.P. (ed.) CP 2009. LNCS, vol. 5732, pp. 746–761. Springer, Heidelberg (2009). https://doi.org/10.1007/978-3-642-04244-7_58
32. Shtrichman, O.: Pruning techniques for the SAT-based bounded model checking problem. In: Margaria, T., Melham, T. (eds.) CHARME 2001. LNCS, vol. 2144, pp. 58–70. Springer, Heidelberg (2001). https://doi.org/10.1007/3-540-44798-9_4
33. Whittemore, J., Kim, J., Sakallah, K.: Satire: a new incremental satisfiability engine. In: Proceedings of the 38th annual Design Automation Conference, pp. 542–545. ACM (2001)
34. Zha, A., Koshimura, M., Fujita, H.: N-level modulo-based CNF encodings of pseudo-boolean constraints for MaxSAT. Constraints **24**(2), 133–161 (2019)

# Author Index

Printed in the United States
By Bookmasters